STUDENT'S SOLUTIONS MANUAL

BEVERLY FUSFIELD

COLLEGE ALGEBRA
THIRD EDITION

J.S. Ratti
University of South Florida

Marcus McWaters
University of South Florida

PEARSON

Boston   Columbus   Indianapolis   New York   San Francisco   Upper Saddle River
Amsterdam   Cape Town   Dubai   London   Madrid   Milan   Munich   Paris   Montreal   Toronto
Delhi   Mexico City   São Paulo   Sydney   Hong Kong   Seoul   Singapore   Taipei   Tokyo

The author and publisher of this book have used their best efforts in preparing this book. These efforts include the development, research, and testing of the theories and programs to determine their effectiveness. The author and publisher make no warranty of any kind, expressed or implied, with regard to these programs or the documentation contained in this book. The author and publisher shall not be liable in any event for incidental or consequential damages in connection with, or arising out of, the furnishing, performance, or use of these programs.

Reproduced by Pearson from electronic files supplied by the author.

Copyright © 2015, 2011, 2007 Pearson Education, Inc.
Publishing as Pearson, 75 Arlington Street, Boston, MA 02116.

ISBN-13: 978-0-321-91743-0
ISBN-10: 0-321-91743-X

www.pearsonhighered.com

PEARSON

CONTENTS

Chapter P Basic Concepts of Algebra

P.1 The Real Numbers and Their Properties

P.1 Practice Problems

1. Let $x = 2.132132132....$
   Then, $1000x = 2132.132132...$
   $1000x = 2132.132132...$
   $\underline{x = 2.132132...}$
   $999x = 2130$
   $x = \dfrac{2130}{999} = \dfrac{710}{333}$

2. a. True   b. True
   c. False

3. $A = \{-3, -1, 0, 1, 3\}$, $B = \{-4, -2, 0, 2, 4\}$
   $A \cap B = \{0\}$,
   $A \cup B = \{-4, -3, -2, -1, 0, 1, 2, 3, 4\}$

4. a. $I_1 \cup I_2 = (-\infty, \infty)$
   b. $I_1 \cap I_2 = [-2, 5)$

5. a. $|-10| = 10$
   b. $|3 - 4| = |-1| = 1$
   c. $|2(-3) + 7| = |1| = 1$

6.
   $d(-7, 2) = |-7 - 2| = |-9| = 9$

7. a. $(-3) \cdot 5 + 20 = -15 + 20 = 5$
   b. $5 - 12 \div 6 \cdot 2 = 5 - 2 \cdot 2 = 5 - 4 = 1$
   c. $\dfrac{9-1}{4} - 5 \cdot 7 = \dfrac{8}{4} - 5 \cdot 7 = 2 - 35 = -33$
   d. $-3 + (6-4)^2 = -3 + 2^2 = -3 + 4 = 1$

8. a. $\dfrac{7}{4} + \dfrac{3}{8} = \dfrac{14}{8} + \dfrac{3}{8} = \dfrac{17}{8}$
   b. $\dfrac{8}{3} - \dfrac{2}{5} = \dfrac{40}{15} - \dfrac{6}{15} = \dfrac{34}{15}$

c. $\dfrac{9}{14} \cdot \dfrac{7}{3} = \dfrac{\overset{3}{\cancel{9}}}{\underset{2}{\cancel{14}}} \cdot \dfrac{\overset{1}{\cancel{7}}}{\underset{1}{\cancel{3}}} = \dfrac{3}{2}$

d. $\dfrac{\frac{5}{8}}{\frac{15}{16}} = \dfrac{5}{8} \div \dfrac{15}{16} = \dfrac{5}{8} \cdot \dfrac{16}{15} = \dfrac{\overset{1}{\cancel{5}}}{\underset{1}{\cancel{8}}} \cdot \dfrac{\overset{2}{\cancel{16}}}{\underset{3}{\cancel{15}}} = \dfrac{2}{3}$

9. a. $(3-2) \div 3 + 3 = 1 \div 3 + 3 = \dfrac{1}{3} + 3 = \dfrac{10}{3}$
   b. $7 - \dfrac{-1}{|3|} = 7 + \dfrac{1}{3} = \dfrac{22}{3}$

10. $°C = \dfrac{39}{3} + 4 = 13 + 4 = 17°C$

P.1 Basic Concepts and Skills

1. Whole numbers are formed by adding the number <u>zero</u> to the set of natural numbers.

3. If $a < b$, then a is to the <u>left</u> of b on the number line.

5. True

7. $0.\overline{3}$, repeating

9. -0.8, terminating

11. $0.\overline{27}$, repeating

13. $3.1\overline{6}$, repeating

15. $3.75 = \dfrac{375}{100} = \dfrac{15}{4}$

17. $10x = 53.\overline{3}$
    $\underline{x = 5.\overline{3}}$
    $9x = 48$
    $x = \dfrac{48}{9} = \dfrac{16}{3}$
    $-5.\overline{3} = -\dfrac{16}{3}$

19. $100x = 213.\overline{13}$
    $\underline{x = 2.\overline{13}}$
    $99x = 211$
    $x = \dfrac{211}{99}$

21. $100x = 452.3\overline{23}$
    $\underline{x = 4.5\overline{23}}$
    $99x = 447.8$
    $x = \dfrac{447.8}{99} = \dfrac{4478}{990} = \dfrac{2239}{495}$

23. Rational   25. Rational

27. Rational   **29.** Irrational

31. $3 > -2$ **33.** $\dfrac{1}{2} \geq \dfrac{1}{2}$

35. $5 \leq 2x$ **37.** $-x > 0$

39. $2x + 7 \leq 14$ **41.** $4 = \dfrac{24}{6}$

43. $-4 < 0$

45. $A \cup B = \{-4, -3, -2, 0, 1, 2, 3, 4\}$

47. $A \cap C = \{-4, -2, 0, 2\}$

49. $(B \cap C) \cup A = \{-3, 0, 2\} \cup A$
$ = \{-4, -3, -2, 0, 2, 4\}$

51. $(A \cup B) \cap C = \{-4, -3, -2, 0, 1, 2, 3, 4\} \cap C$
$ = \{-4, -3, -2, 0, 2\}$

53. $I_1 \cup I_2 = (-2, 5); \ \ I_1 \cap I_2 = [1, 3]$

55. $I_1 \cup I_2 = (-6, 10); \ \ I_1 \cap I_2 = \varnothing$

57. $I_1 \cup I_2 = (-\infty, \infty); \ \ I_1 \cap I_2 = [2, 5)$

59. $I_1 \cup I_2 = (-\infty, \infty); \ \ I_1 \cap I_2 = [-1, 3) \cup [5, 7]$

61. $|-4| = 4$ **63.** $\left| \dfrac{5}{-7} \right| = \dfrac{5}{7}$

65. $\left| 5 - \sqrt{2} \right| = 5 - \sqrt{2}$ **67.** $\left| 2 - \sqrt{3} \right| = 2 - \sqrt{3}$

69. $\dfrac{8}{|-8|} = \dfrac{8}{8} = 1$

71. $\big| 5 - |-7| \big| = |5 - 7| = |-2| = 2$

73.

$d(3, 8) = |3 - 8| = |-5| = 5$

75.

$d(-6, 9) = |-6 - 9| = |-15| = 15$

77.

$d(-20, -6) = |-20 - (-6)| = |-14| = 14$

79.

$d\left(\dfrac{22}{7}, -\dfrac{4}{7} \right) = \left| \dfrac{22}{7} - \left(-\dfrac{4}{7} \right) \right| = \left| \dfrac{26}{7} \right| = \dfrac{26}{7}$

81.

$-3 < x \leq 1$

83.

$x \geq -3$

85.

$x \leq 5$

87.

$-\dfrac{3}{4} < x < \dfrac{9}{4}$

89. $4(x + 1) = 4x + 4$

91. $5(x - y + 1) = 5x - 5y + 5$

93. Additive inverse: -5; reciprocal: $\dfrac{1}{5}$

95. Additive inverse: 0; no reciprocal

97. Additive inverse

99. Multiplicative identity

101. Associative property of multiplication

103. Multiplicative inverse

105. Additive identity

107. Associative property of addition

109. $\dfrac{3}{5} + \dfrac{4}{3} = \dfrac{9}{15} + \dfrac{20}{15} = \dfrac{29}{15}$

111. $\dfrac{6}{5} + \dfrac{5}{7} = \dfrac{42}{35} + \dfrac{25}{35} = \dfrac{67}{35}$

113. $\dfrac{5}{6} + \dfrac{3}{10} = \dfrac{25}{30} + \dfrac{9}{30} = \dfrac{34}{30} = \dfrac{17}{15}$

115. $\dfrac{5}{8} - \dfrac{9}{10} = \dfrac{25}{40} - \dfrac{36}{40} = -\dfrac{11}{40}$

117. $\dfrac{5}{9} - \dfrac{7}{11} = \dfrac{55}{99} - \dfrac{63}{99} = -\dfrac{8}{99}$

119. $\dfrac{2}{5}-\dfrac{1}{2}=\dfrac{4}{10}-\dfrac{5}{10}=-\dfrac{1}{10}$

121. $\dfrac{3}{4}\cdot\dfrac{8}{27}=\dfrac{2}{9}$

123. $\dfrac{\frac{8}{5}}{\frac{16}{15}}=\dfrac{8}{5}\div\dfrac{16}{15}=\dfrac{8}{5}\cdot\dfrac{15}{16}=\dfrac{3}{2}$

125. $\dfrac{\frac{7}{8}}{\frac{21}{16}}=\dfrac{7}{8}\div\dfrac{21}{16}=\dfrac{7}{8}\cdot\dfrac{16}{21}=\dfrac{2}{3}$

127. $5\cdot\dfrac{3}{10}-\dfrac{1}{2}=\dfrac{3}{2}-\dfrac{1}{2}=\dfrac{2}{2}=1$

129. $3\cdot\dfrac{2}{15}-\dfrac{1}{3}=\dfrac{2}{5}-\dfrac{1}{3}=\dfrac{6}{15}-\dfrac{5}{15}=\dfrac{1}{15}$

131. $2(x+y)-3y=2(3+(-5))-3(-5)$
$=2(-2)-(-15)=-4+15=11$

133. $3|x|-2|y|=3|3|-2|-5|=3(3)-2(5)$
$=9-10=-1$

135. $\dfrac{x-3y}{2}+xy=\dfrac{3-3(-5)}{2}+3(-5)$
$=\dfrac{3-(-15)}{2}+(-15)$
$=\dfrac{18}{2}+(-15)=9+(-15)=-6$

137. $\dfrac{2(1-2x)}{y}-(-x)y=\dfrac{2(1-2(3))}{-5}-(-3)(-5)$
$=\dfrac{2(-5)}{-5}-15=2-15=-13$

139. $\dfrac{\frac{14}{x}+\frac{1}{2}}{\frac{-y}{4}}=\dfrac{\frac{14}{3}+\frac{1}{2}}{\frac{-(-5)}{4}}=\dfrac{\frac{31}{6}}{\frac{4}{5}}\cdot=\dfrac{31}{6}\cdot\dfrac{4}{5}=\dfrac{62}{15}$

141. $\dfrac{x}{y}+\dfrac{x}{3}=\dfrac{3x+xy}{3y}$

143. $5(x+3)=5x+5(3)=5x+15$.

145. $x-(3y+2)=x-3y-2$

147. $\dfrac{x+y}{x}=\dfrac{x}{x}+\dfrac{y}{x}=1+\dfrac{y}{x}$

149. $(x+1)(y+1)=xy+x+y+1$

151. $d(P,M)=\left|\dfrac{a+b}{2}-a\right|=\left|\dfrac{a+b-2a}{2}\right|=\left|\dfrac{-a+b}{2}\right|$
$d(Q,M)=\left|b-\dfrac{a+b}{2}\right|=\left|\dfrac{2b-a+b}{2}\right|=\left|\dfrac{b-a}{2}\right|$
Since
$d(P,M)=\left|\dfrac{-a+b}{2}\right|=\left|\dfrac{b-a}{2}\right|=d(Q,M)$, M is
the midpoint of the line segment PQ.

P.1 Applying the Concepts

153. a. people who own either MP3 players or people who own DVD players.

b. people who own both MP3 players and DVD players.

155. $119.5\le x\le134.5$

157. a. $|124-120|=4$

b. $|137-120|=17$

c. $|114-120|=|-6|=6$

159. Let $x=$ the number of calories from broccoli. Then we have
$522.5-55x=0\Rightarrow522.5=55x\Rightarrow9.5=x$
The number of grams of broccoli is
$9.5\times100=950$ grams.

P.2 Integer Exponents and Scientific Notation

P.2 Practice Exercises

1. a. $2^3=8$

b. $(3a)^2=(3a)(3a)=9a^2$

c. $\left(\dfrac{1}{2}\right)^4=\dfrac{1}{2}\cdot\dfrac{1}{2}\cdot\dfrac{1}{2}\cdot\dfrac{1}{2}=\dfrac{1}{16}$

2. a. $2^{-1}=\dfrac{1}{2}$

b. $\left(\dfrac{4}{5}\right)^0=1$

c. $\left(\dfrac{3}{2}\right)^{-2}=\left(\dfrac{2}{3}\right)^2=\dfrac{4}{9}$

3. a. $x^2\cdot3x^7=3x^9$

b. $\left(2^2 x^3\right)\left(4x^{-3}\right) = 4x^3 \cdot 4x^{-3} = 16x^{3+(-3)}$
$$= 16x^0 = 16 \cdot 1 = 16$$

4. a. $\dfrac{3^4}{3^0} = 3^{4-0} = 3^4 = 81$

b. $\dfrac{5}{5^{-2}} = 5^{1-(-2)} = 5^3 = 125$

c. $\dfrac{2x^3}{3x^{-4}} = \dfrac{2x^{3-(-4)}}{3} = \dfrac{2x^7}{3}$

5. a. $\left(7^{-5}\right)^0 = 7^{(-5)(0)} = 7^0 = 1$

b. $\left(7^0\right)^{-5} = 7^{(0)(-5)} = 7^0 = 1$

c. $\left(x^{-1}\right)^8 = x^{(-1)(8)} = x^{-8} = \dfrac{1}{x^8}$

d. $\left(x^{-2}\right)^{-5} = x^{(-2)(-5)} = x^{10}$

6. a. $\left(\dfrac{1}{2}x\right)^{-1} = \left(\dfrac{1}{2}\right)^{-1}(x)^{-1} = 2\left(\dfrac{1}{x}\right) = \dfrac{2}{x}$

b. $\left(5x^{-1}\right)^2 = 5^2\left(x^{-1}\right)^2 = 5^2 x^{(-1)(2)}$
$$= 25x^{-2} = \dfrac{25}{x^2}$$

c. $\left(xy^2\right)^3 = x^3 y^{2(3)} = x^3 y^6$

d. $\left(x^{-2}y\right)^{-3} = x^{(-2)(-3)} y^{-3} = \dfrac{x^6}{y^3}$

7. a. $\left(\dfrac{1}{3}\right)^2 = \dfrac{1^2}{3^2} = \dfrac{1}{9}$

b. $\left(\dfrac{10}{7}\right)^{-2} = \left(\dfrac{7}{10}\right)^2 = \dfrac{7^2}{10^2} = \dfrac{49}{100}$

8. a. $\left(2x^4\right)^{-2} = \left(\dfrac{1}{2x^4}\right)^2 = \dfrac{1}{4x^8}$

b. $\dfrac{x^2\left(-y\right)^3}{\left(xy^2\right)^3} = \dfrac{-x^2 y^3}{x^3 y^6} = -\dfrac{1}{xy^3}$

9. $732,000 = 7.32 \times 10^5$

10. $\dfrac{2.9 \times 10^{10}}{3.25 \times 10^8} = \dfrac{290}{325} \times 10^2 \approx 0.89 \times 10^2 \approx \$89$
per person

P.2 Basic Concepts and Skills

1. In the expression 7^2, the number 2 is called the <u>exponent</u>.

3. The number $\dfrac{1}{4^{-2}}$ simplifies to be the positive integer <u>16</u>.

5. False. $(-11)^{10} = 11^{10}$

7. Exponent: 3, base: 17

9. Exponent: 0, base: 9

11. Exponent: 5, base: −5

13. Exponent: 3, base: 10

15. Exponent: 2, base: a

17. $6^1 = 6$

19. $7^0 = 1$

21. $\left(2^3\right)^2 = 2^{3 \cdot 2} = 2^6 = 64$

23. $\left(3^2\right)^{-2} = 3^{2(-2)} = 3^{-4} = \dfrac{1}{3^4} = \dfrac{1}{81}$

25. $(5^{-2})^3 = \left(\dfrac{1}{5^2}\right)^3 = \left(\dfrac{1}{25}\right)^3 = \dfrac{1}{15,625}$

27. $(4^{-3}) \cdot (4^5) = 4^{(-3+5)} = 4^2 = 16$

29. $3^0 + 10^0 = 1 + 1 = 2$

31. $3^{-2} + \left(\dfrac{1}{3}\right)^2 = \left(\dfrac{1}{3}\right)^2 + \left(\dfrac{1}{3}\right)^2 = \dfrac{1}{9} + \dfrac{1}{9} = \dfrac{2}{9}$

33. $-2^{-3} = -\dfrac{1}{2^3} = -\dfrac{1}{8}$

35. $(-3)^{-2} = \dfrac{1}{(-3)^2} = \dfrac{1}{9}$

37. $\dfrac{2^{11}}{2^{10}} = 2^{(11-10)} = 2^1 = 2$

39. $\dfrac{(5^3)^4}{5^{12}} = \dfrac{5^{12}}{5^{12}} = 1$

41. $\dfrac{2^5 \cdot 3^{-2}}{2^4 \cdot 3^{-3}} = 2^{(5-4)} \cdot 3^{(-2-(-3))} = 2^1 \cdot 3^1 = 6$

43. $\dfrac{-5^{-2}}{2^{-1}} = -\dfrac{1}{5^2} \cdot 2^1 = -\dfrac{1}{25} \cdot 2 = -\dfrac{2}{25}$

45. $\left(\dfrac{11}{7}\right)^{-2} = \dfrac{1}{\left(\dfrac{11}{7}\right)^2} = \dfrac{1}{\dfrac{121}{49}} = \dfrac{49}{121}$

47. $x^4 y^0 = x^4 \cdot 1 = x^4$

49. $x^{-1}y = \dfrac{1}{x} \cdot y = \dfrac{y}{x}$

51. $x^{-1}y^{-2} = \dfrac{1}{xy^2}$

53. $\left(x^{-3}\right)^4 = x^{-12} = \dfrac{1}{x^{12}}$

55. $\left(x^{-11}\right)^{-3} = x^{(-11)\cdot(-3)} = x^{33}$

57. $-3(xy)^5 = -3x^5 y^5$

59. $4\left(xy^{-1}\right)^2 = 4x^2 y^{-2} = \dfrac{4x^2}{y^2}$

61. $3\left(x^{-1}y\right)^{-5} = 3x^{(-1)\cdot(-5)}y^{-5} = \dfrac{3x^5}{y^5}$

63. $\dfrac{\left(x^3\right)^2}{\left(x^2\right)^5} = \dfrac{x^{3\cdot2}}{x^{2\cdot5}} = \dfrac{x^6}{x^{10}} = x^{6-10} = x^{-4} = \dfrac{1}{x^4}$

65. $\left(\dfrac{2xy}{x^2}\right)^3 = \dfrac{2^3 x^3 y^3}{x^{2\cdot3}} = \dfrac{8x^3 y^3}{x^6} = 8x^{3-6}y^3$
$= 8x^{-3}y^3 = \dfrac{8y^3}{x^3}$

67. $\left(\dfrac{-3x^2 y}{x}\right)^5 = \dfrac{(-3)^5 x^{2\cdot5}y^5}{x^5} = \dfrac{-243x^{10}y^5}{x^5}$
$= -243x^{10-5}y^5 = -243x^5 y^5$

69. $\left(\dfrac{-3x}{5}\right)^{-2} = \dfrac{(-3)^{-2}x^{-2}}{5^{-2}} = \dfrac{\dfrac{1}{9}\cdot\dfrac{1}{x^2}}{\dfrac{1}{25}} = \dfrac{\dfrac{1}{9x^2}}{\dfrac{1}{25}} = \dfrac{25}{9x^2}$

71. $\left(\dfrac{4x^{-2}}{xy^5}\right)^3 = \dfrac{4^3 x^{-2\cdot3}}{x^3 y^{5\cdot3}} = \dfrac{64x^{-6}}{x^3 y^{15}} = \dfrac{64}{x^9 y^{15}}$

73. $\dfrac{x^3 y^{-3}}{x^{-2}y} = x^{3-(-2)}y^{-3-1} = x^5 y^{-4} = \dfrac{x^5}{y^4}$

75. $\dfrac{27x^{-3}y^5}{9x^{-4}y^7} = 3x^{-3-(-4)}y^{5-7} = 3x^1 y^{-2} = \dfrac{3x}{y^2}$

77. $\dfrac{1}{x^3}\left(x^2\right)^3 x^{-4} = x^{-3}\left(x^{2\cdot3}\right)x^{-4} = x^{-3}\left(x^6\right)x^{-4}$
$= x^{-3+6-4} = x^{-1} = \dfrac{1}{x}$

79. $\left(-xy^2\right)^3\left(-2x^2 y^2\right)^{-4} = \dfrac{\left(-xy^2\right)^3}{\left(-2x^2 y^2\right)^4}$
$= \dfrac{(-x)^3\left(y^2\right)^3}{(-2)^4\left(x^2\right)^4\left(y^2\right)^4}$
$= \dfrac{-x^3 \cdot y^{2\cdot3}}{16x^{2\cdot4}y^{2\cdot4}} = \dfrac{-x^3 y^6}{16x^8 y^8}$
$= -\dfrac{1}{16}\left(x^{3-8}\right)\left(y^{6-8}\right)$
$= -\dfrac{1}{16}x^{-5}y^{-2} = -\dfrac{1}{16x^5 y^2}$

81. $\left(4x^3 y^2 z\right)^2\left(x^3 y^2 z\right)^{-7} = \dfrac{\left(4x^3 y^2 z\right)^2}{\left(x^3 y^2 z\right)^7}$
$= \dfrac{4^2\left(x^3\right)^2\left(y^2\right)^2 z^2}{\left(x^3\right)^7\left(y^2\right)^7 z^7}$
$= \dfrac{16x^{3\cdot2}y^{2\cdot2}z^2}{x^{3\cdot7}y^{2\cdot7}z^7}$
$= \dfrac{16x^6 y^4 z^2}{x^{21}y^{14}z^7}$
$= 16x^{6-21}y^{4-14}z^{2-7}$
$= 16x^{-15}y^{-10}z^{-5}$
$= \dfrac{16}{x^{15}y^{10}z^5}$

83. $\dfrac{5a^{-2}bc^2}{a^4 b^{-3}c^2} = 5a^{-2-4}b^{1-(-3)}c^{2-2}$
$= 5a^{-6}b^4 \cdot 1 = \dfrac{5b^4}{a^6}$

85. $\left(\dfrac{xy^{-3}z^{-2}}{x^2y^{-4}z^3}\right)^{-3} = \dfrac{x^{-3}y^{(-3)(-3)}z^{(-2)(-3)}}{x^{2(-3)}y^{(-4)(-3)}z^{3(-3)}}$

$\qquad = \dfrac{x^{-3}y^9z^6}{x^{-6}y^{12}z^{-9}}$

$\qquad = x^{-3-(-6)}y^{9-12}z^{6-(-9)}$

$\qquad = x^3y^{-3}z^{15} = \dfrac{x^3z^{15}}{y^3}$

87. $125 = 1.25 \times 10^2$

89. $850,000 = 8.5 \times 10^5$

91. $0.007 = 7.0 \times 10^{-3}$

93. $0.00000275 = 2.75 \times 10^{-6}$

P.2 Applying the Concepts

95. $135 \text{ ft}^3 \times 2^3 = 1080 \text{ ft}^3$

97. a. $(2x)^2 = 4x^2 = 4A = 2^2A$

    b. $(3x)^2 = 9x^2 = 9A = 3^2A$

99. $F = Sw^2 = 25,000(0.25)^2 = 1562.5 \text{ lb}$

101.

Celestial Body	Equatorial Diameter (km)	Scientific Notation
Earth	12,700	1.27×10^4 km
Moon	3480	3.48×10^3 km
Sun	1,390,000	1.39×10^6 km
Jupiter	134,000	1.34×10^5 km
Mercury	4800	4.8×10^3 km

103. $602,000,000,000,000,000,000 =$
$\qquad\qquad\qquad 6.02 \times 10^{20}$ atoms

105. $0.00000000000000000000167 =$
$\qquad\qquad\qquad 1.67 \times 10^{-21}$ kg

107. $5,980,000,000,000,000,000,000,000 =$
$\qquad\qquad\qquad 5.98 \times 10^{24}$ kg

P.3 Polynomials

P.3 Practice Problems

1. $16(7)^2 + 15(7) = 889 \text{ ft}$

2. $\left(7x^3 + 2x^2 - 5\right) + \left(-2x^3 + 3x^2 + 2x + 1\right)$
$\qquad = 5x^3 + 5x^2 + 2x - 4$

3. $\left(3x^4 - 5x^3 + 2x^2 + 7\right) - \left(-2x^4 + 3x^2 + x - 5\right)$
$\qquad\qquad = 5x^4 - 5x^3 - x^2 - x + 12$

4. $-2x^3\left(4x^2 + 2x - 5\right) = -8x^5 - 4x^4 + 10x^3$

5. $\left(5x^2 + 2x\right)\left(-2x^2 + x - 7\right)$
$\qquad = 5x^2\left(-2x^2 + x - 7\right) + 2x\left(-2x^2 + x - 7\right)$
$\qquad = -10x^4 + 5x^3 - 35x^2 - 4x^3 + 2x^2 - 14x$
$\qquad = -10x^4 + x^3 - 33x^2 - 14x$

6. a. $(4x - 1)(x + 7) = 4x^2 + 28x - x - 7$
$\qquad\qquad\qquad = 4x^2 + 27x - 7$

    b. $(3x - 2)(2x - 5) = 6x^2 - 15x - 4x + 10$
$\qquad\qquad\qquad = 6x^2 - 19x + 10$

7. $(3x + 2)^2 = (3x)^2 + 2(3x)(2) + 2^2$
$\qquad\qquad = 9x^2 + 12x + 4$

8. $(1 - 2x)(1 + 2x) = 1 - 4x^2$

9. $x^2(y + x) = x^2y + x^3$

P.3 Basic Concepts and Skills

1. The polynomial $-3x^7 + 2x^2 - 9x + 4$ has leading coefficient $\underline{-3}$ and degree $\underline{7}$.

3. When a polynomial in x of degree 3 is added to a polynomial in x of degree 4, the resulting polynomial has degree $\underline{4}$.

5. True

7. A polynomial; $x^2 + 2x + 1$

9. Not a polynomial

11. Not a polynomial

13. A polynomial; in standard form

15. Degree: 1; terms: $7x, 3$

17. Degree: 4; terms: $-x^4, x^2, 2x, -9$

19. $\left(x^3 + 2x^2 - 5x + 3\right) + \left(-x^3 + 2x - 4\right)$
$\qquad = \left(x^3 + \left(-x^3\right)\right) + 2x^2 + (-5x + 2x) + (3 - 4)$
$\qquad = 2x^2 - 3x - 1$

21. $\left(2x^3 - x^2 + x - 5\right) - \left(x^3 - 4x + 3\right)$
$\qquad = \left(2x^3 - x^3\right) - x^2 + (x - (-4x)) + (-5 - 3)$
$\qquad = x^3 - x^2 + 5x - 8$

23. $\left(-2x^4 + 3x^2 - 7x\right) - \left(8x^4 + 6x^3 - 9x^2 - 17\right)$
$= \left(-2x^4 - 8x^4\right) - 7x - 6x^3$
$\qquad + \left(3x^2 - \left(-9x^2\right)\right) - \left(-17\right)$
$= -10x^4 - 6x^3 + 12x^2 - 7x + 17$

25. $-2\left(3x^2 + x + 1\right) + 6\left(-3x^2 - 2x - 2\right)$
$= -6x^2 - 2x - 2 - 18x^2 - 12x - 12$
$= -24x^2 - 14x - 14$

27. $\left(3y^3 - 4y + 2\right) + \left(2y + 1\right) - \left(y^3 - y^2 + 4\right)$
$= 3y^3 - 4y + 2 + 2y + 1 - y^3 + y^2 - 4$
$= 2y^3 + y^2 - 2y - 1$

29. $6x(2x + 3) = 12x^2 + 18x$

31. $(x + 1)\left(x^2 + 2x + 2\right)$
$= x\left(x^2 + 2x + 2\right) + 1\left(x^2 + 2x + 2\right)$
$= x^3 + 2x^2 + 2x + x^2 + 2x + 2$
$= x^3 + 3x^2 + 4x + 2$

33. $(3x - 2)\left(x^2 - x + 1\right)$
$= 3x\left(x^2 - x + 1\right) - 2\left(x^2 - x + 1\right)$
$= 3x^3 - 3x^2 + 3x - 2x^2 + 2x - 2$
$= 3x^3 - 5x^2 + 5x - 2$

35. $(x + 1)(x + 2) = x(x + 2) + 1(x + 2)$
$\qquad = x^2 + 2x + x + 2 = x^2 + 3x + 2$

37. $(3x + 2)(3x + 1) = 9x^2 + 3x + 6x + 2$
$\qquad = 9x^2 + 9x + 2$

39. $(-4x + 5)(x + 3) = -4x^2 - 12x + 5x + 15$
$\qquad = -4x^2 - 7x + 15$

41. $(3x - 2)(2x - 1) = 6x^2 - 3x - 4x + 2$
$\qquad = 6x^2 - 7x + 2$

43. $(2x - 3a)(2x + 5a) = 4x^2 + 10ax - 6ax - 15a^2$
$\qquad = 4x^2 + 4ax - 15a^2$

45. $(x + 2)^2 - x^2 = x^2 + 4x + 4 - x^2 = 4x + 4$

47. $(x + 3)^3 - x^3 = x^3 + 9x^2 + 27x + 27 - x^3$
$\qquad = 9x^2 + 27x + 27$

49. $(4x + 1)^2 = 16x^2 + 8x + 1$

51. $(3x + 1)^3 = (3x)^3 + 3(3x)^2(1) + 3(3x)(1^2) + 1^3$
$\qquad = 27x^3 + 27x^2 + 9x + 1$

53. $(5 - 2x)(5 + 2x) = 25 - 4x^2 = -4x^2 + 25$

55. $\left(x + \dfrac{3}{4}\right)^2 = x^2 + 2\left(\dfrac{3}{4}\right)x + \left(\dfrac{3}{4}\right)^2$
$\qquad = x^2 + \dfrac{3}{2}x + \dfrac{9}{16}$

57. $(2x - 3)\left(x^2 - 3x + 5\right)$
$= 2x\left(x^2 - 3x + 5\right) - 3\left(x^2 - 3x + 5\right)$
$= 2x^3 - 6x^2 + 10x - 3x^2 + 9x - 15$
$= 2x^3 - 9x^2 + 19x - 15$

59. $(1 + y)\left(1 - y + y^2\right)$
$= 1\left(1 - y + y^2\right) + y\left(1 - y + y^2\right)$
$= 1 - y + y^2 + y - y^2 + y^3 = y^3 + 1$

61. $(x - 6)\left(x^2 + 6x + 36\right)$
$= x\left(x^2 + 6x + 36\right) - 6\left(x^2 + 6x + 36\right)$
$= x^3 + 6x^2 + 36x - 6x^2 - 36x - 216$
$= x^3 - 216$

63. $(x + 2y)(3x + 5y) = 3x^2 + 5xy + 6xy + 10y^2$
$\qquad\qquad = 3x^2 + 11xy + 10y^2$

65. $(2x - y)(3x + 7y) = 6x^2 + 14xy - 3xy - 7y^2$
$\qquad\qquad = 6x^2 + 11xy - 7y^2$

67. $(x - y)^2(x + y)^2 = (x - y)(x - y)(x + y)(x + y)$
$\qquad = \left[(x - y)(x + y)\right]\left[(x - y)(x + y)\right]$
$\qquad = \left(x^2 - y^2\right)\left(x^2 - y^2\right)$
$\qquad = x^4 - 2x^2y^2 + y^4$

69. $(x + y)(x - 2y)^2 = (x + y)\left(x^2 - 4xy + 4y^2\right)$
$= x\left(x^2 - 4xy + 4y^2\right) + y\left(x^2 - 4xy + 4y^2\right)$
$= x^3 - 4x^2y + 4xy^2 + x^2y - 4xy^2 + 4y^3$
$= x^3 - 3x^2y + 4y^3$

71. $(x-2y)^3(x+2y)$

$= \left(x^3 + 3x^2(-2y) + 3x(-2y)^2 + (-2y)^3\right)$
$\qquad\qquad\qquad\qquad \cdot (x+2y)$

$= \left(x^3 - 6x^2y + 12xy^2 - 8y^3\right)(x+2y)$

$= x\left(x^3 - 6x^2y + 12xy^2 - 8y^3\right)$
$\qquad + 2y\left(x^3 - 6x^2y + 12xy^2 - 8y^3\right)$

$= x^4 - 6x^3y + 12x^2y^2 - 8xy^3$
$\qquad + 2x^3y - 12x^2y^2 + 24xy^3 - 16y^4$

$= x^4 - 4x^3y + 16xy^3 - 16y^4$

For exercises 73–77, use
$a^2 + b^2 = (a+b)^2 - 2ab = (a-b)^2 + 2ab.$

73. $x + y = 4,\ xy = 3$

$x^2 + y^2 = (x+y)^2 - 2xy$

$\qquad = 4^2 - 2\cdot 3 = 16 - 6 = 10$

75. a. $x + \dfrac{1}{x} = 3$

Let $a = x$ and $b = \dfrac{1}{x}$. Then $ab = x\left(\dfrac{1}{x}\right) = 1.$

$x^2 + \dfrac{1}{x^2} = \left(x + \dfrac{1}{x}\right)^2 - 2x\left(\dfrac{1}{x}\right)$

$\qquad = 3^2 - 2 = 9 - 2 = 7$

b. Let $a = x^2$ and $b = \dfrac{1}{x^2}$. Using the result from part a, we have

$x^4 + \dfrac{1}{x^4} = \left(x^2 + \dfrac{1}{x^2}\right)^2 - 2x^2\left(\dfrac{1}{x^2}\right)$

$\qquad = 7^2 - 2 = 49 - 2 = 47$

77. Let $a = 3x$ and $b = 2y$. Then $ab = 6xy$.

$9x^2 + 4y^2 = (3x+2y)^2 - 2(6xy)$

$\qquad = 12^2 - 2\cdot 6\cdot 6 = 144 - 72 = 72$

P.3 Applying the Concepts

79. $-0.025\left(6^2\right) + 0.44(6) + 4.28 = \$6.02$ in 2012
(six years after 2006)

81. $0.1\left(40^2\right) + 40 + 50 = \$250.00$

83. $d = 16\left(5^2\right) + 20(5) = 16(25) + 100 = 500$ feet

85. a. $-x + 22.50$

b. $(30+10x)(22.50-x)$

$= 30(22.50) - 30x + 225x - 10x^2$

$= 675 + 195x - 10x^2$

$= -10x^2 + 195x + 675$

P.4 Factoring Polynomials

P.4 Practice Problems

1. a. $6x^5 + 14x^3 = 2x^3\left(3x^2 + 7\right)$

b. $7x^5 + 21x^4 + 35x^2 = 7x^2\left(x^3 + 3x^2 + 5\right)$

2. a $x^2 + 6x + 8 = (x+4)(x+2)$

b. $x^2 - 3x - 10 = (x-5)(x+2)$

3. a. $x^2 + 4x + 4 = (x+2)^2$

b. $9x^2 - 6x + 1 = (3x-1)^2$

4. a. $x^2 - 16 = (x-4)(x+4)$

b. $4x^2 - 25 = (2x-5)(2x+5)$

5. $x^4 - 81 = \left(x^2 - 9\right)\left(x^2 + 9\right)$

$\qquad = (x-3)(x+3)\left(x^2 + 9\right)$

6. a. $x^3 - 125 = (x-5)\left(x^2 + 5x + 25\right)$

b. $27x^3 + 8 = (3x+2)\left(9x^2 - 6x + 4\right)$

7. Following the reasoning in Example 7, in four years, the company will have invested the intial 12 million dollars, plus an additional 4 million dollars. Thus, the total investment is 16 million dollars. To find the profit or loss with 16 million dollars invested, let $x = 16$ in the profit-loss polynomial:

$(0.012)(x-14)\left(x^2 + 14x + 196\right)$

$= (0.012)(16-14)\left(16^2 + 14\cdot 16 + 196\right)$

$= 16.224$

The company will have made a profit of 16.224 million dollars in four years.

8. a. $x^3 + 3x^2 + x + 3 = x^2(x+3) + 1(x+3)$

$\qquad\qquad = (x+3)\left(x^2 + 1\right)$

b. $28x^3 - 20x^2 - 7x + 5$
$$= 4x^2(7x - 5) - (7x - 5)$$
$$= (7x - 5)(4x^2 - 1)$$
$$= (7x - 5)(2x + 1)(2x - 1)$$

c. $x^2 - y^2 - 2y - 1 = x^2 - (y^2 + 2y + 1)$
$$= x^2 - (y + 1)^2$$
$$= (x - (y + 1))(x + (y + 1))$$
$$= (x - y - 1)(x + y + 1)$$

9. $x^4 + 5x^2 + 9 = x^4 + 6x^2 - x^2 + 9$
$$= x^4 + 6x^2 + 9 - x^2$$
$$= (x^2 + 3)^2 - x^2$$
$$= \left[(x^2 + 3) + x\right]\left[(x^2 + 3) - x\right]$$
$$= (x^2 + x + 3)(x^2 - x + 3)$$

10. a. $5x^2 + 11x + 2 = 5x^2 + 10x + x + 2$
$$= (5x^2 + 10x) + (x + 2)$$
$$= 5x(x + 2) + 1(x + 2)$$
$$= (5x + 1)(x + 2)$$

b. $4x^2 + x - 3 = 4x^2 + 4x - 3x - 3$
$$= (4x^2 + 4x) - (3x + 3)$$
$$= 4x(x + 1) - 3(x + 1)$$
$$= (4x - 3)(x + 1)$$

P.4 Basic Concepts and Skills

1. The polynomials $x + 2$ and $x - 2$ are called factors of the polynomial $x^2 - 4$.

3. The GCF of the polynomial $10x^3 + 30x^2$ is $10x^2$.

5. True

7. $8x - 24 = 8(x - 3)$

9. $-6x^2 + 12x = -6x(x - 2)$

11. $7x^2 + 14x^3 = 7x^2(1 + 2x)$

13. $x^4 + 2x^3 + x^2 = x^2(x^2 + 2x + 1)$

15. $3x^3 - x^2 = x^2(3x - 1)$

17. $8ax^3 + 4ax^2 = 4ax^2(2x + 1)$

19. $x^3 + 3x^2 + x + 3 = x^2(x + 3) + 1(x + 3)$
$$= (x + 3)(x^2 + 1)$$

21. $x^3 - 5x^2 + x - 5 = x^2(x - 5) + 1(x - 5)$
$$= (x - 5)(x^2 + 1)$$

23. $6x^3 + 4x^2 + 3x + 2 = 2x^2(3x + 2) + 1(3x + 2)$
$$= (3x + 2)(2x^2 + 1)$$

25. $12x^7 + 4x^5 + 3x^4 + x^2$
$$= 4x^5(3x^2 + 1) + x^2(3x^2 + 1)$$
$$= (3x^2 + 1)(4x^5 + x^2)$$
$$= x^2(3x^2 + 1)(4x^3 + 1)$$

27. $x^2 + 7x + 12 = (x + 3)(x + 4)$

29. $x^2 - 6x + 8 = (x - 4)(x - 2)$

31. $x^2 - 3x - 4 = (x - 4)(x + 1)$

33. Irreducible

35. $2x^2 + x - 36 = (2x + 9)(x - 4)$

37. $6x^2 + 17x + 12 = (2x + 3)(3x + 4)$

39. $3x^2 - 11x - 4 = (3x + 1)(x - 4)$

41. Irreducible

43. $x^2 + 6x + 9 = (x + 3)^2$

45. $9x^2 + 6x + 1 = (3x + 1)^2$

47. $25x^2 - 20x + 4 = (5x - 2)^2$

49. $49x^2 + 42x + 9 = (7x + 3)^2$

51. $x^2 - 64 = (x - 8)(x + 8)$

53. $4x^2 - 1 = (2x - 1)(2x + 1)$

55. $16x^2 - 9 = (4x - 3)(4x + 3)$

57. $x^4 - 1 = (x^2 - 1)(x^2 + 1) = (x - 1)(x + 1)(x^2 + 1)$

59. $20x^4 - 5 = 5(4x^4 - 1) = 5(2x^2 - 1)(2x^2 + 1)$

61. $x^3 + 64 = x^3 + 4^3 = (x + 4)(x^2 - 4x + 16)$

63. $x^3 - 27 = x^3 - 3^3 = (x-3)(x^2 + 3x + 9)$

65. $8 - x^3 = 2^3 - x^3 = (2-x)(4 + 2x + x^2)$

67. $8x^3 - 27 = (2x)^3 - 3^3 = (2x-3)(4x^2 + 6x + 9)$

69. $40x^3 + 5 = 5(8x^3 + 1) = 5((2x)^3 + 1^3)$
$\qquad = 5(2x+1)(4x^2 - 2x + 1)$

71. $x^4 + x^2 + 25 = x^4 + 10x^2 - 9x^2 + 25$
$\qquad = (x^4 + 10x^2 + 25) - 9x^2$
$\qquad = (x^2 + 5)^2 - 9x^2$
$\qquad = [(x^2 + 5) + 3x][(x^2 + 5) - 3x]$
$\qquad = (x^2 + 3x + 5)(x^2 - 3x + 5)$

73. $x^4 + 15x^2 + 64 = x^4 + 16x^2 - x^2 + 64$
$\qquad = (x^4 + 16x^2 + 64) - x^2$
$\qquad = (x^2 + 8)^2 - x^2$
$\qquad = [(x^2 + 8) + x][(x^2 + 8) - x]$
$\qquad = (x^2 + x + 8)(x^2 - x + 8)$

75. $x^4 - x^2 + 16 = x^4 + 8x^2 - 9x^2 + 16$
$\qquad = (x^4 + 8x^2 + 16) - 9x^2$
$\qquad = (x^2 + 4)^2 - 9x^2$
$\qquad = [(x^2 + 4) + 3x][(x^2 + 4) - 3x]$
$\qquad = (x^2 + 3x + 4)(x^2 - 3x + 4)$

77. $x^4 + 4 = x^4 + 4x^2 - 4x^2 + 4$
$\qquad = (x^4 + 4x^2 + 4) - 4x^2$
$\qquad = (x^2 + 2)^2 - 4x^2$
$\qquad = [(x^2 + 2) + 2x][(x^2 + 2) - 2x]$
$\qquad = (x^2 + 2x + 2)(x^2 - 2x + 2)$

79. $1 - 16x^2 = (1 + 4x)(1 - 4x)$

81. $x^2 - 6x + 9 = (x-3)^2$

83. $4x^2 + 4x + 1 = (2x+1)^2$

85. $2x^2 - 8x - 10 = 2(x^2 - 4x - 5)$
$\qquad = 2(x-5)(x+1)$

87. $2x^2 + 3x - 20 = (2x-5)(x+4)$

89. $x^2 - 24x + 36$ is irreducible.

91. $3x^5 + 12x^4 + 12x^3 = 3x^3(x^2 + 4x + 4)$
$\qquad = 3x^3(x+2)^2$

93. $9x^2 - 1 = (3x-1)(3x+1)$

95. $16x^2 + 24x + 9 = (4x+3)^2$

97. $x^2 + 15$ is irreducible.

99. $45x^3 + 8x^2 - 4x = x(45x^2 + 8x - 4)$
$\qquad = x(5x+2)(9x-2)$

101. $ax^2 - 7a^2x - 8a^3 = a(x^2 - 7ax - 8a^2)$
$\qquad = a(x-8a)(x+a)$

103. $x^2 - 16a^2 + 8x + 16 = (x^2 + 8x + 16) - 16a^2$
$\qquad = (x+4)^2 - 16a^2$
$\qquad = (x+4-4a)(x+4+4a)$

105. $3x^5 + 12x^4y + 12x^3y^2 = 3x^3(x^2 + 4xy + 4y^2)$
$\qquad = 3x^3(x+2y)^2$

107. $x^2 - 25a^2 + 4x + 4 = (x^2 + 4x + 4) - 25a^2$
$\qquad = (x+2)^2 - 25a^2$
$\qquad = (x+2-5a)(x+2+5a)$

109. $18x^6 + 12x^5y + 2x^4y^2 = 2x^4(9x^2 + 6xy + y^2)$
$\qquad = 2x^4(3x+y)^2$

P.4 Applying the Concepts

111.

If one side of the garden is x feet, and the perimeter is 16 feet, then $\dfrac{16 - 2x}{2} = 8 - x$ gives the other dimension of the rectangle. So, the area of the garden is $x(8-x)$.

113.

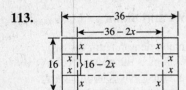

If x = the length of the cut corner, then
$36 - 2x$ = the length of the box, and
$16 - 2x$ = the width of the box. The height of the
box is x. So,
$$v = x(36 - 2x)(16 - 2x) = x(2(18 - x))(2(8 - x))$$
$$= 4x(18 - x)(8 - x)$$

115. The area of the outside circle = 4π cm^2. The
area of the inside circle = πx^2 cm^2. So the
area of the disk = area of outside circle − area
of inside circle = $4\pi - \pi x^2 = \pi(4 - x^2) =$
$\pi(2 - x)(2 + x)$ cm^2.

117. If one side of the fence is x feet, and the
rancher needs a total of 2800 feet of fencing,
then the width of the fence is $2800 - 2x$. So,
the area of the pen is
$$x(2800 - 2x) = 2800x - 2x^2 = 2x(1400 - x)\,\text{ft}^2.$$

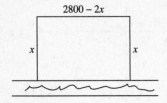

P.5 Rational Expressions

P.5 Practice Problems

1. $\dfrac{0.05(1 - 0.1)}{0.95(0.1) + (0.05)(1 - 0.1)} \approx 0.32$

The likelihood that a student who tests
positive is a nonuser when the test that is used
is 95% accurate is about 32%.

2. a. $\dfrac{2x^3 + 8x^2}{3x^2 + 12x} = \dfrac{2x^2(x + 4)}{3x(x + 4)} = \dfrac{2x}{3}$

   b. $\dfrac{x^2 - 4}{x^2 + 4x + 4} = \dfrac{(x - 2)(x + 2)}{(x + 2)^2} = \dfrac{x - 2}{x + 2}$

3. $\dfrac{\dfrac{x^2 - 2x - 3}{7x^3 + 28x^2}}{\dfrac{4x + 4}{2x^4 + 8x^3}} = \dfrac{x^2 - 2x - 3}{7x^3 + 28x^2} \div \dfrac{4x + 4}{2x^4 + 8x^3}$

$$= \dfrac{x^2 - 2x - 3}{7x^3 + 28x^2} \cdot \dfrac{2x^4 + 8x^3}{4x + 4}$$

$$= \dfrac{(x - 3)(x + 1)}{7x^2(x + 4)} \cdot \dfrac{2x^3(x + 4)}{4(x + 1)}$$

$$= \dfrac{x(x - 3)}{14}$$

4. a. $\dfrac{5x + 22}{x^2 - 36} + \dfrac{2(x + 10)}{x^2 - 36} = \dfrac{5x + 22 + 2x + 20}{x^2 - 36}$

$$= \dfrac{7x + 42}{x^2 - 36}$$

$$= \dfrac{7(x + 6)}{(x - 6)(x + 6)} = \dfrac{7}{x - 6}$$

   b. $\dfrac{4x + 1}{x^2 + x - 12} - \dfrac{3x + 4}{x^2 + x - 12} = \dfrac{(4x + 1) - (3x + 4)}{x^2 + x - 12}$

$$= \dfrac{x - 3}{x^2 + x - 12}$$

$$= \dfrac{x - 3}{(x + 4)(x - 3)}$$

$$= \dfrac{1}{x + 4}$$

5. a. $\dfrac{2x}{x + 2} + \dfrac{3x}{x - 5}$

$$= \dfrac{2x(x - 5)}{(x + 2)(x - 5)} + \dfrac{3x(x + 2)}{(x + 2)(x - 5)}$$

$$= \dfrac{2x^2 - 10x + 3x^2 + 6x}{(x + 2)(x - 5)} = \dfrac{5x^2 - 4x}{(x + 2)(x - 5)}$$

$$= \dfrac{x(5x - 4)}{(x + 2)(x - 5)}$$

   b. $\dfrac{5x}{x - 4} - \dfrac{2x}{x + 3}$

$$= \dfrac{5x(x + 3)}{(x - 4)(x + 3)} - \dfrac{2x(x - 4)}{(x - 4)(x + 3)}$$

$$= \dfrac{5x^2 + 15x - 2x^2 + 8x}{(x - 4)(x + 3)} = \dfrac{3x^2 + 23x}{(x - 4)(x + 3)}$$

$$= \dfrac{x(3x + 23)}{(x - 4)(x + 3)}$$

6. a. $\dfrac{x^2 + 3x}{x^2(x + 2)^2(x - 2)}, \dfrac{4x^2 + 1}{3x(x - 2)^2}$

LCD $= 3x^2(x + 2)^2(x - 2)^2$

b. $\dfrac{2x-1}{x^2-25},\ \dfrac{3-7x^2}{x^2+4x-5}$

$x^2-25=(x-5)(x+5)$

$x^2+4x-5=(x-1)(x+5)$

$LCD=(x-1)(x-5)(x+5)$

7. a. $\dfrac{4}{x^2-4x+4}+\dfrac{x}{x^2-4}$

$=\dfrac{4}{(x-2)^2}+\dfrac{x}{(x-2)(x+2)}$

$=\dfrac{4(x+2)}{(x-2)^2(x+2)}+\dfrac{x(x-2)}{(x-2)^2(x+2)}$

$=\dfrac{(4x+8)+\left(x^2-2x\right)}{(x-2)^2(x+2)}=\dfrac{x^2+2x+8}{(x-2)^2(x+2)}$

b. $\dfrac{2x}{3(x-5)^2}-\dfrac{6x}{2\left(x^2-5x\right)}$

$=\dfrac{2x}{3(x-5)^2}-\dfrac{6x}{2x(x-5)}$

$=\dfrac{2x(2x)}{3\cdot 2x(x-5)^2}-\dfrac{6x(3)(x-5)}{3\cdot 2x(x-5)^2}$

$=\dfrac{4x^2-18x^2+90x}{6x(x-5)^2}=\dfrac{-14x^2+90x}{6x(x-5)^2}$

$=\dfrac{2x(-7x+45)}{6x(x-5)^2}=\dfrac{-7x+45}{3(x-5)^2}$

8. $\dfrac{\dfrac{5}{3x}+\dfrac{1}{3}}{\dfrac{x^2-25}{3x}}=\dfrac{\left(\dfrac{5}{3x}+\dfrac{1}{3}\right)\cdot 3x}{\left(\dfrac{x^2-25}{3x}\right)\cdot 3x}=\dfrac{5+x}{x^2-25}$

$=\dfrac{x+5}{(x-5)(x+5)}=\dfrac{1}{x-5}$

9. $\dfrac{5x}{3x-\dfrac{4}{2-\dfrac{1}{3}}}=\dfrac{5x}{3x-\dfrac{4}{\dfrac{5}{3}}}=\dfrac{5x}{3x-4\left(\dfrac{3}{5}\right)}=\dfrac{5x}{3x-\dfrac{12}{5}}$

$=\dfrac{5x\cdot 5}{\left(3x-\dfrac{12}{5}\right)\cdot 5}=\dfrac{25x}{15x-12}$

P.5 Basic Concepts and Skills

1. The least common denominator for two rational expressions is the polynomial of least degree that contains <u>each denominator</u> as a factor.

3. If the denominators for two rational expressions are x^2-2x and x^2-x-2, then the LCD is $\underline{x(x-2)(x+1)}$.

5. False

7. $\dfrac{2x+2}{x^2+2x+1}=\dfrac{2(x+1)}{(x+1)^2}=\dfrac{2}{x+1},x\neq -1$

9. $\dfrac{3x+3}{x^2-1}=\dfrac{3(x+1)}{(x-1)(x+1)}=\dfrac{3}{x-1},x\neq -1,x\neq 1$

11. $\dfrac{2x-6}{9-x^2}=\dfrac{2(x-3)}{(3-x)(3+x)}=\dfrac{-2(3-x)}{(3-x)(3+x)}$

$=-\dfrac{2}{3+x},x\neq -3,x\neq 3$

13. $\dfrac{2x-1}{1-2x}=\dfrac{-1(1-2x)}{1-2x}=-1,x\neq \dfrac{1}{2}$

15. $\dfrac{x^2-6x+9}{4x-12}=\dfrac{(x-3)^2}{4(x-3)}=\dfrac{x-3}{4},x\neq 3$

17. $\dfrac{7x^2+7x}{x^2+2x+1}=\dfrac{7x(x+1)}{(x+1)^2}=\dfrac{7x}{x+1},x\neq -1$

19. $\dfrac{x^2-11x+10}{x^2+6x-7}=\dfrac{(x-10)(x-1)}{(x+7)(x-1)}$

$=\dfrac{x-10}{x+7},x\neq -7,x\neq 1$

21. $\dfrac{6x^4+14x^3+4x^2}{6x^4-10x^3-4x^2}=\dfrac{2x^2\left(3x^2+7x+2\right)}{2x^2\left(3x^2-5x-2\right)}$

$=\dfrac{2x^2(3x+1)(x+2)}{2x^2(3x+1)(x-2)}$

$=\dfrac{x+2}{x-2},x\neq -\dfrac{1}{3},x\neq 0,x\neq 2$

23. $\dfrac{x-3}{2x+4}\cdot\dfrac{10x+20}{5x-15}=\dfrac{x-3}{2(x+2)}\cdot\dfrac{10(x+2)}{5(x-3)}=1$

25. $\dfrac{2x+6}{4x-8}\cdot\dfrac{x^2+x-6}{x^2-9}=\dfrac{2(x+3)}{4(x-2)}\cdot\dfrac{(x+3)(x-2)}{(x+3)(x-3)}$

$=\dfrac{x+3}{2(x-3)}$

27. $\dfrac{x^2-7x}{x^2-6x-7} \cdot \dfrac{x^2-1}{x^2} = \dfrac{x(x-7)}{(x+1)(x-7)} \cdot \dfrac{(x+1)(x-1)}{x^2} = \dfrac{x-1}{x}$

29. $\dfrac{x^2-x-6}{x^2+3x+2} \cdot \dfrac{x^2-1}{x^2-9} = \dfrac{(x-3)(x+2)}{(x+1)(x+2)} \cdot \dfrac{(x-1)(x+1)}{(x-3)(x+3)} = \dfrac{x-1}{x+3}$

31. $\dfrac{2-x}{x+1} \cdot \dfrac{x^2+3x+2}{x^2-4} = \dfrac{-1(x-2)}{x+1} \cdot \dfrac{(x+1)(x+2)}{(x-2)(x+2)} = -1$

33. $\dfrac{x+2}{6} \div \dfrac{4x+8}{9} = \dfrac{x+2}{6} \cdot \dfrac{9}{4(x+2)} = \dfrac{3}{8}$

35. $\dfrac{x^2-9}{x} \div \dfrac{2x+6}{5x^2} = \dfrac{(x-3)(x+3)}{x} \cdot \dfrac{5x^2}{2(x+3)} = \dfrac{5x(x-3)}{2}$

37. $\dfrac{x^2+2x-3}{x^2+8x+16} \div \dfrac{x-1}{3x+12} = \dfrac{(x-1)(x+3)}{(x+4)^2} \cdot \dfrac{3(x+4)}{x-1} = \dfrac{3(x+3)}{x+4}$

39. $\left(\dfrac{x^2-9}{x^3+8} \div \dfrac{x+3}{x^3+2x^2-x-2} \right) \left(\dfrac{1}{x^2-1} \right) = \dfrac{(x-3)(x+3)}{(x+2)(x^2-2x+4)} \cdot \dfrac{(x^2-1)(x+2)}{x+3} \cdot \dfrac{1}{x^2-1} = \dfrac{x-3}{x^2-2x+4}$

41. $\dfrac{x}{5} + \dfrac{3}{5} = \dfrac{x+3}{5}$

43. $\dfrac{x}{2x+1} + \dfrac{4}{2x+1} = \dfrac{x+4}{2x+1}$

45. $\dfrac{x^2}{x+1} - \dfrac{x^2-1}{x+1} = \dfrac{x^2-(x^2-1)}{x+1} = \dfrac{1}{x+1}$

47. $\dfrac{4}{3-x} + \dfrac{2x}{x-3} = \dfrac{4}{-1(x-3)} + \dfrac{2x}{x-3} = \dfrac{-4+2x}{x-3} = \dfrac{2(x-2)}{x-3}$

49. $\dfrac{5x}{x^2+1} + \dfrac{2x}{x^2+1} = \dfrac{7x}{x^2+1}$

51. $\dfrac{7x}{2(x-3)} + \dfrac{x}{2(x-3)} = \dfrac{8x}{2(x-3)} = \dfrac{4x}{x-3}$

53. $\dfrac{x}{x^2-4} - \dfrac{2}{x^2-4} = \dfrac{x-2}{(x-2)(x+2)} = \dfrac{1}{x+2}$

55. $\dfrac{x-2}{2x+1} - \dfrac{x}{2x-1} = \dfrac{(x-2)(2x-1)}{(2x+1)(2x-1)} - \dfrac{x(2x+1)}{(2x+1)(2x-1)} = \dfrac{(2x^2-5x+2)-(2x^2+x)}{(2x+1)(2x-1)}$

$\qquad\qquad = \dfrac{-6x+2}{(2x+1)(2x-1)} = \dfrac{2(1-3x)}{(2x+1)(2x-1)}$

57.
$$\frac{-x}{x+2}+\frac{x-2}{x}-\frac{x}{x-2}=\frac{-x^2(x-2)}{x(x+2)(x-2)}+\frac{(x-2)^2(x+2)}{x(x+2)(x-2)}-\frac{x^2(x+2)}{x(x+2)(x-2)}$$

$$=\frac{(-x^3+2x^2)}{x(x+2)(x-2)}+\frac{(x^3-2x^2-4x+8)}{x(x+2)(x-2)}+\frac{-(x^3+2x^2)}{x(x+2)(x-2)}$$

$$=\frac{-x^3-2x^2-4x+8}{x(x+2)(x-2)}=-\frac{x^3+2x^2+4x-8}{x(x+2)(x-2)}$$

In exercises 59–65, to find the LCD, first factor each denominator and then multiply each prime factor the greatest number of times it appears as a factor.

59. $3x-6=3(x-2)$ and $4x-8=4(x-2) \Rightarrow \text{LCD}=3\cdot4(x-2)=12(x-2)$

61. $4x^2-1=(2x+1)(2x-1)$ and $(2x+1)^2=(2x+1)(2x+1) \Rightarrow \text{LCD}=(2x-1)(2x+1)^2$

63. $x^2+3x+2=(x+1)(x+2)$ and $x^2-1=(x-1)(x+1) \Rightarrow \text{LCD}=(x-1)(x+1)(x+2)$

65. $x^2-5x+4=(x-1)(x-4)$ and $x^2+x-2=(x-1)(x+2) \Rightarrow \text{LCD}=(x-1)(x-4)(x+2)$

67.
$$\frac{5}{x-3}+\frac{2x}{x^2-9}=\frac{5}{x-3}+\frac{2x}{(x-3)(x+3)}=\frac{5(x+3)}{(x-3)(x+3)}+\frac{2x}{(x-3)(x+3)}=\frac{5x+15+2x}{(x-3)(x+3)}=\frac{7x+15}{(x-3)(x+3)}$$

69.
$$\frac{2x}{x^2-4}-\frac{x}{x+2}=\frac{2x}{(x-2)(x+2)}-\frac{x}{x+2}=\frac{2x}{(x-2)(x+2)}-\frac{x(x-2)}{(x+2)(x-2)}$$

$$=\frac{2x-(x^2-2x)}{(x-2)(x+2)}=\frac{-x^2+4x}{(x-2)(x+2)}=\frac{x(4-x)}{(x-2)(x+2)}$$

71.
$$\frac{x-2}{x^2+3x-10}+\frac{x+3}{x^2+x-6}=\frac{x-2}{(x-2)(x+5)}+\frac{x+3}{(x-2)(x+3)}=\frac{1}{x+5}+\frac{1}{x-2}$$

$$=\frac{x-2}{(x+5)(x-2)}+\frac{x+5}{(x+5)(x-2)}=\frac{2x+3}{(x-2)(x+5)}$$

73.
$$\frac{2x-3}{9x^2-1}+\frac{4x-1}{(3x-1)^2}=\frac{2x-3}{(3x-1)(3x+1)}+\frac{4x-1}{(3x-1)^2}=\frac{(2x-3)(3x-1)}{(3x+1)(3x-1)^2}+\frac{(4x-1)(3x+1)}{(3x+1)(3x-1)^2}$$

$$=\frac{(6x^2-11x+3)+(12x^2+x-1)}{(3x+1)(3x-1)^2}=\frac{18x^2-10x+2}{(3x+1)(3x-1)^2}=\frac{2(9x^2-5x+1)}{(3x+1)(3x-1)^2}$$

75.
$$\frac{x-3}{x^2-25}-\frac{x-3}{x^2+9x+20}=\frac{x-3}{(x-5)(x+5)}-\frac{x-3}{(x+4)(x+5)}=\frac{(x-3)(x+4)}{(x-5)(x+5)(x+4)}-\frac{(x-3)(x-5)}{(x+4)(x+5)(x-5)}$$

$$=\frac{(x^2+x-12)-(x^2-8x+15)}{(x-5)(x+5)(x+4)}=\frac{9x-27}{(x-5)(x+5)(x+4)}=\frac{9(x-3)}{(x-5)(x+5)(x+4)}$$

77.
$$\frac{3}{x^2-4}+\frac{1}{2-x}-\frac{1}{2+x}=-\frac{3}{4-x^2}+\frac{1}{2-x}-\frac{1}{2+x}=-\frac{3}{(2-x)(2+x)}+\frac{1}{2-x}-\frac{1}{2+x}$$

$$=\frac{-3}{(2-x)(2+x)}+\frac{1(2+x)}{(2-x)(2+x)}-\frac{1(2-x)}{(2+x)(2-x)}=\frac{-3+(2+x)-(2-x)}{(2-x)(2+x)}=\frac{2x-3}{(2-x)(2+x)}$$

79.
$$\frac{x+3a}{x-5a}-\frac{x+5a}{x-3a}=\frac{(x+3a)(x-3a)}{(x-5a)(x-3a)}-\frac{(x+5a)(x-5a)}{(x-3a)(x-5a)}=\frac{(x^2-9a^2)-(x^2-25a^2)}{(x-5a)(x-3a)}=\frac{16a^2}{(x-5a)(x-3a)}$$

81.
$$\frac{1}{x+h}-\frac{1}{x}=\frac{x}{x(x+h)}-\frac{1(x+h)}{x(x+h)}=\frac{x-(x+h)}{x(x+h)}=-\frac{h}{x(x+h)}$$

83. $\dfrac{\dfrac{2}{x}}{\dfrac{3}{x^2}} = \dfrac{2}{x} \cdot \dfrac{x^2}{3} = \dfrac{2x}{3}$

85. $\dfrac{\dfrac{1}{x}}{1-\dfrac{1}{x}} = \dfrac{1}{x} \div \left(1-\dfrac{1}{x}\right) = \dfrac{1}{x} \div \left(\dfrac{x-1}{x}\right)$

$\qquad = \dfrac{1}{x} \cdot \dfrac{x}{x-1} = \dfrac{1}{x-1}$

We use Method 2 for exercises 87– 99.

87. $\dfrac{\dfrac{1}{x}-1}{\dfrac{1}{x}+1} = \dfrac{\left(\dfrac{1}{x}-1\right)x}{\left(\dfrac{1}{x}+1\right)x} = \dfrac{1-x}{1+x}$

89. $\dfrac{\dfrac{1}{x}-x}{1-\dfrac{1}{x^2}} = \dfrac{\left(\dfrac{1}{x}-x\right)x^2}{\left(1-\dfrac{1}{x^2}\right)x^2} = \dfrac{x-x^3}{x^2-1} = \dfrac{x\left(1-x^2\right)}{x^2-1}$

$\qquad = -x,\ x \ne 0$

91. $x - \dfrac{x}{x+\dfrac{1}{2}} = x - \dfrac{x}{\dfrac{2x+1}{2}} = x - \left(x \cdot \dfrac{2}{2x+1}\right)$

$\qquad = x - \dfrac{2x}{2x+1} = \dfrac{x(2x+1)}{2x+1} - \dfrac{2x}{2x+1}$

$\qquad = \dfrac{2x^2+x-2x}{2x+1} = \dfrac{2x^2-x}{2x+1}$

$\qquad = \dfrac{x(2x-1)}{2x+1}$

93. $\dfrac{\dfrac{1}{x+h}-\dfrac{1}{x}}{h} = \left(\dfrac{1}{x+h}-\dfrac{1}{x}\right) \div h$

$\qquad = \left(\dfrac{x}{x(x+h)}-\dfrac{x+h}{x(x+h)}\right) \div h$

$\qquad = \dfrac{x-(x+h)}{x(x+h)} \div h = \dfrac{-h}{x(x+h)} \div h$

$\qquad = \dfrac{-h}{x(x+h)} \cdot \dfrac{1}{h} = -\dfrac{1}{x(x+h)}$

95. $\dfrac{\dfrac{1}{x-a}+\dfrac{1}{x+a}}{\dfrac{1}{x-a}-\dfrac{1}{x+a}}$

$\qquad = \left(\dfrac{1}{x-a}+\dfrac{1}{x+a}\right) \div \left(\dfrac{1}{x-a}-\dfrac{1}{x+a}\right)$

$\qquad = \dfrac{(x+a)+(x-a)}{(x-a)(x+a)} \div \dfrac{(x+a)-(x-a)}{(x-a)(x+a)}$

$\qquad = \dfrac{2x}{(x-a)(x+a)} \cdot \dfrac{(x-a)(x+a)}{2a} = \dfrac{x}{a}$

P.5 Applying the Concepts

97. $d = 4 \Rightarrow r = 2$. Substituting 2 for r in the formula gives $\dfrac{125.6}{(3.14)(2^2)} = \dfrac{125.6}{12.56} = 10$ cm.

99. a. Originally the citrus extract is 3 out of 100 gallons or $\dfrac{3}{100}$. When x gallons of water are added to the mixture, the total number of gallons in the mixture is $100 + x$. There are still 3 gallons of citrus extract, so the fraction is $\dfrac{3}{100+x}$.

b. If $x = 50$, then $\dfrac{3}{100+x} = \dfrac{3}{150} = 0.02 = 2\%$

101. a. The volume of a cylinder is given by $V = 120 = \pi r^2 h \Rightarrow h = \dfrac{120}{\pi r^2}$. The cost of each base $= 0.05\pi x^2$; since there are two bases, the total cost of the bases is $0.10\pi x^2$.
The cost of the side is given by $0.01(2\pi rh) = 0.02\pi rh$. Since $r = x$ and $h = \dfrac{120}{\pi r^2}$, this gives $0.02\pi x\left(\dfrac{120}{\pi x^2}\right) = \dfrac{2.4}{x}$.
So the total cost for the container is $0.1\pi x^2 + \dfrac{2.4}{x} = \dfrac{0.1\pi x^3 + 2.4}{x}$.

b. $\dfrac{0.1(3.14)\left(2^3+240\right)}{2} \approx \$2.46$

P.6 Rational Exponents and Radicals

P.6 Practice Problems

1. a. $\sqrt{144} = \sqrt{12^2} = 12$

b. $\sqrt{\dfrac{1}{49}} = \sqrt{\left(\dfrac{1}{7}\right)^2} = \dfrac{1}{7}$

c. $\sqrt{\dfrac{4}{64}} = \sqrt{\dfrac{2^2}{8^2}} = \sqrt{\left(\dfrac{2}{8}\right)^2} = \dfrac{2}{8} = \dfrac{1}{4}$

2. a. $\sqrt{20} = \sqrt{4 \cdot 5} = \sqrt{4}\sqrt{5} = 2\sqrt{5}$

b. $\sqrt{6}\sqrt{8} = \sqrt{6 \cdot 8} = \sqrt{48} = \sqrt{16 \cdot 3}$
$= \sqrt{16}\sqrt{3} = 4\sqrt{3}$

c. $\sqrt{12x^2} = \sqrt{4 \cdot 3 \cdot x^2} = \sqrt{4}\sqrt{3}\sqrt{x^2} = 2|x|\sqrt{3}$

d. $\sqrt{\dfrac{20y^3}{27x^2}} = \dfrac{\sqrt{20y^3}}{\sqrt{27x^2}} = \dfrac{\sqrt{4 \cdot 5 \cdot y^2 \cdot y}}{\sqrt{9 \cdot 3 \cdot x^2}}$
$= \dfrac{\sqrt{4}\sqrt{5}\sqrt{y^2}\sqrt{y}}{\sqrt{9}\sqrt{3}\sqrt{x^2}} = \dfrac{2y\sqrt{5y}}{3|x|\sqrt{3}}$

3. Using Bernoulli's equation, the velocity of the water is
$$v = \sqrt{2gh} \approx \sqrt{2\left(10\,\text{m/s}^2\right)\left(45\,\text{m}\right)}$$
$$= \sqrt{900\ \text{m}^2/\text{s}^2} = 30\,\text{m/s}$$
$30\,\text{m/s} \times 60\,\text{s/min} \times 60\,\text{min/hr} = 108{,}000\,\text{m/hr}$
$108\,\text{km/hr} \times 0.6214\,\text{mi/km} \approx 67\,\text{mi/hr}$

4. a. $\sqrt[3]{-8} = -2$ **b.** $\sqrt[5]{32} = 2$

c. $\sqrt[4]{81} = 3$

d. $\sqrt[6]{-4}$ is not a real number

5. a. $\sqrt[3]{72} = \sqrt[3]{8 \cdot 9} = \sqrt[3]{8} \cdot \sqrt[3]{9} = 2\sqrt[3]{9}$

b. $\sqrt[4]{48a^2} = \sqrt[4]{16 \cdot 3a^2} = 2\sqrt[4]{3a^2}$

6. a. $3\sqrt{12} + 7\sqrt{3} = 3\sqrt{4 \cdot 3} + 7\sqrt{3}$
$= 6\sqrt{3} + 7\sqrt{3} = 13\sqrt{3}$

b. $2\sqrt[3]{135x} - 3\sqrt[3]{40x} = 2\sqrt[3]{27 \cdot 5x} - 3\sqrt[3]{8 \cdot 5x}$
$= 2 \cdot 3\sqrt[3]{5x} - 3 \cdot 2\sqrt[3]{5x}$
$= 6\sqrt[3]{5x} - 6\sqrt[3]{5x} = 0$

7. $\sqrt{3}\sqrt[5]{2} = \sqrt[2]{3}\sqrt[5]{2} = \sqrt[2 \cdot 5]{3^5} \cdot \sqrt[5 \cdot 2]{2^2} = \sqrt[10]{3^5}\sqrt[10]{2^2}$
$= \sqrt[10]{2^2 \cdot 3^5} = \sqrt[10]{972}$

8. a. $\dfrac{7}{\sqrt{8}} = \dfrac{7}{2\sqrt{2}} = \dfrac{7\sqrt{2}}{2\sqrt{2}\sqrt{2}} = \dfrac{7\sqrt{2}}{2 \cdot 2} = \dfrac{7\sqrt{2}}{4}$

b. $\dfrac{\sqrt[3]{3}}{\sqrt[3]{4}} = \dfrac{\sqrt[3]{3} \cdot \sqrt[3]{2}}{\sqrt[3]{4} \cdot \sqrt[3]{2}} = \dfrac{\sqrt[3]{6}}{\sqrt[3]{8}} = \dfrac{\sqrt[3]{6}}{2}$

9. a. $\dfrac{2}{\sqrt{7} - \sqrt{5}} = \dfrac{2\left(\sqrt{7} + \sqrt{5}\right)}{\left(\sqrt{7} - \sqrt{5}\right)\left(\sqrt{7} + \sqrt{5}\right)}$
$= \dfrac{2\left(\sqrt{7} + \sqrt{5}\right)}{7 - 5} = \dfrac{2\left(\sqrt{7} + \sqrt{5}\right)}{2}$
$= \sqrt{7} + \sqrt{5}$

b. $\dfrac{x - 2}{\sqrt{x} + \sqrt{2}} = \dfrac{(x - 2)\left(\sqrt{x} - \sqrt{2}\right)}{\left(\sqrt{x} + \sqrt{2}\right)\left(\sqrt{x} - \sqrt{2}\right)}$
$= \dfrac{(x - 2)\left(\sqrt{x} - \sqrt{2}\right)}{x - 2}$
$= \sqrt{x} - \sqrt{2}$

10. a. $\left(\dfrac{1}{4}\right)^{1/2} = \sqrt{\dfrac{1}{4}} = \dfrac{1}{2}$

b. $(125)^{1/3} = \sqrt[3]{125} = 5$

c. $(-32)^{1/5} = \sqrt[5]{-32} = -2$

11. a. $(25)^{2/3} = \sqrt[3]{25^2} = \sqrt[3]{5^4} = 5\sqrt[3]{5}$

b. $-36^{3/2} = -\left(\sqrt{36}\right)^3 = -(6)^3 = -216$

c. $16^{-5/2} = \dfrac{1}{\left(\sqrt{16}\right)^5} = \dfrac{1}{4^5} = \dfrac{1}{1024}$

d. $(-36)^{-1/2} = \dfrac{1}{\sqrt{-36}}$
Not a real number

12. a. $4x^{1/2} \cdot 3x^{1/5} = 12x^{1/2 + 1/5} = 12x^{7/10}$

b. $\dfrac{25x^{-1/4}}{5x^{1/3}} = 5x^{-1/4 - 1/3} = 5x^{-7/12} = \dfrac{5}{x^{7/12}}$

c. $\left(x^{2/3}\right)^{-1/5} = x^{-2/15} = \dfrac{1}{x^{2/15}}$

13. a. $\sqrt[6]{x^4} = x^{4/6} = x^{2/3} = \sqrt[3]{x^2}$

b. $\sqrt[4]{25}\sqrt{5} = \sqrt[4]{5^2}\sqrt{5} = 5^{2/4} \cdot 5^{1/2} = 5^{1/2} \cdot 5^{1/2} = 5$

c. $\sqrt{\sqrt[3]{x^{12}}} = \sqrt{x^{12/3}} = \sqrt{x^4} = \left(x^4\right)^{1/2} = x^2$

14. $x(x+3)^{-1/2} + (x+3)^{1/2}$

$= \dfrac{x}{(x+3)^{1/2}} + (x+3)^{1/2}$

$= \dfrac{x}{(x+3)^{1/2}} + \dfrac{(x+3)^{1/2}(x+3)^{1/2}}{(x+3)^{1/2}}$

$= \dfrac{x+(x+3)}{(x+3)^{1/2}} = \dfrac{2x+3}{(x+3)^{1/2}}$

$= \dfrac{(2x+3)(x+3)^{1/2}}{(x+3)^{1/2}(x+3)^{1/2}} = \dfrac{(2x+3)(x+3)^{1/2}}{x+3}$

P.6 Basic Concepts and Skills

1. Any positive number has <u>two</u> square roots.

3. Radicals that have the same index and the same radicand are called <u>like radicals</u>.

5. False. For all real x, $\sqrt{x^2} = |x|$.

7. $\sqrt{64} = \sqrt{8^2} = 8$ **9.** $\sqrt[3]{64} = \sqrt[3]{4^3} = 4$

11. $\sqrt[3]{-27} = \sqrt[3]{(-3)^3} = -3$

13. $\sqrt[3]{-\dfrac{1}{8}} = \sqrt[3]{\left(-\dfrac{1}{2}\right)^3} = -\dfrac{1}{2}$

15. $\sqrt{(-3)^2} = \sqrt{9} = 3$

17. There is no real number x such that $x^4 = -16$.

19. $-\sqrt[5]{-1} = -\sqrt[5]{(-1)^5} = -(-1) = 1$

21. $\sqrt[5]{(-7)^5} = -7$

23. $\sqrt{32} = \sqrt{16 \cdot 2} = \sqrt{16}\sqrt{2} = 4\sqrt{2}$

25. $\sqrt{18x^2} = \sqrt{9 \cdot 2x^2} = \sqrt{9}\sqrt{2}\sqrt{x^2} = 3x\sqrt{2}$

27. $\sqrt{9x^3} = \sqrt{9 \cdot x^2 \cdot x} = \sqrt{9}\sqrt{x^2}\sqrt{x} = 3x\sqrt{x}$

29. $\sqrt{6x}\sqrt{3x} = \sqrt{18x^2} = \sqrt{9 \cdot 2 \cdot x^2}$
$= \sqrt{9}\sqrt{2}\sqrt{x^2} = 3x\sqrt{2}$

31. $\sqrt{15x}\sqrt{3x^2} = \sqrt{45x^3} = \sqrt{9 \cdot 5 \cdot x^2 \cdot x}$
$= \sqrt{9}\sqrt{5}\sqrt{x^2}\sqrt{x} = 3x\sqrt{5x}$

33. $\sqrt[3]{-8x^3} = \sqrt[3]{-8}\sqrt[3]{x^3} = -2x$

35. $\sqrt[3]{-x^6} = -x^2$

37. $\sqrt{\dfrac{5}{32}} = \dfrac{\sqrt{5}}{\sqrt{32}} = \dfrac{\sqrt{5}}{\sqrt{16 \cdot 2}} = \dfrac{\sqrt{5}}{4\sqrt{2}}$
$= \dfrac{\sqrt{5}\sqrt{2}}{4\sqrt{2}\sqrt{2}} = \dfrac{\sqrt{10}}{8}$

39. $\sqrt[3]{\dfrac{8}{x^3}} = \dfrac{\sqrt[3]{8}}{\sqrt[3]{x^3}} = \dfrac{2}{x}$

41. $\sqrt[4]{\dfrac{2}{x^5}} = \dfrac{\sqrt[4]{2}}{\sqrt[4]{x^5}} = \dfrac{\sqrt[4]{2}}{\sqrt[4]{x^4 \cdot x}} = \dfrac{\sqrt[4]{2}}{\sqrt[4]{x^4}\sqrt[4]{x}} = \dfrac{\sqrt[4]{2}}{x\sqrt[4]{x}}$
$= \dfrac{\sqrt[4]{2}\sqrt[4]{x^3}}{x\sqrt[4]{x}\sqrt[4]{x^3}} = \dfrac{\sqrt[4]{2x^3}}{x\sqrt[4]{x^4}} = \dfrac{\sqrt[4]{2x^3}}{x^2}$

43. $\sqrt{\sqrt{x^8}} = \sqrt{x^4} = x^2$

45. $\sqrt{\dfrac{9x^6}{y^4}} = \dfrac{\sqrt{9x^6}}{\sqrt{y^4}} = \dfrac{3x^3}{y^2}$

47. $2\sqrt{3} + 5\sqrt{3} = 7\sqrt{3}$

49. $6\sqrt{5} - \sqrt{5} + 4\sqrt{5} = 9\sqrt{5}$

51. $\sqrt{98x} - \sqrt{32x} = 7\sqrt{2x} - 4\sqrt{2x} = 3\sqrt{2x}$

53. $\sqrt[3]{24} - \sqrt[3]{81} = \sqrt[3]{8 \cdot 3} - \sqrt[3]{27 \cdot 3}$
$= 2\sqrt[3]{3} - 3\sqrt[3]{3} = -\sqrt[3]{3}$

55. $\sqrt[3]{3x} - 2\sqrt[3]{24x} + \sqrt[3]{375x}$
$= \sqrt[3]{3x} - 2\sqrt[3]{8 \cdot 3x} + \sqrt[3]{125 \cdot 3x}$
$= \sqrt[3]{3x} - 2\sqrt[3]{8}\sqrt[3]{3x} + \sqrt[3]{125}\sqrt[3]{3x}$
$= \sqrt[3]{3x} - 2 \cdot 2\sqrt[3]{3x} + 5\sqrt[3]{3x}$
$= \sqrt[3]{3x} - 4\sqrt[3]{3x} + 5\sqrt[3]{3x} = 2\sqrt[3]{3x}$

57. $\sqrt{2x^5} - 5\sqrt{32x} + \sqrt{18x^3}$
$= \sqrt{2x^4 x} - 5\sqrt{16 \cdot 2x} + \sqrt{9 \cdot 2x^2 x}$
$= x^2\sqrt{2x} - 5 \cdot 4\sqrt{2x} + 3x\sqrt{2x}$
$= x^2\sqrt{2x} - 20\sqrt{2x} + 3x\sqrt{2x}$
$= \sqrt{2x}\left(x^2 + 3x - 20\right)$

59. $\sqrt{48x^5y} - 4y\sqrt{3x^3y} + y\sqrt{3xy^3}$

$= \sqrt{16\cdot3\,x^4xy} - 4y\sqrt{3x^2xy} + y\sqrt{3xy^2y}$

$= 4x^2\sqrt{3xy} - 4xy\sqrt{3xy} + y^2\sqrt{3xy}$

$= \sqrt{3xy}\left(4x^2 - 4xy + y^2\right) = \sqrt{3xy}(2x - y)^2$

61. $\dfrac{2}{\sqrt{3}} = \dfrac{2\sqrt{3}}{\sqrt{3}\sqrt{3}} = \dfrac{2\sqrt{3}}{3}$

63. $\dfrac{7}{\sqrt{15}} = \dfrac{7\sqrt{15}}{\sqrt{15}\sqrt{15}} = \dfrac{7\sqrt{15}}{15}$

65. $\dfrac{1}{\sqrt{2}+x} = \dfrac{1\left(\sqrt{2}-x\right)}{\left(\sqrt{2}+x\right)\left(\sqrt{2}-x\right)} = \dfrac{\sqrt{2}-x}{2-x^2}$

67. $\dfrac{3}{2-\sqrt{3}} = \dfrac{3\left(2+\sqrt{3}\right)}{\left(2-\sqrt{3}\right)\left(2+\sqrt{3}\right)} = \dfrac{3\left(2+\sqrt{3}\right)}{2^2-\sqrt{3}^2}$

$= \dfrac{3\left(2+\sqrt{3}\right)}{4-3} = 3\left(2+\sqrt{3}\right)$

69. $\dfrac{1}{\sqrt{3}+\sqrt{2}} = \dfrac{1\left(\sqrt{3}-\sqrt{2}\right)}{\left(\sqrt{3}+\sqrt{2}\right)\left(\sqrt{3}-\sqrt{2}\right)}$

$= \dfrac{\sqrt{3}-\sqrt{2}}{\sqrt{3}^2-\sqrt{2}^2} = \dfrac{\sqrt{3}-\sqrt{2}}{3-2}$

$= \sqrt{3}-\sqrt{2}$

71. $\dfrac{\sqrt{5}-\sqrt{2}}{\sqrt{5}+\sqrt{2}} = \dfrac{\left(\sqrt{5}-\sqrt{2}\right)\left(\sqrt{5}-\sqrt{2}\right)}{\left(\sqrt{5}+\sqrt{2}\right)\left(\sqrt{5}-\sqrt{2}\right)}$

$= \dfrac{\sqrt{5}^2 - 2\sqrt{10} + \sqrt{2}^2}{\sqrt{5}^2 - \sqrt{2}^2} = \dfrac{7 - 2\sqrt{10}}{3}$

73. $\dfrac{\sqrt{x+h}-\sqrt{x}}{\sqrt{x+h}+\sqrt{x}} = \dfrac{\left(\sqrt{x+h}-\sqrt{x}\right)\left(\sqrt{x+h}-\sqrt{x}\right)}{\left(\sqrt{x+h}+\sqrt{x}\right)\left(\sqrt{x+h}-\sqrt{x}\right)}$

$= \dfrac{\left(\sqrt{x+h}\right)^2 - 2\sqrt{x}\sqrt{x+h} + \sqrt{x}^2}{\left(\sqrt{x+h}\right)^2 - \sqrt{x}^2}$

$= \dfrac{x+h - 2\sqrt{x(x+h)} + x}{x+h + x}$

$= \dfrac{2x+h - 2\sqrt{x(x+h)}}{h}$

75. $\dfrac{\sqrt{4+h}-2}{h} = \dfrac{\left(\sqrt{4+h}-2\right)\left(\sqrt{4+h}+2\right)}{h\left(\sqrt{4+h}+2\right)}$

$= \dfrac{4+h-4}{h\left(\sqrt{4+h}+2\right)} = \dfrac{h}{h\left(\sqrt{4+h}+2\right)}$

$= \dfrac{1}{\sqrt{4+h}+2}$

77. $\dfrac{2-\sqrt{4-x}}{x} = \dfrac{\left(2-\sqrt{4-x}\right)\left(2+\sqrt{4-x}\right)}{x\left(2+\sqrt{4-x}\right)}$

$= \dfrac{4-(4-x)}{x\left(2+\sqrt{4-x}\right)} = \dfrac{x}{x\left(2+\sqrt{4-x}\right)}$

$= \dfrac{1}{2+\sqrt{4-x}}$

79. $\dfrac{\sqrt{x}-2}{x-4} = \dfrac{\left(\sqrt{x}-2\right)\left(\sqrt{x}+2\right)}{\left(x-4\right)\left(\sqrt{x}+2\right)}$

$= \dfrac{x-4}{\left(x-4\right)\left(\sqrt{x}+2\right)} = \dfrac{1}{\sqrt{x}+2}$

81. $\dfrac{\sqrt{x^2+4x}-x}{2} = \dfrac{\left(\sqrt{x^2+4x}-x\right)\left(\sqrt{x^2+4x}+x\right)}{2\left(\sqrt{x^2+4x}+x\right)}$

$= \dfrac{x^2+4x-x^2}{2\left(\sqrt{x^2+4x}+x\right)}$

$= \dfrac{4x}{2\left(\sqrt{x^2+4x}+x\right)}$

$= \dfrac{2x}{\sqrt{x^2+4x}+x}$

83. $25^{1/2} = \sqrt{25} = 5$

85. $(-8)^{1/3} = \sqrt[3]{-8} = -2$

87. $8^{2/3} = \sqrt[3]{8^2} = \sqrt[3]{(2^3)^2} = \sqrt[3]{2^6} = 2^2 = 4$

89. $-25^{-3/2} = -\dfrac{1}{25^{3/2}} = -\dfrac{1}{\sqrt{25^3}} = -\dfrac{1}{\sqrt{(5^2)^3}}$

$= -\dfrac{1}{\sqrt{5^6}} = -\dfrac{1}{5^3} = -\dfrac{1}{125}$

91. $\left(\dfrac{9}{25}\right)^{-3/2} = \left(\dfrac{25}{9}\right)^{3/2} = \sqrt{\left(\dfrac{25}{9}\right)^3} = \dfrac{\sqrt{25^3}}{\sqrt{9^3}}$

$= \dfrac{\sqrt{(5^2)^3}}{\sqrt{(3^2)^3}} = \dfrac{\sqrt{5^6}}{\sqrt{3^6}} = \dfrac{5^3}{3^3} = \dfrac{125}{27}$

93. $x^{1/2} \cdot x^{2/5} = x^{1/2+2/5} = x^{9/10}$

95. $x^{3/5} \cdot x^{-1/2} = x^{3/5-1/2} = x^{1/10}$

97. $\left(8x^6\right)^{2/3} = 8^{2/3} x^{6(2/3)} = \left(2^3\right)^{2/3} x^{6(2/3)}$
$$= 2^2 x^4 = 4x^4$$

99. $\left(27x^6 y^3\right)^{-2/3} = 27^{-2/3} x^{6(-2/3)} y^{3(-2/3)}$
$$= 3^{3(-2/3)} x^{6(-2/3)} y^{3(-2/3)}$$
$$= 3^{-2} x^{-4} y^{-2}$$
$$= \frac{1}{3^2 x^4 y^2} = \frac{1}{9x^4 y^2}$$

101. $\dfrac{15x^{3/2}}{3x^{1/4}} = 5x^{3/2-1/4} = 5x^{5/4}$

103. $\left(\dfrac{x^{-1/4}}{y^{-2/3}}\right)^{-12} = \dfrac{x^{(-1/4)(-12)}}{y^{(-2/3)(-12)}} = \dfrac{x^3}{y^8}$

105. $\sqrt[4]{3^2} = 3^{2/4} = 3^{1/2}$

107. $\sqrt[3]{x^9} = x^{9/3} = x^3$

109. $\sqrt[3]{x^6 y^9} = x^{6/3} y^{9/3} = x^2 y^3$

111. $\sqrt[4]{9}\sqrt{3} = \sqrt[4]{3^2}\sqrt{3} = 3^{2/4} \cdot 3^{1/2}$
$$= 3^{1/2} \cdot 3^{1/2} = 3^{1/2+1/2} = 3^1 = 3$$

113. $\sqrt{\sqrt[3]{x^{10}}} = \sqrt{x^{10/3}} = \left(x^{10/3}\right)^{1/2}$
$$= x^{(10/3)(1/2)} = x^{5/3}$$

115. $\sqrt[3]{2} \cdot \sqrt[5]{3} = \sqrt[3 \cdot 5]{2^5} \cdot \sqrt[3 \cdot 5]{3^3} = \sqrt[15]{32} \cdot \sqrt[15]{27}$
$$= \sqrt[15]{32 \cdot 27} = \sqrt[15]{864}$$

117. $\sqrt[3]{x^2} \cdot \sqrt[4]{x^3} = \sqrt[3 \cdot 4]{\left(x^2\right)^4} \cdot \sqrt[3 \cdot 4]{\left(x^3\right)^3} = \sqrt[12]{x^8} \cdot \sqrt[12]{x^9} = \sqrt[12]{x^8 \cdot x^9} = \sqrt[12]{x^{17}} = \sqrt[12]{x^{12} \cdot x^5} = \sqrt[12]{x^{12}} \cdot \sqrt[12]{x^5} = x\sqrt[12]{x^5}$

119. $\sqrt[9]{2m^2 n} \cdot \sqrt[3]{5m^5 n^2} = \sqrt[9]{2m^2 n} \cdot \sqrt[3 \cdot 3]{\left(5m^5 n^2\right)^3} = \sqrt[9]{2m^2 n} \cdot \sqrt[9]{5^3 \left(m^5\right)^3 \left(n^2\right)^3} = \sqrt[9]{2m^2 n} \cdot \sqrt[9]{125 m^{15} n^6}$
$$= \sqrt[9]{2m^2 n \cdot 125 m^{15} n^6} = \sqrt[9]{250 m^{2+15} n^{1+6}} = \sqrt[9]{250 m^{17} n^7} = m\sqrt[9]{250 m^8 n^7}$$

121. $\sqrt[5]{x^4 y^3} \cdot \sqrt[6]{x^3 y^5} = \sqrt[5 \cdot 6]{\left(x^4 y^3\right)^6} \cdot \sqrt[5 \cdot 6]{\left(x^3 y^5\right)^5} = \sqrt[30]{\left(x^4\right)^6 \left(y^3\right)^6} \cdot \sqrt[30]{\left(x^3\right)^5 \left(y^5\right)^5} = \sqrt[30]{x^{24} y^{18}} \cdot \sqrt[30]{x^{15} y^{25}}$
$$= \sqrt[30]{x^{24} y^{18} \cdot x^{15} y^{25}} = \sqrt[30]{x^{24+15} y^{18+25}} \qquad = \sqrt[30]{x^{39} y^{43}} = \sqrt[30]{x^{30+9} y^{30+13}}$$
$$= \sqrt[30]{x^{30} x^9 y^{30} y^{13}} = xy\sqrt[30]{x^9 y^{13}}$$

123. $\dfrac{4}{3} x^{1/3} (2x-3) + 2x^{4/3} = 2x^{1/3}\left(\dfrac{2}{3}(2x-3)+x\right) = 2x^{1/3}\left(\dfrac{4x}{3} - 2 + x\right) = 2x^{1/3}\left(\dfrac{7x}{3} - 2\right) = \dfrac{2}{3} x^{1/3} (7x-6)$

125. $4(3x+1)^{1/3}(2x-1) + 2(3x+1)^{4/3} = 2(3x+1)^{1/3}\left(2(2x-1)+(3x+1)\right) = 2(3x+1)^{1/3}(4x-2+3x+1)$
$$= 2(3x+1)^{1/3}(7x-1)$$

127. $3x\left(x^2+1\right)^{1/2}\left(2x^2-x\right) + 2\left(x^2+1\right)^{3/2}(4x-1) = \left(x^2+1\right)^{1/2}\left[3x\left(2x^2-x\right) + 2\left(x^2+1\right)(4x-1)\right]$
$$= \left(x^2+1\right)^{1/2}\left(6x^3 - 3x^2 + 8x^3 - 2x^2 + 8x - 2\right)$$
$$= \left(x^2+1\right)^{1/2}\left(14x^3 - 5x^2 + 8x - 2\right)$$

129. $\dfrac{1}{3-\sqrt{8}} - \dfrac{1}{\sqrt{8}-\sqrt{7}} + \dfrac{1}{\sqrt{7}-\sqrt{6}} - \dfrac{1}{\sqrt{6}-\sqrt{5}} + \dfrac{1}{\sqrt{5}-2}$

$= \dfrac{1\left(3+\sqrt{8}\right)}{\left(3-\sqrt{8}\right)\left(3+\sqrt{8}\right)} - \dfrac{1\left(\sqrt{8}+\sqrt{7}\right)}{\left(\sqrt{8}-\sqrt{7}\right)\left(\sqrt{8}+\sqrt{7}\right)} + \dfrac{1\left(\sqrt{7}+\sqrt{6}\right)}{\left(\sqrt{7}-\sqrt{6}\right)\left(\sqrt{7}+\sqrt{6}\right)} - \dfrac{1\left(\sqrt{6}+\sqrt{5}\right)}{\left(\sqrt{6}-\sqrt{5}\right)\left(\sqrt{6}+\sqrt{5}\right)} + \dfrac{1\left(\sqrt{5}+2\right)}{\left(\sqrt{5}-2\right)\left(\sqrt{5}+2\right)}$

$= \dfrac{3+\sqrt{8}}{9-8} - \dfrac{\sqrt{8}+\sqrt{7}}{8-7} + \dfrac{\sqrt{7}+\sqrt{6}}{7-6} - \dfrac{\sqrt{6}+\sqrt{5}}{6-5} + \dfrac{\sqrt{5}+2}{5-4}$

$= 3+\sqrt{8} - \left(\sqrt{8}+\sqrt{7}\right) + \left(\sqrt{7}+\sqrt{6}\right) - \left(\sqrt{6}+\sqrt{5}\right) + \left(\sqrt{5}+2\right)$

$= 3+\sqrt{8}-\sqrt{8}-\sqrt{7}+\sqrt{7}+\sqrt{6}-\sqrt{6}-\sqrt{5}+\sqrt{5}+2 = 3+2 = 5$

131. a. $\left(2+\sqrt{3}\right) + \dfrac{1}{2+\sqrt{3}} = 2+\sqrt{3} + \dfrac{2-\sqrt{3}}{\left(2+\sqrt{3}\right)\left(2-\sqrt{3}\right)} = 2+\sqrt{3} + \dfrac{2-\sqrt{3}}{4-3} = 2+\sqrt{3}+2-\sqrt{3} = 4$

  b. $\left(2+\sqrt{3}\right)^2 + \dfrac{1}{\left(2+\sqrt{3}\right)^2} = 7+4\sqrt{3} + \dfrac{7-4\sqrt{3}}{\left(7+4\sqrt{3}\right)\left(7-4\sqrt{3}\right)} = 7+4\sqrt{3} + \dfrac{7-4\sqrt{3}}{49-48} = 7+4\sqrt{3}+7-4\sqrt{3} = 14$

P.6 Applying the Concepts

133. Substituting 692 for A in $\sqrt{\dfrac{4A}{\sqrt{3}}}$ gives

  $\sqrt{\dfrac{4(692)}{1.73}} = \sqrt{\dfrac{2768}{1.73}} = \sqrt{1600} = 40$ cm.

135. Substituting 6 for w in $V = \sqrt{\dfrac{w}{1.5}}$ gives

  $V = \sqrt{\dfrac{6}{1.5}} = \sqrt{4} = 2$ cm/sec.

137. Substituting 1058 for W and 2 for R in

  $I = \sqrt{\dfrac{W}{R}}$ gives $I = \sqrt{\dfrac{1058}{2}} = \sqrt{529} = 23$ amp.

139. Substituting 16.25 for h in

  $P = 14.7(0.5)^{h/3.25}$ gives

  $P = 14.7(0.5)^{16.25/3.25} = 14.7(0.5)^5 =$
  $14.7(0.03125) \approx 0.46$ lb/sq in.

Chapter P Review Exercises

Basic Concepts and Skills

1. a. $\{4\}$

  b. $\{0,4\}$

  c. $\{-5,0,4\}$

  d. $\left\{-5,0,0.2,0.\overline{31},\dfrac{1}{2},4\right\}$

  e. $\left\{\sqrt{7},\sqrt{12}\right\}$

  f. $\left\{-5,0,0.2,0.\overline{31},\dfrac{1}{2},\sqrt{7},\sqrt{12},4\right\}$

3. Distributive property

5. Multiplicative identity

7. $x \le 1$

9. $(0,\infty)$

11. $\left|2-|-3|\right| = |2-3| = |-1| = 1$

13. $\left|1-\sqrt{15}\right| = \sqrt{15}-1$

15. $(-4)^3 = -64$

17. $2^4 - 5 \cdot 3^2 = 16 - 5 \cdot 9 = 16 - 45 = -29$

19. $-25^{1/2} = -5$

21. $8^{4/3} = \left(2^3\right)^{4/3} = 2^{3(4/3)} = 2^4 = 16$

23. $\left(\dfrac{25}{36}\right)^{-3/2} = \left(\dfrac{36}{25}\right)^{3/2} = \dfrac{\left(6^2\right)^{3/2}}{\left(5^2\right)^{3/2}}$

  $= \dfrac{6^{2(3/2)}}{5^{2(3/2)}} = \dfrac{6^3}{5^3} = \dfrac{216}{125}$

25. $2 \cdot 5^3 = 2 \cdot 125 = 250$

27. $\dfrac{21\times10^6}{3\times10^7} = \dfrac{7}{10} = 0.7$

29. $\sqrt{5^2} = 5$

31. $\sqrt[3]{-125} = \sqrt[3]{(-5)^3} = -5$

33. $64^{-1/3} = \dfrac{1}{64^{1/3}} = \dfrac{1}{\left(4^3\right)^{1/3}} = \dfrac{1}{4}$

35. $\left(\sqrt{3}+2\right)\left(\sqrt{3}-2\right) = \left(\sqrt{3}\right)^2 - 2^2 = 3 - 4 = -1$

37. $\dfrac{3^{-2}\cdot 7^0}{18^{-1}} = \dfrac{\frac{1}{3^2}\cdot 1}{\frac{1}{18}} = \dfrac{1}{9}\cdot 18 = 2$

39. $\dfrac{2-6\sqrt{4}}{4(2)} = \dfrac{2-6(2)}{8} = \dfrac{2-12}{8} = \dfrac{-10}{8} = -\dfrac{5}{4}$

41. $16^{-1/2} = \dfrac{1}{16^{1/2}} = \dfrac{1}{4}$

43. $\left(x^{-2}\right)^{-5} = x^{(-2)(-5)} = x^{10}$

45. $\left(\dfrac{x^{-3}}{y^{-1}}\right)^{-2} = \dfrac{x^{-3(-2)}}{y^{-1(-2)}} = \dfrac{x^6}{y^2}$

47. $\left(16x^{-2/3}y^{-4/3}\right)^{3/2}$
$= 16^{3/2}x^{(-2/3)(3/2)}y^{(-4/3)(3/2)}$
$= 4^{2(3/2)}x^{-1}y^{-2} = \dfrac{4^3}{xy^2} = \dfrac{64}{xy^2}$

49. $\left(\dfrac{64y^{-9/2}}{x^{-3}}\right)^{-2/3} = \dfrac{4^{3(-2/3)}y^{(-9/2)(-2/3)}}{x^{-3(-2/3)}}$
$= \dfrac{4^{-2}y^3}{x^2} = \dfrac{y^3}{4^2x^2} = \dfrac{y^3}{16x^2}$

51. $\left(7x^{1/4}\right)\left(3x^{3/2}\right) = 21x^{1/4+3/2} = 21x^{7/4}$

53. $\dfrac{x^5(2x)^3}{4x^3} = \dfrac{x^5(2^3)x^3}{4x^3} = \dfrac{8x^{5+3}}{4x^3} = 2x^{8-3} = 2x^5$

55. $\left(\dfrac{x^2y^{4/3}}{x^{1/3}}\right)^6 = \dfrac{x^{2(6)}y^{(4/3)(6)}}{x^{(1/3)(6)}} = \dfrac{x^{12}y^8}{x^2}$
$= x^{12-2}y^8 = x^{10}y^8$

57. $\dfrac{\sqrt{64}}{\sqrt{11}} = \dfrac{8}{\sqrt{11}} = \dfrac{8\sqrt{11}}{\sqrt{11}\cdot\sqrt{11}} = \dfrac{8\sqrt{11}}{11}$

59. $7\sqrt{6} - 3\sqrt{24} = 7\sqrt{6} - 3\sqrt{4\cdot 6}$
$= 7\sqrt{6} - 3\sqrt{4}\sqrt{6} = 7\sqrt{6} - 3(2)\sqrt{6}$
$= 7\sqrt{6} - 6\sqrt{6} = \sqrt{6}$

61. $\sqrt{2x}\sqrt{6x} = \sqrt{12x^2} = \sqrt{4\cdot 3x^2}$
$= \sqrt{4}\sqrt{3}\sqrt{x^2} = 2x\sqrt{3}$

63. $\dfrac{\sqrt{100x^3}}{\sqrt{4x}} = \sqrt{\dfrac{100x^3}{4x}} = \sqrt{25x^2} = 5x$

65. $4\sqrt[3]{135} + \sqrt[3]{40} = 4\sqrt[3]{27\cdot 5} + \sqrt[3]{8\cdot 5}$
$= 4\sqrt[3]{27}\sqrt[3]{5} + \sqrt[3]{8}\sqrt[3]{5}$
$= 4(3)\sqrt[3]{5} + 2\sqrt[3]{5}$
$= 12\sqrt[3]{5} + 2\sqrt[3]{5} = 14\sqrt[3]{5}$

67. $\dfrac{7}{\sqrt{3}} = \dfrac{7\sqrt{3}}{\sqrt{3}\sqrt{3}} = \dfrac{7\sqrt{3}}{3}$

69. $\dfrac{1-\sqrt{3}}{1+\sqrt{3}} = \dfrac{\left(1-\sqrt{3}\right)\left(1-\sqrt{3}\right)}{\left(1+\sqrt{3}\right)\left(1-\sqrt{3}\right)} = \dfrac{1-2\sqrt{3}+3}{1^2-\left(\sqrt{3}\right)^2}$
$= \dfrac{4-2\sqrt{3}}{1-3} = \dfrac{4-2\sqrt{3}}{-2} = -2+\sqrt{3}$

71. $3.7\times\left(6.23\times10^{12}\right) = 23.051\times10^{12}$
$= 2.3051\times10^{13}$

73. $\left(x^3-6x^2+4x-2\right)+\left(3x^3-6x^2+5x-4\right)$
$= 4x^3-12x^2+9x-6$

75. $\left(4x^4+3x^3-5x^2+9\right)+\left(5x^4+8x^3-7x^2+5\right)$
$= 9x^4+11x^3-12x^2+14$

77. $(x-12)(x-3) = x^2-3x-12x+36$
$= x^2-15x+36$

79. $\left(x^5-2\right)\left(x^5+2\right) = \left(x^5\right)^2-2^2 = x^{10}-4$

81. $(2x+5)(3x-11) = 6x^2-22x+15x-55$
$= 6x^2-7x-55$

83. $x^2-3x-10 = (x-5)(x+2)$

85. $24x^2-38x-11 = (6x-11)(4x+1)$

87. $x(x+11)+5(x+11) = (x+11)(x+5)$

89. $x^4 - x^3 + 7x - 7 = x^4 + 7x - x^3 - 7$
$$= x(x^3 + 7) - 1(x^3 + 7)$$
$$= (x^3 + 7)(x - 1)$$

91. $10x^2 + 23x + 12 = (5x + 4)(2x + 3)$

93. $12x^2 + 7x - 12 = (4x - 3)(3x + 4)$

95. $4x^2 - 49 = (2x - 7)(2x + 7)$

97. $x^2 + 12 + 36 = (x + 6)^2$

99. $64x^2 + 48x + 9 = (8x + 3)^2$

101. $8x^3 + 27 = (2x + 3)(4x^2 - 6x + 9)$

103. $x^3 + 5x^2 - 16x - 80 = x^3 - 16x + 5x^2 - 80$
$$= x(x^2 - 16) + 5(x^2 - 16)$$
$$= (x^2 - 16)(x + 5)$$
$$= (x - 4)(x + 4)(x + 5)$$

105. $\dfrac{4}{x - 9} - \dfrac{10}{9 - x} = \dfrac{4}{x - 9} + \dfrac{10}{x - 9} = \dfrac{14}{x - 9}$

---

107. $\dfrac{3x + 5}{x^2 + 14x + 48} - \dfrac{3x - 2}{x^2 + 10x + 16} = \dfrac{3x + 5}{(x + 6)(x + 8)} - \dfrac{3x - 2}{(x + 2)(x + 8)} = \dfrac{(3x + 5)(x + 2)}{(x + 6)(x + 8)(x + 2)} - \dfrac{(3x - 2)(x + 6)}{(x + 2)(x + 6)(x + 8)}$
$$= \dfrac{3x^2 + 11x + 10 - (3x^2 + 16x - 12)}{(x + 2)(x + 6)(x + 8)} = \dfrac{-5x + 22}{(x + 2)(x + 6)(x + 8)}$$

109. $\dfrac{x - 1}{x + 1} - \dfrac{x + 1}{x - 1} = \dfrac{(x - 1)^2}{(x + 1)(x - 1)} - \dfrac{(x + 1)^2}{(x + 1)(x - 1)} = \dfrac{x^2 - 2x + 1 - (x^2 + 2x + 1)}{(x + 1)(x - 1)} = -\dfrac{4x}{(x + 1)(x - 1)}$

111. $\dfrac{x - 1}{2x - 3} \cdot \dfrac{4x^2 - 9}{2x^2 - x - 1} = \dfrac{\cancel{x - 1}}{\cancel{2x - 3}} \cdot \dfrac{(2x + 3)\cancel{(2x - 3)}}{(2x + 1)\cancel{(x - 1)}} = \dfrac{2x + 3}{2x + 1}$

113. $\dfrac{x^2 - 4}{4x^2 - 9} \cdot \dfrac{2x^2 - 3x}{2x + 4} = \dfrac{(x - 2)\cancel{(x + 2)}}{\cancel{(2x - 3)}(2x + 3)} \cdot \dfrac{x\cancel{(2x - 3)}}{2\cancel{(x + 2)}} = \dfrac{x(x - 2)}{2(2x + 3)}$

115. $\dfrac{\dfrac{1}{x^2} - x}{\dfrac{1}{x^2} + x} = \dfrac{\dfrac{1 - x^3}{x^2}}{\dfrac{1 + x^3}{x^2}} = \dfrac{1 - x^3}{x^2} \cdot \dfrac{x^2}{1 + x^3} = \dfrac{1 - x^3}{1 + x^3} = \dfrac{(1 - x)(1 + x + x^2)}{(1 + x)(1 - x + x^2)}$

117. $\dfrac{\dfrac{x}{x - 3} + x}{\dfrac{x}{3 - x} - x} = \dfrac{\dfrac{x + x(x - 3)}{x - 3}}{\dfrac{x - x(3 - x)}{3 - x}} = \dfrac{x^2 - 2x}{x - 3} \cdot \dfrac{3 - x}{x^2 - 2x} = \dfrac{x^2 - 2x}{x - 3} \cdot \left(-\dfrac{x - 3}{x^2 - 2x}\right) = -1$

Applying the Concepts

119. Using the Pythagorean Theorem, we have
$$c^2 = 20^2 + 21^2 \Rightarrow c^2 = 841 \Rightarrow c = 29$$

121. First, find the length of the missing leg:
$$20^2 = 12^2 + x^2 \Rightarrow 400 = 144 + x^2 \Rightarrow$$
$$256 = x^2 \Rightarrow 16 = x.$$
Area $= \dfrac{1}{2}bh \Rightarrow A = \dfrac{1}{2}(12)(16) = 96$.
Perimeter = the sum of the sides $\Rightarrow$
$P = 12 + 16 + 20 = 48$

123. First, find the length of the missing side:
$$10^2 = 6^2 + x^2 \Rightarrow 100 = 36 + x^2 \Rightarrow$$
$64 = x^2 \Rightarrow 8 = x$. So the area $= (6)(8) = 48$ square feet.

125. $\sqrt{7920(0.7) + 0.7^2} = \sqrt{5544 + 0.49}$
$$\approx 74.46 \text{ miles}$$

127. $\dfrac{36 - 3(2^2)}{8(2)} = \dfrac{36 - 3(4)}{16} = \dfrac{36 - 12}{16} = \dfrac{24}{16} = 1.5 \text{ ft}$

Chapter P Practice Test

1. $\big|7 - |-3|\big| = |7 - 3| = |4| = 4$

2. $\big|\sqrt{2} - 100\big| = 100 - \sqrt{2}$

3. $\dfrac{5 - 3(-3)}{2} + (5)(-3) = \dfrac{5 + 9}{2} - 15 = -8$

4. $x \neq -9, x \neq 3 \Rightarrow (-\infty, -9) \cup (-9, 3) \cup (3, \infty)$

5. $\left(\dfrac{-3x^2 y}{x}\right)^3 = \dfrac{(-3)^3 x^{2(3)} y^3}{x^3} = \dfrac{-27 x^6 y^3}{x^3}$
$$= -27 x^{6-3} y^3 = -27 x^3 y^3$$

6. $\sqrt[3]{-8x^6} = \sqrt[3]{(-2)^3 x^6} = -2x^2$

7. $\sqrt{75x} - \sqrt{27x} = \sqrt{3 \cdot 25x} - \sqrt{3 \cdot 9x}$
$$= 5\sqrt{3x} - 3\sqrt{3x} = 2\sqrt{3x}$$

8. $-16^{-3/2} = -\dfrac{1}{16^{3/2}} = -\dfrac{1}{(4^2)^{3/2}}$
$$= -\dfrac{1}{4^{2(3/2)}} = -\dfrac{1}{4^3} = -\dfrac{1}{64}$$

9. $\left(\dfrac{x^{-2} y^4}{25 x^3 y^3}\right)^{-1/2}$
$$= \left(\dfrac{25 x^3 y^3}{x^{-2} y^4}\right)^{1/2} = \dfrac{25^{1/2} \left(x^3\right)^{1/2} \left(y^3\right)^{1/2}}{\left(x^{-2}\right)^{1/2} \left(y^4\right)^{1/2}}$$
$$= \dfrac{5 x^{3/2} y^{3/2}}{x^{-1} y^2} = 5 x^{(3/2)-(-1)} y^{(3/2)-2}$$
$$= 5 x^{5/2} y^{-1/2} = \dfrac{5 x^{5/2}}{y^{1/2}} = \dfrac{5 x^{5/2} y^{1/2}}{y}$$

10. $\dfrac{5}{1 - \sqrt{3}} = \dfrac{5\left(1 + \sqrt{3}\right)}{\left(1 - \sqrt{3}\right)\left(1 + \sqrt{3}\right)}$
$$= \dfrac{5 + 5\sqrt{3}}{1 - 3} = -\dfrac{5 + 5\sqrt{3}}{2}$$

11. $4(x^2 - 3x + 2) + 3(5x^2 - 2x + 1)$
$$= 4x^2 - 12x + 8 + 15x^2 - 6x + 3$$
$$= 19x^2 - 18x + 11$$

12. $(x - 2)(5x - 1) = 5x^2 - 11x + 2$

13. $\left(x^2 + 3y^2\right)^2 = x^4 + 6x^2 y^2 + 9y^4$

14.

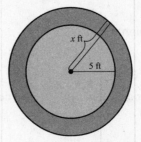

The total area is πx^2 square feet. The area of the rug is 25π square feet. So the area of the border = the total area − the area of the rug = $\pi x^2 - 25\pi = \pi(x^2 - 25) = \pi(x - 5)(x + 5)$.

15. $x^2 - 5x + 6 = (x - 3)(x - 2)$

16. $9x^2 + 12x + 4 = (3x + 2)^2$

17. $8x^3 - 27 = (2x - 3)(4x^2 + 6x + 9)$

18. $\dfrac{6 - 3x}{x^2 - 4} - \dfrac{12}{2x + 4}$
$$= \dfrac{6 - 3x}{(x - 2)(x + 2)} - \dfrac{12}{2(x + 2)}$$
$$= \dfrac{2(6 - 3x)}{2(x - 2)(x + 2)} - \dfrac{12(x - 2)}{2(x - 2)(x + 2)}$$
$$= \dfrac{12 - 6x - (12x - 24)}{2(x - 2)(x + 2)} = \dfrac{-18x + 36}{2(x - 2)(x + 2)}$$
$$= \dfrac{-18(x - 2)}{2(x - 2)(x + 2)} = -\dfrac{9}{x + 2}$$

19. $\dfrac{\dfrac{1}{x^2}}{9 - \dfrac{1}{x^2}} = \dfrac{\dfrac{1}{x^2}}{\dfrac{9x^2 - 1}{x^2}} = \dfrac{1}{x^2} \cdot \dfrac{x^2}{(3x - 1)(3x + 1)}$
$$= \dfrac{1}{(3x - 1)(3x + 1)}$$

20. $c^2 = 35^2 + 12^2 \Rightarrow c^2 = 1369 \Rightarrow c = 37$

Chapter 1 Equations and Inequalities

1.1 Linear Equations in One Variable

1.1 Practice Problems

1. a. Both sides of the equation $\frac{x}{3} - 7 = 5$ are defined for all real numbers, so the domain is $(-\infty, \infty)$.

   b. The left side of the equation $\frac{2}{2-x} = 4$ is not defined if $x = 2$. The right side of the equation is defined for all real numbers, so the domain is $(-\infty, 2) \cup (2, \infty)$.

   c. The left side of the equation $\sqrt{x-1} = 0$ is not defined if $x < 1$. The right side of the equation is defined for all real numbers, so the domain is $[1, \infty)$.

2. $\frac{2}{3} - \frac{3}{2}x = \frac{1}{6} - \frac{7}{3}x$

   To clear the fractions, multiply both sides of the equation by the LCD, 6.
   $$4 - 9x = 1 - 14x$$
   $$4 - 9x + 14x = 1 - 14x + 14x$$
   $$4 + 5x = 1$$
   $$4 + 5x - 4 = 1 - 4$$
   $$5x = -3$$
   $$\frac{5x}{5} = \frac{-3}{5}$$
   $$x = -\frac{3}{5}$$

   Solution set: $\left\{ -\frac{3}{5} \right\}$

3. $3x - \left[2x - 6(x+1) \right] = 7x - 1$
   $$3x - (2x - 6x - 6) = 7x - 1$$
   $$3x - (-4x - 6) = 7x - 1$$
   $$3x + 4x + 6 = 7x - 1$$
   $$7x + 6 = 7x - 1$$
   $$7x + 6 - 7x = 7x - 1 - 7x$$
   $$6 = -1$$

   Since $6 = -1$ is false, no number satisfies this equation. Thus, the equation is inconsistent, and the solution set is $\varnothing$.

4. $2(3x - 6) + 5 = 12 - (19 - 6x)$
   $$6x - 12 + 5 = 12 - 19 + 6x$$
   $$6x - 7 = -7 + 6x$$
   $$6x - 7 - 6x = -7 + 6x - 6x$$
   $$-7 = -7$$
   $$-7 + 7 = -7 + 7$$
   $$0 = 0$$

   The equation $0 = 0$ is always true. Therefore, the original equation is an identity, and the solution set is $(-\infty, \infty)$.

5. $$F = \frac{9}{5}C + 32$$
   $$50 = \frac{9}{5}C + 32$$
   $$50 - 32 = \frac{9}{5}C + 32 - 32$$
   $$18 = \frac{9}{5}C$$
   $$18 \cdot \frac{5}{9} = \frac{5}{9} \cdot \frac{9}{5}C$$
   $$10 = C$$

   Thus, 50°F converts to 10°C.

6. $P = 2l + 2w$

   Subtract $2l$ from both sides.
   $$P - 2l = 2w$$
   Now, divide both sides by 2.
   $$\frac{P - 2l}{2} = w$$

7. Let w = the width of the rectangle.
   Then $2w + 5$ = the length of the rectangle.
   $P = 2l + 2w$, so we have
   $$28 = 2(2w + 5) + 2w$$
   $$28 = 4w + 10 + 2w$$
   $$28 = 6w + 10$$
   $$18 = 6w$$
   $$3 = w$$

   The width of the rectangle is 3 m and the length is $2(3) + 5 = 11$ m

8. Let x = the amount invested in stocks. Then $15{,}000 - x$ = the amount invested in bonds.
   $$x = 3(15{,}000 - x)$$
   $$x = 45{,}000 - 3x$$
   $$4x = 45{,}000$$
   $$x = 11{,}250$$

   Tyrick invested $11,250 in stocks and $15,000 − $11,250 = $3,750 in bonds.

9. Let x = the amount of capital. Then $\dfrac{x}{5}$ = the amount invested at 5%, $\dfrac{x}{6}$ = the amount invested at 8%, and $x - \left(\dfrac{x}{5} + \dfrac{x}{6}\right) = \dfrac{19x}{30}$ = the amount invested at 10%.

Principal	Rate	Time	Interest
$\dfrac{x}{5}$	0.05	1	$0.05\left(\dfrac{x}{5}\right)$
$\dfrac{x}{6}$	0.08	1	$0.08\left(\dfrac{x}{6}\right)$
$\dfrac{19x}{30}$	0.1	1	$0.1\left(\dfrac{19x}{30}\right)$

The total interest is $130, so
$$0.05\left(\dfrac{x}{5}\right) + 0.08\left(\dfrac{x}{6}\right) + 0.1\left(\dfrac{19x}{30}\right) = 130$$
Multiply by the LCD, 30.
$$0.3x + 0.4x + 1.9x = 3900$$
$$2.6x = 3900$$
$$x = 1500$$
The total capital is $1500.

10. Let x = the length of the bridge. Then $x + 130$ = the distance the train travels. $rt = d$, so
$$25(21) = x + 130 \Rightarrow 525 = x + 130 \Rightarrow 395 = x$$
The bridge is 395 m long.

11. Following the reasoning in example 10, we have $x + 2x = 3x$ is the maximum extended length (in feet) of the cord.
$$3x + 7 + 10 = 120$$
$$3x + 17 = 120$$
$$3x + 17 - 17 = 120 - 17$$
$$3x = 103$$
$$\dfrac{3x}{3} = \dfrac{103}{3} \Rightarrow x \approx 34.3$$
The cord should be no longer than 34.3 feet.

1.1 Basic Concepts and Skills

1. The domain of the variable in an equation is the set of all real number for which both sides of the equation are <u>defined</u>.

3. Two equations with the same solution sets are called <u>equivalent</u>.

5. False. The interest $I = (100)(0.05)(3)$.

7. a. Substitute 0 for x in the equation $x - 2 = 5x + 6$:
$$0 - 2 = 5(0) + 6 \Rightarrow -2 \neq 6$$
So, 0 is not a solution of the equation.

b. Substitute -2 for x in the equation $x - 2 = 5x + 6$:
$$-2 - 2 = 5(-2) + 6 \Rightarrow -4 = -10 + 6 \Rightarrow$$
$$-4 = -4$$
So, -2 is a solution of the equation.

9. a. Substitute 4 for x in the equation $\dfrac{2}{x} = \dfrac{1}{3} + \dfrac{1}{x+2}$:
$$\dfrac{2}{4} = \dfrac{1}{3} + \dfrac{1}{4+2} \Rightarrow \dfrac{1}{2} = \dfrac{1}{3} + \dfrac{1}{6} \Rightarrow \dfrac{1}{2} = \dfrac{1}{2}$$
So, 4 is a solution of the equation.

b. Substitute 1 for x in the equation $\dfrac{2}{x} = \dfrac{1}{3} + \dfrac{1}{x+2}$:
$$\dfrac{2}{1} = \dfrac{1}{3} + \dfrac{1}{1+2} \Rightarrow 2 = \dfrac{1}{3} + \dfrac{1}{3} \Rightarrow 2 \neq \dfrac{2}{3}$$
So, 1 is not a solution of the equation.

11. a. The equation $2x + 3x = 5x$ is an identity, so every real number is a solution of the equation. Thus 157 is a solution of the equation. This can be checked by substituting 157 for x in the equation:
$$2(157) + 3(157) = 5(157) \Rightarrow$$
$$314 + 471 = 785 \Rightarrow 785 = 785$$

b. The equation $2x + 3x = 5x$ is an identity, so every real number is a solution of the equation. Thus -2046 is a solution of the equation. This can be checked by substituting -2046 for x in the equation:
$$2(-2046) + 3(-2046) = 5(-2046)$$
$$-4092 - 6138 = -10{,}230$$
$$-10{,}230 = -10{,}230$$

13. The left side of the equation $\dfrac{y}{y-1} = \dfrac{3}{y+2}$ is not defined if $y = 1$, and the right side of the equation is not defined if $y = -2$. The domain is $(-\infty, -2) \cup (-2, 1) \cup (1, \infty)$.

15. The left side of the equation $\dfrac{3x}{(x-3)(x-4)} = 2x + 9$ is not defined if $x = 3$ or $x = 4$. The right side is defined for all real numbers. So, the domain is $(-\infty, 3) \cup (3, 4) \cup (4, \infty)$.

17. Substitute 0 for x in $2x+3=5x+1$. Because $3 \neq 1$, the equation is not an identity.

19. When the terms on the left side of the equation $\dfrac{1}{x}+\dfrac{1}{2}=\dfrac{2+x}{2x}$ are collected, the equation becomes $\dfrac{2+x}{2x}=\dfrac{2+x}{2x}$, which is an identity.

In exercises 21–45, solve the equations using the procedures listed on page 79 in your text: eliminate fractions, simplify, isolate the variable term, combine terms, isolate the variable term, and check the solution.

21.
$$3x+5=14$$
$$3x+5-5=14-5$$
$$3x=9$$
$$\frac{3x}{3}=\frac{9}{3}$$
$$x=3$$
Solution set: $\{3\}$

23.
$$-10x+12=32$$
$$-10x+12-12=32-12$$
$$-10x=20$$
$$\frac{-10x}{-10}=\frac{20}{-10}$$
$$x=-2$$
Solution set: $\{-2\}$

25.
$$3-y=-4$$
$$3-y-3=-4-3$$
$$-y=-7 \Rightarrow y=7$$
Solution set: $\{7\}$

27.
$$7x+7=2(x+1)$$
$$7x+7=2x+2$$
$$7x+7-7=2x+2-7$$
$$7x=2x-5$$
$$7x-2x=2x-5-2x$$
$$5x=-5$$
$$\frac{5x}{5}=\frac{-5}{5} \Rightarrow x=-1$$
Solution set: $\{-1\}$

29.
$$3(2-y)+5y=3y$$
$$6-3y+5y=3y$$
$$6+2y=3y$$
$$6+2y-2y=3y-2y \Rightarrow 6=y$$
Solution set: $\{6\}$

31. $4y-3y+7-y=2-(7-y)$
Distribute -1 to clear the parentheses.
$$7=2-7+y$$
$$7=-5+y$$
$$7+5=-5+y+5 \Rightarrow 12=y$$
Solution set: $\{12\}$

33. $3(x-2)+2(3-x)=1$
$$3x-6+6-2x=1 \Rightarrow x=1$$
Solution set: $\{1\}$

35.
$$2x+3(x-4)=7x+10$$
$$2x+3x-12=7x+10$$
$$5x-12=7x+10$$
$$5x-12+12=7x+10+12$$
$$5x=7x+22$$
$$5x-7x=7x+22-7x$$
$$-2x=22$$
$$\frac{-2x}{-2}=\frac{22}{-2} \Rightarrow x=-11$$
Solution set: $\{-11\}$

37. $4[x+2(3-x)]=2x+1$
Distribute 2 to clear the inner parentheses.
$$4[x+6-2x]=2x+1$$
Combine like terms within the brackets.
$$4[6-x]=2x+1$$
Distribute 4 to clear the brackets.
$$24-4x=2x+1$$
$$24-4x-24=2x+1-24$$
$$-4x=2x-23$$
$$-4x-2x=-23$$
$$-6x=-23$$
$$\frac{-6x}{-6}=\frac{-23}{-6} \Rightarrow x=\frac{23}{6}$$
Solution set: $\left\{\dfrac{23}{6}\right\}$

39. $3(4y-3)=4[y-(4y-3)]$
Distribute 3 on the left side and -1 on the right side to clear parentheses.
$$12y-9=4[y-4y+3]$$
Combine like terms in the brackets.
$$12y-9=4[-3y+3]$$
Distribute 4 to clear the brackets.
$$12y-9=-12y+12$$
$$12y-9+9=-12y+12+9$$
$$12y=-12y+21$$
$$12y+12y=-12y+21+12y$$
$$24y=21$$
$$\frac{24y}{24}=\frac{21}{24} \Rightarrow y=\frac{21}{24}=\frac{7}{8}$$
Solution set: $\left\{\dfrac{7}{8}\right\}$

41. $2x-3(2-x)=(x-3)+2x+1$
Distribute -3 on the left to clear the parentheses.
$$2x-6+3x=x-3+2x+1$$
$$5x-6=3x-2$$
(continued on next page)

(continued)

$$5x - 6 + 6 = 3x - 2 + 6$$
$$5x = 3x + 4$$
$$5x - 3x = 3x + 4 - 3x$$
$$2x = 4$$
$$\frac{2x}{2} = \frac{4}{2} \Rightarrow x = 2$$

Solution set: $\{2\}$

43. $\dfrac{2x+1}{9} - \dfrac{x+4}{6} = 1$

To clear the fractions, multiply both sides of the equation by the least common denominator, 36.

$$36\left(\frac{2x+1}{9} - \frac{x+4}{6}\right) = 36(1)$$
$$4(2x+1) - 6(x+4) = 36$$
$$8x + 4 - 6x - 24 = 36$$
$$2x - 20 = 36$$
$$2x - 20 + 20 = 36 + 20$$
$$2x = 56$$
$$\frac{2x}{2} = \frac{56}{2} \Rightarrow x = 28$$

Solution set: $\{28\}$

45. $\dfrac{1-x}{4} + \dfrac{5x+1}{2} = 3 - \dfrac{2(x+1)}{8}$

To clear the fractions, multiply both sides by the least common denominator, 8.

$$8\left(\frac{1-x}{4} + \frac{5x+1}{2}\right) = 8\left(3 - \frac{2(x+1)}{8}\right)$$

Distribute the 8 on both sides.

$$8\left(\frac{1-x}{4}\right) + 8\left(\frac{5x+1}{2}\right) = 8(3) - 8\left(\frac{2(x+1)}{8}\right)$$
$$2(1-x) + 4(5x+1) = 8(3) - 2(x+1)$$

Simplify by collecting like terms and combining constants.

$$2 - 2x + 20x + 4 = 24 - 2x - 2$$
$$18x + 6 = 22 - 2x$$
$$18x + 6 + 2x = 22 - 2x + 2x$$
$$20x + 6 = 22$$
$$20x + 6 - 6 = 22 - 6$$
$$20x = 16$$
$$\frac{20x}{20} = \frac{16}{20}$$
$$x = \frac{16}{20} = \frac{4}{5}$$

Solution set: $\left\{\dfrac{4}{5}\right\}$

47. To solve $d = rt$ for r, divide both sides of the equation by t. $r = \dfrac{d}{t}$.

49. To solve $C = 2\pi r$ for r, divide both sides of the equation by 2π. $r = \dfrac{C}{2\pi}$.

51. To solve $I = \dfrac{E}{R}$ for R, multiply both sides by R.

$$RI = R\left(\frac{E}{R}\right) \Rightarrow RI = E$$

Divide both sides by I.

$$\frac{RI}{I} = \frac{E}{I} \Rightarrow R = \frac{E}{I}$$

53. To solve $A = \dfrac{(a+b)h}{2}$ for h, multiply both sides by 2.

$$2A = (a+b)h$$

Divide both sides by $(a+b)$.

$$\frac{2A}{a+b} = \frac{(a+b)h}{a+b} \Rightarrow \frac{2A}{a+b} = h$$

55. To solve $\dfrac{1}{f} = \dfrac{1}{u} + \dfrac{1}{v}$ for u, clear the fractions by multiplying both sides by the least common denominator, fuv.

$$fuv\left(\frac{1}{f}\right) = fuv\left(\frac{1}{u} + \frac{1}{v}\right)$$
$$fuv\left(\frac{1}{f}\right) = fuv\left(\frac{1}{u}\right) + fuv\left(\frac{1}{v}\right)$$

Simplify.

$$uv = fv + fu$$

Subtract fu from both sides.

$$uv - fu = fv + fu - fu$$
$$uv - fu = fv$$

Factor the left side.

$$u(v - f) = fv$$

Divide both sides by $v - f$.

$$\frac{u(v-f)}{v-f} = \frac{fv}{v-f} \Rightarrow u = \frac{fv}{v-f}$$

57. To solve $y = mx + b$ for m, subtract b from both sides.

$$y - b = mx + b - b \Rightarrow y - b = mx$$

Divide both sides by x.

$$\frac{y-b}{x} = \frac{mx}{x} \Rightarrow \frac{y-b}{x} = m$$

59. $0.065x$

61. $\$22,000 - x$

1.1 Applying the Concepts

63. The formula for volume is $V = lwh$.
Substitute 2808 for V, 18 for l, and 12 for h.
Solve for w.

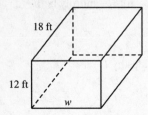

$$2808 = 18 \cdot 12 \cdot w$$
$$2808 = 216w$$
$$\frac{2808}{216} = \frac{216w}{216}$$
$$13 = w$$

The width of the pool is 13 ft.

65. Let w = the width of the rectangle.
Then $2w - 5$ = the length of the rectangle.
$$2w + 2(2w - 5) = 80$$
$$2w + 4w - 10 = 80$$
$$6w - 10 = 80$$
$$6w = 90$$
$$w = 15, \; 2w - 5 = 25$$

The width of the rectangle is 15 ft and its length is 25 feet.

67. The formula for circumference of a circle is
$C = 2\pi r$. Substitute 114π for C. Solve for r.
$$114\pi = 2\pi r \Rightarrow \frac{114\pi}{2\pi} = \frac{2\pi r}{2\pi} \Rightarrow 57 = r$$
The radius is 57 cm.

69. The formula for surface area of a cylinder is
$S = 2\pi rh + 2\pi r^2$. Substitute 6π for S and 1 for r. Solve for h.

$$6\pi = 2\pi(1)h + 2\pi(1^2)$$
$$6\pi = 2\pi h + 2\pi$$
$$6\pi - 2\pi = 2\pi h + 2\pi - 2\pi$$
$$4\pi = 2\pi h$$
$$\frac{4\pi}{2\pi} = \frac{2\pi h}{2\pi} \Rightarrow 2 = h$$

The height is 2 m.

71. The formula for area of a trapezoid is
$A = \frac{1}{2}h(b_1 + b_2)$. Substitute 66 for A, 6 for h,
and 3 for b_1. Solve for b_2.

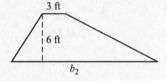

$$66 = \frac{1}{2} \cdot 6(3 + b_2)$$
$$66 = 3(3 + b_2)$$
$$66 = 9 + 3b_2$$
$$66 - 9 = 9 + 3b_2 - 9$$
$$57 = 3b_2$$
$$\frac{57}{3} = \frac{3b_2}{3}$$
$$19 = b_2$$

The length of the second base is 19 ft.

73. Let x = the cost of the less expensive land.
Then $x + 23,000$ = the cost of the more expensive land. Together they cost \$147,000, so
$$x + (x + 23,000) = 147,000$$
$$2x + 23,000 = 147,000$$
$$2x = 124,000 \Rightarrow x = 62,000$$
The less expensive piece of land costs \$62,000 and the more expensive piece of land costs
\$62,000 + \$23,000 = \$85,000.

75. Let x = the lottery ticket sales in July. Then
$1.10x$ = the lottery ticket sales in August.
A total of 1113 tickets were sold, so
$$x + 1.10x = 1113$$
$$2.10x = 1113 \Rightarrow x = 530$$
530 tickets were sold in July, and
$1.10(530) = 583$ tickets were sold in August.

77. Let x = the amount the younger son receives.
Then $4x$ = the amount the older son receives.
Together they receive \$225,000, so
$$x + 4x = 225,000 \Rightarrow 5x = 225,000 \Rightarrow$$
$$x = 45,000$$
The younger son will received \$45,000, and the older son will receive
$4(\$45,000) = \$180,000$.

79. a. Let x = the number of points needed to average 75.
$$\frac{87 + 59 + 73 + x}{4} = 75$$
$$219 + x = 300$$
$$x = 81$$
You need to score 81 in order to average 75.

b. $\dfrac{87 + 59 + 73 + 2x}{5} = 75$
$$219 + 2x = 375$$
$$2x = 156$$
$$x = 78$$
You need to score 78 in order to average 75 if the final carries double weight.

81. Let $x =$ the amount invested in a tax shelter. Then $7000 - x =$ the amount invested in a bank.

Investment	Principal	Rate	Time	Interest
Tax shelter	x	0.09	1	$0.09x$
Bank	$7000 - x$	0.06	1	$0.06(7000 - x)$

The total interest was $540, so
$$0.09x + 0.06(7000 - x) = 540$$
$$0.09x + 420 - 0.06x = 540$$
$$0.03x + 420 = 540$$
$$0.03x = 120 \Rightarrow x = 4000$$
Mr. Mostafa invested $4000 in a tax shelter and $7000 - 4000 = \$3000$ in a bank.

83. Let $x =$ the amount to be invested at 8%.

Principal	Rate	Time	Interest
5000	0.05	1	250
x	0.08	1	$0.08x$
$5000 + x$	0.06	1	$0.06(5000 + x)$

The amount of interest for the total investment is the sum of the interest earned on the individual investments, so
$$0.06(5000 + x) = 250 + 0.08x$$
$$300 + 0.06x = 250 + 0.08x$$
$$50 + 0.06x = 0.08x$$
$$50 = 0.02x \Rightarrow 2500 = x$$
So, $2500 must be invested at 8%.

85. There is a profit of $2 on each shaving set. They want to earn $40,000 + \$30,000 = \$70,000$. Let $x =$ the number of shaving sets to be sold. Then $2x =$ the amount of profit for x shaving sets. So, $2x = 70,000 \Rightarrow x = 35,000$
They must sell 35,000 shaving sets.

87. Let $x =$ the time the second car travels. Then $1 + x =$ the time the first car travels. So,

	Rate	Time	Distance
First car	50	$1 + x$	$50(1 + x)$
Second car	70	x	$70x$

The distances are equal, so
$$50(1 + x) = 70x$$
$$50 + 50x = 70x$$
$$50 = 20x \Rightarrow 2.5 = x$$
So, it will take the second car 2.5 hours to overtake the first car.

89. At 20 miles per hour, it will take Lucas two minutes to bike the remaining 2/3 of a mile.
$$\left(\frac{20\,\text{mi}}{1\,\text{hr}} = \frac{20\,\text{mi}}{60\,\text{min}} = \frac{1\,\text{mi}}{3\,\text{min}} \right)$$ So his brother
will have to bike 1 mile in 2 minutes:
$$\frac{1\,\text{mi}}{2\,\text{min}} = \frac{30\,\text{mi}}{60\,\text{min}} = \frac{30\,\text{mi}}{1\,\text{hr}}$$

91. Let $x =$ the rate the slower car travels. Then $x + 5 =$ the rate the faster car travels. So,

	Rate	Time	Distance
First car	x	3	$3x$
Second car	$x + 5$	3	$3(x + 5)$

The cars are 405 miles apart, so
$$3x + 3(x + 5) = 405$$
$$3x + 3x + 15 = 405$$
$$6x + 15 = 405 \Rightarrow 6x = 390 \Rightarrow x = 65$$
One car is traveling at 65 miles per hour, and the other car is traveling at 70 miles per hour.

93. Substitute 170 for P into the formula $P = 200 - 0.02q$. Solve for q.
$$170 = 200 - 0.02q$$
$$170 - 200 = 200 - 0.02q - 200$$
$$-30 = -0.02q$$
$$\frac{-30}{-0.02} = \frac{-0.02q}{-0.02} \Rightarrow 1500 = q$$
Note that the solution must fall between 100 and 2000 cameras. 1500 cameras must be ordered.

95. Let $x =$ one number. Then $3x =$ the other number. $x + 3x = 28 \Rightarrow 4x = 28 \Rightarrow x = 7, 3x = 3(7) = 21$
The numbers are 7 and 21.

97. Let $x =$ one number. Then $2x$ is the second number.
$$2x - x = 14$$
$$x = 14$$
The numbers are 14 and 28.

1.1 Beyond the Basics

99. a. The solution set of $x^2 = x$ is $\{0,1\}$, while the solution set of $x = 1$ is $\{1\}$. Therefore, the equations are not equivalent.

b. The solution set of $x^2 = 9$ is $\{-3,3\}$, while the solution set of $x = 3$ is $\{3\}$. Therefore, the equations are not equivalent.

c. The solution set of $x^2 - 1 = x - 1$ is $\{0, 1\}$, while the solution set of $x = 0$ is $\{0\}$. Therefore, the equations are not equivalent.

d. The equation $\dfrac{x}{x-2} = \dfrac{2}{x-2}$ is an inconsistent equation, so is solution set is $\varnothing$. The solution set of $x = 2$ is $\{2\}$. Therefore, the equations are not equivalent.

101. Let x = the average speed for the second half of the trip.

	Rate	Distance	Time
1st half	75	D	$\dfrac{D}{75}$
2nd half	x	D	$\dfrac{D}{x}$
Entire trip	60	$2D$	$\dfrac{2D}{60}$

$$\frac{D}{75} + \frac{D}{x} = \frac{2D}{60}$$
$$300x\left(\frac{D}{75} + \frac{D}{x}\right) = 300x\left(\frac{2D}{60}\right)$$
$$4Dx + 300D = 5x(2D)$$
$$4Dx + 300D = 10Dx$$
$$300D = 6Dx \Rightarrow 50 = x$$

The average speed for the second half of the drive is 50 mph.

103. Let x = the number of liters of water in the original mixture. Then $5x$ = the number of liters of alcohol in the original mixture, and $6x$ = the total number of liters in the original mixture. $x + 5$ = the number of liters of water in the new mixture. Then $6x + 5$ = the total number of liters in the new mixture. Since the ratio of alcohol to water in the new mixture is 5:2, then the amount of alcohol in the new mixture is 5/7 of the total mixture or $\dfrac{5}{7}(6x + 5)$.

There was no alcohol added, so the amount of alcohol in the original mixture equals the amount of alcohol in the new mixture. This gives
$$\frac{5}{7}(6x + 5) = 5x$$
$$5(6x + 5) = 35x$$
$$30x + 25 = 35x \Rightarrow 25 = 5x \Rightarrow 5 = x$$

So, there were 5 liters of water in the original mixture and 25 liters of alcohol.

105. Let x = Democratus' age now. Then $x/6$ = the number of years as a boy, $x/8$ = the number of years as a youth, and $x/2$ = the number of years as a man. He has spent 15 years as a mature adult. So,
$$\frac{x}{6} + \frac{x}{8} + \frac{x}{2} + 15 = x$$
$$24\left(\frac{x}{6} + \frac{x}{8} + \frac{x}{2} + 15\right) = 24x$$
$$4x + 3x + 12x + 360 = 24x$$
$$19x + 360 = 24x$$
$$360 = 5x \Rightarrow 72 = x$$
Democratus is 72 years old.

107. There are 180 minutes from 3 p.m. to 6 p.m. So, the number of minutes before 6 p.m. plus 50 minutes plus 4 × the number of minutes before 6 p.m. equals 180 minutes. Let x = the number of minutes before 6 p.m. So,
$x + 50 + 4x = 180 \Rightarrow 5x + 50 = 180 \Rightarrow$
$5x = 130 \Rightarrow x = 26$
So it is 26 minutes before 6 p.m. or 5:34 p.m. Check this by verifying that $26 + 50 = 76$ minutes before 6 p.m. is the same time as $4(26) = 104$ minutes after 3 p.m. Seventy-six minutes before 6 p.m. is 4:44 p.m., while 104 minutes after 3 p.m. is also 4:44 p.m.

109. a. Because of the head wind, the plane flies at 140 mph from Atlanta to Washington and 160 mph from Washington to Atlanta. Let x = the distance the plane flew before turning back. So,

	Rate	Distance	Time
to	140	x	$x/140$
from	160	x	$x/160$

$$\frac{x}{140} + \frac{x}{160} = 1.5$$
$$160x + 140x = 1.5(140)(160)$$
$$300x = 33,600 \Rightarrow x = 112$$
The plane flew 112 miles before turning back.

b. The plane traveled 224 miles in 1.5 hours, so the average speed is $\dfrac{224}{1.5} = 149.33$ mph.

1.1 Critical Thinking/Discussion/Writing

111. If x represents the amount the pawn shop owner paid for the first watch and the owner made a profit of 10%, then $1.1x = 499$, so $x = 453.64$. If y represents the amount the pawn shop owner paid for the second watch and the owner lost 10%, then $0.9y = 499$, so $y = 554.44$. Together the two watches cost $\$453.64 + \$554.44 = \$1008.08$. But the pawn shop owner sold the two watches for $998, so there was a loss. The amount of loss is $(1008.08 - 998)/1008.08 = 10.08/1008.08 \approx 0.01 = 1\%$. The answer is (C).

112. Let x represent the amount of gasoline used in July. Then $0.8x$ represents the amount of gasoline used in August. Let y represent the price of gasoline in July. Then $1.2y$ represents the cost of gasoline in August. The cost of gasoline used in July is xy (amount × price), and the cost of gasoline used in August is $0.8x \times 1.2y = 0.96xy$. So the cost of gasoline used in August is 96% of the cost of gasoline used in July, which is a decrease of 4%. The answer is (D).

1.1 Maintaining Skills

113. $\sqrt{8} = \sqrt{4 \cdot 2} = \sqrt{4}\sqrt{2} = 2\sqrt{2}$

115. $\sqrt{12} = \sqrt{4 \cdot 3} = \sqrt{4}\sqrt{3} = 2\sqrt{3}$

117. $\dfrac{2 + 3\sqrt{8}}{2} = \dfrac{2 + 3 \cdot 2\sqrt{2}}{2} = \dfrac{2}{2} + \dfrac{3 \cdot 2\sqrt{2}}{2} = 1 + 3\sqrt{2}$

119. $\dfrac{15 - \sqrt{75}}{5} = \dfrac{15 - 5\sqrt{3}}{5} = \dfrac{15}{5} - \dfrac{5\sqrt{3}}{5} = 3 - \sqrt{3}$

121. $x^2 + x = x(x + 1)$

123. $x^2 - 4 = (x - 2)(x + 2)$

125. $x^2 + 4x + 4 = (x + 2)^2$

127. $x^2 - 8x + 7 = (x - 1)(x - 7)$

129. $6x^2 - x - 1 = (3x + 1)(2x - 1)$

131. $-5x^2 + 3x + 2 = (5x + 2)(-x + 1)$
$\qquad\qquad\qquad = (5x + 2)(1 - x)$

1.2 Quadratic Equations

1.2 Practice Problems

1. $\quad x^2 + 25x = -84$
$x^2 + 25x + 84 = 0$
$(x + 4)(x + 21) = 0$
$x + 4 = 0 \ \big| \ x + 21 = 0$
$\ \ x = -4 \ \big| \quad\ x = -21$
Solution set: $\{-21, -4\}$

2. $\quad 2m^2 = 5m$
$2m^2 - 5m = 0$
$m(2m - 5) = 0$
$m = 0 \ \big| \ 2m - 5 = 0$
$\quad\quad\ \big| \quad\ 2m = 5$
$\quad\quad\ \big| \quad\quad\ m = \dfrac{5}{2}$
Solution set: $\left\{0, \dfrac{5}{2}\right\}$

3. $\quad x^2 - 6x = -9$
$x^2 - 6x + 9 = 0$
$(x - 3)^2 = 0$
$x - 3 = 0$
$x = 3$
Solution set: $\{3\}$

4. $(x + 2)^2 = 5$
$x + 2 = \pm\sqrt{5}$
$x = -2 \pm \sqrt{5}$
Solution set: $\left\{-2 - \sqrt{5}, \ -2 + \sqrt{5}\right\}$

5. $x^2 - 6x + 7 = 0$
$x^2 - 6x = -7$
$x^2 - 6x + 9 = -7 + 9$
$(x - 3)^2 = 2$
$x - 3 = \pm\sqrt{2}$
$x = 3 \pm \sqrt{2}$
Solution set: $\left\{3 - \sqrt{2}, \ 3 + \sqrt{2}\right\}$

6. $4x^2 - 24x + 25 = 0$

$$4x^2 - 24x = -25$$

$$x^2 - 6x = -\frac{25}{4}$$

$$x^2 - 6x + 9 = -\frac{25}{4} + 9$$

$$(x-3)^2 = \frac{11}{4}$$

$$x - 3 = \pm\frac{\sqrt{11}}{2} \Rightarrow x = 3 \pm \frac{\sqrt{11}}{2}$$

Solution set: $\left\{ 3 - \dfrac{\sqrt{11}}{2}, \ 3 + \dfrac{\sqrt{11}}{2} \right\}$

7. $6x^2 - x - 2 = 0$
$a = 6, \ b = -1, \ c = -2$

$$x = \frac{-b \pm \sqrt{b^2 - 4ac}}{2a}$$

$$= \frac{-(-1) \pm \sqrt{(-1)^2 - 4(6)(-2)}}{2(6)}$$

$$= \frac{1 \pm \sqrt{49}}{12} = \frac{1 \pm 7}{12} = \frac{-6}{12} = -\frac{1}{2} \text{ or } \frac{8}{12} = \frac{2}{3}$$

Solution set: $\left\{ -\dfrac{1}{2}, \dfrac{2}{3} \right\}$

8. Let x = the frontage of the building. Then $5x$ = the depth of the building and $5x - 45$ = the depth of the rear portion.

$$x(5x - 45) = 2100$$

$$5x^2 - 45x = 2100$$

$$5x^2 - 45x - 2100 = 0$$

$a = 5, \ b = -45, \ c = -2100$

$$x = \frac{-(-45) \pm \sqrt{(-45)^2 - 4(5)(-2100)}}{2(5)}$$

$$= \frac{45 \pm \sqrt{44,025}}{10} \approx -16.48 \text{ or } 25.48$$

Reject the negative solution.
$5x = 5 \cdot 25.482 = 127.41$
The building is approximately 25.48 ft by 127.41 ft.

9. $\Phi = \dfrac{\text{length}}{\text{width}} \Rightarrow \dfrac{1 + \sqrt{5}}{2} = \dfrac{x}{36}$

$$x = 36\left(\frac{1 + \sqrt{5}}{2} \right) = 18 + 18\sqrt{5} \approx 58.25 \text{ ft}$$

1.2 Basic Concepts and Skills

1. Any equation of the form $ax^2 + bx + c = 0$ with $a \neq 0$, is called a <u>quadratic</u> equation.

3. From the Square Root Property, we know that if $u^2 = 5$, then $\underline{u = \pm\sqrt{5}}$.

5. True

7. $(-6)^2 + 4(-6) - 12 = 36 - 24 - 12 = 0$
-6 is a solution of the equation.

9. $3\left(\dfrac{2}{3}\right)^2 + 7\left(\dfrac{2}{3}\right) - 6 = 3\left(\dfrac{4}{9}\right) + \dfrac{14}{3} - 6$

$$= \frac{4}{3} + \frac{14}{3} - 6 = 0$$

$2/3$ is a solution of the equation.

11. $\left(2 - \sqrt{3}\right)^2 - 4\left(2 - \sqrt{3}\right) + 1$

$$= \left(4 - 4\sqrt{3} + 3\right) - 8 + 4\sqrt{3} + 1 = 0$$

$2 - \sqrt{3}$ is a solution of the equation.

13. $4\left(2 + \sqrt{3}\right)^2 - 8\left(2 + \sqrt{3}\right) + 13$

$$= 4\left(7 + 4\sqrt{3}\right) - 8\left(2 + \sqrt{3}\right) + 13$$

$$= 28 + 16\sqrt{3} - 16 - 8\sqrt{3} + 13 = 25 + 8\sqrt{3} \neq 0$$

$2 + \sqrt{3}$ is not a solution of the equation.

15. $k(1)^2 + 1 - 3 = 0 \Rightarrow k - 2 = 0 \Rightarrow k = 2$

17. $x^2 - 5x = 0 \Rightarrow x(x - 5) = 0 \Rightarrow$
$x = 0 \text{ or } x - 5 = 0 \Rightarrow x = 0 \text{ or } x = 5$

19. $x^2 + 5x = 14$
$x^2 + 5x - 14 = 0$
$(x + 7)(x - 2) = 0$
$x + 7 = 0 \text{ or } x - 2 = 0$
$x = -7 \text{ or } x = 2$

21. $x^2 = 5x + 6$
$x^2 - 5x - 6 = 0$
$(x - 6)(x + 1) = 0$
$x - 6 = 0 \text{ or } x + 1 = 0$
$x = 6 \text{ or } x = -1$

23. $3x^2 = 48 \Rightarrow x^2 = 16 \Rightarrow x = \pm 4$

25. $x^2 + 1 = 5 \Rightarrow x^2 = 4 \Rightarrow x = \pm 2$

27. $(x - 1)^2 = 16$
$x - 1 = -4 \text{ or } x - 1 = 4 \Rightarrow x = -3 \text{ or } x = 5$

29. To complete the square, find $1/2$ of the coefficient of the x-term, $4/2 = 2$, and then square the answer. $2^2 = 4$.

31. To complete the square, find $1/2$ of the coefficient of the x-term, $6/2 = 3$, and then square the answer. $3^2 = 9$.

33. To complete the square, find $1/2$ of the coefficient of the x-term and then square the answer. $\left(\dfrac{7}{2}\right)^2 = \dfrac{49}{4}$.

35. To complete the square, find $1/2$ of the coefficient of the x-term, $\dfrac{1}{2}\cdot\dfrac{1}{3} = \dfrac{1}{6}$ and then square the answer. $\left(\dfrac{1}{6}\right)^2 = \dfrac{1}{36}$.

37. To complete the square, find $1/2$ of the coefficient of the x-term and then square the answer. $(a/2)^2 = a^2/4$.

39. $x^2 + 2x - 5 = 0 \Rightarrow x^2 + 2x = 5$
Now, complete the square.
$x^2 + 2x + 1 = 5 + 1 \Rightarrow (x+1)^2 = 6 \Rightarrow$
$x + 1 = \pm\sqrt{6} \Rightarrow x = -1 \pm \sqrt{6}$

41. $x^2 - 3x - 1 = 0$
$\quad x^2 - 3x = 1$
Now, complete the square.
$x^2 - 3x + \dfrac{9}{4} = 1 + \dfrac{9}{4}$
$\left(x - \dfrac{3}{2}\right)^2 = \dfrac{13}{4}$
$x - \dfrac{3}{2} = \pm\dfrac{\sqrt{13}}{2}$
$x = \dfrac{3}{2} \pm \dfrac{\sqrt{13}}{2} = \dfrac{3 \pm \sqrt{13}}{2}$

43.
$2r^2 + 3r = 9$
$r^2 + \dfrac{3}{2}r = \dfrac{9}{2}$
$r^2 + \dfrac{3}{2}r + \dfrac{9}{16} = \dfrac{9}{2} + \dfrac{9}{16}$
$\left(r + \dfrac{3}{4}\right)^2 = \dfrac{81}{16}$
$r + \dfrac{3}{4} = \pm\dfrac{9}{4} \Rightarrow r = \dfrac{-3 \pm 9}{4} = \dfrac{3}{2}$ or -3

In exercises 45–49, use the quadratic formula
$$x = \frac{-b \pm \sqrt{b^2 - 4ac}}{2a}.$$

45. $x^2 + 2x - 4 = 0 \Rightarrow a = 1, b = 2, c = -4$
$x = \dfrac{-2 \pm \sqrt{2^2 - 4(1)(-4)}}{2(1)} = \dfrac{-2 \pm \sqrt{4 + 16}}{2}$
$= \dfrac{-2 \pm \sqrt{20}}{2} = \dfrac{-2 \pm 2\sqrt{5}}{2} = -1 \pm \sqrt{5}$

47. $6x^2 = 7x + 5 \Rightarrow 6x^2 - 7x - 5 = 0 \Rightarrow$
$a = 6, b = -7, c = -5$
$x = \dfrac{-(-7) \pm \sqrt{(-7)^2 - 4(6)(-5)}}{2(6)}$
$= \dfrac{7 \pm \sqrt{49 + 120}}{12} = \dfrac{7 \pm \sqrt{169}}{12} = \dfrac{7 \pm 13}{12}$
$x = \dfrac{7 + 13}{12} = \dfrac{20}{12} = \dfrac{5}{3}$ or
$x = \dfrac{7 - 13}{12} = \dfrac{-6}{12} = -\dfrac{1}{2}$

49. $3z^2 - 2z = 7 \Rightarrow 3z^2 - 2z - 7 = 0 \Rightarrow$
$a = 3, b = -2, c = -7$
$z = \dfrac{-(-2) \pm \sqrt{(-2)^2 - 4(3)(-7)}}{2(3)}$
$= \dfrac{2 \pm \sqrt{4 + 84}}{6} = \dfrac{2 \pm \sqrt{88}}{6} = \dfrac{2 \pm 2\sqrt{22}}{6}$
$= \dfrac{1 \pm \sqrt{22}}{3}$

51. $2x^2 + 5x - 3 = 0$
$(2x - 1)(x + 3) = 0$
$\quad 2x - 1 = 0$ or $x + 3 = 0$
$\quad\quad x = \dfrac{1}{2}$ or $x = -3$

53. $(3x - 2)^2 - 16 = 0$
$(3x - 2)^2 = 16$
$3x - 2 = \pm 4$
$3x - 2 = -4$ or $3x - 2 = 4$
$3x = -2 \quad\quad\quad 3x = 6$
$x = -\dfrac{2}{3}$ or $x = 2$

55.
$$5x^2 - 6x = 4x^2 + 6x - 3$$
$$x^2 - 12x = -3$$
Now, complete the square.
$$x^2 - 12x + 36 = -3 + 36$$
$$(x - 6)^2 = 33$$
$$x - 6 = \pm\sqrt{33} \Rightarrow x = 6 \pm \sqrt{33}$$

57. $3p^2 + 8p + 4 = 0 \Rightarrow a = 3, b = 8, c = 4$
$$p = \frac{-8 \pm \sqrt{8^2 - 4(3)(4)}}{2(3)}$$
$$= \frac{-8 \pm \sqrt{64 - 48}}{6} = \frac{-8 \pm \sqrt{16}}{6} = \frac{-8 \pm 4}{6}$$
$$p = \frac{-4}{6} = -\frac{2}{3} \text{ or } p = \frac{-12}{6} = -2$$

59.
$$3y^2 + 5y + 2 = 0$$
$$(3y + 2)(y + 1) = 0$$
$$3y + 2 = 0 \text{ or } y + 1 = 0$$
$$y = -\frac{2}{3} \text{ or } y = -1$$

61. $5x^2 + 12x + 4 = 0$
$$(5x + 2)(x + 2) = 0$$
$$5x + 2 = 0 \text{ or } x + 2 = 0$$
$$x = -\frac{2}{5} \text{ or } x = -2$$

63. $5y^2 + 10y + 4 = 2y^2 + 3y + 1$
$$3y^2 + 7y = -3$$
$$y^2 + \frac{7}{3}y = -1$$
Now, complete the square.
$$y^2 + \frac{7}{3}y + \frac{49}{36} = -1 + \frac{49}{36}$$
$$\left(y + \frac{7}{6}\right)^2 = \frac{13}{36}$$
$$y + \frac{7}{6} = \pm\sqrt{\frac{13}{36}} = \pm\frac{\sqrt{13}}{6}$$
$$y = -\frac{7}{6} \pm \frac{\sqrt{13}}{6} = \frac{-7 \pm \sqrt{13}}{6}$$

65.
$$2x^2 + x = 15$$
$$2x^2 + x - 15 = 0$$
$$(2x - 5)(x + 3) = 0$$
$$2x - 5 = 0 \text{ or } x + 3 = 0 \Rightarrow x = \frac{5}{2} \text{ or } x = -3$$

67.
$$12x^2 - 10x = 12$$
$$12x^2 - 10x - 12 = 0$$
$$2(6x^2 - 5x - 6) = 0$$
$$6x^2 - 5x - 6 = 0$$
$$(3x + 2)(2x - 3) = 0$$
$$3x + 2 = 0 \text{ or } 2x - 3 = 0$$
$$x = -\frac{2}{3} \text{ or } x = \frac{3}{2}$$

69. $(x + 13)(x + 5) = -2 \Rightarrow x^2 + 18x + 65 = -2 \Rightarrow$
$$x^2 + 18x + 67 = 0 \Rightarrow a = 1, b = 18, c = 67$$
$$x = \frac{-18 \pm \sqrt{18^2 - 4(1)(67)}}{2(1)}$$
$$= \frac{-18 \pm \sqrt{324 - 268}}{2} = \frac{-18 \pm \sqrt{56}}{2}$$
$$= \frac{-18 \pm 2\sqrt{14}}{2} = -9 \pm \sqrt{14}$$

71.
$$18x^2 - 45x = -7$$
$$18x^2 - 45x + 7 = 0$$
$$(3x - 7)(6x - 1) = 0$$
$$3x - 7 = 0 \text{ or } 6x - 1 = 0 \Rightarrow x = \frac{7}{3} \text{ or } x = \frac{1}{6}$$

73. $2t^2 - 5 = 0 \Rightarrow a = 2, b = 0, c = -5$
$$t = \frac{0 \pm \sqrt{0^2 - 4(2)(-5)}}{2(2)} = \frac{\pm\sqrt{40}}{4}$$
$$= \pm\frac{2\sqrt{10}}{4} = \pm\frac{\sqrt{10}}{2}$$

75.
$$4x^2 - 10x - 750 = 0$$
$$2(2x^2 - 5x - 375) = 0$$
$$2x^2 - 5x - 375 = 0$$
$$(2x + 25)(x - 15) = 0$$
$$2x + 25 = 0 \text{ or } x - 15 = 0$$
$$x = -\frac{25}{2} \text{ or } x = 15$$

77. $\Phi = \frac{\text{length}}{\text{width}} \Rightarrow \frac{1 + \sqrt{5}}{2} = \frac{x}{14.72}$
$$x = 14.72\left(\frac{1 + \sqrt{5}}{2}\right) \approx 23.82 \text{ in.}$$

79. $\Phi = \frac{\text{length}}{\text{width}} \Rightarrow \frac{1 + \sqrt{5}}{2} = \frac{8.46}{x}$
$$x = 8.46\left(\frac{2}{1 + \sqrt{5}}\right) \approx 5.23 \text{ cm}$$

1.2 Applying the Concepts

81. Let x = the width of the plot. Then $3x$ = the length of the plot. So, $3x^2 = 10,800 \Rightarrow$

$x^2 = 3600 \Rightarrow x = 60$.

The plot is 60 ft by 180 ft.

83. Let x = the first integer. Then $28 - x$ = the second integer. So, $x(28 - x) = 147 \Rightarrow$

$28x - x^2 = 147 \Rightarrow 0 = x^2 - 28x + 147 \Rightarrow$
$0 = (x - 7)(x - 21) \Rightarrow x = 7$ or $x = 21$.

The sum of the two numbers is 28. So, the numbers are 7 and 21.

85. Let x = one number. Then $57 - x$ = the other number.

$$x(57 - x) = 782$$
$$-x^2 + 57x = 782$$
$$x^2 - 57x = -782$$
$$x^2 - 57x + 782 = 0$$
$$(x - 23)(x - 34) = 0$$
$$x - 23 = 0 \text{ or } x - 34 = 0$$
$$x = 23 \text{ or } x = 34$$

87. Let x = the width of the rectangle. Then $x + 5$ = the length of the rectangle. So,
$x(x + 5) = 500$

$\Rightarrow x^2 + 5x = 500 \Rightarrow x^2 + 5x - 500 = 0 \Rightarrow$
$(x + 25)(x - 20) = 0 \Rightarrow x = -25$ or $x = 20$.

Since the length cannot be negative, we reject that solution. $x = 20 \Rightarrow x + 5 = 25$. So the rectangle is 25 cm by 20 cm.

89. Let x = one piece of the wire. Then $16 - x$ = the other piece of the wire. Each piece is bent into a square, so the sides of the squares are

$\dfrac{x}{4}$ and $\dfrac{16 - x}{4}$, respectively.

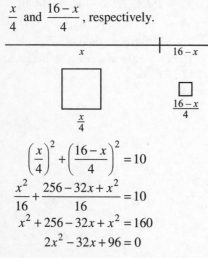

$$\left(\frac{x}{4}\right)^2 + \left(\frac{16 - x}{4}\right)^2 = 10$$
$$\frac{x^2}{16} + \frac{256 - 32x + x^2}{16} = 10$$
$$x^2 + 256 - 32x + x^2 = 160$$
$$2x^2 - 32x + 96 = 0$$

$$x^2 - 16x + 48 = 0$$
$$(x - 12)(x - 4) = 0 \Rightarrow x = 12 \text{ or } x = 4$$

If one piece of the wire is 12, then the other piece is $16 - 12 = 4$. So the pieces are 12 in. and 4 in.

91. Let r = the radius of the can. Then

$$32\pi = 2\pi r(6) + 2\pi r^2$$
$$32\pi = 12\pi r + 2\pi r^2$$

Divide both sides by 2π

$$16 = 6r + r^2$$
$$0 = r^2 + 6r - 16$$
$$0 = (r - 2)(r + 8)$$
$$r = 2 \text{ or } r = -8$$

The answer cannot be negative, so we reject −8. The radius of the can is 2 inches.

93. Let x = the length of the piece of tin. Then $x - 10$ = the length of the box.

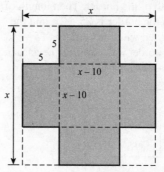
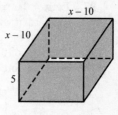

$$(x - 10)(x - 10)(5) = 480$$
$$5(x^2 - 20x + 100) = 480$$
$$x^2 - 20x + 100 = 96$$
$$x^2 - 20x + 4 = 0$$

Solve using the quadratic formula with $a = 1$, $b = -20$, and $c = 4$.

$$x = \frac{-(-20) \pm \sqrt{(-20)^2 - 4(1)(4)}}{2(1)}$$
$$= \frac{20 \pm \sqrt{400 - 16}}{2} = \frac{20 \pm \sqrt{384}}{2}$$
$$\approx \frac{20 + 19.6}{2} \text{ or } \frac{20 - 19.6}{2} \approx 19.8 \text{ or } 0.18$$

The length cannot be 0.18 inches, so we reject that solution. The tin is 19.8 in. by 19.8 in.

95. Let x = the time the buses travel. So the distance the first bus travels = $52x$ mi, and the distance the second bus travels = $39x$ mi.

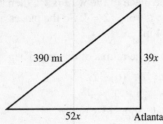

Using the Pythagorean theorem, we have
$$(52x)^2 + (39x)^2 = 390^2$$
$$2704x^2 + 1521x^2 = 152,100$$
$$4225x^2 = 152,100$$
$$x^2 = 36 \Rightarrow x = \pm 6$$
Time cannot be negative, so we reject –6. The buses will be 390 miles apart after 6 hours.

97. Let x = the width of the border. Then length of the garden with the border is $25 + 2x$, and the width of the garden with the border is $15 + 2x$. The area of the border = the area of the garden with the border – the area of the garden.

$$A_b = 624 = (25 + 2x)(15 + 2x) - (15)(25)$$
$$= 375 + 80x + 4x^2 - 375$$
$$= 80x + 4x^2$$
$$624 = 80x + 4x^2$$
$$0 = 4x^2 + 80x - 624$$
$$0 = x^2 + 20x - 156$$
$$0 = (x + 26)(x - 6)$$
$$x = -26 \text{ or } x = 6$$
We reject the negative solution. The width of the border is 6 feet.

99. a. $h = -16(2^2) + 112(2) = 160$ feet

b. $96 = -16t^2 + 112t \Rightarrow 16t^2 - 112t + 96 = 0$
$\Rightarrow t^2 - 7t + 6 = 0 \Rightarrow (t - 1)(t - 6) = 0 \Rightarrow$
$t = 1$ or $t = 6$
The ball will be at a height of 96 feet at 1 second and at 6 seconds.

c. $0 = -16t^2 + 112t \Rightarrow 16t^2 - 112t = 0 \Rightarrow$
$t^2 - 7t = 0 \Rightarrow t(t - 7) = 0 \Rightarrow t = 0$ or
$t = 7$
The ball will return to the ground at 7 seconds.

101. a. $-16t^2 + 96t + 480 = 592$
$$-16t^2 + 96t - 112 = 0$$
$$a = -16, b = 96, c = -112$$
$$x = \frac{-96 \pm \sqrt{96^2 - 4(-16)(-112)}}{2(-16)}$$
$$\approx 1.59 \text{ or } 4.41$$
The projectile will be at a height of 592 ft at about 1.59 sec and 4.41 sec.

b. $-16t^2 + 96t + 480 = 0$
$$a = -16, b = 96, c = 480$$
$$x = \frac{-96 \pm \sqrt{96^2 - 4(-16)(480)}}{2(-16)}$$
$$\approx -3.24 \text{ (reject this) or } 9.24$$
The projectile will crash on the ground after approximately 9.24 seconds.

103. a. $-16t^2 + 256t + 480 = 592$
$$-16t^2 + 256t - 112 = 0$$
$$a = -16, b = 256, c = -112$$
$$x = \frac{-256 \pm \sqrt{256^2 - 4(-16)(-112)}}{2(-16)}$$
$$\approx 0.45 \text{ or } 15.55$$
The projectile will be at a height of 592 ft at about 0.45 sec and 15.55 sec.

b. $-16t^2 + 256t + 480 = 0$
$$a = -16, b = 256, c = 480$$
$$x = \frac{-256 \pm \sqrt{256^2 - 4(-16)(480)}}{2(-16)}$$
$$\approx -1.70 \text{ (reject this) or } 17.70$$
The projectile will crash on the ground after 17.70 seconds.

1.2 Beyond the Basics

In exercises 105–109, the equation has equal roots if it can be written in the form $(ax+b)^2 = 0$ or $(ax-b)^2 = 0$. Then we equate the coefficients.

105. $x^2 - kx + 3 = (x-b)^2 = x^2 - 2bx + b^2$

So, $b^2 = 3 \Rightarrow b = \pm\sqrt{3}$.

Then, $k = 2b = 2(\pm\sqrt{3}) = \pm 2\sqrt{3}$.

107. $2x^2 + kx + k = 2(x-b)^2 = 2x^2 - 4bx + 2b^2$

So, $k = 2b^2$. Then, $k = -4b \Rightarrow 2b^2 = -4b \Rightarrow$
$b^2 + 2b = 0 \Rightarrow b(b+2) = 0 \Rightarrow b = 0$ or $b = -2$.

Therefore, $k = 2(0)^2 = 0$ or $k = 2(-2)^2 = 8$.

109. $x^2 + k^2 = 2(k+1)x \Rightarrow$
$x^2 - 2(k+1)x + k^2 = (x-b)^2 = x^2 - 2bx + b^2$

So, $k^2 = b^2 \Rightarrow k = \pm b$.
$2(k+1) = 2b \Rightarrow k + 1 = b$

If $k = b$, we have $b + 1 = b$, which is false, so we disregard this solution. If $k = -b$, we have

$-b + 1 = b \Rightarrow 1 = 2b \Rightarrow b = \dfrac{1}{2}$. Therefore,

$k = \dfrac{1}{2}$.

111. Since r and s are roots of the equation, then
$$ar^2 + br + c = 0 = as^2 + bs + c \Rightarrow$$
$$ar^2 + br + c - c = as^2 + bs + c - c$$
$$ar^2 + br = as^2 + bs$$
$$ar^2 - as^2 = bs - br$$
$$a(r^2 - s^2) = b(s - r)$$
$$a(r^2 - s^2) = -b(r - s)$$
$$\frac{r^2 - s^2}{r - s} = -\frac{b}{a} \Rightarrow r + s = -\frac{b}{a}$$

To find $r \cdot s$, first divide both sides of

$ar^2 + br + c = 0$ by a. We have

$r^2 + \dfrac{b}{a}r + \dfrac{c}{a} = 0$. Now use the results from

the first part of the problem and substitute

$-(r+s)$ for $\dfrac{b}{a}$. This gives

$$r^2 - (r+s)r + \frac{c}{a} = 0$$
$$r^2 - r^2 - rs + \frac{c}{a} = 0$$
$$\frac{c}{a} = rs$$

113. Using the results from exercise 111, we have

$-\dfrac{k-3}{2} = \dfrac{3k-5}{2} \Rightarrow -k + 3 = 3k - 5 \Rightarrow$
$-4k = -8 \Rightarrow k = 2$

115. From exercise 111, we have

$r + s = -\dfrac{b}{a} \Rightarrow b = -a(r+s)$ and

$rs = \dfrac{c}{a} \Rightarrow c = ars$. Substitute these values

into the equation:
$$ax^2 + bx + c = ax^2 - a(r+s)x + ars$$
$$= a(x^2 - (r+s)x + rs)$$
$$= a(x^2 - rx - sx + rs)$$
$$= a[x(x-r) - s(x-r)]$$
$$= a(x-r)(x-s)$$

117. a. $(x-(-3))(x-4) = 0 \Rightarrow (x+3)(x-4) = 0 \Rightarrow$
$x^2 - x - 12 = 0$

b. $(x-5)(x-5) = 0 \Rightarrow x^2 - 10x + 25 = 0$

c. $(x-(3+\sqrt{2}))(x-(3-\sqrt{2})) = 0 \Rightarrow$
$x^2 - 3x - x\sqrt{2} - 3x + x\sqrt{2} + 7 = 0 \Rightarrow$
$x^2 - 6x + 7 = 0$

119. To show that a rectangle is a golden rectangle, we need to show that the ratio of the longer side to the shorter side equals the golden ratio,

Φ, or $\dfrac{1+\sqrt{5}}{2}$.

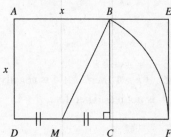

Note that $DM = CM = \dfrac{x}{2}$. Use the

Pythagorean theorem to find the length of MB.

(continued on next page)

(*continued*)

$$MB^2 = CM^2 + BC^2 = \left(\frac{x}{2}\right)^2 + x^2 = \frac{x^2}{4} + x^2$$

$$= \frac{5}{4}x^2 \Rightarrow MB = \frac{\sqrt{5}}{2}x.$$

Then

$$DF = DM + MF = \frac{x}{2} + \frac{\sqrt{5}}{2}x = \frac{1+\sqrt{5}}{2}x, \text{ and}$$

$$\frac{DF}{AD} = \frac{\left(\dfrac{1+\sqrt{5}}{2}\right)x}{x} = \frac{1+\sqrt{5}}{2} = \Phi.$$

So, *AEFD* is a golden rectangle.

121. Using the results from exercises 111 and 115, if r, s, and t are three distinct roots, we know

that $rs = \dfrac{c}{a}, st = \dfrac{c}{a},$ and $rt = \dfrac{c}{a}$. Then

$rs = st \Rightarrow r = s$, and $rs = rt \Rightarrow s = t$. So, $r = s = t$, which contradicts our assumption. Therefore, there cannot be three distinct roots of a quadratic equation.

123. Amar's solutions of 8 and 2 give the equation $(x-8)(x-2) = x^2 - 10x + 16 = 0$. Akbar's solutions of –9 and –1 give the equation $(x+9)(x+1) = x^2 + 10x + 9 = 0$. We know that Akbar misread the coefficient of the x term, and that Amar misread the constant term, so the correct equation must be $x^2 - 10x + 9 = 0$. Anthony correctly solved the equation:

$x^2 - 10x + 9 = 0 \Rightarrow (x-9)(x-1) = 0 \Rightarrow$
$x = 9$ or $x = 1$

125. B. If $k > 4$, then we have

$$x^2 + 4x = -k$$
$$x^2 + 4x + 4 = -k + 4$$
$$(x+2)^2 = -k + 4$$
$$x + 2 = \pm\sqrt{-k+4}$$

However, if $k > 4$, then $-k + 4$ is negative and $\sqrt{-k+4}$ is not real. Therefore, $k \le 4$.

1.2 Critical Thinking/Discussion/Writing

126. $(x-a)(x-b) = k^2 \Rightarrow$
$x^2 - (a+b)x + ab - k^2 = 0$

$$\left[-(a+b)\right]^2 - 4(1)\left(ab - k^2\right)$$
$$= a^2 + 2ab + b^2 - 4ab + 4k^2$$
$$= a^2 - 2ab + b^2 + 4k^2$$
$$= (a-b)^2 + 4k^2 > 0$$

Thus, the solutions of the equation are real.

127. $ax(1-x) = 1 \Rightarrow ax - ax^2 = 1 \Rightarrow$
$-ax^2 + ax - 1 = 0$
Examining the discriminant, we have
$a^2 - 4(-a)(-1) = a^2 - 4a$

$a^2 - 4a < 0 \Rightarrow ax(1-x) = 1$ has no real solutions for $0 < a < 4$.

1.2 Maintaining Skills

129. $\left(1 - 2\sqrt{7}\right)\left(1 + 2\sqrt{7}\right) = 1^2 - \left(2\sqrt{7}\right)^2 = 1 - 4 \cdot 7$
$\qquad\qquad = 1 - 28 = -27$

131. $\left(\sqrt{2} - \sqrt{3}\right)^2 = \left(\sqrt{2}\right)^2 - 2 \cdot \sqrt{2} \cdot \sqrt{3} + \left(\sqrt{3}\right)^2$
$\qquad\qquad = 2 - 2\sqrt{6} + 3 = 5 - 2\sqrt{6}$

133. $\dfrac{6 + \sqrt{27}}{3} = 2 + \dfrac{\sqrt{9 \cdot 3}}{3} = 2 + \dfrac{3\sqrt{3}}{3} = 2 + \sqrt{3}$

135. $\dfrac{18 - \sqrt{108}}{12} = \dfrac{18 - \sqrt{36 \cdot 3}}{12} = \dfrac{18 - 6\sqrt{3}}{12} = \dfrac{3 - \sqrt{3}}{2}$

137. $(5x - 9) + (6 - x) = 4x - 3$

139. $(6x - 5) - (3x + 4) = 6x - 5 - 3x - 4 = 3x - 9$

141. $(2x - 5)(3x + 4) = 6x^2 + 8x - 15x - 20$
$\qquad\qquad = 6x^2 - 7x - 20$

143. $(5x + 2)(5x - 2) = (5x)^2 - 2^2 = 25x^2 - 4$

Solve each equation in exercises 145–147 using the

quadratic formula, $x = \dfrac{-b \pm \sqrt{b^2 - 4ac}}{2a}$.

145. $x^2 + 5x + 5 = 0$
$a = 1, b = 5, c = 5$

$$x = \frac{-5 \pm \sqrt{5^2 - 4 \cdot 1 \cdot 5}}{2 \cdot 1} = \frac{-5 \pm \sqrt{5}}{2}$$

147. $-2x^2 + 5x - 1 = 0$

$a = -2, b = 5, c = -1$

$$x = \frac{-5 \pm \sqrt{5^2 - 4(-2)(-1)}}{2(-2)} = \frac{-5 \pm \sqrt{17}}{-4}$$

$$= \frac{5 \pm \sqrt{17}}{4}$$

1.3 Complex Numbers: Quadratic Equations with Complex Solutions

1.3 Practice Problems

1. a. $-1 + 2i$

real part: -1, imaginary part: 2

b. $-\frac{1}{3} - 6i$

real part: $-\frac{1}{3}$, imaginary part: -6

c. $8 = 8 + 0i$

real part: 8, imaginary part: 0

2. Let $z = (1 - 2a) + 3i$ and let $w = 5 - (2b - 5)i$.
Then

$\text{Re}(z) = \text{Re}(w)$ and $\text{Im}(z) = \text{Im}(w)$

$1 - 2a = 5$	$3 = -(2b - 5)$
$-2a = 4$	$3 = -2b + 5$
$a = -2$	$-2 = -2b$
	$1 = b$

3. a. $(1 - 4i) + (3 + 2i) = 4 - 2i$

b. $(4 + 3i) - (5 - i) = -1 + 4i$

c. $(3 - \sqrt{-9}) - (5 - \sqrt{-64}) = (3 - 3i) - (5 - 8i)$

$$= -2 + 5i$$

4. a. $(2 - 6i)(1 + 4i) = 2 + 8i - 6i - 24i^2$

$$= 2 + 2i + 24 = 26 + 2i$$

b. $-3i(7 - 5i) = -21i + 15i^2 = -15 - 21i$

5. a. $(-3 + \sqrt{-4})^2 = (-3 + 2i)^2$

$$= (-3)^2 + 2(-3)(2i) + (2i)^2$$

$$= 9 - 12i + 4i^2 = 9 - 12i - 4$$

$$= 5 - 12i$$

b. $(5 + \sqrt{-2})(4 + \sqrt{-8}) = (5 + i\sqrt{2})(4 + 2i\sqrt{2})$

$$= 20 + 10\sqrt{2}i + 4\sqrt{2}i + 4i^2$$

$$= 20 + 14\sqrt{2}i - 4$$

$$= 16 + 14\sqrt{2}i$$

6. a. $z = 1 + 6i \Rightarrow \bar{z} = 1 - 6i$

$z\bar{z} = (1 + 6i)(1 - 6i) = 1 - 36i^2 = 1 + 36 = 37$

b. $z = -2i \Rightarrow \bar{z} = 2i$

$z\bar{z} = (-2i)(2i) = -4i^2 = 4$

7. a. $\dfrac{2}{1-i} = \dfrac{2}{1-i} \cdot \dfrac{1+i}{1+i} = \dfrac{2+2i}{1-i^2} = \dfrac{2+2i}{1+1}$

$$= \frac{2+2i}{2} = 1 + i$$

b. $\dfrac{-3i}{4 + \sqrt{-25}} = \dfrac{-3i}{4 + 5i} = \dfrac{-3i}{4 + 5i} \cdot \dfrac{4 - 5i}{4 - 5i}$

$$= \frac{-12i + 15i^2}{16 - 25i^2} = \frac{-15 - 12i}{16 + 25}$$

$$= \frac{-15 - 12i}{41} = -\frac{15}{41} - \frac{12i}{41}$$

8. $Z_t = \dfrac{Z_1 Z_2}{Z_1 + Z_2} = \dfrac{(1 + 2i)(2 - 3i)}{(1 + 2i) + (2 - 3i)}$

$$= \frac{2 - 3i + 4i - 6i^2}{3 - i} = \frac{2 + i + 6}{3 - i} = \frac{8 + i}{3 - i}$$

$$= \frac{8 + i}{3 - i} \cdot \frac{3 + i}{3 + i} = \frac{24 + 8i + 3i + i^2}{9 - i^2}$$

$$= \frac{24 + 11i - 1}{9 + 1} = \frac{23 + 11i}{10} = \frac{23}{10} + \frac{11}{10}i$$

9. a. $4x^2 + 9 = 0$

$$4x^2 = -9$$

$$x^2 = -\frac{9}{4} \Rightarrow x = \pm\sqrt{-\frac{9}{4}} = \pm\frac{3}{2}i$$

Solution set: $\left\{ -\dfrac{3}{2}i, \dfrac{3}{2}i \right\}$

b. $x^2 = 4x - 13$

$$x^2 - 4x + 13 = 0$$

$$x = \frac{-(-4) \pm \sqrt{(-4)^2 - 4(1)(13)}}{2(1)}$$

$$= \frac{4 \pm \sqrt{-36}}{2} = \frac{4 \pm 6i}{2} = 2 \pm 3i$$

Solution set: $\{2 - 3i, 2 + 3i\}$

10. a. $9x^2 - 6x + 1 = 0 \Rightarrow a = 9, b = -6, c = 1$

So, $D = (-6)^2 - 4(9)(1) = 36 - 36 = 0$.

Therefore, there is one real root.

b. $x^2 - 5x + 3 = 0 \Rightarrow a = 1, b = -5, c = 3$

So, $D = (-5)^2 - 4(1)(3) = 25 - 12 = 13 > 0$

Therefore, there are two unequal real roots.

c. $2x^2 - 3x + 4 = 0 \Rightarrow a = 2, b = -3, c = 4$

So, $D = (-3)^2 - 4(2)(4) = 9 - 32 = -23 < 0$

Therefore, there are two nonreal complex solutions.

1.3 Basic Concepts and Skills

1. We define $i = \sqrt{-1}$, so that $i^2 = \underline{-1}$.

3. For $b > 0$, $\sqrt{-b} = i\sqrt{b}$.

5. True

In exercises 7–10, to find the real numbers x and y that make the equation true, set the real parts of the equation equal to each other and then set the imaginary parts of the equation equal to each other.

7. $2 + xi = y + 3i$, so $x = 3$ and $y = 2$.

9. $x - \sqrt{-16} = 2 + yi$. $\sqrt{-16} = 4i$, so the equation becomes $x - 4i = 2 + yi$.
$x = 2$ and $y = -4$.

11. $(5 + 2i) + (3 + i) = (5 + 3) + (2 + 1)i = 8 + 3i$

13. $(4 - 3i) - (5 + 3i) = (4 - 5) + (-3 - 3)i$
$= -1 - 6i$

15. $(-2 - 3i) + (-3 - 2i) = [-2 + (-3)] + (-3 - 2)i$
$= -5 - 5i$

17. $3(5 + 2i) = 3(5) + 3(2i) = 15 + 6i$

19. $-4(2 - 3i) = -4(2) - 4(-3i) = -8 + 12i$

21. $3i(5 + i) = 3i(5) + 3i(i) = 15i + 3i^2$

Because $i^2 = -1$, $3i^2 = -3$.

So, $15i + 3i^2 = 15i - 3 = -3 + 15i$.

23. $4i(2 - 5i) = 4i(2) + 4i(-5i) = 8i - 20i^2$.

Because $i^2 = -1$, $-20i^2 = (-20)(-1) = 20$.

So, $8i - 20i^2 = 8i + 20 = 20 + 8i$.

25. $(3 + i)(2 + 3i) = 3 \cdot 2 + 3 \cdot 3i + i \cdot 2 + i \cdot 3i$
$= 6 + 9i + 2i + 3i^2$
$= 6 + 11i + 3(-1)$
$= 3 + 11i$

27. $(2 - 3i)(2 + 3i)$
$= 2 \cdot 2 + 2 \cdot 3i + (-3i) \cdot 2 + (-3i) \cdot 3i$
$= 4 + 6i - 6i - 9i^2$
$= 4 - 9(-1) = 4 + 9 = 13$

29. $(3 + 4i)(4 - 3i)$
$= 3 \cdot 4 + 3 \cdot (-3i) + 4i \cdot 4 + (4i) \cdot (-3i)$
$= 12 - 9i + 16i - 12i^2$
$= 12 + 7i - 12(-1)$
$= 12 + 7i + 12 = 24 + 7i$

31. $\left(\sqrt{3} - 12i\right)^2$
$= \left(\sqrt{3} - 12i\right)\left(\sqrt{3} - 12i\right)$
$= \sqrt{3} \cdot \sqrt{3} + \sqrt{3} \cdot (-12i)$
$\quad - 12i \cdot \sqrt{3} + (-12i) \cdot (-12i)$
$= 3 - 12\sqrt{3}i - 12\sqrt{3}i + 144i^2$
$= 3 - 24\sqrt{3}i + 144(-1) = 3 - 24\sqrt{3}i - 144$
$= -141 - 24\sqrt{3}i$

33. $\left(2 - \sqrt{-16}\right)(3 + 5i)$
$= (2 - 4i)(3 + 5i)$
$= 2 \cdot 3 + 2 \cdot 5i - 4i \cdot 3 - 4i \cdot 5i$
$= 6 + 10i - 12i - 20i^2$
$= 6 - 2i - 20(-1)$
$= 6 - 2i + 20$
$= 26 - 2i$

35. If $z = 2 - 3i$ then $\overline{z} = 2 + 3i$, and
$z\overline{z} = (2 - 3i)(2 + 3i) = 4 - 9i^2 = 4 + 9 = 13$.

37. If $z = \frac{1}{2} - 2i$ then $\overline{z} = \frac{1}{2} + 2i$, and
$z\overline{z} = \left(\frac{1}{2} - 2i\right)\left(\frac{1}{2} + 2i\right) = \frac{1}{4} - 4i^2 = \frac{1}{4} + 4 = \frac{17}{4}$

39. If $z = \sqrt{2} - 3i$ then $\overline{z} = \sqrt{2} + 3i$, and
$z\overline{z} = \left(\sqrt{2} - 3i\right)\left(\sqrt{2} + 3i\right) = 2 - 9i^2 = 2 + 9 = 11$

41. The denominator is $-i$, so its conjugate is i. Multiply the numerator and denominator by i.
$\frac{5}{-i} = \frac{5i}{-i \cdot i} = \frac{5i}{1} = 5i$

43. The denominator is $1 + i$, so its conjugate is $1 - i$. Multiply the numerator and denominator by $1 - i$.
$\frac{-1}{1 + i} = \frac{-1(1 - i)}{(1 + i)(1 - i)} = \frac{-1 + i}{1 + 1}$
$= \frac{-1 + i}{2} = -\frac{1}{2} + \frac{1}{2}i$

45. The denominator is $2+i$, so its conjugate is $2-i$. Multiply the numerator and denominator by $2-i$.

$$\frac{5i}{2+i} = \frac{5i(2-i)}{(2+i)(2-i)} = \frac{10i - 5i^2}{4+1}$$
$$= \frac{10i + 5}{5} = 1 + 2i$$

47. The denominator is $1+i$, so its conjugate is $1-i$. Multiply the numerator and denominator by $1-i$.

$$\frac{2+3i}{1+i} = \frac{(2+3i)(1-i)}{(1+i)(1-i)} = \frac{2 - 2i + 3i - 3i^2}{1+1}$$
$$= \frac{2 + i + 3}{2} = \frac{5+i}{2} = \frac{5}{2} + \frac{1}{2}i$$

49. The denominator is $4-7i$, so its conjugate is $4+7i$. Multiply the numerator and denominator by $4+7i$.

$$\frac{2-5i}{4-7i} = \frac{(2-5i)(4+7i)}{(4-7i)(4+7i)} = \frac{8 + 14i - 20i - 35i^2}{16+49}$$
$$= \frac{8 - 6i + 35}{65} = \frac{43 - 6i}{65} = \frac{43}{65} - \frac{6}{65}i$$

51. The denominator is $1\ i$, so its conjugate is $1 - i$. Multiply the numerator and denominator by $1 - i$.

$$\frac{2+\sqrt{-4}}{1+i} = \frac{(2+2i)}{(1+i)} = \frac{(2+2i)(1-i)}{(1+i)(1-i)}$$
$$= \frac{2 - 2i + 2i - 2i^2}{1+1} = \frac{2 - 2i^2}{2} = \frac{2+2}{2} = 2$$

53. The denominator is $2-3i$, so its conjugate is $2+3i$. Multiply the numerator and denominator by $2+3i$.

$$\frac{-2+\sqrt{-25}}{2-3i} = \frac{-2+5i}{2-3i} = \frac{(-2+5i)(2+3i)}{(2-3i)(2+3i)}$$
$$= \frac{-4 - 6i + 10i + 15i^2}{4+9}$$
$$= \frac{-4 + 4i - 15}{13} = \frac{-19 + 4i}{13}$$
$$= -\frac{19}{13} + \frac{4}{13}i$$

55. $x^2 + 5 = 1 \Rightarrow x^2 = -4 \Rightarrow x = \pm\sqrt{-4} = \pm 2i$

57. $z^2 - 2z + 2 = 0$
$$z^2 - 2z = -2$$
Now, complete the square.
$$z^2 - 2z + 1 = -2 + 1$$
$$(z-1)^2 = -1 \Rightarrow z - 1 = \pm i \Rightarrow z = 1 \pm i$$

59. $2x^2 - 20x + 49 = -7$
$$2x^2 - 20x = -56$$
$$x^2 - 10x = -28$$
Now, complete the square.
$$x^2 - 10x + 25 = -28 + 25 \Rightarrow (x-5)^2 = -3$$
$$x - 5 = \pm i\sqrt{3} \Rightarrow x = 5 \pm i\sqrt{3}$$

61. $8(x^2 - x) = x^2 - 3 \Rightarrow 8x^2 - 8x = x^2 - 3 \Rightarrow$
$7x^2 - 8x + 3 = 0 \Rightarrow a = 7, b = -8, c = 3$
$$x = \frac{-(-8) \pm \sqrt{(-8)^2 - 4(7)(3)}}{2(7)}$$
$$= \frac{8 \pm \sqrt{64 - 84}}{14} = \frac{8 \pm \sqrt{-20}}{14}$$
$$= \frac{8 \pm 2i\sqrt{5}}{14} = \frac{4 \pm i\sqrt{5}}{7}$$
$$x = \frac{4}{7} + \frac{\sqrt{5}}{7}i \text{ or } x = \frac{4}{7} - \frac{\sqrt{5}}{7}i$$

63. $9k^2 + 25 = 0 \Rightarrow a = 9, b = 0, c = 25$
$$t = \frac{0 \pm \sqrt{0^2 - 4(9)(25)}}{2(9)} = \frac{\pm\sqrt{-900}}{18}$$
$$= \pm\frac{30i}{18} = \pm\frac{5}{3}i$$

1.3 Applying the Concepts

65. $Z_1 = 4 + 3i$ and $Z_2 = 5 - 2i$.
So, $Z_1 + Z_2 = (4 + 3i) + (5 - 2i) = 9 + i$.

67. $Z = \frac{V}{I}$, $I = 7 + 5i$, $V = 35 + 70i$. Then,

$Z = \frac{35 + 70i}{7 + 5i}$. Simplify the fraction by

multiplying the numerator and denominator by $7 - 5i$.
$$\frac{35 + 70i}{7 + 5i} = \frac{(35 + 70i)(7 - 5i)}{(7 + 5i)(7 - 5i)}$$
$$= \frac{245 - 175i + 490i - 350i^2}{49 + 25}$$
$$= \frac{245 + 315i + 350}{74} = \frac{595 + 315i}{74}$$
$$= \frac{595}{74} + \frac{315}{74}i$$

69. $Z = \frac{V}{I}$, $Z = 5 - 7i$, $I = 2 + 5i$. Then,
$V = ZI = (5 - 7i)(2 + 5i)$
$$= 10 + 25i - 14i - 35i^2$$
$$= 10 + 11i + 35 = 45 + 11i$$

71. $Z = \dfrac{V}{I}$, $V = 12 + 10i$, $Z = 12 + 6i$. Then,

$I = \dfrac{V}{Z} = \dfrac{12 + 10i}{12 + 6i}$. Simplify the fraction by

multiplying the numerator and denominator by $12 - 6i$.

$\dfrac{12 + 10i}{12 + 6i} = \dfrac{(12 + 10i)(12 - 6i)}{(12 + 6i)(12 - 6i)}$

$= \dfrac{144 - 72i + 120i - 60i^2}{144 + 36}$

$= \dfrac{144 + 48i + 60}{180}$

$= \dfrac{204 + 48i}{180} = \dfrac{17}{15} + \dfrac{4}{15}i$

1.3 Beyond the Basics

73. To find i^{17}, first divide 17 by 4. The remainder is 1, so $i^{17} = i^1 = i$.

75. To find i^{-7}, first rewrite it as $\dfrac{1}{i^7}$. Then

divide 7 by 4. The remainder is 3, so $\dfrac{1}{i^7} = \dfrac{1}{i^3}$.

Simplify the fraction by multiplying the numerator and denominator by i.

$\dfrac{1}{i^3} \cdot \dfrac{i}{i} = \dfrac{i}{i^4} = i$.

77. To find i^{10}, first divide 10 by 4. The remainder is 2, so $i^{10} = i^2 = -1$. So $i^{10} + 7 = -1 + 7 = 6$.

79. To find i^5, first divide 5 by 4. The remainder is 1, so $i^5 = i$. So $3i^5 = 3i$. $i^3 = -i$, so $-2i^3 = 2i$. Then, $3i^5 - 2i^3 = 3i + 2i = 5i$.

81. $i^3 = -i$, so $2i^3 = -2i$. $i^4 = 1$, so $1 + i^4 = 1 + 1 = 2$. Then $2i^3(1 + i^4) = -2i(2) = -4i$.

83. $\dfrac{1}{a + bi} = \dfrac{1(a - bi)}{(a + bi)(a - bi)} = \dfrac{a - bi}{a^2 + b^2}$

$= \dfrac{a}{a^2 + b^2} - \dfrac{b}{a^2 + b^2}i$

85. $z = a + bi$, so $\text{Im}(z) = b$.

$\dfrac{z - \overline{z}}{2i} = \dfrac{(a + bi) - (a - bi)}{2i} = \dfrac{2bi}{2i} = b = \text{Im}(z)$

87. $z\overline{z} = (a + bi)(a - bi) = a^2 + b^2$

$a^2 + b^2 = 0$ if and only if $a = 0$ and $b = 0$. So, $z = 0 + 0i = 0$.

89. $x - \dfrac{y}{i} = 4i + 1 \Rightarrow x = 1$

$-\dfrac{y}{i} = 4i \Rightarrow -y = 4i^2 \Rightarrow -y = -4 \Rightarrow y = 4$

91. $\dfrac{5x + yi}{2 - i} = 2 + i \Rightarrow 5x + yi = (2 + i)(2 - i)$

$\Rightarrow 5x + yi = 5 \Rightarrow x = 1, y = 0$.

93. $\dfrac{1 + i}{1 - i} \div \dfrac{2 + i}{1 + 2i} = \dfrac{1 + i}{1 - i} \cdot \dfrac{1 + 2i}{2 + i} = \dfrac{1 + 2i + i + 2i^2}{2 + i - 2i - i^2}$

$= \dfrac{1 + 3i - 2}{2 - i + 1} = \dfrac{-1 + 3i}{3 - i}$

$= \dfrac{-1 + 3i}{3 - i} \cdot \dfrac{3 + i}{3 + i} = \dfrac{-3 - i + 9i + 3i^2}{9 - i^2}$

$= \dfrac{-3 + 8i - 3}{9 + 1} = \dfrac{-6 + 8i}{10} = -\dfrac{3}{5} + \dfrac{4}{5}i$

95. $z = 2 - 3i$, $w = 1 + 2i$

a. $\overline{(zw)} = \overline{((2 - 3i)(1 + 2i))} = \overline{(2 + 4i - 3i - 6i^2)}$

$= \overline{(8 + i)} = 8 - i$

$(\overline{z})(\overline{w}) = (2 + 3i)(1 - 2i) = 2 - 4i + 3i - 6i^2$

$= 8 - i$

b. $\dfrac{z}{w} = \dfrac{2 - 3i}{1 + 2i} = \dfrac{2 - 3i}{1 + 2i} \cdot \dfrac{1 - 2i}{1 - 2i} = \dfrac{2 - 4i - 3i + 6i^2}{1 - 4i^2}$

$= \dfrac{-4 - 7i}{5} = -\dfrac{4}{5} - \dfrac{7}{5}i$

$\overline{\left(\dfrac{z}{w}\right)} = -\dfrac{4}{5} + \dfrac{7}{5}i$

$\dfrac{\overline{z}}{\overline{w}} = \dfrac{2 + 3i}{1 - 2i} = \dfrac{2 + 3i}{1 - 2i} \cdot \dfrac{1 + 2i}{1 + 2i}$

$= \dfrac{2 + 4i + 3i + 6i^2}{1 - 4i^2} = \dfrac{-4 + 7i}{5} = -\dfrac{4}{5} + \dfrac{7}{5}i$

1.3 Critical Thinking/Discussion/Writing

96. a. True. Every real number a can be written as a complex number $a + 0i$.

b. False.

c. False. A complex number with the form $a + 0i$ does not have an imaginary component.

d. True

e. True. $(a+bi)(a-bi) = a^2 + b^2$. There is no imaginary component.

f. True

97. $\dfrac{1+i}{1-i} = \dfrac{1+i}{1-i} \cdot \dfrac{1+i}{1+i} = \dfrac{1+2i-1}{2} = \dfrac{2i}{2} = i$

$i^n = 1 \Rightarrow n = 4$

98. If $i = 0$, then $i^2 = 0^2 \Rightarrow -1 = 0$, which is a contradiction. If $i < 0$, then $i \cdot i > 0 \cdot i$ (since i is negative) $\Rightarrow i^2 > 0 \Rightarrow -1 > 0$, a contradiction.

If $i > 0$, then $i \cdot i > 0 \cdot i \Rightarrow i^2 > 0 \Rightarrow -1 > 0$, a contradiction. Thus, the set of complex numbers does not have the ordering properties of the set of real numbers.

1.3 Maintaining Skills

99. $x^3 - x^2 - 9x + 9 = \left(x^3 - x^2\right) - (9x - 9)$

$\qquad = x^2(x-1) - 9(x-1)$

$\qquad = \left(x^2 - 9\right)(x-1)$

$\qquad = (x-3)(x+3)(x-1)$

101. $\dfrac{1}{x-1}, \dfrac{1}{x}$

LCD $= x(x-1)$

103. $\dfrac{15}{x+3}, \dfrac{2x+1}{x-3}$

LCD $= (x+3)(x-3)$

105. $\dfrac{x+1}{(x-3)(2-x)}, \dfrac{12}{(x-2)(x+1)}$

Note that $x-2 = -(2-x)$.

The LCD is $(x-3)(x-2)(x+1)$.

107. $27^{2/3} = \left(\sqrt[3]{27}\right)^2 = 3^2 = 9$

109. $4^{5/2} = \left(\sqrt{4}\right)^5 = 2^5 = 32$

111. $\left(\sqrt{3+x^2}\right)^2 = 3 + x^2$

113. $\left(\left(3x^2 + 7\right)^{1/3}\right)^3 = \left(3x^2 + 7\right)^{(1/3)(3)} = 3x^2 + 7$

1.4 Solving Other Types of Equations

1.4 Practice Problems

1. $x^3 + 2x^2 - x - 2 = 0$

$x^2(x+2) - (x+2) = 0$

$\left(x^2 - 1\right)(x+2) = 0$

$(x-1)(x+1)(x+2) = 0$

$x-1 = 0 \Rightarrow x = 1$ or $x+1 = 0 \Rightarrow x = -1$ or $x+2 = 0 \Rightarrow x = -2$

Solution set: $\{-2, -1, 1\}$

2. $x^4 = 4x^2$

$x^4 - 4x^2 = 0$

$x^2\left(x^2 - 4\right) = 0$

$x^2(x-2)(x+2) = 0$

$x^2 = 0 \Rightarrow x = 0$ or $x-2 = 0 \Rightarrow x = 2$ or $x+2 = 0 \Rightarrow x = -2$

Solution set: $\{-2, 0, 2\}$

3. $x^3 - 5x^2 = 4x - 20$

$x^3 - 5x^2 - 4x + 20 = 0$

$x^2(x-5) - 4(x-5) = 0$

$\left(x^2 - 4\right)(x-5) = 0$

$(x-2)(x+2)(x-5) = 0$

$x-2 = 0 \Rightarrow x = 2$ or $x+2 = 0 \Rightarrow x = -2$ or $x-5 = 0 \Rightarrow x = 5$

Solution set: $\{-2, 2, 5\}$

4. $\dfrac{1}{x} - \dfrac{12}{5x+10} = \dfrac{1}{5}$

$\dfrac{1}{x} - \dfrac{12}{5(x+2)} = \dfrac{1}{5}$

Multiply by the LCD, $5x(x+2)$.

$5(x+2) - 12x = x(x+2)$

$5x + 10 - 12x = x^2 + 2x$

$0 = x^2 + 9x - 10$

$0 = (x+10)(x-1)$

$x+10 = 0 \Rightarrow x = -10$ or $x-1 = 0 \Rightarrow x = 1$

Be sure to check the solutions in the original equation.

Solution set: $\{-10, 1\}$

5. $\dfrac{x}{x-2} - \dfrac{2}{x+2} = \dfrac{4x}{x^2-4}$

$\dfrac{x}{x-2} - \dfrac{2}{x+2} = \dfrac{4x}{(x-2)(x+2)}$

Multiply by the common denominator
$(x-2)(x+2)$.

$x(x+2) - 2(x-2) = 4x$

$x^2 + 2x - 2x + 4 = 4x$

$x^2 - 4x + 4 = 0$

$(x-2)^2 = 0 \Rightarrow x = 2$

Since neither $\dfrac{x}{x-2}$ nor $\dfrac{4x}{x^2-4}$ is defined for

$x = 2$, this is an extraneous solution.
Solution set: $\varnothing$

6. $x = \sqrt{x^3 - 6x}$

$x^2 = \left(\sqrt{x^3-6x}\right)^2 = x^3 - 6x$

$0 = x^3 - x^2 - 6x = x\left(x^2 - x - 6\right)$

$\quad = x(x-3)(x+2)$

$x = 0$ or $x - 3 = 0 \Rightarrow x = 3$ or
$x + 2 = 0 \Rightarrow x = -2$

Now check to see if –2, 0, or 3 are extraneous
solutions. If $x = -2$, then

$\sqrt{(-2)^3 - 6(-2)} = \sqrt{4} = 2 \neq -2$, so –2 is an

extraneous solution. If $x = 0$, then

$\sqrt{0^3 - 6(0)} = \sqrt{0} = 0$, so 0 is a solution.

If $x = 3$, then $\sqrt{3^3 - 6(3)} = \sqrt{9} = 3$, so 3 is a
solution.
Solution set: {0, 3}

7. $\sqrt{6x+4} + 2 = x$

$\sqrt{6x+4} = x - 2$

$\left(\sqrt{6x+4}\right)^2 = (x-2)^2$

$6x + 4 = x^2 - 4x + 4$

$0 = x^2 - 10x = x(x-10)$

$x = 0$ or $x - 10 = 0 \Rightarrow x = 10$

Now check to see if 0 or 10 are extraneous
solutions. If $x = 0$, then

$\sqrt{6(0)+4} + 2 = \sqrt{4} + 2 = 2 + 2 = 4 \neq 0$, so 0 is

an extraneous solution. If $x = 10$, then

$\sqrt{6(10)+4} + 2 = \sqrt{64} + 2 = 8 + 2 = 10$, so 10

is a solution.
Solution set: {10}

8. $\sqrt{x-5} + \sqrt{x} = 5$

$\sqrt{x-5} = 5 - \sqrt{x}$

$\left(\sqrt{x-5}\right)^2 = \left(5 - \sqrt{x}\right)^2$

$x - 5 = 25 - 10\sqrt{x} + x$

$-30 = -10\sqrt{x}$

$3 = \sqrt{x} \Rightarrow 9 = x$

Now check to see if 9 is an extraneous
solution.

$\sqrt{9-5} + \sqrt{9} = \sqrt{4} + \sqrt{9} = 2 + 3 = 5$, so 9 is a

solution.
Solution set: {9}

9. a. $3(x-2)^{3/5} + 4 = 7$

$3(x-2)^{3/5} = 3$

$(x-2)^{3/5} = 1$

$\left[(x-2)^{3/5}\right]^{5/3} = 1^{5/3}$

$x - 2 = 1$

$x = 3$

Be sure to check that 3 satisfies the original
equation.
Solution set: {3}

b. $(2x+1)^{4/3} - 7 = 9$

$(2x+1)^{4/3} = 16$

$\left[(2x+1)^{4/3}\right]^{3/4} = \pm\left(16^{3/4}\right)$

$2x + 1 = \pm 8$

$2x+1 = -8$	$2x+1 = 8$
$2x = -9$	$2x = 7$
$x = -\dfrac{9}{2}$	$x = \dfrac{7}{2}$

Check:

$\left[2\left(-\dfrac{9}{2}\right)+1\right]^{4/3} - 7 \overset{?}{=} 9$

$(-9+1)^{4/3} - 7 \overset{?}{=} 9$

$(-8)^{4/3} - 7 \overset{?}{=} 9$

$16 - 7 = 9$

$\left[2\left(\dfrac{7}{2}\right)+1\right]^{4/3} - 7 \overset{?}{=} 9$

$(7+1)^{4/3} - 7 \overset{?}{=} 9$

$(8)^{4/3} - 7 \overset{?}{=} 9$

$16 - 7 = 9$

Solution set: $\left\{-\dfrac{9}{2}, \dfrac{7}{2}\right\}$

10. $x^{2/3} - 7x^{1/3} + 6 = 0$

Let $u = x^{1/3}$. Then the original equation becomes

$u^2 - 7u + 6 = 0 \Rightarrow (u-1)(u-6) = 0 \Rightarrow$
$u = 1, 6$

Now solve for x:

$1 = x^{1/3} \Rightarrow 1^3 = \left(x^{1/3}\right)^3 \Rightarrow 1 = x$

$6 = x^{1/3} \Rightarrow 6^3 = \left(x^{1/3}\right)^3 \Rightarrow 216 = x$

Be sure to check that both solutions satisfy the original equation.
Solution set: $\{1, 216\}$

11. $\left(1 + \dfrac{1}{x}\right)^2 - 6\left(1 + \dfrac{1}{x}\right) + 8 = 0$

Let $u = \left(1 + \dfrac{1}{x}\right)$. Then the original equation becomes

$u^2 - 6u + 8 = 0 \Rightarrow (u-2)(u-4) = 0 \Rightarrow u = 2, 4$

Now solve for x:

$1 + \dfrac{1}{x} = 2 \Rightarrow x + 1 = 2x \Rightarrow x = 1$

$1 + \dfrac{1}{x} = 4 \Rightarrow x + 1 = 4x \Rightarrow 1 = 3x \Rightarrow \dfrac{1}{3} = x$

Be sure to check that both solutions satisfy the original equation.

Solution set: $\left\{\dfrac{1}{3}, 1\right\}$

12. $t_0 = t\sqrt{1 - \dfrac{v^2}{c^2}}, \, t_0 = 20, \, t = 25$

$20 = 25\sqrt{1 - \dfrac{v^2}{c^2}} \Rightarrow \dfrac{4}{5} = \sqrt{1 - \dfrac{v^2}{c^2}} \Rightarrow$

$\dfrac{16}{25} = 1 - \dfrac{v^2}{c^2} \Rightarrow \dfrac{v^2}{c^2} = 1 - \dfrac{16}{25} = \dfrac{9}{25} \Rightarrow$

$\dfrac{v}{c} = \dfrac{3}{5} \Rightarrow v = 0.6c$

The spacecraft must have been traveling at 60% of the speed of light.

1.4 Basic Concepts and Skills

1. If an apparent solution does not satisfy the equation, it is called an <u>extraneous</u> solution.

3. If $x^{3/4} = 8$, then $x = \underline{16}$.

5. False

7. $x^3 = 2x^2 \Rightarrow x^3 - 2x^2 = 0 \Rightarrow x^2(x-2) = 0 \Rightarrow$
$x^2 = 0$ or $x - 2 = 0 \Rightarrow x = 0$ or $x = 2$

9. $\left(\sqrt{x}\right)^3 = \sqrt{x} \Rightarrow \left(\sqrt{x}\right)^2 \sqrt{x} = \sqrt{x} \Rightarrow$
$x\sqrt{x} = \sqrt{x} \Rightarrow x\sqrt{x} - \sqrt{x} = 0 \Rightarrow$
$\sqrt{x}(x-1) = 0 \Rightarrow \sqrt{x} = 0$ or $(x-1) = 0 \Rightarrow$
$x = 0$ or $x = 1$

11. $x^3 + x = 0 \Rightarrow x(x^2 + 1) = 0 \Rightarrow$
$x = 0$ or $x^2 + 1 = 0 \Rightarrow x = 0$ or $x^2 = -1 \Rightarrow$
$x = 0$ or $x = i$
We are looking for real roots only, so we reject $x = i$. Solution set: $\{0\}$

13. $\qquad x^4 - x^3 = x^2 - x$
$x^4 - x^3 - x^2 + x = 0$
Factor by grouping.
$\qquad x^3(x-1) - x(x-1) = 0$
$\qquad (x^3 - x)(x-1) = 0$
$\qquad x(x^2 - 1)(x-1) = 0$
$\qquad x(x+1)(x-1)(x-1) = 0$
$\qquad x = 0$ or $x = -1$ or $x = 1$

15. $\qquad x^4 = 27x$
$\qquad x^4 - 27x = 0$
$\qquad x(x^3 - 27) = 0$
$\qquad x(x^3 - 3^3) = 0$
$\qquad x(x-3)(x^2 + 3x + 9) = 0$
$\qquad x = 0$ or $x = 3$ or $x^2 + 3x + 9 = 0$
Solving $x^2 + 3x + 9 = 0$ gives

$x = -\dfrac{3}{2} \pm \dfrac{3i\sqrt{3}}{2}$. However, we are looking for

the real roots only, so we reject these roots.
Solution set: $\{0, 3\}$

17. $\qquad \dfrac{x+1}{3x-2} = \dfrac{5x-4}{3x+2}$
$\qquad (x+1)(3x+2) = (5x-4)(3x-2)$
$\qquad 3x^2 + 5x + 2 = 15x^2 - 22x + 8$
$\qquad -12x^2 + 27x - 6 = 0$
$\qquad -3(4x^2 - 9x + 2) = 0$
$\qquad 4x^2 - 9x + 2 = 0 \Rightarrow (4x-1)(x-2) = 0$

$4x - 1 = 0$ or $x - 2 = 0 \Rightarrow x = \dfrac{1}{4}$ or $x = 2$

19.
$$\frac{1}{x}+\frac{2}{x+1}=1$$
$$(x)(x+1)\left(\frac{1}{x}+\frac{2}{x+1}\right)=(x)(x+1)(1)$$
$$(x+1)+2x=x^2+x$$
$$3x+1=x^2+x$$
$$0=x^2-2x-1$$
Solve using the quadratic formula.
$$x=\frac{-(-2)\pm\sqrt{(-2)^2-4(1)(-1)}}{2}$$
$$x=\frac{2\pm\sqrt{4+4}}{2}=\frac{2\pm\sqrt{8}}{2}=\frac{2\pm2\sqrt{2}}{2}=1\pm\sqrt{2}$$

21.
$$\frac{6x-7}{x}-\frac{1}{x^2}=5$$
$$x^2\left(\frac{6x-7}{x}-\frac{1}{x^2}\right)=5x^2$$
$$x(6x-7)-1=5x^2$$
$$6x^2-7x-1=5x^2$$
$$x^2-7x-1=0$$
Solve using the quadratic formula.
$$x=\frac{-(-7)\pm\sqrt{(-7)^2-4(1)(-1)}}{2}$$
$$x=\frac{7\pm\sqrt{49+4}}{2}=\frac{7\pm\sqrt{53}}{2}=\frac{7}{2}\pm\frac{\sqrt{53}}{2}$$

23.
$$\frac{1}{x}+\frac{1}{x-3}=\frac{7}{3x-5}$$
$$x(x-3)(3x-5)\left(\frac{1}{x}+\frac{1}{x-3}=\frac{7}{3x-5}\right)$$
$$(x-3)(3x-5)+x(3x-5)=7x(x-3)$$
$$3x^2-14x+15+3x^2-5x=7x^2-21x$$
$$6x^2-19x+15=7x^2-21x$$
$$-x^2+2x+15=0$$
$$-1(x^2-2x-15)=0$$
$$x^2-2x-15=0$$
$$(x-5)(x+3)=0\Rightarrow$$
$$x-5=0 \text{ or } x+3=0\Rightarrow x=5 \text{ or } x=-3$$

25.
$$\frac{5}{x+1}-\frac{4}{2x+2}+\frac{2}{2x-1}=\frac{13}{18}$$
$$\frac{5}{x+1}-\frac{4}{2(x+1)}+\frac{2}{2x-1}=\frac{13}{18}$$
$$18(x+1)(2x-1)\left(\frac{5}{x+1}-\frac{4}{2(x+1)}+\frac{2}{2x-1}\right)$$
$$=18(x+1)(2x-1)\left(\frac{13}{18}\right)$$

$$90(2x-1)-36(2x-1)+36(x+1)$$
$$=13(x+1)(2x-1)$$
$$180x-90-72x+36+36x+36$$
$$=13(2x^2+x-1)$$
$$144x-18=26x^2+13x-13$$
$$0=26x^2-131x+5$$
$$0=(26x-1)(x-5)$$
$$26x-1=0 \text{ or } x-5=0$$
$$x=\frac{1}{26}\text{ or }x=5$$

27.
$$\frac{x}{x-5}-\frac{5}{x+5}=\frac{10x}{x^2-25}$$
$$x(x+5)-5(x-5)=10x$$
$$x^2+5x-5x+25=10x$$
$$x^2-10x+25=0$$
$$(x-5)^2=0\Rightarrow x=5$$
Since $x=5$ makes the denominator in the first fraction equal zero, there is no solution. Solution set: $\varnothing$

29.
$$\frac{x}{x-3}+\frac{3}{x+3}=\frac{6x}{x^2-9}$$
$$x(x+3)+3(x-3)=6x$$
$$x^2+3x+3x-9=6x$$
$$x^2-9=0$$
$$(x+3)(x-3)=0\Rightarrow x=-3,\,3$$
Since $x=-3$ makes the denominator in the second fraction equal zero and $x=3$ makes the denominator in the first fraction equal zero, there is no solution. Solution set: $\varnothing$

31.
$$\frac{1}{x-1}+\frac{x}{x+3}=\frac{4}{x^2+2x-3}$$
$$\frac{1}{x-1}+\frac{x}{x+3}=\frac{4}{(x-1)(x+3)}$$
$$(x+3)+x(x-1)=4$$
$$x+3+x^2-x=4\Rightarrow x^2=1\Rightarrow x=\pm1$$
If $x=1$, then the denominator in the first fraction equals zero, so 1 is an extraneous solution. Solution set: $\{-1\}$

33.
$$\frac{2x}{x+3} - \frac{x}{x-1} = \frac{14}{x^2+2x-3}$$
$$\frac{2x}{x+3} - \frac{x}{x-1} = \frac{14}{(x+3)(x-1)}$$
$$2x(x-1) - x(x+3) = 14$$
$$2x^2 - 2x - x^2 - 3x = 14$$
$$x^2 - 5x - 14 = 0$$
$$(x-7)(x+2) = 0 \Rightarrow x = 7, -2$$
Solution set: $\{-2, 7\}$

35. $\sqrt[3]{3x-1} = 2 \Rightarrow \left(\sqrt[3]{3x-1}\right)^3 = 2^3 \Rightarrow$
$3x - 1 = 8 \Rightarrow 3x = 9 \Rightarrow x = 3$

37. There is no solution for $\sqrt{x-1} = -2$ because the square root is not negative. The solution set is $\varnothing$.

39. $x + \sqrt{x+6} = 0 \Rightarrow x = -\sqrt{x+6} \Rightarrow$
$x^2 = \left(-\sqrt{x+6}\right)^2 \Rightarrow x^2 = x + 6 \Rightarrow$
$x^2 - x - 6 = 0 \Rightarrow (x-3)(x+2) = 0 \Rightarrow$
$x = 3$ or $x = -2$
Now check to see if 3 or -2 are extraneous roots. If $x = 3$, then $x + \sqrt{x+6} = 3 + \sqrt{3+6} =$
$3 + 3 \neq 0$. So 3 is extraneous. If $x = -2$,
then $x + \sqrt{x+6} = -2 + \sqrt{-2+6} = -2 + \sqrt{4} =$
$-2 + 2 = 0$. Solution set: $\{-2\}$

41. $\sqrt{y+6} = y \Rightarrow \left(\sqrt{y+6}\right)^2 = y^2 \Rightarrow y + 6 = y^2 \Rightarrow$
$0 = y^2 - y - 6 \Rightarrow 0 = (y-3)(y+2) \Rightarrow$
$y = 3$ or $y = -2$
Now check to see if 3 or -2 are extraneous roots. If $y = 3$, then $\sqrt{y+6} = y \Rightarrow$
$\sqrt{3+6} = 3 \Rightarrow \sqrt{9} = 3 \Rightarrow 3 = 3$. So 3 is a solution. If $y = -2$, then $\sqrt{y+6} = y \Rightarrow$
$\sqrt{-2+6} = -2 \Rightarrow \sqrt{4} = -2 \Rightarrow 2 \neq -2$. So -2 is extraneous. Solution set: $\{3\}$

43.
$$\sqrt{6y-11} = 2y - 7$$
$$\left(\sqrt{6y-11}\right)^2 = (2y-7)^2$$
$$6y - 11 = 4y^2 - 28y + 49$$
$$0 = 4y^2 - 34y + 60$$
$$0 = 2y^2 - 17y + 30$$
$$0 = (2y-5)(y-6) \Rightarrow y = \frac{5}{2} \text{ or } y = 6$$

Now check to see if 5/2 or 6 are extraneous roots. If $y = \frac{5}{2}$, then $\sqrt{6y-11} = 2y - 7 \Rightarrow$
$\sqrt{6\left(\frac{5}{2}\right) - 11} = 2\left(\frac{5}{2}\right) - 7 \Rightarrow \sqrt{15 - 11} = 5 - 7 \Rightarrow$
$\sqrt{4} = -2 \Rightarrow 2 \neq -2$. So 5/2 is extraneous.
If $y = 6$, then $\sqrt{6y-11} = 2y - 7 \Rightarrow$
$\sqrt{6(6) - 11} = 2(6) - 7 \Rightarrow \sqrt{36 - 11} = 12 - 7 \Rightarrow$
$\sqrt{25} = 5 \Rightarrow 5 = 5$. Solution set: $\{6\}$

45.
$$t - \sqrt{3t+6} = -2$$
$$t + 2 = \sqrt{3t+6}$$
$$(t+2)^2 = \left(\sqrt{3t+6}\right)^2$$
$$t^2 + 4t + 4 = 3t + 6$$
$$t^2 + t - 2 = 0$$
$$(t+2)(t-1) = 0 \Rightarrow t = -2 \text{ or } t = 1$$
Now check to see if -2 or 1 are extraneous roots. If $t = -2$, then $t - \sqrt{3t+6} = -2 \Rightarrow$
$-2 - \sqrt{3(-2)+6} = -2 \Rightarrow -2 - \sqrt{-6+6} \Rightarrow$
$-2 - 0 = -2$. So -2 is a solution.
If $t = 1$, then $t - \sqrt{3t+6} = -2 \Rightarrow$
$1 - \sqrt{3(1)+6} = -2 \Rightarrow 1 - \sqrt{9} = -2 \Rightarrow$
$1 - 3 = -2 \Rightarrow -2 = -2$. So 1 is a solution.
Solution set: $\{-2, 1\}$

47.
$$\sqrt{x-3} = \sqrt{2x-5} - 1$$
$$\left(\sqrt{x-3}\right)^2 = \left(\sqrt{2x-5} - 1\right)^2$$
$$x - 3 = (2x-5) - 2\sqrt{2x-5} + 1$$
$$x - 3 = 2x - 4 - 2\sqrt{2x-5}$$
$$2\sqrt{2x-5} = x - 1 \Rightarrow \left(2\sqrt{2x-5}\right)^2 = (x-1)^2$$
$$4(2x-5) = x^2 - 2x + 1$$
$$8x - 20 = x^2 - 2x + 1$$
$$0 = x^2 - 10x + 21$$
$$0 = (x-7)(x-3) \Rightarrow x = 7 \text{ or } x = 3$$
Now check to see if 3 or 7 are extraneous roots. If $x = 3$, then $\sqrt{x-3} = \sqrt{2x-5} - 1 \Rightarrow$
$\sqrt{3-3} = \sqrt{2(3)-5} - 1 \Rightarrow 0 = 0$. So 3 is a solution.
If $x = 7$, then $\sqrt{x-3} = \sqrt{2x-5} - 1 \Rightarrow$
$\sqrt{7-3} = \sqrt{2(7)-5} - 1 \Rightarrow \sqrt{4} = \sqrt{9} - 1 \Rightarrow$
$2 = 2$. So 7 is a solution.
Solution set: $\{3, 7\}$

49.
$$\sqrt{2y+9} = 2 + \sqrt{y+1}$$
$$\left(\sqrt{2y+9}\right)^2 = \left(2+\sqrt{y+1}\right)^2$$
$$2y+9 = 4 + 4\sqrt{y+1} + y + 1$$
$$y+4 = 4\sqrt{y+1}$$
$$(y+4)^2 = \left(4\sqrt{y+1}\right)^2$$
$$y^2 + 8y + 16 = 16(y+1)$$
$$y^2 + 8y + 16 = 16y + 16$$
$$y^2 - 8y = 0$$
$$y(y-8) = 0 \Rightarrow y = 0 \text{ or } y = 8$$

Now check to see if 0 or 8 are extraneous roots. If $y = 0$, then $\sqrt{2y+9} = 2 + \sqrt{y+1} \Rightarrow$ $\sqrt{2(0)+9} = 2 + \sqrt{0+1} \Rightarrow \sqrt{9} = 2 + 1 \Rightarrow 3 = 3$. So 0 is a solution. If $y = 8$, then $\sqrt{2y+9} = 2 + \sqrt{y+1} \Rightarrow \sqrt{2(8)+9} =$ $2 + \sqrt{8+1} \Rightarrow \sqrt{25} = 2 + \sqrt{9} \Rightarrow 5 = 5$. So 8 is a solution. Solution set: $\{0, 8\}$

51.
$$\sqrt{7z+1} - \sqrt{5z+4} = 1$$
$$\sqrt{7z+1} = \sqrt{5z+4} + 1$$
$$\left(\sqrt{7z+1}\right)^2 = \left(\sqrt{5z+4}+1\right)^2$$
$$7z+1 = 5z+4 + 2\sqrt{5z+4} + 1$$
$$2z-4 = 2\sqrt{5z+4}$$
$$(2z-4)^2 = \left(2\sqrt{5z+4}\right)^2$$
$$4z^2 - 16z + 16 = 4(5z+4)$$
$$4z^2 - 16z + 16 = 20z + 16$$
$$4z^2 - 36z = 0 \Rightarrow 4z(z-9) = 0 \Rightarrow$$
$$z = 0 \text{ or } z = 9$$

Now check to see if 0 or 9 are extraneous roots. If $z = 0$, then $\sqrt{7z+1} - \sqrt{5z+4} = 1 \Rightarrow$ $\sqrt{7(0)+1} - \sqrt{5(0)+4} = 1 \Rightarrow \sqrt{1} - \sqrt{4} = 1 \Rightarrow$ $1 - 2 \neq 1$. So 0 is extraneous. If $z = 9$, then $\sqrt{7z+1} - \sqrt{5z+4} = 1 \Rightarrow \sqrt{7(9)+1} - \sqrt{5(9)+4}$ $= 1 \Rightarrow \sqrt{64} - \sqrt{49} \Rightarrow 8 - 7 = 1 \Rightarrow 1 = 1$. So 9 is a solution. Solution set: $\{9\}$

53.
$$\sqrt{2x+5} + \sqrt{x+6} = 3$$
$$\sqrt{2x+5} = 3 - \sqrt{x+6}$$
$$\left(\sqrt{2x+5}\right)^2 = \left(3-\sqrt{x+6}\right)^2$$
$$2x+5 = 9 - 6\sqrt{x+6} + x + 6$$
$$x-10 = -6\sqrt{x+6}$$
$$(x-10)^2 = \left(-6\sqrt{x+6}\right)^2$$

$$x^2 - 20x + 100 = 36x + 216$$
$$x^2 - 56x - 116 = 0$$
$$(x-58)(x+2) = 0 \Rightarrow x = 58 \text{ or } x = -2$$

Now check to see if 58 or –2 are extraneous roots. If $x = 58$, then $\sqrt{2x+5} + \sqrt{x+6} = 3 \Rightarrow$ $\sqrt{2(58)+5} + \sqrt{58+6} = 3 \Rightarrow \sqrt{121} + \sqrt{64} = 3 \Rightarrow$ $11 + 8 \neq 3$. So 58 is extraneous. If $x = -2$, then $\sqrt{2x+5} + \sqrt{x+6} = 3 \Rightarrow \sqrt{2(-2)+5} +$ $\sqrt{-2+6} = 3 \Rightarrow \sqrt{1} + \sqrt{4} \Rightarrow 1 + 2 = 3$. So –2 is a solution. Solution set: $\{-2\}$

55.
$$\sqrt{2x-5} - \sqrt{x-3} = 1$$
$$\sqrt{2x-5} = 1 + \sqrt{x-3}$$
$$\left(\sqrt{2x-5}\right)^2 = \left(1+\sqrt{x-3}\right)^2$$
$$2x-5 = 1 + 2\sqrt{x-3} + x - 3$$
$$x-3 = 2\sqrt{x-3}$$
$$(x-3)^2 = \left(2\sqrt{x-3}\right)^2$$
$$x^2 - 6x + 9 = 4x - 12$$
$$x^2 - 10x + 21 = 0$$
$$(x-7)(x-3) = 0$$
$$x = 7 \text{ or } x = 3$$

Now check to see if 7 or 3 are extraneous roots. If $x = 7$, then $\sqrt{2x-5} - \sqrt{x-3} = 1 \Rightarrow$ $\sqrt{2(7)-5} - \sqrt{7-3} = 1 \Rightarrow \sqrt{9} - \sqrt{4} = 1 \Rightarrow$ $3 - 2 = 1$. So 7 is a solution. If $x = 3$, then $\sqrt{2x-5} - \sqrt{x-3} = 1 \Rightarrow \sqrt{2(3)-5} -$ $\sqrt{3-3} = 1 \Rightarrow \sqrt{1} - \sqrt{0} = 1 \Rightarrow 1 = 1$. So 3 is a solution. Solution set: $\{3, 7\}$

57.
$$(x-4)^{3/2} = 27$$
$$\left[(x-4)^{3/2}\right]^{2/3} = 27^{2/3}$$
$$x-4 = 9 \Rightarrow x = 13$$

Verify that 13 satisfies the original equation. Solution set: $\{13\}$

59.
$$(5x-3)^{2/3} - 5 = 4$$
$$(5x-3)^{2/3} = 9$$
$$\left[(5x-3)^{2/3}\right]^{3/2} = \pm 9^{3/2}$$
$$5x-3 = \pm 27$$

$$\begin{array}{c|c} 5x-3 = -27 & 5x-3 = 27 \\ 5x = -24 & 5x = 30 \\ x = -\dfrac{24}{5} & x = 6 \end{array}$$

(continued on next page)

(*continued*)

If $x = -24/5$, then

$$\left(5\left(-\frac{24}{5}\right) - 3\right)^{2/3} - 5 = (-24 - 3)^{2/3} - 5$$

$$= (-27)^{2/3} - 5$$

$$= 9(-1)^{2/3} - 5 = 4,$$

so $x = -\dfrac{24}{5}$ is a solution. Verify that 6 satisfies the original equation.

Solution set: $\left\{-\dfrac{24}{5}, 6\right\}$

61. Let $u = \sqrt{x}$. Then $u^2 = x$. So the equation becomes $u^2 - 5u + 6 = 0 \Rightarrow$
$(u - 3)(u - 2) = 0 \Rightarrow u = 3$ or $u = 2$. Solve for x: $3 = \sqrt{x} \Rightarrow x = 9$ or $2 = \sqrt{x} \Rightarrow 4 = x$. Check to make sure that neither solution is extraneous. If $x = 9$, then $9 - 5\sqrt{9} + 6 = 0 \Rightarrow$
$9 - 15 + 6 = 0 \Rightarrow 0 = 0$. If $x = 4$, then
$4 - 5\sqrt{4} + 6 = 0 \Rightarrow 4 - 10 + 6 = 0 \Rightarrow 0 = 0$.
Solution set: $\{4, 9\}$

63. Let $u = \sqrt{y}$. Then $u^2 = y$. So the equation becomes $2u^2 - 15u = -7 \Rightarrow$
$2u^2 - 15u + 7 = 0 \Rightarrow (2u - 1)(u - 7) = 0 \Rightarrow$
$u = \dfrac{1}{2}$ or $u = 7$. Solve for y: $\dfrac{1}{2} = \sqrt{y} \Rightarrow$
$\dfrac{1}{4} = y$ or $7 = \sqrt{y} \Rightarrow 49 = y$. Now check to make sure that neither solution is extraneous.

If $y = \dfrac{1}{4}$, then $2\left(\dfrac{1}{4}\right) - 15\sqrt{\dfrac{1}{4}} = -7 \Rightarrow$

$\dfrac{1}{2} - \dfrac{15}{2} = -7 \Rightarrow -7 = -7$.

If $y = 49$, then

$2(49) - 15\sqrt{49} = -7 \Rightarrow 98 - 105 = -7 \Rightarrow$

$-7 = -7$. Solution set: $\left\{\dfrac{1}{4}, 49\right\}$

65. $x^{-2} - x^{-1} - 42 = 0$

Let $u = x^{-1}$. Then the equation becomes
$u^2 - u - 42 = 0 \Rightarrow (u - 7)(u + 6) = 0 \Rightarrow$
$u = 7, -6$. Now solve for x:

$7 = x^{-1} \Rightarrow x = \dfrac{1}{7}; -6 = x^{-1} \Rightarrow x = -\dfrac{1}{6}$

Be sure to check that neither of the solutions are extraneous.

Solution set: $\left\{-\dfrac{1}{6}, \dfrac{1}{7}\right\}$

67. $x^{2/3} - 6x^{1/3} + 8 = 0$

Let $u = x^{1/3}$. Then the equation becomes
$u^2 - 6x + 8 = 0 \Rightarrow (u - 4)(u - 2) = 0 \Rightarrow$
$u = 4, 2$. Now solve for x:
$x^{1/3} = 4 \Rightarrow x = 64; x^{1/3} = 2 \Rightarrow x = 8$
Be sure to check that neither of the solutions are extraneous. Solution set: $\{8, 64\}$

69. $2x^{1/2} + 3x^{1/4} - 2 = 0$

Let $u = x^{1/4}$. Then the equation becomes
$2u^2 + 3u - 2 = 0 \Rightarrow (2u - 1)(u + 2) = 0 \Rightarrow$

$u = \dfrac{1}{2}, -2$. Now solve for x:

$x^{1/4} = \dfrac{1}{2} \Rightarrow x = \dfrac{1}{16}; x^{1/4} = -2 \Rightarrow x$ is not a

real number. Be sure to check that $\dfrac{1}{16}$ is not

extraneous. Solution set: $\left\{\dfrac{1}{16}\right\}$

71. Let $u = x^2$. Then $x^4 = u^2$. So the equation becomes $u^2 - 13u + 36 = 0 \Rightarrow$
$(u - 4)(u - 9) = 0 \Rightarrow u = 4$ or $u = 9$. Now

solve for x: $4 = x^2 \Rightarrow \pm 2 = x$ or

$9 = x^2 \Rightarrow \pm 3 = x$. Now check to make sure that none of the solutions are extraneous.

If $x = \pm 2$, then $(\pm 2)^4 - 13(\pm 2)^2 + 36 = 0 \Rightarrow$
$16 - 52 + 36 = 0 \Rightarrow 0 = 0$. If $x = \pm 3$, then
$(\pm 3)^4 - 13(\pm 3)^2 + 36 = 0 \Rightarrow 81 - 117 + 36 = 0$
$\Rightarrow 0 = 0$. The solutions are ± 2 and ± 3.

73. Let $u = t^2$. Then $t^4 = u^2$. So the equation becomes $2u^2 + u - 1 = 0 \Rightarrow (2u - 1)(u + 1) = 0$

$\Rightarrow u = \dfrac{1}{2}$ or $u = -1$. Now solve for t:

$\dfrac{1}{2} = t^2 \Rightarrow \pm\sqrt{\dfrac{1}{2}} = \pm\dfrac{\sqrt{2}}{2} = t$ or $-1 = t^2 \Rightarrow$

$\pm i = t$.

(*continued on next page*)

(continued)

Now check to make sure that none of the solutions are extraneous.

If $t = \pm\dfrac{\sqrt{2}}{2}$, then $2\left(\pm\dfrac{\sqrt{2}}{2}\right)^4 + \left(\dfrac{\sqrt{2}}{2}\right)^2 - 1 = 0 \Rightarrow$

$\dfrac{1}{2} + \dfrac{1}{2} - 1 = 0 \Rightarrow 0 = 0.$ If $x = \pm i$, then

$2(\pm i)^4 + (\pm i)^2 - 1 = 0 \Rightarrow 2 - 1 - 1 = 0 \Rightarrow$

$0 = 0.$ The solutions are $\pm i$ and $\pm\dfrac{\sqrt{2}}{2}$.

75. Let $u = \sqrt{p^2 - 3}$. Then $p^2 - 3 = u^2$. So the

equation becomes $u^2 + 4u - 5 = 0 \Rightarrow$

$(u + 5)(u - 1) = 0 \Rightarrow u = -5$ or $u = 1$. Now

solve for p: $-5 = \sqrt{p^2 - 3}$, which is not

possible, or $1 = \sqrt{p^2 - 3} \Rightarrow$

$1 = p^2 - 3 \Rightarrow 1 = p^2 - 3 \Rightarrow 4 = p^2 \Rightarrow \pm 2 = p.$

Now check to make sure that none of the solutions are extraneous. If $p = \pm 2$, then

$(\pm 2)^2 - 3 + 4\sqrt{(\pm 2)^2 - 3} - 5 = 0 \Rightarrow$

$4 - 3 + 4\sqrt{1} - 5 = 0 \Rightarrow 0 = 0.$

Solution set: $\{\pm 2\}$.

77. Let $u = 3t + 1$. Then the equation becomes

$u^2 - 3u + 2 = 0 \Rightarrow (u - 2)(u - 1) = 0 \Rightarrow$

$u = 2$ or $u = 1$. Now solve for t: $2 = 3t + 1 \Rightarrow$

$\dfrac{1}{3} = t$ or $1 = 3t + 1 \Rightarrow 0 = t.$ Now check to

make sure that neither solution is extraneous.

If $t = 0$,

$(3 \cdot 0 + 1)^2 - 3(3 \cdot 0 + 1) + 2 = 0 \Rightarrow$

$1 - 3 + 2 = 0 \Rightarrow 0 = 0,$ so 0 is a solution. If

$t = \dfrac{1}{3}$, then $\left(3\left(\dfrac{1}{3}\right) + 1\right)^2 - 3\left(3\left(\dfrac{1}{3}\right) + 1\right) + 2 = 0 \Rightarrow$

$4 - 6 + 2 = 0 \Rightarrow 0 = 0.$ So 1/3 is a solution.

Solution set: $\left\{0, \dfrac{1}{3}\right\}$.

79. Let $u = \sqrt{y} + 5$. Then the equation becomes

$u^2 - 9u + 20 = 0 \Rightarrow (u - 5)(u - 4) = 0 \Rightarrow$

$u = 5$ or $u = 4$. Now solve for y:

$5 = \sqrt{y} + 5 \Rightarrow y = 0$ or $4 = \sqrt{y} + 5 \Rightarrow$

$-1 = \sqrt{y}$, which is not defined. Now check to

see if $y = 0$ is extraneous.

$\left(\sqrt{0} + 5\right)^2 - 9\left(\sqrt{0} + 5\right) + 20 = 0 \Rightarrow$

$25 - 45 + 20 = 0 \Rightarrow 0 = 0.$

Solution set: $\{0\}$

81. Let $u = x^2 - 4$. So the equation becomes

$u^2 - 3u - 4 = 0 \Rightarrow (u - 4)(u + 1) = 0 \Rightarrow$

$u = 4$ or $u = -1$. Now solve for x:

$4 = x^2 - 4 \Rightarrow x = \pm 2\sqrt{2}$ or

$-1 = x^2 - 4 \Rightarrow x = \pm\sqrt{3}$. Now check to make

sure that neither solution is extraneous. If

$x = \pm 2\sqrt{2}$, then

$\left[(\pm 2\sqrt{2})^2 - 4\right]^2 - 3\left[(\pm 2\sqrt{2})^2 - 4\right] - 4 = 0 \Rightarrow$

$16 - 12 - 4 = 0 \Rightarrow 0 = 0.$ So $\pm 2\sqrt{2}$ are

solutions.

If $x = \pm\sqrt{3}$, then

$\left[(\pm\sqrt{3})^2 - 4\right]^2 - 3\left[(\pm\sqrt{3})^2 - 4\right] - 4 = 0 \Rightarrow$

$1 + 3 - 4 = 0 \Rightarrow 0 = 0.$ So $\pm\sqrt{3}$ are solutions.

Solution set: $\left\{\pm\sqrt{3}, \pm 2\sqrt{2}\right\}$.

83. Let $u = x^2 - 3x$. So the equation becomes

$u^2 - 2u - 8 = 0 \Rightarrow (u + 2)(u - 4) = 0 \Rightarrow$

$u = -2$ or $u = 4$. Now solve for x:

$-2 = x^2 - 3x \Rightarrow x^2 - 3x + 2 = 0 \Rightarrow$

$(x - 2)(x - 1) = 0 \Rightarrow x = 2$ or $x = 1$ or

$4 = x^2 - 3x \Rightarrow x^2 - 3x - 4 = 0 \Rightarrow$

$(x - 4)(x + 1) = 0 \Rightarrow x = 4$ or $x = -1$.

Now check to make sure that none of the solutions are extraneous. If $x = 2$, then

$(2^2 - 3(2))^2 - 2(2^2 - 3(2)) - 8 = 0 \Rightarrow$

$4 - (-4) - 8 = 0 \Rightarrow 0 = 0.$ So 2 is a solution.

If $x = 1$, then

$(1^2 - 3(1))^2 - 2(1^2 - 3(1)) - 8 = 0 \Rightarrow$

$4 + 4 - 8 = 0 \Rightarrow 0 = 0.$ So 1 is a solution. If

$x = 4$, then

$(4^2 - 3(4))^2 - 2(4^2 - 3(4)) - 8 = 0 \Rightarrow$

$16 - 8 - 8 = 0 \Rightarrow 0 = 0.$ So 4 is a solution.

If $x = -1$, then

$\left[(-1)^2 - 3(-1)\right]^2 - 2\left[(-1)^2 - 3(-1)\right] - 8 = 0 \Rightarrow$

$16 - 8 - 8 = 0 \Rightarrow 0 = 0.$ So -1 is a solution.

Solution set: $\{-1, 1, 2, 4\}$.

1.4 Applying the Concepts

85. Let $x =$ the denominator. Then $x - 2 =$ the numerator.

$$\frac{x-2}{x} + \frac{x}{x-2} = \frac{25}{12}$$

$$12x(x-2)\left(\frac{x-2}{x} + \frac{x}{x-2}\right) = 12x(x-2)\left(\frac{25}{12}\right)$$

$$12(x-2)^2 + 12x^2 = 25x(x-2)$$

$$24x^2 - 48x + 48 = 25x^2 - 50x$$

$$0 = x^2 - 2x - 48$$

$$0 = (x-8)(x+6)$$

$$x = 8 \text{ or } x = -6$$

The solution must be positive. So the fraction is $\frac{6}{8}$.

87. Let $x =$ the number of shares of stock. Then $1800/x =$ the price of each share of stock.

$$\left(\frac{1800}{x} - 18\right)(x+5) = 1800$$

$$\frac{1800(x+5)}{x} - 18(x+5) = 1800$$

$$\frac{1800x + 9000}{x} - 18x - 90 = 1800$$

$$1800x + 9000 - 18x^2 - 90x = 1800x$$

$$-18x^2 - 90x + 9000 = 0$$

$$-18(x^2 + 5x - 500) = 0$$

$$(x-20)(x+25) = 0$$

$$x = 20 \text{ or } x = -25$$

The answer must be positive, so we reject –25.

a. Latasha bought 20 shares of stock.

b. She paid $1800/20 = \$90$ for each share.

89. Let $d =$ the depth of the well and let $t_1 =$ the time it takes for the stone to hit the water. Using Galileo's formula, we have

$$d = 16t_1^2 \Rightarrow \frac{\sqrt{d}}{4} = t_1 . \text{ Let } t_2 = \text{ the time it}$$

takes for the sound of the splash to return to the top of the well. Then $t_2 = \dfrac{d}{1100}$, and

$$t_1 + t_2 = \frac{\sqrt{d}}{4} + \frac{d}{1100} = 4$$

$$1100\left(\frac{\sqrt{d}}{4} + \frac{d}{1100}\right) = 1100(4)$$

$$275\sqrt{d} + d = 4400$$

Now let $u = \sqrt{d}$ and $u^2 = d$. This gives

$$u^2 + 275u = 4400 \Rightarrow u^2 + 275u - 4400 = 0 .$$

Now solve for u using the quadratic formula with $a = 1$, $b = 275$, and $c = -4400$:

$$u = \frac{-275 \pm \sqrt{275^2 - 4(1)(-4400)}}{2(1)} \approx 15.2 .$$

Note that only the positive answer has meaning, so we reject the negative answer. $u^2 = d \approx 230$. So the well is approximately 230 feet deep.

91. Let $x =$ the speed of the current. Then $10 + x =$ the speed of the motorboat going downstream and $10 - x =$ the speed of the motorboat going upstream. $\dfrac{12}{10+x} =$ the time the boat took to go downstream and $\dfrac{12}{10-x} =$ the time the boat took to go upstream. So, we have

$$\frac{12}{10-x} = \frac{12}{10+x} + \frac{1}{2}$$

$$2(10+x)(10-x)\left(\frac{12}{10-x} = \frac{12}{10+x} + \frac{1}{2}\right)$$

$$240 + 24x = 240 - 24x + 100 - x^2$$

$$x^2 + 48x - 100 = 0$$

$$(x+50)(x-2) = 0 \Rightarrow x = -50 \text{ or } x = 2$$

Only the positive answer has meaning, so we reject –50. The rate of the current was 2 mph.

93. Let $x =$ the time it took the wife to wash the car alone. Then $x + 20 =$ the time it took the husband to wash the car alone. So we have

$$\frac{1}{x} + \frac{1}{x+20} = \frac{1}{24}$$

$$24x(x+20)\left(\frac{1}{x} + \frac{1}{x+20} = \frac{1}{24}\right)$$

$$24x + 480 + 24x = x^2 + 20x$$

$$0 = x^2 - 28x - 480$$

$$0 = (x-40)(x+12)$$

$$x = 40 \text{ or } x = -12$$

Only the positive answer has meaning, so we reject –12. The wife took 40 minutes to wash the car alone.

95. Let $x =$ the length of the shorter side. Then $100 + x =$ the length of the longer side. Use the Pythagorean theorem to find the length of the diagonal $= \sqrt{x^2 + (x+100)^2}$. So we have

$$\sqrt{x^2 + (x+100)^2} + (x+100) = 3x \Rightarrow$$

$$\sqrt{x^2 + (x+100)^2} = 2x - 100 \Rightarrow$$

(continued on next page)

(*continued*)

$$\left(\sqrt{x^2 + (x+100)^2}\right)^2 = (2x-100)^2 \Rightarrow$$

$$x^2 + (x+100)^2 = 4x^2 - 400x + 10,000 \Rightarrow$$

$$2x^2 + 200x + 10,000 = 4x^2 - 400x + 10,000 \Rightarrow$$

$$0 = 2x^2 - 600x \Rightarrow 0 = 2x(x-300) \Rightarrow$$

$$x = 0 \text{ or } x = 300$$

The shorter side is 300 feet and the longer side is 400 feet. So the area is 120,000 square feet.

97. Use the Pythagorean theorem to find the lengths of AE and BE. $AE = \sqrt{64 + x^2}$ and $BE = \sqrt{(18-x)^2 + 25}$.

So we have $AE + BE = 23$ or

$$\sqrt{64 + x^2} + \sqrt{(18-x)^2 + 25} = 23 \Rightarrow$$

$$\sqrt{(18-x)^2 + 25} = 23 - \sqrt{64 + x^2} \Rightarrow$$

$$\left(\sqrt{(18-x)^2 + 25}\right)^2 = \left(23 - \sqrt{64 + x^2}\right)^2 \Rightarrow$$

$$(18-x)^2 + 25 = 529 - 46\sqrt{64 + x^2} + 64 + x^2 \Rightarrow$$

$$324 - 36x + x^2 + 25 = 593 - 46\sqrt{64 + x^2} + x^2 \Rightarrow$$

$$-244 - 36x = -46\sqrt{64 + x^2} \Rightarrow$$

$$(244 + 36x)^2 = \left(46\sqrt{64 + x^2}\right)^2$$

$$1296x^2 + 17,568x + 59,536 =$$
$$135,424 + 2116x^2$$

$$-820x^2 + 17,568x - 75,888 = 0$$

$$-4(205x^2 - 4392x + 18,972) = 0$$

$$(x-6)(205x - 3162) = 0 \Rightarrow$$

$$x = 6 \text{ or } x = \frac{3162}{205} \approx 15.4$$

Check for extraneous solutions. Neither is extraneous, so $CE = 6$ mi or $CE \approx 15.4$ mi.

1.4 Beyond the Basics

99.

$$\left(\frac{x^2-3}{x}\right)^2 - 6\left(\frac{x^2-3}{x}\right) + 8 = 0$$

Let $u = \dfrac{x^2-3}{x}$, so we have

$$u^2 - 6u + 8 - 0 \Rightarrow (u-4)(u-2) = 0 \Rightarrow$$
$$u = 4 \text{ or } u = 2.$$

Now solve for x.

$$\frac{x^2-3}{x} = 4 \Rightarrow x^2 - 3 = 4x \Rightarrow$$

$$x^2 - 4x - 3 = 0 \Rightarrow$$

$$x = 2 \pm 2\sqrt{7} \text{ (using the quadratic formula) or}$$

$$\frac{x^2-3}{x} = 2 \Rightarrow x^2 - 3 = 2x \Rightarrow$$

$$x^2 - 2x - 3 = 0 \Rightarrow$$

$$(x-3)(x+1) = 0 \Rightarrow x = 3 \text{ or } x = -1$$

Be sure to check to make sure that none of the solutions are extraneous. None are extraneous.

Solution set: $\left\{-1, \, 2 \pm 2\sqrt{7}, \, 3\right\}$.

101.

$$\sqrt{\sqrt{3x-2} + 2\sqrt{4x+1}} = 2\sqrt{2}$$

$$\sqrt{3x-2} + 2\sqrt{4x+1} = 8$$

$$\sqrt{3x-2} = 8 - 2\sqrt{4x+1}$$

$$3x - 2 = 64 - 32\sqrt{4x+1} + 4(4x+1)$$

$$3x - 2 = 16x + 68 - 32\sqrt{4x+1}$$

$$-13x - 70 = -32\sqrt{4x+1}$$

$$169x^2 + 1820x + 4900 = 4096x + 1024$$

$$169x^2 - 2276x + 3876 = 0$$

Solve for x using the quadratic formula.

$$x = \frac{2276 \pm \sqrt{2276^2 - 4(169)(3876)}}{2(169)}$$

$$x = 2 \text{ or } x \approx 11.47$$

Check to make sure that neither solution is extraneous. 11.47 is extraneous.

Solution set: {2}.

103.

$$\frac{x+2}{x} + \frac{2x+6}{x^2+4x} = -\frac{1}{x+4}$$

$$x(x+4)\left(\frac{x+2}{x} + \frac{2x+6}{x^2+4x}\right) = -\frac{1}{x+4}\right)$$

$$(x+4)(x+2) + 2x + 6 = -x$$

$$x^2 + 9x + 14 = 0$$

$$(x+7)(x+2) = 0$$

$$x = -7 \text{ or } x = -2$$

Solution set: $\{-7, -2\}$

105.

$$\left(\frac{t}{t+2}\right)^2 + \frac{2t}{t+2} - 15 = 0$$

Let $u = \dfrac{t}{t+2}$. This gives

$$u^2 + 2u - 15 = 0 \Rightarrow (u+5)(u-3) = 0 \Rightarrow$$

$$u = -5 \text{ or } u = 3.$$

Now solve for t.

$$-5 = \frac{t}{t+2} \Rightarrow -5t - 10 = t \Rightarrow t = -\frac{5}{3} \text{ or}$$

$$3 = \frac{t}{t+2} \Rightarrow 3t + 6 = t \Rightarrow t = -3.$$

Be sure to check to make sure that neither solution is extraneous. Neither is extraneous.

Solution set: $\left\{-3, \, -\dfrac{5}{3}\right\}$.

107. $2x^{1/3} + 2x^{-1/3} - 5 = 0$

Let $u = x^{1/3}$. So $\dfrac{1}{u} = x^{-1/3}$.

$2u + \dfrac{2}{u} - 5 = 0 \Rightarrow 2u^2 - 5u + 2 = 0 \Rightarrow$

$(2u - 1)(u - 2) = 0 \Rightarrow u = 1/2$ or $u = 2$.

Now solve for x:

$\dfrac{1}{2} = x^{1/3} \Rightarrow \dfrac{1}{8} = x$ or $2 = x^{1/3} \Rightarrow 8 = x$

Make sure that neither solution is extraneous.
Neither is extraneous.

Solution set: $x = \left\{ \dfrac{1}{8}, 8 \right\}$.

109. $x^{4/3} - 5x^{2/3} + 6 = 0$

Let $u = x^{2/3}$. The equation becomes

$u^2 - 5u + 6 = 0 \Rightarrow (u - 2)(u - 3) = 0 \Rightarrow$

$u = 2, 3$

$$
\begin{array}{c|c}
x^{2/3} = 2 & x^{2/3} = 3 \\
\left(x^{2/3}\right)^{3/2} = \pm\left(2^{3/2}\right) & \left(x^{2/3}\right)^{3/2} = \pm\left(3^{3/2}\right) \\
x = \pm\sqrt{8} & x = \pm\sqrt{27} \\
= \pm 2\sqrt{2} & = \pm 3\sqrt{3}
\end{array}
$$

Verify that none of the solutions are
extraneous. Solution set: $\left\{ \pm 2\sqrt{2}, \pm 3\sqrt{3} \right\}$.

111. a. *Rationalization method*:

$x + \sqrt{x} - 42 = 0 \Rightarrow \sqrt{x} = 42 - x \Rightarrow$

$x = 1764 - 84x + x^2 \Rightarrow$

$x^2 - 85x + 1764 = 0 \Rightarrow$

$(x - 36)(x - 49) = 0 \Rightarrow x = 36$ or $x = 49$.

Note that $x = 49$ is extraneous.
Solution set: $\{36\}$.

b. *Substitution method*:

Let $u = \sqrt{x}$. Then $u^2 = x$. So,

$u^2 + u - 42 = 0 \Rightarrow (u + 7)(u - 6) = 0 \Rightarrow$

$u = -7$ or $u = 6$. Now solve for x.

$-7 = \sqrt{x}$, which is not defines, or

$6 = \sqrt{x} \Rightarrow 36 = x$.

Be sure to check to make sure that $x = 36$ is
not extraneous. Solution set: $\{36\}$.

113. $\dfrac{x + b}{x - b} = \dfrac{x - 5b}{2x - 5b}$

$(2x - 5b)(x + b) = (x - b)(x - 5b)$

$2x^2 - 3bx - 5b^2 = x^2 - 6bx + 5b^2$

$x^2 + 3bx - 10b^2 = 0$

$(x + 5b)(x - 2b) = 0$

$\qquad\qquad x = -5b$ or $x = 2b$

Solution set: $\{-5b, 2b\}$

115. $\qquad\qquad 3x - 2a = \sqrt{a(3x - 2a)}$

$9x^2 - 12ax + 4a^2 = 3ax - 2a^2$

$9x^2 - 15ax + 6a^2 = 0$

$3(3x - 2a)(x - a) = 0$

$\qquad\qquad x = \dfrac{2a}{3}$ or $x = a$

Solution set: $\left\{ \dfrac{2a}{3}, a \right\}$.

1.4 Critical Thinking/Discussion/Writing

117. To find the value of x, we can let $x = \sqrt{1 + x}$
because the square root contains a copy of
itself. Now solve the equation:

$x = \sqrt{1 + x} \Rightarrow x^2 = 1 + x \Rightarrow$

$x^2 - x - 1 = 0$. Now use the quadratic

formula. $x = \dfrac{1 \pm \sqrt{1 + 4}}{2} = \dfrac{1 \pm \sqrt{5}}{2}$.

The original expression is positive, so

$x = \dfrac{1 - \sqrt{5}}{2}$ cannot be the solution.

Solution set: $\left\{ \dfrac{1 + \sqrt{5}}{2} \right\}$.

118. To find the value of x, we can let $x = \sqrt{20 + x}$
because the square root contains a copy of
itself. Now solve the equation:

$x = \sqrt{20 + x} \Rightarrow x^2 = 20 + x \Rightarrow$

$x^2 - x - 20 = 0$. Now use the quadratic

formula. $x = \dfrac{1 \pm \sqrt{1 + 80}}{2} = \dfrac{1 \pm 9}{2} \Rightarrow x = 5$

or $x = -4$. The original expression is positive,
so $x = -4$ cannot be a solution.
Solution set: $\{5\}$.

119. To find the value of x, we can let $x = \sqrt{n + x}$ because the square root contains a copy of itself. Now solve the equation:

$x = \sqrt{n + x} \Rightarrow x^2 = n + x \Rightarrow$

$x^2 - x - n = 0$. Now use the quadratic

formula. $x = \dfrac{1 \pm \sqrt{1 + 4n}}{2} = \dfrac{1}{2} \pm \dfrac{\sqrt{1 + 4n}}{2}$.

The original expression is positive, so we reject the negative root.

Solution set: $\left\{ \dfrac{1}{2} + \dfrac{\sqrt{1 + 4n}}{2} \right\}$.

120. $\sqrt{x + 2\sqrt{x - 1}} + \sqrt{x - 2\sqrt{x - 1}} = 4$

Note that $x + 2\sqrt{x - 1} = (x - 1) + 1 + 2\sqrt{x - 1}$

$= \left(\sqrt{x - 1} + 1 \right)^2$. Similarly, $x - 2\sqrt{x - 1} =$

$(x - 1) - 2\sqrt{x - 1} + 1 = \left(\sqrt{x - 1} - 1 \right)^2$.

Substitute these expressions into the original equation:

$\sqrt{\left(\sqrt{x - 1} + 1 \right)^2} + \sqrt{\left(\sqrt{x - 1} - 1 \right)^2} = 4 \Rightarrow$

$\sqrt{x - 1} + 1 + \sqrt{x - 1} - 1 = 4 \Rightarrow$

$2\sqrt{x - 1} = 4 \Rightarrow \sqrt{x - 1} = 2 \Rightarrow x - 1 = 4 \Rightarrow$

$x = 5$.

Solution set: $\{5\}$

1.4 Maintaining Skills

121. $12x - 4 = 28 - 4x$

$16x = 32$

$x = 2$

Solution set: $\{2\}$

123. $2x + 3(x - 4) = 7x + 10$

$2x + 3x - 12 = 7x + 10$

$5x - 12 = 7x + 10$

$-2x = 22$

$x = -11$

Solution set: $\{-11\}$

125. $\dfrac{-2x}{3} = x + \dfrac{x}{2}$

$6\left(\dfrac{-2x}{3} \right) = 6x + 6\left(\dfrac{x}{2} \right)$

$-4x = 6x + 3x$

$-4x = 9x$

$0 = 13x \Rightarrow 0 = x$

Solution set: $\{0\}$

127. $6x^2 - 102 = 0$

$6x^2 = 102$

$x^2 = 17$

$x = \pm\sqrt{17}$

Solution set: $\left\{ \pm\sqrt{17} \right\}$

129. $25x^2 + 1 = 10x$

$25x^2 - 10x + 1 = 0$

$(5x - 1)^2 = 0$

$5x - 1 = 0$

$5x = 1$

$x = \dfrac{1}{5}$

Solution set: $\left\{ \dfrac{1}{5} \right\}$

131. $3x^2 + 3x - 8 = 10$

$3x^2 + 3x - 18 = 0$

$3\left(x^2 + x - 6 \right) = 0$

$x^2 + x - 6 = 0$

$(x + 3)(x - 2) = 0 \Rightarrow x = -3,\ 2$

Solution set: $\{-3, 2\}$

133. $\dfrac{5}{2x} + 3 = \dfrac{5 + 3(2x)}{2x} = \dfrac{5 + 6x}{2x}$

135. $\dfrac{3}{x + 1} - \dfrac{x - 2}{x + 3} = \dfrac{3(x + 3) - (x + 1)(x - 2)}{(x + 1)(x + 3)}$

$= \dfrac{3x + 9 - \left(x^2 - x - 2 \right)}{(x + 1)(x + 3)}$

$= \dfrac{-x^2 + 4x + 11}{(x + 1)(x + 3)}$

1.5 Inequalities

1.5 Practice Problems

1. a. $4x + 9 > 2(x + 6) + 1 \Rightarrow 4x + 9 > 2x + 13 \Rightarrow$

$2x > 4 \Rightarrow x > 2$

Solution set: $(2, \infty)$

b. $7 - 2x \geq -3 \Rightarrow -2x \geq -10 \Rightarrow x \leq 5$

Solution set: $(-\infty, 5]$

2. Let t = time elapsed since the plane went on autopilot. Then $340t$ = distance the plane has flown in t hours, and $185 + 340t$ = the plane's distance from Miami after t hours.
$185 + 340t \geq 1035 \Rightarrow 340t \geq 850 \Rightarrow t \geq 2.5$
The tower will suspect trouble if the plane has not arrived in 2.5 hours.

3. a. $2(4-x)+6x < 4(x+1)+7$
$8-2x+6x < 4x+4+7$
$8+4x < 4x+11$
$0 < 3$
The last inequality is always true, so the solution set is $(-\infty, \infty)$.

b. $3(x-2)+5 \geq 7(x-1)-4(x-2)$
$3x-6+5 \geq 7x-7-4x+8$
$3x-1 \geq 3x+1 \Rightarrow -1 \geq 1$
The last inequality is always false, so the solution set is $\varnothing$.

4. $3x-5 \geq 7$ or $5-2x \geq 1$
$3x \geq 12$ or $-2x \geq -4$
$x \geq 4$ or $x \leq 2$

Solution set: $(-\infty, 2] \cup [4, \infty)$

5. $2(3-x)-3 < 5$ and $2(x-5)+7 \leq 3$
$6-2x-3 < 5$ and $2x-10+7 \leq 3$
$3-2x < 5$ and $2x-3 \leq 3$
$-2x < 2$ and $2x \leq 6$
$x > -1$ and $x \leq 3$

Solution set: $(-1, 3]$

6. $-6 \leq 4x-2 < 4 \Rightarrow -4 \leq 4x < 6 \Rightarrow -1 \leq x < \frac{3}{2}$

Solution set: $\left[-1, \frac{3}{2}\right)$

7. Start with the interval for x.
$-3 \leq x \leq 2$
$3(-3) \leq 3x \leq 3(2)$
$-9 \leq 3x \leq 6$
$-9+5 \leq 3x+5 \leq 6+5$
$-4 \leq 3x+5 \leq 11$
Therefore, $a = -4$ and $b = 11$.

8. $15 < C < 25$
$\left(\frac{9}{5}\right)15 < \frac{9}{5}C < \left(\frac{9}{5}\right)25$
$27 < \frac{9}{5}C < 45$
$27+32 < \frac{9}{5}C+32 < 45+32$
$59 < \frac{9}{5}C+32 < 77$

The temperature range from $15°C$ to $25°C$ corresponds to a range from $59°F$ to $77°F$.

9. $x^2 + 2 < 3x + 6$
$x^2 - 3x - 4 < 0 \Rightarrow (x+1)(x-4) < 0$
The factors equal 0 at $x = -1$ and $x = 4$.

Interval	Test point	Value of x^2-3x-4	Result
$(-\infty, -1)$	-2	6	+
$(-1, 4)$	0	-4	−
$(4, \infty)$	5	6	+

The solution set is $(-1, 4)$.

10. $\frac{2x+5}{x-1} \leq 1$
$\frac{2x+5}{x-1} - 1 \leq 0$
$\frac{2x+5-x+1}{x-1} \leq 0 \Rightarrow \frac{x+6}{x-1} \leq 0$
$x+6 = 0 \Rightarrow x = -6$
$x-1 = 0 \Rightarrow x = 1$
The test points are -6 and 1.

Interval	Test point	Value of $\frac{x+6}{x-1}$	Result
$(-\infty, -6)$	-9	$\frac{3}{10}$	+
$(-6, 1)$	0	-6	−
$(1, \infty)$	2	8	+

Note that the expression is undefined for $x = 1$. The solution set is $[-6, 1)$.

1.5 Basic Concepts and Skills

1. < **3.** $\geq$ **5.** < **7.** $\geq$

9. <

11. $(-2,5)$

13. $(0, 4]$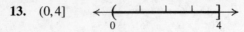

15. $[-1, \infty)$

17. $(-\infty, -2]$

19. $x + 3 < 6 \Rightarrow x < 3$
$(-\infty, 3)$

21. $1 - x \le 4 \Rightarrow -x \le 3 \Rightarrow x \ge -3$
$[-3, \infty)$

23. $2x + 5 < 9 \Rightarrow 2x < 4 \Rightarrow x < 2$
$(-\infty, 2)$

25. $3 - 3x > 15 \Rightarrow -3x > 12 \Rightarrow x < -4$
$(-\infty, -4)$

27. $3(x + 2) < 2x + 5 \Rightarrow 3x + 6 < 2x + 5 \Rightarrow x < -1$
$(-\infty, -1)$

29. $3(x - 3) \le 3 - x \Rightarrow 3x - 9 \le 3 - x \Rightarrow$
$4x \le 12 \Rightarrow x \le 3$
$(-\infty, 3]$

31. $6x + 4 > 3x + 10 \Rightarrow 3x > 6 \Rightarrow x > 2$
$(2, \infty)$

33. $8(x - 1) - x \le 7x - 12$
$8x - 8 - x \le 7x - 12$
$7x - 8 \le 7x - 12 \Rightarrow -8 \le -12$ False
The solution set is $\varnothing$.

35. $5(x + 2) \le 3(x + 1) + 10 \Rightarrow$
$5x + 10 \le 3x + 3 + 10 \Rightarrow 5x + 10 \le 3x + 13 \Rightarrow$
$2x \le 3 \Rightarrow x \le \dfrac{3}{2}$
$\left(-\infty, \dfrac{3}{2}\right]$

37. $2(x + 1) + 3 \ge 2(x + 2) - 1$
$2x + 2 + 3 \ge 2x + 4 - 1$
$2x + 5 \ge 2x + 3 \Rightarrow 5 \ge 3$ True
$(-\infty, \infty)$

39. $2(x + 1) - 2 \le 3(2 - x) + 9$
$2x + 2 - 2 \le 6 - 3x + 9$
$2x \le 15 - 3x$
$5x \le 15 \Rightarrow x \le 3$
$(\infty, 3]$

41. $9x - 6 \ge \dfrac{3}{2}x + 9$
$18x - 12 \ge 3x + 18$
$15x \ge 30 \Rightarrow x \ge 2$
Solution set: $[2, \infty)$

43. $\dfrac{x - 3}{3} \le 2 + \dfrac{x}{2} \Rightarrow 2x - 6 \le 12 + 3x \Rightarrow$
$-18 \le x \Rightarrow x \ge -18$
Solution set: $[-18, \infty)$

45. $\dfrac{3x + 1}{2} < x - 1 + \dfrac{x}{2}$
$3x + 1 < 2x - 2 + x$
$3x + 1 < 3x - 2 \Rightarrow 1 < -2$ False
Solution set: $\varnothing$

47. $\dfrac{x - 3}{2} \ge \dfrac{x}{3} + 1$
$3x - 9 \ge 2x + 6 \Rightarrow x \ge 15$
Solution set: $[15, \infty)$

49. $\dfrac{3x + 1}{3} - \dfrac{x}{2} \le \dfrac{x + 2}{2}$
$6x + 2 - 3x \le 3x + 6$
$3x + 2 \le 3x + 6 \Rightarrow 2 \le 6$ True
Solution set: $(-\infty, \infty)$

51. $2x + 5 < 1$ or $2 + x > 4$
$2x < -4$ or $x > 2$
$x < -2$ or
Solution set: $(-\infty, -2) \cup (2, \infty)$

53. $\dfrac{2x - 3}{4} \le 2$ or $\dfrac{4 - 3x}{2} \ge 2$
$2x - 3 \le 8$ or $4 - 3x \ge 4$
$2x \le 11$ or $-3x \ge 0$
$x \le \dfrac{11}{2}$ or $x \le 0$
Solution set: $\left(-\infty, \dfrac{11}{2}\right]$

55. $\dfrac{2x+1}{3} \geq x+1 \quad$ or $\quad \dfrac{x}{2}-1 > \dfrac{x}{3}$

$2x+1 \geq 3x+3 \quad$ or $\quad 3x-6 > 2x$

$\quad -2 \geq x \qquad$ or $\qquad -6 > -x$

$\quad x \leq -2 \qquad$ or $\qquad 6 < x \Rightarrow x > 6$

Solution set: $(-\infty,\,-2] \cup (6,\,\infty)$

57. $\dfrac{x-1}{2} > \dfrac{x}{3}-1 \quad$ or $\quad \dfrac{2x+5}{3} \leq \dfrac{x+1}{6}$

$3x-3 > 2x-6 \quad$ or $\quad 4x+10 \leq x+1$

$\quad x > -3 \qquad$ or $\qquad 3x \leq -9 \Rightarrow x \leq -3$

Solution set: $(-\infty,\,\infty)$

59. $3-2x \leq 7 \quad$ and $\quad 2x-3 \leq 7$

$\quad -2x \leq 4 \quad$ and $\qquad 2x \leq 10$

$\quad x \geq -2 \quad$ and $\qquad x \leq 5$

Solution set: $[-2,\,5]$

61. $2(x+1)+3 \geq 1 \quad$ and $\quad 2(2-x) > -6$

$\quad 2x+2+3 \geq 1 \quad$ and $\quad 4-2x > -6$

$\qquad 2x+5 \geq 1 \quad$ and $\qquad -2x > -10$

$\qquad\quad 2x \geq -4 \quad$ and $\qquad\quad x < 5$

$\qquad\qquad x \geq -2$

Solution set: $[-2,\,5)$

63. $2(x+1)-3 > 7 \quad$ and $\quad 3(2x+1)+1 < 10$

$\quad 2x+2-3 > 7 \quad$ and $\qquad 6x+3+1 < 10$

$\qquad 2x-1 > 7 \quad$ and $\qquad\quad 6x+4 < 10$

$\qquad\quad 2x > 8 \quad$ and $\qquad\qquad 6x < 6$

$\qquad\quad x > 4 \quad$ and $\qquad\qquad x < 1$

Since x cannot be less than 1 and greater than 4 at the same time, the solution set is $\varnothing$.

65. $5+3(x-1) < 3+3(x+1) \quad$ and $\quad 3x-7 \leq 8$

$\quad 5+3x-3 < 3+3x+3 \quad$ and $\qquad 3x \leq 15$

$\qquad 2+3x < 6+3x \qquad$ and $\qquad\quad x \leq 5$

$\qquad\qquad 2 < 6 \;$ True

Solution set: $(-\infty,\,5]$

67. $3 < x+5 < 4 \Rightarrow -2 < x < -1$ or $(-2,-1)$

69. $-4 \leq x-2 < 2 \Rightarrow -2 \leq x < 4$ or $[-2,4)$

71. $-9 \leq 2x+3 \leq 5 \Rightarrow -12 \leq 2x \leq 2 \Rightarrow$

$-6 \leq x \leq 1$ or $[-6,1]$

73. $0 \leq 1-\dfrac{x}{3} < 2 \Rightarrow 0 \leq 3-x < 6 \Rightarrow$

$-3 \leq -x < 3 \Rightarrow 3 \geq x > -3$ or $(-3,3]$

75. $-1 < \dfrac{2x-3}{5} \leq 0 \Rightarrow -5 < 2x-3 \leq 0 \Rightarrow$

$-2 < 2x \leq 3 \Rightarrow -1 < x \leq \dfrac{3}{2}$ or $\left(-1,\dfrac{3}{2}\right]$

77. $5x \leq 3x+1 < 4x-2 \Rightarrow 2x \leq 1 < x-2 \Rightarrow$

$x \leq \dfrac{1}{2}$ and $3 < x \Rightarrow x > 3$

x cannot be less than or equal to $\dfrac{1}{2}$ at the same time that 3 is less than x, so the solution set is $\varnothing$.

79. $-2 < x < 1 \Rightarrow -2+7 < x+7 < 1+7 \Rightarrow$

$5 < x+7 < 8 \Rightarrow a = 5, b = 8$

81. $-1 < x < 1 \Rightarrow 1 > -x > -1 \Rightarrow$

$2+1 > 2-x > 2-1 \Rightarrow 3 > 2-x > 1 \Rightarrow$

$1 < 2-x < 3 \Rightarrow a = 1, b = 3$

83. $0 < x < 4 \Rightarrow 0 < 5x < 20 \Rightarrow -1 < 5x-1 < 19 \Rightarrow$

$a = -1, b = 19$

85. $x^2 + 4x - 12 \leq 0 \Rightarrow (x+6)(x-2) \leq 0$.

Now solve the associated equation:

$(x+6)(x-2) = 0 \Rightarrow x = -6$ or $x = 2$.

So, the intervals are

$(-\infty,-6], [-6,2],$ and $[2,\infty)$.

Interval	Test point	Value of $x^2+4x-12$	Result
$(-\infty,-6]$	-10	48	$+$
$[-6,2]$	0	-12	$-$
$[2,\infty)$	3	9	$+$

The solution set is $[-6,2]$.

87. $6x^2 + 7x - 3 \geq 0 \Rightarrow (3x-1)(2x+3) \geq 0$.

Now solve the associated equation:

$(3x-1)(2x+3) = 0 \Rightarrow x = 1/3$ or $x = -3/2$.

The intervals are $(-\infty,-3/2], [-3/2,1/3],$

and $[1/3,\infty)$.

Interval	Test point	Value of $6x^2+7x-3$	Result
$(-\infty,-3/2]$	-2	7	$+$
$[-3/2,1/3]$	0	-3	$-$
$[1/3,\infty)$	1	10	$+$

The solution set is $(-\infty,-3/2] \cup [1/3,\infty)$.

89. $(x+3)(x+1)(x-1) \geq 0$.

Now solve the associated equation:

$(x+3)(x+1)(x-1) = 0 \Rightarrow$

$x = -3, x = -1$ or $x = 1$.

The intervals are

$(-\infty, -3], [-3, -1], [-1, 1]$, and $[1, \infty)$.

Interval	Test point	Value of $(x+3)(x+1)(x-1)$	Result
$(-\infty, -3]$	-10	-693	$-$
$[-3, -1]$	-2	3	$+$
$[-1, 1]$	0	-3	$-$
$[1, \infty)$	10	1287	$+$

The solution set is $[-3, -1] \cup [1, \infty)$.

91. $x^3 - 4x^2 - 12x > 0$

Solve the associated equation

$x^3 - 4x^2 - 12x = 0$.

$x^3 - 4x^2 - 12x = 0 \Rightarrow x(x^2 - 4x - 12) = 0 \Rightarrow$

$x(x+2)(x-6) = 0 \Rightarrow x = 0, -2, 6$

The intervals are $(-\infty, -2), (-2, 0), (0, 6)$,

and $(6, \infty)$.

Interval	Test point	Value of $x^3 - 4x^2 - 12x$	Result
$(-\infty, -2)$	-10	-1280	$-$
$(-2, 0)$	-1	7	$+$
$(0, 6)$	1	-15	$-$
$(6, \infty)$	10	480	$+$

The solution set is $(-2, 0) \cup (6, \infty)$.

93. $x^2 + 2x < -1 \Rightarrow x^2 + 2x + 1 < 0$.

Now solve the associated equation:

$x^2 + 2x + 1 = 0 \Rightarrow (x+1)^2 = 0 \Rightarrow x = -1$.

The intervals are $(-\infty, -1)$ and $(-1, \infty)$.

Interval	Test point	Value of $x^2 + 2x + 1$	Result
$(\infty, -1)$	-2	1	$+$
$(-1, \infty)$	0	1	$+$

Neither interval has a negative solution, so the solution set is $\varnothing$.

95. $x^3 - x^2 \geq 0$.

Now solve the associated equation:

$x^3 - x^2 = 0 \Rightarrow x^2(x-1) = 0 \Rightarrow$

$x = 0$ or $x = 1$.

The intervals are $(-\infty, 0], [0, 1]$ and $[1, \infty)$.

Interval	Test point	Value of $x^3 - x^2$	Result
$(-\infty, 0]$	-1	-2	$-$
$[0, 1]$	$\dfrac{1}{2}$	$-\dfrac{1}{8}$	$-$
$[1, \infty)$	2	4	$+$

Note that 0 is a solution. The solution set is $\{0\} \cup [1, \infty)$.

97. $x^2 \geq 1 \Rightarrow x^2 - 1 \geq 0$.

Now solve the associated equation:

$x^2 - 1 = 0 \Rightarrow (x-1)(x+1) = 0 \Rightarrow$

$x = -1$ or $x = 1$.

The intervals are $(-\infty, -1], [-1, 1]$, and $[1, \infty)$.

Interval	Test point	Value of $x^2 - 1$	Result
$(-\infty, -1]$	-2	3	$+$
$[-1, 1]$	0	-1	$-$
$[1, \infty)$	2	3	$+$

The solution set is $(-\infty, -1] \cup [1, \infty)$.

99. $x^3 < -8 \Rightarrow x^3 + 8 < 0$.

Now solve the associated equation:

$x^3 + 8 = 0 \Rightarrow (x+2)(x^2 - 2x + 4) = 0 \Rightarrow$

$x = -2$ or $x^2 - 2x + 4 = 0$.

Solve $x^2 - 2x + 4 = 0$ using the quadratic formula:

$x = \dfrac{-(-2) \pm \sqrt{(-2)^2 - 4(1)(4)}}{2(1)} = 1 \pm i\sqrt{3}$.

We cannot use complex intervals, so the only intervals we examine are $(-\infty, -2)$ and $(-2, \infty)$.

(continued on next page)

(continued)

Interval	Test point	Value of $x^3 + 8$	Result
$(-\infty, -2)$	-10	-992	$-$
$(-2, \infty)$	10	1008	$+$

The solution set is $(-\infty, -2)$.

101. $\dfrac{x+2}{x-5} < 0$.

Now solve $x + 2 = 0 \Rightarrow x = -2$ and
$x - 5 = 0 \Rightarrow x = 5$.
So the intervals are
$(-\infty, -2), (-2, 5),$ and $(5, \infty)$.

Interval	Test point	Value of $\dfrac{x+2}{x-5}$	Result
$(-\infty, -2)$	-10	$8/15$	$+$
$(-2, 5)$	0	$-2/5$	$-$
$(5, \infty)$	10	$12/5$	$+$

Note that the fraction is undefined if $x = 5$.
The solution set is $(-2, 5)$.

103. $\dfrac{x+4}{x} < 0$.

Solve $x + 4 = 0 \Rightarrow x = -4$ and $x = 0$.
The fraction is undefined if $x = 0$. So the
intervals are $(-\infty, -4), (-4, 0),$ and $(0, \infty)$.

Interval	Test point	Value of $\dfrac{x+4}{x}$	Result
$(-\infty, -4)$	-10	$3/5$	$+$
$(-4, 0)$	-2	-1	$-$
$(0, \infty)$	2	3	$+$

The solution set is $(-4, 0)$.

105. $\dfrac{x+1}{x+2} \le 3 \Rightarrow \dfrac{x+1}{x+2} - 3 \le 0 \Rightarrow$

$\dfrac{x+1-3(x+2)}{x+2} \le 0 \Rightarrow \dfrac{-2x-5}{x+2} \le 0$.
Now solve $-2x - 5 = 0 \Rightarrow x = -5/2$ and
$x + 2 = 0 \Rightarrow x = -2$. So the intervals are
$(-\infty, -5/2], [-5/2, -2),$ and $(-2, \infty)$. The
original fraction is not defined if $x = -2$, so -2
is not included in the intervals.

Interval	Test point	Value of $\dfrac{-2x-5}{x+2}$	Result
$(-\infty, -5/2]$	-3	-1	$-$
$[-5/2, -2)$	$-9/4$	2	$+$
$(-2, \infty)$	3	$-11/5$	$-$

The solution set is $(-\infty, -5/2] \cup (-2, \infty)$.

107. $\dfrac{(x-2)(x+2)}{x} > 0$.

Now we have $x = 0$ and
$(x-2)(x+2) = 0 \Rightarrow x = 2$ and $x = -2$.
So the intervals are $(-\infty, -2), (-2, 0), (0, 2),$
and $(2, \infty)$. Note that the original fraction is
not defined if $x = 0$, so 0 is not included in the
intervals.

Interval	Test point	Value of $\dfrac{(x-2)(x+2)}{x}$	Result
$(-\infty, -2)$	-3	$-5/3$	$-$
$(-2, 0)$	-1	3	$+$
$(0, 2)$	1	-3	$-$
$(2, \infty)$	3	$5/3$	$+$

The solution set is $(-2, 0) \cup (2, \infty)$.

109. $\dfrac{(x-2)(x+1)}{(x-3)(x+5)} \ge 0$.

Set the numerator and denominator equal to
zero and solve for x.
$(x-2)(x+1) = 0 \Rightarrow x = 2, -1$
$(x-3)(x+5) = 0 \Rightarrow x = 3, -5$
The intervals are $(-\infty, -5), (-5, -1],$
$[-1, 2], [2, 3),$ and $(3, \infty)$.

Interval	Test point	Value of $\dfrac{(x-2)(x+1)}{(x-3)(x+5)}$	Result
$(-\infty, -5)$	-6	$40/9$	$+$
$(-5, -1]$	-2	$-4/15$	$-$

(continued on next page)

(continued)

Interval	Test point	Value of $\dfrac{(x-2)(x+1)}{(x-3)(x+5)}$	Result
$[-1, 2]$	0	$2/15$	+
$[2, 3)$	$5/2$	$-5/15$	−
$(3, \infty)$	5	$9/10$	+

The solution set is
$(-\infty, -5) \cup [-1, 2] \cup (3, \infty)$.

111. $\dfrac{x^2-1}{x^2-4} \le 0.$

Set the numerator and denominator equal to zero and solve for x.

$x^2 - 1 = 0 \Rightarrow x = \pm 1;\ x^2 - 4 = 0 \Rightarrow x = \pm 2$

The intervals are $(-\infty, -2), (-2, -1],$
$[-1, 1], [1, 2),$ and $[2, \infty).$

Interval	Test point	Value of $\dfrac{x^2-1}{x^2-4}$	Result
$(-\infty, -2)$	-3	$8/5$	+
$(-2, -1]$	$-3/2$	$-5/7$	−
$[-1, 1]$	0	$1/4$	+
$[1, 2)$	$3/2$	$-5/7$	−
$[2, \infty)$	3	$8/5$	+

The solution set is $(-2, -1] \cup [1, 2).$

113. $\dfrac{x+4}{3x-2} \ge 1 \Rightarrow \dfrac{x+4}{3x-2} - 1 \ge 0 \Rightarrow$
$\dfrac{x+4-3x+2}{3x-2} \ge 0 \Rightarrow \dfrac{-2x+6}{3x-2} \ge 0.$

Now we have $-2x + 6 = 0 \Rightarrow x = 3$ and
$3x - 2 = 0 \Rightarrow x = 2/3$. So the intervals are
$(-\infty, 2/3), (2/3, 3],$ and $[3, \infty)$. Note that the
original fraction is not defined if $x = 2/3$, so
$2/3$ is not included in the intervals.

Interval	Test point	Value of $\dfrac{x+4}{3x-2} - 1$	Result
$(-\infty, 2/3)$	0	-3	−
$(2/3, 3]$	1	4	+
$[3, \infty)$	4	$-1/5$	−

The solution set is $(2/3, 3].$

115. $3 \le \dfrac{2x+6}{2x+1} \Rightarrow \dfrac{2x+6}{2x+1} - 3 \ge 0 \Rightarrow$
$\dfrac{2x+6-3(2x+1)}{2x+1} \ge 0 \Rightarrow \dfrac{-4x+3}{2x+1} \ge 0.$
This gives $-4x + 3 = 0 \Rightarrow x = 3/4$ and
$2x + 1 = 0 \Rightarrow x = -1/2$. The intervals are
$(-\infty, -1/2), (-1/2, 3/4],$ and $[3/4, \infty)$. The
original fraction is not defined if $x = -1/2$, so
$-1/2$ is not included in the intervals.

Interval	Test point	Value of $\dfrac{2x+6}{2x+1} - 3$	Result
$(-\infty, -1/2)$	-1	-7	−
$(-1/2, 3/4]$	0	3	+
$[3/4, \infty)$	1	$-1/3$	−

The solution set is $(-1/2, 3/4].$

117. $\dfrac{x+2}{x-3} \ge \dfrac{x-1}{x+3} \Rightarrow \dfrac{x+2}{x-3} - \dfrac{x-1}{x+3} \ge 0 \Rightarrow$
$\dfrac{(x+2)(x+3)-(x-1)(x-3)}{(x-3)(x+3)} \ge 0 \Rightarrow$
$\dfrac{(x^2+5x+6)-(x^2-4x+3)}{(x-3)(x+3)} \ge 0 \Rightarrow$
$\dfrac{9x+3}{(x-3)(x+3)} \ge 0$

Set the numerator and the denominator equal
to zero and solve for x. $9x + 3 = 0 \Rightarrow x = -1/3$
and $(x-3)(x+3) = 0 \Rightarrow x = 3$ or $x = -3$. The
intervals are $(-\infty, -3), (-3, -1/3],$
$[-1/3, 3),$ and $(3, \infty)$. Note that the original
fractions are not defined if $x = -3$ or $x = 3$, so
-3 and 3 are not included in the intervals.

(continued on next page)

(continued)

Interval	Test point	Value of $\dfrac{x+2}{x-3}-\dfrac{x-1}{x+3}$	Result
$(-\infty,\,-3)$	-6	$-17/9$	$-$
$(-3,\,-1/3]$	-1	$3/4$	$+$
$[-1/3,\,3)$	1	$-3/2$	$-$
$(3,\,\infty)$	6	$19/9$	$+$

The solution set is $(-3,\,-1/3]\cup(3,\,\infty)$.

119. $\dfrac{x-1}{x+1}\le\dfrac{x+2}{x-3}\Rightarrow\dfrac{x-1}{x+1}-\dfrac{x+2}{x-3}\le0\Rightarrow$

$\dfrac{(x-1)(x-3)-(x+2)(x+1)}{(x+1)(x-3)}\le0\Rightarrow$

$\dfrac{\left(x^2-4x+3\right)-\left(x^2+3x+2\right)}{(x+1)(x-3)}\le0\Rightarrow$

$\dfrac{-7x+1}{(x+1)(x-3)}\le0$

Set the numerator and the denominator equal to zero and solve for x. $-7x+1=0\Rightarrow x=1/7$ and $(x+1)(x-3)=0\Rightarrow x=-1$ or $x=3$. The intervals are $(-\infty,\,-1),\,(-1,\,1/7],\,[1/7,\,3),$ and $(3,\,\infty)$. Note that the original fractions are not defined if $x=-1$ or $x=3$, so -1 and 3 are not included in the intervals.

Interval	Test point	Value of $\dfrac{x-1}{x+1}-\dfrac{x+2}{x-3}$	Result
$(-\infty,\,-1)$	-2	3	$+$
$(-1,\,1/7]$	0	$-1/3$	$-$
$[1/7,\,3)$	1	$3/2$	$+$
$(3,\,\infty)$	4	$-27/5$	$-$

The solution set is $(-1,\,1/7]\cup(3,\,\infty)$.

1.5 Applying the Concepts

121. Let $x=$ the selling price of the refrigerator. Then
$1750+0.15(1750)\le x\le1750+0.20(1750)\Rightarrow$
$2012.50\le x\le2100$
The refrigerator's selling price ranges from \$2012.50 to \$2100.

123. Let $x=$ the amount of gasoline in the car at the start of the trip. Then
$300\le40x\le480\Rightarrow7.5\le x\le12$
The car had between 7.5 gallons and 12 gallons of gas at the start of the trip.

125. Let $x=$ the amount of cream.
Then $270-x=$ the amount of milk. So,
$0.3x+0.03(270-x)\ge0.045(270)\Rightarrow$
$0.3x+8.1-0.03x\ge12.15\Rightarrow$
$0.27x\ge4.05\Rightarrow x\ge15$
At least 15 quarts of cream must be added.

127. Let $x=$ the number of 4-door sedans.
Then $3x=$ the number of SUV's and $2x=$ the number of convertibles.
$x+3x+2x\ge48\Rightarrow6x\ge48\Rightarrow x\ge8$
There are at least 8 four-door sedans.

129. $132t-t^2\ge3200\Rightarrow-t^2+132t-3200\ge0\Rightarrow$
$t^2-132t+3200\le0\Rightarrow(t-100)(t-32)\le0$
Solving the associated equation gives $t=100$ or $t=32$. The intervals to be checked are $(-\infty,\,32],[32,\,100],$ and $[100,\,\infty)$. Checking a number in each interval shows that the temperature must fall in the range $[32°F,\,100°F]$.

131. Probability must be less than 1, so we have
$0.5<\dfrac{64-0.2x}{208-x}\le1\Rightarrow$
$0.5<\dfrac{64-0.2x}{208-x}$ and $\dfrac{64-0.2x}{208-x}\le1$
Solving each inequality independently gives
$0.5<\dfrac{64-0.2x}{208-x}\Rightarrow0<\dfrac{64-0.2x}{208-x}-0.5\Rightarrow$
$0<\dfrac{64-0.2x-0.5(208-x)}{208-x}\Rightarrow$
$0<\dfrac{-40+0.3x}{208-x}$
Now we have $-40+0.3x=0\Rightarrow x=400/3$ and $208-x=0\Rightarrow208=x$. The original fraction is not defined if $x=208$, so 208 is not in the solution set. So there need to be more than 133 cards. (Note that the value of x is rounded up to account for the partial value.)
$\dfrac{64-0.2x}{208-x}\le1\Rightarrow\dfrac{64-0.2x}{208-x}-1\le0\Rightarrow$
$\dfrac{64-0.2x-208+x}{208-x}\le0\Rightarrow\dfrac{0.8x-144}{208-x}\le0\Rightarrow$
$0.8x-144\le0\Rightarrow0.8x\le144\Rightarrow x\le180$
So, the likelihood that the next card dealt would be a jack, queen, king, or ace is greater than 50% if more than 133 cards and less than 180 cards are dealt.

1.5 Beyond the Basics

133. $2x^2 + kx + 2 = 0$ has two real solutions if the discriminant is greater than zero. So $k^2 - 4(2)(2) > 0 \Rightarrow k^2 - 16 > 0$. Solving the associated equation gives $k = -4$ or $k = 4$. The intervals to be tested are $(-\infty, -4), (-4, 4)$, and $(4, \infty)$. $k^2 - 16 > 0$ for $(-\infty, -4) \cup (4, \infty)$.

135. $x^2 + kx + k = 0$ has two real solutions if the discriminant is greater than zero. So $k^2 - 4k > 0 \Rightarrow k(k - 4) > 0$. Solving the associated equation gives $k = 0$ or $k = 4$. The intervals to be tested are $(-\infty, 0)$, $(0, 4)$, and $(4, \infty)$. $k^2 - 4k > 0$ for $(-\infty, 0) \cup (4, \infty)$.

137. $\dfrac{x}{2x+1} \geq \dfrac{1}{4}$ and $\dfrac{6x}{4x-1} < \dfrac{1}{2}$

We will solve each inequality independently and then determine where the solution sets intersect.

$$\frac{x}{2x+1} \geq \frac{1}{4} \Rightarrow \frac{x}{2x+1} - \frac{1}{4} \geq 0 \Rightarrow$$
$$\frac{4x - (2x+1)}{4(2x+1)} \geq 0 \Rightarrow \frac{2x-1}{8x+4} \geq 0$$

Now, we have $2x - 1 = 0 \Rightarrow x = 1/2$ and $8x + 4 = 0 \Rightarrow x = -1/2$. Note that the original fraction $\dfrac{x}{2x+1}$ is not defined if $x = -1/2$, so the intervals to be tested are $(-\infty, -1/2)$, $(-1/2, 1/2]$, and $[1/2, \infty)$.

Interval	Test point	Value of $\dfrac{x}{2x+1} - \dfrac{1}{4}$	Result
$(-\infty, -1/2)$	-1	$3/4$	$+$
$(-1/2, 1/2]$	0	$-1/4$	$-$
$[1/2, \infty)$	1	$1/12$	$+$

The solution set is $(-\infty, -1/2) \cup [1/2, \infty)$.

$$\frac{6x}{4x-1} < \frac{1}{2} \Rightarrow \frac{6x}{4x-1} - \frac{1}{2} < 0 \Rightarrow$$
$$\frac{12x - (4x-1)}{2(4x-1)} < 0 \Rightarrow \frac{8x+1}{8x-2} < 0$$

The original fractions $\dfrac{6x}{4x-1}$ and $\dfrac{8x+1}{8x-2}$ are not defined if $x = 1/4$. Also, $8x + 1 = 0 \Rightarrow x = -1/8$. The intervals to be tested are $(-\infty, -1/8)$, $(-1/8, 1/4)$, and $(1/4, \infty)$.

Interval	Test point	Value of $\dfrac{6x}{4x-1} - \dfrac{1}{2}$	Result
$(-\infty, -1/8)$	-1	$7/10$	$+$
$(-1/8, 1/4)$	0	$-1/2$	$-$
$(1/4, \infty)$	1	$3/2$	$+$

The solution set is $(-1/8, 1/4)$.

The two solution sets do not intersect, so the solution set of $\dfrac{x}{2x+1} \geq \dfrac{1}{4}$ and $\dfrac{6x}{4x-1} < \dfrac{1}{2}$ is $\varnothing$.

139. Let x = one number. Then $c - x$ = the other number. So we have
$$x(c - x) = 36 \Rightarrow cx - x^2 = 36 \Rightarrow$$
$$x^2 - cx + 36 = 0$$
$x^2 - cx + 36 = 0$ has one or two real solutions if the discriminant is greater than or equal to zero. So
$$(-c)^2 - 4(1)(36) \geq 0 \Rightarrow c^2 - 144 \geq 0 \Rightarrow$$
$$c^2 > 144 \Rightarrow c \geq 12 \text{ or } c \leq -12$$
Solution set: $(-\infty, -12] \cup [12, \infty)$

141. Let x = the total number of radios imported. Then $x - 1000$ = the number of radios subject to the penalty tax. So we have
$$10x + 0.05x(x - 1000) \leq 640,000$$
$$10x + 0.05x^2 - 50x \leq 640,000$$
$$0.05x^2 - 40x - 640,000 \leq 0$$
$$x^2 - 800x - 12,800,000 \leq 0$$
$$(x - 4000)(x + 3200) \leq 0$$
Solving the associated equation gives $x = 4000$ or $x = -3200$. Checking the intervals $[0, 4000]$ and $[4000, \infty)$, we find that the solution is in the range $[0, 4000]$. So they can import no more than 4000 radios.

1.5 Critical Thinking/Discussion/Writing

In Exercises 143 and 144, answers may vary. Sample responses are given.

143. a. $(x+4)(x-5) < 0$

   b. $(x+2)(x-6) \leq 0$

   c. $x^2 \geq 0$ **d.** $x^2 < 0$

   e. $(x-3)^2 \leq 0$ **f.** $(x-2)^2 > 0$

144. a. $\dfrac{x-4}{x+2} \leq 0$ **b.** $\dfrac{x-3}{x-5} \leq 0$

   c. It is not possible to have a quadratic inequality with solution set (2, 5]. If we try $(x-2)(x-5) < 0$, then the solution set will be (2, 5). If we try $(x-2)(x-5) \leq 0$, then the solution set will be [2, 5].

1.5 Maintaining Skills

145. $|-3| = 3$ **147.** $|6-4| = |2| = 2$

149. $|0| = 0$

151. $d = |5-(-2)| = |7| = 7$

153. $d = |5.7 - 2.3| = |3.4| = 3.4$

155. $|x-(-2)| = 5$ or $|x+2| = 5$

157. $|x-4| \leq 2$ **159.** $|x-5| \leq 3$

1.6 Equations and Inequalities Involving Absolute Value

1.6 Practice Problems

1. a. $|x-2| = 0 \Rightarrow x-2 = 0 \Rightarrow x = 2$
Solution set: {2}

   b. $|6x-3| - 8 = 1$
       $|6x-3| = 9$

    $6x-3 = 9$ or $6x-3 = -9$
      $6x = 12$ $6x = -6$
        $x = 2$ $x = -1$
   Solution set: {−1, 2}

2. $|x+2| = |x-3|$

   $x+2 = x-3$ or $x+2 = -(x-3)$
     $0 = -5$ False $x+2 = -x+3$
                     $2x = 1 \Rightarrow x = \dfrac{1}{2}$

   Solution set: $\left\{\dfrac{1}{2}\right\}$

3. $|3x-4| = 2(x-1)$

   $3x-4 = 2(x-1)$ or $3x-4 = -2(x-1)$
   $3x-4 = 2x-2$ $3x-4 = -2x+2$
     $x = 2$ $5x = 6 \Rightarrow x = \dfrac{6}{5}$

   Solution set: $\left\{\dfrac{6}{5}, 2\right\}$

4. $|3x+3| \leq 6 \Rightarrow -6 \leq 3x+3 \leq 6 \Rightarrow$
   $-9 \leq 3x \leq 3 \Rightarrow -3 \leq x \leq 1$
   Solution set: $[-3, 1]$

5. Let x = the actual speed of the search plane, in miles per hour. Then,
$|x-115| \leq 25 \Rightarrow -25 \leq x-115 \leq 25 \Rightarrow$
$90 \leq x \leq 140$
Thus, the actual speed of the plane is between 90 and 140 miles per hour. Since the plane uses 10 gallons of fuel per hour and has 30 gallons of fuel, it can fly for 3 hours. The actual number of miles the search plane can fly is $3x$:
$3(90) \leq 3x \leq 3(140) \Rightarrow 270 \leq x \leq 420$
The plane can fly between 270 miles and 420 miles.

6. $|2x+3| \geq 6 \Rightarrow 2x+3 \leq -6$ or $2x+3 \geq 6$

$2x+3 \leq -6$	$2x+3 \geq 6$
$2x \leq -9$	$2x \geq 3$
$x \leq -\dfrac{9}{2}$	$x \geq \dfrac{3}{2}$

   Solution set: $\left(-\infty, -\dfrac{9}{2}\right] \cup \left[\dfrac{3}{2}, \infty\right)$

7. a. $|5-9x| > -3$
   Since absolute value is always nonnegative, $|5-9x| > -3$ is true for all real numbers.
   Solution set: $(-\infty, \infty)$

b. $|7x - 4| \leq -1$

Since absolute value is always nonnegative, $|7x - 4| \leq -1$ is false for all real numbers.

Solution set: $\varnothing$

8. $\left|\dfrac{x-2}{x+4}\right| < 4 \Rightarrow -4 < \dfrac{x-2}{x+4} < 4$

$$0 < \dfrac{x-2}{x+4} + 4 \qquad\qquad \dfrac{x-2}{x+4} - 4 < 0$$

$$0 < \dfrac{x-2+4(x+4)}{x+4} \qquad \dfrac{x-2-4(x+4)}{x+4} < 0$$

$$0 < \dfrac{x-2+4x+16}{x+4} \qquad \dfrac{x-2-4x-16}{x+4} < 0$$

$$0 < \dfrac{5x+14}{x+4} \qquad\qquad \dfrac{-3x-18}{x+4} < 0$$

$5x + 14 = 0 \Rightarrow x = -\dfrac{14}{5}; \quad x + 4 = 0 \Rightarrow x = -4$

$-3x - 18 = 0 \Rightarrow x = -6$

Interval	Test point	Value of $\dfrac{5x+14}{x+4}$	Result
$(-\infty, -6)$	-10	6	$+$
$\left(-4, -\frac{14}{5}\right)$	-3	-1	$-$
$\left(-\frac{14}{5}, \infty\right)$	-1	3	$+$

Note that the expression is undefined for $x = -4$. The solution set is for this part of the original inequality is $(-\infty, -6) \cup \left(-\frac{14}{5}, \infty\right)$.

Interval	Test point	Value of $\dfrac{-3x-18}{x+4}$	Result
$(-\infty, -6)$	-7	-1	$-$
$(-6, -4)$	-5	3	$+$
$(-4, \infty)$	-1	-5	$-$

Note that the expression is undefined for $x = 4$. The solution set is for this part of the original inequality is $(-\infty, -6) \cup (-4, \infty)$.

The figure shows that both inequalities are true on $(-\infty, -6) \cup \left(-\frac{14}{5}, \infty\right)$, so the solution set is $(-\infty, -6) \cup \left(-\frac{14}{5}, \infty\right)$.

1.6 Basic Concepts and Skills

1. The solution set of the equation $|x| = a$ is $\underline{\{-a, a\}}$.

3. The solution set of the inequality $|x| \geq a$ is $\underline{(-\infty, -a] \cup [a, \infty)}$.

5. True

7. $|3x| = 9 \Rightarrow 3x = 9$ or $3x = -9 \Rightarrow$ $x = 3$ or $x = -3$

9. $|-2x| = 6 \Rightarrow -2x = 6$ or $-2x = -6 \Rightarrow$ $x = -3$ or $x = 3$

11. $|x + 3| = 2 \Rightarrow x + 3 = 2$ or $x + 3 = -2 \Rightarrow$ $x = -1$ or $x = -5$

13. $|6 - 2x| = 8 \Rightarrow 6 - 2x = 8$ or $6 - 2x = -8 \Rightarrow$ $x = -1$ or $x = 7$

15. $|6x - 2| = 9 \Rightarrow 6x - 2 = 9$ or $6x - 2 = -9 \Rightarrow$ $x = \dfrac{11}{6}$ or $x = -\dfrac{7}{6}$

17. $|2x + 3| - 1 = 0 \Rightarrow |2x + 3| = 1 \Rightarrow 2x + 3 = 1$ or $2x + 3 = -1 \Rightarrow x = -1$ or $x = -2$

19. $\dfrac{1}{2}|x| = 3 \Rightarrow |x| = 6 \Rightarrow x = -6$ or $x = 6$

21. $\left|\dfrac{1}{4}x + 2\right| = 3 \Rightarrow \dfrac{1}{4}x + 2 = -3$ or $\dfrac{1}{4}x + 2 = 3 \Rightarrow$ $x = -20$ or $x = 4$

23. $6|1 - 2x| - 8 = 10 \Rightarrow 6|1 - 2x| = 18 \Rightarrow$ $|1 - 2x| = 3 \Rightarrow 1 - 2x = 3$ or $1 - 2x = -3 \Rightarrow$ $x = -1$ or $x = 2$

25. $2|3x - 4| + 9 = 7 \Rightarrow 2|3x - 4| = -2 \Rightarrow$ $|3x - 4| = -1$

The solution set is $\varnothing$ because an absolute value cannot be negative.

27. $|2x + 1| = -1$

The solution set is $\varnothing$ because an absolute value cannot be negative.

29. $\left| x^2 - 4 \right| = 0 \Rightarrow x^2 - 4 = 0 \Rightarrow x = \pm 2$
Solution set: $\{-2, 2\}$

31. $\left| 1 - 2x \right| = 3 \Rightarrow 1 - 2x = 3$ or $1 - 2x = -3 \Rightarrow$
$x = -1$ or $x = 2$
Solution set: $\{-1, 2\}$

33. $\left| \dfrac{1}{3} - x \right| = \dfrac{2}{3} \Rightarrow \dfrac{1}{3} - x = \dfrac{2}{3}$ or $\dfrac{1}{3} - x = -\dfrac{2}{3} \Rightarrow$
$x = -\dfrac{1}{3}$ or $x = 1$
Solution set: $\left\{ -\dfrac{1}{3}, 1 \right\}$

In exercises 35–44, be sure to check answers to eliminate extraneous solutions.

35. $\left| x + 3 \right| = \left| x + 5 \right| \Rightarrow x + 3 = x + 5$ (impossible)
or $x + 3 = -(x + 5) \Rightarrow x + 3 = -x - 5 \Rightarrow x = -4$
The solution set is $\{-4\}$.

37. $\left| 3x - 2 \right| = \left| 6x + 7 \right| \Rightarrow 3x - 2 = 6x + 7 \Rightarrow$
$x = -3$ or $3x - 2 = -(6x + 7) \Rightarrow$
$3x - 2 = -6x - 7 \Rightarrow x = -\dfrac{5}{9}$
The solution set is $\left\{ -3, -\dfrac{5}{9} \right\}$.

39. $\left| 2x - 1 \right| = x + 1 \Rightarrow$
$2x - 1 = x + 1$ or $2x - 1 = -(x + 1)$
$\quad x = 2 \qquad\qquad 2x - 1 = -x - 1$
$\qquad\qquad\qquad\qquad\quad 3x = 0 \Rightarrow x = 0$
Solution set: $\{0, 2\}$

41. $\left| 4 - 3x \right| = x - 1 \Rightarrow$
$4 - 3x = x - 1$ or $4 - 3x = -(x - 1)$
$\quad 5 = 4x \qquad\qquad 4 - 3x = -x + 1$
$\quad \dfrac{5}{4} = x \qquad\qquad 3 = 2x \Rightarrow \dfrac{3}{2} = x$
Solution set: $\left\{ \dfrac{5}{4}, \dfrac{3}{2} \right\}$.

43. $\left| 3x + 2 \right| = 2(x - 1) \Rightarrow$
$3x + 2 = 2(x - 1)$ or $3x + 2 = -2(x - 1)$
$3x + 2 = 2x - 2 \qquad 3x + 2 = -2x + 2$
$\quad x = -4 \qquad\qquad\quad 5x = 0 \Rightarrow x = 0$
If $x = -4$, then $|3x + 2| = |3(-4) + 2| = |-10| = 10$, while $2(x - 1) = 2(-4 - 1) = 2(-5) = -10$.
Therefore, -4 is not a solution.
If $x = 0$, then $|3x + 2| = |3(0) + 2| = |2| = 2$, while $2(x - 1) = 2(0 - 1) = 2(-1) = -2$.
Therefore, 0 is not a solution.
Solution set: $\varnothing$

45. $\left| 3x \right| < 12 \Rightarrow -12 < 3x < 12 \Rightarrow -4 < x < 4$
The solution set is $(-4, 4)$.

47. $\left| 4x \right| > 16 \Rightarrow 4x < -16$ or $4x > 16 \Rightarrow$
$x < -4$ or $x > 4$.
The solution set is $(-\infty, -4) \cup (4, \infty)$.

49. $\left| x + 1 \right| < 3 \Rightarrow -3 < x + 1 < 3 \Rightarrow -4 < x < 2$.
The solution set is $(-4, 2)$.

51. $\left| x \right| + 2 \geq -1 \Rightarrow \left| x \right| \geq -3$
Since absolute value is always nonnegative, the inequality is true for all real numbers.
Solution set: $(-\infty, \infty)$

53. $\left| 2x - 3 \right| < 4 \Rightarrow -4 < 2x - 3 < 4 \Rightarrow$
$-1 < 2x < 7 \Rightarrow -\dfrac{1}{2} < x < \dfrac{7}{2}$.
The solution set is $\left(-\dfrac{1}{2}, \dfrac{7}{2} \right)$.

55. $\left| 5 - 2x \right| > 3 \Rightarrow 5 - 2x < -3$ or $5 - 2x > 3 \Rightarrow$
$x > 4$ or $x < 1$.
The solution set is $(-\infty, 1) \cup (4, \infty)$.

57. $\left| 3x + 4 \right| \leq 19 \Rightarrow -19 \leq 3x + 4 \leq 19 \Rightarrow$
$-23 \leq 3x \leq 15 \Rightarrow -\dfrac{23}{3} \leq x \leq 5$.
The solution set is $\left[-\dfrac{23}{3}, 5 \right]$.

59. $\left| 2x - 15 \right| < 0$. The solution set is $\varnothing$ because an absolute value cannot be negative.

61. $\left| \dfrac{x - 2}{x + 3} \right| < 1 \Rightarrow -1 < \dfrac{x - 2}{x + 3} < 1$

$0 < \dfrac{x - 2}{x + 3} + 1 \qquad\qquad \dfrac{x - 2}{x + 3} - 1 < 0$

$0 < \dfrac{x - 2 + (x + 3)}{x + 3} \quad \dfrac{x - 2 - (x + 3)}{x + 3} < 0$

$0 < \dfrac{2x + 1}{x + 3} \qquad\qquad\quad \dfrac{-5}{x + 3} < 0$

$2x + 1 = 0 \Rightarrow x = -\dfrac{1}{2}$; $x + 3 = 0 \Rightarrow x = -3$

(continued on next page)

(continued)

Interval	Test point	Value of $\dfrac{2x+1}{x+3}$	Result
$(-\infty, -3)$	-4	7	$+$
$\left(-3, -\frac{1}{2}\right)$	-2	-3	$-$
$\left(-\frac{1}{2}, \infty\right)$	2	1	$+$

Note that the expression is undefined for $x = -3$. The solution set is for this part of the original inequality is $(-\infty, -3) \cup \left(-\frac{1}{2}, \infty\right)$.

Interval	Test point	Value of $\dfrac{-5}{x+3}$	Result
$(-\infty, -3)$	-4	5	$+$
$\left(-3, -\frac{1}{2}\right)$	-2	-5	$-$
$\left(-\frac{1}{2}, \infty\right)$	2	-1	$-$

Note that the expression is undefined for $x = -3$. The solution set is for this part of the original inequality is $\left(-3, -\frac{1}{2}\right) \cup \left(-\frac{1}{2}, \infty\right)$.

Both inequalities are true on $\left(-\frac{1}{2}, \infty\right)$, so the solution set is $\left(-\frac{1}{2}, \infty\right)$.

63. $\left|\dfrac{2x-3}{x+1}\right| \le 3 \Rightarrow -3 \le \dfrac{2x-3}{x+1} \le 3$

$$0 \le \dfrac{2x-3}{x+1} + 3 \qquad \dfrac{2x-3}{x+1} - 3 \le 0$$

$$0 \le \dfrac{2x-3+3(x+1)}{x+1} \quad \dfrac{2x-3-3(x+1)}{x+1} \le 0$$

$$0 \le \dfrac{5x}{x+1} \qquad\qquad \dfrac{-x-6}{x+1} \le 0$$

$5x = 0 \Rightarrow x = 0; \ x+1 = 0 \Rightarrow x = -1$
$-x - 6 = 0 \Rightarrow x = -6$

Interval	Test point	Value of $\dfrac{5x}{x+1}$	Result
$(-\infty, -6]$	-11	$\frac{11}{2}$	$+$
$[-6, -1)$	-2	10	$+$
$(-1, 0]$	$-\frac{1}{2}$	-5	$-$
$[0, \infty)$	4	4	$+$

Note that the expression is undefined for $x = -1$. The solution set is for this part of the original inequality is
$(-\infty, -6] \cup [-6, -1) \cup [0, \infty)$.

Interval	Test point	Value of $\dfrac{-x-6}{x+1}$	Result
$(-\infty, -6]$	-11	$-\frac{1}{2}$	$-$
$[-6, -1)$	-2	4	$+$
$(-1, 0]$	$-\frac{1}{2}$	-11	$-$
$[0, \infty)$	4	-2	$-$

Note that the expression is undefined for $x = -1$. The solution set is for this part of the original inequality is
$(-\infty, -6] \cup (-1, 0] \cup [0, \infty)$.

Both inequalities are true on
$(-\infty, -6] \cup [0, \infty)$, so the solution set is
$(-\infty, -6] \cup [0, \infty)$.

65. $\left|\dfrac{x-1}{x+2}\right| \ge 2 \Rightarrow \dfrac{x-1}{x+2} \le -2$ or $\dfrac{x-1}{x+2} \ge 2$

$$\dfrac{x-1}{x+2} \le -2 \qquad\qquad \dfrac{x-1}{x+2} \ge 2$$

$$\dfrac{x-1}{x+2} + 2 \le 0 \qquad\qquad \dfrac{x-1}{x+2} - 2 \ge 0$$

$$\dfrac{x-1+2(x+2)}{x+2} \le 0 \quad \dfrac{x-1-2(x+2)}{x+2} \ge 0$$

$$\dfrac{3x+3}{x+2} \le 0 \qquad\qquad \dfrac{-x-5}{x+2} \ge 0$$

$3x + 3 = 0 \Rightarrow x = -1; \ x + 2 = 0 \Rightarrow x = -2$
$-x - 5 = 0 \Rightarrow x = -5$

Interval	Test point	Value of $\dfrac{3x+3}{x+2}$	Result
$(-\infty, -5]$	-6	$\frac{15}{4}$	$+$
$[-5, -2)$	-3	6	$+$
$(-2, -1]$	$-\frac{3}{2}$	-3	$-$
$[-1, \infty)$	1	2	$+$

Note that the expression is undefined for $x = -2$. The solution set is for this part of the original inequality is $(-2, -1]$.

(continued on next page)

(continued)

Interval	Test point	Value of $\dfrac{-x-5}{x+2}$	Result
$(-\infty, -5]$	-6	$-\dfrac{1}{4}$	$-$
$[-5, -2)$	-3	2	$+$
$(-2, -1]$	$-\dfrac{3}{2}$	-7	$-$
$[-1, \infty)$	1	-2	$-$

Note that the expression is undefined for $x = -2$. The solution set is for this part of the original inequality is $[-5, -2)$. Since the original inequality is an "or" inequality, the solution set of the original inequality is the union of the two solution sets.
The solution set is $(-2, -1] \cup [-5, -2)$.

Interval	Test point	Value of $\dfrac{-2x+5}{x-1}$	Result
$\left(-\infty, \frac{1}{2}\right)$	0	-5	$-$
$\left(\frac{1}{2}, 1\right)$	$\frac{3}{4}$	-14	$-$
$\left(1, \frac{5}{2}\right)$	2	1	$+$
$\left(\frac{5}{2}, \infty\right)$	4	-1	$-$

Note that the expression is undefined for $x = -2$. The solution set is for this part of the original inequality is $\left(1, \frac{5}{2}\right)$. Since the original inequality is an "or" inequality, the solution set of the original inequality is the union of the two solution sets. The solution set is $\left(\frac{1}{2}, 1\right) \cup \left(1, \frac{5}{2}\right)$.

67. $\left|\dfrac{2x+1}{x-1}\right| > 4 \Rightarrow \dfrac{2x+1}{x-1} < -4 \ \text{ or } \ \dfrac{2x+1}{x-1} > 4$

$$\begin{array}{c|c} \dfrac{2x+1}{x-1} < -4 & \dfrac{2x+1}{x-1} > 4 \\[2mm] \dfrac{2x+1}{x-1} + 4 < 0 & \dfrac{2x+1}{x-1} - 4 > 0 \\[2mm] \dfrac{2x+1+4(x-1)}{x-1} < 0 & \dfrac{2x+1-4(x-1)}{x-1} > 0 \\[2mm] \dfrac{6x-3}{x-1} < 0 & \dfrac{-2x+5}{x-1} > 0 \end{array}$$

$6x - 3 = 0 \Rightarrow x = \dfrac{1}{2}; \ x - 1 = 0 \Rightarrow x = 1$

$-2x + 5 = 0 \Rightarrow x = \dfrac{5}{2}$

Interval	Test point	Value of $\dfrac{6x-3}{x-1}$	Result
$\left(-\infty, \frac{1}{2}\right)$	0	3	$+$
$\left(\frac{1}{2}, 1\right)$	$\frac{3}{4}$	-6	$-$
$\left(1, \frac{5}{2}\right)$	2	9	$+$
$\left(\frac{5}{2}, \infty\right)$	4	7	$+$

Note that the expression is undefined for $x = -2$. The solution set is for this part of the original inequality is $\left(\frac{1}{2}, 1\right)$.

69. $|x-1| \le 2|2x-5| \Rightarrow \left|\dfrac{x-1}{2x-5}\right| \le 2 \Rightarrow$

$-2 \le \dfrac{x-1}{2x-5} \le 2$

$$\begin{array}{c|c} 0 \le \dfrac{x-1}{2x-5} + 2 & \dfrac{x-1}{2x-5} - 2 \le 0 \\[2mm] 0 \le \dfrac{x-1+2(2x-5)}{2x-5} & \dfrac{x-1-2(2x-5)}{2x-5} \le 0 \\[2mm] 0 \le \dfrac{x-1+4x-10}{2x-5} & \dfrac{x-1-4x+10}{2x-5} \le 0 \\[2mm] 0 \le \dfrac{5x-11}{2x-5} & \dfrac{-3x+9}{2x-5} \le 0 \end{array}$$

$5x - 11 = 0 \Rightarrow x = \dfrac{11}{5}; \ 2x - 5 = 0 \Rightarrow x = \dfrac{5}{2}$

$-3x + 9 = 0 \Rightarrow x = 3$

Interval	Test point	Value of $\dfrac{5x-11}{2x-5}$	Result
$\left(-\infty, \frac{11}{5}\right]$	0	$\frac{11}{5}$	$+$
$\left[\frac{11}{5}, \frac{5}{2}\right)$	$\frac{23}{10}$	$-\frac{5}{4}$	$-$
$\left(\frac{5}{2}, \infty\right)$	3	4	$+$

Note that the expression is undefined for $x = \frac{5}{2}$. The solution set is for this part of the original inequality is $S_1 = \left(-\infty, \frac{11}{5}\right] \cup \left(\frac{5}{2}, \infty\right)$.

(continued on next page)

(continued)

Interval	Test point	Value of $\dfrac{-3x+9}{2x-5}$	Result
$\left(-\infty, \frac{5}{2}\right)$	0	$-\frac{9}{5}$	$-$
$\left(\frac{5}{2}, 3\right]$	$\frac{11}{4}$	$\frac{3}{2}$	$+$
$[3, \infty)$	5	$-\frac{6}{5}$	$-$

Note that the expression is undefined for $x = \frac{5}{2}$. The solution set is for this part of the original inequality is $S_2 = \left(-\infty, \frac{5}{2}\right) \cup [3, \infty)$.

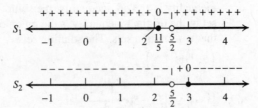

The figure shows that both inequalities are true on $\left(-\infty, \frac{11}{5}\right] \cup [3, \infty)$, so the solution set is $\left(-\infty, \frac{11}{5}\right] \cup [3, \infty)$.

1.6 Applying the Concepts

71. $|T - 75| = 20 \Rightarrow T - 75 = -20$ or
$T - 75 = 20 \Rightarrow T = 55$ or $T = 95$
The temperatures in Tampa during December are between 55°F and 95°F.

73. $|x - 700| \le 50$ **75.** $|x - 120| \le 6.75$

77. Let $x =$ the number of people at a party. Then
$|120 - x| \le 15 \Rightarrow -15 \le 120 - x \le 15 \Rightarrow$
$-135 \le -x \le -105 \Rightarrow 105 \le x \le 135$. So, between 105 and 135 people will be at the party. The total spent on food will be between $48(105) = \$5040$ and $48(135) = \$6480$.

79. Let $x =$ the actual weight of Sarah's catch.
Then $|32 - x| \le \dfrac{1}{2} \Rightarrow -\dfrac{1}{2} \le 32 - x \le \dfrac{1}{2} \Rightarrow$
$-1 \le 64 - 2x \le 1 \Rightarrow -65 \le -2x \le -63 \Rightarrow$.
$32.5 \ge x \ge 31.5$.
So, her catch is between 31.5 pounds and 32.5 pounds. She will be paid between
$0.60(31.5) = \$18.90$ and $0.60(32.5) = \$19.50$.

1.6 Beyond the Basics

81. a. $\left|x^2 - 9\right| = x - 3 \Rightarrow x^2 - 9 = -(x - 3)$ or
$x^2 - 9 = x - 3$
$x^2 - 9 = -(x - 3) \Rightarrow x^2 - 9 = -x + 3 \Rightarrow$
$x^2 + x - 12 = 0 \Rightarrow (x + 4)(x - 3) = 0 \Rightarrow$
$x = -4, 3$
$x^2 - 9 = x - 3 \Rightarrow x^2 - x - 6 = 0 \Rightarrow$
$(x + 2)(x - 3) = 0 \Rightarrow x = -2, 3$
Checking $x = -4$, $x = -2$, and $x = 3$ in the original equation shows that $x = -4$ and $x = -2$ are extraneous solutions.
The solution set is $\{3\}$.

b. $\left|x^2 - 8\right| = -2x \Rightarrow x^2 - 8 = -(-2x)$ or
$x^2 - 8 = -2x$
$x^2 - 8 = -(-2x) \Rightarrow x^2 - 8 = 2x \Rightarrow$
$x^2 - 2x + 8 = 0 \Rightarrow (x + 2)(x - 4) = 0 \Rightarrow$
$x = -2, 4$
$x^2 - 8 = -2x \Rightarrow x^2 + 2x - 8 = 0 \Rightarrow$
$(x + 4)(x - 2) = 0 \Rightarrow x = -4, 2$
Checking $x = -4$, $x = -2$, $x = 2$, and $x = 4$ in the original equation shows that $x = 2$, and $x = 4$ are extraneous solution.
The solution set is $\{-4, -2\}$.

c. $\left|x^2 - 5x\right| = 6 \Rightarrow x^2 - 5x = -6$ or
$x^2 - 5x = 6$
$x^2 - 5x = -6 \Rightarrow x^2 - 5x + 6 = 0 \Rightarrow$
$(x - 2)(x - 3) = 0 \Rightarrow x = 2, 3$
$x^2 - 5x = 6 \Rightarrow x^2 - 5x - 6 = 0 \Rightarrow$
$(x + 1)(x - 6) = 0 \Rightarrow x = -1, 6$
Checking $x = -1$, $x = 2$, $x = 3$, and $x = 6$ in the original equation shows that all values are solutions.
The solution set is $\{-1, 2, 3, 6\}$.

d. $\left|x^2 + 3x - 2\right| = 2 \Rightarrow x^2 + 3x - 2 = -2$ or

$x^2 + 3x - 2 = 2$. If $x^2 + 3x - 2 = -2$, then

$x^2 + 3x = 0 \Rightarrow x(x+3) = 0 \Rightarrow x = 0$ or

$x = -3$.

If $x^2 + 3x - 2 = 2$, then $x^2 + 3x - 4 = 0 \Rightarrow$

$(x+4)(x-1) = 0 \Rightarrow x = 1$ or $x = -4$.

Checking $x = -4$, $x = -3$, $x = 0$, and $x = 1$ in the original equation shows that all values are solutions.

The solution set is $\{-4, -3, 0, 1\}$.

83. $\left|2x - 3\right| + \left|x - 2\right| = 4$

The points $x = \frac{3}{2}$ and $x = 2$ are the points where the absolute value expressions equal zero. Since these expressions must be negative or positive for other x-values, then these points divide the number line into intervals each of which should be considered separately. Thus, we will consider the intervals $\left(-\infty, \frac{3}{2}\right]$, $\left(\frac{3}{2}, 2\right]$, and $(2, \infty)$.

Interval	Test point	Sign of $2x - 3$	Sign of $x - 2$
$\left(-\infty, \frac{3}{2}\right]$	0	$-$	$-$
$\left(\frac{3}{2}, 2\right]$	$\frac{7}{4}$	$+$	$-$
$(2, \infty)$	3	$+$	$+$

On the first interval, both absolute-value expressions will have negative values, so change the signs on both of them when taking the bars off.

$-(2x - 3) - (x - 2) = 4$
$-2x + 3 - x + 2 = 4$
$-3x + 5 = 4$
$-3x = -1 \Rightarrow x = \frac{1}{3}$

Since $x = \frac{1}{3}$ lies in the interval $\left(-\infty, \frac{3}{2}\right]$, this is a valid solution.

On the second interval, $\left|2x - 3\right|$ is positive, so just take the bars off. But $\left|x - 2\right|$ is negative, change the sign when taking the bars off.

$(2x - 3) - (x - 2) = 4$
$2x - 3 - x + 2 = 4$
$x - 1 = 4 \Rightarrow x = 5$

Since $x = 5$ does not lie in the interval $\left(\frac{3}{2}, 2\right]$, it is not a valid solution of the original equation.

On the third interval, the arguments of both absolute values expressions are positive, so just remove the absolute value bars.

$(2x - 3) + (x - 2) = 4$
$2x - 3 + x - 2 = 4$
$3x - 5 = 4$
$3x = 9 \Rightarrow x = 3$

Since $x = 3$ lies in the interval $(2, \infty)$, the solution is valid. Thus, the solution set for the original equation is $\left\{\frac{1}{3}, 3\right\}$.

85. Let $u = \left|x\right|$. Then $\left|x\right|^2 - 4\left|x\right| - 7 = 5 \Rightarrow$

$u^2 - 4u - 7 = 5 \Rightarrow u^2 - 4u - 12 = 0 \Rightarrow$

$(u - 6)(u + 2) = 0 \Rightarrow u = -2, 6$.

Now solve for x: $-2 = \left|x\right|$ is not possible.

$6 = \left|x\right| \Rightarrow x = 6$ or $x = -6$

Solution set: $\{-6, 6\}$.

87. $0 < a < b \Rightarrow b - a > 0$ and

$0 < c < d \Rightarrow d - c > 0$. The product of two positive numbers is positive, so

$(b - a)c > 0 \Leftrightarrow bc > ac$ (1) and

$(d - c)a > 0 \Leftrightarrow ad > ac$ (2) . We know that

$bc > ac$ and $bd > ad$, so

$bd - bc > ad - ac \Rightarrow bd + ac > ad + bc$.

Substituting (1) and (2) into the inequality, we have $bd + ac > ad + bc > ac + ac \Rightarrow bd > ac$.

89. $x^2 < a \Rightarrow x^2 - a < 0 \Rightarrow \left(x - \sqrt{a}\right)\left(x + \sqrt{a}\right) < 0$

$\left(x - \sqrt{a}\right)\left(x + \sqrt{a}\right) < 0 \Rightarrow$

$\left(x - \sqrt{a}\right) > 0$ and $\left(x + \sqrt{a}\right) < 0 \Rightarrow$

$x > \sqrt{a}$ and $x < -\sqrt{a} \Rightarrow \sqrt{a} < x < -\sqrt{a} \Rightarrow$

$\sqrt{a} < -\sqrt{a}$, a contradiction

or

$\left(x - \sqrt{a}\right) < 0$ and $\left(x + \sqrt{a}\right) > 0 \Rightarrow$

$x > -\sqrt{a}$ and $x < \sqrt{a} \Rightarrow -\sqrt{a} < x < \sqrt{a}$

Thus, the solution set is $\left(-\sqrt{a}, \sqrt{a}\right)$.

In Exercises 91–102, answers may vary. Sample responses are given.

91. $\left|x - 4\right| < 3$ **93.** $\left|x - 4\right| \le 6$

95. $\left|x - 7\right| > 4$ **97.** $\left|2x - 5\right| \ge 15$

99. $\left|x - \dfrac{a + b}{2}\right| < \dfrac{a + b}{2} \Rightarrow \left|2x - a - b\right| < a + b$

101. $|2x - a - b| > b - a$

103. $|x - 39| \le 31$

105. $1 \le |x - 2| \le 3 \Rightarrow$

$$
\begin{array}{ll}
1 \le x - 2 \le 3 \quad \text{or} & 1 \le -(x - 2) \le 3 \\
3 \le x \le 5 & 1 \le -x + 2 \le 3 \\
& -1 \le -x \le 1 \\
& 1 \ge x \ge -1 \Rightarrow -1 \le x \le 1
\end{array}
$$

The union of the two solution sets is the solution set of the original inequality.

Solution set: $[-1, 1] \cup [3, 5]$

107. $|x - 2| + |x - 4| \ge 8$.

Solve the equation $|x - 2| + |x - 4| = 8$ to find the critical values for the inequality. The points $x = 2$ and $x = 4$ are the points where the absolute value expressions equal zero. Since these expressions must be negative or positive for other x-values, then these points divide the number line into intervals each of which should be considered separately. Thus, we will consider the intervals $(-\infty, 2]$, $(2, 4]$, and $(4, \infty)$.

Interval	Test point	Sign of $x - 2$	Sign of $x - 4$
$(-\infty, 2]$	0	−	−
$(2, 4]$	3	+	−
$(4, \infty)$	5	+	+

On the first interval, both absolute-value expressions will have negative values, so change the signs on both of them when taking the bars off.

$$
\begin{array}{c}
-(x - 2) + [-(x - 4)] = 8 \\
-2x + 6 = 8 \\
-2x = 2 \Rightarrow x = -1
\end{array}
$$

Since $x = -1$ lies in the interval $(-\infty, 2]$, $x = -1$ is a valid solution.

On the second interval, $x - 2$ is positive, so just take the bars off. But $x - 4$ is negative, change the sign when taking the bars off.

$$
\begin{array}{c}
(x - 2) + [-(x - 4)] = 8 \\
2 = 8
\end{array}
$$

Since this is a false statement, no values of x in the interval $(2, 4]$ are valid solutions of the equation.

On the third interval, the arguments of both absolute values expressions are positive, so

just remove the absolute value bars.

$$
\begin{array}{c}
x - 2 + x - 4 = 8 \\
2x - 6 = 8 \\
2x = 14 \Rightarrow x = 7
\end{array}
$$

Since 7 lies in the interval $(2, \infty)$, $x = 7$ is a valid solution. So $x = -1, 2, 4,$ and 7 are critical values. Test values in the intervals $(-\infty, -1]$, $[-1, 2]$, $[2, 4]$, $[4, 7]$, and $[7, \infty)$ to see where the original inequality is true.

Interval	Test point	Value of $|x - 2| + |x - 4|$	Value ≥ 8?
$(-\infty, -1]$	−2	10	Yes
$[-1, 2]$	0	6	No
$[2, 4]$	3	2	No
$[4, 7]$	5	4	No
$[7, \infty)$	10	14	Yes

Thus, the solution set is $(-\infty, -1] \cup [7, \infty)$.

Verify this by graphing $Y_1 = |x - 2| + |x - 4|$ and $Y_2 = 8$.

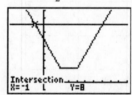 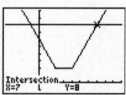

109. $\dfrac{|x - 1| - (x + 2)}{x + 2} \ge 0$.

The value of $x - 1$ changes from negative to positive at $x = 1$, where it is 0. So, we will consider two cases,

$$
\begin{cases}
\dfrac{(x - 1) - (x + 2)}{x + 2} \ge 0 & \text{if } x \ge 1 \\[2ex]
\dfrac{-(x - 1) - (x + 2)}{x + 2} \ge 0 & \text{if } x < 1
\end{cases}
$$

Test the case for $x \ge 1$.

$$
\dfrac{(x - 1) - (x + 2)}{x + 2} \ge 0 \Rightarrow \dfrac{-3}{x + 2} \ge 0
$$

$$
x + 2 = 0 \Rightarrow x = -2
$$

Since this test case is for $x \ge 1$, it is only necessary to test the interval $[1, \infty)$.

(continued on next page)

(*continued*)

Interval	Test point	Value of $\dfrac{-3}{x+2}$	Result
$[1, \infty)$	2	$-\dfrac{3}{4}$	−

$-\dfrac{3}{4}$ is not ≥ 1, so there is no solution in this interval.

Now test the case $x < 1$.

$$\dfrac{-(x-1)-(x+2)}{x+2} \geq 0 \Rightarrow \dfrac{-2x-1}{x+2} \geq 0$$

$$-2x-1 = 0 \Rightarrow x = -\tfrac{1}{2}; \ x+2 = 0 \Rightarrow x = -2$$

The intervals to be tested are $(-\infty, -2)$, $\left(-2, -\tfrac{1}{2}\right]$, and $\left[-\tfrac{1}{2}, 1\right]$.

Interval	Test point	Value of $\dfrac{-2x-1}{x+2}$	Result
$(-\infty, -2)$	−3	−5	−
$\left(-2, -\tfrac{1}{2}\right]$	−1	1	+
$\left[-\tfrac{1}{2}, 1\right]$	0	$-\tfrac{1}{2}$	−

The only interval where the value of $\dfrac{-2x-1}{x+2} \geq 0$ is $\left(-2, -\tfrac{1}{2}\right]$, so the solution set is $\left(-2, -\tfrac{1}{2}\right]$. Verify this by graphing

$$Y_1 = \dfrac{|x-1|-(x+2)}{x+2}.$$

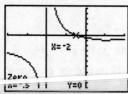

1.6 Critical Thinking/Discussion/Writing

110. To find what values make $\sqrt{(x-3)^2} = x-3$ true, solve $x-3 = 0 \Rightarrow x = 3$, so we check the intervals $(-\infty, 3]$ and $[3, \infty)$ to see which makes the equation true. The equation is true for $[3, \infty)$.

111. To find what values make

$$\sqrt{\left(x^2 - 6x + 8\right)^2} = x^2 - 6x + 8 \text{ true, solve}$$

$x^2 - 6x + 8 = 0 \Rightarrow (x-2)(x-4) = 0 \Rightarrow x = 2$ or $x = 4$. Then check the intervals $(-\infty, 2]$, $[2, 4]$, and $[4, \infty)$ to see which make the equation true. The equation is true for $(-\infty, 2] \cup [4, \infty)$.

112. To solve $|x-3|^2 - 7|x-3| + 10 = 0$, let $u = |x-3|$. So we have $u^2 - 7u + 10 = 0 \Rightarrow$ $(u-5)(u-2) = 0 \Rightarrow u = 5$ or $u = 2$. Now solve for x.

$5 = |x-3| \Rightarrow -5 = x-3$ or $5 = x-3 \Rightarrow$ $x = -2$ or $x = 8$. $2 = |x-3| \Rightarrow -2 = x-3$ or $2 = x-3 \Rightarrow x = 1$ or $x = 5$. So the solution set is $\{-2, 1, 5, 8\}$.

1.6 Maintaining Skills

113. $\dfrac{2+5}{2} = \dfrac{7}{2} = 3.5$

115. $\dfrac{-3-7}{2} = \dfrac{-10}{2} = -5$

117. $\sqrt{(5-2)^2 + (3-7)^2} = \sqrt{3^2 + (-4)^2}$
$$= \sqrt{9+16} = \sqrt{25} = 5$$

119. $\sqrt{(2-5)^2 + (8-6)^2} = \sqrt{(-3)^2 + 2^2} = \sqrt{9+4}$
$$= \sqrt{13}$$

121. $x^2 + 4x + \underline{4} = (x+2)^2$

123. $x^2 - 5x + \left(\dfrac{-5}{2}\right)^2 = x^2 - 5x + \dfrac{25}{4}\left(x - \dfrac{5}{2}\right)^2$

125. $x^2 + \dfrac{3}{2}x + \left(\dfrac{\frac{3}{2}}{2}\right)^2 = x^2 + \dfrac{3}{2}x + \left(\dfrac{3}{4}\right)^2$
$$= x^2 + \dfrac{3}{2}x + \dfrac{9}{16} = \left(x + \dfrac{3}{4}\right)^2$$

Chapter 1 Review Exercises

Basic Concepts and Skills

1. $5x - 4 = 11 \Rightarrow 5x = 15 \Rightarrow x = 3$

3. $3(2x - 4) = 9 - (x + 7) \Rightarrow 6x - 12 = 2 - x \Rightarrow$
   $7x = 14 \Rightarrow x = 2$

5. $3x + 8 = 3(x + 2) + 2 \Rightarrow 3x + 8 = 3x + 8$ This is
   an identity, so the solution set is $(-\infty, \infty)$.

7. $x - (5x - 2) = 7(x - 1) - 2 \Rightarrow$
   $-4x + 2 = 7x - 9 \Rightarrow -11x = -11 \Rightarrow x = 1$

9. $\dfrac{2}{x+3} = \dfrac{5}{11x-1} \Rightarrow 2(11x - 1) = 5(x + 3) \Rightarrow$
   $22x - 2 = 5x + 15 \Rightarrow 17x = 17 \Rightarrow x = 1$

11. $\dfrac{y+5}{2} + \dfrac{y-1}{3} = \dfrac{7y+3}{8} + \dfrac{4}{3} \Rightarrow$
    $12(y + 5) + 8(y - 1) = 3(7y + 3) + 8(4) \Rightarrow$
    $20y + 52 = 21y + 41 \Rightarrow 11 = y$

13. $|2x - 3| = |4x + 5| \Rightarrow 2x - 3 = 4x + 5 \Rightarrow$
    $-2x = 8 \Rightarrow x = -4$
    or $2x - 3 = -(4x + 5) \Rightarrow$
    $2x - 3 = -4x - 5 \Rightarrow 6x = -2 \Rightarrow x = -1/3$
    The solution set is $\left\{-4, -\dfrac{1}{3}\right\}$.

15. $|2x - 1| = |2x + 7|$

    $2x - 1 = 2x + 7$ or $2x - 1 = -(2x + 7)$
    $-2 = 7$ False $\qquad 2x - 1 = -2x - 7$
    $\qquad\qquad\qquad\qquad 4x = -6 \Rightarrow x = -\dfrac{3}{2}$

    Solution set: $\left\{-\dfrac{3}{2}\right\}$

17. $|3x - 2| = 2x + 1$

    $3x - 2 = 2x + 1$ or $3x - 2 = -(2x + 1)$
    $\quad x = 3 \qquad\qquad 3x - 2 = -2x - 1$
    $\qquad\qquad\qquad\qquad 5x = 1$
    $\qquad\qquad\qquad\qquad x = \dfrac{1}{5}$

    Solution set: $\left\{\dfrac{1}{5}, 3\right\}$

19. $p = k + gt \Rightarrow p - k = gt \Rightarrow \dfrac{p-k}{t} = g$

21. $T = \dfrac{2B}{B-1} \Rightarrow TB - T = 2B \Rightarrow TB - 2B = T \Rightarrow$
    $B(T - 2) = T \Rightarrow B = \dfrac{T}{T-2}$

23. $x^2 - 7x = 0 \Rightarrow x(x - 7) = 0 \Rightarrow$
    $x = 0$ or $x = 7$

25. $x^2 - 3x - 10 = 0 \Rightarrow (x - 5)(x + 2) = 0 \Rightarrow$
    $x = 5$ or $x = -2$

27. $(x - 1)^2 = 2x^2 + 3x - 5 \Rightarrow$
    $x^2 - 2x + 1 = 2x^2 + 3x - 5 \Rightarrow$
    $0 = x^2 + 5x - 6 \Rightarrow 0 = (x + 6)(x - 1) \Rightarrow$
    $x = -6$ or $x = 1$

29. $\dfrac{x^2}{4} + x = \dfrac{5}{4} \Rightarrow x^2 + 4x = 5 \Rightarrow$
    $x^2 + 4x - 5 = 0 \Rightarrow (x + 5)(x - 1) = 0 \Rightarrow$
    $x = -5$ or $x = 1$

31. $3x(x + 1) = 2x + 2 \Rightarrow 3x^2 + 3x = 2x + 2 \Rightarrow$
    $3x^2 + x - 2 = 0 \Rightarrow (3x - 2)(x + 1) = 0 \Rightarrow$
    $x = \dfrac{2}{3}$ or $x = -1$

33. Use the quadratic formula to solve
    $x^2 - 3x - 1 = 0$:
    $$x = \frac{-(-3) \pm \sqrt{(-3)^2 - 4(1)(-1)}}{2(1)}$$
    $$= \frac{3 \pm \sqrt{13}}{2} = \frac{3}{2} \pm \frac{\sqrt{13}}{2}$$

35. $2x^2 + x - 1 = 0 \Rightarrow (2x - 1)(x + 1) = 0 \Rightarrow$
    $x = \dfrac{1}{2}$ or $x = -1$

37. First factor 3 from $3x^2 - 12x - 24 = 0$ to get
    $x^2 - 4x - 8 = 0$. Now use the quadratic
    formula:
    $$x = \frac{-(-4) \pm \sqrt{(-4)^2 - 4(1)(-8)}}{2(1)}$$
    $$= \frac{4 \pm \sqrt{48}}{2} = \frac{4 \pm 4\sqrt{3}}{2} = 2 \pm 2\sqrt{3}$$

39. Use the quadratic formula to solve
    $2x^2 - x - 2 = 0$:
    $$x = \frac{-(-1) \pm \sqrt{(-1)^2 - 4(2)(-2)}}{2(2)}$$
    $$= \frac{1 \pm \sqrt{17}}{4} = \frac{1}{4} \pm \frac{\sqrt{17}}{4}$$

41. Use the quadratic formula to solve
$x^2 - x + 1 = 0$:

$$x = \frac{-(-1) \pm \sqrt{(-1)^2 - 4(1)(1)}}{2(1)}$$

$$= \frac{1 \pm \sqrt{-3}}{2} = \frac{1 \pm i\sqrt{3}}{2} = \frac{1}{2} \pm \frac{\sqrt{3}}{2}i$$

43. Use the quadratic formula to solve
$x^2 - 6x + 13 = 0$:

$$x = \frac{-(-6) \pm \sqrt{(-6)^2 - 4(1)(13)}}{2(1)}$$

$$= \frac{6 \pm \sqrt{-16}}{2} = \frac{6 \pm 4i}{2} = 3 \pm 2i$$

45. Use the quadratic formula to solve
$4x^2 - 8x + 13 = 0$:

$$x = \frac{-(-8) \pm \sqrt{(-8)^2 - 4(4)(13)}}{2(4)}$$

$$= \frac{8 \pm \sqrt{-144}}{8} = \frac{8 \pm 12i}{8} = 1 \pm \frac{3}{2}i$$

47. The discriminant is $(-11)^2 - 4(3)(6) = 49 > 0$, so there are 2 real unequal roots.

49. The discriminant is $2^2 - 4(5)(1) = -16 < 0$, so there are 2 unequal complex roots.

51. $\sqrt{x^2 - 16} = 0 \Rightarrow x^2 - 16 = 0 \Rightarrow$
$(x + 4)(x - 4) = 16 \Rightarrow x = \pm 4$

53. $\sqrt{4 - 7x} = \sqrt{2}x \Rightarrow 4 - 7x = 2x^2 \Rightarrow$
$2x^2 + 7x - 4 = 0 \Rightarrow (2x - 1)(x + 4) = 0 \Rightarrow$
$x = \frac{1}{2}$ or $x = -4$
If $x = -4$, then the equation becomes
$\sqrt{4 - 7(-4)} = \sqrt{2}(-4) \Rightarrow \sqrt{32} = -4\sqrt{2}$, which
is not true, so we reject that root. The solution
set is $\left\{\frac{1}{2}\right\}$.

55. $y - 2\sqrt{y} - 3 = 0$. Let $u = \sqrt{y}$, so the equation
becomes $u^2 - 2u - 3 = 0 \Rightarrow$
$(u - 3)(u + 1) = 0 \Rightarrow u = 3$ or $u = -1$.
Now solve for y. $3 = \sqrt{y} \Rightarrow 9 = y$ or
$-1 = \sqrt{y}$ (reject this). The solution set is $\{9\}$.

57. $\sqrt{x} - 1 = \sqrt{5 + \sqrt{x}}$. Let $u = \sqrt{x}$. The equation
becomes $u - 1 = \sqrt{5 + u} \Rightarrow (u - 1)^2 = 5 + u \Rightarrow$
$u^2 - 2u + 1 = 5 + u \Rightarrow u^2 - 3u - 4 \Rightarrow$
$(u - 4)(u + 1) = 0 \Rightarrow u = 4$ or $u = -1$.
Now solve for x.
$4 = \sqrt{x} \Rightarrow 16 = x$ or $-1 = \sqrt{x}$
(not possible). The solution set is $\{16\}$.

59. $(7x + 5)^2 + 2(7x + 5) - 15 = 0$. Let $u = 7x + 5$.
Then the equation becomes
$u^2 + 2u - 15 = 0 \Rightarrow$
$(u + 5)(u - 3) = 0 \Rightarrow u = -5$ or $u = 3$.
Now solve for x: $-5 = 7x + 5 \Rightarrow x = -\frac{10}{7}$ or
$3 = 7x + 5 \Rightarrow x = -\frac{2}{7}$.
The solution set is $\left\{-10/7, -2/7\right\}$.

61. $x^{2/3} + 3x^{1/3} - 4 = 0$. Let $u = x^{1/3}$. So the
equation becomes $u^2 + 3u - 4 = 0 \Rightarrow$
$(u + 4)(u - 1) = 0 \Rightarrow u = -4$ or $u = 1$. Now
solve for u: $-4 = x^{1/3} \Rightarrow -64 = x$ or
$1 = x^{1/3} \Rightarrow 1 = x$. The solution set is $\{-64, 1\}$.

63. $\left(\sqrt{t} + 5\right)^2 - 9\left(\sqrt{t} + 5\right) + 20 = 0$.
Let $u = \sqrt{t} + 5$, so the equation becomes
$u^2 - 9u + 20 = 0 \Rightarrow (u - 5)(u - 4) = 0 \Rightarrow$
$u = 5$ or $u = 4$. Now solve for t. $5 = \sqrt{t} + 5 \Rightarrow$
$0 = t$ or $4 = \sqrt{t} + 5 \Rightarrow -1 = \sqrt{t}$. (reject this)
The solution set is $\{0\}$.

65. $4x^4 - 37x^2 + 9 = 0$. Let $u = x^2$. So the
equation becomes $4u^2 - 37u + 9 = 0 \Rightarrow$
$(4u - 1)(u - 9) = 0 \Rightarrow u = \frac{1}{4}$ or $u = 9$. Now
solve for x: $\frac{1}{4} = x^2 \Rightarrow x = \pm\frac{1}{2}$ or $9 = x^2 \Rightarrow$
$x = \pm 3$. The solution set is $\left\{-3, -\frac{1}{2}, \frac{1}{2}, 3\right\}$.

67.
$$\frac{2x + 1}{2x - 1} = \frac{x - 1}{x + 1}$$
$$(2x + 1)(x + 1) = (2x - 1)(x - 1)$$
$$2x^2 + 3x + 1 = 2x^2 - 3x + 1$$
$$6x = 0 \Rightarrow x = 0$$
The solution set is $\{0\}$.

69. $\left(\dfrac{7x}{x+1}\right)^2 - 3\left(\dfrac{7x}{x+1}\right) = 18$. Let $u = \dfrac{7x}{x+1}$. So

the equation becomes $u^2 - 3u = 18 \Rightarrow$

$u^2 - 3u - 18 = 0 \Rightarrow (u-6)(u+3) = 0 \Rightarrow$

$u = 6$ or $u = -3$. Now solve for x:

$-3 = \dfrac{7x}{x+1} \Rightarrow -3x - 3 = 7x \Rightarrow x = -\dfrac{3}{10}$ or

$6 = \dfrac{7x}{x+1} \Rightarrow 6x + 6 = 7x \Rightarrow x = 6$.

The solution set is $\left\{-\dfrac{3}{10}, 6\right\}$.

71. $x^2 + 2yx - 3y^2 = 0$

$(x+3y)(x-y) = 0$

$x + 3y = 0 \Rightarrow x = -3y$

$x - y = 0 \Rightarrow x = y$

Solution set: $\{-3y, y\}$

73. $x^2 + (3-2y)x + y^2 - 3y + 2 = 0$

Use the quadratic formula with $a = 1$,

$b = 3 - 2y$ and $c = y^2 - 3y + 2$.

$x = \dfrac{-(3-2y) \pm \sqrt{(3-2y)^2 - 4(1)\left(y^2 - 3y + 2\right)}}{2(1)}$

$x = \dfrac{2y - 3 \pm \sqrt{4y^2 - 12y + 9 - \left(4y^2 - 12y + 8\right)}}{2}$

$= \dfrac{(2y-3) \pm 1}{2} = \dfrac{2y-4}{2} = y - 2$ or

$\dfrac{2y-2}{2} = y - 1$

Solution set: $\{y - 2, y - 1\}$

75. $x + 5 < 3 \Rightarrow x < -2$

The solution set is $(-\infty, -2)$.

77. $3(x-3) \le 8 \Rightarrow 3x - 9 \le 8 \Rightarrow 3x \le 17 \Rightarrow x \le \dfrac{17}{3}$

The solution set is $\left(-\infty, \dfrac{17}{3}\right]$.

79. $x + 2 \ge \dfrac{2}{3}x - 2x \Rightarrow 3x + 6 \ge 2x - 6x \Rightarrow$

$3x + 6 \ge -4x \Rightarrow 6 \ge -7x \Rightarrow -\dfrac{6}{7} \le x$

The solution set is $\left[-\dfrac{6}{7}, \infty\right)$.

81. $\dfrac{1}{6} > \dfrac{4-3x}{3} \Rightarrow 1 > 2(4-3x) \Rightarrow 1 > 8 - 6x \Rightarrow$

$-7 > -6x \Rightarrow \dfrac{7}{6} < x$

Solution set: $\left(\dfrac{7}{6}, \infty\right)$

83. $\dfrac{x-3}{3} - 2 \le \dfrac{x}{6} + \dfrac{1}{2} \Rightarrow 2(x-3) - 2(6) \le x + 3 \Rightarrow$

$2x - 6 - 12 \le x + 3 \Rightarrow 2x - 18 \le x + 3 \Rightarrow x \le 21$

Solution set: $(-\infty, 21]$

85. $\dfrac{3-2x}{4} + 1 > \dfrac{x-5}{3}$

$3(3-2x) + 12 > 4(x-5)$

$9 - 6x + 12 > 4x - 20$

$-6x + 21 > 4x - 20$

$-10x > -41 \Rightarrow x < \dfrac{41}{10}$

Solution set: $\left(-\infty, \dfrac{41}{10}\right)$

87. $3x - 1 < 2$ or $11 - 2x < 5$

$3x < 3$ or $\quad -2x < -6$

$x < 1$ or $\qquad x > 3$

Solution set: $(-\infty, 1) \cup (3, \infty)$

89. $4x - 5 < 7$ and $7 - 3x < 1$

$4x < 12$ and $\quad -3x < -6$

$x < 3$ and $\qquad x > 2$

Solution set: $(2, 3)$

91. $-3 \le 2x + 1 < 7 \Rightarrow -4 \le 2x < 6 \Rightarrow -2 \le x < 3$

Solution set: $[-2, 3)$

93. $-3 < 3 - 2x \le 97 \Rightarrow -6 < -2x \le 94 \Rightarrow$

$3 > x \ge -47 \Rightarrow -47 \le x < 3$

Solution set: $[-47, 3)$

95. $x^2 + x - 6 \ge 0 \Rightarrow (x+3)(x-2) \ge 0$

Solve the associated equation:

$(x+3)(x-2) = 0 \Rightarrow x = -3$ or $x = 2$.

So, the intervals are $(-\infty, -3]$, $[-3, 2]$, and

and $[2, \infty)$.

(continued on next page)

(continued)

Interval	Test point	Value of x^2+x-6	Result
$(-\infty,-3]$	-4	6	$+$
$[-3,2]$	0	-6	$-$
$[2,\infty)$	3	6	$+$

The solution set is $(-\infty,-3]\cup[2,\infty)$.

97. $\dfrac{(x-1)(x+3)}{(x+2)(x+5)}\geq 0$

Set the numerator and denominator equal to zero and solve for x.

$(x-1)(x+3)=0\Rightarrow x=1,-3$

$(x+2)(x+5)=0\Rightarrow x=-2,-5$

The intervals are $(-\infty,-5)$, $(-5,-3]$, $[-3,-2)$, $(-2,1]$, and $(1,\infty)$.

Interval	Test point	Value of $\dfrac{(x-1)(x+3)}{(x+2)(x+5)}$	Result
$(-\infty,-5)$	-6	$\frac{21}{4}$	$+$
$(-5,-3]$	-4	$-\frac{5}{2}$	$-$
$[-3,-2)$	$-\frac{5}{2}$	$\frac{7}{5}$	$+$
$(-2,1]$	0	$-\frac{3}{10}$	$-$
$(1,\infty)$	2	$\frac{5}{28}$	$+$

The solution set is $(-\infty,-5)\cup[-3,-2)\cup(1,\infty)$.

99. $|3x+2|\leq 7\Rightarrow -7\leq 3x+2\leq 7\Rightarrow$

$-9\leq 3x\leq 5\Rightarrow -3\leq x\leq\dfrac{5}{3}$

The solution set is $\left[-3,\dfrac{5}{3}\right]$.

101. $4|x-2|+8>12\Rightarrow 4|x-2|>4\Rightarrow$

$|x-2|>1\Rightarrow x-2<-1$ or $x-2>1\Rightarrow$

$x<1$ or $x>3$

The solution set is $(-\infty,1)\cup(3,\infty)$.

103. $\left|\dfrac{4-x}{5}\right|\geq 1\Rightarrow \dfrac{4-x}{5}\leq -1$ or $\dfrac{4-x}{5}\geq 1\Rightarrow$

$4-x\leq -5\Rightarrow x\geq 9$ or $4-x\geq 5\Rightarrow x\leq -1$

The solution set is $(-\infty,-1]\cup[9,\infty)$.

105. $\left|\dfrac{x-1}{x+2}\right|\leq 3\Rightarrow -3\leq\dfrac{x-1}{x+2}\leq 3$

$0<\dfrac{x-1}{x+2}+3 \quad\Big|\quad \dfrac{x-1}{x+2}-3<0$

$0<\dfrac{x-1+3(x+2)}{x+2} \quad\Big|\quad \dfrac{x-1-3(x+2)}{x+2}<0$

$0<\dfrac{4x+5}{x+2} \quad\Big|\quad \dfrac{-2x-7}{x+2}<0$

$4x+5=0\Rightarrow x=-\dfrac{5}{4};\; x+2=0\Rightarrow x=-2$

Interval	Test point	Value of $\dfrac{4x+5}{x+2}$	Result
$(-\infty,-2)$	-4	$\frac{11}{2}$	$+$
$\left(-2,-\frac{5}{4}\right]$	$-\frac{3}{2}$	-2	$-$
$\left[-\frac{5}{4},\infty\right)$	0	$\frac{5}{2}$	$+$

Note that the expression is undefined for $x=-2$. The solution set is for this part of the original inequality is $(-\infty,-2)\cup\left[-\frac{5}{4},\infty\right)$.

$-2x-7=0\Rightarrow x=-\dfrac{7}{2};\; x+2=0\Rightarrow x=-2$

Interval	Test point	Value of $\dfrac{-2x-7}{x+2}$	Result
$\left(-\infty,-\frac{7}{2}\right]$	-4	$-\frac{1}{2}$	$-$
$\left[-\frac{7}{2},-2\right)$	-3	1	$+$
$(-2,\infty)$	0	$-\frac{7}{2}$	$-$

Note that the expression is undefined for $x=-2$. The solution set is for this part of the original inequality is $\left(-\infty,-\frac{7}{2}\right]\cup(-2,\infty)$.

Both inequalities are true on $\left(-\infty,-\frac{7}{2}\right]$ and $\left[-\frac{5}{4},\infty\right)$, so the solution set is $\left(-\infty,-\frac{7}{2}\right]\cup\left[-\frac{5}{4},\infty\right)$.

107. $\left|\dfrac{2x-3}{x+2}\right| \geq 2 \Rightarrow \dfrac{2x-3}{x+2} \leq -2$ or $\dfrac{2x-3}{x+2} \geq 2$

$$\dfrac{2x-3}{x+2} \leq -2 \qquad\qquad \dfrac{2x-3}{x+2} \geq 2$$

$$\dfrac{2x-3}{x+2}+2 \leq 0 \qquad\qquad \dfrac{2x-3}{x+2}-2 \geq 0$$

$$\dfrac{2x-3+2(x+2)}{x+2} \leq 0 \qquad \dfrac{2x-3-2(x+2)}{x+2} \geq 0$$

$$\dfrac{4x+1}{x+2} \leq 0 \qquad\qquad \dfrac{-7}{x+2} \geq 0$$

$$4x+1 = 0 \Rightarrow x = -\dfrac{1}{4};\ x+2 = 0 \Rightarrow x = -2$$

Interval	Test point	Value of $\dfrac{4x+1}{x+2}$	Result
$(-\infty, -2)$	-3	11	$+$
$\left(-2, -\frac{1}{4}\right]$	-1	-3	$-$
$\left[-\frac{1}{4}, \infty\right)$	0	$\frac{1}{2}$	$+$

Note that the expression is undefined for $x = -2$. The solution set is for this part of the original inequality is $\left(-2, -\frac{1}{4}\right]$.

Interval	Test point	Value of $\dfrac{-7}{x+2}$	Result
$(-\infty, -2)$	-3	$\frac{7}{5}$	$+$
$(-2, \infty)$	0	$-\frac{7}{2}$	$-$

Note that the expression is undefined for $x = -2$. The solution set is for this part of the original inequality is $(-\infty, -2)$. Since the original inequality is an "or" inequality, the solution set of the original inequality is the union of the two solution sets.

The solution set is $(-\infty, -2) \cup \left(-2, -\frac{1}{4}\right]$.

Applying the Concepts

109. $C = 2\pi r \Rightarrow 22 = 2\pi r \Rightarrow \dfrac{11}{\pi} = r$

The radius is $11/\pi$ cm.

111. $A = \dfrac{1}{2}h(b_1 + b_2) \Rightarrow 32 = \dfrac{1}{2}(8)(5 + b_2) \Rightarrow$
$32 = 4(5 + b_2) \Rightarrow 8 = 5 + b_2 \Rightarrow 3 = b_2$
The other base is 3 meters.

113. $V = lwh \Rightarrow 4212 = 27(12)h \Rightarrow 13 = h$
The box is 13 cm high.

115. $V = \pi r^2 h \Rightarrow 8750\pi = 5^2 \pi h \Rightarrow 350 = h$
The height is 350 cm.

117. Let x = measure or the base angle.
Then $40 + 2x$ = the measure of the third angle. So,
$x + x + 40 + 2x = 180 \Rightarrow 4x + 40 = 180 \Rightarrow$
$x = 35$.
The three angles are 35°, 35°, and 110°.

119. Let x = amount invested at 6%.
Then $30{,}000 - x$ = amount invested at 8%. So, we have $0.06x + 0.08(30{,}000 - x) = 2160 \Rightarrow$
$-0.02x + 2400 = 2160 \Rightarrow -0.02x = -240 \Rightarrow$
$x = 12{,}000$. $12,000 was invested at 6%, and $18,000 was invested at 8%.

121. Let x = amount of 4 1/2% solution.
Then $10 - x$ = the amount of 12% solution. So, we have $0.045x + 0.12(10 - x) = 0.06(10) \Rightarrow$
$-0.075x + 1.2 = 0.6 \Rightarrow -0.075x = -0.6 \Rightarrow$
$x = 8$. There are 8 liters of the 4 1/2% solution and 2 liters of the 12% solution.

123. Let x = the number of people at the party. Each person shook $x - 1$ hands. So, there are $x(x - 1)$ handshakes. However, each handshake is counted twice (if A shakes hands with B, that is the same as B shaking hands with A).

So, we have $\dfrac{x(x-1)}{2} = 28 \Rightarrow x^2 - x = 56 \Rightarrow$

$x^2 - x - 56 = 0 \Rightarrow (x - 8)(x + 7) = 0 \Rightarrow$
$x = 8$ or $x = -7$. We reject the negative answer. There were 8 people at the party.

125. Let x = the number of shares that Lavina bought. She spent $18,040 for the stock, or $18{,}040/x$ per share.

She sold $x - 20$ shares at $\dfrac{18{,}040}{x} + 18$ per share for a total of $20,000. Therefore,

$$(x - 20)\left(\dfrac{18{,}040}{x} + 18\right) = 20{,}000$$

$$\dfrac{18{,}040 + 18x}{x} = \dfrac{20{,}000}{x - 20}$$

$$(x - 20)(18{,}040 + 18x) = 20{,}000x$$

$$18x^2 + 17{,}680x - 360{,}800 = 20{,}000x$$

$$18x^2 - 2320x - 360{,}800 = 0$$

$$2(9x + 820)(x - 220) = 0$$

(continued on next page)

(continued)

$9x + 820 = 0 \Rightarrow x = -\dfrac{820}{9}$ (reject this)

$x - 220 = 0 \Rightarrow x = 220$

Lavina bought 220 shares of stock.

127. Let x = the number of horses in the herd. Then $x/4$ = the number of horses in the forest, and $2\sqrt{x}$ = the number of horses in the mountains.

$\dfrac{x}{4} + 2\sqrt{x} + 15 = x \Rightarrow x + 8\sqrt{x} + 60 = 4x \Rightarrow$

$8\sqrt{x} = 3x - 60 \Rightarrow 64x = (3x - 60)^2 \Rightarrow$

$64x = 9x^2 - 360x + 3600 \Rightarrow$

$9x^2 - 424x + 3600 = 0 \Rightarrow$

$(x - 36)(9x - 100) = 0 \Rightarrow x = 36$ or $x = 100/9$

Reject the fractional solution. There were 36 horses in the herd.

129. Let x = the original number of members going on the trip. Then $x + 4$ = the final number of members going on the trip. The cost per member for the original number going on the trip was $\dfrac{324}{x}$, and the cost for the final number going on the trip is $\dfrac{324}{x} - 0.9$. Therefore,

$\dfrac{324}{x + 4} = \dfrac{324}{x} - 0.9$

$324x = 324(x + 4) - 0.9x(x + 4)$

$324x = 324x + 1296 - 0.9x^2 - 3.6x$

$0 = -0.9x^2 - 3.6x + 1296$

$0 = x^2 + 4x - 1440$

$0 = (x - 36)(x + 40) \Rightarrow x = 36$ or

$x = -40$ (reject this). Thus, 36 people originally signed up for the trip, and 40 people went on the trips.

Chapter 1 Practice Test A

1. $5x - 9 = 3x - 5 \Rightarrow 2x = 4 \Rightarrow x = 2$

2. $\dfrac{7}{24} = \dfrac{x}{8} + \dfrac{1}{6} \Rightarrow 7 = 3x + 4 \Rightarrow x = 1$

3. $\dfrac{1}{x - 2} - 5 = \dfrac{1}{x + 2} \Rightarrow \dfrac{1 - 5x + 10}{x - 2} = \dfrac{1}{x + 2} \Rightarrow$

$\dfrac{-5x + 11}{x - 2} = \dfrac{1}{x + 2} \Rightarrow$

$(-5x + 11)(x + 2) = x - 2$

$-5x^2 + x + 22 = x - 2$

$-5x^2 + 24 = 0 \Rightarrow -5x^2 = -24 \Rightarrow$

$x^2 = \dfrac{24}{5} \Rightarrow x = \pm\dfrac{2\sqrt{6}}{\sqrt{5}} = \pm\dfrac{2\sqrt{30}}{5}$

4. Let x = the length of the rectangle. Then $x - 3$ = the width of the rectangle.

$x(x - 3) = 54 \Rightarrow x^2 - 3x - 54 = 0 \Rightarrow$

$(x + 6)(x - 9) = 0 \Rightarrow x = -6$ or $x = 9$

We reject the negative answer. The rectangle is 9 cm by 6 cm.

5. $x^2 + 36 = -13x \Rightarrow x^2 + 13x + 36 = 0 \Rightarrow$

$(x + 9)(x + 4) = 0 \Rightarrow x = -9$ or $x = -4$

6. Let x = the amount to be invested at 8%. Then the total amount invested is $x + 8200$.

$0.06(8200) + 0.08x = 0.07(8200 + x)$

$492 + 0.08x = 574 + 0.07x$

$0.01x = 82 \Rightarrow x = 8200$

Fran must invest \$8200 at 7%.

7. Let x = the width of the border. Then the length of the border is $2x + 25$, and the width of the border is $2x + 11$.

$(2x + 25)(2x + 11) = 351$

$4x^2 + 72x + 275 = 351$

$4x^2 + 72x - 76 = 0$

$4(x + 19)(x - 1) = 0 \Rightarrow x = -19$ or $x = 1$

The border is 1 inch wide.

8. $\qquad -6x - 15 = (2x + 5)^2$

$\qquad -6x - 15 = 4x^2 + 20x + 25$

$4x^2 + 26x + 40 = 0$

$2(2x + 5)(x + 4) = 0 \Rightarrow x = -\dfrac{5}{2}$ or $x = -4$

The solution set is $\left\{-4, -5/2\right\}$.

9. To find the constant term, find 1/2 of 2/3 = 1/3 and then square the answer: $(1/3)^2 = 1/9$.

The trinomial is $x^2 + \dfrac{2}{3}x + \dfrac{1}{9}$, which factors

into $\left(x + \dfrac{1}{3}\right)^2$.

10. $x = \dfrac{-(-5) \pm \sqrt{(-5)^2 - 4(3)(-1)}}{2(3)}$

$= \dfrac{5 \pm \sqrt{37}}{6} = \dfrac{5}{6} \pm \dfrac{\sqrt{37}}{6}$

11. Let x = the length of the side of the original piece of cardboard. Then $x - 4$ = the length of the side of the box.

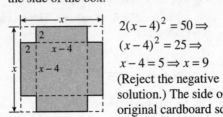

$2(x-4)^2 = 50 \Rightarrow$
$(x-4)^2 = 25 \Rightarrow$
$x - 4 = 5 \Rightarrow x = 9$
(Reject the negative solution.) The side of the original cardboard square is 9 inches.

12. $x = \dfrac{-12 \pm \sqrt{12^2 - 4(1)(33)}}{2(1)}$

$= \dfrac{-12 \pm \sqrt{12}}{2} = \dfrac{-12 \pm 2\sqrt{3}}{2} = -6 \pm \sqrt{3}$

13. $3x^4 - 75x^2 = 0 \Rightarrow 3x^2(x^2 - 25) = 0 \Rightarrow$
$3x^2(x-5)(x+5) = 0 \Rightarrow 3x^2 = 0 \Rightarrow x = 0$
or $x - 5 = 0 \Rightarrow x = 5$
or $x + 5 = 0 \Rightarrow x = -5$
The solution set is $\{-5, 0, 5\}$.

14. Let $u = \sqrt{x}$. The equation becomes
$3u^2 - 2 - 5u = 0 \Rightarrow 3u^2 - 5u - 2 = 0 \Rightarrow$
$(3u + 1)(u - 2) = 0 \Rightarrow u = -\dfrac{1}{3}$ or $u = 2$

Now solve for x: $2 = \sqrt{x} \Rightarrow x = 4$ (We reject the negative solution). The solution set is $\{4\}$.

15. $\left|\dfrac{1}{3}x + 5\right| = \left|\dfrac{2}{3}x + 7\right| \Rightarrow \dfrac{1}{3}x + 5 = \dfrac{2}{3}x + 7$

or $\dfrac{1}{3}x + 5 = -\left(\dfrac{2}{3}x + 7\right)$.

$\dfrac{1}{3}x + 5 = \dfrac{2}{3}x + 7 \Rightarrow x + 15 = 2x + 21 \Rightarrow x = -6$.

$\dfrac{1}{3}x + 5 = -\left(\dfrac{2}{3}x + 7\right) \Rightarrow \dfrac{1}{3}x + 5 = -\dfrac{2}{3}x - 7 \Rightarrow$
$x + 15 = -2x - 21 \Rightarrow 3x = -36 \Rightarrow x = -12$
The solution set is $\{-12, -6\}$.

16. $\dfrac{x}{2} - 5 \geq \dfrac{4x}{9} \Rightarrow 9x - 90 \geq 8x \Rightarrow x \geq 90$
The solution set is $[90, \infty)$.

17. $-4 < 2x - 3 < 4 \Rightarrow -1 < 2x < 7 \Rightarrow$
$-\dfrac{1}{2} < x < \dfrac{7}{2}$
The solution set is $\left(-\dfrac{1}{2}, \dfrac{7}{2}\right)$.

18. $\left|\dfrac{2}{3}x - 1\right| - 2 > \dfrac{1}{3} \Rightarrow \left|\dfrac{2}{3}x - 1\right| > \dfrac{7}{3} \Rightarrow$
$\dfrac{2}{3}x - 1 < -\dfrac{7}{3}$ or $\dfrac{2}{3}x - 1 > \dfrac{7}{3}$
$\dfrac{2}{3}x - 1 < -\dfrac{7}{3} \Rightarrow 2x - 3 < -7 \Rightarrow 2x < -4 \Rightarrow$
$x < -2$
$\dfrac{2}{3}x - 1 > \dfrac{7}{3} \Rightarrow 2x - 3 > 7 \Rightarrow 2x > 10 \Rightarrow x > 5$
The solution set is $(-\infty, -2) \cup (5, \infty)$.

19. $\dfrac{2}{5}y - 13 \leq -\left(7 + \dfrac{13}{5}y\right)$
$\dfrac{2}{5}y - 13 \leq -7 - \dfrac{13}{5}y$
$2y - 65 \leq -35 - 13y \Rightarrow 15y \leq 30 \Rightarrow y \leq 2$
The solution set is $(-\infty, 2]$.

20. $0 \leq 5x - 2 \leq 8 \Rightarrow 2 \leq 5x \leq 10 \Rightarrow \dfrac{2}{5} \leq x \leq 2$
The solution set is $\left[\dfrac{2}{5}, 2\right]$.

Chapter 1 Practice Test B

1. $2x - 2 = 5x + 34 \Rightarrow -3x = 36 \Rightarrow x = -12$.
The answer is D.

2. $\dfrac{z}{2} = 2z + 35 \Rightarrow z = 4z + 70 \Rightarrow -3z = 70 \Rightarrow$
$z = -\dfrac{70}{3}$. The answer is D.

3. $\dfrac{1}{t-2} - \dfrac{1}{2} = \dfrac{-2t}{4t-1} \Rightarrow$
$2(4t-1) - (t-2)(4t-1) = -2t(2)(t-2) \Rightarrow$
$8t - 2 - 4t^2 + 9t - 2 = -4t^2 + 8t \Rightarrow$
$9t - 4 = 0 \Rightarrow 9t = 4 \Rightarrow t = \dfrac{4}{9}$.

The answer is A.

4. Let w = the width of the rectangle.
Then $w + 4$ = the length of the rectangle.
$w(w + 4) = 77 \Rightarrow w^2 + 4w - 77 = 0 \Rightarrow$
$(w + 11)(w - 7) = 0 \Rightarrow w = -11$ or $w = 7$
We reject the negative solution. The rectangle is 7 cm by 11 cm. The answer is C.

5. $x^2 + 12 = -7x \Rightarrow x^2 + 7x + 12 = 0 \Rightarrow$
$(x+4)(x+3) = 0 \Rightarrow x = -4 \text{ or } x = -3$
The answer is B.

6. Let x = the amount to be invested at 12%.
Then the total amount invested is $7500 + x$.
$0.07(7500) + 0.12x = 0.10(7500 + x)$
$$525 + 0.12x = 750 + 0.10x$$
$$0.02x = 225 \Rightarrow x = 11,250$$
Rena must invest \$11,250 at 12%.
The answer is A.

7. Let x = the width of the border. Then the length of the border is $2x + 20$, and the width of the border is $2x + 13$.

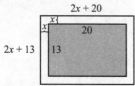

$(2x+20)(2x+13) = 368$
$$4x^2 + 66x + 260 = 368$$
$$4x^2 + 66x - 108 = 0$$
$$2(2x-3)(x+18) = 0 \Rightarrow x = \frac{3}{2} \text{ or } x = -18$$
The border is 1.5 inch wide. The answer is D.

8. $$-6x - 2 = (3x+1)^2$$
$$-6x - 2 = 9x^2 + 6x + 1$$
$$9x^2 + 12x + 3 = 0$$
$$3x^2 + 4x + 1 = 0$$
$$(3x+1)(x+1) = 0 \Rightarrow x = -\frac{1}{3} \text{ or } x = -1$$
The answer is D.

9. To find the constant term, find
$1/2$ of $1/6 = 1/12$ and then square the answer:
$(1/12)^2 = 1/144$.

The trinomial is $x^2 + \dfrac{1}{6}x + \dfrac{1}{144}$, which

factors into $\left(x + \dfrac{1}{12}\right)^2$. The answer is B.

10. $x = \dfrac{-10 \pm \sqrt{10^2 - 4(7)(2)}}{2(7)}$

$= \dfrac{-10 \pm \sqrt{44}}{14} = \dfrac{-10 \pm 2\sqrt{11}}{14} = -\dfrac{5}{7} \pm \dfrac{\sqrt{11}}{7}$
The answer is A.

11. Let x = the length of the side of the original piece of cardboard. Then $x - 6$ = the length of the side of the box.

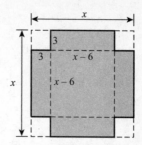

$3(x-6)^2 = 675 \Rightarrow (x-6)^2 = 225 \Rightarrow$
$x - 6 = 15 \Rightarrow x = 21$ (Note that we reject the negative solution.) The side of the original cardboard square is 21 inches.
The answer is B.

12. $x = \dfrac{-14 \pm \sqrt{14^2 - 4(1)(38)}}{2(1)}$

$= \dfrac{-14 \pm \sqrt{44}}{2} = \dfrac{-14 \pm 2\sqrt{11}}{2}$

$= -7 \pm \sqrt{11}$
The answer is D.

13. $5x^4 - 45x^2 = 0 \Rightarrow 5x^2(x^2 - 9) = 0 \Rightarrow$
$5x^2(x-3)(x+3) = 0 \Rightarrow 5x^2 = 0 \Rightarrow x = 0$
or $x - 3 = 0 \Rightarrow x = 3$
or $x + 3 = 0 \Rightarrow x = -3$
The solution set is $\{-3, 0, 3\}$.
The answer is A.

14. Let $u = \sqrt{x}$. The equation becomes
$u^2 - 2048 - 32u = 0 \Rightarrow$
$u^2 - 32u - 2048 = 0 \Rightarrow$
$(u-64)(u+32) = 0 \Rightarrow u = 64 \text{ or } u = -32$
Now solve for x: $64 = \sqrt{x} \Rightarrow x = 4096$ (We reject the negative solution). The answer is A.

15. $\left|\dfrac{1}{2}x + 2\right| = \left|\dfrac{3}{4}x - 2\right| \Rightarrow \dfrac{1}{2}x + 2 = \dfrac{3}{4}x - 2$

or $\dfrac{1}{2}x + 2 = -\left(\dfrac{3}{4}x - 2\right)$.

$\dfrac{1}{2}x + 2 = \dfrac{3}{4}x - 2 \Rightarrow 2x + 8 = 3x - 8 \Rightarrow x = 16$

$\dfrac{1}{2}x + 2 = -\left(\dfrac{3}{4}x - 2\right) \Rightarrow \dfrac{1}{2}x + 2 = -\dfrac{3}{4}x + 2 \Rightarrow$

$2x + 8 = -3x + 8 \Rightarrow 5x = 0 \Rightarrow x = 0$
The answer is D.

16. $\dfrac{x}{6} - \dfrac{1}{3} \le \dfrac{x}{3} + 1 \Rightarrow x - 2 \le 2x + 6 \Rightarrow -8 \le x$
The answer is D.

17. $-13 \le -3x + 2 < -4 \Rightarrow -15 \le -3x < -6 \Rightarrow$
$5 \ge x > 2$
The answer is B.

18. $8 + \left| 1 - \dfrac{x}{2} \right| \ge 10 \Rightarrow \left| 1 - \dfrac{x}{2} \right| \ge 2 \Rightarrow$
$1 - \dfrac{x}{2} \le -2 \Rightarrow -\dfrac{x}{2} \le -3 \Rightarrow x \ge 6$ or
$1 - \dfrac{x}{2} \ge 2 \Rightarrow -\dfrac{x}{2} \ge 1 \Rightarrow x \le -2$
The solution set is $(-\infty, -2] \cup [6, \infty)$.
The answer is C.

19. $\dfrac{2}{3}x - 2 < \dfrac{5}{3}x \Rightarrow 2x - 6 < 5x \Rightarrow -6 < 3x \Rightarrow$
$-2 < x$
The solution set is $(-2, \infty)$. The answer is A.

20. $0 \le 7x - 1 \le 13 \Rightarrow 1 \le 7x \le 14 \Rightarrow \dfrac{1}{7} \le x \le 2$
The solution set is $\left[\dfrac{1}{7}, 2 \right]$. The answer is A.

Chapter 2 Graphs and Functions

2.1 The Coordinate Plane

2.1 Practice Problems

1.

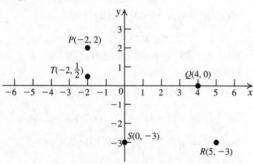

2. (2005, 17.9), (2006, 17.8), (2007, 17.8), (2008, 17.6), (2009, 17.6), (2010, 16.3), (2011, 16.0), (2012, 18.0)

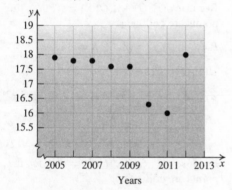

3. $(x_1, y_1) = (-5, 2); \ (x_2, y_2) = (-4, 1)$

$$d = \sqrt{(x_2 - x_1)^2 + (y_2 - y_1)^2}$$
$$= \sqrt{(-4 - (-5))^2 + (1 - 2)^2}$$
$$= \sqrt{1^2 + (-1)^2} = \sqrt{2} \approx 1.4$$

4. $(x_1, y_1) = (6, 2); \ (x_2, y_2) = (-2, 0)$
$(x_3, y_3) = (1, 5)$

$$d_1 = \sqrt{(x_2 - x_1)^2 + (y_2 - y_1)^2}$$
$$= \sqrt{(-2 - 6)^2 + (0 - 2)^2}$$
$$= \sqrt{(-8)^2 + (-2)^2} = \sqrt{68}$$

$$d_2 = \sqrt{(x_3 - x_1)^2 + (y_3 - y_1)^2}$$
$$= \sqrt{(1 - 6)^2 + (5 - 2)^2}$$
$$= \sqrt{(-5)^2 + (3)^2} = \sqrt{34}$$

$$d_3 = \sqrt{(x_3 - x_2)^2 + (y_3 - y_2)^2}$$
$$= \sqrt{(1 - (-2))^2 + (5 - 0)^2}$$
$$= \sqrt{(3)^2 + (5)^2} = \sqrt{34}$$

Yes, the triangle is an isosceles triangle.

5.

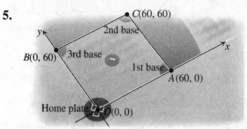

We are asked to find the distance between the points $A(60, 0)$ and $B(0, 60)$.

$$d(A, \ B) = \sqrt{(60 - 0)^2 + (0 - 60)^2}$$
$$= \sqrt{(60)^2 + (-60)^2} = \sqrt{2(60)^2}$$
$$= 60\sqrt{2} \approx 84.85$$

6. $M = \left(\dfrac{5 + 6}{2}, \dfrac{-2 + (-1)}{2} \right) = \left(\dfrac{11}{2}, -\dfrac{3}{2} \right)$

2.1 Basic Concepts and Skills

1. A point with a negative first coordinate and a positive second coordinate lies in the <u>second</u> quadrant.

3. The distance between the points $P(x_1, y_1)$ and $Q(x_2, y_2)$ is given by the formula <u>$d(P, Q) = \sqrt{(x_2 - x_1)^2 + (y_2 - y_1)^2}$</u>.

5. True

7.

(2, 2): Q1; (3, −1): Q4; (−1, 0): *x*-axis
(−2, −5): Q3; (0, 0): origin; (−7, 4): Q2
(0, 3): *y*-axis; (−4, 2): Q2

9. a. If the *x*-coordinate of a point is 0, the point lies on the *y*-axis.

b.

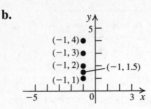

The set of all points of the form $(-1, y)$ is a vertical line that intersects the *x*-axis at -1.

11. a. $y > 0$ **b.** $y < 0$

   c. $x < 0$ **d.** $x > 0$

In Exercises 13–22, use the distance formula, $d = \sqrt{(x_2 - x_1)^2 + (y_2 - y_1)^2}$ and the midpoint formula, $(x, y) = \left(\dfrac{x_1 + x_2}{2}, \dfrac{y_1 + y_2}{2} \right)$.

13. a. $d = \sqrt{(2-2)^2 + (5-1)^2} = \sqrt{4^2} = 4$

   b. $M = \left(\dfrac{2+2}{2}, \dfrac{1+5}{2} \right) = (2, 3)$

15. a. $d = \sqrt{(2 - (-1))^2 + (-3 - (-5))^2}$
$= \sqrt{3^2 + 2^2} = \sqrt{13}$

   b. $M = \left(\dfrac{-1+2}{2}, \dfrac{-5 + (-3)}{2} \right) = (0.5, -4)$

17. a. $d = \sqrt{(3 - (-1))^2 + (-6.5 - 1.5)^2}$
$= \sqrt{4^2 + (-8)^2} = \sqrt{80} = 4\sqrt{5}$

   b. $M = \left(\dfrac{-1+3}{2}, \dfrac{1.5 + (-6.5)}{2} \right) = (1, -2.5)$

19. a. $d = \sqrt{\left(\sqrt{2} - \sqrt{2} \right)^2 + (5 - 4)^2} = \sqrt{1^2} = 1$

   b. $M = \left(\dfrac{\sqrt{2} + \sqrt{2}}{2}, \dfrac{4 + 5}{2} \right) = \left(\sqrt{2}, 4.5 \right)$

21. a. $d = \sqrt{(k - t)^2 + (t - k)^2}$
$= \sqrt{\left(k^2 - 2tk + t^2 \right) + \left(t^2 - 2kt + k^2 \right)}$
$= \sqrt{2t^2 - 4tk + 2k^2} = \sqrt{2 \left(t^2 - 2tk + k^2 \right)}$
$= \sqrt{2(t - k)^2} = |t - k| \sqrt{2}$

b. $M = \left(\dfrac{t + k}{2}, \dfrac{k + t}{2} \right)$

23. $P = (-1, -2), Q = (0, 0), R = (1, 2)$
$d(P, Q) = \sqrt{(0 - (-1))^2 + (0 - (-2))^2} = \sqrt{5}$
$d(Q, R) = \sqrt{(1 - 0)^2 + (2 - 0)^2} = \sqrt{5}$
$d(P, R) = \sqrt{(1 - (-1))^2 + (2 - (-2))^2}$
$= \sqrt{2^2 + 4^2} = \sqrt{20} = 2\sqrt{5}$
Because $d(P, Q) + d(Q, R) = d(P, R)$, the points are collinear.

25. $P = (4, -2), Q = (1, 3), R = (-2, 8)$
$d(P, Q) = \sqrt{(1 - 4)^2 + (3 - (-2))^2} = \sqrt{34}$
$d(Q, R) = \sqrt{(-2 - 1)^2 + (8 - 3)^2} = \sqrt{34}$
$d(P, R) = \sqrt{(-2 - 4)^2 + (8 - (-2))^2}$
$= \sqrt{(-6)^2 + 10^2} = \sqrt{136} = 2\sqrt{34}$
Because $d(P, Q) + d(Q, R) = d(P, R)$, the points are collinear.

27. $P = (-1, 4), Q = (3, 0), R = (11, -8)$
$d(P, Q) = \sqrt{(3 - (-1))^2 + (0 - 4)^2} = 4\sqrt{2}$
$d(Q, R) = \sqrt{(11 - 3)^2 + ((-8) - 0)^2} = 8\sqrt{2}$
$d(P, R) = \sqrt{(11 - (-1))^2 + (-8 - 4)^2}$
$= \sqrt{(12)^2 + (-12)^2} = \sqrt{288} = 12\sqrt{2}$
Because $d(P, Q) + d(Q, R) = d(P, R)$, the points are collinear.

29. It is not possible to arrange the points in such a way so that $d(P, Q) + d(Q, R) = d(P, R)$, so the points are not collinear.

31. First, find the midpoint *M* of *PQ*.
$M = \left(\dfrac{-4 + 0}{2}, \dfrac{0 + 8}{2} \right) = (-2, 4)$
Now find the midpoint *R* of *PM*.
$R = \left(\dfrac{-4 + (-2)}{2}, \dfrac{0 + 4}{2} \right) = (-3, 2)$
Finally, find the midpoint *S* of *MQ*.
$S = \left(\dfrac{-2 + 0}{2}, \dfrac{4 + 8}{2} \right) = (-1, 6)$
Thus, the three points are $(-3, 2)$, $(-2, 4)$, and $(-1, 6)$.

33. $d(P,Q) = \sqrt{(-1-(-5))^2 + (4-5)^2} = \sqrt{17}$

$d(Q,R) = \sqrt{(-4-(-1))^2 + (1-4)^2} = 3\sqrt{2}$

$d(P,R) = \sqrt{(-4-(-5))^2 + (1-5)^2} = \sqrt{17}$

The triangle is isosceles.

35. $d(P,Q) = \sqrt{(0-(-4))^2 + (7-8)^2} = \sqrt{17}$

$d(Q,R) = \sqrt{(-3-0)^2 + (5-7)^2} = \sqrt{13}$

$d(P,R) = \sqrt{(-3-(-4))^2 + (5-8)^2} = \sqrt{10}$

The triangle is scalene.

37. $d(P,Q) = \sqrt{(9-0)^2 + (-9-(-1))^2} = \sqrt{145}$

$d(Q,R) = \sqrt{(5-9)^2 + (1-(-9))^2} = 2\sqrt{29}$

$d(P,R) = \sqrt{(5-0)^2 + (1-(-1))^2} = \sqrt{29}$

The triangle is scalene.

39. $d(P,Q) = \sqrt{(-1-1)^2 + (1-(-1))^2} = 2\sqrt{2}$

$d(Q,R) = \sqrt{(-\sqrt{3}-(-1))^2 + (-\sqrt{3}-1)^2}$

$= \sqrt{(3-2\sqrt{3}+1)+(3+2\sqrt{3}+1)}$

$= \sqrt{8} = 2\sqrt{2}$

$d(P,R) = \sqrt{(-\sqrt{3}-1)^2 + (-\sqrt{3}-(-1))^2}$

$= \sqrt{(3+2\sqrt{3}+1)+(3-2\sqrt{3}+1)}$

$= \sqrt{8} = 2\sqrt{2}$

The triangle is equilateral.

41. First find the lengths of the sides:

$d(P,Q) = \sqrt{(-1-7)^2 + (3-(-12))^2} = 17$

$d(Q,R) = \sqrt{(14-(-1))^2 + (11-3)^2} = 17$

$d(R,S) = \sqrt{(22-14)^2 + (-4-11)^2} = 17$

$d(S,P) = \sqrt{(22-7)^2 + (-4-(-12))^2} = 17$

All the sides are equal, so the quadrilateral is either a square or a rhombus. Now find the length of the diagonals:

$d(P,R) = \sqrt{(14-7)^2 + (11-(-12))^2} = 17\sqrt{2}$

$d(Q,S) = \sqrt{(22-(-1))^2 + (-4-3)^2} = 17\sqrt{2}$

The diagonals are equal, so the quadrilateral is a square.

43. $5 = \sqrt{(x-2)^2 + (2-(-1))^2}$

$= \sqrt{x^2 - 4x + 4 + 9} \Rightarrow$

$5 = \sqrt{x^2 - 4x + 13} \Rightarrow 25 = x^2 - 4x + 13 \Rightarrow$

$0 = x^2 - 4x - 12 \Rightarrow 0 = (x-6)(x+2) \Rightarrow$

$x = -2$ or $x = 6$

45. $P = (-5, 2)$, $Q = (2, 3)$, $R = (x, 0)$ (R is on the x-axis, so the y-coordinate is 0).

$d(P,R) = \sqrt{(x-(-5))^2 + (0-2)^2}$

$d(Q,R) = \sqrt{(x-2)^2 + (0-3)^2}$

$\sqrt{(x-(-5))^2 + (0-2)^2} = \sqrt{(x-2)^2 + (0-3)^2}$

$(x+5)^2 + (0-2)^2 = (x-2)^2 + (0-3)^2$

$x^2 + 10x + 25 + 4 = x^2 - 4x + 4 + 9$

$10x + 29 = -4x + 13$

$14x = -16$

$x = -\dfrac{8}{7}$

The coordinates of R are $\left(-\dfrac{8}{7}, 0\right)$.

2.1 Applying the Concepts

47.

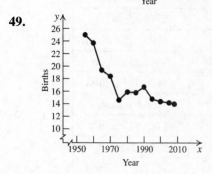

49.

51.

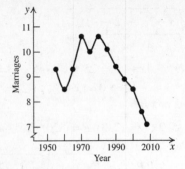

53. $M = \dfrac{16,929 + 14,612}{2} = 15,770.5$

There were about 15,771 murders in 2009.

55. 2008 is the midpoint of the initial range, so

$M_{2008} = \dfrac{548 + 925}{2} = 736.5$.

2006 is the midpoint of the range [2004, 2008], so

$M_{2006} = \dfrac{548 + 736.5}{2} = 642.25$.

2005 is the midpoint of the range [2004, 2006], so

$M_{2005} = \dfrac{548 + 642.25}{2} = 595.125$.

2007 is the midpoint of the range [2006, 2008], so

$M_{2007} = \dfrac{642.25 + 736.5}{2} = 689.375$.

Use similar reasoning to find the amounts for 2009, 2010, and 2011. Defense spending was as follows:

Year	Amount spent
2004	$548 billion
2005	$595.125 billion
2006	$642.25 billion
2007	$689.375 billion
2008	$736.5 billion
2009	$783.625 billion
2010	$830.75 billion
2011	$877.875 billion
2012	$925 billion

57. a.

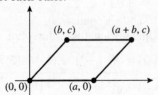

b. $d(D, M) = \sqrt{(800 - 200)^2 + (1200 - 400)^2}$
$ = 1000$

$d(M, P) = \sqrt{(2000 - 800)^2 + (300 - 1200)^2}$
$ = 1500$

The distance traveled by the pilot
$= 1000 + 1500 = 2500$ miles.

c. $d(D, P) = \sqrt{(2000 - 200)^2 + (300 - 400)^2}$
$ = \sqrt{3,250,000} = 500\sqrt{13}$
$ \approx 1802.78$ miles

2.1 Beyond the Basics

59. The midpoint of the diagonal connecting $(0, 0)$ and $(a + b, c)$ is $\left(\dfrac{a+b}{2}, \dfrac{c}{2}\right)$. The midpoint of the diagonal connecting $(a, 0)$ and (b, c) is also $\left(\dfrac{a+b}{2}, \dfrac{c}{2}\right)$. Because the midpoints of the two diagonals are the same, the diagonals bisect each other.

61. The midpoint of the diagonal connecting $(0, 0)$ and (x, y) is $\left(\dfrac{x}{2}, \dfrac{y}{2}\right)$. The midpoint of the diagonal connecting $(a, 0)$ and (b, c) is $\left(\dfrac{a+b}{2}, \dfrac{c}{2}\right)$. Because the diagonals bisect each other, the midpoints coincide. So

$\dfrac{x}{2} = \dfrac{a+b}{2} \Rightarrow x = a + b$, and $\dfrac{y}{2} = \dfrac{c}{2} \Rightarrow y = c$.

Therefore, the quadrilateral is a parallelogram.

63. Let $P(0, 0)$, $Q(a, 0)$, $R(a + b, c)$, and $S(b, c)$ be the vertices of the parallelogram.

$$PQ = RS = \sqrt{(a-0)^2 + (0-0)^2} = a. \quad QR = PS = \sqrt{((a+b)-a)^2 + (c-0)^2} = \sqrt{b^2 + c^2}.$$

The sum of the squares of the lengths of the sides $= 2(a^2 + b^2 + c^2)$.

$$d(P, R) = \sqrt{(a+b)^2 + c^2}. \quad d(Q, S) = \sqrt{(a-b)^2 + (0-c)^2}.$$

The sum of the squares of the lengths of the diagonals is

$$\left((a+b)^2 + c^2\right) + \left((a-b)^2 + c^2\right) = a^2 + 2ab + b^2 + c^2 + a^2 - 2ab + b^2 + c^2$$
$$= 2a^2 + 2b^2 + 2c^2 = 2(a^2 + b^2 + c^2)$$

65. Let $P(0, 0)$, $Q(a, 0)$, and $R(0, a)$ be the vertices of the triangle.

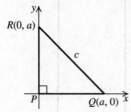

Using the Pythagorean theorem, we have $c^2 = a^2 + a^2 \Rightarrow c^2 = 2a^2 \Rightarrow c = \sqrt{2}a \Rightarrow a = \dfrac{1}{\sqrt{2}}c = \dfrac{\sqrt{2}}{2}c$

67. a. $d(A, C) = \sqrt{\left(x_1 - \dfrac{2x_1 + x_2}{3}\right)^2 + \left(y_1 - \dfrac{2y_1 + y_2}{3}\right)^2} = \sqrt{\left(\dfrac{x_1 - x_2}{3}\right)^2 + \left(\dfrac{y_1 - y_2}{3}\right)^2}$

$d(C, B) = \sqrt{\left(\dfrac{2x_1 + x_2}{3} - x_2\right)^2 + \left(\dfrac{2y_1 + y_2}{3} - y_2\right)^2} = \sqrt{\left(\dfrac{2x_1 - 2x_2}{3}\right)^2 + \left(\dfrac{2y_1 - 2y_2}{3}\right)^2}$

$d(A, C) + d(C, B) = \dfrac{\sqrt{(x_1 - x_2)^2 + (y_1 - y_2)^2}}{3} + \dfrac{\sqrt{(2x_1 - 2x_2)^2 + (2y_1 - 2y_2)^2}}{3}$

$\qquad = \dfrac{\sqrt{(x_1 - x_2)^2 + (y_1 - y_2)^2}}{3} + \dfrac{2\sqrt{(x_1 - x_2)^2 + (y_1 - y_2)^2}}{3} = \sqrt{(x_1 - x_2)^2 + (y_1 - y_2)^2}$

$d(A, B) = \sqrt{(x_1 - x_2)^2 + (y_1 - y_2)^2}$

So A, B, and C are collinear.

$d(A, C) = \sqrt{\left(\dfrac{x_1 - x_2}{3}\right)^2 + \left(\dfrac{y_1 - y_2}{3}\right)^2} = \dfrac{\sqrt{(x_1 - x_2)^2 + (y_1 - y_2)^2}}{3} = \dfrac{1}{3}d(A, B).$

b. $d(A, D) = \sqrt{\left(x_1 - \dfrac{x_1 + 2x_2}{3}\right)^2 + \left(y_1 - \dfrac{y_1 + 2y_2}{3}\right)^2} = \sqrt{\left(\dfrac{2x_1 + 2x_2}{3}\right)^2 + \left(\dfrac{2y_1 + 2y_2}{3}\right)^2}$

$d(D, B) = \sqrt{\left(\dfrac{x_1 + 2x_2}{3} - x_2\right)^2 + \left(\dfrac{y_1 + 2y_2}{3} - y_2\right)^2} = \sqrt{\left(\dfrac{x_1 - x_2}{3}\right)^2 + \left(\dfrac{y_1 - y_2}{3}\right)^2}$

$d(A, D) + d(D, B) = \dfrac{\sqrt{(2x_1 + 2x_2)^2 + (2y_1 + 2y_2)^2}}{3} + \dfrac{\sqrt{(x_1 - x_2)^2 + (y_1 - y_2)^2}}{3}$

$\qquad = \dfrac{2\sqrt{(x_1 - x_2)^2 + (y_1 - y_2)^2}}{3} + \dfrac{\sqrt{(x_1 - x_2)^2 + (y_1 - y_2)^2}}{3}$

$\qquad = \sqrt{(x_1 - x_2)^2 + (y_1 - y_2)^2} = d(A, B)$

(continued on next page)

(*continued*)

So A, B, and C are collinear.

$$d(A, D) = \sqrt{\left(\frac{2x_1 + 2x_2}{3}\right)^2 + \left(\frac{2y_1 - 2y_2}{3}\right)^2}$$

$$= \frac{2\sqrt{(x_1 + x_2)^2 + (y_1 + y_2)^2}}{3}$$

$$= \frac{2}{3} d(A, B).$$

c. $\dfrac{2x_1 + x_2}{3} = \dfrac{2(-1) + 4}{3} = \dfrac{2}{3}$

$\dfrac{2y_1 + y_2}{3} = \dfrac{2(2) + 1}{3} = \dfrac{5}{3}$

$\dfrac{x_1 + 2x_2}{3} = \dfrac{-1 + 2(4)}{3} = \dfrac{7}{3}$

$\dfrac{y_1 + 2y_2}{3} = \dfrac{2 + 2(1)}{3} = \dfrac{4}{3}$

The points of trisection are $\left(\dfrac{2}{3}, \dfrac{5}{3}\right)$ and

$\left(\dfrac{7}{3}, \dfrac{4}{3}\right)$.

2.1 Critical Thinking/Discussion/Writing

69. a. y-axis

   b. x-axis

70. a. The union of the x- and y-axes

   b. The plane without the x- and y-axes

71. a. Quadrants I and III

   b. Quadrants II and IV

72. a. The origin

   b. The plane without the origin

73. a. Right half-plane

   b. Upper half-plane

74. Let (x, y) be the point.

The point lies in	if
Quadrant I	$x > 0$ and $y > 0$
Quadrant II	$x < 0$ and $y > 0$
Quadrant III	$x < 0$ and $y < 0$
Quadrant IV	$x > 0$ and $y < 0$

$(-, +)$ | $(+, +)$

$(-, -)$ | $(+, -)$

2.1 Maintaining Skills

75. a. $x^2 + y^2 = \left(\dfrac{1}{2}\right)^2 + \left(\dfrac{1}{2}\right)^2 = \dfrac{1}{4} + \dfrac{1}{4} = \dfrac{2}{4} = \dfrac{1}{2}$

   b. $x^2 + y^2 = \left(\dfrac{\sqrt{2}}{2}\right)^2 + \left(\dfrac{\sqrt{2}}{2}\right)^2 = \dfrac{2}{4} + \dfrac{2}{4} = 1$

77. a. $\dfrac{x}{|x|} + \dfrac{|y|}{y} = \dfrac{2}{|2|} + \dfrac{|-3|}{-3} = \dfrac{2}{2} + \dfrac{3}{-3} = 1 - 1 = 0$

   b. $\dfrac{x}{|x|} + \dfrac{|y|}{y} = \dfrac{-4}{|-4|} + \dfrac{|3|}{3} = \dfrac{-4}{4} + \dfrac{3}{3} = -1 + 1 = 0$

79. $x^2 - 6x + \left(\dfrac{-6}{2}\right)^2 = x^2 - 6x + 3^2$

$= x^2 - 6x + \underline{9}$

81. $y^2 + 3y = y^2 + 3y + \left(\dfrac{3}{2}\right)^2 = y^2 + 3y + \underline{\dfrac{9}{4}}$

83. $x^2 - ax + \left(\dfrac{-a}{2}\right)^2 = x^2 - ax + \underline{\dfrac{a^2}{4}}$

Section 2.2 Graphs of Equations

2.2 Practice Problems

1. $y = -x^2 + 1$

x	$y = -x^2 + 1$	(x, y)
-2	$y = -(-2)^2 + 1$	$(-2, -3)$
-1	$y = -(-1)^2 + 1$	$(-1, 0)$
0	$y = -(0)^2 + 1$	$(0, 1)$
1	$y = -(1)^2 + 1$	$(1, 0)$
2	$y = -(2)^2 + 1$	$(2, -3)$

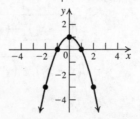

2. To find the *x*-intercept, let $y = 0$, and solve the equation for *x*: $0 = 2x^2 + 3x - 2 \Rightarrow$

   $0 = (2x - 1)(x + 2) \Rightarrow x = \dfrac{1}{2}$ or $x = -2$. To

   find the *y*-intercept, let $x = 0$, and solve the equation for *y*:

   $y = 2(0)^2 + 3(0) - 2 \Rightarrow y = -2$.

   The *x*-intercepts are $\dfrac{1}{2}$ and -2; the *y*-intercept

   is -2.

3. To test for symmetry about the *y*-axis, replace *x* with $-x$ to determine if $(-x, y)$ satisfies the equation.

   $(-x)^2 - y^2 = 1 \Rightarrow x^2 - y^2 = 1$, which is the

   same as the original equation. So the graph is symmetric about the *y*-axis.

4. *x*-axis: $x^2 = (-y)^3 \Rightarrow x^2 = -y^3$, which is not

   the same as the original equation, so the equation is not symmetric with respect to the *x*-axis.

   y-axis: $(-x)^2 = y^3 \Rightarrow x^2 = y^3$, which is the

   same as the original equation, so the equation is symmetric with respect to the *y*-axis.

   origin: $(-x)^2 = (-y)^3 \Rightarrow x^2 = -y^3$, which is

   not the same as the original equation, so the equation is not symmetric with respect to the origin.

5. $y = -t^4 + 77t^2 + 324$

   a. First, find the intercepts. If $t = 0$, then $y = 324$, so the *y*-intercept is $(0, 324)$. If $y = 0$, then we have

   $$0 = -t^4 + 77t^2 + 324$$

   $$t^4 - 77t^2 - 324 = 0$$

   $$(t^2 - 81)(t^2 + 4) = 0$$

   $$(t + 9)(t - 9)(t^2 + 4) = 0 \Rightarrow t = -9, 9, \pm 2i$$

   So, the *t*-intercepts are $(-9, 0)$ and $(9, 0)$. Next, check for symmetry.

   t-axis: $-y = -t^4 + 77t^2 + 324$ is not the

   same as the original equation, so the equation is not symmetric with respect to the *t*-axis.

y-axis: $y = -(-t)^4 + 77(-t)^2 + 324 \Rightarrow$

$y = -t^4 + 77t^2 + 324$, which is the same as

the original equation. So the graph is symmetric with respect to the *y*-axis.

origin: $-y = -(-t)^4 + 77(-t)^2 + 324 \Rightarrow$

$-y = -t^4 + 77t^2 + 324$, which is not the

same as the original equation. So the graph is not symmetric with respect to the origin. Now, make a table of values. Since the graph is symmetric with respect to the *y*-axis, if (t, y) is on the graph, then so is $(-t, y)$. However, the graph pertaining to the physical aspects of the problem consists only of those values for $t \geq 0$.

t	$y = -t^4 + 77t^2 + 324$	(t, y)
0	324	(0, 324)
1	400	(1, 400)
2	616	(2, 616)
3	936	(3, 936)
4	1300	(4, 1300)
5	1624	(5, 1624)
6	1800	(6, 1800)
7	1696	(7, 1696)
8	1156	(8, 1156)
9	0	(9, 0)

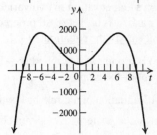

b.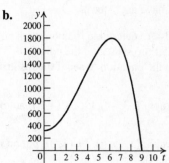

c. The population becomes extinct after 9 years.

6. The standard form of the equation of a circle
is $(x-h)^2 + (y-k)^2 = r^2$
$(h, k) = (3, -6)$ and $r = 10$
The equation of the circle is
$(x-3)^2 + (y+6)^2 = 100$.

7. $(x-2)^2 + (y+1)^2 = 36 \Rightarrow (h,k) = (2,-1)$, $r = 6$
This is the equation of a circle with center $(2, -1)$ and radius 6.

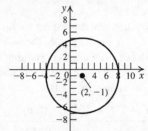

8. $x^2 + y^2 + 4x - 6y - 12 = 0 \Rightarrow$
$x^2 + 4x + y^2 - 6y = 12$
Now complete the square:
$x^2 + 4x + 4 + y^2 - 6y + 9 = 12 + 4 + 9 \Rightarrow$
$(x+2)^2 + (y-3)^2 = 25$
This is a circle with center $(-2, 3)$ and radius 5.

2.2 Basic Concepts and Skills

1. The graph of an equation in two variables, such as x and y, is the set of all ordered pairs (a, b) <u>that satisfy the equation</u>.

3. If $(0, -5)$ is a point of a graph, then -5 is a <u>y-</u> intercept of the graph.

5. False. The equation of a circle has both an x^2-term and a y^2-term. The given equation does not have a y^2-term.

In exercises 7–14, to determine if a point lies on the graph of the equation, substitute the point's coordinates into the equation to see if the resulting statement is true.

7. on the graph: $(-3, -4)$, $(1, 0)$, $(4, 3)$; not on the graph: $(2, 3)$

9. on the graph: $(3, 2)$, $(0, 1)$, $(8, 3)$; not on the graph: $(8, -3)$

11. on the graph: $(1, 0)$, $(2, \sqrt{3})$, $(2, -\sqrt{3})$; not on the graph: $(0, -1)$

13. a. x-intercepts: -3, 3; no y-intercepts

b. Symmetric about the x-axis, y-axis, and origin.

15. a. x-intercepts: -3, 3; y-intercepts: -2, 2

b. Symmetric about the x-axis, y-axis, and origin.

17. a. x-intercept: 0; y-intercept: 0

b. Symmetric about the x-axis.

19. a. x-intercepts: -3, 3; y-interceps: 2

b. Symmetric about the y-axis.

21. a. No x-intercept; y-intercept: 1

b. No symmetries

23. a. x-intercepts: 0, 3; y-intercept: 0

b. Symmetric about the x-axis.

25.

27.

29.

31.

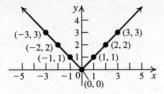

33.

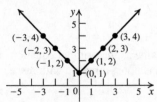

35.

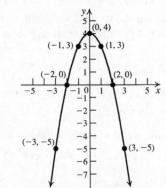

37.

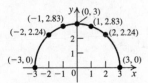

39.

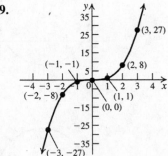

41.

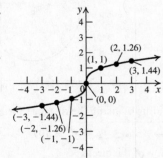

43.

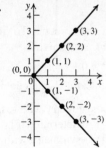

45.

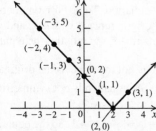

47. To find the x-intercept, let $y = 0$, and solve the equation for x: $3x + 4(0) = 12 \Rightarrow x = 4$. To find the y-intercept, let $x = 0$, and solve the equation for y: $3(0) + 4y = 12 \Rightarrow y = 3$. The x-intercept is 4; the y-intercept is 3.

49. To find the x-intercept, let $y = 0$, and solve the equation for x: $0 = x^2 - 6x + 8 \Rightarrow x = 4$ or $x = 2$. To find the y-intercept, let $x = 0$, and solve the equation for y: $y = 0^2 - 6(0) + 8 \Rightarrow y = 8$. The x-intercepts are 2 and 4; the y-intercept is 8.

51. To find the x-intercept, let $y = 0$, and solve the equation for x: $x^2 + 0^2 = 4 \Rightarrow x = \pm 2$. To find the y-intercept, let $x = 0$, and solve the equation for y: $0^2 + y^2 = 4 \Rightarrow y = \pm 2$. The x-intercepts are –2 and 2; the y-intercepts are –2 and 2.

53. To find the x-intercept, let $y = 0$, and solve the equation for x: $0 = \sqrt{x^2 - 1} \Rightarrow x = \pm 1$. To find the y-intercept, let $x = 0$, and solve the equation for y: $y = \sqrt{0^2 - 1} \Rightarrow$ no solution. The x-intercepts are –1 and 1; there is no y-intercept.

55. To find the x-intercept, let $y = 0$, and solve the equation for x: $0 = x^2 + x + 1 \Rightarrow$

$$x = \frac{-1 \pm \sqrt{1^2 - 4(1)(1)}}{2(1)} = \frac{-1 \pm \sqrt{-3}}{2}, \text{ which}$$

are not real solutions.

(*continued on next page*)

(continued)

Therefore, there are no *x*-intercepts. To find the *y*-intercept, let $x = 0$, and solve the equation for *y*: $y = 0^2 + 0 + 1 = 1$.

There are no *x*-intercepts; the *y*-intercept is 1.

In exercises 57–66, to test for symmetry with respect to the *x*-axis, replace *y* with $-y$ to determine if $(x, -y)$ satisfies the equation. To test for symmetry with respect to the *y*-axis, replace *x* with $-x$ to determine if $(-x, y)$ satisfies the equation. To test for symmetry with respect to the origin, replace *x* with $-x$ and *y* with $-y$ to determine if $(-x, -y)$ satisfies the equation.

57. $-y = x^2 + 1$ is not the same as the original equation, so the equation is not symmetric with respect to the *x*-axis.

$y = (-x)^2 + 1 \Rightarrow y = x^2 + 1$, so the equation is symmetric with respect to the *y*-axis.

$-y = (-x)^2 + 1 \Rightarrow -y = x^2 + 1$, is not the same as the original equation, so the equation is not symmetric with respect to the origin.

59. $-y = x^3 + x$ is not the same as the original equation, so the equation is not symmetric with respect to the *x*-axis.

$y = (-x)^3 - x \Rightarrow y = -x^3 - x \Rightarrow$

$y = -(x^3 + x)$ is not the same as the original equation, so the equation is not symmetric with respect to the *y*-axis.

$-y = (-x)^3 - x \Rightarrow -y = -x^3 - x \Rightarrow$

$-y = -(x^3 + x) \Rightarrow y = x^3 + x$, so the equation is symmetric with respect to the origin.

61. $-y = 5x^4 + 2x^2$ is not the same as the original equation, so the equation is not symmetric with respect to the *x*-axis.

$y = 5(-x)^4 + 2(-x)^2 \Rightarrow y = 5x^4 + 2x^2$, so the equation is symmetric with respect to the *y*-axis.

$-y = 5(-x)^4 + 2(-x) \Rightarrow -y = 5x^4 + 2x^2$ is not the same as the original equation, so the equation is not symmetric with respect to the origin.

63. $-y = -3x^5 + 2x^3$ is not the same as the original equation, so the equation is not symmetric with respect to the *x*-axis.

$y = -3(-x)^5 + 2(-x)^3 \Rightarrow y = 3x^5 - 2x^3$ is not the same as the original equation, so the

equation is not symmetric with respect to the *y*-axis.

$-y = -3(-x)^5 + 2(-x)^3 \Rightarrow -y = 3x^5 - 2x^3 \Rightarrow$

$-y = -(-3x^5 + 2x^3) \Rightarrow y = -3x^5 + 2x^3$, so the equation is symmetric with respect to the origin.

65. $x^2(-y)^2 + 2x(-y) = 1 \Rightarrow x^2 y^2 - 2xy = 1$ is not the same as the original equation, so the equation is not symmetric with respect to the *x*-axis.

$(-x)^2 y^2 + 2(-x)y = 1 \Rightarrow x^2 y^2 - 2xy = 1$ is not the same as the original equation, so the equation is not symmetric with respect to the *y*-axis.

$(-x)^2(-y)^2 + 2(-x)(-y) = 1 \Rightarrow$

$x^2 y^2 + 2xy = 1$, so the equation is symmetric with respect to the origin.

For exercises 67–80, use the standard form of the equation of a circle, $(x - h)^2 + (y - k)^2 = r^2$.

67. Center (2, 3); radius = 6

69. Center (–2, –3); radius = $\sqrt{11}$

71. $x^2 + (y - 1)^2 = 4$

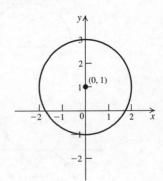

73. $(x + 1)^2 + (y - 2)^2 = 2$

75. Find the radius by using the distance formula:

$d = \sqrt{(-1 - 3)^2 + (5 - (-4))^2} = \sqrt{97}$.

The equation of the circle is

$(x - 3)^2 + (y + 4)^2 = 97$.

(continued on next page)

(*continued*)

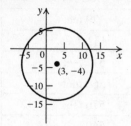

77. The circle touches the *x*-axis, so the radius is 2. The equation of the circle is
$(x-1)^2 + (y-2)^2 = 4$.

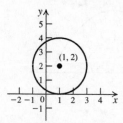

79. Find the diameter by using the distance formula:
$$d = \sqrt{(-3-7)^2 + (6-4)^2} = \sqrt{104} = 2\sqrt{26}\,.$$
So the radius is $\sqrt{26}$. Use the midpoint formula to find the center:
$$M = \left(\frac{7+(-3)}{2}, \frac{4+6}{2}\right) = (2,5).$$ The equation
of the circle is $(x-2)^2 + (y-5)^2 = 26$.

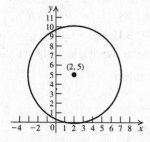

81. a. $x^2 + y^2 - 2x - 2y - 4 = 0 \Rightarrow$
$x^2 - 2x + y^2 - 2y = 4$
Now complete the square:
$x^2 - 2x + 1 + y^2 - 2y + 1 = 4 + 1 + 1 \Rightarrow$
$(x-1)^2 + (y-1)^2 = 6$. This is a circle with
center (1, 1) and radius $\sqrt{6}$.

b. To find the *x*-intercepts, let $y = 0$ and solve for *x*:
$(x-1)^2 + (0-1)^2 = 6 \Rightarrow (x-1)^2 + 1 = 6 \Rightarrow$
$(x-1)^2 = 5 \Rightarrow x-1 = \pm\sqrt{5} \Rightarrow x = 1 \pm \sqrt{5}$

Thus, the *x*-intercepts are $\left(1+\sqrt{5},\, 0\right)$ and
$\left(1-\sqrt{5},\, 0\right)$.
To find the *y*-intercepts, let $x = 0$ and solve for *y*:
$(0-1)^2 + (y-1)^2 = 6 \Rightarrow 1 + (y-1)^2 = 6 \Rightarrow$
$(y-1)^2 = 5 \Rightarrow y-1 = \pm\sqrt{5} \Rightarrow y = 1 \pm \sqrt{5}$
Thus, the *y*-intercepts are $\left(0,\, 1+\sqrt{5}\right)$ and
$\left(0,\, 1-\sqrt{5}\right)$.

83. a. $2x^2 + 2y^2 + 4y = 0 \Rightarrow$
$2(x^2 + y^2 + 2y) = 0 \Rightarrow x^2 + y^2 + 2y = 0$.
Now complete the square:
$x^2 + y^2 + 2y + 1 = 0 + 1 \Rightarrow x^2 + (y+1)^2 = 1$.
This is a circle with center $(0, -1)$ and
radius 1.

b. To find the *x*-intercepts, let $y = 0$ and solve
for *x*: $x^2 + (0+1)^2 = 1 \Rightarrow x^2 = 0 \Rightarrow x = 0$
Thus, the *x*-intercept is $(0, 0)$.
To find the *y*-intercepts, let $x = 0$ and solve
for *y*:
$0^2 + (y+1)^2 = 1 \Rightarrow y+1 = \pm 1 \Rightarrow y = 0, -2$
Thus, the *y*-intercepts are $(0, 0)$ and $(0, -2)$.

85. a. $x^2 + y^2 - x = 0 \Rightarrow x^2 - x + y^2 = 0$.
Now complete the square:
$$x^2 - x + \frac{1}{4} + y^2 = 0 + \frac{1}{4} \Rightarrow$$
$\left(x - \frac{1}{2}\right)^2 + y^2 = \frac{1}{4}$. This is a circle with
center $\left(\frac{1}{2}, 0\right)$ and radius $\frac{1}{2}$.

b. To find the *x*-intercepts, let $y = 0$ and solve
for *x*: $\left(x - \frac{1}{2}\right)^2 + 0^2 = \frac{1}{4} \Rightarrow x - \frac{1}{2} = \pm\frac{1}{2} \Rightarrow$
$x = 0, 1$. Thus, the *x*-intercepts are $(0, 0)$
and $(1, 0)$. To find the *y*-intercepts, let $x = 0$
and solve for *y*:
$\left(0 - \frac{1}{2}\right)^2 + y^2 = \frac{1}{4} \Rightarrow y^2 + \frac{1}{4} = \frac{1}{4} \Rightarrow$
$y^2 = 0 \Rightarrow y = 0$.
Thus, the *y*-intercept is $(0, 0)$.

2.2 Applying the Concepts

87. The distance from $P(x, y)$ to the x-axis is $|x|$ while the distance from P to the y-axis is $|y|$. So the equation of the graph is $|x| = |y|$.

89. The distance from $P(x, y)$ to $(2, 0)$ is $\sqrt{(x-2)^2 + y^2}$ while the distance from P to the y-axis is $|x|$. So the equation of the graph is

$$\sqrt{(x-2)^2 + y^2} = |x| \Rightarrow (x-2)^2 + y^2 = x^2 \Rightarrow$$
$$x^2 - 4x + 4 + y^2 = x^2 \Rightarrow y^2 = 4x - 4 \Rightarrow$$
$$\frac{y^2 + 4}{4} = \frac{y^2}{4} + 1 = x$$

91. a. Since July 2012 is represented by $t = 0$, March 2012 is represented by $t = -4$. So the monthly profit for March is determined by $P = -0.5(-4)^2 - 3(-4) + 8 = \$12$ million.

b. Since July 2012 is represented by $t = 0$, October 2012 is represented by $t = 3$. So the monthly profit for October is determined by $P = -0.5(3)^2 - 3(3) + 8 = -\$5.5$ million. This is a loss.

c.

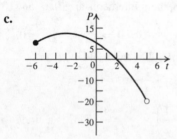

d. To find the t-intercept, set $P = 0$ and solve for t: $0 = -0.5t^2 - 3t + 8 \Rightarrow$
$$t = \frac{3 \pm \sqrt{(-3)^2 - 4(-0.5)(8)}}{2(-0.5)} = \frac{3 \pm \sqrt{25}}{-1}$$
$$= 2 \text{ or} -8$$
The t-intercepts represent the months with no profit and no loss. In this case, $t = -8$ makes no sense in terms of the problem, so we disregard this solution. $t = 2$ represents Sept 2012.

e. To find the P-intercept, set $t = 0$ and solve to P: $P = -0.5(0)^2 - 3(0) + 8 \Rightarrow P = 8$. The P-intercept represents the profit in July 2012.

93. a.

t	Height $= -16t^2 + 128t + 320$
0	320 feet
1	432 feet
2	512 feet
3	560 feet
4	576 feet
5	560 feet
6	512 feet

b.

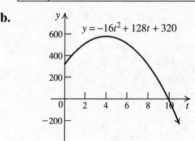

c. $0 \le t \le 10$

d. To find the t-intercept, set $y = 0$ and solve for t:
$$0 = -16t^2 + 128t + 320 \Rightarrow$$
$$0 = -16(t^2 - 8t - 20) \Rightarrow$$
$$0 = (t - 10)(t + 2) \Rightarrow t = 10 \text{ or } t = -2.$$
The graph does not apply if $t < 0$, so the t-intercept is 10. This represents the time when the object hits the ground. To find the y-intercept, set $t = 0$ and solve for y:
$$y = -16(0)^2 + 128(0) + 320 \Rightarrow y = 320.$$
This represents the height of the building.

2.2 Beyond the Basics

95.

97.

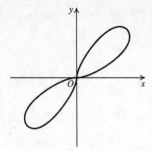

99. $x^2 + y^2 - 4x + 2y - 20 = 0 \Rightarrow$

$x^2 - 4x + y^2 + 2y = 20 \Rightarrow$

$x^2 - 4x + 4 + y^2 + 2y + 1 = 20 + 4 + 1 \Rightarrow$

$(x-2)^2 + (y+1)^2 = 25$

So this is the graph of a circle with center (2, –1) and radius 5. The area of this circle is 25π. $x^2 + y^2 - 4x + 2y - 31 = 0 \Rightarrow$

$x^2 - 4x + y^2 + 2y = 31 \Rightarrow$

$x^2 - 4x + 4 + y^2 + 2y + 1 = 31 + 4 + 1 \Rightarrow$

$(x-2)^2 + (y+1)^2 = 36$

So, this is the graph of a circle with center (2, –1) and radius 6. The area of this circle is 36π. Both circles have the same center, so the area of the region bounded by the two circles equals $36\pi - 25\pi = 11\pi$.

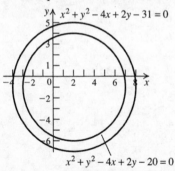

2.2 Critical Thinking/Discussion/Writing

101. The graph of $y^2 = 2x$ is the union of the graphs of $y = \sqrt{2x}$ and $y = -\sqrt{2x}$.

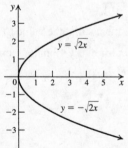

102. Let (x, y) be a point on the graph. Since the graph is symmetric with regard to the x-axis, then the point $(x, -y)$ is also on the graph. Because the graph is symmetric with regard to the y-axis, the point $(-x, y)$ is also on the graph. Therefore the point $(-x, -y)$ is on the graph, and the graph is symmetric with respect to the origin. The graph of $y = x^3$ is an example of a graph that is symmetric with respect to the origin but is not symmetric with respect to the x- and y-axes.

103. a. First find the radius of the circle:

$$d(A, B) = \sqrt{(6-0)^2 + (8-1)^2} = \sqrt{85} \Rightarrow$$

$$r = \frac{\sqrt{85}}{2}.$$

The center of the circle is

$$\left(\frac{6+0}{2}, \frac{1+8}{2} \right) = \left(3, \frac{9}{2} \right).$$

So the equation of the circle is

$$(x-3)^2 + \left(y - \frac{9}{2} \right)^2 = \frac{85}{4}.$$

To find the x-intercepts, set $y = 0$, and solve for x:

$$(x-3)^2 + \left(0 - \frac{9}{2} \right)^2 = \frac{85}{4} \Rightarrow$$

$$(x-3)^2 + \frac{81}{4} = \frac{85}{4} \Rightarrow x^2 - 6x + 9 = 1 \Rightarrow$$

$$x^2 - 6x + 8 = 0$$

The x-intercepts are the roots of this equation.

b. First find the radius of the circle:

$$d(A, B) = \sqrt{(a-0)^2 + (b-1)^2}$$

$$= \sqrt{a^2 + (b-1)^2} \Rightarrow$$

$$r = \frac{\sqrt{a^2 + (b-1)^2}}{2}.$$

The center of the circle is

$$\left(\frac{a+0}{2}, \frac{b+1}{2} \right) = \left(\frac{a}{2}, \frac{b+1}{2} \right)$$

So the equation of the circle is

$$\left(x - \frac{a}{2} \right)^2 + \left(y - \frac{b+1}{2} \right)^2 = \frac{a^2 + (b-1)^2}{4}.$$

(continued on next page)

(continued)

To find the x-intercepts, set $y = 0$ and solve for x:

$$\left(x - \frac{a}{2}\right)^2 + \left(0 - \frac{b+1}{2}\right)^2 = \frac{a^2 + (b-1)^2}{4}$$

$$x^2 - ax + \frac{a^2}{4} + \frac{(b+1)^2}{4} = \frac{a^2 + (b-1)^2}{4}$$

$$4x^2 - 4ax + a^2 + b^2 + 2b + 1 = a^2 + b^2 - 2b + 1$$

$$4x^2 - 4ax + 4b = 0$$

$$x^2 - ax + b = 0$$

The x-intercepts are the roots of this equation.

c. $a = 3$ and $b = 1$. Approximate the roots of the equation by drawing a circle whose diameter has endpoints $A(0, 1)$ and $B(3, 1)$.

The center of the circle is $\left(\frac{3}{2}, 1\right)$ and the

radius is $\frac{3}{2}$. The roots are approximately $(0.4, 0)$ and $(2.6, 0)$.

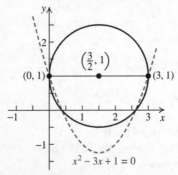

104. a. The coordinates of the center of each circle are (r, r) and $(3r, r)$.

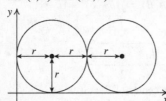

b. To find the area of the shaded region, first find the area of the rectangle shown in the figure below, and then subtract the sum of the areas of the two sectors, A and B.

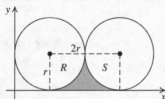

$$A_{\text{rect}} = r(2r) = 2r^2$$

$$A_{\text{sector }R} = A_{\text{sector }S} = \frac{1}{4}\pi r^2$$

$$A_{\text{shaded region}} = 2r^2 - \left(\frac{1}{4}\pi r^2 + \frac{1}{4}\pi r^2\right)$$

$$= 2r^2 - \frac{1}{2}\pi r^2 = \left(2 - \frac{\pi}{2}\right)r^2$$

2.2 Maintaining Skills

105. $\dfrac{5 - 3}{6 - 2} = \dfrac{2}{4} = \dfrac{1}{2}$

107. $\dfrac{2 - (-3)}{3 - 13} = \dfrac{5}{-10} = -\dfrac{1}{2}$

109. $\dfrac{\frac{1}{2} - \frac{1}{4}}{\frac{3}{8} - \left(-\frac{1}{4}\right)} = \dfrac{\frac{1}{4}}{\frac{5}{8}} = \dfrac{1}{4} \cdot \dfrac{8}{5} = \dfrac{2}{5}$

111. $A = P + Prt = P(1 + rt) \Rightarrow P = \dfrac{A}{1 + rt}$

113. $\dfrac{x}{2} - \dfrac{y}{5} = 3 \Rightarrow \dfrac{x}{2} - 3 = \dfrac{y}{5} \Rightarrow \dfrac{5}{2}x - 15 = y$

115. $ax + by + c = 0 \Rightarrow by = -ax - c \Rightarrow$

$$y = -\frac{a}{b}x - \frac{c}{b}$$

117. $-\dfrac{1}{2}$ **119.** $\dfrac{3}{2}$

121. $-\dfrac{2 + \frac{3}{4}}{1 - \frac{1}{2}} = -\dfrac{\frac{11}{4}}{\frac{1}{2}} = -\dfrac{11}{4} \cdot 2 = -\dfrac{11}{2}$

2.3   Lines

2.3 Practice Problems

1. $m = \dfrac{5 - (-3)}{-7 - 6} = -\dfrac{8}{13}$

A slope of $-\frac{8}{13}$ means that the value of y decreases 8 units for every 13 units increase in x.

2. $P(-2,-3)$, $m = -\dfrac{2}{3}$

$$y-(-3) = -\frac{2}{3}\big[x-(-2)\big] \Rightarrow$$

$$y+3 = -\frac{2}{3}(x+2)$$

$$y+3 = -\frac{2}{3}x - \frac{4}{3} \Rightarrow y = -\frac{2}{3}x - \frac{13}{3}$$

3. $m = \dfrac{-4-6}{-3-(-1)} = \dfrac{-10}{-2} = 5$

Use either point to determine the equation of the line. Using $(-3, -4)$, we have

$$y-(-4) = 5\big[x-(-3)\big] \Rightarrow y+4 = 5(x+3) \Rightarrow$$
$$y+4 = 5x+15 \Rightarrow y = 5x+11$$

Using $(-1, 6)$, we have

$$y-6 = 5\big[x-(-1)\big] \Rightarrow y-6 = 5(x+1) \Rightarrow$$
$$y-6 = 5x+5 \Rightarrow y = 5x+11$$

4. $y - y_1 = m(x - x_1) \Rightarrow y-(-3) = 2(x-0)$
$$\text{point-slope form}$$

$$y-(-3) = 2(x-0) \Rightarrow$$
$$y+3 = 2x \Rightarrow y = 2x-3$$

5. The slope is $-2/3$ and the y-intercept is 4.

The line goes through $(0, 4)$, so locate a second point by moving two units down and three units right. Thus, the line goes through $(3, 2)$.

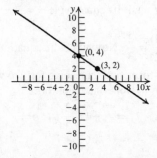

6. $x = -3$. The slope is undefined, and there is no y-intercept. The x-intercept is -3.
$y = 7$. The slope is 0, and the y-intercept is 7.

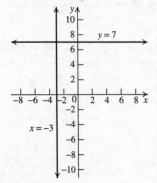

7. First, solve for y to write the equation in slope-intercept form:
$$3x+4y = 24 \Rightarrow 4y = -3x+24 \Rightarrow$$

$y = -\dfrac{3}{4}x + 6$. The slope is $-\dfrac{3}{4}$, and the

y-intercept is 6. Find the x-intercept by setting $y = 0$ and solving the equation for x:

$0 = -\dfrac{3}{4}x + 6 \Rightarrow 6 = \dfrac{3}{4}x \Rightarrow 8 = x$. Thus, the

graph passes through the points $(0, 6)$ and $(8, 0)$.

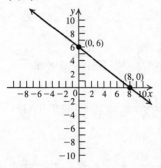

8. Use the equation $H = 2.6x+65$.
$$H_1 = 2.6(43)+65 = 176.8$$
$$H_2 = 2.6(44)+65 = 179.4$$
The person is between 176.8 cm and 179.4 cm tall, or 1.768 m and 1.794 m.

9. a. Parallel lines have the same slope, so the
slope of the line is $m = \dfrac{3-7}{2-5} = \dfrac{-4}{-3} = \dfrac{4}{3}$.
Using the point-slope form, we have

$$y-5 = \frac{4}{3}\big[x-(-2)\big] \Rightarrow 3y-15 = 4(x+2) \Rightarrow$$
$$3y-15 = 4x+8 \Rightarrow 4x-3y+23 = 0$$

b. The slopes of perpendicular lines are negative reciprocals. Write the equation $4x+5y+1=0$ in slope-intercept form to find its slope: $4x+5y+1=0 \Rightarrow$

$5y = -4x-1 \Rightarrow y = -\dfrac{4}{5}x - \dfrac{1}{5}$. The slope of

a line perpendicular to this lines is $\dfrac{5}{4}$.

Using the point-slope form, we have

$$y-(-4) = \frac{5}{4}(x-3) \Rightarrow 4(y+4) = 5(x-3) \Rightarrow$$
$$4y+16 = 5x-15 \Rightarrow 5x-4y-31 = 0$$

10. Since 2014 is 8 years after 2006, set $x = 8$.
Then $y = 0.44(8) + 6.70 = 10.22$
There were 10.22 million registered motorcycles in the U.S. in 2014.

2.3 Basic Concepts and Skills

1. The number that measure the "steepness" of a line is called the <u>slope</u>.

3. If two line have the same slope, then they are <u>parallel</u>.

5. A line perpendicular to a line with slope $-\dfrac{1}{5}$ has slope <u>5</u>.

7. False. The slope of the line through $(1, 2)$ and $(2, 4)$ is $\dfrac{4-2}{2-1} = 2$.

9. $m = \dfrac{7-3}{4-1} = \dfrac{4}{3}$; the graph is rising.

11. $m = \dfrac{-2-(-2)}{-2-6} = \dfrac{0}{-8} = 0$; the graph is horizontal.

13. $m = \dfrac{-3.5-2}{3-0.5} = \dfrac{-5.5}{2.5} = -2.2$; the graph is falling.

15. $m = \dfrac{5-1}{\left(1+\sqrt{2}\right)-\sqrt{2}} = \dfrac{4}{1} = 4$; the graph is rising.

17. a. ℓ_3 b. ℓ_2 c. ℓ_4 d. ℓ_1

19. a. $y = 0$ b. $x = 0$

21. $y = \dfrac{1}{2}x + 4$

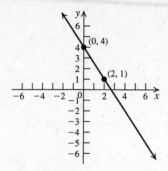

23. $y - 1 = -\dfrac{3}{2}(x-2) \Rightarrow y - 1 = -\dfrac{3}{2}x + 3 \Rightarrow$

$y = -\dfrac{3}{2}x + 4$

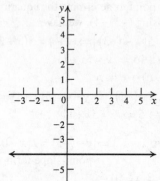

25. $y + 4 = 0(x-5) \Rightarrow y + 4 = 0 \Rightarrow y = -4$

27. $m = \dfrac{0-1}{1-0} = -1$. The y-intercept is $(0, 1)$, so the equation is $y = -x + 1$.

29. $m = \dfrac{3-3}{3-(-1)} = 0$ Because the slope $= 0$, the line is horizontal. Its equation is $y = 3$.

31. $m = \dfrac{1-(-1)}{1-(-2)} = \dfrac{2}{3}$. Now write the equation in point-slope form, and then solve for y to write the equation in slope-intercept form.

$y + 1 = \dfrac{2}{3}(x+2) \Rightarrow y + 1 = \dfrac{2}{3}x + \dfrac{4}{3} \Rightarrow$

$y = \dfrac{2}{3}x + \dfrac{1}{3}$

33. $m = \dfrac{2 - \dfrac{1}{4}}{0 - \dfrac{1}{2}} = \dfrac{\dfrac{7}{4}}{-\dfrac{1}{2}} = -\dfrac{7}{2}$. Now write the equation in point-slope form, and then solve for y to write the equation in slope-intercept form. $y - 2 = -\dfrac{7}{2}x \Rightarrow y = -\dfrac{7}{2}x + 2$

35. $x = 5$ 37. $y = 0$

39. $y = 14$ 41. $y = -\dfrac{2}{3}x - 4$

43. $m = \dfrac{4-0}{0-(-3)} = \dfrac{4}{3}; \quad y = \dfrac{4}{3}x + 4$

45. $y = 7$ **47.** $y = -5$

49. Two lines are parallel if their slopes are equal. The lines are perpendicular if the slope of one is the negative reciprocal of the slope of the other.

   a. $m_{\ell_2} = \dfrac{1-5}{3-1} = \dfrac{-4}{2} = -2 \Rightarrow \ell_2 \parallel \ell_1$

   b. $m_{\ell_2} = \dfrac{4-3}{5-7} = -\dfrac{1}{2}$. The slope of ℓ_2 is neither equal to the slope of ℓ_1 nor the negative reciprocal of the slope of ℓ_1. Therefore, the lines are neither parallel nor perpendicular.

   c. $m_{\ell_2} = \dfrac{4-3}{4-2} = \dfrac{1}{2} \Rightarrow \ell_2 \perp \ell_1$

51. $x + 2y - 4 = 0 \Rightarrow 2y = -x + 4 \Rightarrow y = -\dfrac{1}{2}x + 2$.

The slope is $-1/2$, and the y-intercept is $(0, 2)$. To find the x-intercept, set $y = 0$ and solve for x: $x + 2(0) - 4 = 0 \Rightarrow x = 4$.

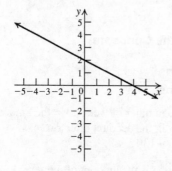

53. $3x - 2y + 6 = 0 \Rightarrow 3x + 6 = 2y \Rightarrow \dfrac{3}{2}x + 3 = y$.

The slope is $3/2$, and the y-intercept is $(0, 3)$. To find the x-intercept, set $y = 0$ and solve for x: $3x - 2(0) + 6 = 0 \Rightarrow 3x = -6 \Rightarrow x = -2$.

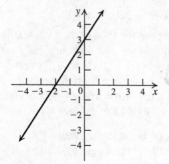

55. $x - 5 = 0 \Rightarrow x = 5$
The slope is undefined, and there is no y-intercept. The x-intercept is 5.

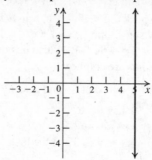

57. $x = 0$
The slope is undefined, and the y-intercepts are the y-axis. This is a vertical line whose x-intercept is 0.

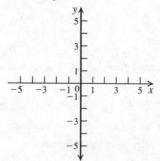

59. The slope of the line through $(a, 0)$ and $(0, b)$ is $\dfrac{b-0}{0-a} = -\dfrac{b}{a}$. The equation of the line can be written as

$$y - b = -\dfrac{b}{a}x \Rightarrow ay - ab = -bx \Rightarrow$$

$$ay + bx = ab \Rightarrow \dfrac{ay}{ab} + \dfrac{bx}{ab} = \dfrac{ab}{ab} \Rightarrow \dfrac{y}{b} + \dfrac{x}{a} = 1$$

61. $2x + 3y = 6 \Rightarrow \dfrac{2x}{6} + \dfrac{3y}{6} = \dfrac{6}{6} \Rightarrow \dfrac{x}{3} + \dfrac{y}{2} = 1$; x-intercept $= 3$; y-intercept $= 2$

63. Let the intercepts be $(a, 0)$ and $(0, a)$. Then the equation of the line is $\dfrac{x}{a} + \dfrac{y}{a} = 1$. Now substitute $x = 3$ and $y = -5$ into the equation to solve for a: $\dfrac{3}{a} - \dfrac{5}{a} = 1 \Rightarrow 3 - 5 = a \Rightarrow -2 = a$.

So the equation of the line is $-\dfrac{x}{2} - \dfrac{y}{2} = 1 \Rightarrow$

$-x - y = 2 \Rightarrow y = -x - 2$.

65. $m = \dfrac{9-4}{7-2} = \dfrac{5}{5} = 1$. The equation of the line through (2, 4) and (7, 9) is $y - 4 = 1(x - 2) \Rightarrow$ $y = x + 2$. Check to see if (–1, 1) satisfies the equation by substituting $x = -1$ and $y = 1$: $1 = -1 + 2 \Rightarrow 1 = 1$. So (–1, 1) also lies on the line. The points are collinear.

67. The line with x-intercept –4 and passing through $\left(\frac{1}{2}, 9\right)$ has slope

$m = \dfrac{9-0}{\frac{1}{2} - (-4)} = \dfrac{9}{9/2} = 2$. Then the line

perpendicular to this line has slope $-\dfrac{1}{2}$. Since

the line also passes through $\left(\frac{1}{2}, 9\right)$, its

equation is

$y - 9 = -\dfrac{1}{2}\left(x - \dfrac{1}{2}\right) \Rightarrow y = -\dfrac{1}{2}x + \dfrac{37}{4}$.

69. Both lines are vertical lines. The lines are parallel.

71. The slope of $2x + 3y = 7$ is $-2/3$, while $y = 2$ is a horizontal line. The lines are neither parallel nor perpendicular.

73. The slope of $x = 4y + 8$ is $1/4$. The slope of $y = -4x + 1$ is -4, so the lines are perpendicular.

75. The slope of $3x + 8y = 7$ is $-3/8$, while the slope of $5x - 7y = 0$ is $5/7$. The lines are neither parallel nor perpendicular.

77. a. Use the point-slope form.
$y - (-3) = 3(x - 2) \Rightarrow y + 3 = 3x - 6 \Rightarrow$
$y = 3x - 9$

b. The slope of the line we are seeking is 2. Using the point-slope form, we have
$y - 2 = 2(x - (-1)) \Rightarrow y - 2 = 2(x + 1) \Rightarrow$
$y - 2 = 2x + 2 \Rightarrow y = 2x + 4$

79. The slope of $x + y = 1$ is –1. The lines are parallel, so they have the same slope. The equation of the line through (1, 1) with slope –1 is $y - 1 = -(x - 1) \Rightarrow y - 1 = -x + 1 \Rightarrow$ $y = -x + 2$.

81. The slope of $3x - 9y = 18$ is $1/3$. The lines are perpendicular, so the slope of the new line is -3. The equation of the line through (–2, 4) with slope -3 is $y - 4 = -3(x - (-2)) \Rightarrow$ $y - 4 = -3x - 6 \Rightarrow y = -3x - 2$.

83. The slope of the line $y = 6x + 5$ is 6. The lines are perpendicular, so the slope of the new line is $-1/6$. The equation of the line with slope $-1/6$ and y-intercept 4 is

$y = -\dfrac{1}{6}x + 4$.

85. The slope of AB is $\dfrac{3-5}{2-(-1)} = -\dfrac{2}{3}$, so the

slope of its perpendicular bisector is $\dfrac{3}{2}$. The

midpoint of AB is

$\left(\dfrac{2 + (-1)}{2}, \dfrac{3 + 5}{2}\right) = \left(\dfrac{1}{2}, 4\right)$.

Using the point-slope form, the equation of the perpendicular bisector is

$y - 4 = \dfrac{3}{2}\left(x - \dfrac{1}{2}\right) \Rightarrow y - 4 = \dfrac{3}{2}x - \dfrac{3}{4} \Rightarrow$

$y = \dfrac{3}{2}x + \dfrac{13}{4}$

2.3 Applying the Concepts

87. slope $= \dfrac{\text{rise}}{\text{run}} \Rightarrow \dfrac{4}{40} = \dfrac{1}{10}$

89. a. $x =$ the number of weeks; $y =$ the amount of money in the account after x weeks; $y = 7x + 130$

b. The slope is the amount of money deposited each week; the y-intercept is the initial deposit.

91. a. $x =$ the number of hours worked per week; $y =$ the amount earned per week;
$y = \begin{cases} 11x & x \le 40 \\ 16.5x - 220 & x > 40 \end{cases}$

To compute the salary when $x > 40$, use the following steps: For 40 hours, Judy earns $40(11) = \$440$. The number of overtime hours is $x - 40$. For those hours, she earns $(1.5)(11)(x - 40) = 16.5x - 660$. So her total wage is $440 + 16.50x - 660 = 16.5x - 220$.

b. The slope is the hourly wage; the y-intercept is the wage for 0 hours of work.

93. a. x = the number of rupees; y = the number of dollars equal to x rupees.

$$y = \frac{1}{53.87}x \approx .0186x.$$

b. The slope is the number of dollars per rupee. When $x = 0$, $y = 0$.

95. a. The y-intercept represents the initial value of the machine, $9000.

b. The point $(10, 1)$ gives the value of the machine after 10 years as $1000.

c. The value of the machine decreased from $9000 to $1000 over 10 years. This is a decrease of $\dfrac{9000 - 1000}{10} = \dfrac{8000}{10} = \$800$ per year.

d. Using the points $(0, 9)$ and $(10, 1)$, the slope is $\dfrac{9-1}{0-10} = -0.8$.

$$y - 9 = -0.8(x - 0) \Rightarrow y = -0.8x + 9$$

e. The slope gives the machine's yearly depreciation, $-0.8(1000) = -800$.

97. $y = 5x + 40,000$

99. a. The year 2005 is represented by $t = 0$, and the year 2011 is represented by $t = 6$. The points are $(0, 2425)$ and $(6, 4026)$. So the slope is $\dfrac{4026 - 2425}{6} \approx 266.8$

The equation is $y - 2425 = 266.8(t - 0) \Rightarrow$
$y = 266.8t + 2425$

b.

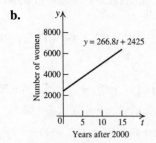

c. The year 2008 is represented by $t = 3$. So $y = 266.8(3) + 2425 \Rightarrow y = 3225.4$.

Note that there cannot be a fraction of a person, so. there were 3225 women prisoners in 2008.

d. The year 2017 is represented by $t = 12$. So $y = 266.8(12) + 2425 \Rightarrow y = 5626.6$.

There will be 5627 women prisoners in 2017.

101. a. The two points are $(4, 210.20)$ and $(10, 348.80)$. So the slope is

$$\frac{348.80 - 210.20}{10 - 4} = \frac{138.6}{6} = 23.1.$$

The equation is $y - 348.8 = 23.1(x - 10) \Rightarrow$
$y = 23.1x + 117.8$

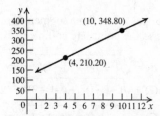

b. The slope represents the cost of producing one modem. The y-intercept represents the fixed cost.

c. $y = 23.1(12) + 117.8 \Rightarrow y = \$395$

103. The independent variable t represents the number of years after 2005, with $t = 0$ representing 2005. The two points are $(0, 12.7)$ and $(3, 11.68)$. So the slope is $\dfrac{12.7 - 11.68}{-3} = -0.34$. The equation is

$p - 12.7 = -0.34(t - 0) \Rightarrow p = -0.34t + 12.7.$

The year 2013 is represented by $t = 8$.
$p = -0.34(8) + 12.7 \Rightarrow p = 9.98\%.$

105. a.

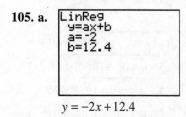

$y = -2x + 12.4$

b.

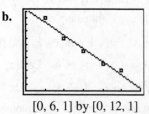

$[0, 6, 1]$ by $[0, 12, 1]$

c. The price in the table is given as the number of nickels. $35¢ = 7$ nickels, so let $x = 7$. $y = -2(7) + 12.4 = -1.6$

Thus, no newspapers will be sold if the price per copy is $35¢$. Note that this is also clear from the graph, which appears to cross the x-axis at approximately $x = 6$.

2.3 Beyond the Basics

107. a. Let $A = (0, 1)$, $B = (1, 3)$, $C = (-1, -1)$.

$$m_{AB} = \frac{3-1}{1-0} = 2; m_{BC} = \frac{-1-3}{-1-1} = \frac{-4}{-2} = 2$$

$$m_{AC} = \frac{-1-1}{-1-0} = 2$$

The slopes of the three segments are the same, so the points are collinear.

b. $d(A, B) = \sqrt{(1-0)^2 + (3-1)^2} = \sqrt{5}$

$d(B, C) = \sqrt{(-1-1)^2 + (-1-3)^2} = 2\sqrt{5}$

$d(A, C) = \sqrt{(-1-0)^2 + (-1-1)^2} = \sqrt{5}$

Because $d(B, C) = d(A, B) + d(A, C)$, the three points are collinear.

109. Since the points are collinear, the slope is the same no matter which two points are used to determine the slope. So we have

$$\frac{c-1}{1-(-5)} = \frac{-2-1}{4-(-5)} \Rightarrow \frac{c-1}{6} = \frac{-3}{9} \Rightarrow$$

$$9(c-1) = -18 \Rightarrow 9c - 9 = -18 \Rightarrow$$

$$9c = -9 \Rightarrow c = -1$$

111. a. $m_{AB} = \frac{4-1}{-1-1} = -\frac{3}{2}; m_{BC} = \frac{8-4}{5-(-1)} = \frac{2}{3}$.

The product of the slopes $= -1$, so $AB \perp BC$, and the triangle is a right triangle.

b. $d(A, B) = \sqrt{(-1-1)^2 + (4-1)^2} = \sqrt{13}$

$d(B, C) = \sqrt{(5-(-1))^2 + (8-4)^2} = \sqrt{52}$

$d(A, C) = \sqrt{(5-1)^2 + (8-1)^2} = \sqrt{65}$

$(d(A, B))^2 + (d(B, C))^2 = (d(A, C))^2$, so the triangle is a right triangle.

113. The equation of ℓ_1 is $y = m_1 x + b_1$ and the equation of ℓ_2 is $y = m_2 x + b_2$. Let (x_1, y_1) and (x_2, y_2) be on ℓ_1. If $\ell_1 \parallel \ell_2$, then the distance between them is $b_1 - b_2$. In other words, $(x_1, y_1 - (b_1 - b_2))$ and $(x_2, y_2 - (b_1 - b_2))$ are on ℓ_2. So,

$y_1 - (b_1 - b_2) = m_2 x_1 + b_2 \Rightarrow y_1 - b_1 = m_2 x_1 \Rightarrow$

$y_1 = m_2 x_1 + b_1$. However, (x_1, y_1) lies on ℓ_1.

So $y_1 = m_2 x_1 + b_1 = m_1 x_1 + b_1 \Rightarrow m_2 = m_1$.

115. Let (x, y) be the coordinates of point B. Then

$$d(A, B) = 12.5 = \sqrt{(x-2)^2 + (y-2)^2} \Rightarrow$$

$$(x-2)^2 + (y-2)^2 = 156.25 \text{ and}$$

$$m_{AB} = \frac{4}{3} = \frac{y-2}{x-2} \Rightarrow 4(x-2) = 3(y-2) \Rightarrow$$

$y = \frac{4}{3}x - \frac{2}{3}$. Substitute this into the first equation and solve for x:

$$(x-2)^2 + \left(\left(\frac{4}{3}x - \frac{2}{3}\right) - 2\right)^2 = 156.25$$

$$(x-2)^2 + \left(\frac{4}{3}x - \frac{8}{3}\right)^2 = 156.25$$

$$x^2 - 4x + 4 + \frac{16}{9}x^2 - \frac{64}{9}x + \frac{64}{9} = 156.25$$

$$9x^2 - 36x + 36 + 16x^2 - 64x + 64 = 1406.25$$

$$25x^2 - 100x - 1306.25 = 0$$

Solve this equation using the quadratic formula:

$$x = \frac{100 \pm \sqrt{100^2 - 4(25)(-1306.25)}}{2(25)}$$

$$= \frac{100 \pm \sqrt{10,000 + 130,625}}{50}$$

$$= \frac{100 \pm \sqrt{140,625}}{50} = \frac{100 \pm 375}{50} = 9.5 \text{ or } -5.5$$

Now find y by substituting the x-values into

the slope formula: $\frac{4}{3} = \frac{y-2}{9.5-2} \Rightarrow y = 12$ or

$\frac{4}{3} = \frac{y-2}{-5.5-2} \Rightarrow y = -8$. So the coordinates of B are $(9.5, 12)$ or $(-5.5, -8)$.

117. Write the equations of each circle in standard form to find the centers.

$$\left(x^2 + 6x + 9\right) + \left(y^2 - 14y + 49\right) = 1 + 9 + 49 \Rightarrow$$

$$(x+3)^2 + (y-7)^2 = 59$$

$$\left(x^2 - 4x + 4\right) + \left(y^2 + 10y + 25\right) = 2 + 4 + 25$$

$$(x-2)^2 + (y+5)^2 = 31$$

The centers are $(-3, 7)$ and $(2, -5)$.

Using the result from exercise 116, we have

$$(x-(-3))(x-2) + (y-7)(y-(-5)) = 0 \Rightarrow$$

$$(x+3)(x-2) + (y-7)(y+5) = 0 \Rightarrow$$

$$x^2 + x - 6 + y^2 - 2y - 35 = 0 \Rightarrow$$

$$x^2 + y^2 + x - 2y - 41 = 0$$

119. The tangent line at a point is perpendicular to the radius drawn to that point. The center of $x^2 + y^2 = 25$ is $(0, 0)$, so the slope of the radius is $-\dfrac{3}{4}$ and the slope of the tangent is $\dfrac{4}{3}$. Using the point-slope form, the equation of the tangent is

$$y - (-3) = \frac{4}{3}(x - 4) \Rightarrow y + 3 = \frac{4}{3}x - \frac{16}{3} \Rightarrow$$
$$y = \frac{4}{3}x - \frac{25}{3}.$$

121.

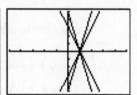

The family of lines has slope 2. The lines have different y-intercepts.

123.

The lines pass through $(1, 0)$. The lines have different slopes.

2.3 Critical Thinking/Discussion/Writing

125. a.

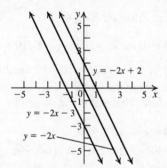

This is a family of lines parallel to the line $y = -2x$. They all have slope -2.

b.

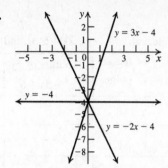

This is a family of lines that passes through the point $(0, -4)$.

126. $\left.\begin{array}{l} y = m_1 x + b_1 \\ y = m_2 x + b_2 \end{array}\right\} \Rightarrow m_1 x + b_1 = m_2 x + b_2 \Rightarrow$

$m_1 x - m_2 x = b_2 - b_1 \Rightarrow x(m_1 - m_2) = b_2 - b_1 \Rightarrow$

$x = \dfrac{b_2 - b_1}{m_1 - m_2}$

a. If $m_1 > m_2 > 0$ and $b_1 > b_2$, then

$$x = \frac{b_2 - b_1}{m_1 - m_2} = -\frac{b_1 - b_2}{m_1 - m_2}.$$

b. If $m_1 > m_2 > 0$ and $b_1 < b_2$, then

$$x = \frac{b_2 - b_1}{m_1 - m_2}.$$

c. If $m_1 < m_2 < 0$ and $b_1 > b_2$, then

$$x = \frac{b_2 - b_1}{m_1 - m_2} = \frac{b_1 - b_2}{m_2 - m_1}.$$

d. If $m_1 < m_2 < 0$ and $b_1 < b_2$, then

$$x = \frac{b_2 - b_1}{m_1 - m_2} = -\frac{b_2 - b_1}{m_2 - m_1}.$$

2.3 Maintaining Skills

127. $x^2 - x - 2 = 0 \Rightarrow (x - 2)(x + 1) = 0 \Rightarrow$
$x = 2, -1$
Solution set: $\{-1, 2\}$

129. $\dfrac{x^2 + 5x + 6}{x - 1} = 0 \Rightarrow x^2 + 5x + 6 = 0 \Rightarrow$
$(x + 2)(x + 3) = 0 \Rightarrow x = -2, -3$
Solution set: $\{-3, -2\}$

131. $x^2 - 5x + 6 \ge 0 \Rightarrow (x - 3)(x - 2) \ge 0$
Now solve the associated equation:
$(x - 3)(x - 2) = 0 \Rightarrow x = 3$ or $x = 2$.

(continued on next page)

(*continued*)

So, the intervals are $(-\infty, 2], [2, 3],$ and $[3, \infty)$.

Interval	Test point	Value of $x^2 - 5x + 6$	Result
$(-\infty, 3]$	0	6	$+$
$[2, 3]$	$\frac{5}{2}$	$-\frac{1}{4}$	$-$
$[3, \infty)$	5	6	$+$

The solution set is $(-\infty, 2] \cup [3, \infty)$.

133. $\dfrac{1}{h}\left[\dfrac{1}{x+h} - \dfrac{1}{x}\right] = \dfrac{1}{h}\left[\dfrac{x}{x(x+h)} - \dfrac{x+h}{x(x+h)}\right]$

$= \dfrac{1}{h}\left[\dfrac{x-(x+h)}{x(x+h)}\right]$

$= \dfrac{1}{h}\left[\dfrac{-h}{x(x+h)}\right] = -\dfrac{1}{x(x+h)}$

135. $\dfrac{\sqrt{5}-\sqrt{2}}{3} \cdot \dfrac{\sqrt{5}+\sqrt{2}}{\sqrt{5}+\sqrt{2}} = \dfrac{5-2}{3\left(\sqrt{5}+\sqrt{2}\right)}$

$= \dfrac{3}{3\left(\sqrt{5}+\sqrt{2}\right)} = \dfrac{1}{\sqrt{5}+\sqrt{2}}$

2.4 Functions

2.4 Practice Problems

1.a. The domain of R is $\{2, -2, 3\}$ and its range is $\{1, 2\}$. The relation R is a function because no two ordered pairs in R have the same first component.

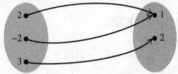

b. The domain of S is $\{2, 3\}$ and its range is $\{5, -2\}$. The relation S is not a function because the ordered paired $(3, -2)$ and $(3, 5)$ have the same first component.

2. Solve each equation for y.

a. $2x^2 - y^2 = 1 \Rightarrow 2x^2 - 1 = y^2 \Rightarrow$

$\pm\sqrt{2x^2 - 1} = y$; not a function

b. $x - 2y = 5 \Rightarrow x - 5 = 2y \Rightarrow \dfrac{1}{2}(x-5) = y$; a function

3.a. $g(0) = -2(0)^2 + 5(0) = 0$

b. $g(-1) = -2(-1)^2 + 5(-1) = -7$

c. $g(x+h) = -2(x+h)^2 + 5(x+h)$

$= -2\left(x^2 + 2xh + h^2\right) + 5x + 5h$

$= -2x^2 - 4hx + 5x - 2h^2 + 5h$

4. $A_{TLMS} = (\text{length})(\text{height}) = \left(|3-1|\right)(22)$

$= (2)(22) = 44$ sq. units

5.a. $f(x) = \dfrac{1}{\sqrt{1-x}}$ is not defined when

$1 - x = 0 \Rightarrow x = 1$ or when $1 - x < 0 \Rightarrow 1 < x$.. Thus, the domain of f is $(-\infty, 1)$.

b. $f(x) = \sqrt{\dfrac{x+2}{x-3}}$ is not defined when the

denominator equals 0 or when $\dfrac{x+2}{x-3} < 0$.

$\dfrac{x+2}{x-3} < 0$ for $(-2, 3)$, so this interval is not in the domain of f. The denominator $x - 3 \leq 0 \Rightarrow x \leq 3$, so all numbers less than or equal to 3 are not in the domain of f. Thus, the domain of f is $(-\infty, -2] \cup (3, \infty)$.

6. $f(x) = x^2$, domain $X = [-3, 3]$

a. $f(x) = 10 \Rightarrow x^2 = 10 \Rightarrow x = \pm\sqrt{10} \approx \pm 3.16$

Since $\sqrt{10} > 3$ and $-\sqrt{10} < -3$, neither solution is in the interval $X = [-3, 3]$. Therefore, 10 is not in the range of f.

b. $f(x) = 4 \Rightarrow x^2 = 4 \Rightarrow x = \pm 2$

Since $-3 < -2 < 2 < 3$, 4 is in the range of f.

c. The range of f is the interval $[0, 9]$ because for each number y in this interval, the number $x = \sqrt{y}$ is in the interval $[-3, 3]$.

7.

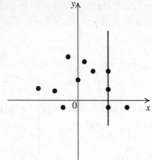

The graph is not a function because a vertical line can be drawn through three points, as shown.

8. $y = f(x) = x^2 + 4x - 5$

a. Check whether the ordered pair (2, 7) satisfies the equation:

$7 \overset{?}{=} 2^2 + 4(2) - 5$
$7 = 7 \checkmark$

The point (2, 7) is on the graph.

b. Let $y = -8$, then solve for x:

$-8 = x^2 + 4x - 5 \Rightarrow 0 = x^2 + 4x + 3 \Rightarrow$
$0 = (x+3)(x+1) \Rightarrow x = -3$ or $x = -1$

The points (−3, −8) and (−1, −8) lie on the graph.

c. Let $x = 0$, then solve for y:

$y = 0^2 + 4(0) - 5 = -5$

The y-intercept is −5.

d. Let $y = 0$, then solve for x:

$0 = x^2 + 4x - 5 \Rightarrow 0 = (x+5)(x-1) \Rightarrow$
$x = -5$ or $x = 1$

The x-intercepts are −5 and 1.

9.

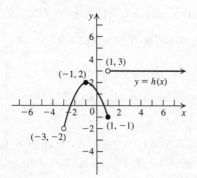

Domain: $(-3, \infty)$; range: $(-2, 2] \cup \{3\}$

10. The range of $C(t)$ is [6, 12).

$C(11) = \frac{1}{2}C(10) + 6 = \frac{1}{2}(11.988) + 6 = 11.994$.

11. From Example 11, we have $AP = \sqrt{500^2 + x^2}$ and $\overline{PD} = 1200 - x$ feet. If c = the cost on land, the total cost C is given by

$C = 1.3c(PD) + c(AP)$
$= 1.3c\sqrt{500^2 + x^2} + c(1200 - x)$

12.a. $C(x) = 1200x + 100,000$

b. $R(x) = 2500x$

c. $P(x) = R(x) - C(x)$
$= 2500x - (1200x + 100,000)$
$= 1300x - 100,000$

d. The break-even point occurs when $C(x) = R(x)$.

$1200x + 100,000 = 2500x$
$100,000 = 1300x \Rightarrow x \approx 77$

Metro needs 77 shows to break even.

2.4 Basic Concepts and Skills

1. In the functional notation $y = f(x)$, x is the <u>independent</u> variable.

3. If the point (9, −14) is on the graph of a function f, then $\underline{f(9) = -14}$.

5. To find the x-intercepts of the graph of an equation in x and y, we solve the equation $\underline{y = 0}$.

7. True. $-x = 7$ and the square root function is defined for all positive numbers.

9. Domain: $\{a, b, c\}$; range: $\{d, e\}$; function

11. Domain: $\{a, b, c\}$; range: $\{1, 2\}$; function

13. Domain: $\{0, 3, 8\}$; range: $\{-3, -2, -1, 1, 2\}$; not a function

15. $x + y = 2 \Rightarrow y = -x + 2$; a function

17. $y = \dfrac{1}{x}$; a function

19. $y^2 = x^2 \Rightarrow y = \pm\sqrt{x^2} \Rightarrow y = \pm x$; not a function

21. $y = \dfrac{1}{\sqrt{2x-5}}$; a function

23. $2 - y = 3x \Rightarrow y = 2 - 3x$; a function

25. $x + y^2 = 8 \Rightarrow y = \pm\sqrt{8-x}$; not a function

27. $x^2 + y^3 = 5 \Rightarrow y = \sqrt[3]{5 - x^2}$; a function

In exercises 29–32, $f(x) = x^2 - 3x + 1$, $g(x) = \dfrac{2}{\sqrt{x}}$,

and $h(x) = \sqrt{2 - x}$.

29. $f(0) = 0^2 - 3(0) + 1 = 1$

$g(0) = \dfrac{2}{\sqrt{0}} \Rightarrow g(0)$ is undefined

$; h(0) = \sqrt{2 - 0} = \sqrt{2}$

$f(a) = a^2 - 3a + 1$

$f(-x) = (-x)^2 - 3(-x) + 1 = x^2 + 3x + 1$

31. $f(-1) = (-1)^2 - 3(-1) + 1 = 5;$

$g(-1) = \dfrac{2}{\sqrt{-1}} \Rightarrow g(-1)$ is undefined;

$h(-1) = \sqrt{2 - (-1)} = \sqrt{3}; h(c) = \sqrt{2 - c};$

$h(-x) = \sqrt{2 - (-x)} = \sqrt{2 + x}$

33.a. $f(0) = \dfrac{2(0)}{\sqrt{4 - 0^2}} = 0$

b. $f(1) = \dfrac{2(1)}{\sqrt{4 - 1^2}} = \dfrac{2}{\sqrt{3}} = \dfrac{2\sqrt{3}}{3}$

c. $f(2) = \dfrac{2(2)}{\sqrt{4 - 2^2}} = \dfrac{4}{0} \Rightarrow f(2)$ is undefined

d. $f(-2) = \dfrac{2(-2)}{\sqrt{4 - (-2)^2}} = \dfrac{-4}{0} \Rightarrow f(-2)$ is

undefined

e. $f(-x) = \dfrac{2(-x)}{\sqrt{4 - (-x)^2}} = \dfrac{-2x}{\sqrt{4 - x^2}}$

35. The width of each rectangle is 1. The height of the left rectangle is $f(1) = 1^2 + 2 = 3$. The height of the right rectangle is $f(2) = 2^2 + 2 = 6.$

$A = (1)(f(1)) + (1)(f(2))$
$= 1(3) + (1)(6) = 9$ sq. units

37. $(-\infty, \infty)$

39. The denominator is not defined for $x = 9$. The domain is $(-\infty, 9) \cup (9, \infty)$

41. The denominator is not defined for $x = -1$ or $x = 1$. The domain is $(-\infty, -1) \cup (-1, 1) \cup (1, \infty)$.

43. The numerator is not defined for $x < 3$, and the denominator is not defined for $x = -2$. The domain is $[3, \infty)$

45. The denominator equals 0 if $x = -1$ or $x = -2$. The domain is $(-\infty, -2) \cup (-2, -1) \cup (-1, \infty)$.

47. The denominator is not defined for $x = 0$. The domain is $(-\infty, 0) \cup (0, \infty)$

49. a function

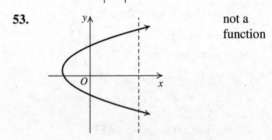

51. a function

53. not a function

55. $f(-4) = -2; f(-1) = 1; f(3) = 5; f(5) = 7$

57. $h(-2) = -5; h(-1) = 4; h(0) = 3; h(1) = 4$

59. $h(x) = 7$, so solve the equation $7 = x^2 - x + 1$.

$x^2 - x - 6 = 0 \Rightarrow (x - 3)(x + 2) = 0 \Rightarrow x = -2$ or $x = 3.$

61.a. $1 = -2(1 + 1)^2 + 7 \Rightarrow 1 = -1$, which is false. Therefore, (1, 1) does not lie on the graph of f.

b. $1 = -2(x + 1)^2 + 7 \Rightarrow 2(x + 1)^2 = 6 \Rightarrow$
$(x + 1)^2 = 3 \Rightarrow x + 1 = \pm\sqrt{3} \Rightarrow x = -1 \pm \sqrt{3}$
The points $\left(-1 - \sqrt{3}, 1\right)$ and $\left(-1 + \sqrt{3}, 1\right)$ lie on the graph of f.

c. $y = -2(0+1)^2 + 7 \Rightarrow y = 5$

The y-intercept is $(0, 5)$.

d. $0 = -2(x+1)^2 + 7 \Rightarrow -7 = -2(x+1)^2 \Rightarrow$

$\dfrac{7}{2} = (x+1)^2 \Rightarrow \pm\sqrt{\dfrac{7}{2}} = \pm\dfrac{\sqrt{14}}{2} = x+1 \Rightarrow$

$x = -1 \pm \dfrac{\sqrt{14}}{2}$

The x-intercepts are $\left(-1 - \dfrac{\sqrt{14}}{2}, 0\right)$ and

$\left(-1 + \dfrac{\sqrt{14}}{2}, 0\right)$.

63. Domain: $[-3, 2]$; range: $[-3, 3]$

65. Domain: $[-4, \infty)$; range: $[-2, 3]$

67. Domain: $[-3, \infty)$; range: $[-1, 4] \cup \{-3\}$

69. Domain: $(-\infty, 4] \cup [-2, 2] \cup [4, \infty)$

Range: $[-2, 2] \cup \{3\}$

71. $[-9, \infty)$ 　　　　**73.** 　$-3, 4, 7, 9$

75. $f(-7) = 4$, $f(1) = 5$, $f(5) = 2$

77. $\{-3.75, -2.25, 3\} \cup [12, \infty)$

79. $[-9, \infty)$

81. $g(-4) = -1$, $g(1) = 3$, $g(3) = 4$

83. $[-9, -5)$

2.4 Applying the Concepts

85. A function because there is only one high temperature per day.

87. Not a function because there are several states that begin with N (i.e., New York, New Jersey, New Mexico, Nevada, North Carolina, North Dakota); there are also several states that begin with T and S.

89. $A(x) = x^2$; $A(4) = 16$; $A(4)$ represents the area of a tile with side 4.

91. It is a function. $S(x) = 6x^2$; $S(3) = 54$

93.a. The domain is $[0, 8]$.

b. $h(2) = 128(2) - 16(2^2) = 192$

$h(4) = 128(4) - 16(4^2) = 256$

$h(6) = 128(6) - 16(6^2) = 192$

c. $0 = 128t - 16t^2 \Rightarrow 0 = 16t(8 - t) \Rightarrow$

$t = 0$ or $t = 8$. It will take 8 seconds for the stone to hit the ground.

d.

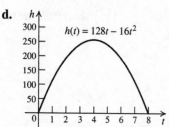

95. $x + y = 28 \Rightarrow y = 28 - x$

$P = x(28 - x) = 28x - x^2$

97. Note that the length of the base = the width of the base = x.

$V = lwh = x^2h = 64 \Rightarrow h = \dfrac{64}{x^2}$

$S = 2lw + 2lh + 2wh$

$= 2x^2 + 2x\left(\dfrac{64}{x^2}\right) + 2x\left(\dfrac{64}{x^2}\right)$

$= 2x^2 + \dfrac{128}{x} + \dfrac{128}{x} = 2x^2 + \dfrac{256}{x}$

99. The piece with length x is formed into a circle, so $C = x = 2\pi r \Rightarrow r = \dfrac{x}{2\pi}$. Thus, the area of

the circle is $A = \pi r^2 = \pi\left(\dfrac{x}{2\pi}\right)^2 = \dfrac{x^2}{4\pi}$.

The piece with length $20 - x$ is formed into a

square, so $P = 20 - x = 4s \Rightarrow s = \dfrac{1}{4}(20 - x)$.

Thus, the area of the square is

$s^2 = \left[\dfrac{1}{4}(20 - x)\right]^2 = \dfrac{1}{16}(20 - x)^2$.

The sum of the areas is $A = \dfrac{x^2}{4\pi} + \dfrac{1}{16}(20 - x)^2$

101. The volume of the pool is

$V = 288 = x^2h \Rightarrow h = \dfrac{288}{x^2}$.

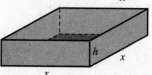

(continued on next page)

(continued)

The total area to be tiled is

$$4xh = 4x\left(\frac{288}{x^2}\right) = \frac{1152}{x}$$

The cost of the tile is $6\left(\frac{1152}{x}\right) = \frac{6912}{x}$.

The area of the bottom of the pool is x^2, so the cost of the cement is $2x^2$. Therefore, the total

cost is $C = 2x^2 + \frac{6912}{x}$.

103.

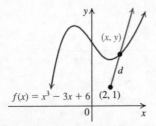

Using the distance formula we have

$$d = \sqrt{(x-2)^2 + (y-1)^2}$$
$$= \sqrt{(x-2)^2 + \left[\left(x^3 - 3x + 6\right) - 1\right]^2}$$
$$= \sqrt{(x-2)^2 + \left(x^3 - 3x + 5\right)^2}$$
$$= \left[(x-2)^2 + \left(x^3 - 3x + 5\right)^2\right]^{1/2}$$

105.a. $p(5) = 1275 - 25(5) = 1150$. If 5000 TVs can be sold, the price per TV is \$1150.
$p(15) = 1275 - 25(15) = 900$. If 15,000 TVs can be sold, the price per TV is \$900.
$p(30) = 1275 - 25(30) = 525$. If 30,000 TVs can be sold, the price per TV is \$525.

b.

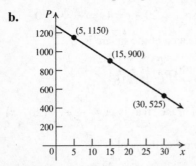

c. $650 = 1275 - 25x \Rightarrow -625 = -25x \Rightarrow x = 25$
25,000 TVs can be sold at \$650 per TV.

107.a. $C(x) = 5.5x + 75,000$

b. $R(x) = 0.6(15)x = 9x$

c. $P(x) = R(x) - C(x) = 9x - (5.5x + 75,000)$
$= 3.5x - 75,000$

d. The break-even point is when the profit is zero: $3.5x - 75,000 = 0 \Rightarrow x = 21,429$

e. $P(46,000) = 3.5(46,000) - 75,000$
$= \$86,000$
The company's profit is \$86,000 when 46,000 copies are sold.

2.4 Beyond the Basics

109. $x = \frac{2}{y-4} \Rightarrow xy - 4x = 2 \Rightarrow xy = 2 + 4x \Rightarrow$

$y = \frac{4x+2}{x} \Rightarrow f(x) = \frac{4x+2}{x}$;

Domain: $(-\infty, 0) \cup (0, \infty)$. $f(4) = \frac{9}{2}$.

111. $\left(x^2 + 1\right)y + x = 2 \Rightarrow y = \frac{2-x}{x^2+1} \Rightarrow$

$f(x) = \frac{2-x}{x^2+1}$; Domain: $(-\infty, \infty)$; $f(4) = -\frac{2}{17}$

113. $f(x) \neq g(x)$ because they have different domains.

115. $f(x) \neq g(x)$ because they have different domains. $g(x)$ is not defined for $x = -1$, while $f(x)$ is defined for all real numbers.

117. $f(x) = g(x)$ because $f(3) = 10 = g(3)$ and $f(5) = 26 = g(5)$.

119. $f(2) = 15 = a(2^2) + 2a - 3 \Rightarrow 15 = 6a - 3 \Rightarrow a = 3$.

121. $h(6) = 0 = \frac{3(6) + 2a}{2(6) - b} \Rightarrow 0 = 18 + 2a \Rightarrow a = -9$

$h(3)$ is undefined $\Rightarrow \frac{3(3) + 2(-9)}{2(3) - b}$ has a zero in

the denominator. So $6 - b = 0 \Rightarrow b = 6$.

123. $g(x) = x^2 - \frac{1}{x^2} \Rightarrow g\left(\frac{1}{x}\right) = \frac{1}{x^2} - \frac{1}{\frac{1}{x^2}} = \frac{1}{x^2} - x^2$

$g(x) + g\left(\frac{1}{x}\right) = \left(x^2 - \frac{1}{x^2}\right) + \left(\frac{1}{x^2} - x^2\right) = 0$

125. $f(x) = \dfrac{x+3}{4x-5} \Rightarrow$

$$f(t) = \frac{\dfrac{3+5x}{4x-1}+3}{4\left(\dfrac{3+5x}{4x-1}\right)-5} = \frac{\dfrac{(3+5x)+3(4x-1)}{4x-1}}{\dfrac{(12+20x)-(5(4x-1))}{4x-1}}$$

$$= \frac{(3+5x)+(12x-3)}{(12+20x)-(20x-5)} = \frac{17x}{17} = x$$

2.4 Critical Thinking/Discussion/Writing

126. Answers may vary. Sample answers are given

   a. $y = \sqrt{x-2}$ **b.** $y = \dfrac{1}{\sqrt{x-2}}$

   c. $y = \sqrt{2-x}$ **d.** $y = \dfrac{1}{\sqrt{2-x}}$

127.a. $ax^2 + bx + c = 0$

   b. $y = c$

   c. The equation will have no x-intercepts if $b^2 - 4ac < 0$.

   d. It is not possible for the equation to have no y-intercepts because $y = f(x)$.

128.a. $f(x) = |x|$ **b.** $f(x) = 0$

   c. $f(x) = x$

   d. $f(x) = \sqrt{-x^2}$ (Note: the point is the origin.)

   e. $f(x) = 1$

   f. A vertical line is not a function.

129.a. $\{(a, 1), (b, 1)\}$ $\{(a, 2), (b, 1)\}$
         $\{(a, 1), (b, 2)\}$ $\{(a, 2), (b, 2)\}$
         $\{(a, 1), (b, 3)\}$ $\{(a, 2), (b, 3)\}$

         $\{(a, 3), (b, 1)\}$
         $\{(a, 3), (b, 2)\}$
         $\{(a, 3), (b, 3)\}$
    There are nine functions from X to Y.

   b. $\{(1, a)\}, \{(2, a)\}, \{(3, a)\}$
      $\{(1, a)\}, \{(2, a)\}, \{(3, b)\}$
      $\{(1, a)\}, \{(2, b)\}, \{(3, a)\}$
      $\{(1, b)\}, \{(2, a)\}, \{(3, a)\}$
      $\{(1, b)\}, \{(2, a)\}, \{(3, a)\}$
      $\{(1, b)\}, \{(2, b)\}, \{(3, a)\}$
      $\{(1, b)\}, \{(2, a)\}, \{(3, b)\}$
      $\{(1, b)\}, \{(2, b)\}, \{(3, b)\}$
    There are eight functions from Y to X.

130. If a set X has m elements and a set of Y has n elements, there are n^m functions that can be defined from X to Y. This is true since a function assigns each element of X to an element of Y. There are m possibilities for each element of X, so there are

$$\underbrace{n \cdot n \cdot n \cdots n}_{m} = n^m \text{ possible functions.}$$

2.4 Maintaining Skills

131. $m = \dfrac{-2-0}{2-0} = -1$

$$y - 0 = -1(x-0) \Rightarrow y = -x$$

133. $m = \dfrac{4-2}{1-(-3)} = \dfrac{2}{4} = \dfrac{1}{2}$

$$y - 2 = \frac{1}{2}\left(x-(-3)\right) \Rightarrow y - 2 = \frac{1}{2}(x+3) \Rightarrow$$

$$y - 2 = \frac{1}{2}x + \frac{3}{2} \Rightarrow y = \frac{1}{2}x + \frac{7}{2}$$

135. $f(x) = 2x^2 - 3x$

   a. $f(-x) = 2(-x)^2 - 3(-x) = 2x^2 + 3x$

   b. $-f(x) = -\left(2x^2 - 3x\right) = -2x^2 + 3x$

137. $f(x) = x^3 - 2x$

   a. $f(-x) = (-x)^3 - 2(-x) = -x^3 + 2x$

   b. $-f(x) = -\left(x^3 - 2x\right) = -x^3 + 2x$

2.5   Properties of Functions

2.5 Practice Problems

1. f is increasing on $(2, \infty)$, decreasing on $(-\infty, -3)$, and constant on $(-3, 2)$.

2.

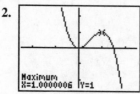

Relative minimum: $(0, 0)$
Relative maximum: $(1, 1)$

3. $v = (11 - r)r^2$

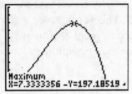

[0, 13, 1] by [0, 250, 25]

Mrs. Osborn's windpipe should be contracted to a radius of 7.33 mm for maximizing the airflow velocity.

4. $f(x) = -x^2$

Replace x with $-x$:

$f(-x) = -(-x)^2 = -x^2 = f(x)$

Thus, the function is even.

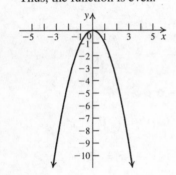

5. $f(x) = -x^3$

Replace x with $-x$:

$f(-x) = -(-x)^3 = x^3 = -f(x)$

Thus, the function is odd.

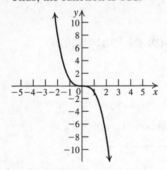

6.a. $g(-x) = 3(-x)^4 - 5(-x)^2$
$\qquad = 3x^4 - 5x^2 = f(x) \Rightarrow$
$g(x)$ is even.

b. $f(-x) = 4(-x)^5 + 2(-x)^3 = -4x^5 - 2x^3$
$\qquad = -\left(4x^5 + 2x^3\right) = -f(x) \Rightarrow$
$f(x)$ is odd.

c. $h(-x) = 2(-x) + 1 = -2x + 1$
$\qquad \neq h(x)$
$\qquad \neq h(-x) \Rightarrow h$ is neither even nor odd.

7. $f(x) = 1 - x^2; \ a = 2, b = 4$
$f(2) = 1 - 2^2 = -3; \ f(4) = 1 - 4^2 = -15$
$\dfrac{f(b) - f(a)}{b - a} = \dfrac{-15 - (-3)}{4 - 2} = \dfrac{-12}{2} = -6$
The average rate of change is -6.

8. $f(t) = 1 - t; \ a = 2, \ b = x, \ x \neq 2$
$f(a) = f(2) = 1 - 2 = -1$
$f(b) = f(x) = 1 - x$
$\dfrac{f(b) - f(a)}{b - a} = \dfrac{(1 - x) - (-1)}{x - 2} = \dfrac{2 - x}{x - 2}$
$\qquad\qquad = \dfrac{-1(x - 2)}{x - 2} = -1$
The average rate of change is -1.

9. $f(x) = -x^2 + x - 3$
$f(x + h) = -(x + h)^2 + (x + h) - 3$
$\qquad\qquad = -x^2 - 2xh - h^2 + x + h - 3$
$\dfrac{f(x + h) - f(x)}{h}$
$= \dfrac{\left(-x^2 - 2xh - h^2 + x + h - 3\right) - \left(-x^2 + x - 3\right)}{h}$
$= \dfrac{-2xh - h^2 + h}{h} = -2x - h + 1$

2.5 Basic Concepts and Skills

1. A function f is decreasing if $x_1 < x_2$ implies that $\underline{f(x_1) > f(x_2)}$.

3. A function f is even if $\underline{f(-x) = f(x)}$ for all x in the $\underline{\text{domain of } f}$.

5. True

7. Increasing on $(-\infty, \infty)$

9. Increasing on $(-\infty, 2)$, decreasing on $(2, \infty)$

11. Increasing on $(-\infty, -2)$, constant on $(-2, 2)$, increasing on $(2, \infty)$

13. Increasing on $(-\infty, -3)$ and $\left(-\frac{1}{2}, 2\right)$, decreasing on $\left(-3, -\frac{1}{2}\right)$ and $(2, \infty)$

15. No relative extrema

17. (2, 10) is a relative maximum point and a turning point.

19. Any point on $(x, 2)$ is a relative maximum and a relative minimum point on the interval $(-2, 2)$. Relative maximum at $(-2, 2)$; relative minimum at $(2, 2)$. None of these points are turning points.

21. $(-3, 4)$ and $(2, 5)$ are relative maxima points and turning points. $\left(-\frac{1}{2}, -2\right)$ is a relative minimum and a turning point.

For exercises 23–32, recall that the graph of an even function is symmetric about the y-axis, and the graph of an odd function is symmetric about the origin.

23. The graph is symmetric with respect to the origin. The function is odd.

25. The graph has no symmetries, so the function is neither odd nor even.

27. The graph is symmetric with respect to the origin. The function is odd.

29. The graph is symmetric with respect to the y-axis. The function is even.

31. The graph is symmetric with respect to the origin. The function is odd.

For exercises 33–46, $f(-x) = f(x) \Rightarrow f(x)$ is even and $f(-x) = -f(x) \Rightarrow f(x)$ is odd.

33. $f(-x) = 2(-x)^4 + 4 = 2x^4 + 4 = f(x) \Rightarrow$
$f(x)$ is even.

35. $f(-x) = 5(-x)^3 - 3(-x) = -5x^3 + 3x$
$= -(5x^3 - 3x) = -f(x) \Rightarrow$
$f(x)$ is odd.

37. $f(-x) = 2(-x) + 4 = -2x + 4$
$\neq -f(x) \neq f(x) \Rightarrow$
$f(x)$ is neither even nor odd.

39. $f(-x) = \dfrac{1}{(-x)^2 + 4} = \dfrac{1}{x^2 + 4} = f(x) \Rightarrow$
$f(x)$ is even.

41. $f(-x) = \dfrac{(-x)^3}{(-x)^2 + 1} = -\dfrac{x^3}{x^2 + 1} = -f(x) \Rightarrow$
$f(x)$ is odd.

43. $f(-x) = \dfrac{-x}{(-x)^5 - 3(-x)^3} = \dfrac{-x}{-x^5 + 3x^3}$
$= \dfrac{(-1)(-x)}{(-1)\left(-x^5 + 3x^3\right)} = \dfrac{x}{x^5 - 3x^3} = f(x)$
Thus, $f(x)$ is even.

45. $f(-x) = \dfrac{(-x)^2 - 2(-x)}{5(-x)^4 + 4(-x)^2 + 7} = \dfrac{x^2 + 2x}{5x^4 + 4x^2 + 7}$
$\neq -f(x) \neq f(x)$
Thus, $f(x)$ is neither even nor odd.

47.a. domain: $(-\infty, \infty)$; range: $(-\infty, 3]$

b. x-intercepts: $(-3, 0)$, $(3, 0)$
y-intercept: $(0, 3)$

c. increasing on $(-\infty, 0)$, decreasing on $(0, \infty)$

d. relative maximum at $(0, 3)$

e. even

49.a. domain: $(-3, 4)$; range: $[-2, 2]$

b. x-intercept: $(1, 0)$; y-intercept: $(0, -1)$

c. constant on $(-3, -1)$ and $(3, 4)$
increasing on $(-1, 3)$

d. Since the function is constant on $(-3, -1)$, any point $(x, -2)$ is both a relative maximum and a relative minimum on that interval. Since the function is constant on $(3, 4)$, any point $(x, 2)$ is both a relative maximum and a relative minimum on that interval.

e. neither even not odd

51.a. domain: $(-2, 4)$; range: $(-2, 3)$

b. x-intercept: $(0, 0)$; y-intercept: $(0, 0)$

c. decreasing on $(-2, -1)$ and $(3, 4)$
increasing on $(-1, 3)$

d. relative maximum: $(3, 3)$
relative minimum: $(-1, -2)$

e. neither even nor odd

53.a. domain: $(-\infty, \infty)$; range: $(0, \infty)$

b. no x-intercept; y-intercept: $(0, 1)$

c. increasing on $(-\infty, \infty)$

d. no relative minimum or relative maximum

e. neither even nor odd

55. $f(x) = -2x + 7;\ a = -1,\ b = 3$
$f(3) = -2(3) + 7 = 1;\ f(-1) = -2(-1) + 7 = 9$

average rate of change $= \dfrac{f(3) - f(-1)}{3 - (-1)}$

$\qquad\qquad\qquad = \dfrac{1 - 9}{4} = -2$

57. $f(x) = 3x + c;\ a = 1,\ b = 5$
$f(5) = 3 \cdot 5 + c = 15 + c;\ f(1) = 3 \cdot 1 + c = 3 + c$

average rate of change $= \dfrac{f(5) - f(1)}{5 - 1}$

$\qquad\qquad\qquad = \dfrac{15 + c - (3 + c)}{4}$

$\qquad\qquad\qquad = \dfrac{12}{4} = 3$

59. $h(x) = x^2 - 1;\ a = -2,\ b = 0$
$h(0) = 0^2 - 1 = -1;\ h(-2) = (-2)^2 - 1 = 3$

average rate of change $= \dfrac{h(0) - h(-2)}{0 - (-2)}$

$\qquad\qquad\qquad = \dfrac{-1 - 3}{2} = -2$

61. $f(x) = (3 - x)^2;\ a = 1,\ b = 3$
$f(4) = (3 - 3)^2 = 0;\ f(1) = (3 - 1)^2 = 4$

average rate of change $= \dfrac{f(3) - f(1)}{3 - 1}$

$\qquad\qquad\qquad = \dfrac{0 - 4}{2} = -2$

63. $g(x) = x^3;\ a = -1,\ b = 3$
$g(3) = 3^3 = 27;\ g(-1) = (-1)^3 = -1$

average rate of change $= \dfrac{g(3) - g(-1)}{3 - (-1)}$

$\qquad\qquad\qquad = \dfrac{27 - (-1)}{4} = 7$

65. $h(x) = \dfrac{1}{x};\ a = 2,\ b = 6$

$h(2) = \dfrac{1}{2};\ h(6) = \dfrac{1}{6}$

average rate of change $= \dfrac{h(6) - h(2)}{6 - 2}$

$\qquad\qquad\qquad = \dfrac{\frac{1}{6} - \frac{1}{2}}{4} = -\dfrac{1}{12}$

67. $f(x) = 2x,\ a = 3 \Rightarrow f(a) = 2 \cdot 3 = 6$

$\dfrac{f(x) - f(a)}{x - a} = \dfrac{2x - 6}{x - 3} = \dfrac{2(x - 3)}{x - 3} = 2$

69. $f(x) = -x^2,\ a = 1 \Rightarrow f(a) = -1$

$\dfrac{f(x) - f(a)}{x - a} = \dfrac{-x^2 - (-1)}{x - 1} = \dfrac{-x^2 + 1}{x - 1}$

$\qquad\qquad\qquad = \dfrac{-(x - 1)(x + 1)}{x - 1} = -x - 1$

71. $f(x) = 3x^2 + x,\ a = 2 \Rightarrow$

$f(a) = 3(2)^2 + 2 = 14$

$\dfrac{f(x) - f(a)}{x - a} = \dfrac{3x^2 + x - 14}{x - 2} = \dfrac{(3x + 7)(x - 2)}{x - 2}$

$\qquad\qquad\qquad = 3x + 7$

73. $f(x) = \dfrac{4}{x},\ a = 1 \Rightarrow f(a) = 4$

$\dfrac{f(x) - f(a)}{x - a} = \dfrac{\frac{1}{x} - 4}{x - 1} = \dfrac{\frac{4 - 4x}{x}}{x - 1} = \dfrac{-4(x - 1)}{x(x - 1)}$

$\qquad\qquad\qquad = -\dfrac{4}{x}$

75. $f(x + h) = x + h$
$f(x + h) - f(x) = x + h - x = h$

$\dfrac{f(x + h) - f(x)}{h} = \dfrac{h}{h} = 1$

77. $f(x + h) = -2(x + h) + 3 = -2x - 2h + 3$
$f(x + h) - f(x) = -2x - 2h + 3 - (-2x + 3)$
$\qquad\qquad\qquad = -2h$

$\dfrac{f(x + h) - f(x)}{h} = \dfrac{-2h}{h} = -2$

79. $f(x + h) = m(x + h) + b = mx + mh + b$
$f(x + h) - f(x) = mx + mh + b - (mx + b)$
$\qquad\qquad\qquad = mh$

$\dfrac{f(x + h) - f(x)}{h} = \dfrac{mh}{h} = m$

81. $f(x + h) = (x + h)^2 = x^2 + 2xh + h^2$
$f(x + h) - f(x) = x^2 + 2xh + h^2 - x^2$
$\qquad\qquad\qquad = 2xh + h^2$

$\dfrac{f(x + h) - f(x)}{h} = \dfrac{2xh + h^2}{h} = 2x + h$

83. $f(x + h) = 2(x + h)^2 + 3(x + h)$
$\qquad\quad = 2x^2 + 4xh + 2h^2 + 3x + 3h$
$\qquad\quad = 2x^2 + 4xh + 3x + 2h^2 + 3h$
$f(x + h) - f(x)$
$\qquad = 2x^2 + 4xh + 3x + 2h^2 + 3h - (2x^2 + 3x)$
$\qquad = 4xh + 2h^2 + 3h$

(continued on next page)

(continued)

$$\frac{f(x+h)-f(x)}{h} = \frac{4xh+2h^2+3h}{h}$$
$$= 4x+2h+3$$

85. $f(x+h) = 4$

$f(x+h) - f(x) = 4 - 4 = 0$

$\dfrac{f(x+h)-f(x)}{h} = \dfrac{0}{h} = 0$

87. $f(x+h) = \dfrac{1}{x+h}$

$f(x+h) - f(x) = \dfrac{1}{x+h} - \dfrac{1}{x}$

$= \dfrac{x}{x(x+h)} - \dfrac{x+h}{x(x+h)}$

$= -\dfrac{h}{x(x+h)}$

$\dfrac{f(x+h)-f(x)}{h} = \dfrac{-\dfrac{h}{x(x+h)}}{h} = -\dfrac{1}{x(x+h)}$

2.5 Applying the Concepts

89. domain: $[0, \infty)$

The particle's motion is tracked indefinitely from time $t = 0$.

91. The graph is above the t-axis on the intervals $(0, 9)$ and $(21, 24)$. This means that the particle was moving forward between 0 and 9 seconds and between 21 and 24 seconds.

93. The function is increasing on $(0, 3)$, $(5, 6)$, $(16, 19)$, and $(21, 23)$. However, the speed $|v|$ of the particle is increasing on $(0, 3)$, $(5, 6)$, $(11, 15)$, and $(21, 23)$. Note that the particle is moving forward on $(0, 3)$, $(5, 6)$, and $(21, 23)$, and moving backward on $(11, 15)$.

95. The maximum speed is between times $t = 15$ and $t = 16$.

97. The particle is moving forward with increasing velocity.

99.

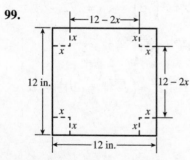

a. $V = lwh = (12-2x)(12-2x)x$
$$= \left(144 - 48x + 4x^2\right)x$$
$$= 4x^3 - 48x^2 + 144x$$

b. The length of the squares in the corners must be greater than 0 and less than 6, so the domain of V is $(0, 6)$.

c.

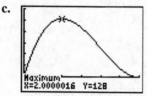

[0, 6, 1] by [−25, 150, 25]
range: [0, 128]

d. V is at its maximum when $x = 2$.

101.a. $C(x) = 210x + 10,500$

b. $C(50) = 210(50) + 10,500 = \$21,000$
It costs $21,000 to produce 50 notebooks per day.

c. average cost $= \dfrac{\$21,000}{50} = \$420$

d. $\dfrac{210x+10,500}{x} = 315$
$210x + 10,500 = 315x$
$10,500 = 105x \Rightarrow x = 100$
The average cost per notebook will be $315 when 100 notebooks are produced.

103.a. $f(0) = 0^2 + 3(0) + 4 = 4$
The particle is 4 ft to the right from the origin.

b. $f(4) = 4^2 + 3(4) + 4 = 32$
The particle started 4 ft from the origin, so it traveled $32 - 4 = 28$ ft in four seconds.

c. $f(3) = 3^2 + 3(3) + 4 = 22$
The particle started 4 ft from the origin, so it traveled $22 - 4 = 18$ ft in three seconds. The average velocity is $18/3 = 6$ ft/sec

d. $f(2) = 2^2 + 3(2) + 4 = 14$
$f(5) = 5^2 + 3(5) + 4 = 44$
The particle traveled $44 - 14 = 30$ ft between the second and fifth seconds. The average velocity is $30/(5-2) = 10$ ft/sec

2.5 Beyond the Basics

105. $f(x) = \dfrac{x-1}{x+1}$

$$f(2x) = \dfrac{2x-1}{2x+1}$$

$$\dfrac{3f(x)+1}{f(x)+3} = \dfrac{3\left(\dfrac{x-1}{x+1}\right)+1}{\dfrac{x-1}{x+1}+3} = \dfrac{\dfrac{3x-3}{x+1}+1}{\dfrac{x-1+3(x+1)}{x+1}}$$

$$= \dfrac{\dfrac{3x-3+x+1}{x+1}}{\dfrac{x-1+3(x+1)}{x+1}} = \dfrac{4x-2}{4x+2}$$

$$= \dfrac{2(2x-1)}{2(2x+1)} = \dfrac{2x-1}{2x+1} = f(2x)$$

107. In order to find the relative maximum, first observe that the relative maximum of

$-(x+1)^2 \le 0$. Then $-(x+1)^2 \le 0 \Rightarrow$

$(x+1)^2 \ge 0 \Rightarrow x \ge -1$.

Thus, the x-coordinate of the relative maximum is -1. $f(-1) = -(-1+1)^2 + 5 = 5$

The relative maximum is $(-1, 5)$.
There is no relative minimum.

109. $f(x) = \sqrt{x}$

$$\dfrac{f(x+h)-f(x)}{h} = \dfrac{\sqrt{x+h}-\sqrt{x}}{h} \cdot \dfrac{\sqrt{x+h}+\sqrt{x}}{\sqrt{x+h}+\sqrt{x}}$$

$$= \dfrac{x+h-x}{h\left(\sqrt{x+h}+\sqrt{x}\right)}$$

$$= \dfrac{h}{h\left(\sqrt{x+h}+\sqrt{x}\right)}$$

$$= \dfrac{1}{\sqrt{x+h}+\sqrt{x}}$$

111. $f(x) = -\dfrac{1}{\sqrt{x}}$

$$\dfrac{f(x+h)-f(x)}{h}$$

$$= \dfrac{-\dfrac{1}{\sqrt{x+h}}+\dfrac{1}{\sqrt{x}}}{h} = \dfrac{\dfrac{\sqrt{x+h}-\sqrt{x}}{\sqrt{x}\sqrt{x+h}}}{h}$$

$$= \dfrac{\dfrac{\sqrt{x+h}-\sqrt{x}}{\sqrt{x(x+h)}}}{h} = \dfrac{\sqrt{x+h}-\sqrt{x}}{h\sqrt{x(x+h)}}$$

$$= \dfrac{\sqrt{x+h}-\sqrt{x}}{h\sqrt{x(x+h)}} \cdot \dfrac{\sqrt{x+h}+\sqrt{x}}{\sqrt{x+h}+\sqrt{x}}$$

$$= \dfrac{(x+h)-x}{h\sqrt{x(x+h)}\left(\sqrt{x+h}+\sqrt{x}\right)}$$

$$= \dfrac{1}{\sqrt{x(x+h)}\left(\sqrt{x}+\sqrt{x+h}\right)}$$

2.5 Critical Thinking/Discussion/Writing

113. f has a relative maximum at $x = a$ if there is an interval $[a, x_1)$ with $a < x_1 < b$ such that $f(a) \ge f(x)$, or $f(x) \le f(a)$, for every x in the interval $(x_1, b]$.

114. f has a relative minimum at $x = b$ if there is x_1 in $[a, b]$ such that $f(x) \ge f(b)$ for every x in the interval $(x_1, b]$.

115. Answers will vary. Sample answers are given.

   a. $f(x) = x$ on the interval $[-1, 1]$

   b. $f(x) = \begin{cases} x & \text{if } 0 \le x < 1 \\ 0 & \text{if } x = 1 \end{cases}$

   c. $f(x) = \begin{cases} x & \text{if } 0 < x \le 1 \\ 1 & \text{if } x = 0 \end{cases}$

   d. $f(x) = \begin{cases} 0 & \text{if } x = 0 \text{ or } x = 1 \\ 1 & \text{if } 0 < x < 1 \text{ and } x \text{ is rational} \\ -1 & \text{if } 0 < x < 1 \text{ and } x \text{ is irrational} \end{cases}$

2.5 Maintaining Skills

117. $m = \dfrac{-1-2}{7-6} = \dfrac{-3}{1} = -3$

$$y - 2 = -3(x-6) \Rightarrow y - 2 = -3x + 18 \Rightarrow$$
$$y = -3x + 20$$

119. $f(x) = x^{3/2}$

   (i) $f(2) = 2^{3/2} = \left(\sqrt{2}\right)^3 = \sqrt{8} = 2\sqrt{2}$

   (ii) $f(4) = 4^{3/2} = \left(\sqrt{4}\right)^3 = 2^3 = 8$

   (iii) $f(-4) = (-4)^{3/2} = \left(\sqrt{-4}\right)^3 = (2i)^3 = -8i$

2.6   A Library of Functions

2.6 Practice Problems

1. Since $g(-2) = 2$ and $g(1) = 8$, the line passes through the points $(-2, 2)$ and $(1, 8)$.

$$m = \frac{8-2}{1-(-2)} = \frac{6}{3} = 2$$

Use the point-slope form:

$$y - 8 = 2(x-1) \Rightarrow y - 8 = 2x - 2 \Rightarrow$$
$$y = 2x + 6 \Rightarrow g(x) = 2x + 6$$

2. Using the formula
Shark length = (0.96)(tooth height) – 0.22, gives:
Shark length = (0.96)(16.4) – 0.22 = 15.524 m

3.

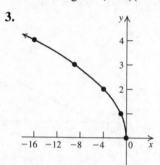

Domain: $(-\infty, 0]$; range: $[0, \infty)$

4.

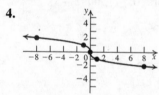

Domain: $(-\infty, \infty)$; range: $(-\infty, \infty)$

5. $f(x) = \begin{cases} x^2 & \text{if } x \le -1 \\ 2x & \text{if } x > -1 \end{cases}$

$f(-2) = (-2)^2 = 4; \quad f(3) = 2(3) = 6$

6.a. $f(x) = \begin{cases} 50 + 4(x-55) & 56 \le x < 75 \\ 200 + 5(x-75) & x \ge 75 \end{cases}$

b. The fine for driving 60 mph is
$50 + 4(60 - 55) = \$70$.

c. The fine for driving 90 mph is
$200 + 5(90 - 75) = \$275$.

7. $f(x) = \begin{cases} -3x & \text{if } x \le -1 \\ 2x & \text{if } x > -1 \end{cases}$

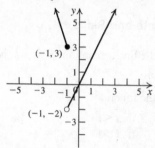

Graph $f(x) = -3x$ on the interval $(-\infty, -1]$,
and graph $f(x) = 2x$ on the interval $(-1, \infty)$.

8.

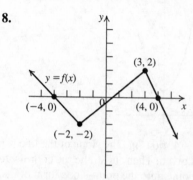

The graph of f is made up of three parts.
For $x \le -2$, the graph is made up of the half-line passing through the points $(-4, 0)$ and $(-2, -2)$.

$$m = \frac{-2-0}{-2-(-4)} = \frac{-2}{2} = -1$$

$$y - 0 = -1(x-(-4)) \Rightarrow y = -x - 4$$

For $-2 < x < 3$, the graph is a line segment passing through the points $(-2, -2)$ and $(3, 2)$.

$$m = \frac{2-(-2)}{3-(-2)} = \frac{4}{5}$$

$$y - 2 = \frac{4}{5}(x-3) \Rightarrow y - 2 = \frac{4}{5}x - \frac{12}{5} \Rightarrow$$

$$y = \frac{4}{5}x - \frac{2}{5}$$

For $x \ge 3$, the graph is a half-line passing through $(3, 2)$ and $(4, 0)$.

$$m = \frac{0-2}{4-3} = -2$$

$$y - 0 = -2(x-4) \Rightarrow y = -2x + 8$$

Combining the three parts, we have

$$f(x) = \begin{cases} -x - 4 & \text{if } x \le -2 \\ \dfrac{4}{5}x - \dfrac{2}{5} & \text{if } -2 < x < 3 \\ -2x + 8 & \text{if } x \ge 3 \end{cases}$$

9. $f(x) = [\![x]\!]$
$f(-3.4) = -4; \ f(4.7) = 4$

2.6 Basic Concepts and Skills

1. The graph of the linear function $f(x) = b$ is a <u>horizontal</u> line.

3. The graph of the function
$$f(x) = \begin{cases} x^2 + 2 & \text{if } x \le 1 \\ ax & \text{if } x > 1 \end{cases}$$
will have a break at $x = 1$ unless $a = \underline{3}$.

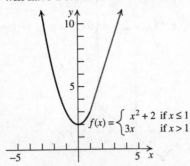

$f(x) = \begin{cases} x^2 + 2 & \text{if } x \le 1 \\ 3x & \text{if } x > 1 \end{cases}$

In exercises 5–14, first find the slope of the line using the two points given. Then substitute the coordinates of one of the points into the point-slope form of the equation to solve for b.

5. The two points are $(0, 1)$ and $(-1, 0)$.
$$m = \frac{0 - 1}{-1 - 0} = 1. \ \ 1 = 1(0) + b \Rightarrow b = 1.$$
$$f(x) = x + 1$$

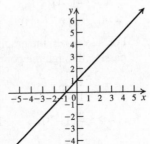

7. The two points are $(-1, 1)$ and $(2, 7)$.
$$m = \frac{7 - 1}{2 - (-1)} = 2. \ \ 1 = 2(-1) + b \Rightarrow 3 = b.$$
$$f(x) = 2x + 3$$

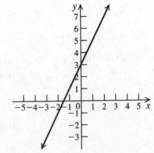

9. The two points are $(1, 1)$ and $(2, -2)$.
$$m = \frac{-2 - 1}{2 - 1} = -3. \ \ 1 = -3(1) + b \Rightarrow b = 4.$$
$$f(x) = -3x + 4.$$

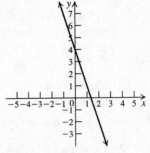

11. The two points are $(-2, 2)$ and $(2, 4)$.
$$m = \frac{4 - 2}{2 - (-2)} = \frac{1}{2}. \ \ 4 = \frac{1}{2}(2) + b \Rightarrow b = 3.$$
$$f(x) = \frac{1}{2}x + 3.$$

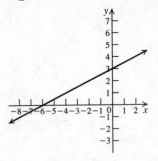

13. The two points are $(0, -1)$ and $(3, -3)$.
$$m = \frac{-3 - (-1)}{3 - 0} = -\frac{2}{3}.$$
$$-1 = -\frac{2}{3}(0) + b \Rightarrow b = -1.$$
$$f(x) = -\frac{2}{3}x - 1.$$

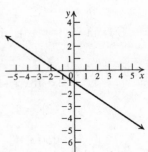

15. $f(x) = \begin{cases} x & \text{if } x \ge 2 \\ 2 & \text{if } x < 2 \end{cases}$

   a. $f(1) = 2; \ f(2) = 2; \ f(3) = 3$

b.

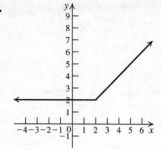

17. $g(x) = \begin{cases} 1 & \text{if } x > 0 \\ -1 & \text{if } x < 0 \end{cases}$

  a. $f(-15) = -1; \ f(12) = 1$

  b.

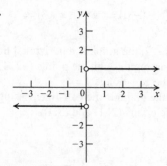

  c. domain: $(-\infty, 0) \cup (0, \infty)$

    range: $\{-1, 1\}$

19. $f(x) = \begin{cases} 2x & \text{if } x < 0 \\ x^2 & \text{if } x \geq 0 \end{cases}$

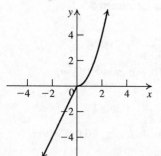

Domain: $(-\infty, \infty)$; range: $(-\infty, \infty)$

21. $g(x) = \begin{cases} \dfrac{1}{x} & \text{if } x < 0 \\ \sqrt{x} & \text{if } x \geq 0 \end{cases}$

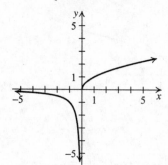

Domain: $(-\infty, \infty)$; range: $(-\infty, \infty)$

23. $f(x) = \begin{cases} [\![x]\!] & \text{if } x < 1 \\ \sqrt[3]{x} & \text{if } x \geq 1 \end{cases}$

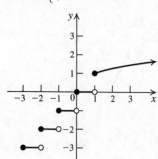

Domain: $(-\infty, \infty)$;

range: $\{\ldots, -3, -2, -1, 0\} \cup [1, \infty)$

25. $g(x) = \begin{cases} |x| & \text{if } x < 1 \\ x^3 & \text{if } x \geq 1 \end{cases}$

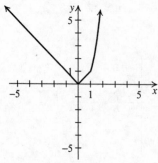

Domain: $(-\infty, \infty)$; range: $[0, \infty)$

27. $f(x) = \begin{cases} |x| & \text{if } -2 \le x < 1 \\ \sqrt{x} & \text{if } x \ge 1 \end{cases}$

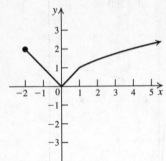

Domain: $[-2, \infty)$; range: $[0, \infty)$

29. $f(x) = \begin{cases} [[x]] & \text{if } x < 1 \\ -2x & \text{if } 1 \le x \le 4 \end{cases}$

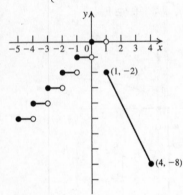

Domain: $(-\infty, 4]$;

range: $\{..., -3, -2, -1, 0\} \cup [-8, -2]$

31. $f(x) = \begin{cases} 2x+3 & \text{if } x < -2 \\ x+1 & -2 \le x < 1 \\ -x+3 & \text{if } x \ge 1 \end{cases}$

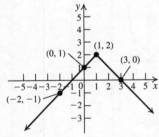

Domain: $(-\infty, \infty)$; range: $(-\infty, 2]$

33. The graph of f is made up of two parts.
For $x < 2$, the graph is made up of the half-line passing through the points $(-1, 0)$ and $(2, 3)$.

$$m = \frac{3-0}{2-(-1)} = \frac{3}{3} = 1$$

$$y - 0 = x - (-1) \Rightarrow y = x + 1$$

For $x \ge 2$, the graph is a line segment passing through the points $(2, 3)$ and $(3, 0)$.

$$m = \frac{0-3}{3-2} = -3$$

$$y - 0 = -3(x-3) \Rightarrow y = -3x + 9$$

Combining the two parts, we have

$$f(x) = \begin{cases} x+1 & \text{if } x < -2 \\ -3x+9 & \text{if } x \ge 2 \end{cases}$$

35. The graph of f is made up of three parts.
For $x < -2$, the graph is the half-line passing through the points $(-2, 2)$ and $(-3, 0)$.

$$m = \frac{0-2}{-3-(-2)} = \frac{-2}{-1} = 2$$

$$y - 0 = 2(x-(-3)) \Rightarrow y = 2(x+3) \Rightarrow$$
$$y = 2x + 6$$

For $-2 \le x < 2$, the graph is a horizontal line segment passing through the points $(-2, 4)$ and $(2, 4)$, so the equation is $y = 4$.
For $x \ge 2$, the graph is the half-line passing through the points $(2, 1)$ and $(3, 0)$.

$$m = \frac{0-1}{3-2} = -1$$

$$y - 0 = -(x-3) \Rightarrow y = -x + 3$$

Combining the three parts, we have

$$f(x) = \begin{cases} 2x+6 & \text{if } x < -2 \\ 4 & \text{if } -2 \le x < 2 \\ -x+3 & \text{if } x \ge 2 \end{cases}$$

2.6 Applying the Concepts

37.a. $f(x) = \dfrac{x}{33.81}$

Domain: $[0, \infty)$; range: $[0, \infty)$.

b. $f(3) = \dfrac{3}{33.81} \approx 0.0887$

This means that 3 oz ≈ 0.0887 liters.

c. $f(12) = \dfrac{12}{33.81} \approx 0.3549$ liters.

39.a. $P(0) = \dfrac{1}{33}(0) + 1 = 1$. The y-intercept is 1.

This means that the pressure at sea level $(d = 0)$ is 1 atm.

$$0 = \frac{1}{33}d + 1 \Rightarrow d = -33.$$

d can't be negative, so there is no d-intercept.

b. $P(0) = 1$ atm; $P(10) = \dfrac{1}{33}(10) + 1 \approx 1.3$ atm;

$P(33) = \dfrac{1}{33}(33) + 1 = 2$ atm;

$P(100) = \dfrac{1}{33}(100) + 1 \approx 4.03$ atm.

c. $5 = \dfrac{1}{33}d + 1 \Rightarrow d = 132$ feet

The pressure is 5 atm at 132 feet.

41.a. $C(x) = 50x + 6000$

b. The y-intercept is the fixed overhead cost.

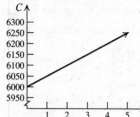

c. $11,500 = 50x + 6000 \Rightarrow 110$

110 printers were manufactures on a day when the total cost was \$11,500.

43.a. $R = 900 - 30x$

b. $R(6) = 900 - 30(6) = 720$

If you move in 6 days after the first of the month, the rent is \$720.

c. $600 = 900 - 30x \Rightarrow x = 10$

You moved in ten days after first of the month.

45. The rate of change (slope) is $\dfrac{100 - 40}{20 - 80} = -1$.

Use the point $(20, 100)$ to find the equation of the line: $100 = -20 + b \Rightarrow b = 120$. The equation of the line is $y = -x + 120$. Now solve $50 = -x + 120 \Rightarrow x = 70$.

Age 70 corresponds to 50% capacity.

47.a. The rate of change (slope) is

$\dfrac{50 - 30}{420 - 150} = \dfrac{2}{27}$.

The equation of the line is

$y - 30 = \dfrac{2}{27}(x - 150) \Rightarrow$

$y = \dfrac{2}{27}(x - 150) + 30$.

b. $y = \dfrac{2}{27}(350 - 150) + 30 \Rightarrow y = \dfrac{1210}{27} \approx 44.8$

There can't be a fractional number of deaths, so round up. There will be about 45 deaths when $x = 350$ milligrams per cubic meter.

c. $70 = \dfrac{2}{27}(x - 150) + 30 \Rightarrow x = 690$

If the number of deaths per month is 70, the concentration of sulfur dioxide in the air is 690 mg/m^3.

49.a.

b.(i) $T(12,000) = 0.04(12,000) = \$480$

(ii) $T(20,000) = 800 + 0.06(20,000 - 20,000)$
$= \$800$

(iii) $T(50,000) = 800 + 0.06(50,000 - 20,000)$
$= \$2600$

c. (i) $600 = 0.04x \Rightarrow x = \$15,000$

(ii) $1200 = 0.04x \Rightarrow x = \$30,000$, which is outside of the domain. Try
$1200 = 800 + 0.06(x - 20,000) \Rightarrow$
$x \approx \$26,667$

(iii) $2300 = 800 + 0.06(x - 20,000) \Rightarrow$
$x = \$45,000$

2.6 Beyond the Basics

51.a. (i) $f(-2) = 3(-2) + 5 = -1$

(ii) $f(-1) = 2(-1) + 1 = -1$

(iii) $f(3) = 2 - 3 = -1$

b. Try the first rule: $2 = 3x + 5 \Rightarrow x = -1$, which is not in the domain for that rule. Now try the second rule: $2 = 2x + 1 \Rightarrow x = \dfrac{1}{2}$, which is in the domain for that rule.

c.

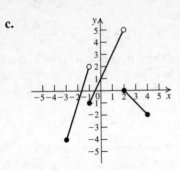

53.a. Domain: $(-\infty, \infty)$; range: $[0, 1)$

b. The function is increasing on $(n, n + 1)$ for every integer n.

c. $f(-x) = -x - [\![-x]\!] \neq -f(x) \neq f(x)$, so the function is neither even nor odd.

55. $|f(x) - f(-x)| = \left| \dfrac{|x|}{x} - \dfrac{|-x|}{-x} \right| = \left| \dfrac{|x|}{x} + \dfrac{|x|}{x} \right|$

$$= |1 + 1| = 2 \text{ or } |-1 + (-1)| = 2$$

Thus, $|f(x) - f(-x)| = 2$

2.6 Critical Thinking/Discussion/Writing

57.a. $C(x) = 17(f(x) - 1) + 44$
$$= -17([\![-x]\!] + 1) + 44$$

b.

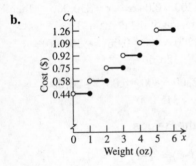

Weight (oz)

c. Domain: $(0, \infty)$

range: $\{17n + 44 : n \text{ a nonnegative integer}\}$

58. $C(x) = 2[\![x]\!] + 4$

59.a. $C(x) = \begin{cases} 150 & \text{if } x \leq 100 \\ -0.2[\![x - 100]\!] + 150 & \text{if } x > 100 \end{cases}$

b.

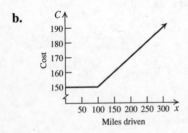

Miles driven

c. $190 = 0.2[\![x - 99]\!] + 150$
$40 = 0.2[\![x - 99]\!] \Rightarrow 200 = [\![x - 99]\!] \Rightarrow$
$x = (299, 300]$ miles

60.a. $f(x) = [\![x]\!] + [\![-x]\!] \Rightarrow$
$$f(x) = \begin{cases} -1 & \text{if } x \text{ is not an integer} \\ 0 & \text{if } x \text{ is an integer} \end{cases}$$

For example, $\left[\!\left[\dfrac{3}{2}\right]\!\right] + \left[\!\left[-\dfrac{3}{2}\right]\!\right] = 1 + (-2) = -1$ and

$[\![3]\!] + [\![-3]\!] = 3 + (-3) = 0$.

b. $g(x) = [\![x]\!] - [\![-x]\!] \Rightarrow$
$$g(x) = \begin{cases} 2[\![x]\!] + 1 & \text{if } x \text{ is not an integer} \\ 2[\![x]\!] & \text{if } x \text{ is an integer} \end{cases}$$

For example, $\left[\!\left[\dfrac{3}{2}\right]\!\right] - \left[\!\left[-\dfrac{3}{2}\right]\!\right] = 1 - (-2) = 3$,

while $2\left[\!\left[\dfrac{3}{2}\right]\!\right] + 1 = 2 + 1 = 3$ and

$[\![3]\!] - [\![-3]\!] = 3 - (-3) = 6 = 2[\![3]\!]$.

c. $h(x) = x - |x| \Rightarrow$
$$h(x) = \begin{cases} 2x & \text{if } x < 0 \\ 0 & \text{if } x \geq 0 \end{cases}$$

For example, $-3 - |-3| = -3 - 3 = -6 = 2(-3)$
and $3 - |3| = 3 - 3 = 0$.

d. $F(x) = x|x| \Rightarrow$
$$F(x) = \begin{cases} -x^2 & \text{if } x < 0 \\ x^2 & \text{if } x \geq 0 \end{cases}$$

For example, $-4|-4| = -4(4) = -16 = -4^2$
and $4|4| = 4(4) = 16 = 4^2$.

e. $G(x) = |x - 1| + |x - 2| \Rightarrow$
$$G(x) = \begin{cases} -2x + 3 & \text{if } x < 1 \\ 1 & \text{if } 1 \leq x < 2 \\ 2x - 3 & \text{if } x \geq 2 \end{cases}$$

For example, if $x = 0$, then
$G(x) = -2(0) + 3 = 3$. If $x = 5$, then
$G(x) = 2(5) - 3 = 7$.

2.6 Maintaining Skills

61. If we add 3 to each y-coordinate of the graph of f, we will obtain the graph of $y = \underline{f(x) + 3}$.

63. If we replace each x-coordinate with its opposite in the graph of f, we will obtain the graph of $y = \underline{f(-x)}$.

65.

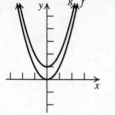

67.

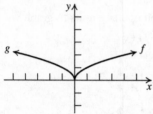

2.7 Transformations of Functions

2.7 Practice Problems

1.

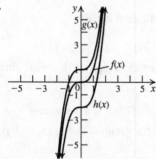

The graph of *g* is the graph of *f* shifted one unit up. The graph of *h* is the graph of *f* shifted two units down.

2.

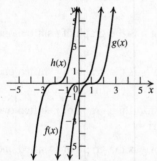

The graph of *g* is the graph of *f* shifted one unit to the right. The graph of *h* is the graph of *f* shifted two units to the left.

3. The graph of $f(x) = \sqrt{x-2} + 3$ is the graph of $g(x) = \sqrt{x}$ shifted two units to the right and three units up.

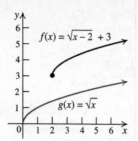

4. The graph of $y = -(x-1)^2 + 2$ can be obtained from the graph of $y = x^2$ by first shifting the graph of $y = x^2$ one unit to the right. Reflect the resulting graph about the *x*-axis, and then shift the graph two units up

5. The graph of $y = 2x - 4$ is obtained from the graph of $y = 2x$ by shifting it down by four units. We know that

$$|y| = \begin{cases} y & \text{if } y \geq 0 \\ -y & \text{if } y < 0. \end{cases}$$

This means that the portion of the graph on or above the *x*-axis $(y \geq 0)$ is unchanged while the portion of the graph below the *x*-axis $(y < 0)$ is reflected above the *x*-axis.

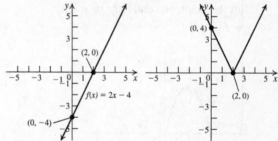

6.

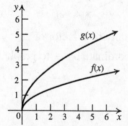

The graph of *g* is the graph of *f* stretched vertically by multiplying each of its *y*-coordinates by 2.

7.a.

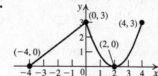

b.

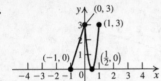

8. Start with the graph of $y = \sqrt{x}$. Shift the graph one unit to the left, then stretch the graph vertically by a factor of three. Shift the resulting graph down two units.

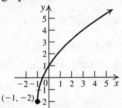

9.

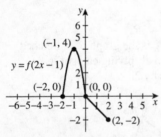

Shift the graph one unit right to graph $y = f(x-1)$.

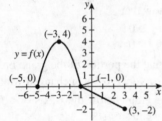

Compress horizontally by a factor of 2. Multiply each x-coordinate by $\frac{1}{2}$ to graph $y = f(2x-1)$.

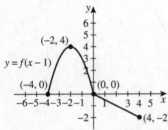

Compress vertically by a factor of $\frac{1}{2}$. Multiply each y-coordinate by $\frac{1}{2}$ to graph $y = \frac{1}{2}f(2x-1)$.

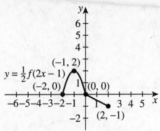

Shift the graph up three units to graph $y = \frac{1}{2}f(2x-1)+3$.

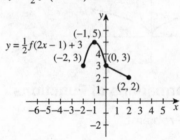

2.7 Basic Concepts and Skills

1. The graph of $y = f(x)-3$ is found by vertically shifting the graph of $y = f(x)$ three units <u>down</u>.

3. The graph of $y = f(bx)$ is a horizontal compression of the graph of $y = f(x)$ is b <u>is greater than 1</u>.

5. False. The graphs are the same if the function is an even function.

7.a. The graph of g is the graph of f shifted two units up.

   b. The graph of h is the graph of f shifted one unit down.

9.a. The graph of g is the graph of f shifted one unit to the left.

   b. The graph of h is the graph of f shifted two units to the right.

11.a. The graph of g is the graph of f shifted one unit left and two units down.

   b. The graph of h is the graph of f shifted one unit right and three units up.

13.a. The graph of g is the graph of f reflected about the x-axis.

   b. The graph of h is the graph of f reflected about the y-axis.

15.a. The graph of g is the graph of f vertically stretched by a factor of 2.

b. The graph of *h* is the graph of *f* horizontally compressed by a factor of 2.

17.a. The graph of *g* is the graph of *f* reflected about the *x*-axis and then shifted one unit up.

b. The graph of *h* is the graph of *f* reflected about the *y*-axis and then shifted one unit up.

19.a. The graph of *g* is the graph of *f* shifted one unit up.

b. The graph of *h* is the graph of *f* shifted one unit to the left.

21. e   **23.** g   **25.** i   **27.** b

29. l   **31.** d

33.

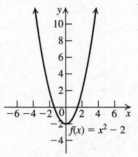

35.

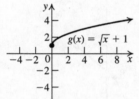

37.

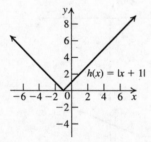

39.

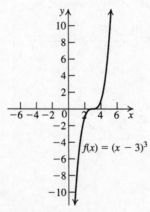

41.

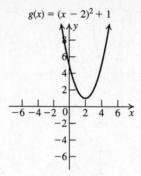

43.

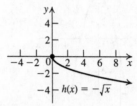

45.

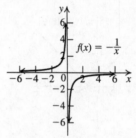

47.

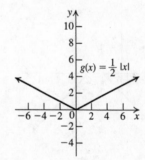

49.

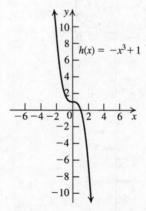

51.

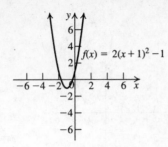

$f(x) = 2(x+1)^2 - 1$

53.

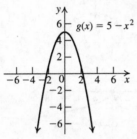

$g(x) = 5 - x^2$

55.

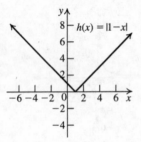

$h(x) = |1-x|$

57.

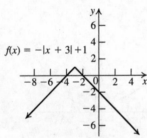

$f(x) = -|x+3| + 1$

59.

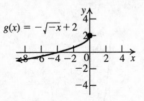

$g(x) = -\sqrt{-x} + 2$

61. $h(x) = 2[[x+1]]$

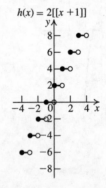

63. $y = x^3 + 2$ **65.** $y = -|x|$

67. $y = (x-3)^2 + 2$ **69.** $y = -\sqrt{x+3} - 2$

71. $y = 3(-x+4)^3 + 2$ **73.** $y = -2|x-4| - 3$

75.

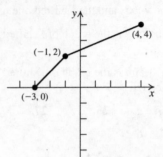

77.

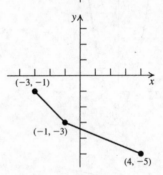

79.

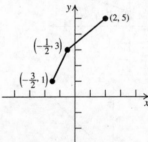

81.

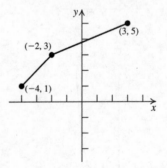

83.

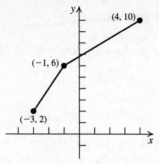

b.

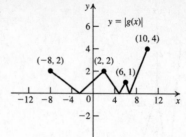

85.

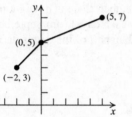

93.a.

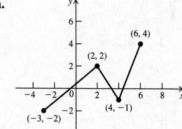

87.

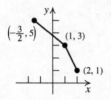

b.

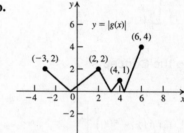

89.a.

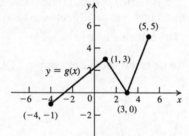

95.a.

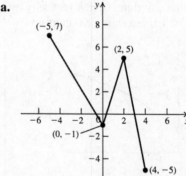

b.

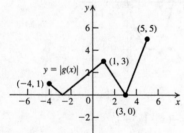

b.

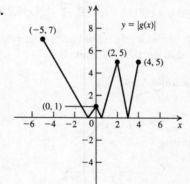

91.a.

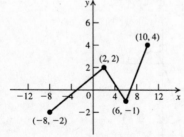

97.a.

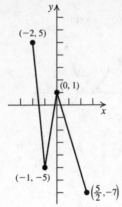

b.

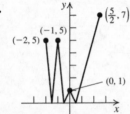

2.7 Applying the Concepts

99. $g(x) = f(x) + 800$

101. $p(x) = 1.02(f(x) + 500)$

103.a. Shift one unit right, stretch vertically by a factor of 10, and shift 5000 units up.

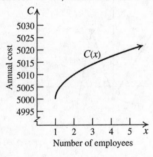

b. $C(400) = 5000 + 10\sqrt{400 - 1} = \$5199.75$

105.a. Shift one unit left, reflect across the x-axis, and shift up 109,561 units.

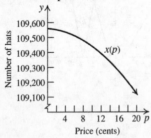

b. $69,160 = 109,561 - (p+1)^2$
$40,401 = (p+1)^2$
$\qquad 201 = p+1 \Rightarrow p = 200\cancel{c} = \$2.00$

c. $\qquad 0 = 109,561 - (p+1)^2$
$109,561 = (p+1)^2$
$\qquad 331 = p+1 \Rightarrow p = 330\cancel{c} = \$3.30$

107. The first coordinate gives the month; the second coordinate gives the hours of daylight. From March to September, there is daylight more than half of the day each day. From September to March, more than half of the day is dark each day.

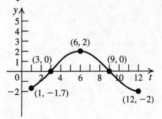

2.7 Beyond the Basics

109. The graph is shifted one unit right then reflected about the x-axis, and finally reflected about the y-axis. The equation is $g(x) = -\sqrt{1 - x}$.

111. Shift two units left and 4 units down.

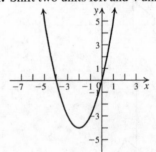

113. $f(x) = -x^2 + 2x = -\left(x^2 - 2x + 1\right) + 1$
$\qquad = -(x-1)^2 + 1$

Shift one unit right, reflect about the x-axis, shift one unit up.

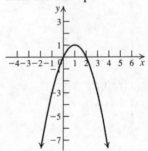

115. $f(x) = 2x^2 - 4x = 2(x^2 - 2x + 1) - 2$

$$= 2(x-1)^2 - 2$$

Shift one unit right, stretch vertically by a factor of 2, shift two units down.

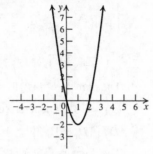

117. $f(x) = -2x^2 - 8x + 3 = -2(x^2 + 4x - 1.5)$

$$= -2(x^2 + 4x - 1.5 + 5.5) + 11$$

$$= -2(x+2)^2 + 11$$

Shift two units left, stretch vertically by a factor of 2, reflect across the x-axis, shift eleven units up.

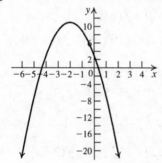

119.

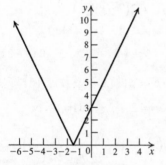

121.

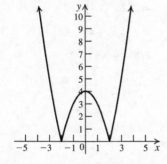

123.

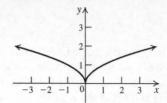

2.7 Critical Thinking/Discussion/Writing

125.a. The function is shifted three units to the right, so the x-intercept is $-2 + 3 = 1$. The y-intercept cannot be determined.

   b. The function is stretched horizontally by a factor of 5, so the x-intercept is not affected and remains at -2, while the y-intercept is $5 \cdot 3 = 15$.

   c. The function is reflected about the x-axis, so the x-intercept is not affected and remains at -2, while the y-intercept is -3.

   d. The function is reflected about the y-axis, so the x-intercept is 2, while the y-intercept is not changed and remains at 3.

126.a. The function is shifted two units to the left, so the x-intercept is $4 - 2 = 2$. The y-intercept cannot be determined.

   b. The function is stretched horizontally by a factor of 2, so the x-intercept is not affected and remains at 4, while the y-intercept is $2(-1) = -2$.

   c. The function is reflected about the x-axis, so the x-intercept is not affected and remains at 4, while the y-intercept is 1.

   d. The function is reflected in the y-axis, so the x-intercept is -4, while the y-intercept is not changed and remains at -1.

127.a. $g(x) = h(x-3) + 3$

   The graph of g is the graph of h shifted three units to the right and three units up.

   b. $g(x) = h(x-1) - 1$.

   The graph of g is the graph of h shifted one unit to the left and one unit down.

   c. $g(x) = 2h\left(\dfrac{1}{2}x\right)$.

   The graph of g is the graph of h stretched horizontally and vertically by a factor of 2.

d. $g(x) = -3h\left(-\dfrac{1}{3}x\right)$.

The graph of g is the graph of h stretched horizontally by a factor of 3, reflected about the y-axis, stretched vertically by a factor of 3, and reflected about the x-axis.

128. $y = f(x) = f\left(-\dfrac{1}{4}(-4x)\right)$.

Stretch the graph of $y = f(-4x)$ horizontally by a factor of 4 and reflect it about the y-axis.

2.7 Maintaining Skills

129. $\left(5x^2 + 5x + 7\right) + \left(x^2 + 9x - 4\right) = 6x^2 + 14x + 3$

131. $\left(5x^2 + 6x - 2\right) - \left(3x^2 - 9x + 1\right) = 2x^2 + 15x - 3$

133. $(x-2)\left(x^2 + 2x + 4\right)$
$$= x^3 + 2x^2 + 4x - 2x^2 - 4x - 8$$
$$= x^3 - 8$$

135. $f(x) = \dfrac{2x - 3}{x^2 - 5x + 6}$

The function is not defined when the denominator is zero.

$x^2 - 5x + 6 = 0 \Rightarrow (x-2)(x-3) = 0 \Rightarrow x = 2, 3$

The domain is $(-\infty, 2) \cup (2, 3) \cup (3, \infty)$.

137. $f(x) = \sqrt{2x - 3}$

The function is defined only if $2x - 3 \geq 0$.

$2x - 3 \geq 0 \Rightarrow 2x \geq 3 \Rightarrow x \geq \dfrac{3}{2}$

The domain is $\left[\dfrac{3}{2}, \infty\right)$.

2.8   Combining Functions; Composite Functions

2.8 Practice Problems

1. $f(x) = 3x - 1$, $g(x) = x^2 + 2$

$(f + g)(x) = f(x) + g(x)$
$$= 3x - 1 + x^2 + 2 = x^2 + 3x + 1$$
$(f - g)(x) = f(x) - g(x)$
$$= (3x - 1) - \left(x^2 + 2\right) = -x^2 + 3x - 3$$

$(fg)(x) = f(x) \cdot g(x)$
$$= (3x - 1)\left(x^2 + 2\right) = 3x^3 - x^2 + 6x - 2$$
$\left(\dfrac{f}{g}\right)(x) = \dfrac{f(x)}{g(x)} = \dfrac{3x - 1}{x^2 + 2}$

2. $f(x) = \sqrt{x - 1}$, $g(x) = \sqrt{3 - x}$

The domain of f is $[1, \infty)$ and the domain of g is $(-\infty, 3]$. The intersection of D_f and D_g,

$D_f \cap D_g = [1, 3]$.

The domain of fg is $[1, 3]$.

The domain of $\dfrac{f}{g}$ is $[1, 3)$.

The domain of $\dfrac{g}{f}$ is $(1, 3]$.

3. $f(x) = -5x$, $g(x) = x^2 + 1$

a. $(f \circ g)(0) = f(g(0))$
$$= f\left(0^2 + 1\right) = f(1) = -5$$

b. $(g \circ f)(0) = g(f(0))$
$$= g(-5 \cdot 0) = g(0) = 1$$

4. $f(x) = 2 - x$, $g(x) = 2x^2 + 1$

a. $(g \circ f)(x) = g(f(x)) = g(2 - x)$
$$= 2(2 - x)^2 + 1$$
$$= 2\left(4 - 4x + x^2\right) + 1$$
$$= 8 - 8x + 2x^2 + 1 = 2x^2 - 8x + 9$$

b. $(f \circ g)(x) = f(g(x)) = f\left(2x^2 + 1\right)$
$$= 2 - \left(2x^2 + 1\right) = 1 - 2x^2$$

c. $(g \circ g)(x) = g(g(x)) = g\left(2x^2 + 1\right)$
$$= 2\left(2x^2 + 1\right)^2 + 1$$
$$= 2\left(4x^4 + 4x^2 + 1\right) + 1$$
$$= 8x^4 + 8x^2 + 3$$

5. $f(x) = \sqrt{x+1}$, $g(x) = \dfrac{2}{x-3}$

Let $A = \{x \mid g(x) \text{ is defined}\}$.

$g(x)$ is not defined if $x = 3$, so

$A = (-\infty, 3) \cup (3, \infty)$.

Let $B = \{x \mid f(g(x)) \text{ is defined}\}$.

$f(g(x)) = \sqrt{\dfrac{2}{x-3} + 1} = \sqrt{\dfrac{2+x-3}{x-3}} = \sqrt{\dfrac{x-1}{x-3}}$

$f(g(x))$ is not defined if $x = 3$ or if $\dfrac{x-1}{x-3} < 0$.

$x - 1 = 0 \Rightarrow x = 1$

Interval	Test point	Value of $\dfrac{x-1}{x-3}$	Result
$(-\infty, 1]$	0	$\frac{1}{3}$	+
$[1, 3)$	2	-1	$-$
$(3, \infty)$	4	3	+

$f(g(x))$ is not defined for $[1, 3)$, so

$B = (-\infty, 1] \cup (3, \infty)$.

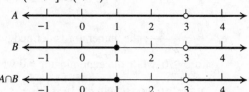

The domain of $f \circ g$ is

$A \cap B = (-\infty, 1] \cup (3, \infty)$.

6. $f(x) = \sqrt{x-1}$, $g(x) = \sqrt{4 - x^2}$

a. $(f \circ g)(x) = f(g(x)) = f\left(\sqrt{4-x^2}\right)$

$= \sqrt{\sqrt{4-x^2} - 1}$

The function $g(x) = \sqrt{4 - x^2}$ is defined for

$4 - x^2 \geq 0 \Rightarrow x^2 \leq 4 \Rightarrow -2 \leq x \leq 2$. So,

$A = [-2, 2]$.

The function $f(g(x))$ is defined for

$\sqrt{4-x^2} - 1 \geq 0 \Rightarrow \sqrt{4-x^2} \geq 1 \Rightarrow$

$4 - x^2 \geq 1 \Rightarrow -x^2 \geq -3 \Rightarrow x^2 \leq 3 \Rightarrow$

$-\sqrt{3} \leq x \leq \sqrt{3}$

So, $B = \left[-\sqrt{3}, \sqrt{3}\right]$.

The domain of $f \circ g$ is

$A \cap B = \left[-\sqrt{3}, \sqrt{3}\right]$.

b. $(g \circ f)(x) = g(f(x)) = \sqrt{4 - \left(\sqrt{x-1}\right)^2}$

$= \sqrt{4 - (x-1)} = \sqrt{5-x}$

The function $f(x) = \sqrt{x-1}$ is defined for

$x - 1 \geq 0 \Rightarrow x \geq 1$. So, $A = [1, \infty)$.

The function $g(f(x))$ is defined for

$5 - x \geq 0 \Rightarrow 5 \geq x$, or $x \leq 5$. So,

$B = (-\infty, 5]$.

The domain of $g \circ f$ is $A \cap B = [1, 5]$.

7. $H(x) = \dfrac{1}{\sqrt{2x^2 + 1}}$, $g(x) = \sqrt{2x^2 + 1}$

If $f(x) = \dfrac{1}{x}$, then

$H(x) = (f \circ g)(x) = f\left(\sqrt{2x^2+1}\right) = \dfrac{1}{\sqrt{2x^2+1}}$.

8.a. $A = f(g(t)) = f(g(3)) = f(3t)$

$= \pi(3t)^2 = 9\pi t^2$

b. $A = 9\pi t^2 = 9\pi(6)^2 = 324\pi$

The area covered by the oil slick is $324\pi \approx 1018$ square miles.

9.a. $r(x) = x - 4500$

b. $d(x) = x - 0.06x = 0.94x$

c. i. $(r \circ d)(x) = r(0.94x) = 0.94x - 4500$

ii. $(d \circ r)(x) = d(x - 4500) = 0.94(x - 4500)$
$= 0.94x - 4230$

d. $(d \circ r)(x) - (r \circ d)(x)$
$= (0.94x - 4230) - (0.94x - 4500)$
$= 270$

2.8 Basic Concepts and Skills

1. $(g - f)(2) = 4 - 12 = -8$

3. $(f \circ g)(x) = f(g(x)) = f(1) = 1$

5. False. $(f \circ g)(x) = f(g(x)) = f\left(\dfrac{1}{2x}\right) = \dfrac{1}{x}$

7.a. $(f + g)(-1) = f(-1) + g(-1)$
$= 2(-1) + -(-1) = -2 + 1 = -1$

b. $(f - g)(0) = f(0) - g(0) = 2(0) - (-0) = 0$

c. $(f \cdot g)(2) = f(2) \cdot g(2) = 2(2) \cdot (-2) = -8$

d. $\left(\dfrac{f}{g}\right)(1) = \dfrac{f(1)}{g(1)} = \dfrac{2(1)}{-1} = -2$

9.a. $(f + g)(-1) = f(-1) + g(-1)$
$= \dfrac{1}{\sqrt{-1+2}} + (2(-1) + 1) = 0$

b. $(f - g)(0) = f(0) - g(0)$
$= \dfrac{1}{\sqrt{0+2}} - (2(0) + 1) = \dfrac{\sqrt{2}}{2} - 1$

c. $(f \cdot g)(2) = f(2) \cdot g(2)$
$= \dfrac{1}{\sqrt{2+2}} \cdot (2(2) + 1) = \dfrac{5}{2}$

d. $\left(\dfrac{f}{g}\right)(1) = \dfrac{f(1)}{g(1)} = \dfrac{\frac{1}{\sqrt{1+2}}}{2(1)+1} = \dfrac{1}{3\sqrt{3}} = \dfrac{\sqrt{3}}{9}$

11.a. $f + g = x^2 + x - 3$; domain: $(-\infty, \infty)$

b. $f - g = x - 3 - x^2 = -x^2 + x - 3$;
domain: $(-\infty, \infty)$

c. $f \cdot g = (x - 3)x^2 = x^3 - 3x^2$;
domain: $(-\infty, \infty)$

d. $\dfrac{f}{g} = \dfrac{x - 3}{x^2}$; domain: $(-\infty, 0) \cup (0, \infty)$

e. $\dfrac{g}{f} = \dfrac{x^2}{x - 3}$; domain: $(-\infty, 3) \cup (3, \infty)$

13.a. $f + g = \left(x^3 - 1\right) + \left(2x^2 + 5\right) = x^3 + 2x^2 + 4$;
domain: $(-\infty, \infty)$

b. $f - g = \left(x^3 - 1\right) - \left(2x^2 + 5\right) = x^3 - 2x^2 - 6$;
domain: $(-\infty, \infty)$

c. $f \cdot g = \left(x^3 - 1\right)\left(2x^2 + 5\right)$
$= 2x^5 + 5x^3 - 2x^2 - 5$;
domain: $(-\infty, \infty)$

d. $\dfrac{f}{g} = \dfrac{x^3 - 1}{2x^2 + 5}$; domain: $(-\infty, \infty)$

e. $\dfrac{g}{f} = \dfrac{2x^2 + 5}{x^3 - 1}$; domain: $(-\infty, 1) \cup (1, \infty)$

15.a. $f + g = 2x + \sqrt{x} - 1$; domain: $[0, \infty)$

b. $f - g = 2x - \sqrt{x} - 1$; domain: $[0, \infty)$

c. $f \cdot g = (2x - 1)\sqrt{x} = 2x\sqrt{x} - \sqrt{x}$;
domain: $[0, \infty)$

d. $\dfrac{f}{g} = \dfrac{2x - 1}{\sqrt{x}}$; domain: $(0, \infty)$

e. $\dfrac{g}{f} = \dfrac{\sqrt{x}}{2x - 1}$; the numerator is defined only
for $x \geq 0$, while the denominator $= 0$ when
$x = \dfrac{1}{2}$, so the domain is $\left[0, \dfrac{1}{2}\right) \cup \left(\dfrac{1}{2}, \infty\right)$.

17.a. $f + g = \dfrac{2}{x + 1} + \dfrac{x}{x + 1} = \dfrac{2 + x}{x + 1}$
Neither f nor g is defined for $x = -1$, so the
domain is $(-\infty, -1) \cup (-1, \infty)$.

b. $f - g = \dfrac{2}{x + 1} - \dfrac{x}{x + 1} = \dfrac{2 - x}{x + 1}$.
domain: $(-\infty, -1) \cup (-1, \infty)$.

c. $f \cdot g = \left(\dfrac{2}{x + 1}\right)\left(\dfrac{x}{x + 1}\right) = \dfrac{2x}{(x + 1)^2}$.
domain: $(-\infty, -1) \cup (-1, \infty)$.

d. $\dfrac{f}{g} = \dfrac{\frac{2}{x + 1}}{\frac{x}{x + 1}} = \dfrac{2}{x}$
Neither f nor g is defined for $x = -1$, and f/g
is not defined for $x = 0$, so the domain is
$(-\infty, -1) \cup (-1, 0) \cup (0, \infty)$.

e. $\dfrac{f}{g} = \dfrac{\frac{x}{x+1}}{\frac{2}{x+1}} = \dfrac{x}{2}$

Neither f nor g is defined for $x = -1$, so the domain is $(-\infty, -1) \cup (-1, \infty)$.

19. $f(x) = \dfrac{x^2}{x+1}$; $g(x) = \dfrac{2x}{x^2-1}$

a. $f + g = \dfrac{x^2}{x+1} + \dfrac{2x}{x^2-1} = \dfrac{x^2(x-1)}{x^2-1} + \dfrac{2x}{x^2-1}$

$= \dfrac{x^3 - x^2 + 2x}{x^2-1}$

f is not defined for $x = -1$, g is not defined for $x = \pm 1$, and $f + g$ is not defined for either -1 or 1, so the domain is

$(-\infty, -1) \cup (-1, 1) \cup (1, \infty)$.

b. $f - g = \dfrac{x^2}{x+1} - \dfrac{2x}{x^2-1} = \dfrac{x^2(x-1)}{x^2-1} - \dfrac{2x}{x^2-1}$

$= \dfrac{x^3 - x^2 - 2x}{x^2-1} = \dfrac{x(x^2 - x - 2)}{x^2-1}$

$= \dfrac{x(x-2)(x+1)}{(x-1)(x+1)} = \dfrac{x^2 - 2x}{x-1}$

f is not defined for $x = -1$, g is not defined for $x = \pm 1$, and $f - g$ is not defined for 1, so the domain is $(-\infty, -1) \cup (-1, 1) \cup (1, \infty)$.

c. $f \cdot g = \dfrac{x^2}{x+1} \cdot \dfrac{2x}{x^2-1} = \dfrac{2x^3}{x^3 + x^2 - x - 1}$

f is not defined for $x = -1$, g is not defined for $x = \pm 1$, and fg is not defined for either -1 or 1, so the domain is

$(-\infty, -1) \cup (-1, 1) \cup (1, \infty)$.

d. $\dfrac{f}{g} = \dfrac{\frac{x^2}{x+1}}{\frac{2x}{x^2-1}} = \dfrac{x^2}{x+1} \cdot \dfrac{x^2-1}{2x} = \dfrac{x(x-1)}{2}$

f is not defined for $x = -1$, g is not defined for $x = \pm 1$, and f/g is not defined for either -1 or 1, so the domain is

$(-\infty, -1) \cup (-1, 1) \cup (1, \infty)$.

e. $\dfrac{g}{f} = \dfrac{\frac{2x}{x^2-1}}{\frac{x^2}{x+1}} = \dfrac{2x}{x^2-1} \cdot \dfrac{x+1}{x^2} = \dfrac{2}{x(x-1)}$

Neither f nor g is defined for $x = -1$ and g/f is not defined for $x = 0$ or $x = 1$, so the domain is $(-\infty, -1) \cup (-1, 0) \cup (0, 1) \cup (1, \infty)$.

21. $f(x) = \dfrac{2x}{x^2-16}$; $g(x) = \dfrac{2x-7}{x^2-7x+12}$

a. $f + g = \dfrac{2x}{x^2-16} + \dfrac{2x-7}{x^2-7x+12}$

$= \dfrac{2x}{(x-4)(x+4)} + \dfrac{2x-7}{(x-4)(x-3)}$

$= \dfrac{2x(x-3) + (2x-7)(x+4)}{(x-4)(x+4)(x-3)}$

$= \dfrac{2x^2 - 6x + 2x^2 + x - 28}{x^3 - 3x^2 - 16x + 48}$

$= \dfrac{4x^2 - 5x - 28}{x^3 - 3x^2 - 16x + 48}$

f is not defined for $x = -4$ and $x = 4$, g is not defined for $x = 3$ and $x = 4$, and $f + g$ is not defined for -4, 4, or 3, so the domain is

$(-\infty, -4) \cup (-4, 3) \cup (3, 4) \cup (4, \infty)$.

b. $f - g = \dfrac{2x}{x^2-16} - \dfrac{2x-7}{x^2-7x+12}$

$= \dfrac{2x}{(x-4)(x+4)} - \dfrac{2x-7}{(x-4)(x-3)}$

$= \dfrac{2x(x-3) - (2x-7)(x+4)}{(x-4)(x+4)(x-3)}$

$= \dfrac{2x^2 - 6x - (2x^2 + x - 28)}{(x-4)(x+4)(x-3)}$

$= \dfrac{-7x + 28}{(x-4)(x+4)(x-3)}$

$= \dfrac{-7(x-4)}{(x-4)(x+4)(x-3)}$

$= -\dfrac{7}{(x+4)(x-3)}$

f is not defined for $x = -4$ and $x = 4$, g is not defined for $x = 3$ and $x = 4$, and $f + g$ is not defined for -4, 4, or 3, so the domain is

$(-\infty, -4) \cup (-4, 3) \cup (3, 4) \cup (4, \infty)$.

c. $f \cdot g = \dfrac{2x}{x^2 - 16} \cdot \dfrac{2x - 7}{x^2 - 7x + 12}$

$= \dfrac{4x^2 - 14x}{x^4 - 7x^3 - 4x^2 + 112x - 192}$

f is not defined for $x = -4$ and $x = 4$, g is not defined for $x = 3$ and $x = 4$, and fg is not defined for -4, 4 or 3, so the domain is
$(-\infty, -4) \cup (-4, 3) \cup (3, 4) \cup (4, \infty)$.

d. $\dfrac{f}{g} = \dfrac{\dfrac{2x}{x^2 - 16}}{\dfrac{2x - 7}{x^2 - 7x + 12}}$

$= \dfrac{2x}{(x - 4)(x + 4)} \cdot \dfrac{(x - 4)(x - 3)}{2x - 7}$

$= \dfrac{2x^2 - 6x}{2x^2 + x - 28}$

f is not defined for $x = -4$ and $x = 4$, g is not defined for $x = 3$ and $x = 4$, and f/g is not defined for $x = -4$ and $x = 7/2$, so the domain is

$(-\infty, -4) \cup (-4, 3) \cup \left(3, \dfrac{7}{2}\right) \cup \left(\dfrac{7}{2}, 4\right)$
$\cup (4, \infty)$.

e. $\dfrac{g}{f} = \dfrac{\dfrac{2x - 7}{x^2 - 7x + 12}}{\dfrac{2x}{x^2 - 16}}$

$= \dfrac{2x - 7}{(x - 4)(x - 3)} \cdot \dfrac{(x - 4)(x + 4)}{2x}$

$= \dfrac{(2x - 7)(x + 4)}{2x(x - 3)} = \dfrac{2x^2 + x - 28}{2x^2 - 6x}$

f is not defined for $x = -4$ and $x = 4$, g is not defined for $x = 3$ and $x = 4$, and g/f is not defined for $x = 0$ and $x = 3$ so the domain is
$(-\infty, -4) \cup (-4, 0) \cup (0, 3) \cup (3, 4)$
$\cup (4, \infty)$.

23. $f(x) = \sqrt{x - 1}$, $g(x) = \sqrt{5 - x}$

$f(x)$ is defined if $x - 1 \geq 0 \Rightarrow x \geq 1$. Thus, $D_1 = [1, \infty)$. $g(x)$ is defined if $5 - x \geq 0 \Rightarrow$
$-x \geq -5 \Rightarrow x \leq 5$. Thus, $D_2 = (-\infty, 5]$.

a. The domain of fg is $D_1 \cap D_2 = [1, 5]$.

b. f/g is not defined for $\sqrt{5 - x} = 0 \Rightarrow x = 5$, so, the domain of f/g is $[1, 5)$.

25. $f(x) = \sqrt{x + 2}$, $g(x) = \sqrt{9 - x^2}$

$f(x)$ is defined if $x + 2 \geq 0 \Rightarrow x \geq -2$. Thus, $D_1 = [-2, \infty)$. $g(x)$ is defined if $9 - x^2 \geq 0 \Rightarrow$
$9 \geq x^2 \Rightarrow -3 \leq x \leq 3$. Thus, $D_2 = [-3, 3]$.

a. The domain of fg is $D_1 \cap D_2 = [-2, 3]$.

b. f/g is not defined for
$\sqrt{9 - x^2} = 0 \Rightarrow x = \pm 3$.
Thus, the domain of f/g is $[-2, 3)$.

27. $(g \circ f)(x) = 2(x^2 - 1) + 3 = 2x^2 + 1;$
$(g \circ f)(2) = 2(2^2 - 1) + 3 = 9;$
$(g \circ f)(-3) = 2((-3)^2 - 1) + 3 = 19$

29. $(f \circ g)(2) = 2(2(2^2) - 3) + 1 = 11$

31. $(f \circ g)(-3) = 2(2(-3)^2 - 3) + 1 = 31$

33. $(g \circ f)(a) = 2(2a + 1)^2 - 3$
$= 2(4a^2 + 4a + 1) - 3$
$= 8a^2 + 8a - 1$

35. $(f \circ f)(1) = 2(2(1) + 1) + 1 = 7$

37. $(f \circ g)(x) = \dfrac{2}{\dfrac{1}{x} + 1} = \dfrac{2}{\dfrac{x + 1}{x}} = \dfrac{2x}{x + 1}$

The domain of g is $(-\infty, 0) \cup (0, \infty)$. Since -1 is not in the domain of f, we must exclude those values of x that make $g(x) = -1$.
$\dfrac{1}{x} = -1 \Rightarrow x = -1$

Thus, the domain of $f \circ g$ is
$(-\infty, -1) \cup (-1, 0) \cup (0, \infty)$.

39. $(f \circ g)(x) = \sqrt{(2 - 3x) - 3} = \sqrt{-1 - 3x}$.
The domain of g is $(-\infty, \infty)$. Since f is not defined for $(-\infty, 3)$, we must exclude those values of x that make $g(x) > 3$:
$2 - 3x < 3 \Rightarrow -3x < 1 \Rightarrow x > -\dfrac{1}{3}$

Thus, the domain of $f \circ g$ is $\left(-\infty, -\dfrac{1}{3}\right]$.

41. $(f \circ g)(x) = |x^2 - 1|$; domain: $(-\infty, \infty)$

43. $f(x) = 2x - 3,\ g(x) = x + 4$

Domain of $f = (-\infty,\ \infty)$

Domain of $g = (-\infty,\ \infty)$

a. $(f \circ g)(x) = 2(x+4) - 3 = 2x + 5;$
Domain: $(-\infty, \infty)$

b. $(g \circ f)(x) = (2x-3) + 4 = 2x + 1;$
Domain: $(-\infty, \infty)$

c. $(f \circ f)(x) = 2(2x-3) - 3 = 4x - 9;$
Domain: $(-\infty, \infty)$

d. $(g \circ g)(x) = (x+4) + 4 = x + 8;$
Domain: $(-\infty, \infty)$

45. $f(x) = 1 - 2x,\ g(x) = 1 + x^2$

Domain of $f = (-\infty,\ \infty)$

Domain of $g = (-\infty,\ \infty)$

a. $(f \circ g)(x) = 1 - 2(1 + x^2) = -2x^2 - 1;$
Domain: $(-\infty, \infty)$

b. $(g \circ f)(x) = 1 + (1 - 2x)^2 = 4x^2 - 4x + 2;$
Domain: $(-\infty, \infty)$

c. $(f \circ f)(x) = 1 - 2(1 - 2x) = 4x - 1;$
Domain: $(-\infty, \infty)$

d. $(g \circ g)(x) = 1 + (1 + x^2)^2 = x^4 + 2x^2 + 2;$
Domain: $(-\infty, \infty)$

47. $f(x) = x^2,\ g(x) = \sqrt{x}$

Domain of $f = (-\infty,\ \infty)$

Domain of $g = [0,\ \infty)$

a. $(f \circ g)(x) = \left(\sqrt{x}\right)^2 = x;$ domain: $[0, \infty)$

b. $(g \circ f)(x) = \sqrt{x^2} = |x|;$ domain: $(-\infty, \infty)$

c. $(f \circ f)(x) = \left(x^2\right)^2 = x^4;$ domain: $(-\infty, \infty)$

d. $(g \circ g)(x) = \sqrt{\sqrt{x}} = \sqrt[4]{x};$ domain: $[0, \infty)$

49. $f(x) = \dfrac{1}{2x-1},\ g(x) = \dfrac{1}{x^2}$

Domain of $f = \left(-\infty, \dfrac{1}{2}\right) \cup \left(\dfrac{1}{2}, \infty\right).$

Domain of $g = (-\infty, 0) \cup (0, \infty).$

a. $(f \circ g)(x) = \dfrac{1}{2\left(\dfrac{1}{x^2}\right) - 1} = \dfrac{1}{\dfrac{2 - x^2}{x^2}}$

$= \dfrac{x^2}{2 - x^2} = -\dfrac{x^2}{x^2 - 2}.$

The domain of g is $(-\infty, 0) \cup (0, \infty)$. Since $\frac{1}{2}$ is not in the domain of f, we must find those values of x that make $g(x) = \frac{1}{2}$.

$\dfrac{1}{x^2} = \dfrac{1}{2} \Rightarrow x^2 = 2 \Rightarrow x = \pm\sqrt{2}$

Thus, the domain of $f \circ g$ is

$\left(-\infty, -\sqrt{2}\right) \cup \left(-\sqrt{2}, 0\right) \cup \left(0, \sqrt{2}\right) \cup \left(\sqrt{2}, \infty\right).$

b. $(g \circ f) = \dfrac{1}{\left(\dfrac{1}{2x-1}\right)^2} = (2x - 1)^2$

The domain of f is $\left(-\infty, \dfrac{1}{2}\right) \cup \left(\dfrac{1}{2}, \infty\right)$. Since 0 is not in the domain of g, we must find those values of x that make $f(x) = 0$. However, there are no such values, so the domain of $g \circ f$ is $\left(-\infty, \dfrac{1}{2}\right) \cup \left(\dfrac{1}{2}, \infty\right)$.

c. $(f \circ f)(x) = \dfrac{1}{2\left(\dfrac{1}{2x-1}\right) - 1} = \dfrac{1}{\dfrac{2 - 2x + 1}{2x - 1}}$

$= \dfrac{2x - 1}{3 - 2x} = -\dfrac{2x - 1}{2x - 3}.$

The domain of f is $\left(-\infty, \dfrac{1}{2}\right) \cup \left(\dfrac{1}{2}, \infty\right)$.

$-\dfrac{2x - 1}{2x - 3}$ is defined for $\left(-\infty, \dfrac{3}{2}\right) \cup \left(\dfrac{3}{2}, \infty\right)$,

so the domain of $f \circ f$ is

$\left(-\infty, \dfrac{1}{2}\right) \cup \left(\dfrac{1}{2}, \dfrac{3}{2}\right) \cup \left(\dfrac{3}{2}, \infty\right).$

d. $(g \circ g)(x) = \dfrac{1}{\left(\dfrac{1}{x^2}\right)^2} = x^4.$

The domain of g is $(-\infty, 0) \cup (0, \infty)$, while $g \circ g = x^4$ is defined for all real numbers. Thus, the domain of $g \circ g$ is $(-\infty, 0) \cup (0, \infty)$.

51. $f(x) = \sqrt{x-1}$, $g(x) = \sqrt{4-x}$

Domain of $f = [1, \infty)$.

Domain of $g = (-\infty, 4]$.

  a. $(f \circ g)(x) = \sqrt{\sqrt{4-x}-1}$; domain: $(-\infty, 3]$

  b. $(g \circ f)(x) = \sqrt{4-\sqrt{x-1}}$; domain: $[1, 17]$

  c. $(f \circ f)(x) = \sqrt{\sqrt{x-1}-1}$; domain: $[2, \infty)$

  d. $(g \circ g)(x) = \sqrt{4-\sqrt{4-x}}$; domain: $[-12, 4]$

53. $f(x) = \sqrt{x^2-1}$, $g(x) = \sqrt{4-x^2}$

Domain of $f = (-\infty, -1] \cup [1, \infty)$.

Domain of $g = [-2, 2]$

  a. $(f \circ g)(x) = \sqrt{\left(\sqrt{4-x^2}\right)^2 - 1}$

          $= \sqrt{4-x^2-1} = \sqrt{3-x^2}$

    Domain: $\left[-\sqrt{3}, \sqrt{3}\right]$

  b. $(g \circ f)(x) = \sqrt{4 - \left(\sqrt{x^2-1}\right)^2}$

          $= \sqrt{4-x^2+1} = \sqrt{5-x^2}$

    Domain: $\left[-\sqrt{5}, -1\right] \cup \left[1, \sqrt{5}\right]$

  c. $(f \circ f)(x) = \sqrt{\left(\sqrt{x^2-1}\right)^2 - 1} = \sqrt{x^2-2}$

    Domain: $f = \left(-\infty, -\sqrt{2}\right] \cup \left[\sqrt{2}, \infty\right)$.

  d. $(g \circ g)(x) = \sqrt{4 - \left(\sqrt{4-x^2}\right)^2}$

          $= \sqrt{4-4+x^2} = \sqrt{x^2} = |x|$

    Domain: $[-2, 2]$

55. $f(x) = 1 + \dfrac{1}{x}$, $g(x) = \dfrac{1+x}{1-x}$

Domain of $f = (-\infty, 0) \cup (0, \infty)$.

Domain of $g = (-\infty, 1) \cup (1, \infty)$.

  a. $(f \circ g)(x) = 1 + \dfrac{1}{\dfrac{1+x}{1-x}} = 1 + \dfrac{1-x}{1+x}$

         $= \dfrac{1+x+1-x}{1+x} = \dfrac{2}{1+x}.$

The domain of g is $(-\infty, 1) \cup (1, \infty)$.

Since 0 is not in the domain of f, we must find those values of x that make $g(x) = 0$.

$\dfrac{1+x}{1-x} = 0 \Rightarrow 1+x = 0 \Rightarrow x = -1$

Thus, the domain of $f \circ g$ is

$(-\infty, -1) \cup (-1, 1) \cup (1, \infty)$.

  b. $(g \circ f)(x) = \dfrac{1 + \left(1 + \dfrac{1}{x}\right)}{1 - \left(1 + \dfrac{1}{x}\right)} = \dfrac{2 + \dfrac{1}{x}}{-\dfrac{1}{x}} \cdot \dfrac{x}{x}$

         $= \dfrac{2x+1}{-1} = -2x-1$

The domain of f is $(-\infty, 0) \cup (0, \infty)$. Since 1 is not in the domain of g, we must find those values of x that make $f(x) = 1$.

$1 + \dfrac{1}{x} = 1 \Rightarrow \dfrac{1}{x} = 0$

There are no values of x that make this true, so there are no additional values to be excluded from the domain of $g \circ f$. Thus, the domain of $g \circ f$ is $(-\infty, 0) \cup (0, \infty)$.

  c. $(f \circ f)(x) = 1 + \dfrac{1}{1 + \dfrac{1}{x}} = 1 + \dfrac{1}{\dfrac{x+1}{x}} = 1 + \dfrac{x}{x+1}$

         $= \dfrac{2x+1}{x+1}$

The domain of f is $(-\infty, 0) \cup (0, \infty)$.

$\dfrac{2x+1}{x+1}$ is defined for $(-\infty, -1) \cup (-1, \infty)$, so the domain of $f \circ f$ is

$(-\infty, -1) \cup (-1, 0) \cup (0, \infty)$.

  d. $(g \circ g)(x) = \dfrac{1 + \dfrac{1+x}{1-x}}{1 - \dfrac{1+x}{1-x}} = \dfrac{\dfrac{2}{1-x}}{\dfrac{-2x}{1-x}} = -\dfrac{1}{x}.$

The domain of g is $(-\infty, 1) \cup (1, \infty)$,

while $-\dfrac{1}{x}$ is defined for $(-\infty, 0) \cup (0, \infty)$.

The domain of $g \circ g$ is

$(-\infty, 0) \cup (0, 1) \cup (1, \infty)$.

In exercises 57–66, sample answers are given. Other answers are possible.

57. $H(x) = \sqrt{x+2} \Rightarrow f(x) = \sqrt{x}, g(x) = x+2$

59. $H(x) = \left(x^2-3\right)^{10} \Rightarrow f(x) = x^{10}, g(x) = x^2 - 3$

61. $H(x) = \dfrac{1}{3x-5} \Rightarrow f(x) = \dfrac{1}{x}, g(x) = 3x-5$

63. $H(x) = \sqrt[3]{x^2-7} \Rightarrow f(x) = \sqrt[3]{x}, g(x) = x^2-7$

65. $H(x) = \dfrac{1}{\left|x^3-1\right|} \Rightarrow f(x) = \dfrac{1}{|x|}, g(x) = x^3-1$

2.8 Applying the Concepts

67.a. $f(x)$ is the cost function.

   b. $g(x)$ is the revenue function.

   c. $h(x)$ is the selling price of x shirts including sales tax.

   d. $P(x)$ is the profit function.

69.a. $P(x) = R(x) - C(x) = 25x - (350+5x)$
$= 20x - 350$

   b. $P(20) = 20(20) - 350 = 50.$ This represents the profit when 20 radios are sold.

   c. $P(x) = 20x - 350; 500 = 20x - 350 \Rightarrow x = 43$

   d. $C = 350 + 5x \Rightarrow x = \dfrac{C-350}{5} = x(C).$

$(R \circ x)(C) = 25\left(\dfrac{C-350}{5}\right) = 5C - 1750.$

This function represents the revenue in terms of the cost C.

71.a. $f(x) = 0.7x$

   b. $g(x) = x - 5$

   c. $(g \circ f)(x) = 0.7x - 5$

   d. $(f \circ g)(x) = 0.7(x-5)$

   e. $(f \circ g) - (g \circ f) = 0.7(x-5) - (0.7x-5)$
$= 0.7x - 3.5 - 0.7x + 5$
$= \$1.50$

73.a. $f(x) = 1.1x; g(x) = x + 8$

   b. $(f \circ g)(x) = 1.1(x+8) = 1.1x + 8.8$
This represents a final test score computed by first adding 8 points to the original score and then increasing the total by 10%.

   c. $(g \circ f)(x) = 1.1x + 8$
This represents a final test score computed by first increasing the original score by 10% and then adding 8 points.

   d. $(f \circ g)(70) = 1.1(70+8) = 85.8;$
$(g \circ f)(70) = 1.1(70) + 8 = 85.0;$

   e. $(f \circ g)(x) \neq (g \circ f)(x)$

   f. **(i)** $(f \circ g)(x) = 1.1x + 8.8 \geq 90 \Rightarrow x \geq 73.82$

     (ii) $(g \circ f)(x) = 1.1x + 8 \geq 90 \Rightarrow x \geq 74.55$

75.a. $f(x) = \pi x^2$

   b. $g(x) = \pi(x+30)^2$

   c. $g(x) - f(x)$ represents the area between the fountain and the fence.

   d. The circumference of the fence is $2\pi(x+30)$.
$10.5(2\pi(x+30)) = 4200 \Rightarrow$
$\pi(x+30) = 200 \Rightarrow$
$\pi x + 30\pi = 200 \Rightarrow \pi x = 200 - 30\pi.$

$g(x) - f(x) = \pi(x+30)^2 - \pi x^2$
$= \pi(x^2 + 60x + 900) - \pi x^2$
$= 60\pi x + 900\pi.$ Now substitute $200 - 30\pi$ for πx to compute the estimate:
$1.75[60(200 - 30\pi) + 900\pi]$
$= 1.75(12,000 - 900\pi) \approx \$16,052.$

77.a. $(f \circ g)(t) = \pi(2t+1)^2$

   b. $A(t) = f(2t+1) = \pi(2t+1)^2$

   c. They are the same.

2.8 Beyond the Basics

79.a. When you are looking for the domain of the sum of two functions that are given as sets, you are looking for the intersection of their domains. Since the x-values that f and g have in common are −2, 1, and 3, the domain of $f + g$ is {−2, 1, 3}. Now add the y-values.
$(f + g)(-2) = 3 + 0 = 3$
$(f + g)(1) = 2 + (-2) = 0$
$(f + g)(3) = 0 + 2 = 2$
Thus, $f + g = \{(-2, 3), (1, 0), (3, 2)\}.$

b. When you are looking for the domain of the product of two functions that are given as sets, you are looking for the intersection of their domains. Since the x-values that f and g have in common are -2, 1, and 3, the domain of $f+g$ is $\{-2, 1, 3\}$. Now multiply the y-values.

$(fg)(-2) = 3 \cdot 0 = 0$

$(fg)(1) = 2 \cdot (-2) = -4$

$(fg)(3) = 0 \cdot 2 = 0$

Thus, $fg = \{(-2, 0), (1, -4), (3, 0)\}$.

c. When you are looking for the domain of the quotient of two functions that are given as sets, you are looking for the intersection of their domains and values of x that do not cause the denominator to equal zero. The x-values that f and g have in common are -2, 1, and 3; however, $g(-2) = 0$, so the domain is $\{1, 3\}$. Now divide the y-values.

$\left(\dfrac{f}{g}\right)(1) = \dfrac{2}{-2} = -1; \quad \left(\dfrac{f}{g}\right)(3) = \dfrac{0}{2} = 0$

Thus, $\dfrac{f}{g} = \{(1, -1), (3, 0)\}$.

d. When you are looking for the domain of the composition of two functions that are given as sets, you are looking for values that come from the domain of the inside function and when you plug those values of x into the inside function, the output is in the domain of the outside function.

$f(g(-2)) = f(0)$, which is undefined

$f(g(0)) = f(2) = 1$,

$f(g(1)) = f(-2) = 3$,

$f(g(3)) = f(2) = 1$

Thus, $f \circ g = \{(0, 1), (1, 3), (3, 1)\}$.

81.a. $f(-x) = h(-x) + h(-(-x)) = h(-x) + h(x)$
$= f(x) \Rightarrow f(x)$ is an even function.

b. $g(-x) = h(-x) - h(-(-x)) = h(-x) - h(x)$
$= -g(x) \Rightarrow g(x)$ is an odd function.

c. $\begin{cases} f(x) = h(x) + h(-x) \\ g(x) = h(x) - h(-x) \Rightarrow \end{cases}$
$f(x) + g(x) = 2h(x) \Rightarrow$
$h(x) = \dfrac{f(x) + g(x)}{2} = \dfrac{f(x)}{2} + \dfrac{g(x)}{2} \Rightarrow$
$h(x)$ is the sum of an even function and an odd function.

83. $f(x) = \sqrt{\dfrac{1 - |x|}{2 - |x|}}$

$f(x)$ is defined if $\dfrac{1 - |x|}{2 - |x|} \geq 0$ and $2 - |x| \neq 0$.

$2 - |x| = 0 \Rightarrow 2 = |x| \Rightarrow x = \pm 2$

Thus, the values -2 and 2 are not in the domain of f.

$\dfrac{1 - |x|}{2 - |x|} \geq 0$ if $1 - |x| \geq 0$ and $2 - |x| > 0$, or if

$1 - |x| \leq 0$ and $2 - |x| < 0$.

Case 1: $1 - |x| \geq 0$ and $2 - |x| > 0$.

$1 - |x| \geq 0 \Rightarrow 1 \geq |x| \Rightarrow -1 \leq x \leq 1$

$2 - |x| > 0 \Rightarrow 2 > |x| \Rightarrow -2 < x < 2$

Thus, $1 - |x| \geq 0$ and $2 - |x| > 0 \Rightarrow -1 \leq x \leq 1$.

Case 2: $1 - |x| \leq 0$ and $2 - |x| < 0$.

$1 - |x| \leq 0 \Rightarrow 1 \leq |x| \Rightarrow (-\infty, -1] \cup [1, \infty)$

$2 - |x| < 0 \Rightarrow 3 \leq |x| \Rightarrow (-\infty, -2) \cup (2, \infty)$

Thus, $1 - |x| \leq 0$ and $2 - |x| < 0 \Rightarrow$

$(-\infty, -2) \cup (2, \infty)$.

The domain of f is $(-\infty, -2) \cup [-1, 1] \cup (2, \infty)$.

2.8 Critical Thinking/Discussion/Writing

85.a. The domain of $f(x)$ is $(-\infty, 0) \cup [1, \infty)$.

b. The domain of $g(x)$ is $[0, 2]$.

c. The domain of $f(x) + g(x)$ is $[1, 2]$.

d. The domain of $\dfrac{f(x)}{g(x)}$ is $[1, 2)$.

86.a. The domain of f is $(-\infty, 0)$. The domain of $f \circ f$ is $\varnothing$ because $f \circ f = \dfrac{1}{\sqrt{-\dfrac{1}{\sqrt{-x}}}}$ and

the denominator is the square root of a negative number.

b. The domain of f is $(-\infty, 1)$. The domain of $f \circ f$ is $(-\infty, 0)$ because

$f \circ f = \dfrac{1}{\sqrt{1 - \dfrac{1}{\sqrt{1 - x}}}}$ and the denominator

must be greater than 0. If $x = 0$, then the denominator $= 0$.

87.a. The sum of two even functions is an even function. $f(x) = f(-x)$ and $g(x) = g(-x) \Rightarrow$

$$(f + g)(x) = f(x) + g(x) = f(-x) + g(-x)$$
$$= (f + g)(-x).$$

b. The sum of two odd functions is an odd function.

$$f(-x) = -f(x) \text{ and } g(-x) = -g(x) \Rightarrow$$
$$(f + g)(-x) = f(-x) + g(-x) = -f(x) - g(x)$$
$$= -(f + g)(x).$$

c. The sum of an even function and an odd function is neither even nor odd.

$$f(x) \text{ even} \Rightarrow f(x) = f(-x) \text{ and } g(x) \text{ odd} \Rightarrow$$
$$g(-x) = -g(x) \Rightarrow f(-x) + g(-x) =$$
$$f(x) + (-g(x)), \text{ which is neither even nor odd.}$$

d. The product of two even functions is an even function. $f(x) = f(-x)$ and $g(x) = g(-x) \Rightarrow$

$$(f \cdot g)(x) = f(x) \cdot g(x) = f(-x) \cdot (g(-x))$$
$$= (f \cdot g)(-x).$$

e. The product of two odd functions is an even function.

$$f(-x) = -f(x) \text{ and } g(-x) = -g(x) \Rightarrow$$
$$(f \cdot g)(-x) = f(-x) \cdot g(-x) = -f(x) \cdot (-g(x))$$
$$= (f \cdot g)(x).$$

f. The product of an even function and an odd function is an odd function.

$$f(x) \text{ even} \Rightarrow f(x) = f(-x) \text{ and } g(x) \text{ odd} \Rightarrow$$
$$g(-x) = -g(x) \Rightarrow$$
$$f(-x) \cdot g(-x) = f(x) \cdot (-g(x)) = -(f \cdot g)(x)$$

88.a. $f(-x) = -f(x)$ and $g(-x) = -g(x) \Rightarrow$

$$(f \circ g)(-x) = f(g(-x)) = f(-g(x)) =$$
$$-f(g(x)) \Rightarrow (f \circ g)(x) \text{ is odd.}$$

b. $f(x) = f(-x)$ and $g(x) = g(-x) \Rightarrow$

$$(f \circ g)(-x) = f(g(-x)) = f(g(x)) \Rightarrow$$
$$(f \circ g)(x) \text{ is even.}$$

c. $f(x) \text{ odd} \Rightarrow f(-x) = -f(x)$ and

$$g(x) \text{ even} \Rightarrow g(x) = g(-x) \Rightarrow (f \circ g)(-x)$$
$$f(g(x)) = f(g(-x)) \Rightarrow (f \circ g)(x) \text{ is even.}$$

d. $f(x) \text{ even} \Rightarrow f(x) = f(-x)$ and $g(x) \text{ odd} \Rightarrow$

$$g(-x) = -g(x) \Rightarrow (f \circ g)(-x) = f(-g(x))$$
$$= f(g(x)) = (f \circ g)(-x) \Rightarrow (f \circ g)(x) \text{ is even.}$$

2.8 Maintaining Skills

89.a. Yes, R defines a function.

b. $S = \{(2, -3), (1, -1), (3, 1), (1, 2)\}$

No, S does not define a function since the first value 1 maps to two different second values, -1 and 2.

91. $x = 2y + 3 \Rightarrow x - 3 = 2y \Rightarrow \dfrac{x - 3}{2} = y$

93. $x^2 + y^2 = 4, \ x \le 0 \Rightarrow x^2 = 4 - y^2 \Rightarrow$

$$x = -\sqrt{4 - y^2}$$

Section 2.9 Inverse Functions

2.9 Practice Problems

1. $f(x) = (x - 1)^2$ is not one-to-one because the horizontal line $y = 1$ intersects the graph at two different points.

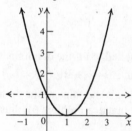

2.a. $f^{-1}(12) = -3$

b. $f(9) = 4$

3. $f(x) = 3x - 1, \ g(x) = \dfrac{x + 1}{3}$

$$(f \circ g)(x) = f\left(\frac{x + 1}{3}\right) = 3\left(\frac{x + 1}{3}\right) - 1 = x$$
$$(g \circ f)(x) = g(3x - 1) = \frac{3x - 1 + 1}{3} = x$$

Since $f(g(x)) = g(f(x)) = x$, the two functions are inverses.

4. The graph of f^{-1} is the reflection of the graph of f about the line $y = x$.

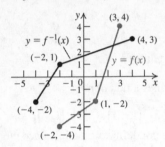

5. $f(x) = -2x + 3$ is a one-to-one function, so the function has an inverse. Interchange the variables and solve for y:
$$f(x) = y = -2x + 3 \Rightarrow x = -2y + 3 \Rightarrow$$
$$\frac{x-3}{-2} = y \Rightarrow y = f^{-1}(x) = \frac{3-x}{2}.$$

6. Interchange the variables and solve for y:
$$f(x) = y = \frac{x}{x+3}, x \neq 3$$
$$x = \frac{y}{y+3} \Rightarrow xy + 3x = y \Rightarrow 3x = y - xy \Rightarrow$$
$$3x = y(1-x) \Rightarrow \frac{3x}{1-x} = y \Rightarrow$$
$$f^{-1}(x) = \frac{3x}{1-x}, x \neq 1$$

7. $f(x) = \frac{x}{x+3}$

The function is not defined if the denominator is zero, so the domain is $(-\infty, -3) \cup (-3, \infty)$. The range of the function is the same as the domain of the inverse (see practice problem 6), thus the range is $(-\infty, 1) \cup (1, \infty)$.

8. G is one-to-one since the domain is restricted, so an inverse exists.

$G(x) = y = x^2 - 1, x \leq 0$. Interchange the variables and solve for y:
$$x = y^2 - 1, y \leq 0 \Rightarrow y = G^{-1}(x) = -\sqrt{x+1}.$$

9. From the text, we have $d = \frac{11p}{5} - 33$.

$$d = \frac{11 \cdot 1650}{5} - 33 = 3597$$

The bell was 3597 feet below the surface when the gauge failed.

2.9 Basic Concepts and Skills

1. If no horizontal line intersects the graph of a function f in more than one point, the f is a <u>one-to-one</u> function.

3. If $f(x) = 3x$, then $f^{-1}(x) = \frac{1}{3}x$.

5. True

7. One-to-one   **9.** Not one-to-one

11. Not one-to-one   **13.** One-to-one

15. $f(2) = 7 \Rightarrow f^{-1}(7) = 2$

17. $f(-1) = 2 \Rightarrow f^{-1}(2) = -1$

19. $f(a) = b \Rightarrow f^{-1}(b) = a$

21. $\left(f^{-1} \circ f\right)(337) = f^{-1}(f(337)) = 337$

23. $\left(f \circ f^{-1}\right)(-1580) = f(f^{-1}(-1580)) = -1580$

25.a. $f(3) = 2(3) - 3 = 3$

  b. Using the result from part (a), $f^{-1}(3) = 3$.

  c. $\left(f \circ f^{-1}\right)(19) = f(f^{-1}(19)) = 19$

  d. $\left(f \circ f^{-1}\right)(5) = f(f^{-1}(5)) = 5$

27.a. $f(1) = 1^3 + 1 = 2$

  b. Using the result from part (a), $f^{-1}(2) = 1$.

  c. $\left(f \circ f^{-1}\right)(269) = f(f^{-1}(269)) = 269$

29. $f(g(x)) = 3\left(\frac{x-1}{3}\right) + 1 = x - 1 + 1 = x$
$$g(f(x)) = \frac{(3x+1)-1}{3} = \frac{3x}{3} = x$$

31. $f(g(x)) = \left(\sqrt[3]{x}\right)^3 = x$
$$g(f(x)) = \sqrt[3]{x^3} = x$$

33. $f(g(x)) = \dfrac{\dfrac{1+2x}{1-x} - 1}{\dfrac{1+2x}{1-x} + 2} = \dfrac{\dfrac{1+2x-(1-x)}{1-x}}{\dfrac{1+2x+2(1-x)}{1-x}} = \dfrac{3x}{3} = x$

(continued on next page)

(*continued*)

$$g\left(f\left(x\right)\right) = \frac{1 + 2\left(\dfrac{x-1}{x+2}\right)}{1 - \dfrac{x-1}{x+2}} = \frac{1 + \dfrac{2x-2}{x+2}}{1 - \dfrac{x-1}{x+2}}$$

$$= \frac{\dfrac{x+2+2x-2}{x+2}}{\dfrac{x+2-(x-1)}{x+2}} = \frac{3x}{3} = x$$

35.

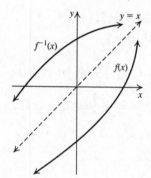

37.

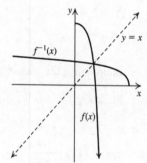

39.

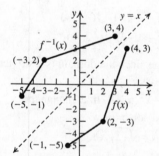

41.a. One-to-one

b. $f(x) = y = 15 - 3x.$ Interchange the variables and solve for $y: x = 15 - 3y \Rightarrow$

$y = f^{-1}(x) = \dfrac{15 - x}{3} = 5 - \dfrac{1}{3}x.$

c.

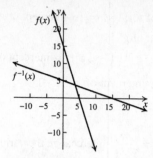

d. Domain of f: $(-\infty, \infty)$; x-intercept of f: 5; y-intercept of f: 15

domain of f^{-1}: $(-\infty, \infty)$; x-intercept of f^{-1}: 15; y-intercept of f^{-1}: 5

43.a. Not one-to-one

45.a. One-to-one

b. $f(x) = y = \sqrt{x} + 3$ Interchange the variables and solve for $y: x = \sqrt{y} + 3 \Rightarrow$

$x - 3 = \sqrt{y} \Rightarrow y = f^{-1}(x) = (x-3)^2.$

c.

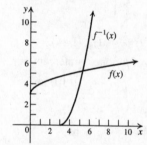

d. Domain of f: $[0, \infty)$; x-intercept of f: none; y-intercept of f: 3

domain of f^{-1}: $[3, \infty)$; x-intercept of f^{-1}: 3; y-intercept of f^{-1}: none

47.a. One-to-one

b. $g(x) = y = \sqrt[3]{x + 1}.$ Interchange the variables and solve for $y: x = \sqrt[3]{y + 1} \Rightarrow$

$x^3 = y + 1 \Rightarrow y = g^{-1}(x) = x^3 - 1$

c.

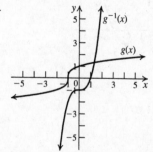

d. Domain of g: $(-\infty, \infty)$; x-intercept of g: -1;
y-intercept of g: 1
domain of g^{-1}: $(-\infty, \infty)$; x-intercept of
g^{-1}: 1; y-intercept of g^{-1}: -1

49.a. One-to-one

b. $f(x) = y = \dfrac{1}{x-1}$. Interchange the variables

and solve for y: $x = \dfrac{1}{y-1} \Rightarrow x(y-1) = 1 \Rightarrow$

$\dfrac{1}{x} = y - 1 \Rightarrow y = f^{-1}(x) = \dfrac{1}{x} + 1 = \dfrac{1+x}{x}$.

c.

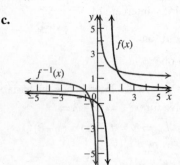

d. Domain of f: $(-\infty, 1) \cup (1, \infty)$
x-intercept of f: none; y-intercept of f: -1
domain of f^{-1}: $(-\infty, 0) \cup (0, \infty)$
x-intercept of f^{-1}: -1
y-intercept of f^{-1}: none

51.a. One-to-one

b. $f(x) = y = 2 + \sqrt{x+1}$. Interchange the
variables and solve for y: $x = 2 + \sqrt{y+1} \Rightarrow$
$x - 2 = \sqrt{y+1} \Rightarrow (x-2)^2 = y + 1 \Rightarrow$
$y = f^{-1}(x) = (x-2)^2 - 1 = x^2 - 4x + 3$

c.

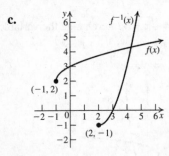

d. Domain of f: $[-1, \infty)$; x-intercept of f: none;
y-intercept of f: 3
Domain of f^{-1}: $[2, \infty)$;
x-intercept of f^{-1}: 3
y-intercept of f^{-1}: none

In exercises 53 and 54, use the fact that the range of f
is the same as the domain of f^{-1}.

53. Domain: $(-\infty, -2) \cup (-2, \infty)$
Range: $(-\infty, 1) \cup (1, \infty)$

55. $f(x) = y = \dfrac{x+1}{x-2}$. Interchange the variables

and solve for y: $x = \dfrac{y+1}{y-2} \Rightarrow xy - 2x = y + 1 \Rightarrow$

$xy - y = 2x + 1 \Rightarrow y(x-1) = 2x + 1 \Rightarrow$

$y = f^{-1}(x) = \dfrac{2x+1}{x-1}$.

Domain of f: $(-\infty, 2) \cup (2, \infty)$
Range of f: $(-\infty, 1) \cup (1, \infty)$.

57. $f(x) = y = \dfrac{1-2x}{1+x}$. Interchange the variables

and solve for y: $x = \dfrac{1-2y}{1+y} \Rightarrow$

$x + xy = 1 - 2y \Rightarrow xy + 2y = 1 - x \Rightarrow$

$y(x+2) = 1 - x \Rightarrow y = f^{-1}(x) = \dfrac{1-x}{x+2}$.

Domain of f: $(-\infty, -1) \cup (-1, \infty)$
Range of f: $(-\infty, -2) \cup (-2, \infty)$.

59. f is one-to-one since the domain is restricted, so
an inverse exists.

$f(x) = y = -x^2, x \geq 0$. Interchange the
variables and solve for y:

$x = -y^2 \Rightarrow y = \sqrt{-x}, x \leq 0$.

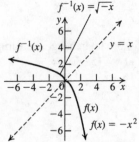

61. f is one-to-one since the domain is restricted, so an inverse exists.

$f(x) = y = |x| = x$, $x \geq 0$. Interchange the variables and solve for y: $y = x$, $x \geq 0$.

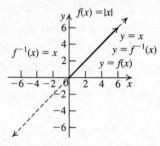

63. f is one-to-one since the domain is restricted, so an inverse exists.

$f(x) = y = x^2 + 1$, $x \leq 0$. Interchange the variables and solve for y:

$x = y^2 + 1 \Rightarrow y = -\sqrt{x - 1}$, $x \geq 1$.

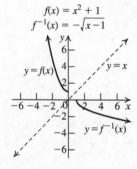

65. f is one-to-one since the domain is restricted, so an inverse exists.

$f(x) = y = -x^2 + 2$, $x \leq 0$. Interchange the variables and solve for y:

$x = -y^2 + 2 \Rightarrow y = -\sqrt{2 - x}$, $x \leq 2$.

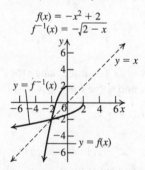

2.9 Applying the Concepts

67.a. $K(C) = C + 273 \Rightarrow$

$C(K) = K - 273 = K^{-1}(C)$.

This represents the Celsius temperature corresponding to a given Kelvin temperature.

b. $C(300) = 300 - 273 = 27°C$

c. $K(22) = 22 + 273 = 295°K$

69.a. $F(K(C)) = \dfrac{9}{5}(C + 273) - \dfrac{2297}{5}$

$= \dfrac{9}{5}C + \dfrac{9(273)}{5} - \dfrac{2297}{5}$

$= \dfrac{9}{5}C + \dfrac{160}{5} = \dfrac{9}{5}C + 32$

b. $C(K(F)) = \dfrac{5}{9}F + \dfrac{2297}{9} - 273$

$= \dfrac{5}{9}F + \dfrac{2297 - 2457}{9}$

$= \dfrac{5}{9}F - \dfrac{160}{9}$

71.a. $E(x) = 0.75x$ where x represents the number of dollars; $D(x) = 1.25x$ where x represents the number of euros.

b. $E(D(x)) = 0.75(1.25x) = 0.9375x \neq x$.
Therefore, the two functions are not inverses.

c. She loses money either way.

73.a. $7 = 4 + 0.05x \Rightarrow x = \$60$. This means that if food sales $\leq \$60$, he will receive the minimum hourly wage. If food sales $> \$60$, his wages will be based on food sales.

$w = \begin{cases} 4 + 0.05x & \text{if } x > 60 \\ 7 & \text{if } x \leq 60 \end{cases}$

b. The function does not have an inverse because it is constant on $(0, 60)$, and it is not one-to-one.

c. If the domain is restricted to $[60, \infty)$, the function has an inverse.

75.a. $V = 8\sqrt{x} \Rightarrow \dfrac{V}{8} = \sqrt{x} \Rightarrow \dfrac{1}{64}V^2 = x = V^{-1}(x)$

This represents the height of the water in terms of the velocity.

b. (i) $x = \dfrac{1}{64}(30^2) = 14.0625$ ft

(ii) $x = \dfrac{1}{64}(20^2) = 6.25$ ft

77.a. The function represents the amount she still owes after x months.

b. $y = 36,000 - 600x$. Interchange the variables and solve for y: $x = 36,000 - 600y \Rightarrow$

$$600y = 36,000 - x \Rightarrow y = 60 - \frac{x}{600} \Rightarrow$$

$f^{-1}(x) = 60 - \dfrac{1}{600}x$. This represents the number of months that have passed from the first payment until the balance due is $\$x$.

c. $y = 60 - \dfrac{1}{600}(22,000) = 23.33 \approx 24$ months

There are 24 months remaining.

2.9 Beyond the Basics

79. $f(g(3)) = f(1) = 3, f(g(5)) = f(3) = 5$, and
$f(g(2)) = f(4) = 2 \Rightarrow f(g(x)) = x$ for each x.
$g(f(1)) = g(3) = 1, g(f(3)) = g(5) = 3$, and
$g(f(4)) = g(2) = 4 \Rightarrow g(f(x)) = x$ for each x.
So, f and g are inverses.

81.a.

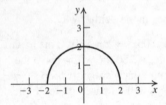

b. f is not one-to-one

c. Domain: $[-2, 2]$; range: $[0, 2]$

83.a. f satisfies the horizontal line test.

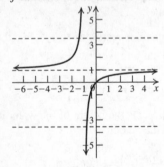

b. $y = 1 - \dfrac{1}{x+1}$. Interchange the variables and solve for y: $x = 1 - \dfrac{1}{y+1} \Rightarrow$

$$\dfrac{1}{y+1} = 1 - x \Rightarrow 1 = y + 1 - xy - x \Rightarrow$$

$$xy - y = -x \Rightarrow y(x-1) = -x \Rightarrow$$

$$y = f^{-1}(x) = -\dfrac{x}{x-1} = \dfrac{x}{1-x}$$

c. Domain of f: $(-\infty, -1) \cup (-1, \infty)$;
range of f: $(-\infty, 1) \cup (1, \infty)$.

85.a. $M = \left(\dfrac{3+7}{2}, \dfrac{7+3}{2}\right) = (5, 5)$.

Since the coordinates of M satisfy the equation $y = x$, it lies on the line.

b. The slope of $y = x$ is 1, while the slope of

$\overline{PQ}$ is $\dfrac{3-7}{7-3} = -1$. So, $y = x$ is perpendicular to $\overline{PQ}$.

87.a. The graph of g is the graph of f shifted one unit to the right and two units up.

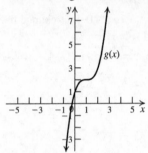

b. $g(x) = y = (x-1)^3 + 2$
Interchange the variables and solve for y.
$x = (y-1)^3 + 2 \Rightarrow y = g^{-1}(x) = \sqrt[3]{x-2} + 1$

c.

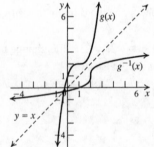

89.a. **(i)** $f(x) = y = 2x + 3$. Interchange the variables and solve for y: $x = 2y + 3 \Rightarrow$

$$y = f^{-1}(x) = \dfrac{1}{2}x - \dfrac{3}{2}$$

(ii) $g(x) = y = x^3 - 1$. Interchange the variables and solve for y: $x = y^3 - 1 \Rightarrow$

$$y = g^{-1}(x) = \sqrt[3]{x+1}$$

(iii) $(f \circ g)(x) = 2(x^3 - 1) + 3 = 2x^3 + 1$

(iv) $(g \circ f)(x) = (2x+3)^3 - 1$
$$= 8x^3 + 36x^2 + 54x + 26$$

(v) $(f \circ g)(x) = y = 2x^3 + 1$. Interchange the variables and solve for y:

$$x = 2y^3 + 1 \Rightarrow (f \circ g)^{-1}(x) = \sqrt[3]{\frac{x-1}{2}}$$

(vi) $(g \circ f)(x) = y$
$$= 8x^3 + 36x^2 + 54x + 26$$

Interchange the variables and solve for y:

$$x = 8y^3 + 36y^2 + 54y + 26 \Rightarrow$$
$$x + 1 = 8y^3 + 36y^2 + 54y + 27 \Rightarrow$$
$$x + 1 = (2y+3)^3 \Rightarrow \sqrt[3]{x+1} = 2y + 3 \Rightarrow$$
$$y = (g \circ f)^{-1}(x) = \frac{1}{2}\sqrt[3]{x+1} - \frac{3}{2}$$

(vii) $\left(f^{-1} \circ g^{-1}\right)(x) = \frac{1}{2}\left(\sqrt[3]{x+1}\right) - \frac{3}{2}$

(viii) $\left(g^{-1} \circ f^{-1}\right)(x) = \sqrt[3]{\frac{1}{2}x - \frac{3}{2} + 1}$
$$= \sqrt[3]{\frac{1}{2}x - \frac{1}{2}} = \sqrt[3]{\frac{x-1}{2}}$$

b. (i) $(f \circ g)^{-1}(x) = \sqrt[3]{\frac{1}{2}x - \frac{1}{2}} = \sqrt[3]{\frac{x-1}{2}}$
$$= \left(g^{-1} \circ f^{-1}\right)(x)$$

(ii) $(g \circ f)^{-1}(x) = \frac{1}{2}\left(\sqrt[3]{x+1}\right) - \frac{3}{2}$
$$= \left(f^{-1} \circ g^{-1}\right)(x)$$

2.9 Critical Thinking/Discussion/Writing

91. No. For example, $f(x) = x^3 - x$ is odd, but it does not have an inverse, because $f(0) = f(1)$, so it is not one-to-one.

92. Yes. The function $f = \{(0,1)\}$ is even, and it has an inverse: $f^{-1} = \{(1,0)\}$.

93. Yes, because increasing and decreasing functions are one-to-one.

94.a. $R = \{(-1,1),(0,0),(1,1)\}$

b. $R = \{(-1,1),(0,0),(1,2)\}$

2.9 Maintaining Skills

95. $x^2 - 7x + 12 = 0 \Rightarrow (x-3)(x-4) = 0 \Rightarrow$
$x = 3, 4$
Solution set: $\{3, 4\}$

97. $12 - 3(x-1)^2 = 0 \Rightarrow 3\left(4 - (x-1)^2\right) = 0 \Rightarrow$
$4 - (x-1)^2 = 0 \Rightarrow$
$[2 + (x-1)][2 - (x-1)] = 0 \Rightarrow$
$(1+x)(3-x) = 0 \Rightarrow x = -1, 3$
Solution set: $\{-1, 3\}$

99. Shift the graph of $y = x^2$ two units left and three units down.

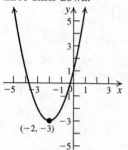

(−2, −3)

Chapter 2 Review Exercises

Basic Concepts and Skills

1. False. The midpoint is $\left(\dfrac{-3+3}{2}, \dfrac{1+11}{2}\right) = (0,6)$.

3. True

5. False.
The slope is 4/3 and the y-intercept is 3.

7. True

9.a. $d(P,Q) = \sqrt{(-1-3)^2 + (3-5)^2} = 2\sqrt{5}$

b. $M = \left(\dfrac{3+(-1)}{2}, \dfrac{5+3}{2}\right) = (1,4)$

c. $m = \dfrac{3-5}{-1-3} = \dfrac{1}{2}$

11.a. $d(P,Q) = \sqrt{(9-4)^2 + (-8-(-3))^2} = 5\sqrt{2}$

b. $M = \left(\dfrac{4+9}{2}, \dfrac{-3+(-8)}{2}\right) = \left(\dfrac{13}{2}, -\dfrac{11}{2}\right)$

c. $m = \dfrac{-8-(-3)}{9-4} = -1$

13.a. $D(P,Q) = \sqrt{(5-2)^2 + (-2-(-7))^2} = \sqrt{34}$

b. $M = \left(\dfrac{2+5}{2}, \dfrac{-7+(-2)}{2}\right) = \left(\dfrac{7}{2}, -\dfrac{9}{2}\right)$

c. $m = \dfrac{-2 - (-7)}{5 - 2} = \dfrac{5}{3}$

15. $d(A,B) = \sqrt{(-2-0)^2 + (-3-5)^2} = \sqrt{68}$

$d(A,C) = \sqrt{(3-0)^2 + (0-5)^2} = \sqrt{34}$

$d(B,C) = \sqrt{(3-(-2))^2 + (0-(-3))^2} = \sqrt{34}$

Using the Pythagorean theorem, we have

$AC^2 + BC^2 = \left(\sqrt{34}\right)^2 + \left(\sqrt{34}\right)^2$

$\qquad = 68 = \left(\sqrt{68}\right)^2 = AB^2$

Alternatively, we can show that AC and CB are perpendicular using their slopes.

$m_{AC} = \dfrac{0-5}{3-0} = -\dfrac{5}{3}; m_{CB} = \dfrac{0-(-3)}{3-(-2)} = \dfrac{3}{5}$

$m_{AC} \cdot m_{CB} = -1 \Rightarrow AC \perp CB$, so $\triangle ABC$ is a right triangle.

17. $A = (-6,3), B = (4,5)$

$d(A,O) = \sqrt{(-6-0)^2 + (3-0)^2} = \sqrt{45}$

$d(B,O) = \sqrt{(4-0)^2 + (5-0)^2} = \sqrt{41}$

$(4, 5)$ is closer to the origin.

19. $A = (-5,3), B = (4,7), C = (x,0)$

$d(A,C) = \sqrt{(x-(-5))^2 + (0-3)^2}$

$\qquad = \sqrt{(x+5)^2 + 9}$

$d(B,C) = \sqrt{(x-4)^2 + (0-7)^2}$

$\qquad = \sqrt{(x-4)^2 + 49}$

$d(A,C) = d(B,C) \Rightarrow$

$\sqrt{(x+5)^2 + 9} = \sqrt{(x-4)^2 + 49}$

$(x+5)^2 + 9 = (x-4)^2 + 49$

$x^2 + 10x + 34 = x^2 - 8x + 65$

$x = \dfrac{31}{18} \Rightarrow$ The point is $\left(\dfrac{31}{18}, 0\right)$.

21. Not symmetric with respect to the x-axis; symmetric with respect to the y-axis; not symmetric with respect to the origin.

23. Symmetric with respect to the x-axis; not symmetric with respect to the y-axis; not symmetric with respect to the origin.

25. x-intercept: 4; y-intercept: 2; not symmetric with respect to the x-axis; not symmetric with respect to the y-axis; not symmetric with respect to the origin.

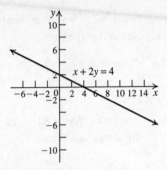

27. x-intercept: 0; y-intercept: 0; not symmetric with respect to the x-axis; symmetric with respect to the y-axis; not symmetric with respect to the origin.

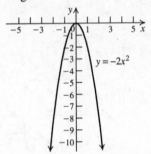

29. x-intercept: 0; y-intercept: 0; not symmetric with respect to the x-axis; not symmetric with respect to the y-axis; symmetric with respect to the origin.

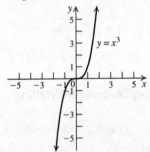

31. No x-intercept; y-intercept: 2; not symmetric with respect to the x-axis; symmetric with respect to the y-axis; not symmetric with respect to the origin.

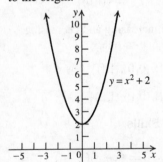

33. x-intercepts: -4, 4; y-intercepts: -4, 4; symmetric with respect to the x-axis; symmetric with respect to the y-axis; symmetric with respect to the origin.

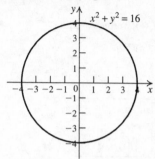

35. $(x-2)^2 + (y+3)^2 = 25$

37. The radius is 2, so the equation of the circle is $(x+2)^2 + (y+5)^2 = 4$.

39. $\dfrac{x}{2} - \dfrac{y}{5} = 1 \Rightarrow 5x - 2y = 10 \Rightarrow \dfrac{5}{2}x - 5 = y$.
Line with slope 5/2 and y-intercept -5.

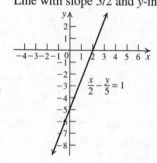

41. $x^2 + y^2 - 2x + 4y - 4 = 0 \Rightarrow$
$x^2 - 2x + 1 + y^2 + 4y + 4 = 4 + 1 + 4 \Rightarrow$
$(x-1)^2 + (y+2)^2 = 9$.
Circle with center $(1, -2)$ and radius 3.

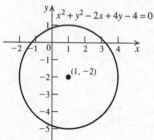

43. $y - 2 = -2(x-1) \Rightarrow y = -2x + 4$

45. $m = \dfrac{7-3}{-1-1} = -2; 3 = -2(1) + b \Rightarrow 5 = b \Rightarrow$
$y = -2x + 5$

47.a. $y = 3x - 2 \Rightarrow m = 3$; $y = 3x + 2 \Rightarrow m = 3$ The slopes are equal, so the lines are parallel.

b. $3x - 5y + 7 \Rightarrow m = 3/5$;
$5x - 3y + 2 = 0 \Rightarrow m = 5/3$
The slopes are neither equal nor negative reciprocals, so the lines are neither parallel nor perpendicular.

c. $ax + by + c = 0 \Rightarrow m = -a/b$;
$bx - ay + d = 0 \Rightarrow m = b/a$
The slopes are negative reciprocals, so the lines are perpendicular.

d. $y + 2 = \dfrac{1}{3}(x-3) \Rightarrow m = \dfrac{1}{3}$;
$y - 5 = 3(x-3) \Rightarrow m = 3$
The slopes are neither equal nor negative reciprocals, so the lines are neither parallel nor perpendicular.

49. Domain: $\{-1, 0, 1, 2\}$; range: $\{-1, 0, 1, 2\}$.
This is a function.

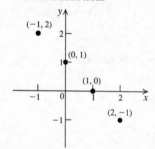

51. Domain: $(-\infty, \infty)$; range: $(-\infty, \infty)$.
This is a function.

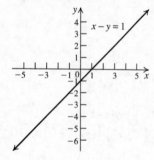

53. Domain: $[-0.2, 0.2]$; range: $[-0.2, 0.2]$.
This is not a function.

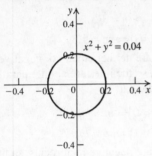

55. Domain: $\{1\}$; range: $(-\infty, \infty)$.
This is not a function.

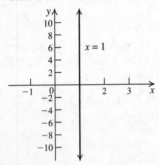

57. Domain: $(-\infty, \infty)$; range: $[0, \infty)$.
This is a function.

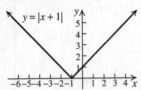

59. $f(-2) = 3(-2) + 1 = -5$

61. $f(x) = 4 \Rightarrow 3x + 1 = 4 \Rightarrow x = 1$

63. $(f + g)(1) = f(1) + g(1)$
$= (3(1) + 1) + (1^2 - 2) = 3$

65. $(f \cdot g)(-2) = f(-2) \cdot g(-2)$
$= (3(-2) + 1) \cdot ((-2)^2 - 2) = -10$

67. $(f \circ g)(3) = 3(3^2 - 2) + 1 = 22$

69. $(f \circ g)(x) = 3(x^2 - 2) + 1 = 3x^2 - 5$

71. $(f \circ f)(x) = 3(3x + 1) + 1 = 9x + 4$

73. $f(a + h) = 3(a + h) + 1 = 3a + 3h + 1$

75. $\dfrac{f(x + h) - f(x)}{h} = \dfrac{(3(x + h) + 1) - (3x + 1)}{h}$
$= \dfrac{3x + 3h + 1 - 3x - 1}{h} = \dfrac{3h}{h} = 3$

77. Domain: $(-\infty, \infty)$; range: $\{-3\}$.
Constant on $(-\infty, \infty)$.

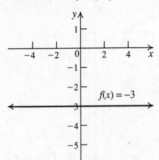

79. Domain: $\left[\dfrac{2}{3}, \infty\right)$; range: $[0, \infty)$

Increasing on $\left(\dfrac{2}{3}, \infty\right)$.

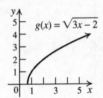

81. Domain: $(-\infty, \infty)$; range: $[1, \infty)$. Decreasing on $(-\infty, 0)$; increasing on $(0, \infty)$.

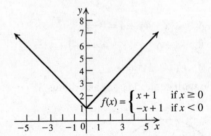

83. The graph of g is the graph of f shifted one unit left.

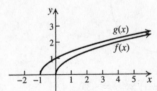

85. The graph of g is the graph of f shifted two units right and then reflected in the x-axis.

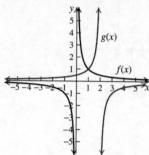

87. $f(-x) = (-x)^2 - (-x)^4 = x^2 - x^4 = f(x) \Rightarrow$

$f(x)$ is even. Not symmetric with respect to the x-axis; symmetric with respect to the y-axis; not symmetric with respect to the origin.

89. $f(-x) = |-x| + 3 = |x| + 3 = f(x) \Rightarrow$

$f(x)$ is even. Not symmetric with respect to the x-axis; symmetric with respect to the y-axis; not symmetric with respect to the origin.

91. $f(-x) = \sqrt{-x} \neq f(x)$ or $f(-x) \Rightarrow f(x)$ is neither even nor odd. Not symmetric with respect to the x-axis; not symmetric with respect to the y-axis; not symmetric with respect to the origin.

93. $f(x) = \sqrt{x^2 - 4} \Rightarrow f(x) = (g \circ h)(x)$ where $g(x) = \sqrt{x}$ and $h(x) = x^2 - 4$.

95. $h(x) = \sqrt{\dfrac{x-3}{2x+5}} \Rightarrow h(x) = (f \circ g)(x)$ where $f(x) = \sqrt{x}$ and $g(x) = \dfrac{x-3}{2x+5}$.

97. $f(x)$ is one-to-one. $f(x) = y = x + 2$.

Interchange the variables and solve for y:

$x = y + 2 \Rightarrow y = x - 2 = f^{-1}(x)$.

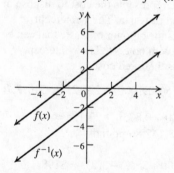

99. $f(x)$ is one-to-one. $f(x) = y = \sqrt[3]{x - 2}$.

Interchange the variables and solve for y:

$x = \sqrt[3]{y-2} \Rightarrow y = x^3 + 2 = f^{-1}(x)$.

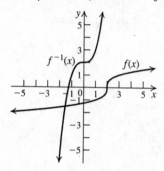

101. $f(x) = y = \dfrac{x-1}{x+2}, x \neq 2$.

Interchange the variables and solve for y.

$x = \dfrac{y-1}{y+2} \Rightarrow xy + 2x = y - 1 \Rightarrow$

$xy - y = -2x - 1 \Rightarrow y(x-1) = -2x - 1 \Rightarrow$

$y = \dfrac{-2x-1}{x-1} \Rightarrow y = f^{-1}(x) = \dfrac{2x+1}{1-x}$

Domain of f: $(-\infty, -2) \cup (-2, \infty)$

Range of f: $(-\infty, 1) \cup (1, \infty)$

103.a. $A = (-3, -3), B = (-2, 0), C = (0, 1), D = (3, 4)$.

Find the equation of each segment:

$m_{AB} = \dfrac{0 - (-3)}{-2 - (-3)} = 3.0 = 3(-2) + b \Rightarrow b = 6$.

The equation of AB is $y = 3x + 6$.

$m_{BC} = \dfrac{1 - 0}{0 - (-2)} = \dfrac{1}{2}; b = 1$.

The equation of BC is $y = \dfrac{1}{2}x + 1$.

$m_{CD} = \dfrac{4 - 1}{3 - 3} = 1; b = 1$.

The equation of CD is $y = x + 1$.

So,

$$f(x) = \begin{cases} 3x + 6 & \text{if } -3 \leq x \leq -2 \\ \dfrac{1}{2}x + 1 & \text{if } -2 < x < 0 \\ x + 1 & \text{if } 0 \leq x \leq 3 \end{cases}$$

b. Domain: $[-3, 3]$; range: $[-3, 4]$

c. x-intercept: -2; y-intercept: 1

d.

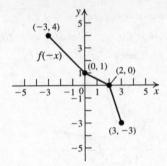

e.

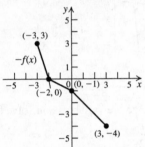

f.

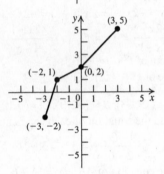

g.

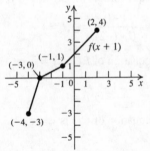

h.

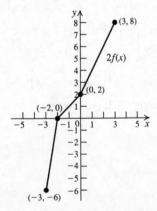

i.

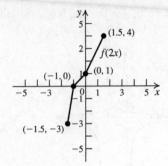

j.

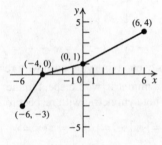

k. *f* is one-to-one because it satisfies the horizontal line test.

l.

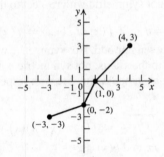

Applying the Concepts

105.a. rate of change $(\text{slope}) = \dfrac{173{,}000 - 54{,}000}{223{,}000 - 87{,}000}$
$= 0.875$

$54{,}000 = 0.875(87{,}000) + b \Rightarrow b = -22{,}125.$
The equation is $C = 0.875w - 22{,}125.$

b. The slope represents the cost to dispose of one pound of waste. The *x*-intercept represents the amount of waste that can be disposed with no cost. The *y*-intercept represents the fixed cost.

c. $C = 0.875(609{,}000) - 22{,}125 = \$510{,}750$

d. $1{,}000{,}000 = 0.875w - 22{,}125 \Rightarrow$
$w = 1{,}168{,}142.86$ pounds

107.a. $f(2) = 100 + 55(2) - 3(2)^2 = \$198.$
She started with \$100, so she won \$98.

b. She was winning at a rate of \$49/hour.

c. $0 = 100 + 55t - 3t^2 \Rightarrow (-t + 20)(3t + 5) \Rightarrow$
$t = 20, t = -5/3$. Since t represents the amount of time, we reject $t = -5/3$.
Chloe will lose all her money after playing for 20 hours.

d. $\$100/20 = \$5/\text{hour}$.

109.a. $(L \circ x)(t) = 0.5\sqrt{\left(1 + 0.002t^2\right)^2 + 4}$

$= 0.5\sqrt{0.000004t^4 + 0.004t^2 + 5}$

b. $(L \circ x)(5) = 0.5\sqrt{\left(1 + 0.002(5^2)\right)^2 + 4}$

$= 0.5\sqrt{(1.05)^2 + 4} = 0.5\sqrt{5.1025}$

≈ 1.13

111.a.

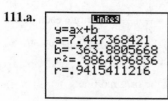

$y \approx 7.4474x - 363.88$

b.

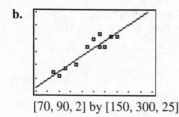

[70, 90, 2] by [150, 300, 25]

c. $y \approx 7.4474(76) - 363.88 \approx 202$

A player whose height is 76 inches weighs about 202 pounds.

Chapter 2 Practice Test A

1. The endpoints of the diameter are $(-2, 3)$ and $(-4, 5)$, so the center of the circle is

$C = \left(\dfrac{-2 + (-4)}{2}, \dfrac{3 + 5}{2}\right) = (-3, 4)$.

The length of the diameter is

$\sqrt{\left(-4 - (-2)\right)^2 + (5 - 3)^2} = \sqrt{8} = 2\sqrt{2}$.

Therefore, the length of the radius is $\sqrt{2}$.
The equation of the circle is

$(x + 3)^2 + (y - 4)^2 = 2$.

2. To test if the graph is symmetric with respect to the y-axis, replace x with $-x$:

$3(-x) + 2(-x)y^2 = 1 \Rightarrow -3x - 2xy^2 = 1$, which is not the same as the original equation, so the graph is not symmetric with respect to the y-axis. To test if the graph is symmetric with respect to the x-axis, replace y with $-y$:

$3x + 2x(-y)^2 = 1 \Rightarrow 3x + 2xy^2 = 1$, which is the same as the original equation, so the graph is symmetric with respect to the x-axis. To test if the graph is symmetric with respect to the origin, replace x with $-x$ and y with $-y$:

$3(-x) + 2(-x)(-y)^2 = 1 \Rightarrow -3x - 2xy^2 = 1$, which is not the same as the original equation, so the graph is not symmetric with respect to the origin.

3. $0 = x^2(x - 3)(x + 1) \Rightarrow x = 0$ or $x = 3$ or $x = -1$

$y = 0^2(0 - 3)(0 + 1) \Rightarrow y = 0$. The x-intercepts are 0, 3, and -1; the y-intercept is 0.

4.

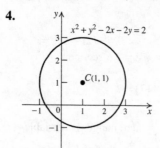

Intercepts:
$y^2 - 2y = 2 \Rightarrow y = 1 \pm \sqrt{3}$
$x^2 - 2x = 2 \Rightarrow x = 1 \pm \sqrt{3}$

5. $7 = -1(2) + b \Rightarrow 9 = b$
The equation is $y = -x + 9$.

6. $8x - 2y = 7 \Rightarrow y = 4x - \dfrac{7}{2} \Rightarrow$ the slope of the line is 4. $-1 = 4(2) + b \Rightarrow b = -9$. So the equation is $y = 4x - 9$.

7. $(fg)(2) = f(2) \cdot g(2)$
$= (-2(2) + 1)\left(2^2 + 3(2) + 2\right)$
$= (-3)(12) = -36$

8. $g(f(2)) = g(2(2) - 3) = g(1) = 1 - 2(1)^2 = -1$

9. $(f \circ f)(x) = (x^2 - 2x)^2 - 2(x^2 - 2x)$
$= x^4 - 4x^3 + 4x^2 - 2x^2 + 4x$
$= x^4 - 4x^3 + 2x^2 + 4x$

10.a. $f(-1) = (-1)^3 - 2 = -3$

b. $f(0) = 0^3 - 2 = -2$

c. $f(1) = 1 - 2(1)^2 = -1$

11. $1 - x > 0 \Rightarrow x < 1; x$ must also be greater than or equal to 0, so the domain is $[0, 1)$.

12. $\dfrac{f(4) - f(1)}{4 - 1} = \dfrac{(2(4) + 7) - (2(1) + 7)}{3} = 2$

13. $f(-x) = 2(-x)^4 - \dfrac{3}{(-x)^2} = 2x^4 - \dfrac{3}{x^2} = f(x) \Rightarrow$

$f(x)$ is even.

14. Increasing on $(-\infty, 0)$ and $(2, \infty)$; decreasing on $(0, 2)$.

15. Shift the graph of $y = \sqrt{x}$ three units to the right, then stretch the graph vertically by a factor of 2, and then shift the resulting graph four units up.

16. $25 = 25 - (2t - 5)^2 \Rightarrow 0 = -(2t - 5)^2 \Rightarrow$
$0 = 2t - 5 \Rightarrow t = 5/2 = 2.5$ seconds

17. $f(2) = 7 \Rightarrow f^{-1}(7) = 2$

18. $f(x) = y = \dfrac{2x}{x - 1}$. Interchange the variables and solve for y: $x = \dfrac{2y}{y - 1} \Rightarrow$

$xy - x = 2y \Rightarrow xy - 2y = x \Rightarrow$

$y(x - 2) = x \Rightarrow y = f^{-1}(x) = \dfrac{x}{x - 2}$

19. $A(x) = 100x + 1000$

20.a. $C(230) = 0.25(230) + 30 = \$87.50$

b. $57.50 = 0.25m + 30 \Rightarrow m = 110$ miles

Chapter 2 Practice Test B

1. To test if the graph is symmetric with respect to the y-axis, replace x with $-x$:

$|-x| + 2|y| = 2 \Rightarrow |x| + 2|y| = 2$, which is the same as the original equation, so the graph is symmetric with respect to the y-axis. To test if the graph is symmetric with respect to the x-axis, replace y with $-y$:

$|x| + 2|-y| = 2 \Rightarrow |x| + 2|y| = 2$, which is the same as the original equation, so the graph is symmetric with respect to the x-axis. To test if the graph is symmetric with respect to the origin, replace x with $-x$, and y with $-y$:

$|-x| + 2|-y| = 2 \Rightarrow |x| + 2|y| = 2$, which is the same as the original equation, so the graph is symmetric with respect to the origin. The answer is D.

2. $0 = x^2 - 9 \Rightarrow x = \pm 3; y = 0^2 - 9 \Rightarrow y = -9$. The x-intercepts are ± 3; the y-intercept is -9. The answer is B.

3. D   **4.** D   **5.** C

6. Suppose the coordinates of the second point are (a, b). Then $-\dfrac{1}{2} = \dfrac{b - 2}{a - 3}$. Substitute each of the points given into this equation to see which makes it true. The answer is C.

7. Find the slope of the original line:

$6x - 3y = 5 \Rightarrow y = 2x - \dfrac{5}{3}$. The slope is 2. The equation of the line with slope 2, passing through $(-1, 2)$ is $y - 2 = 2(x + 1)$.
The answer is D.

8. $(f \circ g)(x) = 3(2 - x^2) - 5 = 1 - 3x^2$.
The answer is B.

9. $(f \circ f)(x) = 2(2x^2 - x)^2 - (2x^2 - x)$
$= 8x^4 - 8x^3 + x$
The answer is A.

10. $g(a - 1) = \dfrac{1 - (a - 1)}{1 + (a - 1)} = \dfrac{2 - a}{a}$.
The answer is C.

11. $1 - x \geq 0 \Rightarrow x \leq 1$; x must also be greater than or equal to 0, so the domain is $[0, 1]$.
The answer is A.

12. $x^2 + 3x - 4 = 6 \Rightarrow x^2 + 3x - 10 = 0 \Rightarrow$
$(x + 5)(x - 2) = 0 \Rightarrow x = -5, 2$
The answer is D.

13. A   **14.** A   **15.** B

16. D   **17.** C

18. $f(x) = y = \dfrac{x}{3x+2}$. Interchange the variables

and solve for y: $x = \dfrac{y}{3y+2} \Rightarrow$

$3xy + 2x = y \Rightarrow 3xy - y = -2x \Rightarrow$

$y(3x - 1) = -2x \Rightarrow y = f^{-1}(x) = -\dfrac{2x}{3x-1} \Rightarrow$

$f^{-1}(x) = \dfrac{2x}{1-3x}$

The answer is C.

19. $w = 5x - 190; w = 5(70) - 190 = 160$.
The answer is B.

20. $50 = 0.2m + 25 \Rightarrow m = 125$. The answer is A.

Cumulative Review Exercises (Chapters P–2)

1.a. $\left(\dfrac{x^3}{y^2}\right)^2 \left(\dfrac{y^2}{x^3}\right)^3 = \left(\dfrac{x^6}{y^4}\right)\left(\dfrac{y^6}{x^9}\right) = \dfrac{y^2}{x^3}$

b. $\dfrac{x^{-1}y^{-1}}{x^{-1}+y^{-1}} = \dfrac{\frac{1}{x}\cdot\frac{1}{y}}{\frac{1}{x}+\frac{1}{y}} = \dfrac{\frac{1}{xy}}{\frac{y+x}{xy}} = \dfrac{1}{x+y}$

2.a. $2x^2 + x - 15 = (2x - 5)(x + 3)$

b. $x^3 - 2x^2 + 4x - 8 = x^2(x - 2) + 4(x - 2)$
$\qquad\qquad\qquad = (x^2 + 4)(x - 2)$

3.a. $\sqrt{75} + \sqrt{108} - \sqrt{192} = 5\sqrt{3} + 6\sqrt{3} - 8\sqrt{3}$
$\qquad\qquad\qquad\qquad = 3\sqrt{3}$

b. $\dfrac{x-1}{x+1} - \dfrac{x-2}{x+2} = \dfrac{(x-1)(x+2) - (x-2)(x+1)}{(x+1)(x+2)}$

$\qquad = \dfrac{(x^2+x-2)-(x^2-x-2)}{(x+1)(x+2)}$

$\qquad = \dfrac{2x}{(x+1)(x+2)}$

4.a. $\dfrac{1}{2+\sqrt{3}} = \dfrac{1}{2+\sqrt{3}}\cdot\dfrac{2-\sqrt{3}}{2-\sqrt{3}} = \dfrac{2-\sqrt{3}}{4-3} = 2-\sqrt{3}$

b. $\dfrac{1}{\sqrt{5}-2} = \dfrac{1}{\sqrt{5}-2}\cdot\dfrac{\sqrt{5}+2}{\sqrt{5}+2} = \dfrac{\sqrt{5}+2}{5-4} = \sqrt{5}+2$

5.a. $3x - 7 = 5 \Rightarrow 3x = 12 \Rightarrow x = 4$

b. $\dfrac{1}{x-1} = \dfrac{3}{x-1} \Rightarrow$ There is no solution.

6.a. $x^2 - 3x = 0 \Rightarrow x(x - 3) = 0 \Rightarrow$
$x = 0$ or $x = 3$

b. $x^2 + 3x - 10 = 0 \Rightarrow (x+5)(x-2) = 0 \Rightarrow$
$x = -5$ or $x = 2$

7.a. $2x^2 - x + 3 = 0 \Rightarrow x = \dfrac{1 \pm \sqrt{1-4(2)(3)}}{2(2)} \Rightarrow$

$x = \dfrac{1 \pm \sqrt{-23}}{4} \Rightarrow x = \dfrac{1 \pm i\sqrt{23}}{4}$

b. $4x^2 - 12x + 9 = 0 \Rightarrow (2x-3)^2 = 0 \Rightarrow x = \dfrac{3}{2}$

8.a. $x - 6\sqrt{x} + 8 = 0 \Rightarrow \left(\sqrt{x}-4\right)\left(\sqrt{x}-2\right) = 0 \Rightarrow$
$\sqrt{x} = 4 \Rightarrow x = 16$ or $\sqrt{x} = 2 \Rightarrow x = 4$

b. $\left(x - \dfrac{1}{x}\right)^2 - 10\left(x - \dfrac{1}{x}\right) + 21 = 0.$

Let $u = x - \dfrac{1}{x}$.

$u^2 - 10u + 21 = 0 \Rightarrow (u-7)(u-3) = 0 \Rightarrow$
$u = 7$ or $u = 3$;

$x - \dfrac{1}{x} = 7 \Rightarrow x^2 - 1 = 7x \Rightarrow x^2 - 7x - 1 = 0 \Rightarrow$

$x = \dfrac{7 \pm \sqrt{7^2 - (4)(-1)}}{2} \Rightarrow x = \dfrac{7 \pm \sqrt{53}}{2};$

$x - \dfrac{1}{x} = 3 \Rightarrow x^2 - 1 = 3x \Rightarrow x^2 - 3x - 1 = 0 \Rightarrow$

$x = \dfrac{3 \pm \sqrt{3^2 - 4(-1)}}{2} \Rightarrow x = \dfrac{3 \pm \sqrt{13}}{2}$

The solution set is

$\left\{\dfrac{7-\sqrt{53}}{2}, \dfrac{7+\sqrt{53}}{2}, \dfrac{3-\sqrt{13}}{2}, \dfrac{3+\sqrt{13}}{2}\right\}.$

9.a. $\sqrt{3x-1} = 2x - 1 \Rightarrow 3x - 1 = (2x-1)^2 \Rightarrow$
$3x - 1 = 4x^2 - 4x + 1 \Rightarrow 4x^2 - 7x + 2 = 0 \Rightarrow$
$x = \dfrac{7 \pm \sqrt{(-7)^2 - 4(4)(2)}}{2(4)} = \dfrac{7 \pm \sqrt{17}}{8}.$ If

$x = \dfrac{7 - \sqrt{17}}{8}, \sqrt{3\left(\dfrac{7-\sqrt{17}}{8}\right)} - 1 \approx 0.281$ while

$2\left(\dfrac{7-\sqrt{17}}{8}\right) - 1 \approx -0.281$, so the solution set

is $\left\{\dfrac{7+\sqrt{17}}{8}\right\}.$

b.
$$\sqrt{1-x} = 2 - \sqrt{2x+1}$$
$$\left(\sqrt{1-x}\right)^2 = \left(2 - \sqrt{2x+1}\right)^2$$
$$1 - x = 4 - 4\sqrt{2x+1} + 2x + 1$$
$$-4 - 3x = -4\sqrt{2x+1}$$
$$\left(-4-3x\right)^2 = \left(-4\sqrt{2x+1}\right)^2$$
$$16 + 24x + 9x^2 = 16(2x+1)$$
$$16 + 24x + 9x^2 = 32x + 16$$
$$9x^2 - 8x = 0 \Rightarrow x(9x-8) = 0$$
$$x = 0 \text{ or } x = \frac{8}{9}.$$

Check to make sure that neither solution is extraneous. The solution set is $\{0, 8/9\}$.

10.a. $2x - 5 < 11 \Rightarrow x < 8 \Rightarrow (-\infty, 8)$

b. $-3x + 4 > -5 \Rightarrow x < 3 \Rightarrow (-\infty, 3)$

11.a. $-3 < 2x - 3 < 5 \Rightarrow 0 < 2x < 8 \Rightarrow 0 < x < 4$.
The solution set is $(0, 4)$.

b. $5 \le 1 - 2x \le 7 \Rightarrow 4 \le -2x \le 6 \Rightarrow -2 \ge x \ge -3$.
The solution set is $[-3, -2]$.

12.a. $|2x-1| \le 7 \Rightarrow 2x - 1 \le 7 \Rightarrow x \le 4$
or $2x - 1 \ge -7 \Rightarrow x \ge -3$.
The solution set is $[-3, 4]$.

b. $|2x-3| \ge 5 \Rightarrow 2x - 3 \ge 5 \Rightarrow x \ge 4$ or
$2x - 3 \le -5 \Rightarrow x \le -1$.
The solution set is $(-\infty, -1] \cup [4, \infty)$.

13. $d(A,C) = \sqrt{(2-5)^2 + (2-(-2))^2} = 5$

$d(B,C) = \sqrt{(2-6)^2 + (2-5)^2} = 5$

Since the lengths of the two sides are equal, the triangle is isosceles.

14.

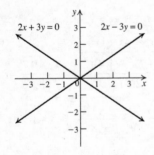

15. First, find the equation of the circle with center $(2, -1)$ and radius determined by $(2, -1)$ and $(-3, -1)$: $r = \sqrt{2 - (-3))^2 + (-1-(-1))^2} = 5$.
The equation is $(x-2)^2 + (y+1)^2 = 5^2$. Now check to see if the other three points satisfy the equation: $(2-2)^2 + (4+1)^2 = 5^2 \Rightarrow 5^2 = 5^2$,
$(5-2)^2 + (3+1)^2 = 5^2 \Rightarrow 3^2 + 4^2 = 5^2$ (true because 3, 4, 5 is a Pythagorean triple), and
$(6-2)^2 + (2+1)^2 = 5^2 \Rightarrow 4^2 + 3^2 = 5^2$.
Since all the points satisfy the equation, they lie on the circle.

16. $x^2 + y^2 - 6x + 4y + 9 = 0 \Rightarrow$
$x^2 - 6x + y^2 + 4y = -9$.
Now complete both squares:
$x^2 - 6x + 9 + y^2 + 4y + 4 = -9 + 9 + 4 \Rightarrow$
$(x-3)^2 + (y+2)^2 = 4$.
The center is $(3, -2)$ and the radius is 2.

17. $y = -3x + 5$

18. The x-intercept is 4, so $(4, 0)$ satisfies the equation. To write the equation in slope-intercept form, find the y-intercept:
$0 = 2(4) + b \Rightarrow -8 = b$
The equation is $y = 2x - 8$.

19. The slope of the perpendicular line is the negative reciprocal of the slope of the original line. The slope of the original line is 2, so the slope of the perpendicular is $-1/2$. Now find the y-intercept of the perpendicular.

$-1 = -\frac{1}{2}(2) + b \Rightarrow b = 0$.

The equation of the perpendicular is $y = -\frac{1}{2}x$.

20. The slope of the parallel line is the same as the slope of the original line, 2. Now find the y-intercept of the parallel line: $-1 = 2(2) + b \Rightarrow$
$b = -5$. The equation of the parallel line is
$y = 2x - 5$.

21. The slope of the perpendicular line is the negative reciprocal of the slope of the original line. The slope of the original line is

$\dfrac{7-(-1)}{5-3}=4$, so the slope of the perpendicular is –1/4. The perpendicular bisector passes through the midpoint of the original segment.

The midpoint is $\left(\dfrac{3+5}{2},\dfrac{-1+7}{2}\right)=(4,3)$. Use this point and the slope to find the y-intercept:

$3=-\dfrac{1}{4}(4)+b \Rightarrow b=4$. The equation of the

perpendicular bisector is $y=-\dfrac{1}{4}x+4$.

22. The slope is undefined because the line is vertical. Because it passes through (5, 7), the equation of the line is $x=5$.

23. Use the slope formula to solve for x:

$2=\dfrac{5-11}{x-5} \Rightarrow 2(x-5)=-6 \Rightarrow 2x-10=-6 \Rightarrow$
$x=2$

24. The line through $(x, 3)$ and $(3, 7)$ has slope –2 because it is perpendicular to a line with slope 1/2. Use the slope formula to solve for x:

$-2=\dfrac{3-7}{x-3} \Rightarrow -2(x-3)=-4 \Rightarrow x-3=2 \Rightarrow$
$x=5$

25.

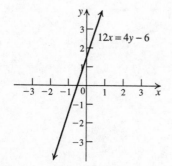

26.

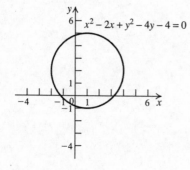

27.

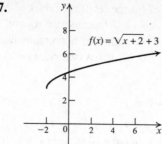

28.

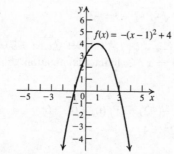

29. Let x = the number of books initially purchased, and $\dfrac{1650}{x}$ = the cost of each book. Then $x-16$ = the number of books sold, and $\dfrac{1650}{x-16}$ = the selling price of each book. The profit = the selling price – the cost, so

$\dfrac{1650}{x-16}-\dfrac{1650}{x}=10 \Rightarrow$
$1650x-1650(x-16)=10x(x-16) \Rightarrow$
$1650x-1650x+26,400=10x^2-160x \Rightarrow$
$10x^2-160x-26,400=0 \Rightarrow$
$x^2-16x-2640=0 \Rightarrow (x-60)(x+44)=0 \Rightarrow$
$x=60,\ x=-44$.

Reject –44 because there cannot be a negative number of books. So she bought 60 books.

30. Let x = the monthly note on the 1.5 year lease, and $1.5(12)x=18x$ = the total expense for the 1.5 year lease. Then $x-250$ = the monthly note on the 2 year lease, and $2(12)(x-250)=24x-6000$ the total expenses for the 2 year lease.

Then $18x+24x-6000=21,000 \Rightarrow$
$42x=27,000 \Rightarrow x=642.86$. So the monthly note for the 1.5 year lease is \$642.86, and the monthly note for the 2 year lease is $\$642.86-250=\$392.86$.

31.a. The domain of *f* is the set of all values of *x* which make $x+1 \geq 0$ (because the square root of a negative number is not a real value.) So $x \geq -1$ or $[-1, \infty)$ in interval notation is the domain.

b. $y = \sqrt{0+1} - 3 \Rightarrow y = -2; 0 = \sqrt{x+1} - 3 \Rightarrow$ $3 = \sqrt{x+1} \Rightarrow 9 = x+1 \Rightarrow 8 = x.$ The *x*-intercept is 8, and the *y*-intercept is –2.

c. $f(-1) = \sqrt{-1+1} - 3 = -3$

d. $f(x) > 0 \Rightarrow \sqrt{x+1} - 3 > 0 \Rightarrow \sqrt{x+1} > 3 \Rightarrow$ $x+1 > 9 \Rightarrow x > 8.$ In interval notation, this is $(8, \infty)$.

32.a. $f(-2) = -(-2) = 2; f(0) = 0^2 = 0;$ $f(2) = 2^2 = 4$

b. *f* decreases on $(-\infty, 0)$ and increases on $(0, \infty)$.

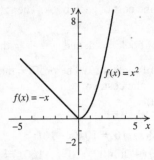

33.a. $(f \circ g)(x) = \dfrac{1}{\dfrac{2}{x} - 2} = \dfrac{1}{\dfrac{2-2x}{x}} = \dfrac{x}{2-2x}.$

Because 0 is not in the domain of *g*, it must be excluded from the domain of $(f \circ g)$.

Because 2 is not in the domain of *f*, any values of *x* for which $g(x) = 2$ must also be excluded from the domain of

$(f \circ g): \dfrac{2}{x} = 2 \Rightarrow x = 1,$ so 1 is excluded

also. The domain of $(f \circ g)$ is

$(-\infty, 0) \cup (0, 1) \cup (1, \infty).$

b. $(g \circ f)(x) = \dfrac{2}{\dfrac{1}{x-2}} = 2(x-2) = 2x - 4.$

Because 2 is not in the domain of *f*, it must be excluded from the domain of $(g \circ f)$.

Because 0 is not in the domain of *g*, any values of *x* for which $f(x) = 0$ must also be excluded from the domain of $(g \circ f)$.

However, there is no value for *x* which makes $f(x) = 0$. So the domain of $(g \circ f)$ is

$(-\infty, 2) \cup (2, \infty).$

Chapter 3 Polynomial and Rational Functions

3.1 Quadratic Functions

3.1 Practice Problems

1. Substitute 1 for h, −5 for k, 3 for x, and 7 for y in the standard form for a quadratic equation to solve for a:

$$7 = a(3-1)^2 - 5 \Rightarrow 7 = 4a - 5 \Rightarrow a = 3$$

The equation is $y = 3(x-1)^2 - 5$.

Since $a = 3 > 0$, f has a minimum value of −5 at $x = 1$.

2. The graph of $f(x) = -2(x+1)^2 + 3$ is a parabola with $a = -2$, $h = -1$ and $k = 3$. Thus, the vertex is (−1, 3), and the maximum value of the function is 3. The parabola opens down because $a < 0$. Now, find the x-intercepts:

$$0 = -2(x+1)^2 + 3 \Rightarrow 2(x+1)^2 = 3 \Rightarrow$$

$$(x+1)^2 = \frac{3}{2} \Rightarrow x + 1 = \pm\sqrt{\frac{3}{2}} \Rightarrow x = \pm\sqrt{\frac{3}{2}} - 1 \Rightarrow$$

$x \approx 0.22$ or $x \approx -2.22$. Next, find the y-intercept: $f(0) = -2(0+1)^2 + 3 = 1$

Plot the vertex, the x-intercepts, and the y-intercept, and join them with a parabola.

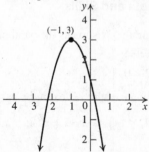

3. The graph of $f(x) = 3x^2 - 3x - 6$ is a parabola with $a = 3$, $b = -3$ and $c = -6$. The parabola opens up because $a > 0$. Now, find the vertex:

$$h = -\frac{b}{2a} = -\frac{-3}{2(3)} = \frac{1}{2}$$

$$k = f(h) = f\left(\frac{1}{2}\right) = 3\left(\frac{1}{2}\right)^2 - 3\left(\frac{1}{2}\right) - 6 = -\frac{27}{4}$$

Thus, the vertex (h, k) is $\left(\frac{1}{2}, -\frac{27}{4}\right)$, and the

minimum value of the function is $-\frac{27}{4}$.

Next, find the x-intercepts:

$$3x^2 - 3x - 6 = 0 \Rightarrow 3\left(x^2 - x - 2\right) = 0 \Rightarrow$$

$$(x-2)(x+1) = 0 \Rightarrow x = 2 \text{ or } x = -1$$

Now, find the y-intercept:

$$f(0) = 3(0)^2 - 3(0) - 6 = -6.$$

Thus, the intercepts are (−1, 0), (2, 0) and (0, −6). Use the fact that the parabola is

symmetric with respect to its axis, $x = \frac{1}{2}$, to

locate additional points. Plot the vertex, the x-intercepts, the y-intercept, and any additional points, and join them with a parabola.

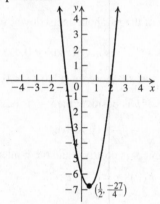

4. The graph of $f(x) = 3x^2 - 6x - 1$ is a parabola with $a = 3$, $b = -6$ and $c = -1$. The parabola opens up because $a > 0$. Complete the square to write the equation in standard form:

$$f(x) = 3x^2 - 6x - 1 = 3\left(x^2 - 2x\right) - 1$$

$$= 3\left(x^2 - 2x + 1\right) - 1 - 3 = 3(x-1)^2 - 4$$

Thus, the vertex is (1, −4). Next, find the x-intercepts:

$$0 = 3(x-1)^2 - 4 \Rightarrow \frac{4}{3} = (x-1)^2 \Rightarrow$$

$$\pm\frac{2\sqrt{3}}{3} = x - 1 \Rightarrow 1 \pm \frac{2\sqrt{3}}{3} = x \Rightarrow x \approx 2.15 \text{ or}$$

$x \approx -0.15$. Now, find the

y-intercept: $f(0) = 3(0)^2 - 6(0) - 1 = -1$. Use

the fact that the parabola is symmetric with respect to its axis, $x = -1$, to locate additional points. Plot the vertex, the x-intercepts, the y-intercept, and any additional points, and join them with a parabola.

(continued on next page)

(*continued*)

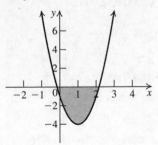

The graph of *f* is below the *x*-axis between the
x-intercepts, so the solution set for

$$f(x) = 3x^2 - 6x - 1 \le 0 \text{ is}$$

$$\left[1 - \frac{2\sqrt{3}}{3}, 1 + \frac{2\sqrt{3}}{3}\right] \text{ or } \left[\frac{3 - 2\sqrt{3}}{3}, \frac{3 + 2\sqrt{3}}{3}\right].$$

5. The graph of the parabola opens down, so
    $a < 0$.
    The vertex $V(h, k)$ is in QII, so $h < 0$. Because

    $h = -\dfrac{b}{2a}$ and $a < 0$, we must have $b < 0$.

    Otherwise, if *b* is positive, then $-b$ is negative,

    and $h = \dfrac{-b}{2a} = -\dfrac{b}{2a}$ is positive.

    The *y*-intercept is positive, and the *y*-intercept

    is $a(0)^2 + b(0) + c = c$, so $c > 0$.

6. $h(t) = -\dfrac{g_E}{2}t^2 + v_0 t + h_0$

    $g_E = 32 \text{ ft/s}^2$, $h_0 = 100 \text{ ft}$, max height = 244 ft

    a. Using the given values, we have

    $$h(t) = -\frac{32}{2}t^2 + v_0 t + 100$$
    $$= -16t^2 + v_0 t + 100$$

    Using the formula for the vertex of a

    parabola gives $t = -\dfrac{v_0}{2(-16)} = \dfrac{v_0}{32}$. This is

    the time at which the maximum height
    $h(t) = 244$ ft is attained. Thus,

    $$h(t) = 244 = h\left(\frac{v_0}{32}\right)$$

    $$= -16\left(\frac{v_0}{32}\right)^2 + v_0\left(\frac{v_0}{32}\right) + 100$$

    $$= -\frac{v_0^2}{64} + \frac{v_0^2}{32} + 100 = \frac{v_0^2}{64} + 100$$

Solving for v_0 yields

$$244 = \frac{v_0^2}{64} + 100 \Rightarrow 144 = \frac{v_0^2}{64} \Rightarrow$$
$$v_0^2 = 144 \cdot 64 \Rightarrow v_0 = \sqrt{144 \cdot 64} = 12 \cdot 8 = 96$$

Thus, $h(t) = -16t^2 + 96t + 100$ feet.

b. Using the formula for the vertex of a

    parabola gives $t = -\dfrac{96}{2(-16)} = \dfrac{96}{32} = 3$.

    The ball reached its highest point 3 seconds
    after it was released.

7. Let *x* = the length of the playground and
    y = the width of the playground.
    Then $2(x + y) = 1000 \Rightarrow x + y = 500 \Rightarrow$
    $y = 500 - x$.
    The area of the playground is

    $$A(x) = xy = x(500 - x) = 500x - x^2.$$

    The vertex for the parabola is (h, k) where

    $$h = -\frac{500}{2(-1)} = 250 \text{ and}$$

    $$k = 500(250) - 250^2 = 62,500.$$

    Thus, the maximum area that can be enclosed
    is 62,500 ft². The playground is a square with
    side length 250 ft.

3.1 Basic Concepts and Skills

1. A point where the axis meets the parabola is
    called the <u>vertex</u>.

3. True. $a = -2 < 0$.

5. The *x*-coordinate of the vertex of the parabola

    in exercise 4, $f(x) = -2 - x + x^2$ is $\dfrac{1}{2}$. Use

    the formula $h = -\dfrac{b}{2a}$, with $b = -1$ and $a = 1$.

7. True. The *x*-coordinate of the vertex of the

    parabola is $h = -\dfrac{b}{2a}$, so the *y*-coordinate of

    the vertex is $f\left(-\dfrac{b}{2a}\right)$.

9. f   11. a   13. h   15. g

17. Substitute -8 for *y* and 2 for *x* to solve for *a*:
     $-8 = a(2)^2 \Rightarrow -2 = a.$

     The equation is $y = -2x^2$.

19. Substitute 20 for *y* and 2 for *x* to solve for *a*:
$20 = a(2)^2 \Rightarrow 5 = a.$

The equation is $y = 5x^2$.

21. Substitute 0 for *h*, 0 for *k*, 8 for *y*, and –2 for *x* in the standard form for a quadratic equation to solve for *a*: $8 = a(-2-0)^2 + 0 \Rightarrow 8 = 4a \Rightarrow$
$2 = a.$ The equation is $y = 2x^2$.

23. Substitute –3 for *h*, 0 for *k*, –4 for *y*, and –5 for *x* in the standard form for a quadratic equation to solve for *a*:
$-4 = a(-5-(-3))^2 + 0 \Rightarrow -1 = a.$

The equation is $y = -(x+3)^2$.

25. Substitute 2 for *h*, 5 for *k*, 7 for *y*, and 3 for *x* in the standard form for a quadratic equation to solve for *a*: $7 = a(3-2)^2 + 5 \Rightarrow 2 = a.$

The equation is $y = 2(x-2)^2 + 5$.

27. Substitute 2 for *h*, –3 for *k*, 8 for *y*, and –5 for *x* in the standard form for a quadratic equation to solve for *a*: $8 = a(-5-2)^2 - 3 \Rightarrow \dfrac{11}{49} = a.$

The equation is $y = \dfrac{11}{49}(x-2)^2 - 3$.

29. Substitute $\dfrac{1}{2}$ for *h*, $\dfrac{1}{2}$ for *k*, $-\dfrac{1}{4}$ for *y*, and $\dfrac{3}{4}$ for *x* in the standard form for a quadratic equation to solve for *a*:
$-\dfrac{1}{4} = a\left(\dfrac{3}{4} - \dfrac{1}{2}\right)^2 + \dfrac{1}{2} \Rightarrow -12 = a.$

The equation is $y = -12\left(x - \dfrac{1}{2}\right)^2 + \dfrac{1}{2}$.

31. The vertex is (–2, 0), and the graph passes through (0, 3). Substitute –2 for *h*, 0 for *k*, 3 for *y*, and 0 for *x* in the standard form for a quadratic equation to solve for *a*:
$3 = a(0-(-2))^2 + 0 \Rightarrow \dfrac{3}{4} = a.$

The equation is $y = \dfrac{3}{4}(x+2)^2$.

33. The vertex is (3, –1), and the graph passes through (5, 2). Substitute 3 for *h*, –1 for *k*, 2 for *y*, and 5 for *x* in the standard form for a quadratic equation to solve for *a*:

$2 = a(5-3)^2 - 1 \Rightarrow \dfrac{3}{4} = a.$

The equation is $y = \dfrac{3}{4}(x-3)^2 - 1$.

35. The graph opens up, so *a* > 0. The vertex is located in QIV, so *h* > 0. Because $h = -\dfrac{b}{2a}$ and *a* > 0, we must have *b* < 0. Otherwise, if *b* is positive, then –*b* is negative, and
$h = \dfrac{-b}{2a} = -\dfrac{b}{2a}$ is negative.
The *y*-intercept is negative, and the *y*-intercept is $a(0)^2 + b(0) + c = c,$ so *c* < 0.

37. The graph opens up, so *a* > 0. The vertex is located on the positive *x*-axis, so *h* > 0 and *k* = 0. Because $h = -\dfrac{b}{2a}$ and *a* > 0, we must have *b* < 0. Otherwise, if *b* is positive, then –*b* is negative, and $h = \dfrac{-b}{2a} = -\dfrac{b}{2a}$ is negative.
The *y*-intercept is positive, and the *y*-intercept is $a(0)^2 + b(0) + c = c,$ so *c* > 0.

39. The graph of the parabola opens down, so *a* < 0.
The vertex *V*(*h*, *k*) is in QII, so *h* < 0. Because $h = -\dfrac{b}{2a}$ and *a* < 0, we must have *b* < 0.
Otherwise, if *b* is positive, then –*b* is negative, and $h = \dfrac{-b}{2a} = -\dfrac{b}{2a}$ is positive.
The *y*-intercept is negative, and the *y*-intercept is $a(0)^2 + b(0) + c = c,$ so *c* < 0.

41. $f(x) = 3x^2$

Stretch the graph of $y = x^2$ vertically by a factor of 3.

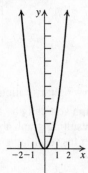

43. $g(x) = (x-4)^2$

Shift the graph of $y = x^2$ right four units.

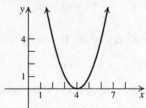

45. $f(x) = -2x^2 - 4$

Stretch the graph of $y = x^2$ vertically by a factor of 2, reflect the resulting graph in the x-axis, then shift the resulting graph down 4 units.

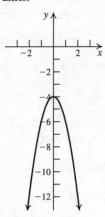

47. $g(x) = (x-3)^2 + 2$

Shift the graph of $y = x^2$ right three units, then shift the resulting graph up two units.

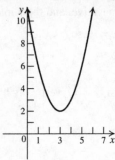

49. $f(x) = -3(x-2)^2 + 4$

Shift the graph of $y = x^2$ right two units, stretch the resulting graph vertically by a factor of 3, reflect the resulting graph about the x-axis, and then shift the resulting graph up four units.

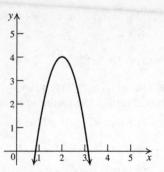

51. Complete the square to write the equation in standard form: $y = x^2 + 4x \Rightarrow$

$y + 4 = x^2 + 4x + 4 \Rightarrow y = (x+2)^2 - 4$. This

is the graph of $y = x^2$ shifted two units left and four units down. The vertex is $(-2, -4)$. The axis of symmetry is $x = -2$. To find the x-intercepts, let $y = 0$ and solve

$0 = (x+2)^2 - 4 \Rightarrow (x+2)^2 = 4 \Rightarrow$

$x + 2 = \pm 2 \Rightarrow x = -4$ or $x = 0$

To find the y-intercept, let $x = 0$ and solve

$y = (0+2)^2 - 4 \Rightarrow y = 0$.

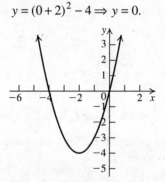

53. Complete the square to write the equation in standard form:

$y = 6x - 10 - x^2 \Rightarrow y = -(x^2 - 6x + 10) \Rightarrow$
$y - 9 = -(x^2 - 6x + 9) - 10 \Rightarrow y = -(x-3)^2 - 1$.

This is the graph of $y = x^2$ shifted three units right, reflected about the x-axis, and then shifted one unit down. The vertex is $(3, -1)$. The axis of symmetry is $x = 3$. To find the x-intercepts, let $y = 0$ and solve

$0 = -(x-3)^2 - 1 \Rightarrow -1 = (x-3)^2 \Rightarrow$ there is no x-intercept. To find the y-intercept, let

$x = 0$ and solve $y = -(0-3)^2 - 1 \Rightarrow y = -10$.

(*continued on next page*)

(*continued*)

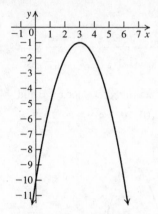

55. Complete the square to write the equation in standard form:

$y = 2x^2 - 8x + 9 \Rightarrow y = 2(x^2 - 4x) + 9 \Rightarrow$

$y + 8 = 2(x^2 - 4x + 4) + 9 \Rightarrow y = 2(x - 2)^2 + 1.$

This is the graph of $y = x^2$ shifted 2 units right, stretched vertically by a factor of 2, and then shifted one unit up. The vertex is (2, 1). The axis of symmetry is $x = 2$. To find the x-intercepts, let $y = 0$ and solve

$0 = 2(x - 2)^2 + 1 \Rightarrow -\dfrac{1}{2} = (x - 2)^2 \Rightarrow$ there is

no x-intercept. To find the y-intercept, let $x = 0$ and solve $y = 2(0 - 2)^2 + 1 \Rightarrow y = 9.$

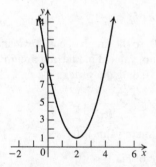

57. Complete the square to write the equation in standard form:

$y = -3x^2 + 18x - 11 \Rightarrow y = -3(x^2 - 6x) - 11 \Rightarrow$

$y - 27 = -3(x^2 - 6x + 9) - 11 \Rightarrow$

$y = -3(x - 3)^2 + 16.$ This is the graph of

$y = x^2$ shifted three units right, stretched vertically by a factor of three, reflected about the x-axis, and then shifted 16 units up. The vertex is (3, 16). The axis of symmetry is $x = 3$. To find the x-intercepts, let $y = 0$ and solve $y = 0 = -3(x - 3)^2 + 16 \Rightarrow$

$\dfrac{16}{3} = (x - 3)^2 \Rightarrow \pm\dfrac{4\sqrt{3}}{3} = x - 3 \Rightarrow$

$3 \pm \dfrac{4}{3}\sqrt{3} = x.$ To find the y-intercept, let $x = 0$

and solve $y = -3(0 - 3)^2 + 16 \Rightarrow y = -11.$

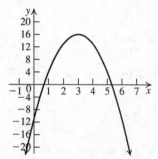

59. a. $a = 1 > 0$, so the graph opens up.

    b. The vertex is

$$\left(-\dfrac{-8}{2(1)}, f\left(-\dfrac{-8}{2(1)}\right) \right) = (4, -1).$$

    c. The axis of symmetry is $x = 4$.

    d. To find the x-intercepts, let $y = 0$ and solve

$0 = x^2 - 8x + 15 \Rightarrow 0 = (x - 3)(x - 5) \Rightarrow$

$x = 3$ or $x = 5$. To find the y-intercept, let

$x = 0$ and solve $y = 0^2 - 8(0) + 15 \Rightarrow$

$y = 15.$

    e.

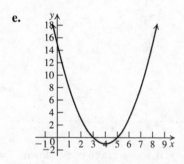

61. a. $a = 1 > 0$, so the graph opens up.

    b. The vertex is

$$\left(-\dfrac{-1}{2(1)}, f\left(-\dfrac{-1}{2(1)}\right) \right) = \left(\dfrac{1}{2}, -\dfrac{25}{4} \right).$$

    c. The axis of symmetry is $x = \dfrac{1}{2}$.

    d. To find the x-intercepts, let $y = 0$ and solve

$0 = x^2 - x - 6 \Rightarrow 0 = (x - 3)(x + 2) \Rightarrow x = 3$

or $x = -2$. To find the y-intercept, let $x = 0$

and solve $y = 0^2 - (0) - 6 \Rightarrow y = -6.$

e.

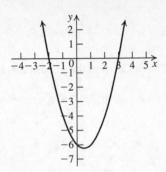

63. a. $a = 1 > 0$, so the graph opens up.

b. The vertex is $\left(-\dfrac{-2}{2(1)}, f\left(-\dfrac{-2}{2(1)}\right)\right) = (1, 3)$.

c. The axis of symmetry is $x = 1$.

d. To find the x-intercepts, let $y = 0$ and solve

$$0 = x^2 - 2x + 4 \Rightarrow$$

$$x = \dfrac{-(-2) \pm \sqrt{(-2)^2 - 4(1)(4)}}{2(1)} \Rightarrow$$

$$x = \dfrac{2 \pm \sqrt{-12}}{2} \Rightarrow \text{ there are no } x\text{-intercepts.}$$

To find the y-intercept, let $x = 0$ and solve

$$y = 0^2 - 2(0) + 4 \Rightarrow y = 4.$$

e.

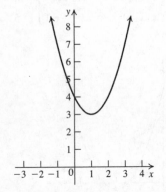

65. a. $a = -1 < 0$, so the graph opens down.

b. The vertex is

$$\left(-\dfrac{-2}{2(-1)}, f\left(-\dfrac{-2}{2(-1)}\right)\right) = (-1, 7).$$

c. The axis of symmetry is $x = -1$.

d. To find the x-intercepts, let $y = 0$ and solve

$$0 = 6 - 2x - x^2 \Rightarrow$$

$$x = \dfrac{-(-2) \pm \sqrt{(-2)^2 - 4(-1)(6)}}{2(-1)} \Rightarrow x = \dfrac{2 \pm \sqrt{28}}{-2} \Rightarrow$$

$$x = -1 \pm \sqrt{7}.$$

To find the y-intercept, let $x = 0$ and solve

$$y = 6 - 2(0) - 0^2 \Rightarrow y = 6.$$

e.

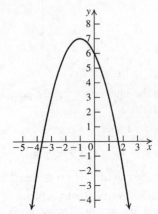

67. a. $a = 1 > 0$, so the graph opens up and has a minimum value. Find the minimum value by finding the vertex:

$$\left(-\dfrac{-4}{2(1)}, f\left(-\dfrac{-4}{2(1)}\right)\right) = (2, -1)$$

The minimum value is -1.

b. The range of f is $[-1, \infty)$.

69. a. $a = -1 < 0$, so the graph opens down and has a maximum value. Find the maximum value by finding the vertex:

$$\left(-\dfrac{4}{2(-1)}, f\left(-\dfrac{4}{2(-1)}\right)\right) = (2, 0)$$

The maximum value is 0.

b. The range of f is $(-\infty, 0]$.

71. a. $a = 2 > 0$, so the graph opens up and has a minimum value. Find the minimum value by finding the vertex:

$$\left(-\dfrac{-8}{2(2)}, f\left(-\dfrac{-8}{2(2)}\right)\right) = (2, -5)$$

The minimum value is -5.

b. The range of f is $[-5, \infty)$.

73. a. $a = -4 < 0$, so the graph opens down and has a maximum value. Find the maximum value by finding the vertex:

$$\left(-\frac{12}{2(-4)}, f\left(-\frac{12}{2(-4)}\right)\right) = \left(\frac{3}{2}, 16\right)$$

The maximum value is 16.

b. The range of f is $(-\infty, 16]$.

In exercises 75–80, the x-intercepts are the boundaries of the intervals.

75. Solution: $[-2, 2]$

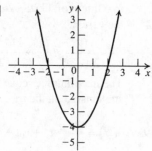

77. Solution:
$(-\infty, 1) \cup (3, \infty)$

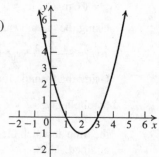

79. Solution:
$\left(-1, \frac{7}{6}\right)$

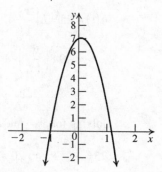

3.1 Applying the Concepts

81. $h(x) = -\frac{32}{132^2}x^2 + x + 3$

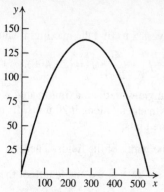

a. Using the graph, we see that the ball traveled approximately 550 ft horizontally.

b. Using the graph, we see that the ball went approximately 140 high.

c. $0 = -\frac{32}{132^2}x^2 + x + 3$

$$x = \frac{-1 \pm \sqrt{1^2 - 4\left(-\frac{32}{132^2}\right)(3)}}{2\left(-\frac{32}{132^2}\right)}$$

≈ -3 or 547

Thus, the ball traveled approximately 547 ft horizontally. The ball reached its maximum height at the vertex of the function,

$$\left(-\frac{b}{2a}, h\left(-\frac{b}{2a}\right)\right).$$

$$-\frac{b}{2a} = -\frac{1}{2\left(-32/\left(132^2\right)\right)} \approx 272.25$$

$$h(272.25) = -\frac{32}{132^2}(272.25)^2 + 272.25 + 3$$

≈ 139

The ball reached approximately 139 ft

83. The vertex of the function is

$$\left(-\frac{114}{2(-3)}, f\left(-\frac{114}{2(-3)}\right)\right) = (19, 1098).$$

The revenue is at its maximum when $x = 19$.

85. $\left(-\frac{-50}{2(1)}, f\left(-\frac{-50}{2(1)}\right)\right) = (25, -425)$.

The total cost is minimum when $x = 25$.

87. Let $x =$ the length of the rectangle. Then
$$\frac{80 - 2x}{2} = 40 - x = \text{ the width of the rectangle.}$$

The area of the rectangle $= x(40 - x)$
$= 40x - x^2$.

(continued on next page)

(continued)

Find the vertex to find the maximum value:

$$\left(-\frac{40}{2(-1)}, f\left(-\frac{40}{2(-1)}\right)\right) = (20, 400).$$

The rectangle with the maximum area is a square with sides of length 20 units. Its area is 400 square units.

89. Let x = the width of the fields. Then,

$$\frac{600 - 3x}{2} = 300 - \frac{3}{2}x = \text{the length of the two}$$

fields together. (Note that there is fencing between the two fields, so there are three "widths.")

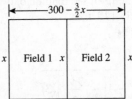

The total area = $x\left(300 - \frac{3}{2}x\right) = 300x - \frac{3}{2}x^2$.

Find the vertex to find the dimensions and maximum value:

$$\left(-\frac{300}{2(-3/2)}, f\left(-\frac{300}{2(-3/2)}\right)\right) = (100, 15{,}000).$$

So the width of each field is 100 meters. The length of the two fields together is $300 - 1.5(100) = 150$ meters, so the length of each field is $150/2 = 75$ meters. The area of each field is $100(75) = 7500$ square meters.

91. The yield per tree is modeled by the equation of a line passing through $(26, 500)$ where the x-coordinate represents the number of trees planted, and the y-coordinate represents the number of apples per tree. The rate of change is -10; that is, for each tree planted the yield decreases by 10. So, the yield per tree is $y - 500 = -10(x - 26) \Rightarrow y = -10x + 760$. Since there are x trees, the total yield = $x(-10x + 760) = -10x^2 + 760x$. Use the vertex to find the number of trees that will maximize the yield:

$$\left(-\frac{760}{2(-10)}, f\left(-\frac{760}{2(-10)}\right)\right) = (38, 14{,}440).$$

So the maximum yield occurs when 38 trees are planted per acre.

93. If 20 students or less go on the trip, the cost is $72 per students. If more than 20 students go on the trip, the cost is reduced by $2 per the

number of students over 20. So the cost per student is a piecewise function based on the number of students, n, going on the trip:

$$f(n) = \begin{cases} 72 & \text{if } n \le 20 \\ 72 - 2(n - 20) = 112 - 2n & \text{if } n > 20 \end{cases}$$

The total revenue is

$$nf(n) = \begin{cases} 72n & \text{if } n \le 20 \\ n(112 - 2n) = 112n - 2n^2 & \text{if } n > 20 \end{cases}$$

The maximum revenue is either 1440 (the revenue if 20 students go on the trip) or the maximum of $112n - 2n^2$. Find this by using the vertex:

$$\left(-\frac{112}{(2)(-2)}, f\left(-\frac{112}{(2)(-2)}\right)\right) = (28, 1568)$$

The maximum revenue is $1568 when 28 students go on the trip.

95. $h(t) = -\dfrac{g_M}{2}t^2 + v_0 t + h_0$

$g_M = 1.6\,\text{m/s}^2$, $h_0 = 5$ m, max height = 25 m

a. Using the given values, we have

$$h(t) = -\frac{1.6}{2}t^2 + v_0 t + 5 = -0.8t^2 + v_0 t + 5$$

Using the formula for the vertex of a parabola gives $t = -\dfrac{v_0}{2(-0.8)} = \dfrac{v_0}{1.6}$. This is the time at which the height $h(t) = 25$ m is attained. Thus,

$$h(t) = 25 = h\left(\frac{v_0}{1.6}\right)$$

$$= -0.8\left(\frac{v_0}{1.6}\right)^2 + v_0\left(\frac{v_0}{1.6}\right) + 5$$

$$= -\frac{v_0^2}{3.2} + \frac{v_0^2}{1.6} + 5 = \frac{v_0^2}{3.2} + 5$$

Solving for v_0 yields

$$25 = \frac{v_0^2}{3.2} + 5 \Rightarrow 20 = \frac{v_0^2}{3.2} \Rightarrow$$

$$v_0^2 = 64 \Rightarrow v_0 = 8$$

Thus, $h(t) = -0.8t^2 + 8t + 5$.

b. Using the formula for the vertex of a parabola gives $t = -\dfrac{8}{2(-0.8)} = \dfrac{8}{1.6} = 5$.

The ball reached its highest point 5 seconds after it was released.

97. a. The maximum height occurs at the vertex:

$$\left(-\frac{64}{2(-16)}, f\left(-\frac{64}{2(-16)}\right)\right) = (2, 64), \text{ so the}$$

maximum height is 64 feet.

b. When the projectile hits the ground, $h = 0$, so solve

$$0 = -16t^2 + 64t \Rightarrow -16t(t-4) = 0 \Rightarrow$$
$t = 0$ or $t = 4$.

The projectile hits the ground at 4 seconds.

99. Let x = the radius of the semicircle. Then the length of the rectangle is $2x$. The circumference of the semicircle is πx, so the perimeter of the rectangular portion of the window is $18 - \pi x$.

The width of the rectangle = $\dfrac{18 - \pi x - 2x}{2}$.

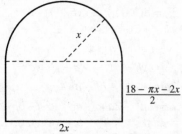

The area of the semicircle is $\pi x^2 / 2$, and the

area of the rectangle is $2x\left(\dfrac{18 - \pi x - 2x}{2}\right) =$

$18x - \pi x^2 - 2x^2$. So the total area is

$$18x - \pi x^2 - 2x^2 + \frac{\pi x^2}{2} = 18x - 2x^2 - \frac{\pi x^2}{2} \Rightarrow$$

$$18x - \left(2 + \frac{\pi}{2}\right)x^2.$$

The maximum area occurs at the x-coordinate of the vertex:

$$-\frac{18}{2\left(-2-\dfrac{\pi}{2}\right)} = -\frac{18}{-4-\pi} = \frac{18}{4+\pi}.$$

This is the radius of the semicircle. The length

of the rectangle is $2\left(\dfrac{18}{4+\pi}\right) = \dfrac{36}{4+\pi}$ ft.

The width of the rectangle is

$$\frac{18 - \pi\left(\dfrac{18}{4+\pi}\right) - 2\left(\dfrac{18}{4+\pi}\right)}{2} = \frac{18}{4+\pi} \text{ ft.}$$

101. a. $V = \left(-\dfrac{1.18}{2(-0.01)}, f\left(-\dfrac{1.18}{2(-0.01)}\right)\right)$
$= (59, 36.81)$

b. $0 = -0.01x^2 + 1.18x + 2 \Rightarrow$

$$x = \frac{-1.18 \pm \sqrt{1.18^2 - 4(-0.01)(2)}}{2(-0.01)}$$

$$= \frac{-1.18 \pm \sqrt{1.4724}}{-0.02} \approx \frac{-1.18 \pm 1.21}{-0.02}$$

$$\approx -1.5 \text{ or } 119.5.$$

Reject the negative answer, and round the positive answer to the nearest whole number. The ball hits the ground approximately 120 feet from the punter.

c. The maximum height occurs at the vertex. Since we know that the ball hits the ground at $x \approx 120$, the vertex occurs at $x \approx 60$.

$f(60) = -0.01(60)^2 + 1.18(60) + 2 = 36.8$.

The maximum height is approximately 37 ft.

d. The player is at $x = 6$ feet.

$f(6) = -0.01(6)^2 + 1.18(6) + 2 = 8.72$.

The player must reach approximately 9 feet to block the ball.

e. $7 = -0.01x^2 + 1.18x + 2$
$0 = -0.01x^2 + 1.18x - 5$

$$x = \frac{-1.18 \pm \sqrt{1.18^2 - 4(-0.01)(-5)}}{2(-0.01)}$$

$$= \frac{-1.18 \pm \sqrt{1.1924}}{-0.02} \Rightarrow$$

$x \approx 4.4$ feet or $x \approx 113.6$ feet

3.1 Beyond the Basics

103. $y = 3(x+2)^2 + 3 \Rightarrow y = 3(x^2 + 4x + 4) + 3 \Rightarrow$
$y = 3x^2 + 12x + 15$

105. The y-intercept is 4, so the graph passes through (0, 4). Substitute the coordinates (0, 4) for x and y and (1, –2) for h and k into the standard form $y = a(x-h)^2 + k$ to solve for a: $4 = a(0-1)^2 - 2 \Rightarrow 6 = a$. The x-coordinate of the vertex is $1 = -\dfrac{b}{2a} = -\dfrac{b}{2(6)} \Rightarrow b = -12$.

The y-intercept = c, so the equation is

$$f(x) = 6x^2 - 12x + 4.$$

107. The x-coordinate of the vertex is 3. Substitute the coordinates of the x-intercept and the x-coordinate of the vertex into the standard form

$y = a(x-h)^2 + k$ to find an expression relating a and k:

$0 = a(7-3)^2 + k \Rightarrow -16a = k$. Now substitute the coordinates of the y-intercept, the x-coordinate of the vertex, and the expression for k into the standard form $y = a(x-h)^2 + k$

to solve for a: $14 = a(0-3)^2 - 16a \Rightarrow -2 = a$.
Use this value to find k: $k = -16(-2) = 32$.
The equation is

$y = -2(x-3)^2 + 32 \Rightarrow y = -2x^2 + 12x + 14$.

109. The x-coordinate of the vertex is $\dfrac{-2+6}{2} = 2$.

Substitute the coordinates of one of the x-intercepts and the x-coordinate of the vertex into the standard form $y = a(x-h)^2 + k$ to find an expression relating a and k:

$0 = a(-2-2)^2 + k \Rightarrow -16a = k$.
Now substitute the coordinates of the other x-intercept, the x-coordinate of the vertex, and the expression for k into the standard form

$y = a(x-h)^2 + k$ to solve for a:

$0 = a(6-2)^2 - 16a \Rightarrow 1 = a$. Use this value to find k: $-16(1) = -16 = k$.
So, one equation is

$y = (x-2)^2 - 16 = x^2 - 4x - 12$. The graph of this equation opens upward. To find the equation of the graph that opens downward, multiply the equation by -1:

$y = -x^2 + 4x + 12$.

111. The x-coordinate of the vertex is $\dfrac{-7-1}{2} = -4$.

Substitute the coordinates of one of the x-intercepts and the x-coordinate of the vertex into the standard form $y = a(x-h)^2 + k$ to find an expression relating a and k:

$0 = a(-7+4)^2 + k \Rightarrow -9a = k$. Now substitute the coordinates of the other x-intercept, the x-coordinate of the vertex, and the expression for k into the standard form $y = a(x-h)^2 + k$

to solve for a: $0 = a(-1+4)^2 - 9a \Rightarrow 1 = a$.
Use this value to find k: $-9(1) = -9 = k$. So

one equation is $y = (x+4)^2 - 9 = x^2 + 8x + 7$.
The graph of this equation opens upward. To find the equation of the graph that opens downward, multiply the equation by -1:

$y = -x^2 - 8x - 7$.

113.
$$f(a+1) = 4(a+1) - (a+1)^2$$
$$= 4a + 4 - (a^2 + 2a + 1) = -a^2 + 2a + 3$$
$$f(a-1) = 4(a-1) - (a-1)^2$$
$$= 4a - 4 - (a^2 - 2a + 1) = -a^2 + 6a - 5$$
$$f(a+1) - f(a-1) = 0 \Rightarrow$$
$$(-a^2 + 2a + 3) - (-a^2 + 6a - 5) = 0 \Rightarrow$$
$$8 - 4a = 0 \Rightarrow a = 2$$

3.1 Critical Thinking/Discussion/Writing

115.
$$f(h+p) = a(h+p)^2 + b(h+p) + c$$
$$f(h-p) = a(h-p)^2 + b(h+p) + c$$
$$f(h+p) = f(h-p) \Leftrightarrow$$
$$a(h+p)^2 + b(h+p) + c$$
$$= a(h-p)^2 + b(h-p) + c \Leftrightarrow$$
$$ah^2 + 2ahp + ap^2 + bh + bp + c$$
$$= ah^2 - 2ahp + ap^2 + bh - bp + c \Leftrightarrow$$
$4ahp = -2bp \Leftrightarrow 2ah = -b$. (We can divide by p because $p \neq 0$.) Since $a \neq 0, 2ah = -b \Rightarrow$

$$h = -\frac{b}{2a}.$$

116. Write $y = 2x^2 - 8x + 9$ in standard form to find the axis of symmetry:

$$y = 2x^2 - 8x + 9 \Rightarrow y + 8 = 2(x^2 - 4x + 4) + 9 \Rightarrow$$
$$y = 2(x-2)^2 + 1.$$

The axis of symmetry is $x = 2$.
Using the results of exercise 115, we know that $f(2+p) = f(-1) = f(2-p)$.
$2 + p = -1 \Rightarrow$
$p = -3$, so $2 - p = 2 - (-3) = 5$.
The point symmetric to the point $(-1, 19)$ across the axis of symmetry is $(5, 19)$.

117. Answers will vary. If $a > 0$, then the parabola $f(x) = ax^2 + bx + c$ opens upward and the solution of the inequality is the portion of the graph that is above the x-axis. If the x-intercepts are represented by x_0 and x_1 with $x_0 \leq x_1$, then the solution of the inequality is $(-\infty, x_0) \cup (x_1, \infty)$.

(continued on next page)

(continued)

If $a < 0$, then the parabola $f(x) = ax^2 + bx + c$ opens downward and the solution of the inequality is the portion of the graph that is below the x-axis. If the x-intercepts are represented by x_0 and x_1 with $x_0 \le x_1$, then the solution of the inequality is (x_0, x_1).

118. a.
$$(f \circ g)(x) = f(mx + b) = a\big[(mx + b) - h\big]^2 + k$$
$$= a\big[(mx + b)^2 - 2h(mx + b) + h^2\big] + k$$
$$= a\big[m^2x^2 + 2bmx + b^2 - 2hmx - 2hb + h^2\big] + k$$
$$= am^2x^2 + 2abmx + ab^2 - 2ahmx - 2ahb + ah^2 + k$$
$$= am^2x^2 + (2abm - 2ahm)x + \left(ab^2 - 2ahb + ah^2 + k\right)$$

This is the equation of a parabola. The x-coordinate of the vertex is
$$-\frac{2abm - 2ahm}{2am^2} = -\frac{2am(b-h)}{2am^2} = -\frac{b-h}{m} \text{ or } \frac{h-b}{m}.$$

The y-coordinate of the vertex is
$$am^2\left(\frac{h-b}{m}\right)^2 + (2abm - 2ahm)\left(\frac{h-b}{m}\right) + \left(ab^2 - 2ahb + ah^2 + k\right)$$
$$= a(h-b)^2 + 2a(b-h)(h-b) + \left(ab^2 - 2ahb + ah^2 + k\right)$$
$$= ah^2 - 2ahb + ab^2 - 2ab^2 + 4abh - 2ah^2 + ab^2 - 2ahb + ah^2 + k$$
$$= k$$

The vertex is $\left(\dfrac{h-b}{m}, k\right)$.

b.
$$(g \circ f)(x) = g\Big[a(x-h)^2 + k\Big]$$
$$= m\Big[a(x-h)^2 + k\Big] + b$$
$$= m\left(ax^2 - 2ahx + ah^2 + k\right) + b$$
$$= max^2 - 2mahx + mah^2 + mk + b$$

This is the equation of a parabola. The x-coordinate of the vertex is $-\dfrac{-2mah}{2ma} = h$.

The y-coordinate of the vertex is
$$mah^2 - 2mah^2 + mah^2 + mk + b = mk + b.$$
Thus, the vertex is $(h, mk + b)$.

119. If the discriminant equals zero, there is exactly one real solution. Thus, the vertex of
$$y = f(x) \text{ lies on the } x\text{-axis at } x = -\frac{b}{2a}.$$
If the discriminant > 0, there are two unequal real solutions. This means that the graph of $y = f(x)$ crosses the x-axis in two places. If $a > 0$, then the vertex lies below the x-axis and the parabola crosses the x-axis; if $a < 0$, then the vertex lies above the x-axis and the parabola crosses the x-axis. If the discriminant < 0, there are two nonreal complex solutions. If $a > 0$, then the vertex lies above the x-axis

and the parabola does not cross the x-axis (it opens upward); if $a < 0$, then the vertex lies below the x-axis and the parabola does not cross the x-axis (it opens downward).

120. Let $g(x) = x^2$ and $k(x) = x - h$. Then
$$f(x) = g(k(x)) = g(x - h) = (x - h)^2, \text{ which}$$
is a horizontal translation of g.

3.1 Maintaining Skills

121. $-5^0 = -\left(5^0\right) = -1$

123. $-4^{-2} = -(4)^{-2} = -\dfrac{1}{(4)^2} = -\dfrac{1}{16}$

125. $x^7 \cdot x^{-7} = x^{7+(-7)} = x^0 = 1$

127. $x^2\left(3 - \dfrac{3}{4}\right) = x^2\left(\dfrac{12}{4} - \dfrac{3}{4}\right) = \dfrac{9}{4}x^2$

129. $4x^2 - 9 = (2x + 3)(2x - 3)$

131. $15x^2 + 11x - 12 = (3x + 4)(5x - 3)$

133. $x^2(x-1)-4(x-1)=(x^2-4)(x-1)$
$$=(x+2)(x-2)(x-1)$$

135. $x^3+4x^2+3x+12=(x^3+4x^2)+(3x+12)$
$$=x^2(x+4)+3(x+4)$$
$$=(x^2+3)(x+4)$$

3.2 Polynomial Functions

3.2 Practice Problems

1. a. $f(x)=\dfrac{x^2+1}{x-1}$ is not a polynomial function
because its domain is not $(-\infty,\infty)$.

b. $g(x)=2x^7+5x^2-17$ is a polynomial
function. Its degree is 7, the leading term is
$2x^7$, and the leading coefficient is 2.

2. $P(x)=4x^3+2x^2+5x-17$
$$=x^3\left[4+\frac{2}{x}+\frac{5}{x^2}-\frac{17}{x^3}\right]$$

When $|x|$ is large, the terms $\dfrac{2}{x}$, $\dfrac{5}{x^2}$, and $-\dfrac{17}{x^3}$
are close to 0. Therefore,
$$P(x)=x^3(4+0+0-0)\approx 4x^3.$$

3. Use the leading-term test to determine the end
behavior of $y=f(x)=-2x^4+5x^2+3$. Here
$n=4$ and $a_n=-2<0$. Thus, Case 2 applies.
The end behavior is described as
$y\to-\infty$ as $x\to-\infty$ and $y\to-\infty$ as $x\to\infty$.

4. First group the terms, then factor and solve
$f(x)=0$:
$$f(x)=2x^3-3x^2+4x-6$$
$$=2x^3+4x-3x^2-6$$
$$=2x(x^2+2)-3(x^2+2)$$
$$=(2x-3)(x^2+2)$$
$$0=(2x-3)(x^2+2)$$
$$0=2x-3\Rightarrow x=\frac{3}{2}\text{ or }$$
$$0=x^2+2\text{ (no real solution)}$$
The only real zero is $\dfrac{3}{2}$.

5. $f(x)=2x^3-3x-6$
$f(1)=-7$ and $f(2)=4$. Since $f(1)$ and
$f(2)$ have opposite signs, by the
Intermediate Value Theorem, f has a real zero
between 1 and 2.

6. $f(x)=(x+1)^2(x-3)(x+5)=0\Rightarrow$
$(x+1)^2=0$ or $x-3=0$ or $x+5=0\Rightarrow$
$x=-1$ or $x=3$ or $x=-5$
$f(x)$ has three distinct zeros.

7. $f(x)=(x-1)^2(x+3)(x+5)=0\Rightarrow$
$(x-1)^2=0$ or $x+3=0$ or $x+5=0\Rightarrow$
$x=1$ (multiplicity 2) or $x=-3$ (multiplicity 1)
or $x=-5$ (multiplicity 1)

8. $f(x)=-x^4+3x^2-2$ has at most three
turning points. Using a graphing calculator,
we see that there are indeed, three turning
points.

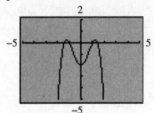

9. $f(x)=-x^4+5x^2-4$
Since the degree, 4, is even and the leading
coefficient is -1, the end behavior is as
shown:

$y\to-\infty$ as $x\to-\infty$ and $y\to-\infty$ as $x\to\infty$.
Now find the zeros of the function:
$$0=-x^4+5x^2-4\Rightarrow 0=-(x^4-5x^2+4)$$
$$=-(x^2-4)(x^2-1)\Rightarrow$$
$$0=x^2-4\text{ or }0=x^2-1\Rightarrow$$
$$x=\pm 2\quad\text{or }x=\pm 1$$
There are four zeros, each of multiplicity 1, so
the graph crosses the x-axis at each zero.

(continued on next page)

(continued)

Next, find the *y*-intercept:

$$f(0) = -x^4 + 5x^2 - 4 = -4$$

Now find the intervals on which the graph lies above or below the *x*-axis. The four zeros divide the *x*-axis into five intervals, $(-\infty, -2)$, $(-2, -1), (-1, 1), (1, 2),$ and $(2, \infty)$. Determine the sign of a test value in each interval

Interval	Test point	Value of $f(x)$	Above/below x-axis
$(-\infty, -2)$	-3	-40	below
$(-2, -1)$	-1.5	2.1875	above
$(-1, 1)$	0	-4	below
$(1, 2)$	1.5	2.1875	above
$(2, \infty)$	3	-40	below

Plot the zeros, *y*-intercepts, and test points, and then join the points with a smooth curve.

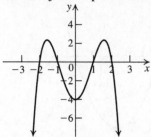

10. $V = \dfrac{\pi}{3\sqrt{3}} x^3 = \dfrac{\pi}{3\sqrt{3}} \cdot 7^3 \approx 207.378 \text{ dm}^3$
    $\approx 207.378 \text{ L}$

3.2 Basic Concepts and Skills

1. The degree of $2x^5 - 3x^4 + x - 6$ is $\underline{5}$, its leading term is $\underline{2x^5}$, its leading coefficient is $\underline{2}$, and its constant term is $\underline{-6}$.

3. The behavior of the function $y = f(x)$ as $x \to \infty$ or $x \to -\infty$ is called the <u>end behavior</u> of the function.

5. A number *c* for which $f(c) = 0$ is called a <u>zero</u> of the function *f*.

7. The graph of a polynomial function of degree *n* has at most $\underline{n-1}$ turning points.

9. Polynomial function; degree: 5; leading term: $2x^5$; leading coefficient: 2

11. Polynomial function; degree: 3; leading term: $\dfrac{2}{3}x^3$; leading coefficient: $\dfrac{2}{3}$

13. Polynomial function; degree: 4; leading term: πx^4; leading coefficient: π

15. Not a polynomial function: the graph has sharp corners; not a smooth curve; presence of $|x|$

17. Not a polynomial function: the domain is not $(-\infty, \infty)$

19. Not a polynomial function: presence of $\sqrt{x}$

21. Not a polynomial function: the domain is not $(-\infty, \infty)$

23. Not a polynomial function: graph is not continuous

25. Not a polynomial function: the graph is not continuous

27. Not a polynomial function: not the graph of a function

29. c   31. a   33. d

35. a. Zeros: $x = -5, 1$

    b. $x = -5$: multiplicity: 1, crosses the *x*-axis; $x = 1$, multiplicity: 1, crosses the *x*-axis

    c. Maximum number of turning points: 1

37. a. Zeros: $x = -1, 1$

    b. $x = -1$: multiplicity: 2, touches but does not cross the *x*-axis; $x = 1$, multiplicity: 3, crosses the *x*-axis

    c. Maximum number of turning points: 4

39. a. Zeros: $x = -\dfrac{2}{3}, \dfrac{1}{2}$

    b. $x = -2/3$: multiplicity: 1, crosses the *x*-axis; $x = 1/2$: multiplicity: 2, touches but does not cross the *x*-axis

    c. Maximum number of turning points: 2

41. a. $f(x) = -x^4 + 6x^3 - 9x^2 = -x^2(x^2 - 6x + 9)$
       $= -x^2(x-3)^2$
       Zeros: $x = 0, 3$

b. $x = 0$: multiplicity: 2, touches but does not cross the x-axis;
$x = 3$, multiplicity: 2, touches but does not cross the x-axis

c. Maximum number of turning points: 3

43. $f(2) = 2^4 - 2^3 - 10 = -2$;

$f(3) = 3^4 - 3^3 - 10 = 44$.

Because the sign changes, there is a real zero between 2 and 3. The zero is approximately 2.09.

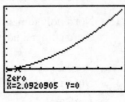

45. $f(2) = 2^5 - 9(2)^2 - 15 = -19$;

$f(3) = 3^5 - 9(3)^2 - 15 = 147$.

Because the sign changes, there is a real zero between 2 and 3. The zero is approximately 2.28.

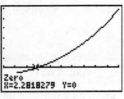

47. a. $y \to \infty$ as $x \to -\infty$; $y \to \infty$ as $x \to \infty$

b. $x^2 + 3 = 0 \Rightarrow x^2 = -3 \Rightarrow x = \pm i\sqrt{3} \Rightarrow$ there are no real zeros.

c. The graph is above the x-axis on $(-\infty, \infty)$.

d. $y = 0^2 + 3 = 3 \Rightarrow$ the y-intercept is 3.

e. $f(-x) = (-x)^2 + 3 = x^2 + 3 \Rightarrow f$ is even. The graph is symmetrical with respect to the y-axis.

f. Maximum number of turning points: 1

g.

49. a. $y \to \infty$ as $x \to -\infty$; $y \to \infty$ as $x \to \infty$

b. $x^2 + 4x - 21 = 0 \Rightarrow (x + 7)(x - 3) = 0 \Rightarrow$
$x = 3$ or $x = -7$.
The graph crosses the x-axis at both zeros.

c. The graph is above the x-axis on $(-\infty, -7) \cup (3, \infty)$ and below the x-axis on $(-7, 3)$.

d. $y = 0^2 + 4(0) - 21 = -21 \Rightarrow$ the y-intercept is -21.

e. $f(-x) = (-x)^2 + 4(-x) - 21$
$= x^2 - 4x - 21 \neq f(x) \Rightarrow f$ is not even.
$f(-x) \neq -f(x) \Rightarrow f$ is not odd.
There are no symmetries with respect to the axes or the origin.

$h = -\dfrac{b}{2a} = -\dfrac{4}{2(1)} = -2,$ so the graph is

symmetric about $x = -2$.

f. Maximum number of turning points: 1

g.

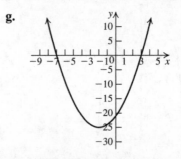

51. a. $y \to \infty$ as $x \to -\infty$; $y \to -\infty$ as $x \to \infty$

b. $-2x^2(x + 1) = 0 \Rightarrow x = 0 \cup x = -1$.
The graph touches but does not cross the x-axis at $x = 0$. The graph crosses the x-axis at $x = -1$.

c. The graph is above the x-axis on $(-\infty, -1)$ and below the x-axis on $(-1, 0) \cup (0, \infty)$.

d. $y = -2(0)^2(0 + 1) = 0 \Rightarrow$ the y-intercept is 0.

e. $f(-x) = 2(-x)^2(-x + 1) = 2x^2(-x + 1)$
$\neq f(x) \Rightarrow f$ is not even.
$f(-x) \neq -f(x) \Rightarrow f$ is not odd.
There are no symmetries.

f. Maximum number of turning points: 2

g.

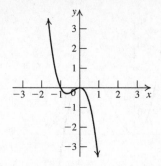

53. a. $y \to \infty$ as $x \to -\infty$; $y \to \infty$ as $x \to \infty$

b. $x^2(x-1)^2 = 0 \Rightarrow x = 0$ or $x = 1$.
The graph touches but does not cross the
x-axis at both zeros.

c. The graph is above the x-axis on
$(-\infty, 0) \cup (0,1) \cup (1, \infty)$.

d. $y = 0^2(0-1)^2 = 0 \Rightarrow$ the y-intercept is 0.

e. $f(-x) = (-x)^2(-x-1)^2 = x^2(x^2 + 2x + 1)$
$\neq f(x) \Rightarrow f$ is not even.
$f(-x) \neq -f(x) \Rightarrow f$ is not odd.
There are no symmetries.

f. Maximum number of turning points: 3

g.

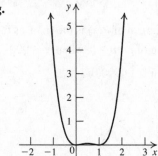

55. a. $y \to \infty$ as $x \to -\infty$; $y \to \infty$ as $x \to \infty$

b. $(x-1)^2(x+3)(x-4) = 0 \Rightarrow x = 1$ or $x = -3$
or $x = 4$. The graph touches but does not
cross the x-axis at $x = 1$. The graph crosses
the x-axis at $x = -3$ and $x = 4$.

c. The graph is above the x-axis on $(-\infty, -3)$
and $(4, \infty)$ and below the x-axis on
$(-3,1) \cup (1,4)$.

d. $y = (0-1)^2(0+3)(0-4) = -12 \Rightarrow$ the
y-intercept is -12.

e. $f(x) = (x-1)^2(x+3)(x-4)$
$= (x^2 - 2x + 1)(x^2 - x - 12)$
$= x^4 - 3x^3 - 9x^2 + 23x - 12.$
$f(-x) = (-x-1)^2(-x+3)(-x-4)$
$= (x^2 + 2x + 1)(x^2 + x - 12)$
$= x^4 + 3x^3 - 9x^2 - 23x - 12.$
$f(x) \neq f(x) \Rightarrow f$ is not even.
$f(-x) \neq -f(x) \Rightarrow f$ is not odd.
There are no symmetries.

f. Maximum number of turning points: 3

g.

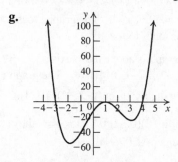

57. a. $y \to \infty$ as $x \to -\infty$; $y \to -\infty$ as $x \to \infty$

b. $-x^2(x-1)^2(x+1) = 0 \Rightarrow x = 0$ or $x = 1$ or
$x = -1$. The graph touches but does not
cross the x-axis at $x = 0$ and $x = -1$. The
graph crosses the x-axis at $x = 1$.

c. The graph is above the x-axis on
$(-\infty, -1) \cup (-1, 0) \cup (0,1)$ and below the
x-axis on $(1, \infty)$.

d. $y = -0^2(0-1)^2(0+1) = 0 \Rightarrow$ the
y-intercept is 0.

e. $f(x) = -x^2(x^2 - 1)(x+1)$
$= -x^5 - x^4 + x^3 + x^2.$
$f(-x) = -(-x)^2((-x)^2 - 1)(-x+1)$
$= x^5 - x^4 - x^3 + x^2.$
$f(x) \neq f(x) \Rightarrow f$ is not even.
$f(-x) \neq -f(x) \Rightarrow f$ is not odd.
There are no symmetries.

f. Maximum number of turning points: 4

g.

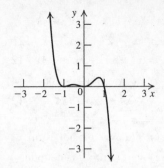

59. a. $y \to \infty$ as $x \to -\infty$; $y \to \infty$ as $x \to \infty$

b. $x(x+1)(x-1)(x+2) = 0 \Rightarrow x = 0$ or $x = -1$
or $x = -2$ or $x = 1$. The graph crosses the
x-axis at all four zeros.

c. The graph is above the x-axis on
$(-\infty, -2), \cup (-1, 0) \cup (1, \infty)$. The graph is
below the x-axis on $(-2, -1) \cup (0, 1)$.

d. $y = 0(0+1)(0-1)(0+2) = 0 \Rightarrow$ the
y-intercept is 0.

e. $f(x) = x(x+1)(x-1)(x+2)$
$= x^4 + 2x^3 - x^2 - 2x.$
$f(-x) = -x(-x+1)(-x-1)(-x+2)$
$= x^4 - 2x^3 - x^2 + 2x.$
$f(x) \neq f(x) \Rightarrow f$ is not even.
$f(-x) \neq -f(x) \Rightarrow f$ is not odd.
There are no symmetries.

f. Maximum number of turning points: 3

g.

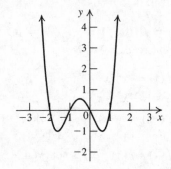

3.2 Applying the Concepts

61. a. $3x^2(4-x) = 0 \Rightarrow x = 0$ or $x = 4$. $x = 0$,
multiplicity 2; $x = 4$, multiplicity 1.

b.

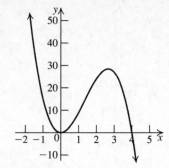

c. There are 2 turning points.

d. Domain: [0, 4]. The portion between the
x-intercepts is the graph of $R(x)$.

63. a. Domain [0, 1]

b.

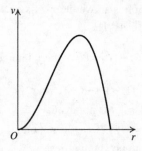

65. a. $R(x) = x\left(27 - \left(\dfrac{x}{300}\right)^2\right) = 27x - \dfrac{x^3}{90,000}$

b. The domain of $R(x)$ is the same as the
domain of p. $p \geq 0$ when $p \leq 900\sqrt{3}$. The
domain is $\left[0, 900\sqrt{3}\right]$.

c.

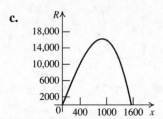

67. a. $N(x) = (x+12)\left(400 - 2x^2\right)$

b. The low end of the domain is 0. (There
cannot be fewer than 0 workers.) The upper
end of the domain is the value where
productivity is 0, so solve $N(x) = 0$ to find
the upper end of the domain.
$(x+12)(400 - 2x^2) = 0 \Rightarrow x = -12$ (reject
this) or $400 - 2x^2 = 0 \Rightarrow 200 = x^2 \Rightarrow$
$x = \pm 10\sqrt{2}$ (reject the negative solution).
The domain is $[0, 10\sqrt{2}]$.

c.

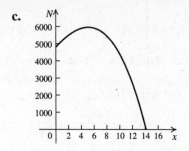

69. a. $V(x) = x(8 - 2x)(15 - 2x)$

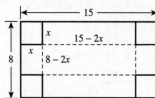

b.

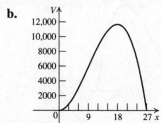

71. a. $V(x) = x^2(108 - 4x)$

b.

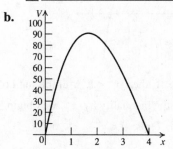

73. a. $V(x) = x^2(62 - 2x)$

b.

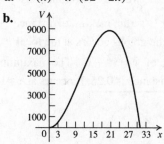

75. a. $V(x) = 2x^2(45 - 3x) = 6x^2(15 - x)$

b.

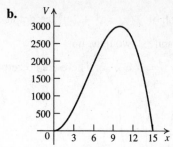

77.

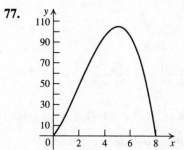

79.

$$D(t) = 0.205996t^3 - 2.526415t^2 + 11.156629t + 17.337104$$

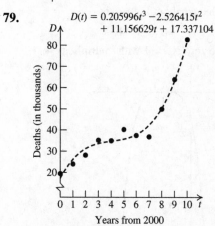

Years from 2000

The model function $D(t)$ fits the actual data very well.

3.2 Beyond the Basics

81. The graph of $f(x) = (x - 1)^4$ is the graph of $y = x^4$ shifted one unit right.

$(x - 1)^4 = 0 \Rightarrow x = 1$. The zero is $x = 1$ with multiplicity 4.

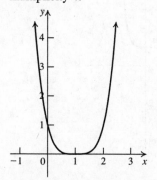

83. The graph of $f(x) = x^4 + 2$ is the graph of

$y = x^4$ shifted two units up.

$x^4 + 2 = 0 \Rightarrow x = \sqrt[4]{-2}$. There are no zeros.

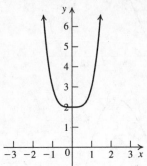

85. The graph of $f(x) = -(x-1)^4$ is the graph of

$y = x^4$ shifted one unit right and then reflected about the x-axis.

$-(x-1)^4 = 0 \Rightarrow x = 1$.

The zero is $x = 1$ with multiplicity 4.

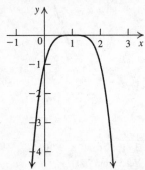

87. The graph of $f(x) = x^5 + 1$ is the graph of

$y = x^5$ shifted one unit up. $x^5 + 1 = 0 \Rightarrow$

$x = \sqrt[5]{-1} \Rightarrow x = -1$.

The zero is -1 with multiplicity 1.

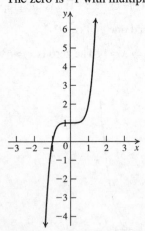

89. The graph of $f(x) = 8 - \dfrac{(x+1)^5}{4}$ is the graph

of $y = x^5$ shifted one unit left, compressed vertically by one-fourth, reflected about the x-axis, and then shifted up eight units.

$8 - \dfrac{(x+1)^5}{4} = 0 \Rightarrow -\dfrac{(x+1)^5}{4} = -8 \Rightarrow$

$(x+1)^5 = 32 \Rightarrow x + 1 = 2 \Rightarrow x = 1.$

The zero is 1 with multiplicity 1.

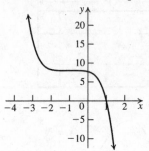

91. If $0 < x < 1$, then $x^2 < 1$. Multiplying both sides of the inequality by x^2, we obtain $x^4 < x^2$. Thus, $0 < x^4 < x^2$.

93.

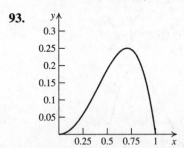

The graph of $y = x^2 - x^4$ represents the difference between the two functions $y = x^2$ and $y = x^4$, so the maximum distance between the graphs occurs at the local maximum of $y = x^2 - x^4$. The maximum vertical distance is 0.25. It occurs at $x \approx 0.71$.

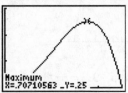

Answers will vary in exercises 95–98. Sample answers are given.

95. $f(x) = x^2(x+1)$

97. $f(x) = 1 - x^4$

99. The smallest possible degree is 5, because the graph has five x-intercepts and four turning points.

101. The smallest possible degree is 6, because the graph has five turning points.

3.2 Critical Thinking/Discussion/Writing

103. It is not possible for a polynomial function to have no y-intercepts because the domain of any polynomial function is $(-\infty, \infty)$, which includes the point $x = 0$.

104. It is possible for a polynomial function to have no x-intercepts because the function can be shifted above the x-axis. An example is the function $y = x^2 + 1$.

105. It is not possible for the graph of a polynomial function of degree 3 to have exactly one local maximum and no local minimum because the graph of a function of degree 3 rises in one direction and falls in the other. This requires an even number of turning points. Since the degree is 3, there can be only zero or two turning points. Therefore, if there is a local maximum, there must also be another turning point, which will be a local minimum.

106. It is not possible for the graph of a polynomial function of degree 4 to have exactly one local maximum and exactly one local minimum because the graph rises in both directions or falls in both directions. This requires an odd number of turning points. Since the degree is 4, there can be only one or three turning points. If there were exactly one local maximum and exactly one local minimum, there would be two turning points.

107. If $P(x) = a_0 x^m + a_1 x^{m-1} + \cdots a_{m-1} x + a_m$ and $Q(x) = b_0 x^n + b_1 x^{n-1} + \cdots b_{n-1} x + b_n$, then

$(P \circ Q)(x)$

$= a_0 \left(b_0 x^n + b_1 x^{n-1} + \cdots b_{n-1} x + b_n \right)^m$

$+ a_1 \left(b_0 x^n + b_1 x^{n-1} + \cdots b_{n-1} x + b_n \right)^{m-1} + \cdots$

$+ a_{m-1} \left(b_0 x^n + b_1 x^{n-1} + \cdots b_{n-1} x + b_n \right) + a_m,$

which is a polynomial of degree mn.

3.2 Maintaining Skills

109. $\dfrac{x^5}{x^7} = x^{5-7} = x^{-2} = \dfrac{1}{x^2}$

111. $\dfrac{x^5 - 3x^2}{x^2} = \dfrac{x^2\left(x^3 - 3\right)}{x^2} = x^3 - 3$

113. a. $7 = 1 \cdot 7$ **b.** $7 = 7 \cdot 1$

    c. $7 = (-1)(-7)$ **d.** $7 = (-7)(-1)$

115. The integer factors of 120 are $-120, -60, -40,$ $-30, -24, -20, -15, -12, -10, -8, -6, -5, -4,$ $-3, -2, -1, 1, 2, 3, 4, 5, 6, 8, 10, 12, 15, 20,$ $24, 30, 40, 60, 120.$

117. $g(x) = x^4 + 2x^3 - 20x - 5$

$g(3) = (3)^4 + 2(3)^3 - 20(3) - 5 = 70$

119. $f(x) = \left(x^8 - 15x^7 + 5x^3\right)(x + 12)$

$f(-12)$

$= \left[(-12)^8 - 15(-12)^7 + 5(-12)^3\right]\left[(-12) + 12\right]$

$= 0$

121. $-2x^2 + x + 1 = -\left(2x^2 - x - 1\right)$

$= -(2x + 1)(x - 1)$

$= (2x + 1)(1 - x)$

123. $(x - 9)\left(6x^2 - 7x - 3\right) = (x - 9)(3x + 1)(2x - 3)$

3.3 Dividing Polynomials

3.3 Practice Problems

1.

$$
\begin{array}{r}
3x + 4 \\
x^2 + 0x + 1 \overline{\smash{\big)}\, 3x^3 + 4x^2 + x + 7} \\
\underline{(-)\,3x^3 + 0x^2 + 3x} \\
4x^2 - 2x + 7 \\
\underline{(-)\,4x^2 + 0x + 4} \\
-2x + 3
\end{array}
$$

Quotient: $3x + 4$; remainder: $-2x + 3$

2.

$$\begin{array}{r} x^2 + x + 3 \\ x^2 - x + 3 \overline{)x^4 + 0x^3 + 5x^2 + 2x + 6} \\ \underline{x^4 - x^3 + 3x^2} \\ x^3 + 2x^2 + 2x \\ \underline{x^3 - x^2 + 3x} \\ 3x^2 - x + 6 \\ \underline{3x^2 - 3x + 9} \\ 2x - 3 \end{array}$$

Quotient: $x^2 + x + 3$; remainder: $2x - 3$ or

$$x^2 + x + 3 + \frac{2x - 3}{x^2 - x + 3}$$

3. $3\rfloor$
$$\begin{array}{rrrr} 2 & -7 & 0 & 5 \\ & 6 & -3 & -9 \\ \hline 2 & -1 & -3 & -4 \end{array}$$

The quotient is $2x^2 - x - 3$ remainder -4 or

$$2x^2 - x - 3 - \frac{4}{x - 3}.$$

4. $-3\rfloor$
$$\begin{array}{rrrr} 2 & 1 & -18 & -7 \\ & -6 & 15 & 9 \\ \hline 2 & -5 & -3 & 2 \end{array}$$

The quotient is $2x^2 - 5x - 3$ remainder 2 or

$$2x^2 - 5x - 3 + \frac{2}{x + 3}.$$

5. $F(1) = 1^{110} - 2 \cdot 1^{57} + 5 = 1 - 2 + 5 = 4$, so the remainder when $F(x) = x^{110} - 2x^{57} + 5$ is divided by $x - 1$ is 4.

6. $-2\rfloor$
$$\begin{array}{rrrrr} 1 & 0 & 10 & 2 & -20 \\ & -2 & 4 & -28 & 52 \\ \hline 1 & -2 & 14 & -26 & 32 \end{array}$$
The remainder is 32, so $f(-2) = 32$.

7. Since -2 is a zero of the function $3x^3 - x^2 - 20x - 12$, $x + 2$ is a factor. Use synthetic division to find the depressed equation.
$$-2\rfloor \begin{array}{rrrr} 3 & -1 & -20 & -12 \\ & -6 & 14 & 12 \\ \hline 3 & -7 & -6 & 0 \end{array}$$
Thus,
$$3x^3 - x^2 - 20x - 12 = (x + 2)(3x^2 - 7x - 6).$$

Now solve $3x^2 - 7x - 6 = 0$ to find the two remaining zeros:

$3x^2 - 7x - 6 = 0 \Rightarrow (x - 3)(3x + 2) = 0 \Rightarrow$

$x - 3 = 0$ or $3x + 2 = 0 \Rightarrow x = 3$ or $x = -\dfrac{2}{3}$

The solution set is $\left\{-2, -\dfrac{2}{3}, 3\right\}$.

8. $C(x) = 0.23x^3 - 4.255x^2 + 0.345x + 41.05$
$C(3) = 10 \Rightarrow C(x) = (x - 3)Q(x) + 10 \Rightarrow$
$C(x) - 10 = (x - 3)Q(x)$
So, 3 is zero of $C(x) - 10 =$
$0.23x^3 - 4.255x^2 + 0.345x + 31.05$.
We need to find another positive zero of $C(x) - 10$. Use synthetic division to find the depressed equation.

$3\rfloor$
$$\begin{array}{rrrr} 0.23 & -4.255 & 0.345 & 31.05 \\ & 0.69 & -10.695 & -31.05 \\ \hline 0.23 & -3.565 & -10.35 & 0 \end{array}$$

Solve the depressed equation

$0.23x^2 - 3.565x - 10.35 = 0$ using the quadratic formula:

$$x = \frac{3.565 \pm \sqrt{(-3.565)^2 - 4(0.23)(-10.35)}}{2(0.23)}$$

≈ 18 or -2.5

The positive zero is 18.
Check by verifying that $C(18) = 10$.

3.3 Basic Concepts and Skills

1. In the division
$$\frac{x^4 - 2x^3 + 5x^2 - 2x + 1}{x^2 - 2x + 3} = x^2 + 2 + \frac{2x - 5}{x^2 - 2x + 3},$$
the dividend is $\underline{x^4 - 2x^3 + 5x^2 - 2x + 1}$, the divisor is $\underline{x^2 - 2x + 3}$, the quotient is $\underline{x^2 + 2}$, and the remainder is $\underline{2x - 5}$.

3. The Remainder Theorem states that if a polynomial $F(x)$ is divided by $(x - a)$, then the remainder $R = \underline{F(a)}$.

5. True

7.
$$\begin{array}{r} 3x - 2 \\ 2x + 1 \overline{)6x^2 - x - 2} \\ \underline{-(6x^2 + 3x)} \\ -4x - 2 \\ \underline{-(-4x - 2)} \\ 0 \end{array}$$

In exercises 9–14, insert zero coefficients for missing terms.

9.
$$
\begin{array}{r}
3x^3 - 3x^2 - 3x + 6 \\
x+1{\overline{\smash{\big)}\,3x^4 + 0x^3 - 6x^2 + 3x - 7}} \\
\underline{-(3x^4 + 3x^3)} \\
-3x^3 - 6x^2 \\
\underline{-(-3x^3 - 3x^2)} \\
-3x^2 + 3x \\
\underline{-(-3x^2 - 3x)} \\
6x - 7 \\
\underline{-(6x + 6)} \\
-13
\end{array}
$$

11.
$$
\begin{array}{r}
2x - 1 \\
2x^2 - x - 5{\overline{\smash{\big)}\,4x^3 - 4x^2 - 9x + 5}} \\
\underline{-(4x^3 - 2x^2 - 10x)} \\
-2x^2 + x + 5 \\
\underline{-(-2x^2 + x + 5)} \\
0
\end{array}
$$

13.
$$
\begin{array}{r}
z^2 + 2z + 1 \\
z^2 - 2z + 1{\overline{\smash{\big)}\,z^4 + 0z^3 - 2z^2 + 0z + 1}} \\
\underline{-(z^4 - 2z^3 + z^2)} \\
2z^3 - 3z^2 + 0z \\
\underline{-(2z^3 - 4z^2 + 2z)} \\
z^2 - 2z + 1 \\
\underline{-(z^2 - 2z + 1)} \\
0
\end{array}
$$

15.

1⌋	1	−1	−7	2
		1	0	−7
	1	0	−7	−5

The quotient is $x^2 - 7$ and the remainder is −5.

17.

−2⌋	1	4	−7	−10
		−2	−4	22
	1	2	−11	12

The quotient is $x^2 + 2x - 11$ and the remainder is 12.

19.

2⌋	1	−3	2	4	5
		2	−2	0	8
	1	−1	0	4	13

The quotient is $x^3 - x^2 + 4$ and the remainder is 13.

21.

$\frac{1}{2}$⌋	2	4	−3	1
		1	$\frac{5}{2}$	$-\frac{1}{4}$
	2	5	$-\frac{1}{2}$	$\frac{3}{4}$

The quotient is $2x^2 + 5x - \frac{1}{2}$ and the remainder is $\frac{3}{4}$.

23.

$-\frac{1}{2}$⌋	2	−5	3	2
		−1	3	−3
	2	−6	6	−1

The quotient is $2x^2 - 6x + 6$ and the remainder is −1.

25.

1⌋	1	1	−7	2	1	−1
		1	2	−5	−3	−2
	1	2	−5	−3	−2	−3

The quotient is $x^4 + 2x^3 - 5x^2 - 3x - 2$ and the remainder is −3.

27.

−1⌋	1	0	0	0	0	1
		−1	1	−1	1	−1
	1	−1	1	−1	1	0

The quotient is $x^4 - x^3 + x^2 - x + 1$ remainder 0.

29.

−1⌋	1	0	2	−1	0	0	5
		−1	1	−3	4	−4	4
	1	−1	3	−4	4	−4	9

The quotient is $x^5 - x^4 + 3x^3 - 4x^2 + 4x - 4$ remainder 9.

31. a.

1⌋	1	3	0	1
		1	4	4
	1	4	4	5

The remainder is 5, so $f(1) = 5$.

b.

−1⌋	1	3	0	1
		−1	−2	2
	1	2	−2	3

The remainder is 3, so $f(-1) = 3$.

c.

$\frac{1}{2}$⌋	1	3	0	1
		$\frac{1}{2}$	$\frac{7}{4}$	$\frac{7}{8}$
	1	$\frac{7}{2}$	$\frac{7}{4}$	$\frac{15}{8}$

The remainder is $\frac{15}{8}$, so $f\left(\frac{1}{2}\right) = \frac{15}{8}$.

d.

$$10 \underline{|\quad 1 \quad 3 \quad 0 \quad 1}$$
$$\underline{\quad\quad 10 \quad 130 \quad 1300}$$
$$1 \quad 13 \quad 130 \quad 1301$$

The remainder is 1301, so $f(10) = 1301$.

33. a.

$$1 \underline{|\quad 1 \quad 5 \quad -3 \quad 0 \quad -20}$$
$$\underline{\quad\quad 1 \quad 6 \quad 3 \quad 3}$$
$$1 \quad 6 \quad 3 \quad 3 \quad -17$$

The remainder is –17, so $f(1) = -17$.

b.

$$-1 \underline{|\quad 1 \quad 5 \quad -3 \quad 0 \quad -20}$$
$$\underline{\quad\quad -1 \quad -4 \quad 7 \quad -7}$$
$$1 \quad 4 \quad -7 \quad 7 \quad -27$$

The remainder is –27, so $f(-1) = -27$.

c.

$$-2 \underline{|\quad 1 \quad 5 \quad -3 \quad 0 \quad -20}$$
$$\underline{\quad\quad -2 \quad -6 \quad 18 \quad -36}$$
$$1 \quad 3 \quad -9 \quad 18 \quad -56$$

The remainder is –56, so $f(-2) = -56$.

d.

$$2 \underline{|\quad 1 \quad 5 \quad -3 \quad 0 \quad -20}$$
$$\underline{\quad\quad 2 \quad 14 \quad 22 \quad 44}$$
$$1 \quad 7 \quad 11 \quad 22 \quad 24$$

The remainder is 24, so $f(2) = 24$.

35. $f(1) = 2(1)^3 + 3(1)^2 - 6(1) + 1 = 0 \Rightarrow x - 1$ is a factor of $2x^3 + 3x^2 - 6x + 1$.
Check as follows:

$$1 \underline{|\quad 2 \quad 3 \quad -6 \quad 1}$$
$$\underline{\quad\quad 2 \quad 5 \quad -1}$$
$$2 \quad 5 \quad -1 \quad 0$$

37. $f(-1) = 5(-1)^4 + 8(-1)^3 + (-1)^2$
$$+ 2(-1) + 4 = 0 \Rightarrow x + 1 \text{ is a}$$
factor of $5x^4 + 8x^3 + x^2 + 2x + 4$.
Check as follows:

$$-1 \underline{|\quad 5 \quad 8 \quad 1 \quad 2 \quad 4}$$
$$\underline{\quad\quad -5 \quad -3 \quad 2 \quad -4}$$
$$5 \quad 3 \quad -2 \quad 4 \quad 0$$

39. $f(2) = 2^4 + 2^3 - 2^2 - 2 - 18 = 0 \Rightarrow x - 2$ is a factor of $x^4 + x^3 - x^2 - x - 18$.
Check as follows:

$$2 \underline{|\quad 1 \quad 1 \quad -1 \quad -1 \quad -18}$$
$$\underline{\quad\quad 2 \quad 6 \quad 10 \quad 18}$$
$$1 \quad 3 \quad 5 \quad 9 \quad 0$$

41. $f(-2) = (-2)^6 - (-2)^5 - 7(-2)^4 + (-2)^3$
$$+ 8(-2)^2 + 5(-2) + 2 = 0 \Rightarrow x + 2$$
is a factor of
$x^6 - x^5 - 7x^4 + x^3 + 8x^2 + 5x + 2$.
Check as follows:

$$-2 \underline{|\quad 1 \quad -1 \quad -7 \quad 1 \quad 8 \quad 5 \quad 2}$$
$$\underline{\quad\quad -2 \quad 6 \quad 2 \quad -6 \quad -4 \quad -2}$$
$$1 \quad -3 \quad -1 \quad 3 \quad 2 \quad 1 \quad 0$$

43. $f(-1) = 0 = (-1)^3 + 3(-1)^2 + (-1) + k \Rightarrow$
$$0 = 1 + k \Rightarrow k = -1$$

45. $f(2) = 0 = 2(2)^3 + (2^2)k - 2k - 2 \Rightarrow$
$$14 + 2k = 0 \Rightarrow k = -7$$

In exercises 47–50, use synthetic division to find the remainder.

47.

$$2 \underline{|\quad -2 \quad 4 \quad -4 \quad 9}$$
$$\underline{\quad\quad -4 \quad 0 \quad -8}$$
$$-2 \quad 0 \quad -4 \quad 1$$

The remainder is 1, so $x - 2$ is not a factor of $-2x^3 + 4x^2 - 4x + 9$.

49.

$$-2 \underline{|\quad 4 \quad 9 \quad 3 \quad 1 \quad 4}$$
$$\underline{\quad\quad -8 \quad -2 \quad -2 \quad 2}$$
$$4 \quad 1 \quad 1 \quad -1 \quad 6$$

The remainder is 6, so $x + 2$ is not a factor of $4x^4 + 9x^3 + 3x^2 + x + 4$.

3.3 Applying the Concepts

51. $A = lw \Rightarrow l = \dfrac{A}{w} \Rightarrow$

$$
\require{enclose}
\begin{array}{r}
2x^2 + 1 \\
x^2 - x + 2 \enclose{longdiv}{2x^4 - 2x^3 + 5x^2 - x + 2} \\
\underline{-(2x^4 - 2x^3 + 4x^2)} \\
x^2 - x + 2 \\
\underline{-(x^2 - x + 2)} \\
0
\end{array}
$$

The width is $2x^2 + 1$.

53. a. $R(40) = 3000 \Rightarrow R(40) - 3000 = 0$ and
$$R(60) = 3000 \Rightarrow R(60) - 3000 = 0. \text{ Thus,}$$
40 and 60 are zeros of $R(x) - 3000$.
Therefore,
$$R(x) - 3000 = a(x - 40)(x - 60)$$
$$= a(x^2 - 100x + 2400)$$

(continued on next page)

(*continued*)

Since (30, 2400) lies on $R(x)$, we have

$$2400 - 3000 = a(30 - 40)(30 - 60) \Rightarrow -600 = a(-10)(-30) \Rightarrow a = -2.$$

Thus $R(x) - 3000 = -2(x - 40)(x - 60) = -2x^2 + 200x - 4800.$

b. $R(x) - 3000 = -2x^2 + 200x - 4800 \Rightarrow R(x) = -2x^2 + 200x - 1800$

c. The maximum weekly revenue occurs at the vertex of the function, $\left(-\dfrac{b}{2a}, \ R\left(-\dfrac{b}{2a} \right) \right).$

$$-\frac{b}{2a} = -\frac{200}{2(-2)} = 50$$

$$R(50) = -2(50)^2 + 200(50) - 1800 = 3200$$

The maximum revenue is \$3200 if the phone is priced at \$50.

55. Since $t = 11$ represents 2002, we have $C(11) = 97.6 \Rightarrow C(x) = (x - 11)Q(x) + 97.6 \Rightarrow$
$C(t) - 97.6 = (t - 11)Q(t).$ So

$$-0.0006t^3 - 0.0613t^2 + 2.0829t + 82.904 - 97.6 = -0.0006t^3 - 0.0613t^2 + 2.0829t - 14.6960$$
$$= (x - 11)Q(x).$$

Use synthetic division to find $Q(x)$:

$$\begin{array}{r|rrrr} 11 & -0.0006 & -0.0613 & 2.0829 & -14.6960 \\ & & -0.0066 & -0.7469 & 14.6960 \\ \hline & -0.0006 & -0.0679 & 1.3360 & 0 \end{array}$$

Now solve the depressed equation to find another zero:

$$-0.0006^2 - 0.0679t + 1.3360 = 0 \Rightarrow t = \frac{0.0679 \pm \sqrt{0.0679^2 - 4(-0.0006)(1.3360)}}{2(-0.0006)} \Rightarrow$$

$$t = \frac{0.0679 \pm \sqrt{0.00781681}}{-0.0012} = \frac{0.0679 \pm 0.0884}{-0.0012} \Rightarrow t \approx -130.26 \text{ or } t \approx 17.0939.$$

Since we must find t greater than 0, $t = 17$, and the year is 1991 + 17 = 2008.

57. $M(t) = -0.0027t^3 + 0.3681t^2 - 5.8645t + 195.2782$

Since $M(2) = 191.736$, $M(t) = (t - 2)Q(t) + 195.2782 \Rightarrow M(t) - 195.2782 = (t - 2)Q(t)$

We must find two other zeros of

$$F(t) = M(t) - 185 = -0.0027t^3 + 0.3681t^2 - 5.8645t + 195.2782 - 185$$

$$= -0.0027t^3 + 0.3681t^2 - 5.8645t + 10.2782$$

Because 2 is a zero of $F(t)$, use synthetic division to find $Q(t)$.

$$\begin{array}{r|rrrr} 2 & -0.0027 & 0.3681 & -5.8645 & 10.2782 \\ & & -0.0054 & 0.7254 & -10.2782 \\ \hline & -0.0027 & 0.3627 & -5.1391 & 0 \end{array}$$

Now use the quadratic formula to solve the depressed equation.

$$t = \frac{-0.3627 \pm \sqrt{0.3627^2 - 4(-0.0027)(-5.1391)}}{2(-0.0027)} = \frac{-0.3627 \pm \sqrt{0.0760}}{-0.0054} \Rightarrow t \approx 16.1 \text{ or } t \approx 118$$

Thus, the model shows that the Marine Corps had about 186,000 when $t \approx 16$, or in the year 1990 + 16 = 2006.

3.3 Beyond the Basics

59. Divide $4x^3 + 8x^2 - 11x + 3$ by $\left(x - \dfrac{1}{2}\right)$:

$$
\begin{array}{r|rrrr}
\frac{1}{2} & 4 & 8 & -11 & 3 \\
 & & 2 & 5 & -3 \\
\hline
 & 4 & 10 & -6 & 0
\end{array}
$$

Now divide $4x^2 + 10x - 6$ by $\left(x - \dfrac{1}{2}\right)$:

$$
\begin{array}{r|rrr}
\frac{1}{2} & 4 & 10 & -6 \\
 & & 2 & 6 \\
\hline
 & 4 & 12 & 0
\end{array}
$$

Since $\left(x - \dfrac{1}{2}\right)$ does not divide $4x + 12$,

$\left(x - \dfrac{1}{2}\right)$ is a root of multiplicity 2 of

$4x^3 + 8x^2 - 11x + 3$.

61. a. $x + a$ is a factor of $x^n + a^n$ if n is an odd integer. The possible rational zeros of $x^n + a^n$ are $\{\pm a, \pm a^2, \pm a^3, \ldots, \pm a^n\}$. Since $x + a$ is a factor means that $-a$ is a root, then $(-a)^n + a^n = 0 \Rightarrow (-a)^n = -a^n$ only for odd values of n.

b. $x + a$ is a factor of $x^n - a^n$ if n is an even integer. The possible rational zeros of $x^n - a^n$ are $\{\pm a, \pm a^2, \pm a^3, \ldots, \pm a^n\}$. Since $x + a$ is a factor means that $-a$ is a root, then $(-a)^n - a^n = 0 \Rightarrow (-a)^n = a^n$ only for even values of n.

c. There is no value of n for which $x - a$ is a factor of $x^n + a^n$. The possible rational zeros of $x^n + a^n$ are $\{\pm a, \pm a^2, \pm a^3, \ldots, \pm a^n\}$. If $x - a$ is a factor, then a is a root, and $a^n + a^n = 0 \Rightarrow a^n = -a^n$, which is not possible.

d. $x - a$ is a factor of $x^n - a^n$ for all positive integers n. The possible rational zeros of $x^n - a^n$ are $\{\pm a, \pm a^2, \pm a^3, \ldots, \pm a^n\}$. If $x - a$ is a factor, then a is a root, and $a^n - a^n = 0$, which is true for all values of n. However, if n is negative, then

$$
x^n - a^n = \frac{1}{x^{-n}} - \frac{1}{a^{-n}} \text{ and } x - a \text{ is not a}
$$

factor.

63. a. Divide the divisor and the dividend by 2 so that the leading coefficient of the divisor is 1:

$$
\frac{2x^3 + 3x^2 + 6x - 2}{2x - 1} = \frac{2x^3 + 3x^2 + 6x - 2}{2\left(x - \dfrac{1}{2}\right)}
$$

$$
= \frac{x^3 + \dfrac{3}{2}x^2 + 3x - 1}{\left(x - \dfrac{1}{2}\right)}
$$

$$
\begin{array}{r|rrrr}
\frac{1}{2} & 1 & \frac{3}{2} & 3 & -1 \\
 & & \frac{1}{2} & 1 & 2 \\
\hline
 & 1 & 2 & 4 & 1
\end{array}
$$

Because the original polynomials were divided by 2 to use synthetic division, multiply the remainder by 2 to find the remainder for the original division, 2.

b. Divide the divisor and the dividend by 2 so that the leading coefficient of the divisor is 1:

$$
\frac{2x^3 - x^2 - 4x + 1}{2x + 3} = \frac{2x^3 - x^2 - 4x + 1}{2\left(x + \dfrac{3}{2}\right)}
$$

$$
= \frac{x^3 - \dfrac{1}{2}x^2 - 2x + \dfrac{1}{2}}{\left(x + \dfrac{3}{2}\right)}
$$

$$
\begin{array}{r|rrrr}
-\frac{3}{2} & 1 & -\frac{1}{2} & -2 & \frac{1}{2} \\
 & & -\frac{3}{2} & 3 & -\frac{3}{2} \\
\hline
 & 1 & -2 & 1 & -1
\end{array}
$$

Because the original polynomials were divided by 2 to use synthetic division, multiply the remainder by 2 to find the remainder for the original division, -2.

65. $f(x) = ax^3 + bx^2 + cx + d$, $a \neq 0$

There are two possibilities for the end behavior of f:

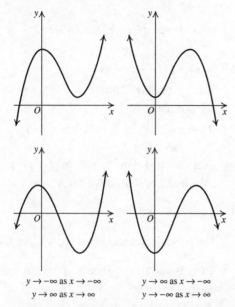

$y \to -\infty$ as $x \to -\infty$ $y \to \infty$ as $x \to -\infty$
$y \to \infty$ as $x \to \infty$ $y \to -\infty$ as $x \to \infty$

Therefore, there is one sign change from the left end of f to the right side of f or there are three sign changes. By the Intermediate Value Theorem, then, there is one zero (if there is one sign change) or there are three zeros (if there are three sign changes.)

67. $dp < 0$ means that d and p have opposite signs, and, thus, are on opposite sides of the x-axis. If $d > 0$ then $p < 0$. Since $kp < 0$, $k > 0$ also. The end behavior is $g(x) \to \infty$ as $x \to -\infty$ and $g(x) \to \infty$ as $x \to \infty$. However, f has end behavior shown in exercise 65. Therefore, the two graphs must intersect. Similar reasoning applies if $d < 0$.
Algebraic proof: Assume $d > 0$ and $p < 0$. Consider the fourth degree polynomial $g - f$. Then $(g - f)(0) = p - d < 0$. Since $k > 0$, $(g - f)(x)$ has upward end behavior on both ends, so $(g - f)(x)$ is eventually > 0. By the Intermediate Value Theorem, there is some c such that $(g - f)(c) = 0 \Rightarrow g(c) = f(c)$. Therefore, the graphs intersect.

69. Answers will vary. One example is
$$f(x) = x^4 + 2$$

3.3 Critical Thinking/Discussion/Writing

70. a. The quotient when $F(x)$ is divided by $2x - 6$ is one-half the quotient when $F(x)$ is divided by $x - 3$, but the remainders are identical.

b. If $\dfrac{F(x)}{x + \frac{b}{a}} = Q(x) + \dfrac{R(x)}{x + \frac{b}{a}}$, then

$$\frac{F(x)}{ax + b} = \frac{1}{a}\left[\frac{F(x)}{x + \frac{b}{a}}\right] = \frac{1}{a}\left[Q(x) + \frac{R(x)}{x + \frac{b}{a}}\right]$$

$$= \frac{1}{a}Q(x) + \frac{R(x)}{ax + b}$$

So, the remainders are again identical and the quotient when $F(x)$ is divided by $ax + b$ is $\dfrac{1}{a}$ times the quotient when $F(x)$ is divided by $x + \dfrac{b}{a}$.

3.3 Maintaining Skills

71.

1⌋	1	−1	2	1	−3
		1	0	2	3
	1	0	2	3	0

The quotient is $x^3 + 2x + 3$ and the remainder is 0.

73.

5⌋	−1	5	0	0	2	−10	3
		−5	0	0	0	10	0
	−1	0	0	0	2	0	3

The quotient is $-x^5 + 2x$ and the remainder is 3.

75. The factors of 11 are $\pm 1, \pm 11$.

77. The factors of 51 are $\pm 1, \pm 3, \pm 17, \pm 51$.

79. $x^2 + 2x - 1$
Degree: 2
Leading coefficient: 1
Constant term: −1

81. $-7x^{10} + 3x^3 - 2x + 4$
Degree: 10
Leading coefficient: −7
Constant term: 4

3.4 The Real Zeros of a Polynomial Function

3.4 Practice Problems

1. $F(x) = 2x^3 + 3x^2 - 6x - 8$

   The factors of the constant term, -8, are $\{\pm 1, \pm 2, \pm 4, \pm 8\}$, and the factors of the leading coefficient, 2, are $\{\pm 1, \pm 2\}$. The possible rational zeros are $\left\{\pm\dfrac{1}{2}, \pm 1, \pm 2, \pm 4, \pm 8\right\}$. Use synthetic division to find one rational root:

   $$\begin{array}{r|rrrr} -2 & 2 & 3 & -6 & -8 \\ & & -4 & 2 & 8 \\ \hline & 2 & -1 & -4 & 0 \end{array}$$

   The remainder is 0, so -2 is a zero of the function.

   $$2x^3 + 3x^2 - 6x - 8 = (x+2)(2x^2 - x - 4)$$

   Now find the zeros of $2x^2 - x - 4$ using the quadratic formula:

   $$x = \frac{1 \pm \sqrt{(-1)^2 - 4(2)(-4)}}{2(1)} = \frac{1 \pm \sqrt{33}}{2}, \text{ which}$$

   are not rational roots.
   The only rational zero is $\{-2\}$.

2. $2x^3 - 9x^2 + 6x - 1 = 0$

   The factors of the constant term, -1, are $\{\pm 1\}$, and the factors of the leading coefficient, 2, are $\{\pm 1, \pm 2\}$. The possible rational zeros are $\left\{\pm\dfrac{1}{2}, \pm 1\right\}$. Use synthetic division to find one rational zero:

   $$\begin{array}{r|rrrr} \frac{1}{2} & 2 & -9 & 6 & -1 \\ & & 1 & -4 & 1 \\ \hline & 2 & -8 & 2 & 0 \end{array}$$

   Thus, $2x^3 - 9x^2 + 6x - 1 = 0 \Rightarrow$

   $$\left(x - \frac{1}{2}\right)(2x^2 - 8x + 2) = 0 \Rightarrow$$

   $$x - \frac{1}{2} = 0 \Rightarrow x = \frac{1}{2} \text{ or}$$

   $$2x^2 - 8x + 2 = 0$$

$$x = \frac{-(-8) \pm \sqrt{(-8)^2 - 4(2)(2)}}{2(2)} = \frac{8 \pm \sqrt{48}}{4}$$

$$= 2 \pm \sqrt{3}$$

Solution set: $\left\{\dfrac{1}{2}, 2 - \sqrt{3}, 2 + \sqrt{3}\right\}$

3. $f(x) = 2x^5 + 3x^2 + 5x - 1$

   There is one sign change in $f(x)$, so there is one positive zero.

   $$f(-x) = 2(-x)^5 + 3(-x)^2 + 5(-x) - 1$$
   $$= -2x^5 + 3x^2 - 5x - 1$$

   There are two sign changes in $f(-x)$, so there are either 2 or 0 negative zeros.

4. $f(x) = 2x^3 + 5x^2 + x - 2$

   The possible rational zeros are $\left\{\pm\dfrac{1}{2}, \pm 1, \pm 2\right\}$.

   There is one sign change in $f(x)$, so there is one positive zero.

   $$f(-x) = 2(-x)^3 + 5(-x)^2 + (-x) - 2$$
   $$= -2x^3 + 5x^2 - x - 2$$

   There are two sign changes in $f(-x)$, so there are either 2 or 0 negative zeros. Try synthetic division by $x - k$ with $k = 1, 2, 3, \ldots$. The first integer that makes each number in the last row a 0 or a positive number is an upper bound on the zeros of $F(x)$. Then use synthetic division by $x - k$ with $k = -1, -2, -3, \ldots$. The first negative integer for which the numbers in the last row alternate in sign is a lower bound on the zeros of $F(x)$. In this case, 1 is an upper bound and -3 is a lower bound.

   $$\begin{array}{r|rrrr} 1 & 2 & 5 & 1 & -2 \\ & & 2 & 7 & 8 \\ \hline & 2 & 7 & 8 & 6 \end{array} \qquad \begin{array}{r|rrrr} -3 & 2 & 5 & 1 & -2 \\ & & -6 & 3 & -12 \\ \hline & 2 & -1 & 4 & -14 \end{array}$$

5. $f(x) = 3x^4 - 11x^3 + 22x - 12$

   Step 1: f has at most 4 real zeros.

   Step 2: $f(x) = 3x^4 - 11x^3 + 0x^2 + 22x - 12$

   There are three sign changes in f, so f has 1 or 3 positive zeros.

   $$f(-x) = 3(-x)^4 - 11(-x)^3 + 0(-x)^2$$
   $$+ 22(-x) - 12$$

   $$= 3x^4 + 11x^3 + 0x^2 - 22x - 12$$

   There is one sign change in $f(-x)$, so f has 1 negative zero.

   (continued on next page)

(*continued*)

Step 3: The factors of the constant term, −12, are $\{\pm1, \pm2, \pm3, \pm4, \pm6, \pm12\}$, and the factors of the leading coefficient, 3, are $\{\pm1, \pm3\}$. The possible rational zeros are

$$\left\{\pm1, \pm2, \pm3, \pm4, \pm6, \pm12, \pm\frac{1}{3}, \pm\frac{2}{3}, \pm\frac{4}{3}\right\}.$$

Steps 4–7: Test for zeros until a zero or an upper bound is found. Try a positive possibility:

$$\begin{array}{r|rrrrr} 3 & 3 & -11 & 0 & 22 & -12 \\ & & 9 & -6 & -18 & 12 \\ \hline & 3 & -2 & -6 & 4 & 0 \end{array}$$

Since 3 is a zero, we have

$$f(x) = (x-3)\left(3x^3 - 2x^2 - 6x + 4\right).$$

Now find the a zero of $Q_1(x) = 3x^3 - 2x^2 - 6x + 4$. The factors of the constant term, 4, are $\{\pm1, \pm2, \pm4\}$, and the factors of the leading coefficient, 3, are $\{\pm1, \pm3\}$. The possible rational zeros are

$$\left\{\pm1, \pm2, \pm4, \pm\frac{1}{3}, \pm\frac{2}{3}, \pm\frac{4}{3}\right\}.$$

Try the positive possibilities first.

$$\begin{array}{r|rrrr} \frac{2}{3} & 3 & -2 & -6 & 4 \\ & & 2 & 0 & -4 \\ \hline & 3 & 0 & -6 & 0 \end{array}$$

Thus, $f(x) = (x-3)\left(x-\frac{2}{3}\right)\left(3x^2 - 6\right).$

Now solve $3x^2 - 6 = 0$.

$$3x^2 - 6 \Rightarrow x^2 = 2 \Rightarrow x = \pm\sqrt{2}$$

Solution set: $\left\{\frac{2}{3}, 3, \sqrt{2}, -\sqrt{2}\right\}$

6. $f(x) = 3x^3 - x^2 - 9x + 3$

   Step 1: Since the degree, 3, is odd, and the leading coefficient, 3, is positive, the end behavior is similar to that of $y = x^3$.

   Step 2: Solve $3x^3 - x^2 - 9x + 3 = 0$ to find the real zeros. There are two sign changes in f, so f has either 2 or 0 positive zeros.

   $$f(-x) = 3(-x)^3 - (-x)^2 - 9(-x) + 3$$
   $$= -3x^3 + x^2 + x + 3$$

   There is one sign change in $f(-x)$, so there is one negative zero.

By the Rational Root Theorem, the possible rational roots are $\left\{\pm1, \pm3, \pm\frac{1}{3}\right\}$. Trying each value, we find that $\frac{1}{3}$ is a rational zero.

$$\begin{array}{r|rrrr} \frac{1}{3} & 3 & -1 & -9 & 3 \\ & & 1 & 0 & -3 \\ \hline & 3 & 0 & -9 & 0 \end{array}$$

Solve the depressed equation $3x^2 - 9 = 0$ to find the remaining zeros.

$$3x^2 - 9 = 0 \Rightarrow 3x^2 = 9 \Rightarrow x^2 = 3 \Rightarrow x = \pm\sqrt{3}$$

Step 3: The three zeros divide the x-axis into four intervals, $\left(-\infty, -\sqrt{3}\right), \left(-\sqrt{3}, \frac{1}{3}\right), \left(\frac{1}{3}, \sqrt{3}\right)$, and $\left(\sqrt{3}, \infty\right)$.

Step 4: Determine the sign of a test value in each interval

Interval	Test point	Value of $f(x)$	Above/below x-axis
$\left(-\infty, -\sqrt{3}\right)$	−2	−7	below
$\left(-\sqrt{3}, \frac{1}{3}\right)$	0	3	above
$\left(\frac{1}{3}, \sqrt{3}\right)$	1	−4	below
$\left(\sqrt{3}, \infty\right)$	2	5	above

Plot the zeros, y-intercepts, and test points, and then join the points with a smooth curve.

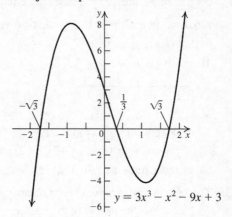

$y = 3x^3 - x^2 - 9x + 3$

3.4 Basic Concepts and Skills

1. Let $P(x) = a_n x^n + \ldots + a_0$ with integer coefficients. If $\dfrac{p}{q}$ is a rational zero of $P(x)$,

   then $\dfrac{p}{q} = \dfrac{\text{possible factors of } \underline{a_0}}{\text{possible factors of } \underline{a_n}}$.

3. The number of positive zeros of a polynomial function $P(x)$, is equal to the number of <u>variations</u> of sign of $P(x)$ or less than that number by <u>an even integer</u>.

5. False. Possible rational zeros of

   $P(x) = 9x^3 - 9x^2 - x + 1$ are $\left\{ \pm\dfrac{1}{9}, \pm\dfrac{1}{3}, \pm 1 \right\}$.

7. The factors of the constant term, 5, are $\{\pm 1, \pm 5\}$, and the factors of the leading coefficient, 3, are $\{\pm 1, \pm 3\}$. The possible

   rational zeros are $\left\{ \pm\dfrac{1}{3}, \pm 1, \pm\dfrac{5}{3}, \pm 5 \right\}$.

9. The factors of the constant term, 6, are $\{\pm 1, \pm 2, \pm 3, \pm 6\}$, and the factors of the leading coefficient, 4, are $\{\pm 1, \pm 2, \pm 4\}$. The possible

   rational zeros are $\left\{ \pm\dfrac{1}{4}, \pm\dfrac{1}{2}, \pm\dfrac{3}{4}, \pm 1, \pm\dfrac{3}{2}, \right.$

   $\left. \pm 2, \pm 3, \pm 6 \right\}$.

11. The factors of the constant term, 4, are $\{\pm 1, \pm 2, \pm 4\}$, and the factors of the leading coefficient, 1, are $\{\pm 1\}$. The possible rational zeros are $\{\pm 1, \pm 2, \pm 4\}$. Use synthetic division to find one rational root:

    $$\begin{array}{r|rrrr} 1 & 1 & -1 & -4 & 4 \\ & & 1 & 0 & -4 \\ \hline & 1 & 0 & -4 & 0 \end{array}$$

    So,
    $$\begin{aligned} x^3 - x^2 - 4x + 4 &= (x-1)(x^2 - 4) \\ &= (x-1)(x-2)(x+2) \Rightarrow \end{aligned}$$
    the rational zeros are $\{-2, 1, 2\}$.

13. The factors of the constant term, 2, are $\{\pm 1, \pm 2\}$, and the factors of the leading coefficient, 1, are $\{\pm 1\}$. The possible rational zeros are $\{\pm 1, \pm 2\}$. Use synthetic division to find one rational root:

$$\begin{array}{r|rrrr} -1 & 1 & 1 & 2 & 2 \\ & & -1 & 0 & -2 \\ \hline & 1 & 0 & 2 & 0 \end{array}$$

So, $x^3 + x^2 + 2x + 2 = (x+1)(x^2 + 2) \Rightarrow$ the rational zero is $\{-1\}$.

15. The factors of the constant term, 6, are $\{\pm 1, \pm 2, \pm 3, \pm 6\}$, and the factors of the leading coefficient, 2, are $\{\pm 1, \pm 2\}$. The possible

    rational zeros are $\left\{ \pm\dfrac{1}{2}, \pm 1, \pm\dfrac{3}{2}, \pm 2, \pm 3, \pm 6 \right\}$.

    Use synthetic division to find one rational root:

    $$\begin{array}{r|rrrr} 2 & 2 & 1 & -13 & 6 \\ & & 4 & 10 & -6 \\ \hline & 2 & 5 & -3 & 0 \end{array}$$

    $$\begin{aligned} 2x^3 + x^2 - 13x + 6 &= (x-2)(2x^2 + 5x - 3) \\ &= (x-2)(x+3)(2x-1) \Rightarrow \end{aligned}$$

    the rational zeros are $\left\{ -3, \dfrac{1}{2}, 2 \right\}$.

17. The factors of the constant term, -2, are $\{\pm 1, \pm 2\}$, and the factors of the leading coefficient, 3, are $\{\pm 1, \pm 3\}$. The possible

    rational zeros are $\left\{ \pm\dfrac{1}{3}, \pm\dfrac{2}{3}, \pm 1, \pm 2 \right\}$. Use

    synthetic division to find one rational root:

    $$\begin{array}{r|rrrr} \frac{2}{3} & 3 & -2 & 3 & -2 \\ & & 2 & 0 & 2 \\ \hline & 3 & 0 & 3 & 0 \end{array}$$

    $3x^3 - 2x^2 + 3x - 2 = \left(x - \dfrac{2}{3} \right)(3x^2 + 3) \Rightarrow$ the

    rational zero is $\left\{ \dfrac{2}{3} \right\}$.

19. The factors of the constant term, 2, are $\{\pm 1, \pm 2\}$, and the factors of the leading coefficient, 3, are $\{\pm 1, \pm 3\}$. The possible

    rational zeros are $\left\{ \pm\dfrac{1}{3}, \pm\dfrac{2}{3}, \pm 1, \pm 2 \right\}$.

    Use synthetic division to find one rational root:

    $$\begin{array}{r|rrrr} -\frac{1}{3} & 3 & 7 & 8 & 2 \\ & & -1 & -2 & -2 \\ \hline & 3 & 6 & 6 & 0 \end{array}$$

    (continued on next page)

(continued)

$$3x^3 + 7x^2 + 8x + 2 = \left(x + \frac{1}{3}\right)(3x^2 + 6x + 6) \Rightarrow$$

the rational zero is $\left\{-\frac{1}{3}\right\}$.

21. The factors of the constant term, –2, are $\{\pm1, \pm2\}$, and the factors of the leading coefficient, 1, are $\{\pm1\}$. The possible rational zeros are $\{\pm1, \pm2\}$. Use synthetic division to find one rational root:

```
−1|  1  −1  −1  −1  −2
        −1   2  −1   2
     1  −2   1  −2   0
```

So, –1 is a rational zero. Use synthetic division again to find another rational zero:

$$x^4 - x^3 - x^2 - x - 2 = (x+1)(x^3 - 2x^2 + x - 2)$$

```
2|  1  −2   1  −2
        2   0   2
     1  0   1   0
```

$$x^4 - x^3 - x^2 - x - 2 = (x+1)(x-2)(x^2+1) \Rightarrow$$

the rational zeros are $\{-1, 2\}$.

23. The factors of the constant term, 12, are $\{\pm1, \pm2, \pm3, \pm4, \pm6, \pm12\}$, and the factors of the leading coefficient, 1, are $\{\pm1\}$. The possible rational zeros are $\{\pm1, \pm2, \pm3, \pm4, \pm6, \pm12\}$. Use synthetic division to find one rational root:

```
−3|  1  −1  −13    1   12
        −3   12    3  −12
     1  −4   −1    4    0
```

So, –3 is a rational zero. Use synthetic division again to find another rational zero:

$$x^4 - x^3 - 13x^2 + x + 12$$
$$= (x+3)(x^3 - 4x^2 - x + 4)$$

```
−1|  1  −4  −1   4
        −1   5  −4
     1  −5   4   0
```

So, –1 is also a rational zero.
$$x^4 - x^3 - 13x^2 + x + 12$$
$$= (x+3)(x+1)(x^2 - 5x + 4)$$
$$= (x+3)(x+1)(x-4)(x-1) \Rightarrow$$
the rational zeros are $\{-3, -1, 1, 4\}$.

25. The factors of the constant term, 1, are $\{\pm1\}$, and the factors of the leading coefficient, 1,

are $\{\pm1\}$. The possible rational zeros are $\{\pm1\}$. Use synthetic division to find one rational root:

```
1|  1  −2  10  −1   1
        1  −1   9   8
    1  −1   9   8   9
```

The remainder is 9 so, 1 is not a rational zero. Try –1:

```
−1|  1  −2  10  −1   1
        −1   3 −13  14
     1  −3  13 −14  15
```

The remainder is 15 so, –1 is not a rational zero. Therefore, there are no rational zeros.

27. $f(x) = 6x^3 + 13x^2 + x - 2$

Zeros: $\left\{-2, -\frac{1}{2}, \frac{1}{3}\right\}$

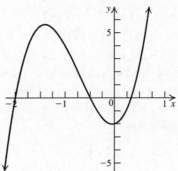

29. $f(x) = 2x^3 - 3x^2 - 14x + 21$

Zeros: $\left\{\pm\sqrt{7}, \frac{3}{2}\right\}$

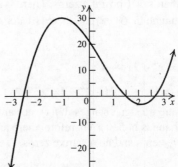

31. $f(x) = 4x^4 + 4x^3 - 3x^2 - 2x + 1$

Zeros: $\left\{-1, \frac{1}{2}\right\}$

33. $f(x) = 5x^3 - 2x^2 - 3x + 4;$
$$f(-x) = 5(-x)^3 - 2(-x)^2 - 3(-x) + 4$$
$$= -5x^3 - 2x^2 + 3x + 4$$
There are two sign changes in $f(x)$, so there are either 2 or 0 positive zeros. There is one sign change in $f(-x)$, so there is one negative zero.

35. $f(x) = 2x^3 + 5x^2 - x + 2;$
$$f(-x) = 2(-x)^3 + 5(-x)^2 - (-x) + 2$$
$$= -2x^3 + 5x^2 + x + 2$$
There are two sign changes in $f(x)$, so there are either 2 or 0 positive zeros. There is one sign change in $f(-x)$, so there is one negative zero.

37. $h(x) = 2x^5 - 5x^3 + 3x^2 + 2x - 1$
$$h(-x) = 2(-x)^5 - 5(-x)^3 + 3(-x)^2 + 2(-x) - 1$$
$$= -2x^5 + 5x^3 + 3x^2 - 2x - 1$$
There are three sign changes in $h(x)$, so there are either 1 or 3 positive zeros. There are two sign changes in $h(-x)$, so there are either 2 or 0 negative zeros.

39. $G(x) = -3x^4 - 4x^3 + 5x^2 - 3x + 7;$
$$G(-x) = -3(-x)^4 - 4(-x)^3 + 5(-x)^2$$
$$- 3(-x) + 7$$
$$= -3x^4 + 4x^3 + 5x^2 + 3x + 7$$
There are three sign changes in $G(x)$, so there are either 3 or 1 positive zeros. There is one sign change in $G(-x)$, so there is 1 negative zero.

41. $f(x) = x^4 + 2x^2 + 4$
$$f(-x) = (-x)^4 + 2(-x)^2 + 4 = x^4 + 2x^2 + 4$$
There are no sign changes in $f(x)$, nor are there sign changes in $f(-x)$. Therefore, there are no positive zeros and no negative zeros.

43. $g(x) = 2x^5 + x^3 + 3x$
$$g(-x) = 2(-x)^5 + (-x)^3 + 3(-x)$$
$$= -2x^5 - x^3 - 3x$$
There are no sign changes in $g(x)$, nor are there sign changes in $g(-x)$. Therefore, there are no positive zeros and no negative zeros.

45. $h(x) = -x^5 - 2x^3 + 4$
$$h(-x) = -(-x)^5 - 2(-x)^3 + 4$$
$$= x^5 + x^3 + 4$$
There is one sign change in $h(x)$, and no changes in $g(-x)$. Therefore, there is one positive zero and there are no negative zeros.

47. The possible rational zeros are $\left\{\pm\dfrac{1}{3}, \pm 1, \pm 3\right\}$.

There are three sign changes in $f(x)$, so there are either 3 or 1 positive zeros. There are no sign changes in $f(-x)$, so there are no negative zeros. Using synthetic division with $k = 1, 2, 3, \ldots$ and $k = -1, -2, -3, \ldots$ gives an upper bound of 1 and a lower bound of -1.

1⌋	3	−1	9	−3		−1⌋	3	−1	9	−3
		3	2	11				−3	4	−13
	3	2	11	8			3	−4	13	−16

49. The possible rational zeros are $\left\{\pm\dfrac{1}{3}, \pm 1, \pm\dfrac{7}{3}, \pm 7\right\}$. There are no sign changes in $F(x)$, so there are no positive zeros. There are three sign changes in $F(-x)$, so there are either 3 or 1 negative zeros. Using synthetic division with $k = 1, 2, 3, \ldots$ and $k = -1, -2, -3, \ldots$ gives an upper bound of 1 and a lower bound of -3.

1⌋	3	2	5	7		−3⌋	3	2	5	7
		3	5	10				−9	21	−75
	3	5	10	17			3	−7	25	−68

51. The possible rational zeros are $\{\pm 1, \pm 31\}$.

There are two sign changes in $h(x)$, so there are either 2 or 0 positive zeros. There are two sign changes in $h(-x)$, so there are either 2 or 0 negative zeros. Using synthetic division with $k = 1, 2, 3, \ldots$ and $k = -1, -2, -3, \ldots$ gives an upper bound of 31 and a lower bound of -31.

31⌋	1	3	−15	−9	31
		31	1054	32,209	998,200
	1	34	1039	32,200	998,231

−31⌋	1	3	−15	−9	31
		−31	868	−26,443	820,012
	1	−28	853	−26,452	820,043

53. The possible rational zeros are
$$\left\{\pm\frac{1}{6}, \pm\frac{1}{3}, \pm\frac{1}{2}, \pm 1, \pm\frac{7}{6}, \pm\frac{7}{3}, \pm\frac{7}{2}, \pm 7\right\}.$$

There are two sign changes in $f(x)$, so there are either 2 or 0 positive zeros. There are two sign changes in $f(-x)$, so there are either 2 or 0 negative zeros. Using synthetic division with $k = 1, 2, 3, \ldots$ and $k = -1, -2, -3, \ldots$ gives an upper bound of 4 and a lower bound of -4.

$$
\begin{array}{r|rrrrr}
4 & 6 & 1 & -43 & -7 & 7 \\
  &   & 24 & 100 & 228 & 884 \\
\hline
  & 6 & 25 & 57 & 221 & 891
\end{array}
$$

$$
\begin{array}{r|rrrrr}
-4 & 6 & 1 & -43 & -7 & 7 \\
   &   & -24 & 92 & -196 & 812 \\
\hline
   & 6 & -23 & 49 & -203 & 819
\end{array}
$$

In exercises 55–74, first check to see the number of possible positive zeros and the number of possible negative zeros.

55. $f(x) = x^3 + 5x^2 - 8x + 2$

The function has degree 3, so there are three zero, zero or two possible positive zeros; one possible negative zero.
The possible rational zeros are $\{\pm 1, \pm 2\}$.

$$
\begin{array}{r|rrrr}
1 & 1 & 5 & -8 & 2 \\
  &   & 1 & 6 & -2 \\
\hline
  & 1 & 6 & -2 & 0
\end{array}
$$

So, 1 is a zero. Now solve the depressed equation $x^2 + 6x - 2 = 0$.
$$x = \frac{-6 \pm \sqrt{36 - 4(1)(-2)}}{2(1)} = \frac{-6 \pm \sqrt{44}}{2}$$
$$= -3 \pm \sqrt{11}$$
The solution set is $\left\{1, -3 \pm \sqrt{11}\right\}$.

57. $f(x) = 2x^3 - x^2 - 6x + 3$

The function has degree 3, so there are three zeros, zero or two possible positive zeros; one possible negative zero. The possible rational zeros are $\left\{\pm 1, \pm 3, \pm\frac{1}{2}, \pm\frac{3}{2}\right\}$.

$$
\begin{array}{r|rrrr}
\frac{1}{2} & 2 & -1 & -6 & 3 \\
  &   & 1 & 0 & -3 \\
\hline
  & 2 & 0 & -6 & 0
\end{array}
$$

So, 1/2 is a zero. Now solve the depressed equation $2x^2 - 6 = 0$.

$2x^2 - 6 = 0 \Rightarrow x^2 = 3 \Rightarrow x = \pm\sqrt{3}$
Solution set: $\left\{-\sqrt{3}, \frac{1}{2}, \sqrt{3}\right\}$

59. $f(x) = 2x^3 - 9x^2 + 6x - 1$

The function has degree 3, so there are three zeros, one or three possible positive zeros; no possible negative zeros.

The possible rational zeros are $\left\{\pm 1, \pm\frac{1}{2}\right\}$.

$$
\begin{array}{r|rrrr}
\frac{1}{2} & 2 & -9 & 6 & -1 \\
  &   & 1 & -4 & 1 \\
\hline
  & 2 & -8 & 2 & 0
\end{array}
$$

So, 1/2 is a zero. Now solve the depressed equation $2x^2 - 8x + 2 = 0$
$$x = \frac{-(-8) \pm \sqrt{(-8)^2 - 4(2)(2)}}{2(2)} = \frac{8 \pm \sqrt{48}}{4}$$
$$= 2 \pm \sqrt{3}$$
Solution set: $\left\{2 - \sqrt{3}, \frac{1}{2}, 2 + \sqrt{3}\right\}$.

61. $f(x) = x^4 + x^3 - 5x^2 - 3x + 6$

The function has degree 4, so there are four zeros, zero or two possible positive zeros; zero or two possible negative zeros.
The possible rational zeros are $\{\pm 1, \pm 2, \pm 3, \pm 6\}$.

$$
\begin{array}{r|rrrrr}
1 & 1 & 1 & -5 & -3 & 6 \\
  &   & 1 & 2 & -3 & -6 \\
\hline
  & 1 & 2 & -3 & -6 & 0
\end{array}
$$

So, 1 is a zero. Now find a zero of the depressed equation $x^3 + 2x^2 - 3x - 6 = 0$.

$$
\begin{array}{r|rrrr}
-2 & 1 & 2 & -3 & -6 \\
   &   & -2 & 0 & 6 \\
\hline
   & 1 & 0 & -3 & 0
\end{array}
$$

So, -2 is a zero. Now solve the depressed equation $x^2 - 3 = 0$.
$x^2 - 3 = 0 \Rightarrow x^2 = 3 \Rightarrow x = \pm\sqrt{3}$
Solution set: $\left\{-2, -\sqrt{3}, 1, \sqrt{3}\right\}$

63. $f(x) = x^4 - 3x^3 + 3x - 1$

The function has degree 4, so there are four zeros, one or three possible positive real zeros; one possible negative real zero. The possible rational zeros are $\{\pm 1\}$.

```
1│  1  -3   0   3  -1
        1  -2  -2   1
   ─────────────────────
     1  -2  -2   1   0
```

So, 1 is a zero. Now find a zero of the depressed equation $g(x) = x^3 - 2x^2 - 2x + 1$.

```
-1│  1  -2  -2   1
        -1   3  -1
   ──────────────────
      1  -3   1   0
```

So, -1 is a zero. Now solve the depressed equation $x^2 - 3x + 1 = 0$

$$x = \frac{3 \pm \sqrt{9 - 4(1)(1)}}{2(1)} = \frac{3 \pm \sqrt{5}}{2}$$

The solution set is $\left\{ \pm 1, \dfrac{3 \pm \sqrt{5}}{2} \right\}$.

65. $f(x) = 2x^4 - 5x^3 - 4x^2 + 15x - 6$

The function has degree 4, so there are four zeros, one or three possible positive real zeros; one possible negative real zero. The possible rational zeros are

$$\left\{ \pm 1, \pm 2, \pm 3, \pm 6, \pm \frac{1}{2}, \pm \frac{3}{2} \right\}.$$

```
2│  2  -5  -4  15  -6
        4  -2 -12   6
   ──────────────────────
     2  -1  -6   3   0
```

So, 2 is a zero. Now find a zero of the depressed equation $g(x) = 2x^3 - x^2 - 6x + 3$.

```
 1 │  2  -1  -6   3
 ─              
 2 │      1   0  -3
   ──────────────────
      2   0  -6   0
```

So, $1/2$ is a zero. Now solve the depressed equation $2x^2 - 6 = 0$.

$2x^2 - 6 = 0 \Rightarrow x^2 = 3 \Rightarrow x = \pm\sqrt{3}$

The solution set is $\left\{ -\sqrt{3}, \dfrac{1}{2}, \sqrt{3}, 2 \right\}$.

67. $f(x) = 6x^4 - x^3 - 13x^2 + 2x + 2$

The function has degree 4, so there are four zeros, zero or two possible positive real zeros; zero or two possible negative real zeros. The

possible rational zeros are $\left\{ \pm 1, \pm 3, \pm \dfrac{1}{6}, \pm \dfrac{1}{3} \right\}$.

```
 1 │  6  -1 -13   2   2
 ─                     
 2 │      3   1  -6  -2
   ──────────────────────
      6   2 -12  -4   0
```

So, $1/2$ is a zero. Now find a zero of the depressed equation

$g(x) = 6x^3 + 2x^2 - 12x - 4$.

```
   1 │  6   2 -12  -4
 - ─                 
   3 │     -2   0   4
     ────────────────
        6   0 -12   0
```

So, $-1/3$ is a zero. Now solve the depressed equation $6x^2 - 12 = 0$.

$6x^2 - 12 = 0 \Rightarrow x^2 = 2 \Rightarrow x = \pm\sqrt{2}$

The solution set is $\left\{ -\sqrt{2}, -\dfrac{1}{3}, \dfrac{1}{2}, \sqrt{2} \right\}$.

69. $f(x) = x^5 - 2x^4 - 4x^3 + 8x^2 + 3x - 6$

The function has degree 5, so there are five zeros. There are three sign changes, so there are either 1, 3, or 5 positive real zeros. There are two sign changes in $f(-x)$, so there are zero or two negative real zeros. The possible rational zeros are $\{\pm 1, \pm 3, \pm 6\}$. Using synthetic division to test the positive values, we find that one zero is 1:

```
1│  1  -2  -4   8   3  -6
        1  -1  -5   3   6
   ──────────────────────────
     1  -1  -5   3   6   0
```

The zeros of the depressed function $x^4 - x^3 - 5x^2 + 3x + 6$ are also zeros of f. Use synthetic division again to find the next zero, 2:

```
2│  1  -1  -5   3   6
        2   2  -6  -6
   ──────────────────────
     1   1  -3  -3   0
```

$x^5 - 2x^4 - 4x^3 + 8x^2 + 3x - 6$

$= (x - 1)(x - 2)(x^3 + x^2 - 3x - 3)$

Now find a zero of the depressed function $x^3 + x^2 - 3x - 3$. Use synthetic division again to find the next zero, -1:

```
-1│  1   1  -3  -3
        -1   0   3
   ──────────────────
     1   0  -3   0
```

(continued on next page)

(continued)

$$x^5 - 2x^4 - x^3 + 8x^2 - 10x + 4$$
$$= (x-1)(x-2)(x+1)\left(x^2 - 3\right)$$

Now solve the depressed equation:

$$x^2 - 3 = 0.$$

$$x^2 - 3 = 0 \Rightarrow x^2 = 3 \Rightarrow x = \pm\sqrt{3}$$

Solution set: $\left\{-\sqrt{3},\ -1,\ 1,\ \sqrt{3},\ 2\right\}$

Using synthetic division to test the positive values, we find that one zero is 2:

$$\begin{array}{r|rrrrrr} 2| & 1 & 1 & -6 & -6 & 8 & 8 \\ & & 2 & 6 & 0 & -12 & -8 \\ \hline & 1 & 3 & 0 & -6 & -4 & 0 \end{array}$$

The zeros of the depressed function

$x^4 + 3x^3 - 6x - 4$ are also zeros of f.
Use synthetic division again to find the next zero, -2:

$$\begin{array}{r|rrrrr} -2| & 1 & 3 & 0 & -6 & -4 \\ & & -2 & -2 & 4 & 4 \\ \hline & 1 & 1 & -2 & -2 & 0 \end{array}$$

$$x^5 + x^4 - 6x^3 - 6x^2 + 8x + 8$$
$$= (x-2)(x+2)\left(x^3 + x^2 - 2x - 2\right).$$

Now find a zero of the depressed function
$x^3 + x^2 - 2x - 2$. Use synthetic division again to find the next zero, -1:

$$\begin{array}{r|rrrr} -1| & 1 & 1 & -2 & -2 \\ & & -1 & 0 & 2 \\ \hline & 1 & 0 & -2 & 0 \end{array}$$

$$x^5 + x^4 - 6x^3 - 6x^2 + 8x + 8$$
$$= (x-2)(x+2)(x+1)\left(x^2 - 2\right)$$

Now solve the depressed equation:

$$x^2 - 2 = 0.$$

$$x^2 - 2 = 0 \Rightarrow x^2 = 2 \Rightarrow x = \pm\sqrt{2}$$

Solution set: $\left\{-2,\ -\sqrt{2},\ -1,\ 2,\ \sqrt{2}\right\}$

71. $f(x) = 2x^5 + x^4 - 11x^3 - x^2 + 15x - 6$

The function has degree 5, so there are five zeros. There are three sign changes, so there are either 1, 3, or 5 positive real zeros. There are two sign changes in $f(-x)$, so there are zero or two negative real zeros. The possible rational zeros are $\left\{\pm1,\ \pm3,\ \pm6,\ \pm\dfrac{1}{2},\ \pm\dfrac{3}{2}\right\}$.

Using synthetic division to test the positive

values, we find that one zero is 1:

$$\begin{array}{r|rrrrrr} 1| & 2 & 1 & -11 & -1 & 15 & -6 \\ & & 2 & 3 & -8 & -9 & 6 \\ \hline & 2 & 3 & -8 & -9 & 6 & 0 \end{array}$$

The zeros of the depressed function

$2x^4 + 3x^3 - 8x^2 - 9x + 6$ are also zeros of f.
Use synthetic division again to find the next zero, -2:

$$\begin{array}{r|rrrrr} -2| & 2 & 3 & -8 & -9 & 6 \\ & & -4 & 2 & 12 & -6 \\ \hline & 2 & -1 & -6 & 3 & 0 \end{array}$$

$$2x^5 + x^4 - 11x^3 - x^2 + 15x - 6$$
$$= (x-1)(x+2)\left(2x^3 - x^2 - 6x + 3\right)$$

Now find a zero of the depressed function
$2x^3 - x^2 - 6x + 3$. Use synthetic division again to find the next zero, $1/2$:

$$\begin{array}{r|rrrr} \dfrac{1}{2}| & 2 & -1 & -6 & 3 \\ & & 1 & 0 & -3 \\ \hline & 2 & 0 & -6 & 0 \end{array}$$

$$2x^5 + x^4 - 11x^3 - x^2 + 15x - 6$$
$$= (x-1)(x+2)\left(x - \dfrac{1}{2}\right)\left(2x^2 - 6\right)$$

Now solve the depressed equation:

$$2x^2 - 6 = 0.$$

$$2x^2 - 6 = 0 \Rightarrow x^2 = 3 \Rightarrow x = \pm\sqrt{3}$$

Solution set: $\left\{-2,\ -\sqrt{3},\ 1,\ \dfrac{1}{2},\ \sqrt{3}\right\}$

73. $f(x) = 2x^5 - 13x^4 + 27x^3 - 17x^2 - 5x + 6$

The function has degree 5, so there are five zeros. There are four sign changes, so there are either 0, 2, or 4 positive real zeros. There is one sign change in $f(-x)$, so there is one possible negative real zero. The possible rational zeros are $\left\{\pm1,\ \pm2,\ \pm3,\ \pm6,\ \pm\dfrac{1}{2},\ \pm\dfrac{3}{2}\right\}$.

Using synthetic division to test the positive values, we find that one zero is 1:

$$\begin{array}{r|rrrrrr} 1| & 2 & -13 & 27 & -17 & -5 & 6 \\ & & 2 & -11 & 16 & -1 & -6 \\ \hline & 2 & -11 & 16 & -1 & -6 & 0 \end{array}$$

The zeros of the depressed function

$2x^4 - 11x^3 + 16x^2 - x - 6$ are also zeros of f.
Use synthetic division again to find the next zero, 1.

(continued on next page)

(*continued*)

$$
\begin{array}{r|rrrrr}
1| & 2 & -11 & 16 & -1 & -6 \\
  &   & 2 & -9 & 7 & 6 \\
\hline
  & 2 & -9 & 7 & 6 & 0
\end{array}
$$

$$2x^5 - 13x^4 + 27x^3 - 17x^2 - 5x + 6$$
$$= (x-1)(x-1)(2x^3 - 9x^2 + 7x + 6)$$

The zeros of the depressed function $2x^3 - 9x^2 + 7x + 6$ are also zeros of P. Use synthetic division again to find the next zero, 2.

$$
\begin{array}{r|rrrr}
2| & 2 & -9 & 7 & 6 \\
  &   & 4 & -10 & -6 \\
\hline
  & 2 & -5 & -3 & 0
\end{array}
$$

$$2x^5 - 13x^4 + 27x^3 - 17x^2 - 5x + 6$$
$$= (x-1)(x-1)(x-2)(2x^2 - 5x - 3)$$

Now solve the depressed equation:
$$2x^2 - 5x - 3 = 0.$$
$$2x^2 - 5x - 3 = 0 \Rightarrow (2x+1)(x-3) = 0 \Rightarrow$$
$$x = -\frac{1}{2}, 3$$

Solution set: $\left\{ -\frac{1}{2}, 1 \text{ (multiplicity 2)}, 2, 3 \right\}$

3.4 Applying the Concepts

75. The length of the rectangle is $x^2 - 2x + 3$ and its width is $x - 2$. Its area is 306 square units, so we have $(x^2 - 2x + 3)(x-2) = 306 \Rightarrow$

$$x^3 - 4x^2 + 7x - 6 = 306 \Rightarrow$$
$$x^3 - 4x^2 + 7x - 312 = 0.$$

There are 3 sign changes in $f(x)$ and no sign changes in $f(-x)$, so there are 1 or 3 possible positive real zeros and no possible negative real zeros. The possible rational zeros are $\{\pm 1, \pm 2, \pm 3, \pm 4, \pm 6, \pm 8, \pm 12, \pm 13, \pm 24, \pm 26, \pm 39, \pm 52, \pm 78, \pm 104, \pm 156, \pm 312\}$

Using synthetic division, we find that 8 is a zero:

$$
\begin{array}{r|rrrr}
8| & 1 & -4 & 7 & -312 \\
  &   & 8 & 32 & 312 \\
\hline
  & 1 & 4 & 39 & 0
\end{array}
$$

Note that the discriminant of the depressed equation $x^2 + 4x + 39 = 0$, $4^2 - 4(1)(39) < 0$, so there are 2 complex solutions to the depressed equation. Therefore, $x = 8$. The width of the rectangle is $8 - 2 = 6$ units and

the length of the rectangle is $8^2 - 2 \cdot 8 + 3 = 51$ units.

77. Use synthetic division to solve the equation $628 = 3x^3 - 6x^2 + 108x + 100 \Rightarrow$ $3x^3 - 6x^2 + 108x - 528 = 0$. The factors of the constant term, -528, are $\{\pm 1, \pm 2, \pm 3, \pm 4, \pm 6, \pm 8, \pm 11, \pm 48, \pm 66, \pm 88, \pm 132, \pm 176, \pm 264, \pm 528\}$. The factors of the leading coefficient, 3, are $\{\pm 1, \pm 3\}$. Only the positive, whole number possibilities make sense for the problem, so the possible rational zeros are $\{1, 2, 3, 4, 6, 8, 11, 16, 22, 44, 48, 66, 88, 132, 176, 264, 528\}$.

$$
\begin{array}{r|rrrr}
4| & 3 & -6 & 108 & -528 \\
  &   & 12 & 24 & 528 \\
\hline
  & 3 & 6 & 132 & 0
\end{array}
$$

Thus, $x = 4$.

79. The cost function gives the result as a number of thousands, so set it equal to 125:
$$x^3 - 15x^2 + 5x + 50 = 125 \Rightarrow$$
$$x^3 - 15x^2 + 5x - 75 = 0 . \text{ The factors of the}$$
constant term, -75, are $\{\pm 1, \pm 3, \pm 5, \pm 15, \pm 25, \pm 75\}$. The factors of the leading coefficient, 1, are $\{\pm 1\}$. Only the positive solutions make sense for the problem, so the possible rational zeros are $\{1, 3, 5, 15, 25, 75\}$. Use synthetic division to find the zero:

$$
\begin{array}{r|rrrr}
15| & 1 & -15 & 5 & -75 \\
   &   & 15 & 0 & 75 \\
\hline
   & 1 & 0 & 5 & 0
\end{array}
$$

The total monthly cost is $125,000 when 1500 units are produced.

81. a. The country exported oil for 5 years, from 2007 to 2012.

   b. There are three zeros, so the minimum degree of the polynomial is 3. The zeros are 0, 3, and 8, so the polynomial is of the form $p(x) = a(x-0)(x-3)(x-8)$. The graph goes through the point $(9, -3)$, so we have $-3 = a(9-0)(9-3)(9-8) \Rightarrow -3 = 54a \Rightarrow$
$$a = -\frac{1}{18}$$
Thus, the equation is
$$p(x) = y = -\frac{1}{18}x(x-3)(x-8).$$

c. $p(5) = -\frac{1}{18}(5)(5-3)(5-8) \approx 1.7$

In 2009, the country exported about 1.7 million barrels of oil.

83. a. There are three zeros, so the minimum degree of the polynomial is 3. The zeros are 0, 6, and 8, so the polynomial is of the form $p(x) = a(x-0)(x-6)(x-8)$. The graph goes through the point (9, 0.5), so we

have $0.5 = a(9-0)(9-6)(9-8) \Rightarrow 0.5 = 27a \Rightarrow a = \frac{1}{54}$

Thus, the equation is $p(x) = y = \frac{1}{54}x(x-6)(x-8)$.

b. $p(4) = \frac{1}{54}(4)(4-6)(4-8) \approx 0.59259$

The profit in 2008 was about \$592.59.

c. Set $Y_1 = \frac{1}{54}x(x-6)(x-8)$ and $Y_2 = 0.6$.

Find the intersection of the two graphs.

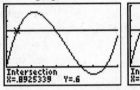 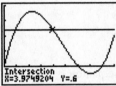

[0, 9, 1] by [−0.25, 1, 0.25]

Ms. Sharpy realized a profit of \$600 in 2004 and 2007.

d.

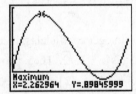

The maximum profit was approximately \$898.46 in 2006.

3.4 Beyond the Basics

85. Let $x = \sqrt{3} \Rightarrow x^2 = 3 \Rightarrow x^2 - 3 = 0$. The only possible rational zeros of this equation are ± 1 and ± 3. Because $\sqrt{3}$ is neither of these, it must be irrational.

87. $x = 3 - \sqrt{2} \Rightarrow 3 - x = \sqrt{2} \Rightarrow 9 - 6x + x^2 = 2 \Rightarrow x^2 - 6x + 7 = 0$. The possible rational zeros of this equation are $\{\pm 1, \pm 7\}$. Since $3 - \sqrt{2}$ is not in the set, it must be irrational.

89. $y = (x-1)(x+1)(x-0) = x^3 - x$

91. $y = \left(x + \frac{1}{2}\right)(x-2)\left(x - \frac{7}{3}\right) = (2x+1)(x-2)(3x-7) = 6x^3 - 23x^2 + 15x + 14$

93. $y = \left(x - \left(1+\sqrt{2}\right)\right)\left(x - \left(1-\sqrt{2}\right)\right)(x-3) = \left(x^2 - 2x - 1\right)(x-3) = x^3 - 5x^2 + 5x + 3$

95. $y = \left(x - \left(1+\sqrt{3}\right)\right)\left(x - \left(1-\sqrt{3}\right)\right) \cdot \left(x - \left(3-\sqrt{2}\right)\right)\left(x - \left(3+\sqrt{2}\right)\right) = \left(x^2 - 2x - 2\right)\left(x^2 - 6x + 7\right)$
$= x^4 - 8x^3 + 17x^2 - 2x - 14$

97. $y = \left(x - \left(\sqrt{2} + \sqrt{3}\right)\right)\left(x - \left(\sqrt{2} - \sqrt{3}\right)\right) = x^2 - 2x\sqrt{2} - 1$

Since we must have integer coefficients, there must be additional zeros. Try $-\left(\sqrt{2} + \sqrt{3}\right)$:

$y = \left(x - \left(\sqrt{2} + \sqrt{3}\right)\right)\left(x - \left(\sqrt{2} - \sqrt{3}\right)\right) \cdot \left(x + \left(\sqrt{2} + \sqrt{3}\right)\right)\left(x + \left(\sqrt{2} - \sqrt{3}\right)\right) = x^4 - 10x^2 + 1$

99. $x = 3 - \sqrt{2} + \sqrt{5} \Rightarrow x - 3 = -\sqrt{2} + \sqrt{5} \Rightarrow (x - 3)^2 = \left(-\sqrt{2} + \sqrt{5}\right)^2 \Rightarrow x^2 - 6x + 9 = 7 - 2\sqrt{10} \Rightarrow$

$x^2 - 6x + 2 = -2\sqrt{10} \Rightarrow \left(x^2 - 6x + 2\right)^2 = \left(-2\sqrt{10}\right)^2 \Rightarrow x^4 - 12x^3 + 40x^2 - 24x + 4 = 40 \Rightarrow$

$y = x^4 - 12x^3 + 40x^2 - 24x - 36$

101. i. Simplifying the fraction if necessary, we can assume that $\dfrac{p}{q}$ is in lowest terms.

Since $\dfrac{p}{q}$ is a zero of F, we have

$F\left(\dfrac{p}{q}\right) = 0.$

ii. Substitute $\dfrac{p}{q}$ for x in the equation $F(x) = 0.$

iii. Multiply the equation in (ii) by q^n.

iv. Subtract $a_0 q^n$ from both sides of the equation.

v. The left side of the equation in (iv) is
$a_n p^n + a_{n-1} p^{n-1} q + \cdots + a_1 p q^{n-1} =$
$p\left(a_n p^{n-1} + a_{n-1} p^{n-2} q + \cdots + a_1 q^{n-1}\right).$
Therefore p is a factor.

vi. $a = b \Leftrightarrow \dfrac{a}{p} = \dfrac{b}{p}.$

vii. Since p and q have no common prime factors, p must be a factor of a_0.

viii. Rearrange the terms of the equation in (iii).

ix. The left side of the equation in (viii) is
$a_{n-1} p^{n-1} q + \cdots + a_1 p q^{n-1} + a_0 q^n =$
$q\left(a_{n-1} p^{n-1} + \cdots + a_1 p q^{n-2} + a_0 q^{n-1}\right).$
Therefore q is a factor.

x. $a = b \Leftrightarrow \dfrac{a}{q} = \dfrac{b}{q}.$

xi. Since p and q have no common prime factors, q must be a factor of a_n.

3.4 Critical Thinking/Discussion/Writing

103. a. False. The factors of the constant term, 3, are $\{\pm 1, \pm 3\}$ and the factors of the leading coefficient, 1, are $\{\pm 1\}$. The possible rational zeros are $\{\pm 1, \pm 3\}$.

b. False. The factors of the constant term, 25, are $\{\pm 1, \pm 5, \pm 25\}$ and the factors of the leading coefficient, 2, are $\{\pm 1, \pm 2\}$. The possible rational zeros are
$\left\{\pm \dfrac{1}{2}, \pm 1, \pm \dfrac{5}{2}, \pm 5, \pm \dfrac{25}{2}, \pm 25\right\}.$

104. Since $f(x)$ has integer coefficients, if $x = \sqrt{2}$ is a zero, then so is $x = -\sqrt{2}$. Then, two of the factors of $f(x)$ are $\left(x + \sqrt{2}\right)$ and $\left(x - \sqrt{2}\right)$.

$\left(x + \sqrt{2}\right)\left(x - \sqrt{2}\right) = x^2 - 2$, so divide $f(x)$ by $x^2 - 2$ to find another factor.

$$
\begin{array}{r}
x^2 - 2x - 2 \\
x^2 + 0x - 2 \overline{)x^4 - 2x^3 - 4x^2 + 4x + 4} \\
\underline{x^4 + 0x^3 - 2x^2} \\
-2x^3 - 2x^2 + 4x \\
\underline{-2x^3 - 0x^2 + 4x} \\
-2x^2 + 0x + 4 \\
\underline{-2x^2 + 0x + 4}
\end{array}
$$

Use the quadratic formula to solve $x^2 - 2x - 2 = 0$.

$x = \dfrac{-(-2) \pm \sqrt{(-2)^2 - 4(1)(-2)}}{2(1)} = \dfrac{2 \pm \sqrt{4 + 8}}{2}$

$= \dfrac{2 \pm 2\sqrt{3}}{2} = 1 \pm \sqrt{3}$

The rational roots of the equation are $x \pm \sqrt{2}$ and $1 \pm \sqrt{3}$.

3.4 Maintaining Skills

105. $(-2i)^5 = (-2)^5 (i)^5 = (-2)^5 (i)^4 i = -32i$

107. $(2x+i)(-2x+i) = -4x^2 + i^2 = -4x^2 - 1$

109. $(5x+2i)(5x-2i) = 25x^2 - 4i^2 = 25x^2 + 4$

111. $(x-1+2i)(x-1-2i)$
$= \big((x-1)+2i\big)\big((x-1)-2i\big) = (x-1)^2 - 4i^2$
$= x^2 - 2x + 1 + 4 = x^2 - 2x + 5$

113. $\big[x-(2+3i)\big]\big[x-(2-3i)\big]$
$= x^2 - (2-3i)x - (2+3i)x + (2+3i)(2-3i)$
$= x^2 - 2x + 3ix - 2x - 3ix + 4 - 9i^2$
$= x^2 - 4x + 4 + 9 = x^2 - 4x + 13$

3.5 The Complex Zeros of a Polynomial Function

3.5 Practice Problems

1. a. $P(x) = 3(x+2)(x-1)\big[x-(1+i)\big]\big[x-(1-i)\big]$
$= 3(x+2)(x-1)(x-1-i)(x-1+i)$

b. $P(x) = 3(x+2)(x-1)\big[x-(1+i)\big]\big[x-(1-i)\big]$
$= 3(x+2)(x-1)\big(x^2 - 2x + 2\big)$
$= 3(x+2)\big(x^3 - 3x^2 + 4x - 2\big)$
$= 3\big(x^4 - x^3 - 2x^2 + 6x - 4\big)$
$= 3x^4 - 3x^3 - 6x^2 + 18x - 12$

2. Since $2 - 3i$ is a zero of multiplicity 2, so is $2 + 3i$. Since i is a zero, so is $-i$. The eight zeros are $-3, -3, 2 - 3i, 2 - 3i, 2 + 3i, 2 + 3i, i,$ and $-i$.

3. The function has degree four, so there are four zeros. Since one zero is $2i$, another zero is $-2i$. So $(x-2i)(x+2i) = x^2 + 4$ is a factor of $P(x)$. Now divide to find the other factors.

$$
\begin{array}{r}
x^2 - 3x + 2 \\
x^2 + 4 \overline{\smash{\big)}\, x^4 - 3x^3 + 6x^2 - 12x + 8} \\
\underline{x^4 \qquad\quad + 4x^2} \\
-3x^3 + 2x^2 - 12x \\
\underline{-3x^3 \qquad\quad - 12x} \\
2x^2 \qquad\quad + 8 \\
\underline{2x^2 \qquad\quad + 8} \\
0
\end{array}
$$

So,
$P(x) = x^4 - 3x^3 + 6x^2 - 12x + 8$
$= (x-2i)(x+2i)(x^2 - 3x + 2)$
$= (x-2i)(x+2i)(x-2)(x-1)$
The zeros of $P(x)$ are $1, 2, 2i,$ and $-2i$.

4. $f(x) = x^4 - 8x^3 + 22x^2 - 28x + 16$
The function has degree 4, so there are four zeros. There are three sign changes, so there are either 3 or 1 positive zeros.
$f(-x) = (-x)^4 - 8(-x)^3 + 22(-x)^2 - 28(-x) + 16$
$= x^4 + 8x^3 + 22x + 28x + 16$
There are no sign changes in $f(-x)$, so there are no negative zeros. The possible rational zeros are $\{\pm 1, \pm 2, \pm 4, \pm 8, \pm 16\}$. Using synthetic division to test the positive values, we find that one zero is 2:

$$
\begin{array}{r|rrrr}
2 & 1 & -8 & 22 & -28 & 16 \\
  &   & 2 & -12 & 20 & -16 \\
\hline
  & 1 & -6 & 10 & -8 & 0
\end{array}
$$

The zeros of the depressed function $x^3 - 6x^2 + 10x - 8$ are also zeros of P. The possible rational zeros of the depressed function are $\{\pm 1, \pm 2, \pm 4, \pm 8\}$. Examine only the positive possibilities and find that 4 is a zero:

$$
\begin{array}{r|rrrr}
4 & 1 & -6 & 10 & -8 \\
  &   & 4 & -8 & 8 \\
\hline
  & 1 & -2 & 2 & 0
\end{array}
$$

So,
$x^4 + 8x^3 + 22x^2 - 28x + 16$
$= (x-2)(x-4)(x^2 - 2x + 2)$

Now find the zeros of $x^2 - 2x + 2$ using the quadratic formula:
$$x = \frac{2 \pm \sqrt{(-2)^2 - 4(1)(2)}}{2(1)} = \frac{2 \pm \sqrt{-4}}{2} = 1 \pm i$$
The zeros are $2, 4, 1 + i,$ and $1 - i$.

3.5 Basic Concepts and Skills

1. The Fundamental Theorem of Algebra states that a polynomial function of degree $n \geq 1$ has at least one <u>complex</u> zero.

3. If P is a polynomial function with real coefficients and if $z = a + bi$ is a zero of P, then $\overline{z} = a - bi$ is also a zero of $P(x)$.

5. False. A polynomial function of degree $n \geq 1$ has at least one complex zero.

7. $x^2 + 25 = 0 \Rightarrow x^2 = -25 \Rightarrow x = \pm 5i$

9. $x^2 + 4x + 4 = -9 \Rightarrow (x+2)^2 = -9 \Rightarrow$
$x + 2 = \pm 3i \Rightarrow x = -2 \pm 3i$

11. $(x-2)(x-3i)(x+3i) = 0 \Rightarrow x = 2$ or $x = \pm 3i$

13. The remaining zero is $3 - i$.

15. The remaining zero is $-5 - i$.

17. The remaining zeros are $-i$ and $-3i$.

19. $P(x) = 2(x-(5-i))(x-(5+i))(x-3i)(x+3i)$
$= 2x^4 - 20x^3 + 70x^2 - 180x + 468$

21. $P(x) = 7(x-5)^2(x-1)(x-(3-i))(x-(3+i))$
$= 7x^5 - 119x^4 + 777x^3 - 2415x^2$
$\qquad + 3500x - 1750$

23. The function has degree four, so there are four zeros. Since one zero is $3i$, another zero is $-3i$. So $(x-3i)(x+3i) = x^2 + 9$ is a factor of $P(x)$. Now divide to find the other factor:

$$
\begin{array}{r}
x^2 + x \\
x^2+9 \overline{\smash{)}\,x^4 + x^3 + 9x^2 + 9x} \\
\underline{x^4 \qquad\;\; + 9x^2} \\
x^3 \qquad\;\; + 9x \\
\underline{x^3 \qquad\;\; + 9x} \\
0
\end{array}
$$

So,
$P(x) = x^4 + x^3 + 9x^2 + 9x$
$= \left(x^2 + 9\right)\left(x^2 + x\right) = x(x+1)\left(x^2 + 9\right)$

25. The function has degree five, so there are five zeros. Since one zero is $3 - i$, another zero is $3 + i$. So
$(x-(3-i))(x-(3+i)) = x^2 - 6x + 10$ is a factor of $P(x)$. Now divide to find the other factor:

$$
\begin{array}{r}
x^3 + x^2 - 2x \\
x^2-6x+10 \overline{\smash{)}\,x^5 - 5x^4 + 2x^3 + 22x^2 - 20x} \\
\underline{x^5 - 6x^4 + 10x^3} \\
x^4 - 8x^3 + 22x^2 \\
\underline{x^4 - 6x^3 + 10x^2} \\
-2x^3 + 12x^2 - 20x \\
\underline{-2x^3 + 12x^2 - 20x} \\
0
\end{array}
$$

$P(x) = x^5 - 5x^4 + 2x^3 + 22x^2 - 20x$
$= \left(x^2 - 6x + 10\right)\left(x^3 + x^2 - 2x\right)$
$= \left(x^2 - 6x + 10\right)\left(x^2 + x - 2\right)x$
$= x(x+2)(x-1)\left(x^2 - 6x + 10\right)$

27. The function has degree 3, so there are three zeros. There are three sign changes, so there are 1 or 3 positive zeros. The possible rational zeros are $\{\pm 1, \pm 17\}$. Using synthetic division we find that one zero is 1:

$$
\begin{array}{r|rrr}
1 & 1 & -9 & 25 & -17 \\
& & 1 & -8 & 17 \\
\hline
& 1 & -8 & 17 & 0
\end{array}
$$

$x^3 - 9x^2 + 25x - 17 = (x-1)(x^2 - 8x + 17)$.
Now solve the depressed equation
$x^2 - 8x + 17 = 0 \Rightarrow$
$x = \dfrac{8 \pm \sqrt{(-8)^2 - 4(1)(17)}}{2(1)} \Rightarrow x = \dfrac{8 \pm \sqrt{-4}}{2} \Rightarrow$
$x = 4 \pm i$. The zeros are $1, 4 \pm i$.

29. The function has degree 3, so there are three zeros. There are two sign changes, so there are either 2 or 0 positive zeros. There is one sign change in $f(-x)$, so there is one negative zero. The possible rational zeros are
$\left\{\pm\dfrac{1}{3}, \pm\dfrac{2}{3}, \pm 1, \pm\dfrac{4}{3}, \pm\dfrac{5}{3}, \pm 2, \pm\dfrac{8}{3}, \pm\dfrac{10}{3}, \pm 4, \right.$
$\left. \pm 5, \pm\dfrac{20}{3}, \pm 8, \pm 10, \pm\dfrac{40}{3}, \pm 20, \pm 40 \right\}$. Using

synthetic division to test the negative values we find that one zero is $-4/3$.

$$
\begin{array}{r|rrr}
-\dfrac{4}{3} & 3 & -2 & 22 & 40 \\
& & -4 & 8 & -40 \\
\hline
& 3 & -6 & 30 & 0
\end{array}
$$

$3x^3 - 2x^2 + 22x + 40 = \left(x + \dfrac{4}{3}\right)(3x^2 - 6x + 30)$
$= 3\left(x + \dfrac{4}{3}\right)(x^2 - 2x + 10)$

Now solve the depressed equation
$x^2 - 2x + 10 = 0 \Rightarrow$
$x = \dfrac{2 \pm \sqrt{(-2)^2 - 4(1)(10)}}{2(1)} \Rightarrow x = \dfrac{2 \pm \sqrt{-36}}{2} \Rightarrow$
$x = 1 \pm 3i$. The zeros are $-\dfrac{4}{3}, 1 \pm 3i$.

31. The function has degree 4, so there are four zeros. There are four sign changes, so there are 4, 2, or 0 positive zeros. There are no sign changes in $f(-x)$, so there are no negative zeros. The possible rational zeros are $\left\{\pm\dfrac{1}{2},\pm1,\pm\dfrac{3}{2},\pm3,\pm\dfrac{9}{2},\pm9\right\}$. Using synthetic division to test the positive values we find that one zero is 1:

$$\begin{array}{r|rrrrr} 1 & 2 & -10 & 23 & -24 & 9 \\ & & 2 & -8 & 15 & -9 \\ \hline & 2 & -8 & 15 & -9 & 0 \end{array}$$

The zeros of the depressed function $2x^3-8x^2+15x-9$ are also zeros of P. The possible rational zeros are $\left\{\pm\dfrac{1}{2},\pm1,\pm\dfrac{3}{2},\pm3,\pm\dfrac{9}{2},\pm9\right\}$.

Using synthetic division to test the positive values we find that one zero is 1:

$$\begin{array}{r|rrrr} 1 & 2 & -8 & 15 & -9 \\ & & 2 & -6 & 9 \\ \hline & 2 & -6 & 9 & 0 \end{array}$$

Thus,
$$\begin{aligned} 2x^4&-10x^3+23x^2-24x+9 \\ &=(x-1)^2\left(2x^2-6x+9\right). \end{aligned}$$

Now solve the depressed equation
$2x^2-6x+9=0$.

$$x=\frac{6\pm\sqrt{(-6)^2-4(2)(9)}}{2(2)}\Rightarrow x=\frac{6\pm\sqrt{-36}}{4}\Rightarrow$$

$$x=\frac{6\pm6i}{4}=\frac{3}{2}\pm\frac{3i}{2}.$$

The zeros of P are 1 (multiplicity 2), $\dfrac{3}{2}\pm\dfrac{3i}{2}$.

33. The function has degree 4, so there are four zeros. There are three sign changes, so there are either 3 or 1 positive zeros. There is one sign change in $f(-x)$, so there is one negative zero. The possible rational zeros are $\{\pm1,\pm2,\pm3,\pm5,\pm6,\pm10,\pm15,\pm30\}$. Using synthetic division to test the negative values, we find that one zero is -3:

$$\begin{array}{r|rrrrr} -3 & 1 & -4 & -5 & 38 & -30 \\ & & -3 & 21 & -48 & 30 \\ \hline & 1 & -7 & 16 & -10 & 0 \end{array}$$

The zeros of the depressed function $x^3-7x^2+16x-10$ are also zeros of P.

The possible rational zeros of the depressed function are $\{\pm1,\pm2,\pm5,\pm10\}$. Since we have already found the negative zero, we examine only the positive possibilities and find that 1 is a zero:

$$\begin{array}{r|rrrr} 1 & 1 & -7 & 16 & -10 \\ & & 1 & -6 & 10 \\ \hline & 1 & -6 & 10 & 0 \end{array}$$

$$\begin{aligned} x^4&-4x^3-5x^2+38x-30 \\ &=(x+3)(x-1)(x^2-6x+10) \end{aligned}$$

Now solve the depressed equation:

$$x^2-6x+10=0\Rightarrow x=\frac{6\pm\sqrt{(-6)^2-4(1)(10)}}{2(1)}\Rightarrow$$

$$x=\frac{6\pm\sqrt{-4}}{2}\Rightarrow x=3\pm i.$$

The zeros are -3, 1, $3\pm i$.

35. The function has degree 5, so there are five zeros. There are five sign changes, so there are either 5, 3, or 1 positive zeros. There are no sign changes in $f(-x)$, so there are no negative zeros. The possible rational zeros are $\left\{\dfrac{1}{2},1,\dfrac{3}{2},2,3,6\right\}$. Using synthetic division to test the positive values, we find that one zero is 1/2.

$$\begin{array}{r|rrrrrr} \frac{1}{2} & 2 & -11 & 19 & -17 & 17 & -6 \\ & & 1 & -5 & 7 & -5 & 6 \\ \hline & 2 & -10 & 14 & -10 & 12 & 0 \end{array}$$

The zeros of the depressed function $2x^4-10x^3+14x^2-10x+12$ are also zeros of P. Use synthetic division again to find the next zero, 2:

$$\begin{array}{r|rrrrr} 2 & 2 & -10 & 14 & -10 & 12 \\ & & 4 & -12 & 4 & -12 \\ \hline & 2 & -6 & 2 & -6 & 0 \end{array}$$

$$\begin{aligned} 2x^5&-11x^4+19x^3-17x^2+17x-6 \\ &=\left(x-\frac{1}{2}\right)(x-2)(2x^3-6x^2+2x-6) \\ &=\left(x-\frac{1}{2}\right)(x-2)(2x^3-6x^2+2x-6) \end{aligned}$$

Use factoring by grouping to factor x^3-3x^2+x-3:

(continued on next page)

(*continued*)

$$x^3 - 3x^2 + x - 3 = x^2(x-3) + 1(x-3)$$
$$= (x^2 + 1)(x-3).$$ So the remaining zeros are 3

and $\pm i$. The zeros are $\dfrac{1}{2}, 2, 3, \pm i$.

37. Since one zero is $3i$, $-3i$ is another zero. There are two zeros, so the degree of the equation is at least 2. Thus, the equation is of the form $f(x) = a(x-3i)(x+3i) = a(x^2 + 9)$. The graph passes through (0, 3), so we have

$3 = a(0^2 + 9) \Rightarrow a = \dfrac{1}{3}$. Thus, the equation of

the function is $f(x) = \dfrac{1}{3}(x^2 + 9)$.

39. Since i and $2i$ are zeros, so are $-i$, and $-2i$. From the graph, we see that 2 is also a zero. There are five zeros, so the degree of the equation is at least 5. Thus, the equation is of the form

$$f(x) = a(x-i)(x+i)(x-2i)(x+2i)(x-2)$$
$$= a(x^2 + 1)(x^2 + 4)(x-2)$$

The graph passes through (0, 4), so we have

$$4 = a(0^2 + 1)(0^2 + 4)(0-2) \Rightarrow a = -\dfrac{1}{2}.$$

Thus, the equation is

$$f(x) = -\dfrac{1}{2}(x^2 + 1)(x^2 + 4)(x-2) \text{ or}$$

$$f(x) = \dfrac{1}{2}(x^2 + 1)(x^2 + 4)(2-x).$$

3.5 Beyond the Basics

41. There are three cube roots. We know that one root is 1. Using synthetic division, we find

$$0 = x^3 - 1 = (x-1)(x^2 + x + 1):$$

$$
\begin{array}{r|rrrr}
1 & 1 & 0 & 0 & -1 \\
  &   & 1 & 1 & 1 \\
\hline
  & 1 & 1 & 1 & 0
\end{array}
$$

Solve the depressed equation

$$x^2 + x + 1 = 0 \Rightarrow x = \dfrac{-1 \pm \sqrt{1^2 - 4(1)(1)}}{2(1)}$$

$$= \dfrac{-1 \pm \sqrt{-3}}{2} = -\dfrac{1}{2} \pm \dfrac{i\sqrt{3}}{2}.$$

So, the cube roots of 1 are 1 and $-\dfrac{1}{2} \pm \dfrac{i\sqrt{3}}{2}$.

43. $P(x) = x^2 + (i-2)x - 2i, \ x = -i$

$$
\begin{array}{r|rrr}
-i & 1 & i-2 & -2i \\
   &   & -i & 2i \\
\hline
   & 1 & -2 & 0
\end{array}
$$

$$x^2 + (i-2)x - 2i = (x+i)(x-2)$$

45. $P(x) = x^3 - (3+i)x^2 - (4-3i)x + 4i, \ x = i$

$$
\begin{array}{r|rrrr}
i & 1 & -(3+i) & -(4-3i) & 4i \\
  &   & i   & -3i   & -4i \\
\hline
  & 1 & -3   & -4   & 0
\end{array}
$$

$$x^3 - (3+i)x^2 - (4-3i)x + 4i$$
$$= (x^2 - 3x - 4)(x-i) = (x-4)(x+1)(x-i)$$

47. Since $1 + 2i$ is a zero, so is $1 - 2i$. The equation is of the form

$$f(x) = a(x-2)(x-(1+2i))(x-(1-2i))$$
$$= a(x-2)(x^2 - 2x + 5)$$

The y-intercept is 40, so we have

$$40 = a(0-2)(0^2 - 2(0) + 5) \Rightarrow a = -4$$

Thus, the equation is

$$f(x) = -4(x-2)(x^2 - 2x + 5).$$

Because $a < 0$, $y \to \infty$ as $x \to -\infty$ and $y \to -\infty$ as $x \to \infty$.

49. Since $3 + i$ is a zero, so is $3 - i$. The equation is of the form

$$f(x) = a(x-1)(x+1)(x-(3+i))(x+(3+i))$$
$$= a(x^2 - 1)(x^2 - 6x + 10)$$

The y-intercept is 20, so we have

$$20 = a(0^2 - 1)(0^2 - 6(0) + 10) \Rightarrow a = -2$$

Thus, the equation is

$$f(x) = -2(x^2 - 1)(x^2 - 6x + 10).$$

Because $a < 0$, $y \to -\infty$ as $x \to -\infty$ and $y \to -\infty$ as $x \to \infty$.

3.5 Critical Thinking/Discussion/Writing

51. Factoring the polynomial we have

$$a_n x^n + a_{n-1} x^{n-1} + \cdots + a_1 x + a_0$$
$$= a_n (x - r_1)(x - r_2) \cdots (x - r_n).$$ Expanding the

right side, we find that the coefficient of x^{n-1} is $-a_n r_1 - a_n r_2 - \cdots - a_n r_n$ and the constant term is $(-1)^n a_n r_1 r_2 \cdots r_n$.

(continued on next page)

(*continued*)

Comparing the coefficients with those on the left side, we obtain

$a_{n-1} = -a_n(r_1 + r_2 + \cdots + r_n)$.

Because $a_n \neq 0$, $-\dfrac{a_{n-1}}{a_n} = r_1 + r_2 + \cdots + r_n$

and $-(1)^n \dfrac{a_0}{a_n} = r_1 r_2 \cdots r_n$.

52. a. $x^3 + 6x = 20 \Rightarrow x^3 + 6x - 20 = 0$. There is one sign change, so there is one positive root. There are no sign changes in $f(-x)$, so there are no negative roots. Therefore, there is only one real solution.

b. Substituting $v - u$ for x and remembering that $v^3 - u^3 = 20$ and $uv = 2$, we have

$(v-u)^3 + 6(v-u)$

$= v^3 - 3v^2u + 3vu^2 - u^3 + 6(v-u)$

$= (v^3 - u^3) - (3v^2u - 3vu^2) + 6(v-u)$

$= (v^3 - u^3) - 3vu(v-u) + 6(v-u)$

$= (v^3 - u^3) - (6 - 3uv)(v-u)$

$= 20 - (6 - 3(2))(v-u) = 20$. Therefore,

$x = v - u$ is the solution.

c. To solve the system $v^3 - u^3 = 20, vu = 2$, solve the second equation for v, and substitute that value into the first equation, keeping in mind that u cannot be zero, so division by u is permitted:

$v^3 - u^3 = 20 \Rightarrow \left(\dfrac{2}{u}\right)^3 - u^3$

$= \dfrac{8}{u^3} - u^3 = 20 \Rightarrow$

$-u^3 + \dfrac{8}{u^3} - 20 = 0$.

Let $-u^3 = a$, so

$a - \dfrac{8}{a} - 20 = 0 \Rightarrow a^2 - 20a - 8 = 0 \Rightarrow$

$a = \dfrac{20 \pm \sqrt{400 + 32}}{2} = 10 \pm 6\sqrt{3} = -u^3 \Rightarrow$

$\sqrt[3]{-10 \pm 6\sqrt{3}} = u \Rightarrow v = \dfrac{2}{\sqrt[3]{-10 \pm 6\sqrt{3}}}$

$= \dfrac{2}{\sqrt[3]{-10 \pm 6\sqrt{3}}} \cdot \dfrac{\sqrt[3]{10 \pm 6\sqrt{3}}}{\sqrt[3]{10 \pm 6\sqrt{3}}} = \sqrt[3]{10 \pm 6\sqrt{3}}$

d. If q is positive, then, according to Descartes's Rule of Signs, the polynomial $x^3 + px - q$ has one positive zero and no negative zeros. If q is negative, then it has no positive zeros and one negative zero. Either way, there is exactly one real solution. From (b), we have $v^3 - u^3 = q$, $vu = \dfrac{p}{3}$, and $x = v - u$. Substituting, we have

$x^3 + px = (v-u)^3 + p(v-u)$

$= v^3 - 3v^2u + 3vu^2 - u^3 + p(v-u)$

$= (v^3 - u^3) - (3v^2u - 3vu^2) + p(v-u)$

$= (v^3 - u^3) - 3vu(v-u) + p(v-u)$

$= (v^3 - u^3) + (p - 3uv)(v-u)$

$= q + \left(p - 3 \cdot \dfrac{p}{3}\right)x = q \Rightarrow x = v - u$ is the

solution.

Solving the system $v^3 - u^3 = q$, $vu = \dfrac{p}{3}$,

we obtain

$v^3 - u^3 = q \Rightarrow \left(\dfrac{p}{3u}\right)^3 - u^3 = \dfrac{p^3}{3^3 u^3} - u^3 = q$

$\Rightarrow -u^3 + \dfrac{p^3}{3^3 u^3} - q = 0$. Let $-u^3 = a$, so

we have $a - \dfrac{p}{3^3 a} - q = 0 \Rightarrow$

$a^2 - qa - \dfrac{p^3}{3^3} = 0 \Rightarrow$

$a = \dfrac{q \pm \sqrt{q^2 + 4\left(\dfrac{p^3}{3^3}\right)}}{2} = \dfrac{q}{2} \pm \sqrt{\dfrac{q^2 + 4\left(\dfrac{p^3}{3^3}\right)}{4}}$

$= \dfrac{q}{2} \pm \sqrt{\left(\dfrac{q}{2}\right)^2 + \left(\dfrac{p}{3}\right)^3} = -u^3 \Rightarrow$

$u = \sqrt[3]{-\dfrac{q}{2} + \sqrt{\left(\dfrac{q}{2}\right)^2 + \left(\dfrac{p}{3}\right)^3}}$ and

$v = \sqrt[3]{\dfrac{q}{2} + \sqrt{\left(\dfrac{q}{2}\right)^2 + \left(\dfrac{p}{3}\right)^3}}$ or

$u = \sqrt[3]{-\dfrac{q}{2} - \sqrt{\left(\dfrac{q}{2}\right)^2 + \left(\dfrac{p}{3}\right)^3}}$ and

$v = \sqrt[3]{\dfrac{q}{2} - \sqrt{\left(\dfrac{q}{2}\right)^2 + \left(\dfrac{p}{3}\right)^3}}$.

(*continued on next page*)

(continued)

The difference $v - u$ is the same in both cases, so

$$x = v - u$$

$$= \sqrt[3]{-\frac{q}{2} + \sqrt{\left(\frac{q}{2}\right)^2 + \left(\frac{p}{3}\right)^3}}$$

$$- \sqrt[3]{\frac{q}{2} + \sqrt{\left(\frac{q}{2}\right)^2 + \left(\frac{p}{3}\right)^3}}.$$

e. Substituting $x = y - \frac{a}{3}$, we have

$$x^3 + ax^2 + bx + c = 0 \Rightarrow$$

$$\left(y - \frac{a}{3}\right)^3 + a\left(y - \frac{a}{3}\right)^2 + b\left(y - \frac{a}{3}\right) + c = 0 \Rightarrow$$

$$\left(y^3 - ay^2 + \frac{a^2}{3}y - \frac{a^3}{27}\right) + \left(ay^2 - \frac{2a^2 y}{3} + \frac{a^3}{9}\right)$$

$$+ by - \frac{ab}{3} + c = 0 \Rightarrow$$

$$y^3 + py = q, \text{ where } p = b - \frac{a^2}{3} \text{ and}$$

$$q = -\frac{2a^3}{27} + \frac{ab}{3} - c$$

$$y^3 - \frac{a^2 y}{3} + by + \frac{2a^3}{27} - \frac{ab}{3} + c = 0 \Rightarrow$$

$$y^3 + \left(b - \frac{a^2}{3}\right)y = -\frac{2a^3}{27} + \frac{ab}{3} - c$$

53. $x^3 + 6x^2 + 10x + 8 = 0 \Rightarrow a = 6, b = 10, c = 8$.

Substituting $x = y - \frac{6}{3} = y - 2$ as in (1e), we have

$$(y - 2)^3 + 6(y - 2)^2 + 10(y - 2) + 8 = 0 \Rightarrow$$

$$(y^3 - 6y^2 + 12y - 8) + (6y^2 - 24y + 24)$$

$$+ 10y - 20 + 8 = 0 \Rightarrow$$

$y^3 - 2y + 4 = 0 \Rightarrow y^3 - 2y = -4$. Then, using the results of (1d), we have

$$y = \sqrt[3]{\frac{-4}{2} + \sqrt{\left(\frac{-4}{2}\right)^2 + \left(\frac{-2}{3}\right)^3}}$$

$$- \sqrt[3]{-\frac{-4}{2} + \sqrt{\left(\frac{-4}{2}\right)^2 + \left(\frac{-2}{3}\right)^3}}.$$

Using a calculator, we find that $y = -2$. So $x = -2 - 2 = -4$. Now use synthetic division to find the depressed equation:

$$\begin{array}{r|rrrr} -4 & 1 & 6 & 10 & 8 \\ & & -4 & -8 & -8 \\ \hline & 1 & 2 & 2 & 0 \end{array}$$

$$x^3 + 6x^2 + 10x + 8 = (x + 4)(x^2 + 2x + 2) = 0.$$

$$x^2 + 2x + 2 = 0 \Rightarrow$$

$$x = \frac{-2 \pm \sqrt{2^2 - 4(1)(2)}}{2(1)} = \frac{-2 \pm \sqrt{-4}}{2}$$

$$= \frac{-2 \pm 2i}{2} = -1 \pm i.$$

The solution set is $\{-4, -1 \pm i\}$.

3.5 Maintaining Skills

55. $2x^2 + x - 3 = 0 \Rightarrow (2x + 3)(x - 1) = 0 \Rightarrow$

$$x = -\frac{3}{2}, x = 1$$

Solution set: $\left\{-\frac{3}{2}, 1\right\}$

57. $x^4 - x^3 - 12x^2 = 0 \Rightarrow x^2\left(x^2 - x - 12\right) = 0 \Rightarrow$

$x^2(x - 4)(x + 3) = 0 \Rightarrow x = 0, x = 4, x = -3$

Solution set: $\{-3, 0, 4\}$

59. $\dfrac{x^2 - x - 4}{x - 2}$

$$\begin{array}{r|rrr} 2 & 1 & -1 & -4 \\ & & 2 & 2 \\ \hline & 1 & 1 & -2 \end{array}$$

$Q(x) = x + 1; \ R(x) = -2$

61.
$$2x^3 + 0x^2 + x \overline{\smash)8x^4 + 6x^3 - 0x^2 + 0x - 5} \quad \begin{array}{r}4x + 3\end{array}$$

$$\underline{8x^4 + 0x^3 + 4x^2}$$

$$6x^3 - 4x^2 + 0x$$

$$\underline{6x^3 + 0x^2 + 3x}$$

$$-4x^2 - 3x - 5$$

$Q(x) = 4x + 3; \ R(x) = -4x^2 - 3x - 5$

63. $\dfrac{7 - (-1)}{2(-1)^2 + 3(-1)} = \dfrac{8}{-1} = -8$

65. $\dfrac{(2)^2 + 4(2) - 1}{9 - (2)^3} = 11$

3.6 Rational Functions

3.6 Practice Problems

1. $f(x) = \dfrac{x-3}{x^2 - 4x - 5}$

The domain of f consists of all real numbers

for which $x^2 - 4x - 5 \neq 0$.

$x^2 - 4x - 5 = 0 \Rightarrow (x-5)(x+1) = 0 \Rightarrow$

$x = 5$ or $x = -1$

Thus, the domain is

$(-\infty, -1) \cup (-1, 5) \cup (5, \infty)$.

2. a. $g(x) = \dfrac{3}{x-2}$

Let $f(x) = \dfrac{1}{x}$. Then

$g(x) = \dfrac{3}{x-2} = 3\left(\dfrac{1}{x-2}\right) = 3f(x-2)$.

The graph of $y = f(x-2)$ is the graph of

$y = f(x)$ shifted two units to the right.

This moves the vertical asymptote two

units to the right. The graph of

$y = 3f(x-2)$ is the graph of

$y = f(x-2)$ stretched vertically three

units.

The domain of g is $(-\infty, 2) \cup (2, \infty)$.

The range of g is $(-\infty, 0) \cup (0, \infty)$.

The vertical asymptote is $x = 2$.

The horizontal asymptote is $y = 0$.

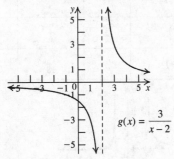

b. $h(x) = \dfrac{2x+5}{x+1}$

$x+1\overline{\smash{\big)}\,2x+5} \quad h(x) = \dfrac{2x+5}{x+1} = 2 + \dfrac{3}{x+1}$
$\underline{2x+2}$
$\quad\quad 3$

Let $f(x) = \dfrac{1}{x}$. Then

$h(x) = \dfrac{2x+5}{x+1} = 2 + \dfrac{3}{x+1} = 2 + 3f(x+1)$.

The graph of $y = h(x)$ is the graph of

$y = f(x)$ shifted one units to the left and

then stretched vertically three units. The

graph is then shifted two units up. This moves

the vertical asymptote one unit to the left. The

horizontal asymptote is shifted two units up.

The domain of h is $(-\infty, -1) \cup (-1, \infty)$.

The range of h is $(-\infty, 2) \cup (2, \infty)$.

The vertical asymptote is $x = -1$.

The horizontal asymptote is $y = 2$.

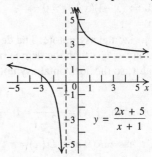

3. $f(x) = \dfrac{x+1}{x^2 + 3x - 10}$

The vertical asymptotes are located at the zeros

of the denominator.

$x^2 + 3x - 10 = 0 \Rightarrow (x+5)(x-2) = 0 \Rightarrow$

$x = -5$ or $x = 2$

The vertical asymptotes are $x = -5$ and $x = 2$.

4. $f(x) = \dfrac{3-x}{x^2 - 9} = \dfrac{-(x-3)}{(x-3)(x+3)} = -\dfrac{1}{x+3}$

$x + 3 = 0 \Rightarrow x = -3$

The vertical asymptote is $x = -3$.

5. a. $f(x) = \dfrac{2x-5}{3x+4}$

Since the numerator and denominator both

have degree 1, the horizontal asymptote is

$y = \dfrac{2}{3}$.

b. $g(x) = \dfrac{x^2 + 3}{x-1}$

Since the degree of the numerator is greater

than the degree of the denominator, there are

no horizontal asymptotes.

c. $h(x) = \dfrac{100x + 57}{0.01x^3 + 8x - 9}$

Since the degree of the numerator is less than the degree of the denominator, the horizontal asymptote is the line $y = 0$.

6. $f(x) = \dfrac{2x}{x^2 - 1}$

There are no common factors of the form $x - a$ between $2x$ and $x^2 - 1$.

First find the intercepts:

$\dfrac{2x}{x^2 - 1} = 0 \Rightarrow x = 0$

The graph passes through the origin.
Find the vertical asymptotes:

$x^2 - 1 = 0 \Rightarrow (x - 1)(x + 1) = 0 \Rightarrow$
$x = 1$ or $x = -1$

Find the horizontal asymptote: The degree of the numerator is less than the degree of the denominator, so the horizontal asymptote is the x-axis.

By long division, $f(x) = \dfrac{2x}{x^2 - 1} = 0 + \dfrac{2x}{x^2 - 1}$.

$R(x) = 2x$ has zero 0 and $D(x) = x^2 - 1$ has zeros -1 and 1. These zeros divide the x-axis into four intervals, $(-\infty, -1)$, $(-1, 0)$, $(0, 1)$, and $(1, \infty)$. Use test values to determine where the graph of f is above and below the x-axis.

Interval	Test point	Value of $f(x)$	Above/below x-axis
$(-\infty, -1)$	-3	$-\frac{3}{4}$	below
$(-1, 0)$	$-\frac{1}{2}$	$\frac{4}{3}$	above
$(0, 1)$	$\frac{1}{2}$	$-\frac{4}{3}$	below
$(1, \infty)$	2	$\frac{4}{3}$	above

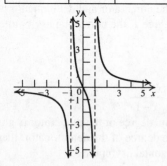

7. $f(x) = \dfrac{2x^2 - 1}{2x^2 + x - 3}$

$f(x)$ is in lowest terms.
Find the intercepts:

$\dfrac{2x^2 - 1}{2x^2 + x - 3} = 0 \Rightarrow x = \pm\dfrac{\sqrt{2}}{2}$; $f(0) = \dfrac{1}{3}$

Find the vertical asymptotes:

$2x^2 + x - 3 = 0 \Rightarrow (2x + 3)(x - 1) = 0 \Rightarrow$
$x = -\dfrac{3}{2}$ or $x = 1$

Find the horizontal asymptote: The degree of the numerator is the same as the degree of the denominator, so the horizontal asymptote is

$y = \dfrac{2}{2} = 1$.

$f(x) = \dfrac{2x^2 - 1}{2x^2 + x - 3} = 1 + \dfrac{2 - x}{2x^2 + x - 3}$

The zero of $R(x) = 2 - x$ is 2 and the zeros of

$D(x) = 2x^2 + x - 3 = (2x + 3)(x - 1)$ are

$x = -\dfrac{3}{2}$ and $x = 1$.

Use test values to determine where the graph of f is above and below the horizontal asymptote $y = 1$.

Interval	Test point	Value of $f(x)$	Above/below $y = 1$
$\left(-\infty, -\frac{3}{2}\right)$	-2	$\frac{7}{3}$	above
$\left(-\frac{3}{2}, 1\right)$	0	$\frac{1}{3}$	below
$(1, 2)$	$\frac{3}{2}$	$\frac{7}{6}$	above
$(2, \infty)$	3	$\frac{17}{18}$	above

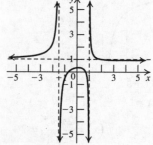

Notice that the graph crosses the horizontal asymptote at $(2, 1)$.

8. $f(x) = \dfrac{x^2+1}{x^2+2}$

$f(x)$ is in lowest terms.

Find the intercepts:

$\dfrac{x^2+1}{x^2+2} = 0 \Rightarrow x = \pm i \Rightarrow$ there is no

x-intercept.

$f(0) = \dfrac{1}{2} \Rightarrow \left(0, \dfrac{1}{2}\right)$ is the y-intercept.

Find the vertical asymptotes:

$x^2+2 = 0 \Rightarrow x = i\sqrt{2} \Rightarrow$ there is no vertical
asymptote.

Find the horizontal asymptote: The degree of
the numerator is the same as the degree of the
denominator, so the horizontal asymptote is

$y = \dfrac{1}{1} = 1$.

$f(x) = \dfrac{x^2+1}{x^2+2} = 1 + \dfrac{-1}{x^2+2}$

Neither $R(x) = -1$ nor $D(x) = x^2+2$ have

real zeros. Since $\dfrac{-1}{x^2+2}$ is negative for all

values of x, the graph of $f(x)$ is always

below the line $y = 1$.

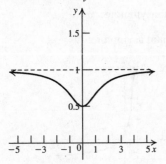

9. $f(x) = \dfrac{x^2+2}{x-1}$

$f(x)$ is in lowest terms.

Find the intercepts:

$\dfrac{x^2+2}{x-1} = 0 \Rightarrow x = \pm i\sqrt{2} \Rightarrow$ there is no

x-intercept. $f(0) = -2 \Rightarrow (0, -2)$ is the

y-intercept.

Find the vertical asymptotes:

$x-1 = 0 \Rightarrow x = 1$

Find the horizontal asymptote: The degree of
the numerator is greater than the degree of the
denominator, so there is no horizontal
asymptote. However, there is an oblique

asymptote.

$\dfrac{x^2+2}{x-1} = x+1 + \dfrac{3}{x-1} \Rightarrow y = x+1$ is the oblique

asymptote.

The graph is above the line $y = x+1$ on $(1, \infty)$

and below the line on $(-\infty, 1)$.

The intervals determined by the zeros of the
numerator and of the denominator of

$f(x) - (x+1) = \dfrac{x^2+2}{x-1} - (x+1) = \dfrac{3}{x-1}$ divide

the x-axis into two intervals, $(-\infty, 1)$ and $(1, \infty)$.

Use test numbers to determine where the graph
of f is above and below the x-axis.

Interval	Test point	Value of $f(x)$	Above/below $y = x+1$
$(-\infty, 1)$	-1	$-\dfrac{3}{2}$	below
$(1, \infty)$	2	6	above

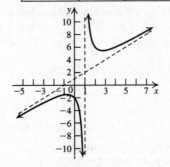

10. $R(x) = \dfrac{x(100-x)}{x+20}$

a. $R(10) = \dfrac{10(100-10)}{10+20} = 30$ billion dollars.

This means that if income is taxed at a rate of
10%, then the total revenue for the
government will be 30 billion dollars.
Similarly, $R(20) = \$40$ billion,
$R(30) = \$42$ billion, $R(40) = \$40$ billion,
$R(50) \approx \$35.7$ billion, $R(60) = \$30$ billion.

b.

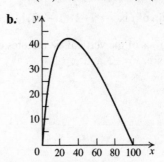

c.

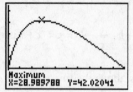

From the graphing calculator screen, we see that a tax rate of about 29% generates the maximum tax revenue of about \$42.02 billion.

3.6 Basic Concepts and Skills

1. A rational function can be expressed in the form $\dfrac{N(x)}{D(x)}$, where $N(x)$ and $D(x)$ are polynomials and $D(x)$ is not the zero polynomial.

3. The line $y = k$ is a horizontal asymptote of f if $f(x) \to k$ as $x \to \infty$ or as $x \to -\infty$.

5. False. A rational function has a vertical asymptote only if the denominator has a real zero and the numerator and denominator have no common factors.

7. $(-\infty, -4) \cup (-4, \infty)$

9. $(-\infty, \infty)$

11. $x^2 - x - 6 = 0 \Rightarrow (x-3)(x+2) = 0 \Rightarrow x = -2, 3$
The domain of the function is
$(-\infty, -2) \cup (-2, 3) \cup (3, \infty)$.

13. $x^2 - 6x + 8 = 0 \Rightarrow (x-4)(x-2) = 0 \Rightarrow x = 2, 4$
The domain of the function is
$(-\infty, 2) \cup (2, 4) \cup (4, \infty)$.

15. As $x \to 1^+$, $f(x) \to \infty$.

17. As $x \to -2^+$, $f(x) \to \infty$.

19. As $x \to \infty$, $f(x) \to 1$.

21. The domain of f is $(-\infty, -2) \cup (-2, 1) \cup (1, \infty)$.

23. The equations of the vertical asymptotes of the graph are $x = -2$ and $x = 1$.

25.

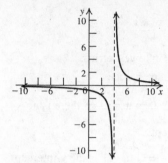

Domain: $(-\infty, 4) \cup (4, \infty)$
Range: $(-\infty, 0) \cup (0, \infty)$
Vertical asymptote: $x = 4$
Horizontal asymptote: $y = 0$

27.

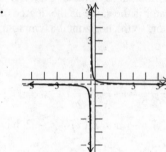

Domain: $\left(-\infty, -\frac{1}{3}\right) \cup \left(-\frac{1}{3}, \infty\right)$
Range: $\left(-\infty, -\frac{1}{3}\right) \cup \left(-\frac{1}{3}, \infty\right)$
Vertical asymptote: $x = -\frac{1}{3}$
Horizontal asymptote: $y = -\frac{1}{3}$

29.

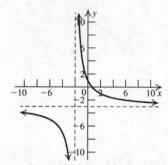

Domain: $(-\infty, -2) \cup (-2, \infty)$
Range: $(-\infty, -3) \cup (-3, \infty)$
Vertical asymptote: $x = -2$
Horizontal asymptote: $y = -3$

31.

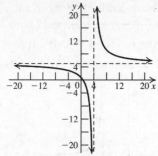

Domain: $(-\infty, 4) \cup (4, \infty)$
Range: $(-\infty, 5) \cup (5, \infty)$
Vertical asymptote: $x = 4$
Horizontal asymptote: $y = 5$

In exercises 33–42, to find the vertical asymptotes, first eliminate any common factors in the numerator and denominator, and then set the denominator equal to zero and solve for x.

33. $x = 1$　　　　**35.** $x = -4, x = 3$

37. $h(x) = \dfrac{x^2 - 1}{x^2 + x - 6} = \dfrac{(x-1)(x+1)}{(x-2)(x+3)}.$

The equations of the vertical asymptotes are $x = -3$ and $x = 2$.

39. $f(x) = \dfrac{x^2 - 6x + 8}{x^2 - x - 12} = \dfrac{(x-4)(x-2)}{(x-4)(x+3)}.$

Disregard the common factor. The vertical asymptote is $x = -3$.

41. There is no vertical asymptote.

For exercises 43–50, locate the horizontal asymptote as follows:

- If the degree of the numerator of a rational function is less than the degree of the denominator, then the x-axis ($y = 0$) if the horizontal asymptote.

- If the degree of the numerator of a rational function equals the degree of the denominator, the horizontal asymptote is the line with the

  equation $y = \dfrac{a_n}{b_m}$, where a_n is the coefficient of

  the leading term of the numerator and b_m is the coefficient of the leading term of the denominator.

- If the degree of the numerator of a rational function is greater than the degree of the denominator, then there is no horizontal asymptote.

43. $y = 0$　　　**45.** $y = \dfrac{2}{3}$

47. There is no horizontal asymptote.

49. $y = 0$

51. d　　　**53.** e　　　**55.** a

57. $0 = \dfrac{2x}{x-3} \Rightarrow x = 0$ is the x-intercept.

$\dfrac{2(0)}{0-3} = 0 \Rightarrow y = 0$ is the y-intercept. The

vertical asymptote is $x = 3$. The horizontal asymptote is $y = 2$. The intervals to be tested are $(-\infty, 3)$ and $(3, \infty)$. The graph is above the horizontal asymptote on $(3, \infty)$ and below the horizontal asymptote on $(-\infty, 3)$.

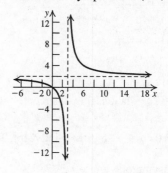

59. $0 = \dfrac{x}{x^2 - 4} \Rightarrow x = 0$ is the x-intercept.

$\dfrac{0}{0^2 - 4} = 0 \Rightarrow y = 0$ is the y-intercept.

The vertical asymptotes are $x = -2$ and $x = 2$. The horizontal asymptote is the x-axis. The intervals to be tested are $(-\infty, -2), (-2, 0), (0, 2)$ and $(2, \infty)$. The graph is above the x-axis on $(-2, 0) \cup (2, \infty)$ and below the x-axis on $(-\infty, -2) \cup (0, 2)$.

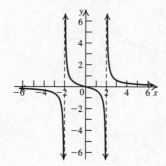

61. $0 = \dfrac{-2x^2}{x^2 - 9} \Rightarrow x = 0$ is the x-intercept.

$\dfrac{0^2}{0^2 - 4} = 0 \Rightarrow y = 0$ is the y-intercept. The vertical asymptotes are $x = -3$ and $x = 3$. The horizontal asymptote is $y = -2$. The intervals to be tested are $(-\infty, -3), (-3, 3),$ and $(3, \infty)$. The graph is above the horizontal asymptote on $(-3, 3)$ and below the horizontal asymptote on $(-\infty, -3) \cup (3, \infty)$.

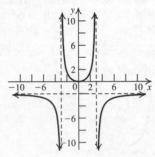

63. $0 = \dfrac{2}{x^2 - 2} \Rightarrow$ there is no x-intercept.

$\dfrac{2}{0^2 - 2} = 0 \Rightarrow y = -1$ is the y-intercept. The vertical asymptotes are $x = \pm\sqrt{2}$. The horizontal asymptote is the x-axis. The intervals to be tested are $(-\infty, -\sqrt{2}), (-\sqrt{2}, \sqrt{2})$ and $(\sqrt{2}, \infty)$. The graph is above the x-axis on $\left(-\infty, -\sqrt{2}\right) \cup \left(\sqrt{2}, \infty\right)$ and below the x-axis on $\left(-\sqrt{2}, \sqrt{2}\right)$.

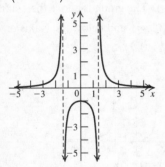

65. $0 = \dfrac{x+1}{(x-2)(x+3)} \Rightarrow x = -1$ is the x-intercept.

$\dfrac{0+1}{(0-2)(0+3)} = -\dfrac{1}{6} \Rightarrow y = -\dfrac{1}{6}$ is the y-intercept. The vertical asymptotes are $x = -3$ and $x = 2$. The horizontal asymptote is the

x-axis. The intervals to be tested are $(-\infty, -3), (-3, -1), (-1, 2),$ and $(2, \infty)$. The graph is above the x-axis on $(-3, -1) \cup (2, \infty)$ and below the x-axis on $(-\infty, -3) \cup (-1, 2)$.

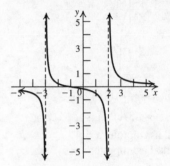

67. $0 = \dfrac{x^2}{x^2 + 1} \Rightarrow x = 0$ is the x-intercept.

$\dfrac{0^2}{0^2 + 1} = 0 \Rightarrow y = 0$ is the y-intercept. There is no vertical asymptote. The horizontal asymptote is $y = 1$. The intervals to be tested are $(-\infty, 0)$ and $(0, \infty)$. The graph is above the x-axis on $(-\infty, 0) \cup (0, \infty)$ and below the horizontal asymptote on $(-\infty, \infty)$.

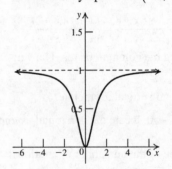

69. $f(x) = \dfrac{x^3 - 4x}{x^3 - 9x} = \dfrac{x(x-2)(x+2)}{x(x-3)(x+3)}$

$\qquad = \dfrac{(x-2)(x+2)}{(x-3)(x+3)}$

$f(x) = 0 \Rightarrow x = \pm 2$ are the x-intercepts.

$\dfrac{x^3 - 4x}{x^3 - 9x} = \dfrac{x(x^2 - 4)}{x(x^2 - 9)} = \dfrac{x^2 - 4}{x^2 - 9} \Rightarrow$

$\dfrac{0^2 - 4}{0^2 - 9} = \dfrac{4}{9} \Rightarrow \dfrac{4}{9}$ is the y-intercept. However,

there is a hole at $\left(0, \dfrac{4}{9}\right)$ since $0^3 - 9(0) = 0$

(continued on next page)

(continued)

$x^2 - 9 = 0 \Rightarrow x(x+3)(x-3) = 0 \Rightarrow x = -3$

and $x = 3$ are the vertical asymptotes. The degree of the numerator is the same as the degree of the denominator, so the horizontal asymptote is $y = 1$. The intervals to be tested are $(-\infty, -3), (-3, 0), (0, 3),$ and $(3, \infty)$. The graph is above the horizontal asymptote on $(-\infty, -3) \cup (3, \infty)$ and below the horizontal asymptote on $(-3, 0) \cup (0, 3)$.

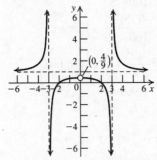

71. $0 = \dfrac{(x-2)^2}{x-2} \Rightarrow$ there is no x-intercept. There

is a hole at $(2, 0)$. $\dfrac{(0-2)^2}{0-2} = -2 \Rightarrow y = -2$ is

the y-intercept. There are no vertical asymptotes. There are no horizontal asymptotes. The intervals to be tested are $(-\infty, 2)$ and $(2, \infty)$. The graph is above the x-axis on $(2, \infty)$ and below the x-axis on $(-\infty, 2)$.

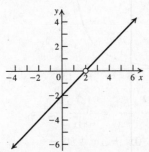

73. The x-intercept is 1 and the vertical asymptote is $x = 2$. The horizontal asymptote is $y = -2$, so the degree of the numerator equals the degree of the denominator, and the ratio of the leading terms of the numerator and the denominator is -2. Thus, the equation is of the form $y = a\left(\dfrac{x-1}{x-2}\right)$. The y-intercept is -1, so

we have $-1 = a\left(\dfrac{0-1}{0-2}\right) \Rightarrow a = -2$.

Thus, the equation is $f(x) = \dfrac{-2(x-1)}{x-2}$.

75. The x-intercepts are 1 and 3, and the vertical asymptotes are $x = 0$ and $x = 2$. The horizontal asymptote is $y = 1$, so the degree of the numerator equals the degree of the denominator, and the ratio of the leading terms of the numerator and the denominator is 1. Thus, the

equation is of the form $y = a\dfrac{(x-1)(x-3)}{x(x-2)}$.

There is no y-intercept, so the equation is

$f(x) = \dfrac{(x-1)(x-3)}{x(x-2)}$.

77. $\dfrac{2x^2 + 1}{x} = 2x + \dfrac{1}{x}$.

The oblique asymptote is $y = 2x$.

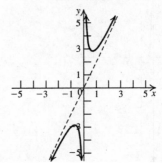

79. $\dfrac{x^3 - 1}{x^2} = x - \dfrac{1}{x^2}$.

The oblique asymptote is $y = x$.

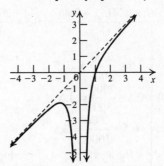

81.
$$
\begin{array}{r}
x - 2 \\
x + 1 \overline{\smash{\big)}\ x^2 - x + 1} \\
\underline{x^2 + x} \\
-2x + 1 \\
\underline{-2x - 2} \\
3
\end{array}
$$

The oblique asymptote is $y = x - 2$.

(continued on next page)

(*continued*)

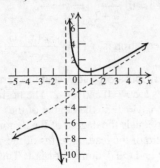

83.

$$x^2 - 1 \overline{\smash{\big)}\, x^3 - 2x^2 + 0x + 1}$$

with quotient $x - 2$

$$\underline{x^3 \qquad - x}$$
$$-2x^2 + x + 1$$
$$\underline{-2x^2 \qquad + 2}$$
$$x - 1$$

The oblique asymptote is $y = x - 2$. Note that there is a hole in the graph at $x = 1$.

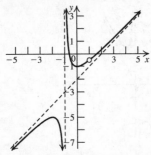

3.6 Applying the Concepts

85. a. $C(x) = 0.5x + 2000$

b. $\bar{C}(x) = \dfrac{C(x)}{x} = \dfrac{0.5x + 2000}{x} = 0.5 + \dfrac{2000}{x}$

c. $\bar{C}(100) = 0.5 + \dfrac{2000}{100} = 20.5$

$\bar{C}(500) = 0.5 + \dfrac{2000}{500} = 4.5$

$\bar{C}(1000) = 0.5 + \dfrac{2000}{1000} = 2.5$

These show the average cost of producing 100, 500, and 1000 trinkets, respectively.

d. The horizontal asymptote of $\bar{C}(x)$ is $y = 0.5$. It means that the average cost approaches the daily fixed cost of producing each trinket as the number of trinkets approaches ∞.

87. a.

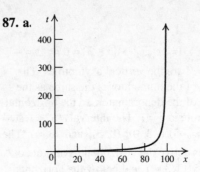

b. $f(50) = \dfrac{4(50) + 1}{100 - 50} \approx 4 \text{ min}$

$f(75) = \dfrac{4(75) + 1}{100 - 75} \approx 12 \text{ min}$

$f(95) = \dfrac{4(95) + 1}{100 - 95} \approx 76 \text{ min}$

$f(99) = \dfrac{4(99) + 1}{100 - 99} \approx 397 \text{ min}$

c. (i) As $x \to 100^-$, $f(x) \to \infty$.

(ii) The statement is not applicable because the domain is $x < 100$.

d. No, the bird doesn't ever collect all the seed from the field.

89. a. $C(50) = \dfrac{3(50^2) + 50}{50(100 - 50)} = \$3.02 \text{ billion}$

b.

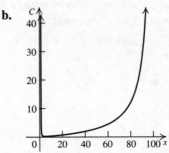

c. $30 = \dfrac{3x^2 + 50}{x(100 - x)} = \dfrac{3x^2 + 50}{100x - x^2} \Rightarrow$

$3000x - 30x^2 = 3x^2 + 50 \Rightarrow$

$-33x^2 + 3000x - 50 = 0 \Rightarrow$

$x = \dfrac{-3000 \pm \sqrt{3000^2 - 4(-33)(-50)}}{2(-33)}$

$= \dfrac{-3000 \pm \sqrt{8,993,400}}{-66} \approx \dfrac{-3000 \pm 2998.9}{-66}$

≈ 90.89 or -0.017. Reject the negative solution. Approximately 90.89% of the impurities can be removed at a cost of $30 billion.

91. a. $P(0) = \dfrac{8(0)+16}{2(0)+1} = 16 \text{ thousand} = 16,000$

b. The horizontal asymptote is $y = 4$. This means that the population will stabilize at 4000.

93. a. $f(x) = \dfrac{10x + 200,000}{x - 2500}$

b. $C(10,000) = \dfrac{10(10,000) + 200,000}{10,000 - 2500} = \$40$

c. $20 > \dfrac{10x + 200,000}{x - 2500} \Rightarrow$
$20x - 50,000 > 10x + 200,000 \Rightarrow$
$10x > 250,000 \Rightarrow x > 25,000$
More than 25,000 books must be sold to bring the average cost under \$20.

d. The vertical asymptote is $x = 2500$. This represents the number of free samples. The horizontal asymptote is $y = 10$. This represents the cost of printing and binding one book.

3.6 Beyond the Basics

95. Stretch the graph of $y = \dfrac{1}{x}$ vertically by a factor of 2, and then reflect the graph about the x-axis.

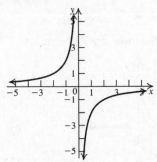

97. Shift the graph of $y = \dfrac{1}{x^2}$ two units right.

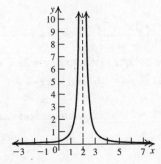

99. Shift the graph of $y = \dfrac{1}{x^2}$ one unit right and two units down.

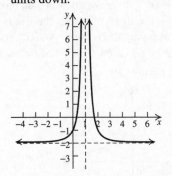

101. Shift the graph of $y = \dfrac{1}{x^2}$ six units left.

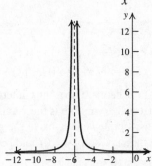

103. $x^2 - 2x + 1 \overline{\smash{\big)}\, x^2 - 2x + 2} \Rightarrow$
$ \underline{x^2 - 2x + 1}$
$ 1$

$\dfrac{x^2 - 2x + 2}{x^2 - 2x + 1} = \dfrac{(x^2 - 2x + 1) + 1}{x^2 - 2x + 1} =$
$\dfrac{x^2 - 2x + 1}{x^2 - 2x + 1} + \dfrac{1}{x^2 - 2x + 1} = 1 + \dfrac{1}{(x-1)^2}$

Shift the graph of $y = \dfrac{1}{x^2}$ one unit right and one unit up.

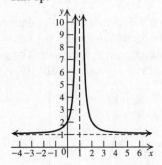

105. a. $f(x) \to 0$ as $x \to -\infty$; $f(x) \to 0$ as $x \to \infty$; $f(x) \to -\infty$ as $x \to 0^-$; $f(x) \to \infty$ as $x \to 0^+$. There are no x- or y-intercepts. The horizontal asymptote is the x-axis. The vertical asymptote is the y-axis. The graph is above the x-axis on $(0, \infty)$ and below the x-axis on $(-\infty, 0)$.

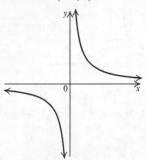

b. $f(x) \to 0$ as $x \to -\infty$; $f(x) \to 0$ as $x \to \infty$; $f(x) \to \infty$ as $x \to 0^-$; $f(x) \to -\infty$ as $x \to 0^+$. There are no x- or y-intercepts. The horizontal asymptote is the x-axis. The vertical asymptote is the y-axis. The graph is above the x-axis on $(-\infty, 0)$ and below the x-axis on $(0, \infty)$.

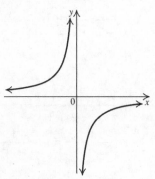

c. $f(x) \to 0$ as $x \to -\infty$; $f(x) \to 0$ as $x \to \infty$; $f(x) \to \infty$ as $x \to 0^-$; $f(x) \to \infty$ as $x \to 0^+$. There are no x- or y-intercepts. The horizontal asymptote is the x-axis. The vertical asymptote is the y-axis. The graph is above the x-axis on $(-\infty, 0) \cup (0, \infty)$.

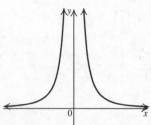

d. $f(x) \to 0$ as $x \to -\infty$; $f(x) \to 0$ as $x \to \infty$; $f(x) \to -\infty$ as $x \to 0^-$; $f(x) \to -\infty$ as $x \to 0^+$. There are no x- or y-intercepts. The horizontal asymptote is the x-axis. The vertical asymptote is the y-axis. The graph is below the x-axis on $(-\infty, 0) \cup (0, \infty)$.

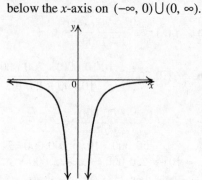

107.

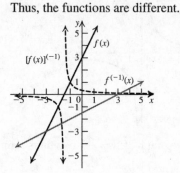

109. $f(x) = 2x + 3 \Rightarrow [f(x)]^{-1} = \dfrac{1}{2x + 3}$ and $y = 2x + 3$ becomes $x = 2y + 3 \Rightarrow$

$$\frac{x-3}{2} = \frac{x}{2} - \frac{3}{2} = y = f^{-1}(x).$$

Thus, the functions are different.

111. $g(x) = \dfrac{2x^3 + 3x^2 + 2x - 4}{x^2 - 1} = 2x + 3 + \dfrac{-1}{x^2 - 1}$

$g(x)$ has the oblique asymptote $y = 2x + 3$.

$g(x) \to -\infty$ as $x \to -\infty$, and $g(x) \to \infty$ as $x \to \infty$.

(continued on next page)

(*continued*)

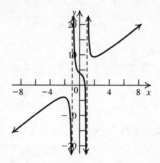

113. The horizontal asymptote is $y = 1$.

$$\frac{x^2 + x - 2}{x^2 - 2x - 3} = 1 \Rightarrow x^2 + x - 2 = x^2 - 2x - 3 \Rightarrow$$

$$3x = -1 \Rightarrow x = -\frac{1}{3}.$$

The point of intersection is $\left(-\frac{1}{3}, 1\right)$.

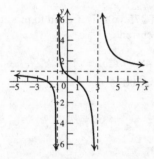

115. Vertical asymptote at $x = 3 \Rightarrow$ the denominator is $x - 3$. Horizontal asymptote at $y = -1 \Rightarrow$ the ratio of the leading coefficients of the numerator and denominator $= -1$. The x–intercept $= 2 \Rightarrow$ the numerator is $x - 2$. So the equation is of the form

$$f(x) = -\frac{x - 2}{x - 3} \text{ or } \frac{2 - x}{x - 3}.$$

117. Vertical asymptote at $x = 0 \Rightarrow$ the denominator is x. The slant asymptote $y = -x \Rightarrow$ the degree of the numerator is one more than the degree of the denominator and the quotient of the numerator and the denominator is $-x$. The x-intercepts -1 and $1 \Rightarrow$ the numerator is $(x - 1)(x + 1)$. So the equation is of the form

$$f(x) = a\frac{(x - 1)(x + 1)}{x} = a\left(\frac{x^2 - 1}{x}\right).$$

$$x\overline{\smash{\big)}x^2 - 1}$$
$$\underline{x^2}$$
$$-1$$

However, the slant asymptote is $y = -x$, so $a = -1$ and the equation is

$$f(x) = -\frac{x^2 - 1}{x} = \frac{-x^2 + 1}{x}.$$

119. Vertical asymptote at $x = 2 \Rightarrow$ the denominator is $x - 2$. Horizontal asymptote at $y = 1 \Rightarrow$ the degrees of the numerator and denominator are the same and the leading coefficients of the numerator and denominator are the same. So the numerator is $x + a$, where a is chosen so that

$$f(0) = -2 \Rightarrow \frac{x + 4}{x - 2} = f(x).$$

121. $f(x) \to -\infty$ as $x \to 1^-$ and $f(x) \to \infty$ as $x \to 1^+ \Rightarrow$ the denominator is zero if $x = 1$. Since $x \to 1$ from both directions, the denominator is $(x - 1)^2$. $f(x) \to 4$ as $x \to \pm\infty \Rightarrow$ the horizontal asymptote is 4. So the leading coefficient of the numerator is 4 and the degree of the numerator is the same as the degree of the denominator. The numerator is $4x^2 + a$, where a is chosen so that

$$f\left(\frac{1}{2}\right) = 0 \Rightarrow \frac{4(1/2)^2 + a}{(1/2 - 1)^2} \Rightarrow a = -1. \text{ So,}$$

$$f(x) = \frac{4x^2 - 1}{(x - 1)^2}.$$

123. Vertical asymptote at $x = 1 \Rightarrow$ the denominator could be $x - 1$. Because the oblique asymptote is $y = 3x + 2$, the numerator is $(3x + 2)(x - 1) + a$, where a can be any number (no x-intercepts or function values are given). Let $a = 1$. So.

$$f(x) = \frac{(3x + 2)(x - 1) + 1}{x - 1} = \frac{3x^2 - x + 1}{x - 1}.$$

3.6 Critical Thinking/Discussion/Writing

125. Answers may vary. Sample answers are given:

a. $f(x) = \frac{1}{x}$ **b.** $f(x) = \frac{x^2 + 2}{x^2 + x - 2}$

c. $f(x) = \frac{x^3 + 2x^2 - 7x - 1}{x^3 + x^2 - 6x + 5}$

126. Answers may vary. Sample answers are given:

a. $f(x) = \frac{x^2 + 1}{x}$

b. $f(x) = \dfrac{x^2 + 2x - 2}{x}$

c. $f(x) = \dfrac{x^5 + x^4 + x^2 - 4}{x^4}$

127. Since $y = x + 1$ is the oblique asymptote, we know that $R(x) = x + 1 + \dfrac{r(x)}{D(x)}$. We are told that the graph of $R(x)$ intersects the asymptote at the points $(2, R(2))$ and $(5, R(5))$, so $R(x) = x + 1$ when $\dfrac{r(x)}{D(x)} = 0$.

Thus, $r(x) = (x - 2)(x - 5)$. Recall that the numerator of a rational function must have degree greater than that of the denominator in order for there to be an oblique asymptote, so the denominator, $D(x)$, can be any polynomial whose degree is greater than or equal to 3. Thus, a possible rational function is

$$R(x) = x + 1 + \dfrac{(x-2)(x-5)}{x^5}$$
$$= \dfrac{(x+1)x^5 + (x-2)(x-5)}{x^5}.$$

128. See exercise 127 for explanation.

$$R(x) = (ax + b) + \dfrac{K(x - c_1)(x - c_2)\cdots(x - c_n)}{D(x)}$$
$$= \dfrac{(ax + b) + D(x) + K(x - c_1)(x - c_2)\cdots(x - c_n)}{D(x)},$$

where $K \neq 0$, $D(x)$ is a polynomial of degree $n + 1$ and none of its zeros are at $c_1, c_2, \ldots, c_n$.

3.6 Maintaining Skills

129. a. $y - 3 = -\dfrac{2}{3}(x - (-5)) \Rightarrow y - 3 = -\dfrac{2}{3}(x + 5) \Rightarrow$

$y = -\dfrac{2}{3}x - \dfrac{2}{3}(5) + 3 \Rightarrow y = -\dfrac{2}{3}x - \dfrac{1}{3}$

b. $y = -\dfrac{2}{3}(1) - \dfrac{1}{3} = -1$

131. $3(1) + 2c = 11 \Rightarrow 2c = 8 \Rightarrow c = 4$

133. $12 = \dfrac{k}{\left(\dfrac{1}{2}\right)^2} \Rightarrow k = 3$

$y = \dfrac{3}{2^2} \Rightarrow y = \dfrac{3}{4}$

3.7 Variation

3.7 Practice Problems

1. $y = kx \Rightarrow 6 = 30k \Rightarrow \dfrac{1}{5} = k$

$y = \left(\dfrac{1}{5}\right)120 = 24$

2. $I = kV \Rightarrow 60 = 220k \Rightarrow \dfrac{3}{11} = k$

$75 = \left(\dfrac{3}{11}\right)V \Rightarrow 275 = V$

A battery of 275 volts is needed to produce 60 amperes of current.

3. $y = kx^2 \Rightarrow 48 = k(2)^2 \Rightarrow 12 = k$

$y = 12(5)^2 = 300$

4. $A = \dfrac{k}{B} \Rightarrow 12 = \dfrac{k}{5} \Rightarrow 60 = k$

$A = \dfrac{60}{3} = 20$

5. $y = \dfrac{k}{\sqrt{x}} \Rightarrow \dfrac{3}{4} = \dfrac{k}{\sqrt{16}} \Rightarrow k = 3$

$2 = \dfrac{3}{\sqrt{x}} \Rightarrow \sqrt{x} = \dfrac{3}{2} \Rightarrow x = \dfrac{9}{4}$

6. $F = G \cdot \dfrac{m_1 m_2}{r^2} \Rightarrow m \cdot g = G \cdot \dfrac{mM_{\text{Mars}}}{R_{\text{Mars}}^2} \Rightarrow$

$g = G \cdot \dfrac{M_{\text{Mars}}}{R_{\text{Mars}}^2}$

Note that the radius of Mars is given in kilometers, which must be converted to meters.

$g = \dfrac{\left(6.67 \times 10^{-11} \text{ m}^3/\text{kg}/\text{sec}^2\right)\left(6.42 \times 10^{23} \text{ kg}\right)}{(3397 \text{ km})^2}$

$= \dfrac{\left(6.67 \times 10^{-11} \text{ m}^3/\text{kg}/\text{sec}^2\right)\left(6.42 \times 10^{23} \text{ kg}\right)}{\left(3.397 \times 10^6 \text{ m}\right)^2}$

$\approx 3.7 \text{ m/sec}^2$

3.7 Basic Concepts and Skills

1. $P = kT$

3. $y = k\sqrt{x}$

5. $V = kx^3 y^4$

7. $z = \dfrac{kxu}{v^2}$

9. $x = ky; 15 = 30k \Rightarrow k = \dfrac{1}{2}; x = \dfrac{1}{2}(28) = 14$

11. $s = kt^2; 64 = 2^2 k \Rightarrow k = 16; s = 5^2(16) = 400$

13. $r = \dfrac{k}{u}; 3 = \dfrac{k}{11} \Rightarrow k = 33; r = \dfrac{33}{1/3} = 99$

15. $B = \dfrac{k}{A^3}; 1 = \dfrac{k}{2^3} \Rightarrow k = 8; B = \dfrac{8}{4^3} = \dfrac{1}{8}$

17. $z = kxy; 42 = (2)(3)k \Rightarrow k = 7; 56 = 2(7)y \Rightarrow y = 4$

19. $z = kx^2; 32 = 4^2 k \Rightarrow k = 2; z = 2(5^2) = 50$

21. $P = kTQ^2; 36 = (17)(6^2)k \Rightarrow k = \dfrac{1}{17};$
$P = \dfrac{1}{17}(4)(9^2) = \dfrac{324}{17}$

23. $z = \dfrac{k\sqrt{x}}{y^2}; 24 = \dfrac{\sqrt{16}k}{3^2} \Rightarrow k = 54$
$27 = \dfrac{54\sqrt{x}}{2^2} \Rightarrow 2 = \sqrt{x} \Rightarrow x = 4$

25. $\dfrac{16}{12} = \dfrac{8}{y} \Rightarrow y = 6$

27. $\dfrac{100}{x_0} = \dfrac{y}{2x_0} \Rightarrow y = 200$

3.7 Applying the Concepts

29. $y = Hx$, where y is the speed of the galaxies and x is the distance between them.

31. a. $y = 30.5x$, where y is the length in centimeters and x is the length in feet.

b. (i) $y = 30.5(8) = 244$ cm

(ii) $y = 30.5\left(5\dfrac{1}{3}\right) \approx 162.67$ cm

c. (i) $57 = 30.5x \Rightarrow x \approx 1.87$ ft
(ii) $124 = 30.5x \Rightarrow x \approx 4.07$ ft

33. $P = kQ; 7 = 20k \Rightarrow \dfrac{7}{20} = k; P = \dfrac{7}{20}(100) = 35$ g

35. $d = kt^2; 64 = 2^2 k \Rightarrow k = 16; 9 = 16t^2 \Rightarrow$
$t = \dfrac{3}{4}$ sec

37. $P = \dfrac{k}{V}; 20 = \dfrac{k}{300} \Rightarrow k = 6000;$
$P = \dfrac{6000}{100} = 60$ lb/in.^{2}

39. a. The astronaut is $6000 + 3960$ miles from the Earth's center.
$W = \dfrac{k}{d^2}; 120 = \dfrac{k}{3960^2} \Rightarrow k = 120(3960^2)$
$W = \dfrac{120(3960^2)}{(6000 + 3960)^2} \approx 18.97$ lb

b. $200 = \dfrac{k}{3960^2} \Rightarrow k = 200(3960^2)$
$W = \dfrac{200(3960^2)}{3950^2} \approx 201.01$ lb

41. 1740 km $= 1,740,000$ m $= 1.740 \times 10^6$ m
$g = G \cdot \dfrac{M}{R^2} = 6.67 \times 10^{-11} \times \dfrac{7.4 \times 10^{22}}{\left(1.740 \times 10^6\right)^2}$
$= \dfrac{(6.67)(7.4)(10^{11})}{1.740^2 \times 10^{12}} \approx 1.63$ m/sec^2

43. a. $I = \dfrac{k}{d^2}; 320 = \dfrac{k}{10^2} \Rightarrow k = 32,000$
$I = \dfrac{32,000}{5^2} = 1280$ candlepower

b. $400 = \dfrac{32,000}{d^2} \Rightarrow d^2 = 80 \Rightarrow d \approx 8.94$ ft from the source.

45. $p = k\sqrt{l} \Rightarrow k = \dfrac{p}{\sqrt{l}} = \dfrac{2p}{2\sqrt{l}} = \dfrac{2p}{\sqrt{4l}} \Rightarrow$ the length is multiplied by 4 if the period is doubled.

47. a. $H = kR^2 N$, where k is a constant

b. $k(2R)^2 N = 4kR^2 N = 4H \Rightarrow$ the horsepower is multiplied by 4 if the radius is doubled.

c. $kR^2(2N) = 2kR^2 N = 2H \Rightarrow$ the horsepower is doubled if the number of pistons is doubled.

d. $k\left(\dfrac{R}{2}\right)^2(2N) = \dfrac{2kR^2N}{4} = \dfrac{kR^2N}{2} = \dfrac{H}{2} \Rightarrow$

the horsepower is halved if the radius is halved and the number of pistons is doubled.

3.7 Beyond the Basics

49. a. $E = kl^2v^3$

b. $1920 = k(10^2)(8^3) \Rightarrow k = \dfrac{3}{80} = 0.0375$

c. $E = 0.0375(8^2)(25^3) = 37{,}500$ watts

d. $kl^2(2v)^3 = 8kl^2v^3 = 8E$

e. $k(2l)^2v^3 = 4kl^2v^3 = 4E$

f. $k(2l)^2(2v)^3 = 4(8)kl^2v^3 = 32E$

51. a. $m = kw^{3/4}; 75 = k(75^{3/4}) \Rightarrow$
$k = \sqrt[4]{3}\sqrt{5} \approx 2.94$

b. $m = 2.94(450^{3/4}) \approx 287.25$ watts

c. $k(4w)^{3/4} = 4^{3/4}kw^{3/4} \approx 2.83kw^{3/4} \approx 2.83\,m$

d. $250 \approx 2.94w^{3/4} \Rightarrow w^{3/4} \approx 85.034 \Rightarrow$
$w \approx 373.93$ kg

53. a. $T^2 = \dfrac{4\pi^2r^3}{G(M_1 + M_2)}$

b. Because gravity is in terms of cubic meters per kilogram per second squared, convert the distance from kilometers to meters:
$1.5 \times 10^8\,\text{km} = 1.5 \times 10^{11}\,\text{m}$.

$(3.15 \times 10^7)^2 \approx \dfrac{4\pi^2(1.5 \times 10^{11})^3}{6.67 \times 10^{-11}M_{\text{sun}}} \Rightarrow$

$9.9225 \times 10^{14} \approx \dfrac{4\pi^2(3.375) \times 10^{33}}{6.67 \times 10^{-11}M_{\text{sun}}} \Rightarrow$

$M_{\text{sun}} \approx \dfrac{133.24 \times 10^{24}}{6.67 \times 10^{-11} \times 9.9225 \times 10^{14}}$
$\approx 2.01 \times 10^{30}$ kg

55. a. $R = kN(P - N)$, where k is the constant of proportionality.

b. $45 = 1000(9000)k \Rightarrow k = 5 \times 10^{-6}$

c. $R = 5 \times 10^{-6}(5000)(5000) = 125$ people per day.

d. $100 = 5 \times 10^{-6}N(10{,}000 - N) \Rightarrow$
$20{,}000{,}000 = 10{,}000N - N^2 \Rightarrow$
$N^2 - 10{,}000N + 20{,}000{,}000 = 0 \Rightarrow$
$N = \dfrac{10{,}000 \pm \sqrt{10{,}000^2 - 4(1)(2 \times 10^7)}}{2(1)}$
$= \dfrac{10{,}000 \pm \sqrt{20{,}000{,}000}}{2} \approx 2764$ or 7236

3.7 Critical Thinking/Discussion/Writing

56. $I = \dfrac{kV}{R} \Rightarrow IR = kV$

$1.3I = \dfrac{kV}{1.2R} \Rightarrow 1.56IR = kV \Rightarrow$ the voltage
must increase by 56%.

57. a. $v = kw^2$. The diamond is cut into two pieces whose weights are $\dfrac{2w}{5}$ and $\dfrac{3w}{5}$. The value of the first piece is $\left(\dfrac{4}{25}\right)(1000) = \$160$, and the value of the second piece is $\left(\dfrac{9}{25}\right)(1000) = \$360$. The two pieces together are valued at \$520, a loss of \$480.

b. The stone is broken into three pieces whose weights are $\dfrac{5w}{25} = \dfrac{w}{5}, \dfrac{9w}{25}$, and $\dfrac{11w}{25}$. So, the values of the three pieces are
$\left(\dfrac{1}{5}\right)^2(25{,}000) = \$1000,$
$\left(\dfrac{9}{25}\right)^2(25{,}000) = \$3240,$ and
$\left(\dfrac{11}{25}\right)^2(25{,}000) = \$4840,$ respectively. The total value is \$9,080, a loss of \$15,920.

c. The weights of the pieces are

$\dfrac{w}{15}, \dfrac{2w}{15}, \dfrac{3w}{15} = \dfrac{w}{5}, \dfrac{4w}{15}$, and $\dfrac{5w}{15} = \dfrac{w}{3}$,

respectively. If the original value is x, then

$$x - 85{,}000 = \left(\dfrac{1}{15}\right)^2 x + \left(\dfrac{2}{15}\right)^2 x + \left(\dfrac{1}{5}\right)^2 x$$
$$+ \left(\dfrac{4}{15}\right)^2 x + \left(\dfrac{1}{3}\right)^2 x \Rightarrow$$

$x - 85{,}000 = \dfrac{11}{45} x \Rightarrow x = \$112{,}500 = $ the

original value of the diamond. A diamond whose weight is twice that of the original diamond is worth 4 times the value of the original diamond = $450,000.

58. $p = kd(n - f)$

$80 = k(40)(30 - f)$ and

$180 = k(60)(35 - f)$. Solving the first

equation for k we have $k = -\dfrac{2}{f - 30}$.

Substitute that value into the second equation and solve for f:

$$180 = \left(\dfrac{2}{30 - f}\right)(60)(35 - f) \Rightarrow$$

$$180 = \dfrac{4200 - 120f}{30 - f} \Rightarrow$$

$5400 - 180f = 4200 - 120f \Rightarrow$
$1200 = 60f \Rightarrow f = 20$

59. $w_{\text{solid}} = kr_o{}^3; w_{\text{hollow}} = kr_o{}^3 - kr_i{}^3 = \dfrac{7}{8} kr_o{}^3$

$kr_o{}^3 - kr_i{}^3 = \dfrac{7}{8} kr_o{}^3 \Rightarrow r_o{}^3 - r_i{}^3 = \dfrac{7}{8} r_o{}^3 \Rightarrow$

$\dfrac{1}{8} r_o{}^3 = r_i{}^3 \Rightarrow \dfrac{1}{8} = \dfrac{r_i{}^3}{r_o{}^3} \Rightarrow \dfrac{1}{2} = \dfrac{r_i}{r_o}$

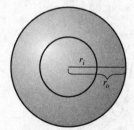

60. $s = 24 - k\sqrt{w}; 20 = 24 - k\sqrt{4} \Rightarrow k = 2$

$0 \le 24 - 2\sqrt{w} \Rightarrow w \le 144$

The greatest number of wagons the engine can move is 144.

3.7 Maintaining Skills

61. $5^0 = 1$ **63.** $3^{-2} = \dfrac{1}{3^2} = \dfrac{1}{9}$

65. $\left(\dfrac{1}{2}\right)^{-4} = 2^4 = 16$

67. $5^{2x-3} \cdot 5^{3-x} = 5^{(2x-3)+(3-x)} = 5^x$

69. $\left(2^{x-1}\right)^x = 2^{(x-1)(x)} = 2^{x^2 - x}$

71. $y = mx + b \Rightarrow y - b = mx \Rightarrow m = \dfrac{y - b}{x}$

73. $A = B(1 + C) \Rightarrow \dfrac{A}{B} = 1 + C \Rightarrow \dfrac{A}{B} - 1 = C$

75. $A = B \cdot 10^{-n} \Rightarrow \dfrac{A}{10^{-n}} = B \Rightarrow A \cdot 10^n = B$

Chapter 3 Review Exercises

Basic Concepts and Skills

1. (i)   Opens up   **(ii)** Vertex: $(1, 2)$
    (iii) Axis: $x = 1$
    (iv) $0 = (x - 1)^2 + 2 \Rightarrow -2 = (x - 1)^2 \Rightarrow$
        there are no x-intercepts.
    (v) $y = (0 - 1)^2 + 2 \Rightarrow y = 3$ is the
        y-intercept.
    (vi) The function is decreasing on $(-\infty, 1)$
        and increasing on $(1, \infty)$.

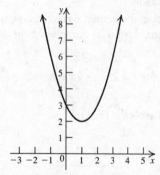

3. (i)   Opens down **(ii)** Vertex: $(3, 4)$
    (iii) Axis: $x = 3$
    (iv) $0 = -2(x - 3)^2 + 4 \Rightarrow 2 = (x - 3)^2 \Rightarrow$
        $x = 3 \pm \sqrt{2}$ are the x-intercepts.
    (v) $y = -2(0 - 3)^2 + 4 \Rightarrow y = -14$ is the
        y-intercept.

(vi) The function is increasing on $(-\infty, 3)$ and decreasing on $(3, \infty)$.

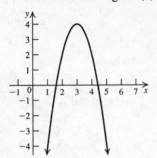

5. **(i)** Opens down **(ii)** Vertex: (0, 3)
   (iii) Axis: $x = 0$ (y-axis)
   (iv) $0 = -2x^2 + 3 \Rightarrow \dfrac{3}{2} = x^2 \Rightarrow$
   $x = \pm\dfrac{\sqrt{6}}{2}$ are the x-intercepts.
   (v) $y = -2(0)^2 + 3 \Rightarrow y = 3$ is the y-intercept.
   (vi) The function is increasing on $(-\infty, 0)$ and decreasing on $(0, \infty)$.

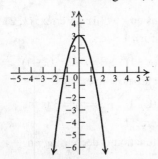

7. **(i)** Opens up
   (ii) To find the vertex, write the equation in standard form by completing the square:
   $$y = 2x^2 - 4x + 3$$
   $$y - 3 + 2 = 2(x^2 - 2x + 1)$$
   $$y = 2(x - 1)^2 + 1$$
   The vertex is (1, 1).
   (iii) Axis: $x = 1$
   (iv) $0 = 2x^2 - 4x + 3 \Rightarrow$
   $x = \dfrac{4 \pm \sqrt{(-4)^2 - 4(2)(3)}}{2(2)} = \dfrac{4 \pm \sqrt{-8}}{4} \Rightarrow$
   there are no x-intercepts.
   (v) $y = 2(0)^2 - 4(0) + 3 \Rightarrow y = 3$ is the y-intercept.

(vi) The function is decreasing on $(-\infty, 1)$ and increasing on $(1, \infty)$.

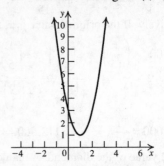

9. **(i)** Opens up
   (ii) To find the vertex, write the equation in standard form by completing the square.
   $$y = 3x^2 - 2x + 1$$
   $$y - 1 + \frac{1}{3} = 3\left(x^2 - \frac{2}{3}x + \frac{1}{9}\right)$$
   $$y = 3\left(x - \frac{1}{3}\right)^2 + \frac{2}{3}$$
   The vertex is $\left(\dfrac{1}{3}, \dfrac{2}{3}\right)$.
   (iii) Axis: $x = \dfrac{1}{3}$
   (iv) $0 = 3x^2 - 2x + 1 \Rightarrow$
   $x = \dfrac{2 \pm \sqrt{(-2)^2 - 4(3)(1)}}{2(3)} = \dfrac{2 \pm \sqrt{-8}}{6} \Rightarrow$
   there are no x-intercepts.
   (v) $y = 3(0)^2 - 3(0) + 1 \Rightarrow y = 1$ is the y-intercept.
   (vi) The function is decreasing on $\left(-\infty, \dfrac{1}{3}\right)$ and increasing on $\left(\dfrac{1}{3}, \infty\right)$.

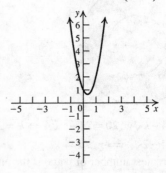

In exercises 11–14, find the vertex using the formula $\left(-\dfrac{b}{2a}, f\left(-\dfrac{b}{2a}\right)\right)$.

11. $a > 0 \Rightarrow$ the graph opens up, so f has a minimum value at its vertex. The vertex is

$$\left(-\frac{-4}{2(1)}, f\left(-\frac{-4}{2(1)}\right)\right) = (2, -1).$$

The minimum value is -1.

13. $a < 0 \Rightarrow$ the graph opens down, so f has a maximum value at its vertex. The vertex is

$$\left(-\frac{-3}{2(-2)}, f\left(-\frac{-3}{2(-2)}\right)\right) = \left(-\frac{3}{4}, \frac{25}{8}\right).$$

The maximum value is $\frac{25}{8}$.

15. Shift the graph of $y = x^3$ one unit left and two units down.

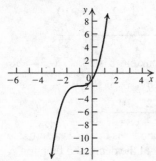

17. Shift the graph of $y = x^3$ one unit right, reflect the resulting graph in the y-axis, and shift it one unit up.

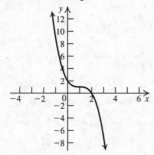

19. **(i)** $f(x) \to -\infty$ as $x \to -\infty$
$f(x) \to \infty$ as $x \to \infty$

(ii) Zeros: $x = -2$, multiplicity 1, crosses the x-axis; $x = 0$, multiplicity 1, crosses the x-axis; $x = 1$, multiplicity 1, crosses the x-axis.

(iii) x-intercepts: $-2, 0, 1$;
$y = 0(0-1)(0+2) \Rightarrow y = 0$ is the y-intercept.

(iv) The intervals to be tested are $(-\infty, -2)$, $(-2, 0), (0, 1), (1, 2),$ and $(2, \infty)$. The graph is above the x-axis on $(-2, 0) \cup (1, \infty)$ and below the x-axis on $(-\infty, -2) \cup (0, 1)$.

(v) $f(-x) = -x(-x-1)(-x+2) \neq f(x) \Rightarrow$ f is not even.
$-f(x) = -(x(x-1)(x+2)) \neq f(-x) \Rightarrow$ f is not odd. There are no symmetries.

(vi)

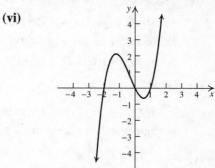

21. **(i)** $f(x) \to -\infty$ as $x \to -\infty$
$f(x) \to -\infty$ as $x \to \infty$

(ii) Zeros: $x = 0$, multiplicity 2, touches but does not cross the x- axis; $x = 1$, multiplicity 2, touches but does not cross the x- axis.

(iii) x-intercepts: 0, 1;
$y = -0^2(0-1)^2 \Rightarrow y = 0$ is the y-intercept.

(iv) The intervals to be tested are $(-\infty, 0)$, $(0, 1),$ and $(1, \infty)$. The graph is below the x-axis on $(-\infty, 0) \cup (0, 1) \cup (1, \infty)$.

(v) $f(-x) = -(-x)^2(-x-1)$
$= -x^2(-x-1) \neq -f(x)$ or $f(x) \Rightarrow$
$= -f(x) \Rightarrow f$ is neither even nor odd. There are no symmetries.

(vi)

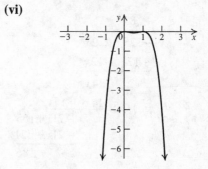

23. **(i)** $f(x) \to -\infty$ as $x \to -\infty$

$f(x) \to -\infty$ as $x \to \infty$

(ii) Zeros: $x = -1$, multiplicity 1, crosses the x-axis; $x = 0$, multiplicity 2, touches but does not cross; $x = 1$, multiplicity 1, crosses the x-axis.

(iii) x-intercepts: $-1, 0, 1$;

$y = -0^2(0^2 - 1) \Rightarrow y = 0$ is the y-intercept.

(iv) The intervals to be tested are $(-\infty, -1)$, $(-1, 0), (0, 1),$ and $(1, \infty)$. The graph is above the x-axis on $(-1, 0) \cup (0, 1)$ and below the x-axis on $(-\infty, -1) \cup (1, \infty)$.

(v) $f(-x) = -(-x)^2((-x)^2 - 1)$

$= -x^2(x^2 - 1) = f(x) \Rightarrow f$ is even, and f is symmetric with respect to the y-axis.

(vi)

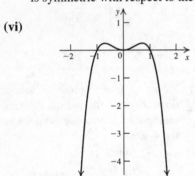

25.

$$\require{enclose}\begin{array}{r} 2x+3 \\ 3x-2 \enclose{longdiv}{6x^2+5x-13} \\ \underline{6x^2-4x} \\ 9x-13 \\ \underline{9x-6} \\ -7 \end{array}$$

27.

$$\begin{array}{r} 8x^3-12x^2+14x-21 \\ x+1\enclose{longdiv}{8x^4-4x^3+2x^2-7x+165} \\ \underline{8x^4+8x^3} \\ -12x^3+2x^2 \\ \underline{-12x^3-12x^2} \\ 14x^2-7x \\ \underline{14x^2+14x} \\ -21x+165 \\ \underline{-21x-21} \\ 186 \end{array}$$

29.

$$\begin{array}{r|rrrr} 3 & 1 & 0 & -12 & 3 \\ & & 3 & 9 & -9 \\ \hline & 1 & 3 & -3 & -6 \end{array}$$

Quotient: $x^2 + 3x - 3$ remainder -6.

31.

$$\begin{array}{r|rrrrr} -1 & 2 & -3 & 5 & -7 & 165 \\ & & -2 & 5 & -10 & 17 \\ \hline & 2 & -5 & 10 & -17 & 182 \end{array}$$

Quotient: $2x^3 - 5x^2 + 10x - 17$ remainder 182.

33. (i) $f(2) = 2^3 - 3(2^2) + 11(2) - 29 = -11$

(ii)

$$\begin{array}{r|rrrr} 2 & 1 & -3 & 11 & -29 \\ & & 2 & -2 & 18 \\ \hline & 1 & -1 & 9 & -11 \end{array}$$

The remainder is -11, so $f(2) = -11$.

35. (i) $f(2) = (-3)^4 - 2(-3)^2 - 5(-3) + 10 = 88$

(ii)

$$\begin{array}{r|rrrrr} -3 & 1 & 0 & -2 & -5 & 10 \\ & & -3 & 9 & -21 & 78 \\ \hline & 1 & -3 & 7 & -26 & 88 \end{array}$$

The remainder is 88, so $f(-3) = 88$.

37.

$$\begin{array}{r|rrrr} 2 & 1 & -7 & 14 & -8 \\ & & 2 & -10 & 8 \\ \hline & 1 & -5 & 4 & 0 \end{array}$$

The remainder is 0, so 2 is a zero. Now find the zeros of the depressed function $x^2 - 5x + 4 = (x - 4)(x - 1) \Rightarrow 4$ and 1 are also zeros. So the zeros of $x^3 - 7x^2 + 14x - 8$ are 1, 2, and 4.

39.

$$\begin{array}{r|rrrr} \frac{1}{3} & 3 & 14 & 13 & -6 \\ & & 1 & 5 & 6 \\ \hline & 3 & 15 & 18 & 0 \end{array}$$

The remainder is 0, so $1/3$ is a zero. Now find the zeros of the depressed function $3x^2 + 15x + 18 = 3(x + 2)(x + 3) \Rightarrow -2$ and -3 are also zeros. So the zeros of $3x^3 + 14x^2 + 13x - 6$ are $-3, -2$, and $1/3$.

41. The factors of the constant term are $\{\pm 1, \pm 2, \pm 3, \pm 6\}$, and the factors of the leading coefficient are $\{\pm 1\}$. So the possible rational zeros are $\{\pm 1, \pm 2, \pm 3, \pm 6\}$.

43. $f(x) = 5x^3 + 11x^2 + 2x;$

$f(-x) = 5(-x)^3 + 11(-x)^2 + 2(-x)$
$\quad = -5x^3 + 11x^2 - 2x$

There are no sign changes in $f(x)$, so there are no positive zeros. There are two sign changes in $f(-x)$, so there are 2 or 0 negative zeros.

$5x^3 + 11x^2 + 2x = x(5x^2 + 11x + 2)$

$= x(5x+1)(2+x) \Rightarrow$ the zeros are $-2, -\dfrac{1}{5}, 0.$

45. $f(x) = x^3 + 3x^2 - 4x - 12$

$f(-x) = (-x)^3 + 3(-x)^2 - 4(-x) - 12$
$\quad = -x^3 + 3x^2 + 4x - 12$

There is one sign change in $f(x)$, so there is one positive zero. There are two sign changes in $f(-x)$, so there are 2 or 0 negative zeros.

The possible rational zeros are $\{\pm1, \pm2, \pm3, \pm4, \pm6, \pm12\}$. Use synthetic division to find one zero:

$$\begin{array}{r|rrrr} -3 & 1 & 3 & -4 & -12 \\ & & -3 & 0 & 12 \\ \hline & 1 & 0 & -4 & 0 \end{array}$$

The remainder is 0, so -3 is a zero. Now find the zeros of the depressed function

$x^2 - 4 = (x-2)(x+2) \Rightarrow -2$ and 2 are also zeros. So the zeros of $x^3 + 3x^2 - 4x - 12$ are $-3, -2,$ and 2.

47. $f(x) = x^3 - 4x^2 - 5x + 14$

$f(-x) = (-x)^3 - 4(-x)^2 - 5(-x) + 14$
$\quad = -x^3 - 4x^2 + 5x + 14$

There are two sign changes in $f(x)$, so there are 0 or 2 real positive zeros. There is one sign change in $f(-x)$, so there is one real negative zero. The possible rational zeros are $\{\pm1, \pm2, \pm7, \pm14\}$. Use synthetic division to find one zero:

$$\begin{array}{r|rrrr} -2 & 1 & -4 & -5 & 14 \\ & & -2 & 12 & -14 \\ \hline & 1 & -6 & 7 & 0 \end{array}$$

The remainder is 0, so -2 is a zero. Now find the zeros of the depressed function

$x^2 - 6x + 7.$

$$x = \frac{-(-6) \pm \sqrt{(-6)^2 - 4(1)(7)}}{2(1)}$$

$$= \frac{6 \pm \sqrt{8}}{2} = \frac{6 \pm 2\sqrt{2}}{2} = 3 \pm \sqrt{2}$$

So, the zeros of $x^3 - 4x^2 - 5x + 14$ are -2, $3 - \sqrt{2}$, and $3 + \sqrt{2}.$

49. $f(x) = 2x^3 - 5x^2 - 2x + 2$

$f(-x) = 2(-x)^3 - 5(-x)^2 - 2(-x) + 2$
$\quad = -2x^3 - 5x^2 + 2x + 2$

There are two sign changes in $f(x)$, so there are 0 or 2 real positive zeros. There is one sign change in $f(-x)$, so there is one real negative zero. The possible rational zeros are $\left\{\pm1, \pm2, \pm\dfrac{1}{2}\right\}$. Use synthetic division to find one zero:

$$\begin{array}{r|rrrr} \frac{1}{2} & 2 & -5 & -2 & 2 \\ & & 1 & -2 & -2 \\ \hline & 2 & -4 & -4 & 0 \end{array}$$

The remainder is 0, so $1/2$ is a zero. Now find the zeros of the depressed function

$2x^2 - 4x - 4.$

$$x = \frac{-(-4) \pm \sqrt{(-4)^2 - 4(2)(-4)}}{2(2)}$$

$$= \frac{4 \pm \sqrt{48}}{4} = \frac{4 \pm 4\sqrt{3}}{4} = 1 \pm \sqrt{3}$$

So, the zeros of $2x^3 - 4x^2 - 2x + 2$ are $1/2$, $1 - \sqrt{3}$, and $1 + \sqrt{3}.$

51. The function has degree three, so there are three zeros. Use synthetic division to find the depressed function:

$$\begin{array}{r|rrrr} 2 & 1 & 0 & -7 & 6 \\ & & 2 & 4 & -6 \\ \hline & 1 & 2 & -3 & 0 \end{array}$$

Now find the zeros of the depressed function

$x^2 + 2x - 3 = (x+3)(x-1) \Rightarrow -3$ and 1 are zeros. The zeros of the original function are $-3, 1, 2.$

53. The function has degree four, so there are four zeros. Use synthetic division twice to find the depressed function:

$$\begin{array}{r} -1\,\underline{|}\;\;1\;\;-2\;\;\;6\;\;-18\;\;-27 \\ \;\;\;\;\;\;\;-1\;\;\;\;3\;\;\;-9\;\;\;\;27 \\ \hline 1\;\;-3\;\;\;\;9\;\;-27\;\;\;\;\;0 \end{array} \qquad \begin{array}{r} 3\,\underline{|}\;\;1\;\;-3\;\;\;\;9\;\;-27 \\ \;\;\;\;\;\;\;\;\;\;\;3\;\;\;\;0\;\;\;\;27 \\ \hline 1\;\;\;\;0\;\;\;\;9\;\;\;\;\;0 \end{array}$$

Alternatively, divide

$x^4 - 2x^3 + 6x^2 - 18x - 27$ by

$(x+1)(x-3) = x^2 - 2x - 3$.

Now find the zeros of the depressed function

$x^2 + 9 \Rightarrow \pm 3i$ are zeros. The zeros of the original function are -1, 3, and $\pm 3i$.

55. The function has degree four, so there are four zeros. Since one zero is $-1 + 2i$, another zero is $-1 - 2i$. Divide $x^4 + 2x^3 + 9x^2 + 8x + 20$ by $(x - (-1 + 2i))(x - (-1 - 2i)) = x^2 + 2x + 5$ to find the depressed function:

$$\begin{array}{r} x^2 + 4 \\ x^2 + 2x + 5 \overline{\big)\,x^4 + 2x^3 + 9x^2 + 8x + 20} \\ \underline{x^4 + 2x^3 + 5x^2 } \\ 4x^2 + 8x + 20 \\ \underline{4x^2 + 8x + 20} \\ 0 \end{array}$$

Now find the zeros of the depressed function:

$x^2 + 4 = 0 \Rightarrow x = \pm 2i$. The zeros of the original function are $\pm 2i$ and $-1 \pm 2i$.

57. The function has degree three, so there are three zeros. The possible rational zeros are $\{\pm 1, \pm 2, \pm 4\}$. Using synthetic division, we find that one zero is 1:

$$\begin{array}{r} 1\,\underline{|}\;\;1\;\;-1\;\;-4\;\;\;\;4 \\ \;\;\;\;\;\;\;\;\;\;1\;\;\;\;0\;\;-4 \\ \hline 1\;\;\;\;0\;\;-4\;\;\;\;0 \end{array}$$

Now find the zeros of the depressed function:

$x^2 - 4 = 0 \Rightarrow x = \pm 2$. The solution set is $\{-2, 1, 2\}$.

59. The function has degree three, so there are three zeros. The possible rational zeros are $\left\{\pm\dfrac{1}{4}, \pm\dfrac{1}{2}, \pm\dfrac{3}{4}, \pm 1, \pm\dfrac{3}{2}, \pm 3\right\}$. Using synthetic division, we find that one zero is -1:

$$\begin{array}{r} -1\,\underline{|}\;\;4\;\;\;\;0\;\;-7\;\;-3 \\ \;\;\;\;\;\;\;\;\;\;\;-4\;\;\;\;4\;\;\;\;3 \\ \hline 4\;\;-4\;\;-3\;\;\;\;0 \end{array}$$

Now find the zeros of the depressed function:

$4x^2 - 4x - 3 = 0 \Rightarrow (2x - 3)(2x + 1) = 0 \Rightarrow$

$x = \dfrac{3}{2}$ or $x = -\dfrac{1}{2}$.

The solution set is $\left\{-1/2, -1, 3/2\right\}$.

61. $x^3 - 8x^2 + 23x - 22 = 0$

The function has degree three, so there are three zeros. The possible rational zeros are $\{\pm 1, \pm 2, \pm 11, \pm 22\}$. Using synthetic division, we find that one zero is 2:

$$\begin{array}{r} 2\,\underline{|}\;\;1\;\;-8\;\;\;\;23\;\;-22 \\ \;\;\;\;\;\;\;\;\;\;2\;\;-12\;\;\;\;22 \\ \hline 1\;\;-6\;\;\;\;11\;\;\;\;\;0 \end{array}$$

Now find the zeros of the depressed function:

$x^2 - 6x + 11 = 0$.

$$x = \frac{-(-6) \pm \sqrt{(-6)^2 - 4(1)(11)}}{2(1)}$$

$$= \frac{6 \pm \sqrt{-8}}{2} = \frac{6 \pm 2i\sqrt{2}}{2} = 3 \pm i\sqrt{2}$$

Solution set: $\left\{2,\, 3 \pm i\sqrt{2}\right\}$

63. $3x^3 - 5x^2 + 16x + 6 = 0$

The function has degree three, so there are three zeros. The possible rational zeros are $\left\{\pm 1, \pm 2, \pm 3, \pm 6, \pm\dfrac{1}{3}, \pm\dfrac{2}{3}\right\}$. Using synthetic division, we find that one zero is $-1/3$:

$$\begin{array}{r} -\dfrac{1}{3}\,\underline{|}\;\;3\;\;-5\;\;\;\;16\;\;\;\;6 \\ \;\;\;\;\;\;\;\;\;\;\;\;-1\;\;\;\;2\;\;-6 \\ \hline 3\;\;-6\;\;\;\;18\;\;\;\;0 \end{array}$$

Now find the zeros of the depressed function:

$3x^2 - 6x + 18 = 0 \Rightarrow 3\left(x^2 - 2x + 6\right) = 0$.

$$x = \frac{-(-2) \pm \sqrt{(-2)^2 - 4(1)(6)}}{2(1)}$$

$$= \frac{2 \pm \sqrt{-20}}{2} = \frac{2 \pm 2i\sqrt{5}}{2} = 1 \pm i\sqrt{5}$$

Solution set: $\left\{-1/3,\, 1 \pm i\sqrt{5}\right\}$

65. $x^4 - x^3 - x^2 - x - 2 = 0$

The function has degree four, so there are four zeros. The possible rational zeros are $\{\pm 1, \pm 2\}$. Using synthetic division, we find that one zero is 2:

$$\begin{array}{r} 2\,\underline{|}\;\;1\;\;-1\;\;-1\;\;-1\;\;-2 \\ \;\;\;\;\;\;\;\;\;\;2\;\;\;\;2\;\;\;\;2\;\;\;\;2 \\ \hline 1\;\;\;\;1\;\;\;\;1\;\;\;\;1\;\;\;\;0 \end{array}$$

(continued on next page)

(*continued*)

Now find the zeros of the depressed function:
$x^3 + x^2 + x + 1 = 0$. We can use synthetic division or factor to find another zero:

$$x^3 + x^2 + x + 1 = 0$$
$$x^2(x+1) + (x+1) = 0$$
$$(x^2 + 1)(x+1) = 0$$

$x^2 + 1 = 0 \Rightarrow x^2 = -1 \Rightarrow x = \pm i$ or
$x + 1 = 0 \Rightarrow x = -1$
Solution set: $\{-1, 2, \pm i\}$

67. $2x^4 - x^3 - 2x^2 + 13x - 6 = 0$
The function has degree four, so there are four zeros. The possible rational zeros are
$\left\{\pm 1, \pm 2, \pm 3, \pm 6, \pm \dfrac{1}{2}, \pm \dfrac{3}{2}\right\}$. Using
synthetic division, we find that one zero is −2:

$$\begin{array}{r|rrrrr} -2 & 2 & -1 & -2 & 13 & -6 \\ & & -4 & 10 & -16 & 6 \\ \hline & 2 & -5 & 8 & -3 & 0 \end{array}$$

Now find the zeros of the depressed function:
$2x^3 - 5x^2 + 8x - 3 = 0$. Again using synthetic division, we find that one zero is 1/2 :

$$\begin{array}{r|rrrr} \frac{1}{2} & 2 & -5 & 8 & -3 \\ & & 1 & -2 & 3 \\ \hline & 2 & -4 & 6 & 0 \end{array}$$

Now find the zeros of the depressed function:
$2x^2 - 4x + 6 = 0 \Rightarrow 2(x^2 - 2x + 3) = 0$.

$$x = \frac{-(-2) \pm \sqrt{(-2)^2 - 4(1)(3)}}{2(1)}$$
$$= \frac{2 \pm \sqrt{-8}}{2} = \frac{2 \pm 2i\sqrt{2}}{2} = 1 \pm i\sqrt{2}$$

Solution set: $\left\{-2, \dfrac{1}{2}, 1 \pm i\sqrt{2}\right\}$.

69. The only possible rational roots are $\{\pm 1, \pm 2\}$. None of these satisfies the equation.

71. $f(1) = 1^3 + 6(1)^2 - 28 = -21;$
$f(2) = 2^3 + 6(2)^2 - 28 = 4$. Because the sign changes, there is a real zero between 1 and 2. The zero is approximately 1.88.

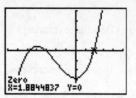

73. $1 + \dfrac{1}{x} = 0 \Rightarrow x = -1$ is the *x*-intercept. There is
no *y*-intercept. The vertical asymptote is the *y*-axis ($x = 0$). The horizontal asymptote is $y = 1$. Testing the intervals
$(-\infty, -1)$, $(-1, 0)$, and $(0, \infty)$, we find that the graph is above the *x*-axis on $(-\infty, -1) \cup (0, \infty)$ and below the *x*-axis on $(-1, 0)$.

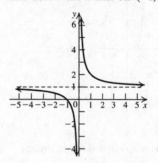

75. $\dfrac{x}{x^2 - 1} = 0 \Rightarrow x = 0$ is the *x*-intercept.

$\dfrac{0}{0^2 - 1} = 0 \Rightarrow y = 0$ is the *y*-intercept. The
vertical asymptotes are $x = 1$ and $x = -1$. The horizontal asymptote is the *x*-axis. Testing the intervals $(-\infty, -1), (-1, 0), (0, 1),$ and $(1, \infty)$, we find that the graph is above the *x*-axis on $(-\infty, -1) \cup (0, \infty)$ and below the *x*-axis on $(-1, 0) \cup (0, 1)$.

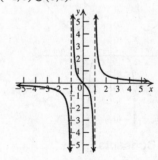

77. $\dfrac{x^3}{x^2-9}=0 \Rightarrow x=0$ is the x-intercept.

$\dfrac{0^3}{0^2-9}=0 \Rightarrow y=0$ is the y-intercept. The vertical asymptotes are $x=3$ and $x=-3$. There is no horizontal asymptote. The oblique asymptote is $y=x$. Testing the intervals $(-\infty,-3),(-3,0),(0,3),$ and $(3,\infty)$, we find that the graph is above the x-axis on $(-3,0)\cup(3,\infty)$ and below the x-axis on $(-\infty,-3)\cup(0,3)$.

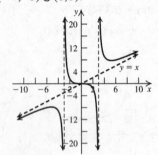

79. $\dfrac{x^4}{x^2-4}=0 \Rightarrow x=0$ is the x-intercept.

$\dfrac{0^4}{0^2-4}=0 \Rightarrow y=0$ is the y-intercept. The vertical asymptotes are $x=2$ and $x=-2$. There is no horizontal asymptote. Testing the intervals $(-\infty,-2),(-2,0),(0,2),$ and $(2,\infty)$, we find that the graph is above the x-axis on $(-\infty,-2)\cup(2,\infty)$ and below the x-axis on $(-2,0)\cup(0,2)$.

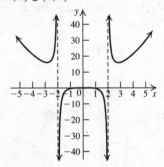

Applying the Concepts

81. $y=kx; 12=4k \Rightarrow k=3; y=3(5)=15$

83. $s=kt^2; 20=2^2 k \Rightarrow 5=k; s=3^2(5)=45$

85. The maximum height occurs at the vertex,

$$\left(-\dfrac{20}{2(-1/10)}, f\left(-\dfrac{20}{2(-1/10)}\right)\right)=(100,1000).$$

The maximum height is 1000. To find where the missile hits the ground, solve

$-\dfrac{1}{10}x^2+20x=0 \Rightarrow x=0$ or $x=200$. The missile hits the ground at $x=200$.

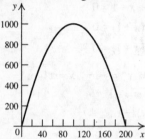

87. The area of each section is 400 square feet. Since the width is x, $y=\dfrac{400}{x}$. The total amount of fencing needed is $4x+\dfrac{2400}{x}$.

Using a graphing calculator, we find that this is a minimum at $x\approx24.5$ feet. So

$y\approx\dfrac{400}{24.5}\approx16.3$ feet.

The dimensions of each pen should be approximately 24.5 ft by 16.3 ft.

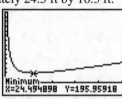

89. a.

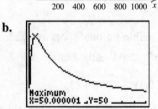

b.

The maximum occurs at $x=50$.

91. a. The revenue is $24x$. Profit = revenue – cost, so

$$P(x) = 24x - \left(150 + 3.9x + \frac{3}{1000}x^2\right)$$

$$= -\frac{3}{1000}x^2 + 20.1x - 150$$

b. The maximum occurs at the x-coordinate of the vertex: $-\dfrac{20.1}{2\left(-\dfrac{3}{1000}\right)} = 3350$

c. $\bar{C}(x) = \dfrac{C(x)}{x} = \dfrac{\dfrac{3}{1000}x^2 + 3.9x + 150}{x}$

$$= \frac{3}{1000}x + 3.9 + \frac{150}{x}$$

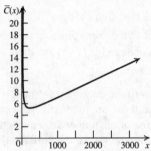

93. $280 = 40k \Rightarrow k = 7; s = 7(35) = \$245$

95. $I = \dfrac{ki}{d^2}$, where I = illumination, i = intensity, and d = distance from the source. The illumination 6 inches from the source is

$I_6 = \dfrac{300k}{6^2}$, while the illumination x inches

from the source is $I_x = \dfrac{300k}{x^2}$. $I_6 = 2I_x \Rightarrow$

$\dfrac{300k}{36} = 2\left(\dfrac{300k}{x^2}\right) \Rightarrow x^2 = 72 \Rightarrow x = 6\sqrt{2}$ in.

97. $I = \dfrac{k}{R}; 30 = \dfrac{k}{300} \Rightarrow k = 9000$

a. $I = \dfrac{9000}{250} = 36$ amp

b. $60 = \dfrac{9000}{R} \Rightarrow R = 150$ ohms

99. $R = ki(p - i)$

a. $255 = k(0.15)(20,000)(0.85)(20,000)$

$$= \frac{1}{200,000}$$

b. $R = \dfrac{1}{200,000}(10,000)(10,000) = 500$
people per day.

c. $95 = \dfrac{1}{200,000}x(20,000 - x) \Rightarrow$

$x^2 - 20,000x + 19,000,000 = 0 \Rightarrow$
$(x - 1000)(x - 19,000) = 0 \Rightarrow x = 1000$ or
$x = 19,000$

Chapter 3 Practice Test A

1. $x^2 - 6x + 2 = 0 \Rightarrow x = \dfrac{6 \pm \sqrt{(-6)^2 - 4(1)(2)}}{2(1)} \Rightarrow$

$x = \dfrac{6 \pm \sqrt{28}}{2} = 3 \pm \sqrt{7}$ are the x-intercepts

2.

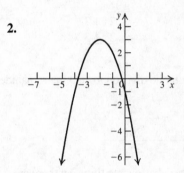

3. The vertex is at $\left(-\dfrac{b}{2a}, f\left(-\dfrac{b}{2a}\right)\right)$

$$= \left(-\frac{14}{2(-7)}, f\left(-\frac{14}{2(-7)}\right)\right) = (1, 10).$$

4. The denominator is 0 when $x = -4$ or $x = 1$. The domain is $(-\infty, -4) \cup (-4, 1) \cup (1, \infty)$.

5. Using either long division or synthetic division, we find that the quotient is $x^2 - 4x + 3$ and the remainder is 0.

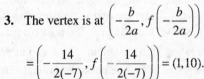

6.

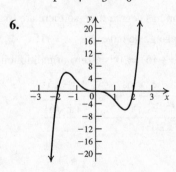

7. The function has degree three, so there are three zeros. Since 2 is a zero, use synthetic division to find the depressed function:

$$2 \underline{|\quad 2 \quad -2 \quad -8 \quad \ 8}$$
$$\underline{\qquad \quad 4 \quad \ \ 4 \quad -8}$$
$$\quad \ \ 2 \quad \ \ 2 \quad -4 \quad \ \ 0$$

Now find the zeros of $2x^2 + 2x - 4$:

$2x^2 + 2x - 4 = 2(x^2 + x - 2) = 2(x+2)(x-1) \Rightarrow$
the zeros are $x = -2$ and $x = 1$. The zeros of the original function are $-2, 1, 2$.

8.
$$\require{enclose}\begin{array}{r} -3x^2 + 5x + 1 \\ 2x+3 \enclose{longdiv}{-6x^3 + \ \ x^2 + 17x + 3} \\ \underline{-6x^3 - 9x^2 \qquad\qquad} \\ 10x^2 + 17x \\ \underline{10x^2 + 15x} \\ 2x + 3 \\ \underline{2x + 3} \\ 0 \end{array}$$

9. Using synthetic division to find the remainder, we have $P(-2) = -53$.

$$-2 \underline{|\quad 1 \quad \ \ 5 \quad -7 \quad \ \ 9 \quad \ \ 17}$$
$$\underline{\qquad \quad -2 \quad -6 \quad 26 \quad -70}$$
$$\quad \ \ 1 \quad \ \ 3 \quad -13 \quad 35 \quad -53$$

10. The function has degree three, so there are three zeros. There are two sign changes in $f(x)$, so there are either 2 or 0 positive zeros. There is one sign change in $f(-x)$, so there is one negative zero. The possible rational zeros are $\{\pm 1, \pm 2, \pm 4, \pm 5, \pm 10, \pm 20\}$. Using synthetic division, we find that -2 is a zero:

$$-2 \underline{|\quad 1 \quad -5 \quad -4 \quad \ \ 20}$$
$$\underline{\qquad \quad -2 \quad 14 \quad -20}$$
$$\quad \ \ 1 \quad -7 \quad 10 \quad \ \ 0$$

Now find the zeros of the depressed function $x^2 - 7x + 10$: $x^2 - 7x + 10 = (x-5)(x-2) \Rightarrow$ the zeros are $x = 2$ and $x = 5$. The zeros of the original function are $-2, 2, 5$.

11. The function has degree four, so there are four zeros. Factoring, we have $x^4 + x^3 - 15x^2$
$= x^2(x^2 + x - 15) \Rightarrow 0$ is a zero of multiplicity 2. Now find the zeros of $x^2 + x - 15$:

$$x = \frac{-1 \pm \sqrt{1^2 - 4(1)(-15)}}{2(1)} = \frac{-1 \pm \sqrt{61}}{2}.$$

The zeros of the original function are 0 and
$\dfrac{-1 \pm \sqrt{61}}{2}$.

12. The factors of the constant term, 9, are $\{\pm 1, \pm 3, \pm 9\}$, while the factors of the leading coefficient are $\{\pm 1, \pm 2\}$. The possible rational zeros are $\left\{\pm \dfrac{1}{2}, \pm 1, \pm \dfrac{3}{2}, \pm 3, \pm \dfrac{9}{2}, \pm 9\right\}$.

13. $f(x) \to -\infty$ as $x \to -\infty$; $f(x) \to \infty$ as $x \to \infty$.

14. $f(x) = (x^2 - 4)(x+2)^2$
$\qquad = (x-2)(x+2)(x+2)^2 \Rightarrow$
the zeros are -2 (multiplicity 3) and 2 (multiplicity 1).

15. There are two sign changes in $f(x)$, so there are either two or zero positive zeros. There is one sign change in $f(-x)$, so there is one negative zero.

16. The zeros of the denominator are $x = 5$ and $x = -4$, so those are the vertical asymptotes. The degree of the numerator equals the degree of the denominator, so the horizontal asymptote is $y = 2$.

17. $y = \dfrac{kx}{t^2}$

18. $6 = \dfrac{8k}{2^2} \Rightarrow k = 3; y = \dfrac{3(12)}{3^2} = 4$

19. The minimum occurs at the x-coordinate of the vertex: $-\dfrac{-30}{2(1)} = 15$ thousand units.

20. $V = x(17 - 2x)(8 - 2x)$

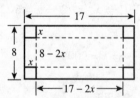

Chapter 3 Practice Test B

1. $x^2 + 5x + 3 = 0 \Rightarrow x = \dfrac{-5 \pm \sqrt{5^2 - 4(1)(3)}}{2(1)}$

$\qquad = \dfrac{-5 \pm \sqrt{13}}{2}$. The answer is B.

2. The graph of $f(x) = 4 - (x-2)^2$ is the graph of $f(x) = x^2$ shifted two units right, reflected across the x-axis, and then shifted 4 units up. The answer is D.

3. $\left(-\dfrac{b}{2a}, f\left(-\dfrac{b}{2a}\right)\right) = \left(-\dfrac{12}{2(6)}, f\left(-\dfrac{12}{2(6)}\right)\right)$.

$\qquad = (-1, -11)$

The answer is A.

4. The denominator is 0 when $x = -3$ or $x = 2$. The answer is B.

5.
$$-3\,\underline{|\ 1 \quad 0 \quad -8 \quad 6}$$
$$\ \underline{\quad -3 \quad 9 \quad -3}$$
$$1 \quad -3 \quad 1 \quad 3$$
The answer is D.

6. $P(x) = x^4 + 2x^3 = x^3(x+2)$. So the zeros are 0 (multiplicity 3) and -2 (multiplicity 1). The only graph with those zeros is C.

7.
$$3\,\underline{|\ 3 \quad -26 \quad 61 \quad -30}$$
$$\ \underline{\quad 9 \quad -51 \quad 30}$$
$$3 \quad -17 \quad 10 \quad 0$$
The zeros of the depressed function $3x^2 - 17x + 10$ are $x = \dfrac{2}{3}$ and $x = 5$.

The answer is C.

8.
$$\begin{array}{r}
-5x^2 + 3x - 4 \\
2x-3\,\overline{\smash{)}\,-10x^3 + 21x^2 - 17x + 12} \\
\underline{-10x^3 + 15x^2} \\
6x^2 - 17x \\
\underline{6x^2 - 9x} \\
-8x + 12 \\
\underline{-8x + 12} \\
0
\end{array}$$
The answer is B.

9.
$$-3\,\underline{|\ 1 \quad 4 \quad 7 \quad 10 \quad 15}$$
$$\ \underline{\quad -3 \quad -3 \quad -12 \quad 6}$$
$$1 \quad 1 \quad 4 \quad -2 \quad 21$$
$P(-3) = 21$. The answer is C.

10. The polynomial has degree three, so there are three zeros. The possible rational zeros are $\{\pm 1, \pm 2, \pm 3, \pm 4, \pm 6, \pm 12\}$. Using synthetic division, we find that one zero is -3:
$$-3\,\underline{|\ -1 \quad 1 \quad 8 \quad -12}$$
$$\ \underline{\quad 3 \quad -12 \quad 12}$$
$$-1 \quad 4 \quad -4 \quad 0$$

The zeros of the depressed function $-x^2 + 4x - 4$ are $x = 2$ (multiplicity 2). The answer is A.

11. The polynomial has degree three, so there are three zeros. Factoring, we have
$$x^3 + x^2 - 30x = x(x^2 + x - 30) = x(x+6)(x-5)$$
The answer is A.

12. The factors of the constant term, 60, are $\{\pm 1, \pm 2, \pm 3, \pm 4, \pm 5, \pm 6, \pm 10, \pm 12, \pm 15, \pm 20, \pm 30, \pm 60\}$. Since the leading coefficient is 1, these are also the possible rational zeros. The answer is D.

13. The answer is C.

14. $(x^2 - 1)(x+1)^2 = (x-1)(x+1)(x+1)^2$. The answer is B.

15. There are three sign changes in $P(x)$, so there are 3 or 1 positive zeros. There are two sign changes in $P(-x)$, so there are 2 or 0 negative zeros. The answer is C.

16. The zeros of the denominator are $x = 3$ and $x = -4$, so those are the vertical asymptotes. The degree of the numerator equals the degree of the denominator, so the horizontal asymptote is $y = 1$. The answer is C.

17. The answer is D.

18. $27 = \dfrac{3^2 k}{1^3} \Rightarrow k = 3; S = \dfrac{3(6^2)}{3^3} = 4$. The answer is B.

19. The minimum occurs at the x-coordinate of the vertex: $-\dfrac{-24}{2(1)} = 12$. The answer is B.

20. $V = x(10 - 2x)(12 - 2x)$. The answer is D.

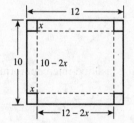

Cumulative Review Exercises
Chapters P–3

1. $d = \sqrt{(x_2 - x_1)^2 + (y_2 - y_1)}$
$= \sqrt{(-1-2)^2 + (3-5)^2} = \sqrt{13}$

2. $M = \left(\dfrac{x_1 + x_2}{2}, \dfrac{y_1 + y_2}{2}\right)$
$= \left(\dfrac{2+(-8)}{2}, \dfrac{-5+(-3)}{2}\right) = (-3, -4)$

3. $0 = x^2 - 2x - 8 \Rightarrow 0 = (x-4)(x+2) \Rightarrow$
$x - 4 = 0 \Rightarrow x = 4$ or $x + 2 = 0 \Rightarrow x = -2$
$f(0) = 0^2 - 2(0) - 8 = -8$
The x-intercepts are $(-2, 0)$ and $(4, 0)$, and the y-intercept is $(0, -8)$.

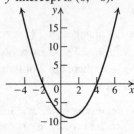

4. Write the equation in slope-intercept form:
$x + 3y - 6 = 0 \Rightarrow 3y = -x + 6 \Rightarrow y = -\dfrac{1}{3}x + 2$

The slope is $-\dfrac{1}{3}$, and the y-intercept is $(0, 2)$.
$x + 3(0) - 6 = 0 \Rightarrow x - 6 = 0 \Rightarrow x = 6$, so the x-intercept is $(6, 0)$.

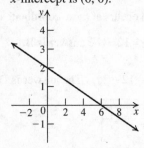

5. Write the equation in slope-intercept form:
$x = 2y - 6 \Rightarrow x + 6 = 2y \Rightarrow \dfrac{1}{2}x + 3 = y$

The slope is $\dfrac{1}{2}$, and the y-intercept is $(0, 3)$.
The x-intercept is $x = 2(0) - 6 = -6$.

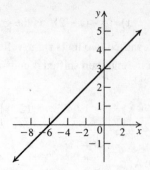

6. $(x-2)^2 + (y+3)^2 = 16$

7. $x^2 + y^2 + 2x - 4y - 4 = 0 \Rightarrow$
$(x^2 + 2x) + (y^2 - 4y) = 4 \Rightarrow$
$(x^2 + 2x + 1) + (y^2 - 4y + 4) = 4 + 1 + 4 \Rightarrow$
$(x+1)^2 + (y-2)^2 = 9$. The center is $(-1, 2)$.
The radius is 3.

8. $y + 2 = 3(x-1) \Rightarrow y = 3x - 5$

9. $2x + 3y = 5 \Rightarrow y = -\dfrac{2}{3}x + \dfrac{5}{3} \Rightarrow$ the slope is $-\dfrac{2}{3}$.
$y - 3 = -\dfrac{2}{3}(x-1) \Rightarrow y = -\dfrac{2}{3}x + \dfrac{11}{3}$.

10. $2x + 3 = 0 \Rightarrow x = -\dfrac{3}{2}$.
The domain is $\left(-\infty, -\dfrac{3}{2}\right) \cup \left(-\dfrac{3}{2}, \infty\right)$.

11. $\sqrt{4 - 2x} = 0 \Rightarrow x = 2$. The domain is $(-\infty, 2)$.

12. $f(-2) = (-2)^2 - 2(-2) + 3 = 11$;
$f(3) = 3^2 - 2(3) + 3 = 6$;
$f(x+h) = (x+h)^2 - 2(x+h) + 3$
$= x^2 + 2xh + h^2 - 2x - 2h + 3$
$= x^2 + 2(h-1)x + h^2 - 2h + 3$

$\dfrac{f(x+h) - f(x)}{h}$
$= \dfrac{(x^2 + 2(h-1)x + h^2 - 2h + 3) - (x^2 - 2x + 3)}{h}$
$= \dfrac{2(h-1)x + 2x + h^2 - 2h}{h}$
$= \dfrac{2hx - 2x + 2x + h^2 - 2h}{h} = \dfrac{h^2 + 2hx - 2h}{h}$
$= h + 2x - 2$

13. a. $f(g(x)) = \sqrt{x^2 + 1}$

b. $g(f(x)) = (\sqrt{x})^2 + 1 = x + 1$

c. $f(f(x)) = \sqrt{\sqrt{x}} = \sqrt[4]{x}$

d. $g(g(x)) = (x^2 + 1)^2 + 1 = x^4 + 2x^2 + 2$

14. a. $f(1) = 3(1) + 2 = 5; f(3) = 4(3) - 1 = 11;$
$f(4) = 6$

b.

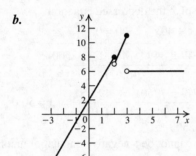

15. $y = 2x - 3.$ Interchange x and y, and then solve for y.

$$x = 2y - 3 \Rightarrow y = \frac{x+3}{2} = \frac{1}{2}x + \frac{3}{2} = f^{-1}(x).$$

16. a. Shift the graph of $y = \sqrt{x}$ two units left.

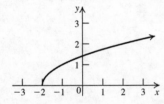

b. Shift the graph of $y = \sqrt{x}$ one unit left, stretch vertically by a factor of two, reflect about the x-axis, and then shift up three units.

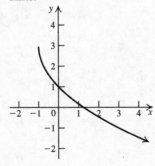

17. The factors of the constant term, -6, are $\{\pm 1, \pm 2, \pm 3, \pm 6\}$, and the factors of the leading coefficient, 2, are $\{\pm 1, \pm 2\}$. The possible rational zeros are $\left\{\pm\dfrac{1}{2}, \pm 1, \pm\dfrac{3}{2}, \pm 2, \pm 3, \pm 6\right\}$.

18.

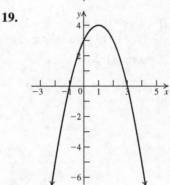

19.

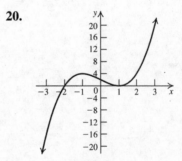

20.

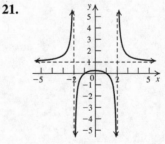

21.

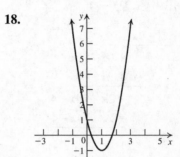

22. Since one zero is $1 + i$, another zero is $1 - i$.

So $(x - (1-i))(x - (1+i)) = x^2 - 2x + 2$ is a factor of $f(x)$. Now divide to find the other factor:

$$x^2 - 2x + 2 \overline{)x^4 - 3x^3 + 2x^2 + 2x - 4}$$

$$\begin{array}{r} x^2 - x - 2 \\ \underline{x^4 - 2x^3 + 2x^2} \\ -x^3 + 0x^2 + 2x \\ \underline{-x^3 + 2x^2 - 2x} \\ -2x^2 + 4x - 4 \\ \underline{-2x^2 + 4x - 4} \\ 0 \end{array}$$

$$\begin{aligned} f(x) &= x^4 - 3x^3 + 2x^2 + 2x - 4 \\ &= (x - (1-i))(x - (1+i))(x^2 - x - 2) \\ &= (x - (1-i))(x - (1+i))(x - 2)(x + 1) \Rightarrow \end{aligned}$$

the zeros of $f(x)$ are $-1, 2, 1 - i, 1 + i$.

23. $y = k\sqrt{x}; 6 = k\sqrt{4} \Rightarrow k = 3; y = 3\sqrt{9} = 9$

24. Profit = revenue − cost

$$150x - (0.02x^2 + 100x + 3000)$$
$$= -0.02x^2 + 50x - 3000.$$

The maximum occurs at the vertex:

$$\left(-\frac{50}{2(-0.02)}, f\left(-\frac{50}{2(-0.02)} \right) \right)$$
$$= (1250, \$28,250).$$

25.

$$\begin{array}{r|rrrr} 10 & 0.02 & 48.8 & -2990 & 25{,}000 \\ & & 0.2 & 490 & -25{,}000 \\ \hline & 0.02 & 49 & -2500 & 0 \end{array}$$

Now solve the depressed equation

$$0.02x^2 + 49x - 2500 = 0.$$

$$x = \frac{-49 \pm \sqrt{49^2 - 4(0.02)(-2500)}}{2(0.02)}$$

$$= \frac{-49 \pm \sqrt{2601}}{0.04} = \frac{-49 \pm 51}{0.04} \Rightarrow x = 50 \text{ or}$$

$$x = -2500$$

There cannot be a negative amount of units sold, so another break-even point is 50.

Chapter 4 Exponential and Logarithmic Functions

4.1 Exponential Functions

4.1 Practice Problems

1. $f(x) = \left(\dfrac{1}{4}\right)^x$

$f(2) = \left(\dfrac{1}{4}\right)^2 = \dfrac{1}{16}$

$f(0) = \left(\dfrac{1}{4}\right)^0 = 1$

$f(-1) = \left(\dfrac{1}{4}\right)^{-1} = 4$

$f\left(\dfrac{5}{2}\right) = \left(\dfrac{1}{4}\right)^{5/2} = \left(\sqrt{\dfrac{1}{4}}\right)^5 = \left(\dfrac{1}{2}\right)^5 = \dfrac{1}{32}$

$f\left(-\dfrac{3}{2}\right) = \left(\dfrac{1}{4}\right)^{-3/2} = 4^{3/2} = \left(\sqrt{4}\right)^3 = 2^3 = 8$

2. a. $3^{\sqrt{8}} \cdot 3^{\sqrt{2}} = 3^{\sqrt{8}+\sqrt{2}} = 3^{2\sqrt{2}+\sqrt{2}} = 3^{3\sqrt{2}} = 27^{\sqrt{2}}$

b. $\left(a^{\sqrt{8}}\right)^{\sqrt{2}} = a^{\sqrt{8}\cdot\sqrt{2}} = a^{\sqrt{16}} = a^4$

3.

x	2^x
-3	$1/8$
-2	$1/4$
-1	$1/2$
0	1
1	2
2	4
3	8

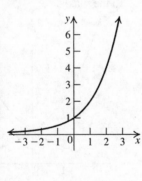

4.

x	$(1/3)^x$
-3	27
-2	9
-1	3
0	1
1	$1/3$
2	$1/9$
3	$1/27$

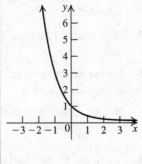

5. a. $(0, 1)$ and $(2, 49)$

$f(0) = 1 \Rightarrow 1 = c \cdot a^0 = c \cdot 1 = c$

Therefore,

$f(x) = 1 \cdot a^x \Rightarrow 49 = 1 \cdot a^x \Rightarrow$

$49 = a^2 \Rightarrow a = 7$

So, $f(x) = 7^x$.

b. $(-2, 16)$ and $\left(3, \dfrac{1}{2}\right)$

$16 = ca^{-2}$

$\dfrac{1}{2} = ca^3$

$\dfrac{16}{1/2} = \dfrac{ca^{-2}}{ca^3}$

$32 = a^{-5} \Rightarrow a = \dfrac{1}{2}$

$16 = c\left(\dfrac{1}{2}\right)^{-2} \Rightarrow 16 = 4c \Rightarrow 4 = c$

So, $f(x) = 4\left(\dfrac{1}{2}\right)^x$.

6. $A = P + I = P + Prt$

$\quad = 10,000 + 10,000(0.075)(2)$

$\quad = 10,000 + 1500 = 11,500$

There will be \$11,500 in the account.

7. a. $A = P(1+r)^t = 8000(1+0.075)^5$

$\quad\quad \approx 11,485.03$

There will be \$11,485.03 in the account.

b. Interest $= A - P = 11,485.03 - 8000$

$\quad\quad = 3485.03$

She will receive \$3485.03

8. The formula for compound interest is

$A = P\left(1 + \dfrac{r}{n}\right)^{nt}$.

For each of the following, $t = 1$.

(i) Annual compounding $(n = 1)$

$A = \$5000\left(1 + \dfrac{0.065}{1}\right)^1 = \$5325.00$

(ii) Semiannual compounding $(n = 2)$

$A = \$5000\left(1 + \dfrac{0.065}{2}\right)^2 \approx \$5330.28$

(iii) Quarterly compounding ($n = 4$)

$$A = \$5000\left(1 + \frac{0.065}{4}\right)^4 \approx \$5333.01$$

(iv) Monthly compounding ($n = 12$)

$$A = \$5000\left(1 + \frac{0.065}{12}\right)^{12} \approx \$5334.86$$

(v) Daily compounding ($n = 365$)

$$A = \$5000\left(1 + \frac{0.065}{365}\right)^{365} \approx \$5335.76$$

9. $\qquad 20,000 = 9000\left(1 + \dfrac{r}{12}\right)^{12\cdot 8}$

$$\left(1 + \frac{r}{12}\right)^{96} = \frac{20,000}{9000} = \frac{20}{9}$$

$$1 + \frac{r}{12} = \left(\frac{20}{9}\right)^{1/96}$$

$$\frac{r}{12} = \left(\frac{20}{9}\right)^{1/96} - 1$$

$$r = 12\left[\left(\frac{20}{9}\right)^{1/96} - 1\right] \approx .10023$$

Carmen needs an interest rate of about 10.023%.

10. $A = Pe^{rt} = \$9000e^{(0.06)(8.25)} \approx \$14,764.48$

11. Shift the graph of $y = e^x$ one unit right, then reflect the graph about the x-axis. Shift the resulting graph two units down.

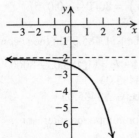

12. Use the exponential growth/decay formula

$$A(t) = A_0 e^{kt}$$

a. $A(30) = 6.08e^{0.016(30)} \approx 9.8257$

The model predicts that if the rate of growth is 1.6% per year, there will be about 9.83 billion people in the world in the year 2030.

b. $A(-10) = 6.08e^{0.016(-10)} \approx 5.1810$

The model predicts that if the rate of growth is 1.6% per year, there were about 5.18 billion people in the world in the year 1990.

13. $A(t) = A_0 e^{-kt}$

$A(6) = 22,000e^{(-0.18)(6)} \approx \$7471.10$

4.1 Basic Concepts and Skills

1. For the exponential function
$f(x) = ca^x, a > 0,\ a \ne 1$, the domain is
$(-\infty, \infty)$ and for $c > 0$, the range is $\underline{(0, \infty)}$.

3. The horizontal asymptote of the graph of
$y = \left(\dfrac{1}{3}\right)^x$ is <u>the x-axis</u>.

5. The formula for compound interest at rate r compounded n times per year is

$$A = P\left(1 + \frac{r}{n}\right)^{nt}.$$

7. False. The graphs are symmetric with respect to the y-axis.

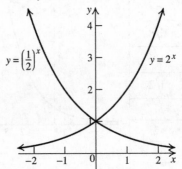

9. True

11. Not an exponential function. The base is not a constant.

13. Exponential function. The base is a constant, 1/2.

15. Not an exponential function. The base is not a constant.

17. Not an exponential function. The base is not a positive constant.

19. $f(0) = 5^{0-1} = 5^{-1} = \dfrac{1}{5}$

21. $g(3.2) = 3^{1-3.2} = 3^{-2.2} = 0.0892$
$g(-1.2) = 3^{1-(-1.2)} = 3^{2.2} = 11.2116$

23. $h(1.5) = \left(\dfrac{2}{3}\right)^{2(1.5)-1} = \left(\dfrac{2}{3}\right)^2 = \dfrac{4}{9}$
$h(-2.5) = \left(\dfrac{2}{3}\right)^{2(-2.5)-1} = \left(\dfrac{2}{3}\right)^{-6} = \dfrac{729}{64}$

25. $3^{\sqrt{2}} \cdot 3^{\sqrt{2}} = 3^{\sqrt{2}+\sqrt{2}} = 3^{2\sqrt{2}}$

27. $8^{\pi} \div 4^{\pi} = \left(2^3\right)^{\pi} \div \left(2^2\right)^{\pi} = 2^{3\pi} \div 2^{2\pi}$
$= 2^{3\pi-2\pi} = 2^{\pi}$

29. $\left(3^{\sqrt{2}}\right)^{\sqrt{3}} = 3^{\sqrt{2}\cdot\sqrt{3}} = 3^{\sqrt{6}}$

31. $\left(a^{\sqrt{3}}\right)^{\sqrt{12}} = a^{\sqrt{3}\cdot\sqrt{12}} = a^{\sqrt{36}} = a^6$

33. a. (0, 1) and (2, 16)
$f(0) = 1 \Rightarrow 1 = ca^0 \Rightarrow 1 = c$
$f(2) = 16 \Rightarrow 16 = 1 \cdot a^2 \Rightarrow a = 4$
$f(x) = 4^x$

b. (0, 1) and $\left(-2, \dfrac{1}{9}\right)$
$f(0) = 1 \Rightarrow 1 = ca^0 \Rightarrow 1 = c$
$f(-2) = \dfrac{1}{9} \Rightarrow \dfrac{1}{9} = 1 \cdot a^{-2} \Rightarrow a = 3$
$f(x) = 3^x$

35. a. (1, 1) and (2, 5)
$f(1) = 1 \Rightarrow 1 = ca^1$
$f(2) = 5 \Rightarrow 5 = ca^2$
$\dfrac{1}{5} = \dfrac{ca^1}{ca^2} \Rightarrow \dfrac{1}{5} = a^{-1} \Rightarrow 5 = a$
$f(1) = 1 \Rightarrow 1 = c \cdot 5^1 \Rightarrow c = \dfrac{1}{5}$

$f(x) = \dfrac{1}{5} \cdot (5)^x$

b. (1, 1) and $\left(2, \dfrac{1}{5}\right)$
$f(1) = 1 \Rightarrow 1 = ca^1$
$f(2) = \dfrac{1}{5} \Rightarrow \dfrac{1}{5} = ca^2$
$\dfrac{1}{1/5} = \dfrac{ca^1}{ca^2} \Rightarrow 5 = a^{-1} \Rightarrow \dfrac{1}{5} = a$
$f(1) = 1 \Rightarrow 1 = c\left(\dfrac{1}{5}\right)^1 \Rightarrow c = 5$

$f(x) = 5 \cdot \left(\dfrac{1}{5}\right)^x$

37.

x	4^x
−2	1/16
−1	1/4
0	1
1	4
2	16

39.

x	$(3/2)^{-x}$
−4	81/16
−2	9/4
−1	3/2
0	1
1	2/3
2	4/9
4	16/81

41.

x	$(1/4)^x$
−2	16
−1	4
0	1
1	1/4
2	1/16

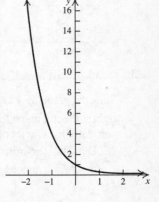

43.

x	1.3^{-x}
-4	≈ 2.86
-3	≈ 2.2
0	1
2	≈ 0.59
4	≈ 0.35

45. c   **47.** a

49. $g(x) = 3^{x-1}$

Shift the graph of $f(x) = 3^x$ one unit right.

Domain: $(-\infty, \infty)$; Range: $(0, \infty)$

Horizontal asymptote: $y = 0$

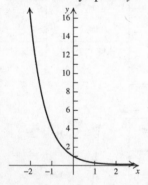

51. $g(x) = 4^{-x}$

Reflect the graph of $f(x) = 4^x$ about the y-axis.

Domain: $(-\infty, \infty)$; Range: $(0, \infty)$

Horizontal asymptote: $y = 0$

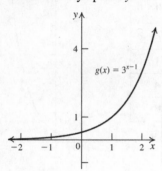

53. $g(x) = -2 \cdot 5^{x-1} + 4$

Shift the graph of $f(x) = 5^x$ one unit to the right, reflect it about the x-axis, stretch vertically by a factor of 2, then shift four units up. Domain: $(-\infty, \infty)$; Range: $(-\infty, 4)$.

Horizontal asymptote: $y = 4$.

55. $g(x) = -e^{x-2} + 3$

Shift the graph of $f(x) = e^x$ two units to the right, then reflect it about the x-axis, then shift the graph three units up.

Domain: $(-\infty, \infty)$; Range: $(-\infty, 3)$

Horizontal asymptote: $y = 3$

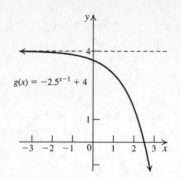

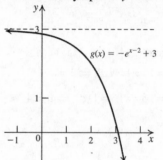

57. The graph passes through the point $(0, 3)$ and $(1, 3.5)$.

$f(0) = 3 \Rightarrow 3 = a^0 + b \Rightarrow 3 = 1 + b \Rightarrow b = 2$

$f(1) = 3.5 \Rightarrow 3.5 = a^1 + 2 \Rightarrow 1.5 = a$

$f(x) = 1.5^x + 2$

$f(2) = 1.5^2 + 2 = 4.25$

59. The graph passes through $(-2, 7)$ and $(-1, 1)$.

$f(-2) = 7 \Rightarrow 7 = a^{-2} + b$ (1)

$f(-1) = 1 \Rightarrow 1 = a^{-1} + b$ (2)

Subtract (2) from (1).

$6 = a^{-2} - a^{-1} \Rightarrow 6 = \dfrac{1}{a^2} - \dfrac{1}{a} \Rightarrow$

$6a^2 = 1 - a \Rightarrow 6a^2 + a - 1 = 0 \Rightarrow$

$(2a+1)(3a-1) = 0 \Rightarrow a = -\dfrac{1}{2}, \dfrac{1}{3}$

If $a = -\dfrac{1}{2}$, then

$1 = \left(-\dfrac{1}{2}\right)^{-1} + b \Rightarrow 1 = -2 + b \Rightarrow b = 3$.

(continued on next page)

(continued)

If $a = \dfrac{1}{3}$, then

$$1 = \left(\dfrac{1}{3}\right)^{-1} + b \Rightarrow 1 = 3 + b \Rightarrow b = -2.$$

The horizontal asymptote of the given graph is $y = -2$, so the equation of the graph is

$$f(x) = \left(\dfrac{1}{3}\right)^x - 2.$$

$$f(2) = \left(\dfrac{1}{3}\right)^2 - 2 = \dfrac{1}{9} - 2 = -\dfrac{17}{9}$$

61. $y = 2^{x+2} + 5$ **63.** $y = 2\left(\dfrac{1}{2}\right)^x - 5$

65. $I = Prt = \$5000 \cdot 0.1 \cdot 5 = \$2500$

67. $I = Prt = \$7800 \cdot 0.06875 \cdot 10.75 = \$5764.69$

69. a. $A = 3500\left(1 + \dfrac{0.065}{1}\right)^{13} = \$7936.21$

    b. interest $= \$7936.21 - \$3500 = \$4436.21$

71. a. $A = 7500e^{0.05(10)} = \$12,365.41$

    b. interest $= \$12,365.41 - \$7500 = \$4865.41$

73. $10,000 = P\left(1 + \dfrac{0.08}{1}\right)^{10} = P(1.08)^{10} \Rightarrow$

    $P = \$4631.93$

75. $10,000 = P\left(1 + \dfrac{0.08}{365}\right)^{365(10)} \Rightarrow P = \$4493.68$

77. Reflect the graph of $y = e^x$ about the y-axis.

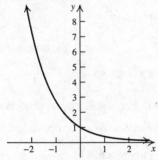

Horizontal asymptote: $y = 0$

79. Shift the graph of $y = e^x$ two units right.

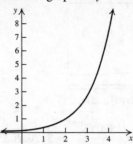

Horizontal asymptote: $y = 0$

81. Shift the graph of $y = e^x$ one unit up.

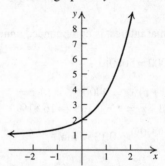

Horizontal asymptote: $y = 1$

83. Shift the graph of $y = e^x$ two units right, reflect the graph about the x-axis, and then shift it three units up.

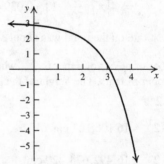

Horizontal asymptote: $y = 3$

4.1 Applying the Concepts

85. a. **(i)** $T = 200 \cdot 4^{-0.1(2)} + 25 = 176.6°C$

    (ii) $T = 200 \cdot 4^{-0.1(3.5)} + 25 = 148.1°C$

    b. $125 = 200 \cdot 4^{-0.1t} + 25 \Rightarrow 100 = 200 \cdot 4^{-0.1t} \Rightarrow$

    $\dfrac{1}{2} = 4^{-0.1t} \Rightarrow 2^{-1} = 2^{-0.2t} \Rightarrow -1 = -0.2t \Rightarrow$

    $t = 5$ hours

c. As $t \to \infty$, $T \to 25$. Verify graphically.

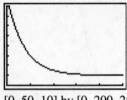

[0, 50, 10] by [0, 200, 25]

87. $A = 190,000\left(1 + \dfrac{0.03}{1}\right)^{1(5)} \approx \$220,262$

89. $A = 80,000\left(1 - \dfrac{0.15}{1}\right)^{1(5)} \approx \$35,496.43$

91. Assume that interest is compounded annually.

$$22,000,000 = 5000\left(1 + \dfrac{r}{1}\right)^{1(54)} \Rightarrow$$

$$4400 = (1+r)^{54} \Rightarrow 4400^{1/54} = 1 + r \Rightarrow$$
$$1.1681 \approx 1 + r \Rightarrow r \approx 0.1681 \approx 16.81\%$$

93. $A = 10e^{-0.43(10)} \approx 0.1357 \text{ mm}^2$

95. $95,600 = 60,000\left(1 + \dfrac{r}{4}\right)^{4(12)} \Rightarrow$

$$\dfrac{239}{150} = \left(1 + \dfrac{r}{4}\right)^{48} \Rightarrow \sqrt[48]{\dfrac{239}{150}} = 1 + \dfrac{r}{4} \Rightarrow$$

$$\sqrt[48]{\dfrac{239}{150}} - 1 = \dfrac{r}{4} \Rightarrow 4\left(\sqrt[48]{\dfrac{239}{150}} - 1\right) \approx 0.039 = r$$

The interest rate on the bond was about 3.9%.

97. The number of pieces of paper is 2^x, where x is the number of tears. The height of the paper is $0.015 \cdot 2^x$.

a. $0.015 \cdot 2^{30} \approx 16,106,127 \text{ cm}$

b. $0.015 \cdot 2^{40} \approx 16,492,674,420 \text{ cm}$

c. $0.015 \cdot 2^{50} \approx 16,888,498,600,000 \text{ cm}$
$$= 1.689 \times 10^{13} \text{ cm}$$

4.1 Beyond the Basics

99. a. $\dfrac{f(x+h) - f(x)}{h} = \dfrac{e^{x+h} - e^x}{h} = \dfrac{e^x(e^h - 1)}{h}$
$$= e^x \dfrac{(e^h - 1)}{h}$$

b. $f(x+y) = e^{x+y} = e^x e^y = f(x)f(y)$

c. $f(-x) = e^{-x} = \dfrac{1}{e^x} = \dfrac{1}{f(x)}$

101. $S_5 = 2 + \dfrac{1}{2!} + \dfrac{1}{3!} + \dfrac{1}{4!} + \dfrac{1}{5!} = \dfrac{163}{60} \approx 2.71667$

$$S_{10} = S_5 + \dfrac{1}{6!} + \dfrac{1}{7!} + \dfrac{1}{8!} + \dfrac{1}{9!} + \dfrac{1}{10!} \approx 2.7182818$$

$$S_{15} = S_{10} + \dfrac{1}{11!} + \dfrac{1}{12!} + \dfrac{1}{13!} + \dfrac{1}{14!} + \dfrac{1}{15!}$$
$$\approx 2.718281828$$

The sum approaches the value of e.

103. $A = P\left(1 + \dfrac{r}{m}\right)^{m(1)} = P(1 + y) \Rightarrow$

$$\left(1 + \dfrac{r}{m}\right)^m = 1 + y \Rightarrow y = \left(1 + \dfrac{r}{m}\right)^m - 1$$

105. $y = e^x \to y = e^{x-1} \to y = e^{2x-1} \to y = 3e^{2x-1}$

107. $y = e^x \to y = e^{2+x} \to y = e^{2+3x} \to$
$$y = e^{2-3x} \to y = 5e^{2-3x} \to y = 5e^{2-3x} + 4$$

4.1 Critical Thinking/Discussion/Writing

109. The base a cannot be 1 because this makes the function a constant, $f(x) = 1$. Similarly, the base a cannot be 0 because this becomes $f(x) = 0^x = 0$, a constant. We rule out negative bases so that the domain can include all real numbers. For example, a cannot be -3 because $f\left(\frac{1}{2}\right) = (-3)^{1/2} = \sqrt{-3}$ is not a real number.

110. $g(x)$ is defined for the set of integers. For example, $g(-2) = (-3)^{-2} = \dfrac{1}{9}$ and
$$g(3) = (-3)^3 = -27.$$

111. If $0 < a < 1$, then, for $x < 0$, the denominator becomes very large, so $y = \dfrac{b}{1 + ca^x}$ approaches 0. For $x \geq 0$, ca^x approaches 0, so the denominator $1 + ca^x$ approaches 1 and $y = \dfrac{b}{1 + ca^x}$ approaches b.

(continued on next page)

(*continued*)

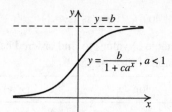

If $a > 1$, then, for $x < 0$, ca^x approaches 0, so the denominator $1 + ca^x$ approaches 1 and $y = \dfrac{b}{1 + ca^x}$ approaches b. For $x \geq 0$, the denominator becomes very large, so $y = \dfrac{b}{1 + ca^x}$ approaches 0.

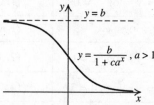

112. There are four possibilities, $0 < a < 1$ with $c < 0$, $0 < a < 1$ with $c > 0$, $a > 1$ with $c < 0$, and $a > 1$ with $c > 0$. They are illustrated below

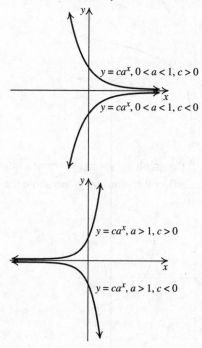

113. The function $y = 2^x$ is an increasing function, so it is one-to-one. The horizontal line $y = k$ for $k > 0$ intersects the graph of $y = 2^x$ in exactly one point, so there is exactly one solution for each value of k.

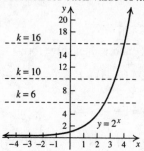

114. The graphs of $y = 2^x$ and $y = x$ intersect in exactly two points: (1, 2) and (2, 4). So $2^x = 2x \Rightarrow x = 1$ or 2.

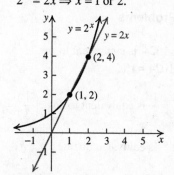

4.1 Maintaining Skills

115. $10^0 = 1$

117. $(-8)^{1/3} = \sqrt[3]{-8} = -2$

119. $\left(\dfrac{1}{7}\right)^{-2} = 7^2 = 49$

121. $18^{1/2} = \sqrt{18} = \sqrt{9 \cdot 2} = 3\sqrt{2}$

123. $f(x) = y = 3x + 4$

Interchange x and y, then solve for y.
$x = 3y + 4 \Rightarrow x - 4 = 3y \Rightarrow$
$\dfrac{x - 4}{3} = y = f^{-1}(x)$

The domain and range of f^{-1} are $(-\infty, \infty)$.

125. $f(x) = y = \sqrt{x},\ x \geq 0$

Interchange x and y, then solve for y.

$x = \sqrt{y} \Rightarrow x^2 = y = f^{-1}(x),\ x \geq 0$

The domain and range of f^{-1} are $[0, \infty)$.

127. $f(x) = y = \dfrac{1}{x-1},\ x \neq 1$

Interchange x and y, then solve for y.

$x = \dfrac{1}{y-1} \Rightarrow x(y-1) = 1 \Rightarrow y - 1 = \dfrac{1}{x} \Rightarrow$

$y = f^{-1}(x) = 1 + \dfrac{1}{x},\ x \neq 0$

The domain of f^{-1} is $(-\infty, 0) \cup (0, \infty)$. The

range of f^{-1} is $(-\infty, 1) \cup (1, \infty)$.

4.2 Logarithmic Functions

4.2 Practice Problems

1. a. $2^{10} = 1024$ is equivalent to

$\log_2 1024 = 10$.

b. $9^{-1/2} = \dfrac{1}{3}$ is equivalent to $\log_9 \left(\dfrac{1}{3}\right) = -\dfrac{1}{2}$.

c. $p = a^q$ is equivalent to $\log_a p = q$

2. a. $\log_2 64 = 6$ is equivalent to $2^6 = 64$.

b. $\log_v u = w$ is equivalent to $v^w = u$.

3. a. $\log_3 9 = y \Rightarrow 3^y = 9 \Rightarrow 3^y = 3^2 \Rightarrow y = 2$

Thus, $\log_3 9 = 2$.

b. $\log_9 \dfrac{1}{3} = y \Rightarrow 9^y = \dfrac{1}{3} \Rightarrow 3^{2y} = 3^{-1} \Rightarrow y = -\dfrac{1}{2}$

Thus, $\log_9 \dfrac{1}{3} = -\dfrac{1}{2}$.

c. $\log_{1/2} 32 = y \Rightarrow \left(\dfrac{1}{2}\right)^y = 32 \Rightarrow 2^{-y} = 2^5 \Rightarrow$

$y = -5$

Thus, $\log_{1/2} 32 = -5$.

4. a. $\log_5 1 = 0$ **b.** $\log_3 3^5 = 5$

c. $7^{\log_7 5} = 5$

5. Since the domain of the logarithmic function

is $(0, \infty)$, the expression $\sqrt{1-x}$ must be

positive. $\sqrt{1-x}$ is defined for $(-\infty, 1)$, so the

domain of $\log_{10} \sqrt{1-x}$ is $(-\infty, 1)$.

6. Create a table of values to find ordered pairs

on the graph of $y = \log_2 x$.

x	$y = \log_2 x$	(x, y)
$\dfrac{1}{8}$	$2^{-3} = \dfrac{1}{8} \Rightarrow y = \log_2 \dfrac{1}{8} = -3$	$\left(\dfrac{1}{8}, -3\right)$
$\dfrac{1}{4}$	$2^{-2} = \dfrac{1}{4} \Rightarrow y = \log_2 \dfrac{1}{4} = -2$	$\left(\dfrac{1}{4}, -2\right)$
$\dfrac{1}{2}$	$2^{-1} = \dfrac{1}{2} \Rightarrow y = \log_2 \dfrac{1}{2} = -1$	$\left(\dfrac{1}{2}, -1\right)$
1	$2^0 = 1 \Rightarrow y = \log_2 1 = 0$	$(1, 0)$
x	$y = \log_2 x$	(x, y)
2	$2^1 = 2 \Rightarrow y = \log_2 2 = 1$	$(2, 1)$
4	$2^2 = 4 \Rightarrow y = \log_2 4 = 2$	$(4, 2)$
8	$2^3 = 8 \Rightarrow y = \log_2 8 = 3$	$(8, 3)$

Plot the ordered pairs and connect them with a

smooth curve.

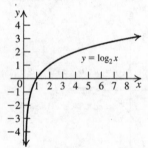

7. Shift the graph of $y = \log_2 x$ three units right,

then reflect the resulting graph about the

x-axis.

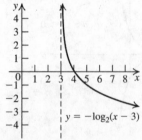

8. $P_2 = \log 3 - \log 2 \approx 0.176$

This means that about 17.6% of the data is

expected to have 2 as the first digit.

9. a. $y = \ln \dfrac{1}{e} \Rightarrow e^y = \dfrac{1}{e} \Rightarrow e^y = e^{-1} \Rightarrow y = -1$

Thus, $\ln \dfrac{1}{e} = -1$.

b. Using a calculator, we have $\ln 2 \approx 0.693$.

10. a. If P dollars are invested, then the amount $A = 3P$.

$A = Pe^{rt} \Rightarrow 3P = Pe^{0.065t} \Rightarrow 3 = e^{0.065t} \Rightarrow$

$\ln 3 = 0.065t \Rightarrow t = \dfrac{\ln 3}{0.065} \approx 16.9$

It will take approximately 17 years to triple your money.

b. $A = Pe^{rt} \Rightarrow 3P = Pe^{5r} \Rightarrow 3 = e^{5r} \Rightarrow$

$\ln 3 = 5r \Rightarrow t = \dfrac{\ln 3}{5} \approx 0.2197$

The investment will triple in 5 years at the rate of 21.97%.

11. Start with equation (2) in Example 11 in the text.

$T = 72 + 108e^{-0.1495317t}$

This equation gives the rate of cooling in a 72°F room when the starting temperature is 180°F. Since the final temperature is 120°F, replace T with 120 and solve for t.

$$120 = 72 + 108e^{-0.1495317t}$$
$$48 = 108e^{-0.1495317t}$$
$$\frac{4}{9} = e^{-0.1495317t}$$
$$\ln\left(\frac{4}{9}\right) = -0.1495317t$$

$-0.8109302 = -0.1495317t \Rightarrow 5.423 = t$

The employees should wait about 5.423 (realistically 5.5 minutes) to deliver the coffee at 120°F.

4.2 Basic Concepts and Skills

1. The domain of the function $y = \log_a x$ is $(0, \infty)$ and it range is $(-\infty, \infty)$.

3. The logarithm with base 10 is called the <u>common</u> logarithm, and the logarithm with base e is called the <u>natural</u> logarithm.

5. False

7. $\log_5 25 = 2$

9. $\log_{1/16} 4 = -\dfrac{1}{2}$

11. $\log_{10} 1 = 0$

13. $\log_{10} 0.1 = -1$

15. $a^2 + 2 = 7 \Rightarrow a^2 = 5 \Rightarrow \log_a 5 = 2$

17. $2a^3 - 3 = 10 \Rightarrow 2a^3 = 13 \Rightarrow a^3 = \dfrac{13}{2} \Rightarrow$

$\log_a \left(\dfrac{13}{2}\right) = 3$

19. $2^5 = 32$

21. $10^2 = 100$

23. $10^0 = 1$

25. $10^{-2} = 0.01$

27. $3\log_8 2 = 1 \Rightarrow \log_8 2 = \dfrac{1}{3} \Rightarrow 8^{1/3} = 2$

29. $e^x = 2$

31. $\log_5 125 = 3$ because $5^3 = 125$

33. $\log 10,000 = 4$ because $10^4 = 10,000$

35. $\log_2 \dfrac{1}{8} = -3$ because $2^{-3} = \dfrac{1}{8}$

37. $\log_3 \sqrt{27} = \dfrac{3}{2}$ because $3^{3/2} = \sqrt{27}$

39. $\log_{16} 2 = \dfrac{1}{4}$ because $16^{1/4} = 2$

41. $\log_3 1 = 0$

43. $\log_7 7 = 1$

45. $\log_6 6^7 = 7$

47. $3^{\log_3 5} = 5$

49. $2^{\log_2 7} + \log_5 5^{-3} = 7 + (-3) = 4$

51. $4^{\log_4 6} - \log_4 4^{-2} = 6 - (-2) = 8$

53. Since the domain of the logarithmic function is $(0, \infty)$, the expression $x + 1$ must be positive. This occurs in the interval $(-1, \infty)$, so the domain of $\log_2(x+1)$ is $(-1, \infty)$.

55. Since the domain of the logarithmic function is $(0, \infty)$, the expression $\sqrt{x-1}$ must be positive. This occurs in the interval $(1, \infty)$, so the domain of $\log_3 \sqrt{x-1}$ is $(1, \infty)$.

57. Since the domain of the logarithmic function is $(0, \infty)$, the expressions $(x - 2)$ and $(2x - 1)$ must be positive. This occurs in the interval $(2, \infty)$ for $x - 2$ and in the interval $\left(\frac{1}{2}, \infty\right)$ for $(2x - 1)$. The intersection of the two intervals is $(2, \infty)$, so the domain of f is $(2, \infty)$.

59. Since the domain of the logarithmic function is $(0, \infty)$, the expressions $(x - 1)$ and $(2 - x)$ must be positive. This occurs in the interval $(1, \infty)$ for $x - 2$ and in the interval $(-\infty, 2)$ for $(2 - x)$. The intersection of the two intervals is $(1, 2)$, so the domain of h is $(1, 2)$.

61. a. F   **b.** A

     c. D   **d.** B

     e. E   **f.** C

63. Shift the graph of $y = \log_4 x$ three units left. Domain: $(-3, \infty)$; range: $(-\infty, \infty)$; asymptote: $x = -3$

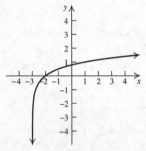

65. Reflect the graph of $y = \log_5 x$ about the x-axis. Domain: $(0, \infty)$; range: $(-\infty, \infty)$; asymptote: $x = 0$

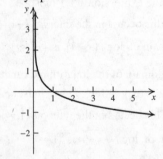

67. Reflect the graph of $y = \log_{1/5} x$ about the y-axis. Domain: $(-\infty, 0)$; range: $(-\infty, \infty)$; asymptote: $x = 0$

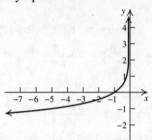

69. On $(0,1)$, reflect the graph of $y = \log_3 x$ about the x-axis. Domain: $(0, \infty)$; range: $[0, \infty)$; asymptote: $x = 0$

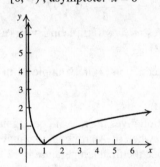

71. Shift the graph of $y = \log_2 x$ one unit right. Domain: $(1, \infty)$; range: $(-\infty, \infty)$; asymptote: $x = 1$

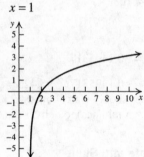

73. Shift the graph of $y = \log_2 x$ three units right and then reflect it about the y-axis. Domain: $(-\infty, 3)$; range: $(-\infty, \infty)$; asymptote: $x = 3$

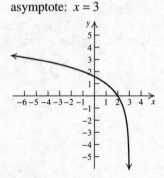

75. Shift the graph of $y = \log_2 x$ three units right, then reflect it about the *y*-axis, and then shift it 2 units up. Domain: $(-\infty, 3)$; range: $(-\infty, \infty)$; asymptote: $x = 3$

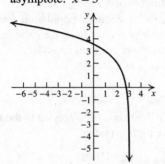

77. On $(-\infty, 0)$, reflect the graph of $y = \log_2 x$ about the *y*-axis. Domain: $(-\infty, 0) \cup (0, \infty)$; range: $(-\infty, \infty)$; asymptote: $x = 0$

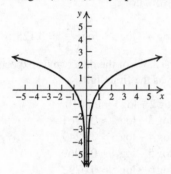

79. $\log_4(\log_3 81) = \log_4 4 = 1$ because $3^4 = 81$ and $4^1 = 4$

81. $\log_{\sqrt{2}} 2 = 2$ because $\sqrt{2}^2 = 2$.

83. $\log_{\sqrt{2}} 4 = 4$ because $\left(\sqrt{2}\right)^4 = 4$.

85. $\log x = 2 \Rightarrow 10^2 = 100 = x$

87. $\ln x = 1 \Rightarrow e^1 = e = x$

89. Shift the graph of $y = \ln x$ two units left.

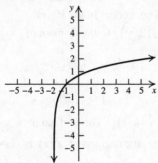

91. Shift the of $y = \ln x$ graph two units right, then reflect the graph about the *y*-axis, and then reflect the graph about the *x*-axis.

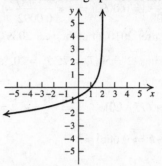

93. Stretch the graph of $y = \ln x$ vertically by a factor of 2, reflect the resulting graph about the *x*-axis, and shift it three units up.

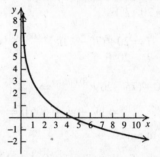

4.2 Applying the Concepts

95. $2P = Pe^{0.08t} \Rightarrow 2 = e^{0.08t} \Rightarrow \ln 2 = 0.08t \Rightarrow$ $t \approx 8.66$ years

97. $2P = Pe^{6k} \Rightarrow 2 = e^{6k} \Rightarrow \ln 2 = 6k \Rightarrow$ $k \approx 0.1155 = 11.55\%$

99. a. The population in 2010 was 109.7% of the population in 2000.
$308 = 1.097 A_0 \Rightarrow A_0 \approx 280.7657247$
The population in 2000 was about 280.8 million.

b. $308 = 280.7657247 e^{10r} \Rightarrow$
$$\frac{308}{280.7657247} = e^{10r} \Rightarrow$$
$$\ln\left(\frac{308}{280.7657247}\right) = 10r \Rightarrow$$
$$r = \frac{\ln\left(\frac{308}{280.7657247}\right)}{10} \approx 0.009257 \approx 0.93\%$$

c. $400 = 308e^{0.0092t} \Rightarrow \dfrac{400}{308} = e^{0.0092t} \Rightarrow$

$\ln\left(\dfrac{400}{308}\right) = 0.0092t \Rightarrow t = \dfrac{\ln\left(\dfrac{400}{308}\right)}{0.0092} \approx 28.4$

years after 2010, or in the year 2039.

101. Find k using $T = 50, T_0 = 75, T_s = 20,$ and $t = 1$ minute.

$50 = 20 + (75 - 20)e^{-k(1)} \Rightarrow \dfrac{30}{55} = e^{-k} \Rightarrow$

$\ln\dfrac{30}{55} = -k \Rightarrow 0.6061 \approx k$

a. **(i)** $T = 20 + (75 - 20)e^{-0.6061(5)} \Rightarrow$

$\quad T \approx 22.66°\text{F}$

(ii) $T = 20 + (75 - 20)e^{-0.6061(10)} \Rightarrow$

$\quad T \approx 20.13°\text{F}$

(iii) $\quad T = 20 + (75 - 20)e^{-0.6061(60)} \Rightarrow$

$\quad\quad T \approx 20°\text{F}$

b. $22 = 20 + (75 - 20)e^{-0.6061t} \Rightarrow$

$\dfrac{2}{55} = e^{-0.6061t} \Rightarrow \ln\dfrac{2}{55} = -0.6061t \Rightarrow$

$t \approx 5.5$ minutes

103. Find k using $T = 160, T_0 = 50, T_s = 400,$ and $t = 10$ minutes.

$160 = 400 + (50 - 400)e^{-10k} \Rightarrow \dfrac{240}{350} = e^{-10k} \Rightarrow$

$\ln\dfrac{24}{35} = -10k \Rightarrow k \approx 0.03773$

Use this value of k to find the time given $T = 220, T_0 = 50,$ and $T_s = 400.$

$220 = 400 + (50 - 400)e^{-0.03773t} \Rightarrow$

$\dfrac{18}{35} = e^{-0.03773t} \Rightarrow \ln\dfrac{18}{35} = -0.03773t \Rightarrow$

$t \approx 17.6$ minutes

105. a. $A(1) = 0.8(0.04) = 0.032$

Using the model $A(t) = A_0e^{rt}$, we have

$0.032 = 0.04e^r \Rightarrow 0.8 = e^r \Rightarrow r = \ln 0.8$
So, in one year (12 months), the concentration of the contaminant will be

$A(12) = 0.04e^{12\ln 0.8} = .0027 = 0.27\%$

b. $0.0001 \le 0.04e^{\ln 0.8t} \Rightarrow 0.0025 \le e^{\ln 0.8t} \Rightarrow$
$\ln 0.0025 \le \ln 0.8t \Rightarrow t \le 26.85$ months

107. a. $3000 = 1500e^{3r} \Rightarrow \ln 2 = 3r \Rightarrow$

$r = \dfrac{\ln 2}{3} \approx 0.231$

Thus, the function is $P \approx 1500e^{0.231t}$.
Alternatively, since the population doubles every three years, we have the geometric progression progression given by

$P = 1500 \cdot 2^{t/3}$.

b. $P = 1500e^{0.231(7)} \approx 7557$
There will be about 7557 sheep in the herd seven years from now.

c. $15{,}000 = 1500e^{0.231t} \Rightarrow 10 = e^{0.231t} \Rightarrow$

$\ln 10 = 0.231t \Rightarrow t = \dfrac{\ln 10}{0.231} \approx 10$

The herd will have 15,000 sheep about 10 years from now.

109. a. $P_3 = \log 4 - \log 3 = \log\left(\dfrac{4}{3}\right) \approx 0.125$

About 12.5% of the data can be expected to have 3 as the first digit.

b. $P_1 = \log 2 - \log 1 \approx 0.3010$
$P_2 = \log 3 - \log 2 \approx 0.1761$
$P_3 = \log 4 - \log 3 \approx 0.1249$
$P_4 = \log 5 - \log 4 \approx 0.0969$
$P_5 = \log 6 - \log 5 \approx 0.0792$
$P_6 = \log 7 - \log 6 \approx 0.0669$
$P_7 = \log 8 - \log 7 \approx 0.0580$
$P_8 = \log 9 - \log 8 \approx 0.0512$
$P_9 = \log 10 - \log 9 \approx 0.0458$
$P_1 + P_2 + P_3 + \cdots P_8 + P_9 = 1$

This means that one of the digits $1\ldots9$ will appear as the first digit.

4.2 Beyond the Basics

111. Since the domain of the logarithmic function is $(0, \infty)$, the expression $\dfrac{x-2}{x+1}$ must be positive. This occurs in the interval $(-\infty, -1) \cup (2, \infty)$, so the domain of

$\log_3\left(\dfrac{x-2}{x+1}\right)$ is $(-\infty, -1) \cup (2, \infty)$.

113. a. $h(x) = \log_3 x$ and $g(x) = \log_2 x$. So, $f(x) = g(h(x))$. The domain of $h(x)$ is $(0, \infty) \Rightarrow$ the domain of $f(x)$ is $(1, \infty)$.

b. $h(x) = \ln(x-1)$ and $g(x) = \log x$. So, $f(x) = g(h(x))$. The domain of $h(x)$ is $(1, \infty) \Rightarrow$ the domain of $f(x)$ is $(2, \infty)$.

c. $h(x) = \log(x-1)$ and $g(x) = \ln x$. So, $f(x) = g(h(x))$. The domain of $h(x)$ is $(1, \infty) \Rightarrow$ the domain of $f(x)$ is $(2, \infty)$.

d. $h(x) = \log(x-1)$ and $g(x) = \log x$. So, $f(x) = g(g(h(x)))$. The domain of $h(x)$ is $(1, \infty)$. The domain of $g(h(x))$ is $(2, \infty)$. (Note that $\log(x-1) = 0 \Rightarrow x = 2$.) So, the domain of $g(g(h(x)))$ is $(11, \infty)$.

115. $y = \log x \rightarrow y = \log(x-2) \rightarrow$

$y = \log\left(\frac{1}{2}x - 2\right) \rightarrow y = 3\log\left(\frac{1}{2}x - 2\right) \rightarrow$

$y = -3\log\left(\frac{1}{2}x - 2\right) \rightarrow y = -3\log\left(\frac{1}{2}x - 2\right) + 4$

117. a. $P = 100,000e^{-0.07(20)} = \$24,659.69$

b. $50,000 = 75,000e^{-10r} \Rightarrow \frac{2}{3} = e^{-10r} \Rightarrow$
$\ln(2/3) = -10r \Rightarrow r \approx 0.0405 = 4.05\%$

4.2 Critical Thinking/Discussion/Writing

119. $2^{\log_2 3} - 3^{\log_3 2} = 3 - 2 = 1$

120. $\log_3 4 = \log_3\left(2^2\right) = 2\log_3 2$
$\log_2 9 = \log_2\left(3^2\right) = 2\log_2 3$
Let $x = \log_3 2$ and let $y = \log_2 3$.
$\left(\log_3 4 + \log_2 9\right)^2 - \left(\log_3 4 - \log_2 9\right)^2$
$= (2x + 2y)^2 - (2x - 2y)^2$
$= \left[(2x+2y) + (2x-2y)\right] \cdot$
$\quad \left[(2x+2y) - (2x-2y)\right]$
$= (4x)(4y) = 16xy$
Note that
$x = \log_3 2 \Rightarrow 3^x = 2$ and $y = \log_2 3 \Rightarrow 2^y = 3$
$3^{xy} = \left(3^x\right)^y = 2^y = 3 \Rightarrow 3^{xy} = 3 \Rightarrow xy = 1$
Alternatively,
$xy = \log_3 2 \cdot \log_2 3 = \frac{\log 2}{\log 3} \cdot \frac{\log 3}{\log 2} = 1.$
Therefore, $16xy = 16 \cdot 1 = 16$, so
$\left(\log_3 4 + \log_2 9\right)^2 - \left(\log_3 4 - \log_2 9\right)^2 = 16.$

121. $\log_3\left[\log_4\left(\log_2 x\right)\right] = 0 \Rightarrow \log_4\left(\log_2 x\right) = 3^0 \Rightarrow$
$\log_4\left(\log_2 x\right) = 1 \Rightarrow \log_2 x = 4 \Rightarrow x = 2^4 = 16$

122. a. $f(x) = |\log x| = \begin{cases} -\log x & \text{if } 0 < x < 1 \\ \log x & \text{if } x \geq 1 \end{cases}$

b. $g(x) = |\ln(x-1)| + |\ln(x-2)|$
$= \begin{cases} \ln(x-1) - \ln(x-2) & \text{if } 2 < x < 3 \\ \ln(x-1) + \ln(x-2) & \text{if } x \geq 3 \end{cases}$

123. a. Yes, the statement is always true.

b. The increasing property is used.

124. There are four possibilities, $0 < a < 1$ with $c < 0$, $0 < a < 1$ with $c > 0$, $a > 1$ with $c < 0$, and $a > 1$ with $c > 0$. They are illustrated below.

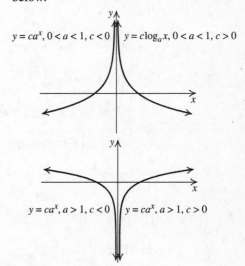

4.2 Maintaining Skills

125. $a^2 \cdot a^7 = a^9$

127. $\sqrt{a^8} = \left(a^8\right)^{1/2} = a^{8 \cdot (1/2)} = a^4$

129. $\left(\frac{243}{32}\right)^{4/5} = \left[\left(\frac{243}{32}\right)^{1/5}\right]^4 = \left(\frac{3}{2}\right)^4$

131. $\left(4.7 \times 10^7\right)\left(8.1 \times 10^5\right) = (4.7 \times 8.1)\left(10^7 \times 10^5\right)$
$= 38.07 \times 10^{12}$
$= 3.807 \times 10^{13}$

133. $\log_3 81 = 4$ because $3^4 = 81$.
$\log_3 3 + \log_3 27 = 1 + 3 = 4$ because $\log_3 3 = 1$ and $3^3 = 27$. Therefore, $\log_3 81 = \log_3 3 + \log_3 27$.

135. $\log_2 16 = 4$ because $2^4 = 16$.

$2\log_2 4 = 2 \cdot 2 = 4$ because $2^2 = 4$.

Therefore, $\log_2 16 = 2\log_2 4$.

4.3 Rules of Logarithms

4.3 Practice Problems

1. Given $\log_5 z = 3$ and $\log_5 y = 2$

a. $\log_5 (y/z) = \log_5 y - \log_5 z = 2 - 3 = -1$

b. $\log_5 \left(y^2 z^3\right) = \log_5 \left(y^2\right) + \log_5 \left(z^3\right)$
$= 2\log_5 y + 3\log_5 z$
$= 2 \cdot 2 + 3 \cdot 3 = 13$

2. a. $\ln \dfrac{2x-1}{x+4} = \ln(2x-1) - \ln(x+4)$

b. $\log \sqrt{\dfrac{4xy}{z}} = \log\left(\dfrac{4xy}{z}\right)^{1/2} = \dfrac{1}{2}\log\left(\dfrac{4xy}{z}\right)$

$= \dfrac{1}{2}\left(\log 4xy - \log z\right)$

$= \dfrac{1}{2}\left(\log 4 + \log x + \log y - \log z\right)$

$= \dfrac{1}{2}\left(\log\left(2^2\right) + \log x + \log y - \log z\right)$

$= \log 2 + \dfrac{1}{2}\log x + \dfrac{1}{2}\log y - \dfrac{1}{2}\log z$

3. $\dfrac{1}{2}\left[\log(x+1) + \log(x-1)\right]$

$= \dfrac{1}{2}\left[\log\left((x+1)(x-1)\right)\right]$

$= \dfrac{1}{2}\log\left(x^2-1\right) = \log\left(x^2-1\right)^{1/2}$

$= \log\sqrt{x^2-1}$

4. $K = 234^{567}$

$\log K = \log\left(234^{567}\right)$

$\log K = 567\log 234 \approx 1343.345391$

Since $\log K$ lies between the integers 1343 and 1344, the number K requires 1344 digits to the left of the decimal point. By definition of the common logarithm, we have

$K = 10^{1343.345391} = 10^{0.345391} \times 10^{1343}$

$\approx 2.215088 \times 10^{1343}$

5. $\log_3 15 = \dfrac{\log 15}{\log 3} \approx 2.46497$

6. a., b., Substitute (3, 3) and (9, 1) in the equation $y = c + b\log x$ to obtain

$3 = c + b\log 3$ (1)
$1 = c + b\log 9$ (2)

Subtract equation (1) from equation (2) and solve the resulting equation for b.

$-2 = b\log 9 - b\log 3 = b\left(\log 9 - \log 3\right)$

$= b\log\dfrac{9}{3} = b\log 3$

$b = -\dfrac{2}{\log 3}$

Substitute the value for b into equation (1) and solve for c.

$3 = c + \left(-\dfrac{2}{\log 3}\right)\log 3 \Rightarrow 3 = c - 2 \Rightarrow c = 5$

Substituting the values for b and c into $y = c + b\log x$ gives

$y = 5 - \dfrac{2}{\log 3}(\log x) = 5 - 2\left(\dfrac{\log x}{\log 3}\right)$

$= 5 - 2\log_3 x$

7. $\ln\left(\dfrac{A(t)}{A_0}\right) = kt \Rightarrow \ln\left(\dfrac{66}{100}\right) = 15k \Rightarrow$

$\ln(0.66) = 15k \Rightarrow k = \dfrac{\ln(0.66)}{15}$

To find the half-life, we use the formula

$h = -\dfrac{\ln 2}{k} = -\dfrac{\ln 2}{\ln(0.66)/15} \approx 25.$

The half-life of strontium-90 is about 25 years.

8. King Tut died in 1346 B.C., so the object was made in 1540 B.C., so the time elapsed between when the object was made and 1960 is $1540 + 1960 = 3500$ years. The decay function for carbon-14 in exponential form is $A(t) = A_0 e^{-0.0001216t}$. (See example 7 in the text.) Let $x =$ the percent of the original amount of carbon-14 in the object remaining after t years. Then

$xA_0 = A_0 e^{-0.0001216t} \Rightarrow x = e^{-0.0001216t}$.

$t = 3500$, so

$x = e^{-0.0001216(3500)} \approx 0.6534 \approx 65.34\%$.

4.3 Basic Concepts and Skills

1. $\log_a MN = \underline{\log_a M} + \underline{\log_a N}$.

3. $\log_a M^r = \underline{r\log_a M}$.

5. False. There is no rule for the logarithm of a sum. $\log_a u + \log_a v = \log_a (uv)$.

7. $\log 6 = \log(2 \cdot 3) = \log 2 + \log 3 = 0.3 + 0.48 = 0.78$

9. $\log 5 = \log\left(\dfrac{10}{2}\right) = \log 10 - \log 2 = 1 - 0.3 = 0.7$

11. $\log\left(\dfrac{2}{x}\right) = \log 2 - \log x = 0.3 - 2 = -1.7$

13. $\log(2x^2 y) = \log 2 + 2\log x + \log y = 0.3 + 2(2) + 3 = 7.3$

15. $\log \sqrt[3]{x^2 y^4} = \log(x^{2/3} y^{4/3}) = \dfrac{2}{3}\log x + \dfrac{4}{3}\log y = \dfrac{2}{3}(2) + \dfrac{4}{3}(3) = \dfrac{16}{3}$

17. $\log \sqrt[3]{48} = \log(2^4 \cdot 3)^{1/3} = \dfrac{4}{3}\log 2 + \dfrac{1}{3}\log 3 = \dfrac{4}{3}(0.3) + \dfrac{1}{3}(0.48) = 0.56$

19. $\ln[x(x-1)] = \ln x + \ln(x-1)$

21. $\log_a \sqrt{x}\, y^3 = \log_a \sqrt{x} + \log_a y^3 = \log_a x^{1/2} + \log_a y^3 = \dfrac{1}{2}\log_a x + 3\log_a y$

23. $\log_a \sqrt[3]{\dfrac{x}{y}} = \log_a \left(\dfrac{x}{y}\right)^{1/3} = \dfrac{1}{3}\log_a \left(\dfrac{x}{y}\right) = \dfrac{1}{3}\log_a x - \dfrac{1}{3}\log_a y$

25. $\log_2 \sqrt[4]{\dfrac{xy^2}{8}} = \log_2 \left(\dfrac{xy^2}{8}\right)^{1/4} = \dfrac{1}{4}\log_2 \left(\dfrac{xy^2}{8}\right) = \dfrac{1}{4}\log_2 \left(xy^2\right) - \dfrac{1}{4}\log_2 8$

$$= \dfrac{1}{4}\log_2 x + \dfrac{1}{4}\log_2 y^2 - \dfrac{1}{4}(3) = \dfrac{1}{4}\log_2 x + 2 \cdot \dfrac{1}{4}\log_2 y - \dfrac{3}{4} = \dfrac{1}{4}\log_2 x + \dfrac{1}{2}\log_2 y - \dfrac{3}{4}$$

27. $\log \dfrac{\sqrt{x^2+1}}{x+3} = \log \sqrt{x^2+1} - \log(x+3) = \dfrac{1}{2}\log(x^2+1) - \log(x+3)$

29. $\log_b x^2 y^3 z = 2\log_b x + 3\log_b y + \log_b z$

31. $\ln\left(\dfrac{x\sqrt{x-1}}{x^2+2}\right) = \ln\left(x\sqrt{x-1}\right) - \ln\left(x^2+2\right) = \ln x + \ln\left((x-1)^{1/2}\right) - \ln\left(x^2+2\right) = \ln x + \dfrac{1}{2}\ln(x-1) - \ln\left(x^2+2\right)$

33. $\ln\left(\dfrac{(x+1)^2}{(x-3)\sqrt{x+4}}\right) = \ln\left((x+1)^2\right) - \ln\left((x-3)\sqrt{x+4}\right) = 2\ln(x+1) - \left(\ln(x-3) + \ln(x+4)^{1/2}\right)$

$$= 2\ln(x+1) - \ln(x-3) - \dfrac{1}{2}\ln(x+4)$$

35. $\ln\left((x+1)\sqrt{\dfrac{x^2+2}{x^2+5}}\right) = \ln(x+1) + \ln\left(\dfrac{x^2+2}{x^2+5}\right)^{1/2} = \ln(x+1) + \dfrac{1}{2}\left(\ln\left(x^2+2\right) - \ln\left(x^2+5\right)\right)$

$$= \ln(x+1) + \dfrac{1}{2}\ln\left(x^2+2\right) - \dfrac{1}{2}\ln\left(x^2+5\right)$$

37. $\ln\left(\dfrac{x^3(3x+1)^4}{\sqrt{x^2+1}\,(x+2)^{-5}(x-3)^2}\right) = \ln\left(\dfrac{x^3(3x+1)^4(x+2)^5}{(x^2+1)^{1/2}(x-3)^2}\right)$

$$= \ln\left[x^3(3x+1)^4(x+2)^5\right] - \ln\left[(x^2+1)^{1/2}(x-3)^2\right]$$

$$= 3\ln x + 4\ln(3x+1) + 5\ln(x+2) - \frac{1}{2}\ln(x^2+1) - 2\ln(x-3)$$

39. $\log_2 x + \log_2 7 = \log_2(7x)$

41. $\dfrac{1}{2}\log x - \log y + \log z = \log\left(\dfrac{z\sqrt{x}}{y}\right)$

43. $\dfrac{1}{5}(\log_2 z + 2\log_2 y) = \log_2\left(y^2 z\right)^{1/5}$

$$= \log_2 \sqrt[5]{y^2 z}$$

45. $\ln x + 2\ln y + 3\ln z = \ln(xy^2 z^3)$

47. $2\ln x - \dfrac{1}{2}\ln(x^2+1) = \ln x^2 - \ln\sqrt{x^2+1}$

$$= \ln\left(\dfrac{x^2}{\sqrt{x^2+1}}\right)$$

49. $K = e^{500}$

$\log K = \log\left(e^{500}\right) = 500\log e = 217.147241 \Rightarrow$

$K = 10^{217.147241} = 10^{0.147241} \times 10^{217}$

$\quad \approx 1.4036 \times 10^{217}$

51. $K = 324^{756}$

$\log K = \log\left(324^{756}\right) = 756\log 324$

$\quad = 1897.972028 \Rightarrow$

$K = 10^{1897.972028} = 10^{0.972028} \times 10^{1897}$

$\quad \approx 9.3762 \times 10^{1897}$

53. $K = 234^{567}$

$\log K = \log\left(234^{567}\right) = 567\log 234$

$\quad = 1343.345$

$M = 567^{234}$

$\log M = \log\left(567^{234}\right) = 234\log 567$

$\quad = 664.338$

Since $\log K > \log M$, $K > M$. Thus, $234^{567} > 567^{234}$.

55. $K = 17^{200} \cdot 53^{67}$

$\log K = \log\left(17^{200} \cdot 53^{67}\right)$

$\quad = \log\left(17^{200}\right) + \log\left(53^{67}\right)$

$\quad = 200\log 17 + 67\log 53$

$\quad = 361.6163$

There are 362 digits in the given product.

57. $\log_2 5 = \dfrac{\log 5}{\log 2} \approx 2.322$

59. $\log_{1/2} 3 \approx -1.585$

61. $\log_{\sqrt{5}} \sqrt{17} = \dfrac{\log\sqrt{17}}{\log\sqrt{5}} \approx 1.760$

63. $\log_2 7 + \log_4 3 = \dfrac{\log 7}{\log 2} + \dfrac{\log 3}{\log 4} \approx 3.6$

65. $\log_3 \sqrt{3} = \log_3 3^{1/2} = \dfrac{1}{2}$

67. $\log_3(\log_2 8) = \log_3(\log_2 2^3) = \log_3 3 = 1$

69. $5^{2\log_5 3 + \log_5 2} = 5^{\log_5 3^2 + \log_5 2} = 5^{\log_5(9\cdot 2)}$

$$= 5^{\log_5 18} = 18$$

71. $\log 4 + 2\log 5 = \log(4\cdot 5^2) = \log 100$

$$= \log 10^2 = 2$$

73. Substitute $(10, 1)$ and $(1, 2)$ in the equation
$y = c + b\log x$ to obtain

$1 = c + b\log 10$ (1)

$2 = c + b\log 1$ (2)

Subtract equation (1) from equation (2) and solve the resulting equation for b.

$1 = b\log 1 - b\log 10 = b\cdot 0 - b\cdot 1 = -b \Rightarrow$

$b = -1$

Substitute the value for b into equation (1) and solve for c.

$1 = c - \log 10 \Rightarrow 1 = c - 1 \Rightarrow c = 2$

Substituting the values for b and c into $y = c + b\log x$ gives $y = 2 - \log x$.

75. Substitute $(e, 1)$ and $(1, 2)$ in the equation
$y = c + b \log x$ to obtain

$1 = c + b \log e$ (1)
$2 = c + b \log 1$ (2)

Since $\log 1 = 0$, equation (2) becomes $c = 2$.
Substitute the value for c into equation (1) and
solve for b.

$$1 = 2 + b \log e \Rightarrow -1 = b \log e \Rightarrow b = -\frac{1}{\log e}$$

Substituting the values for b and c into
$y = c + b \log x$ gives

$$y = 2 - \left(\frac{1}{\log e}\right) \log x = 2 - \left(\frac{\log x}{\log e}\right) = 2 - \ln x.$$

77. Substitute $(5, 4)$ and $(25, 7)$ in the equation
$y = c + b \log x$ to obtain

$4 = c + b \log 5$ (1)
$7 = c + b \log 25$ (2)

Subtract equation (1) from equation (2) and
solve the resulting equation for b.

$$3 = b \log 25 - b \log 5 = b(\log 25 - \log 5)$$
$$= b \log\left(\frac{25}{5}\right) = b \log 5 \Rightarrow$$
$$b = \frac{3}{\log 5}$$

Substitute the value for b into equation (1) and
solve for c.

$$4 = c + \left(\frac{3}{\log 5}\right) \log 5 \Rightarrow 4 = c + 3 \Rightarrow c = 1$$

Substituting the values for b and c into
$y = c + b \log x$ gives

$$y = 1 + \left(\frac{3}{\log 5}\right) \log x = 1 + 3\left(\frac{\log x}{\log 5}\right)$$
$$= 1 + 3 \log_5 x.$$

79. Substitute $(2, 4)$ and $(4, 9)$ in the equation
$y = c + b \log x$ to obtain

$4 = c + b \log 2$ (1)
$9 = c + b \log 4$ (2)

Subtract equation (1) from equation (2) and
solve the resulting equation for b.

$$5 = b \log 4 - b \log 2 = b(\log 4 - \log 2)$$
$$= b \log\left(\frac{4}{2}\right) \Rightarrow 5 = b \log 2 \Rightarrow b = \frac{5}{\log 2}$$

Substitute the value for b into equation (1) and
solve for c.

$$4 = c + \left(\frac{5}{\log 2}\right) \log 2 = c + 5 \Rightarrow -1 = c$$

Substituting the values for b and c into
$y = c + b \log x$ gives

$$y = -1 + \left(\frac{5}{\log 2}\right) \log x = -1 + 5\left(\frac{\log x}{\log 2}\right)$$
$$= -1 + 5 \log_2 x.$$

81. $A(t) = A_0 e^{kt} \Rightarrow 23 = 50 e^{12k} \Rightarrow$
$0.46 = e^{12k} \Rightarrow \ln(0.46) = 12k \Rightarrow$

$$k = \frac{\ln(0.46)}{12}$$

To find the half-life, we use the formula

$$h = -\frac{\ln 2}{k} = -\frac{\ln 2}{\ln(0.46)/12} \approx 10.7.$$

The half-life is about 10.7 years.

83. $A(t) = A_0 e^{kt} \Rightarrow 3.8 = 10.3 e^{15k} \Rightarrow$

$$\frac{3.8}{10.3} = e^{15k} \Rightarrow \ln\left(\frac{3.8}{10.3}\right) = 15k \Rightarrow$$

$$k = \frac{\ln\left(\frac{3.8}{10.3}\right)}{15}$$

To find the half-life, we use the formula

$$h = -\frac{\ln 2}{k} = -\frac{\ln 2}{\ln(3.8/10.3)/15} \approx 10.4.$$

The half-life is about 10.4 hours.

4.3 Applying the Concepts

85. $7.09 = 7 e^{1k} \Rightarrow$
$$\ln\frac{7.09}{7} = k \approx 0.012775 = 1.2775\%$$

87. $20 < 7 e^{100k} \Rightarrow \ln\left(\frac{20}{7}\right) < 100k \Rightarrow$
$0.010498 = 1.0498\% < k$
The maximum rate of growth is 1.0498%.

89. $3500 = 1000 e^{0.1t} \Rightarrow \ln 3.5 = 0.1t \Rightarrow$
$t \approx 12.53$ years

91. Because the half-life is 8 days, there will be
10 grams left after 8 days. Use this to find k:

$$10 = 20 e^{8k} \Rightarrow \frac{1}{2} = e^{8k} \Rightarrow \ln\left(\frac{1}{2}\right) = 8k \Rightarrow$$

$k \approx -0.08664$

$A = 20 e^{-0.08664(5)} \approx 12.969$ grams

93. $\frac{1}{2} = e^{-0.055t} \Rightarrow \ln\left(\frac{1}{2}\right) = -0.055t \Rightarrow t \approx 12.6$ yr.

95. $1.5 = 5 e^{-10k} \Rightarrow 0.3 = e^{-10k} \Rightarrow \ln 0.3 = -10k \Rightarrow$
$k \approx 0.1204$

97. Find k using $A = 8, A_0 = 16,$ and $t = 36$:

$$8 = 16e^{-36k} \Rightarrow \frac{1}{2} = e^{-36k} \Rightarrow$$

$$\ln\left(\frac{1}{2}\right) = 36k \Rightarrow k \approx 0.0193$$

$$A = 16e^{-0.0193(8)} \Rightarrow A \approx 13.7 \text{ grams.}$$

In exercises 99–102, we use the formula

$$P = \frac{r \cdot M}{1 - \left(1 + \frac{r}{n}\right)^{-nt}} \div n,$$

where P = the payment, r = the annual interest rate, M = the mortgage amount, t = the number of years, and n = the number of payments per year.

99. $P = \dfrac{0.06 \cdot 120{,}000}{1 - \left(1 + \frac{0.06}{12}\right)^{-12 \cdot 20}} \div 12 = 859.72$

The monthly payment is \$859.72.
There are 240 payments so the total amount paid is $240 \cdot \$859.72 = \$206{,}332.80$.
The amount of interest paid is
$206{,}332.80 - 120{,}000 = \$86{,}332.80$.

101. $850 = \dfrac{0.085 \cdot M}{1 - \left(1 + \frac{0.085}{12}\right)^{-12 \cdot 30}} \div 12$

$10{,}200 = \dfrac{0.085 \cdot M}{1 - \left(1 + \frac{0.085}{12}\right)^{-12 \cdot 30}}$

$M = \dfrac{10{,}200\left[1 - \left(1 + \frac{0.085}{12}\right)^{-12 \cdot 30}\right]}{0.085}$

$\approx 110{,}545.60$

Andy can afford a mortgage of about \$110,545.60.

4.3 Beyond the Basics

103. $\log_b\left(\sqrt{x^2 + 1} - x\right) + \log_b\left(\sqrt{x^2 + 1} + x\right)$

$= \log_b\left(\left(\sqrt{x^2 + 1} - x\right)\left(\sqrt{x^2 + 1} + x\right)\right)$

$= \log_b\left(x^2 + 1 - x^2\right) = \log_b 1 = 0$

105. $\left(\log_b a\right)\left(\log_a b\right) = \dfrac{\log a}{\log b} \cdot \dfrac{\log b}{\log a} = 1$

107.

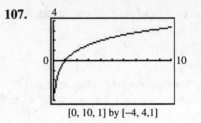

[0, 10, 1] by [−4, 4,1]

109. a. $\log\left(\dfrac{a}{b}\right) + \log\left(\dfrac{b}{a}\right) + \log\left(\dfrac{c}{a}\right) + \log\left(\dfrac{a}{c}\right) = 0$

(See exercise 106.)

b. $\log\left(\dfrac{a^2}{bc}\right) + \log\left(\dfrac{b^2}{ca}\right) + \log\left(\dfrac{c^2}{ab}\right)$

$= 2\log a - (\log b + \log c)$
$\qquad + 2\log b - (\log c + \log a)$
$\qquad + 2\log c - (\log a + \log b) = 0$

c. $\log_2 3 \cdot \log_3 4 = \dfrac{\log 3}{\log 2} \cdot \dfrac{\log 2^2}{\log 3}$

$= \dfrac{\log 3}{\log 2} \cdot \dfrac{2\log 2}{\log 3} = 2$

d. $\log_a b \cdot \log_b c \cdot \log_c a$

$= \dfrac{\log b}{\log a} \cdot \dfrac{\log c}{\log b} \cdot \dfrac{\log a}{\log c} = 1$

111. $f(x) = \log_4\left(\log_5\left(\log_3\left(18x - x^2 - 77\right)\right)\right)$

$18x - x^2 - 77 = -\left(x^2 - 18x + 77\right)$
$\qquad\qquad = -(x - 11)(x - 7)$

$\log_3\left(18x - x^2 - 77\right)$ is defined only for those

values of x which make $18x - x^2 - 77 > 0$.
Thus, the domain of $\log_3\left(18x - x^2 - 77\right)$ is
$(7, 11)$.

$\log_5\left[\log_3\left(18x - x^2 - 77\right)\right]$ is defined only for

those values of x in the interval $(7, 11)$ which
make $\log_3\left(18x - x^2 - 77\right) > 0$. We use a
graphing calculator to solve this. Note that we
use the change of base formula to define the
function.

$Y_1 = \log_3\left(18x - x^2 - 77\right) = \dfrac{\log\left(18x - x^2 - 77\right)}{\log 3}.$

(continued on next page)

(continued)

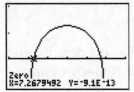 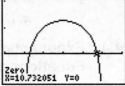

[6, 12] by [−1 ,2]

Thus, the domain of

$\log_5\left[\log_3\left(18x - x^2 - 77\right)\right]$ is approximately

(7.27, 10.73).

$\log_4\left(\log_5\left(\log_3\left(18x - x^2 - 77\right)\right)\right)$ is defined

only for those values of x in the interval
(7.27, 10.73) which make

$\log_5\left[\log_3\left(18x - x^2 - 77\right)\right] > 0.$

We use a graphing calculator to solve this.
Note that we use the change of base formula
to define the function.

$$Y_2 = \log_5\left(\log_3\left(18x - x^2 - 77\right)\right)$$

$$= \log_5\left(Y_1\right) = \frac{Y_1}{\log 5}$$

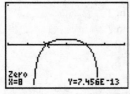 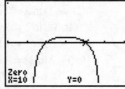

[6, 12] by [−1 ,1]

Thus, the domain of

$\log_4\left(\log_5\left(\log_3\left(18x - x^2 - 77\right)\right)\right)$ is (8, 10).

113. a. False   **b.** True

    c. False   **d.** False

    e. True   **f.** True

    g. False   **h.** True

    i. False   **j.** True

115.

$$\log\left(\frac{a+b}{3}\right) = \frac{1}{2}\left(\log a + \log b\right)$$

$$2\log\left(\frac{a+b}{3}\right) = \log\left(ab\right)$$

$$2\log\left(a+b\right) - 2\log 3 = \log\left(ab\right)$$

$$\log\left(a+b\right)^2 - \log 9 = \log\left(ab\right)$$

$$\log\left(a+b\right)^2 = \log\left(ab\right) + \log 9$$

$$\log\left(a+b\right)^2 = \log\left(9ab\right)$$

$$\left(a+b\right)^2 = 9ab$$

$$a^2 + 2ab + b^2 = 9ab$$

$$a^2 + b^2 - 7ab = 0$$

4.3 Critical Thinking/Discussion/Writing

117. In step 2, $\log\left(\dfrac{1}{2}\right)$ is negative, so $3 < 4 \Rightarrow$

$3\log\left(\dfrac{1}{2}\right) > 4\log\left(\dfrac{1}{2}\right).$

118. The domain of $2\log x$ is $(0, \infty)$, while the

domain of $\log(x^2)$ is $(-\infty, 0) \cup (0, \infty)$.

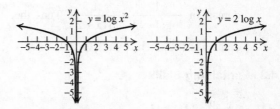

119. If $p = 2^m - 1$ is a prime number, then the

only way p and 2^m have a different number

of digits is if $2^m = 10^k$ or $2^{n-k} = 5^k$.

However, this is impossible because 2^{n-k} is

even and 5^k is odd. Thus, p and 2^m have the
same number of digits.

$$K = 2^{43112609}$$

$$\log K = \log\left(2^{43112609}\right) = 43112609\log 2$$

$$\approx 12978188.5$$

There are 12,978,189 digits in this prime
number.

120. $\dfrac{1}{\log\left(1-x\right)}$ is defined only for those values of

$\log\left(1-x\right) \neq 0$ and for those values of x such

that $\log\left(1-x\right)$ is defined.

$\log\left(1-x\right) = 0 \Rightarrow 1-x = 1 \Rightarrow x = 0$

$\log\left(1-x\right)$ is defined for

$1-x > 0 \Rightarrow 1 > x$ or $x < 1.$

Thus, the domain of $\dfrac{1}{\log\left(1-x\right)}$ is

$(-\infty, 0) \cup (0, 1).$

(continued on next page)

(*continued*)

$\dfrac{1}{\ln(x+2)}$ is defined only for those values of

$\ln(x+2) \neq 0$ and for those values of x such

that $\ln(x+2)$ is defined.

$\ln(x+2) = 0 \Rightarrow x+2 = 1 \Rightarrow x = -1$

$\ln(x+2)$ is defined for

$x+2 > 0 \Rightarrow x > -2$ or $-2 < x$.

Thus, the domain of $\dfrac{1}{\ln(x+2)}$ is

$(-2,-1) \cup (-1, \infty)$.

The intersection of the two domains is the

domain of $\dfrac{1}{\log(1-x)} + \dfrac{1}{\ln(x+2)}$. Thus, the

domain is $(-2,-1) \cup (-1, 0) \cup (0, 1)$.

4.3 Maintaining Skills

121. $11 \cdot 3^0 = 11 \cdot 1 = 11$

123. $4^x \cdot 2^{-2x+1} = \left(2^2\right)^x \cdot 2^{-2x+1} = 2^{2x} \cdot 2^{-2x+1}$

$\qquad = 2^{2x+(-2x+1)} = 2^1 = 2$

For exercises 125–128, let $t = 5^x$. Then $5^{2x} = t^2$.

125. $5^{2x} - 5^x = -1 \Rightarrow t^2 - t = -1 \Rightarrow t^2 - t + 1 = 0$

127. $\dfrac{5^x + 3 \cdot 5^{-x}}{5^x} = \dfrac{1}{4} \Rightarrow \dfrac{t + \dfrac{3}{t}}{t} = \dfrac{1}{4} \Rightarrow \dfrac{\dfrac{t^2+3}{t}}{t} = \dfrac{1}{4} \Rightarrow$

$\dfrac{t^2+3}{t^2} = \dfrac{1}{4} \Rightarrow 4t^2 + 12 = t^2 \Rightarrow 3t^2 + 12 = 0 \Rightarrow$

$t^2 + 4 = 0$

129. $2x - (11 + x) = 8x + (7 + 2x)$

$\qquad x - 11 = 10x + 7$

$\qquad\quad -18 = 9x \Rightarrow x = -2$

Solution set: $\{-2\}$

131. $\quad x^2 + 3x - 1 = 3$

$\quad\; x^2 + 3x - 4 = 0$

$\;\,(x-1)(x+4) = 0$

$x - 1 = 0 \mid x + 4 = 0$

$\quad\; x = 1 \mid \quad x = -4$

Solution set: $\{-4, 1\}$

133. $2x < 7 + x \Rightarrow x < 7$

Solution set: $(-\infty, 7)$

135. $12x > 30 - 3x$

$15x > 30 \Rightarrow x > 2$

Solution set: $(2, \infty)$

4.4 Exponential and Logarithmic Equations and Inequalities

4.4 Practice Problems

1. a. $3^x = 243 \Rightarrow 3^x = 3^5 \Rightarrow x = 5$

b. $8^x = 4 \Rightarrow \left(2^3\right)^x = 2^2 \Rightarrow 2^{3x} = 2^2 \Rightarrow$

$3x = 2 \Rightarrow x = \dfrac{2}{3}$

2. $7 \cdot 3^{x+1} = 11 \Rightarrow 3^{x+1} = \dfrac{11}{7} \Rightarrow$

$\ln\left(3^{x+1}\right) = \ln\left(\dfrac{11}{7}\right) \Rightarrow (x+1)\ln 3 = \ln\left(\dfrac{11}{7}\right) \Rightarrow$

$x + 1 = \dfrac{\ln(11/7)}{\ln 3} \Rightarrow x = \dfrac{\ln(11/7)}{\ln 3} - 1 \approx -0.589$

3. $\qquad\quad 3^{x+1} = 2^{2x}$

$\qquad \ln\left(3^{x+1}\right) = \ln\left(2^{2x}\right)$

$\qquad\quad (x+1)\ln 3 = 2x \ln 2$

$\qquad\quad x \ln 3 + \ln 3 = 2x \ln 2$

$\qquad\quad x \ln 3 - 2x \ln 2 = -\ln 3$

$\qquad\quad x(\ln 3 - 2\ln 2) = -\ln 3$

$\qquad\qquad x = -\dfrac{\ln 3}{\ln 3 - 2\ln 2} \approx 3.819$

4. $\quad e^{2x} - 4e^x - 5 = 0$

$\left(e^x - 5\right)\left(e^x + 1\right) = 0$

$e^x - 5 = 0$ or $e^x + 1 = 0$

$e^x - 5 = 0 \Rightarrow e^x = 5 \Rightarrow \ln\left(e^x\right) = \ln 5 \Rightarrow$

$x \ln e = \ln 5 \Rightarrow x = \ln 5 \approx 1.609$

$e^x + 1 = 0 \Rightarrow e^x = -1$, which is not possible.

5. The given model is $P(t) = P_0 \left(1 + r\right)^t$.

a. In 2020, ten years after the base year, the population of the United States will be

$P(10) = 308(1 + .011)^{10} \approx 343.61$ million.

The population of Pakistan will be

$P(10) = 185(1 + .033)^{10} \approx 255.96$ million.

b. To find when the population of the U.S. will be 350 million, solve for t:

$$350 = 308(1+.011)^t \Rightarrow \frac{350}{308} = 1.011^t \Rightarrow$$

$$\ln\left(\frac{350}{308}\right) = t\ln 1.011 \Rightarrow$$

$$t = \frac{\ln(350/308)}{\ln 1.011} \approx 11.69$$

The population of the U.S. will be 350 million approximately 11.69 years after 2010, sometime in the year 2022.

c. To find when the population of the two countries will be the same, solve for t:

$$308(1+.011)^t = 185(1+.033)^t$$

$$308(1.011)^t = 185(1.033)^t$$

$$\frac{308}{185} = \frac{1.033^t}{1.011^t} = \left(\frac{1.033}{1.011}\right)^t$$

$$\ln\left(\frac{308}{185}\right) = t\ln\left(\frac{1.033}{1.011}\right)$$

$$\frac{\ln(308/185)}{\ln(1.033/1.011)} \approx 23.68 = t$$

The populations of the two countries will be the same about 23.68 years after 2010, sometime in the year 2033.

6. $1 + 2\ln x = 4 \Rightarrow 2\ln x = 3 \Rightarrow \ln x = \dfrac{3}{2} \Rightarrow x = e^{3/2}$

7. $\log_3(x-8) + \log_3 x = 2$

$$\log_3\left[x(x-8)\right] = 2$$

$$x^2 - 8x = 3^2$$

$$x^2 - 8x - 9 = 0$$

$$(x-9)(x+1) = 0$$

$$x - 9 = 0 \quad \text{or} \quad x + 1 = 0$$

$$x = 9 \qquad\qquad x = -1$$

Now check each possible solution in the original equation.

$$\log_3(9-8) + \log_3 9 \overset{?}{=} 2$$

$$\log_3 1 + \log_3 9 \overset{?}{=} 2$$

$$0 + 2 = 2 \checkmark$$

$$\log_3(-1-8) + \log_3(-1) \overset{?}{=} 2$$

This is not possible because logarithms are not defined for negative values.
The solution set is $\{9\}$.

8. $\ln(x+5) + \ln(x+1) = \ln(x-1)$

$$\ln\left[(x+5)(x+1)\right] = \ln(x-1)$$

$$\ln\left(x^2 + 6x + 5\right) = \ln(x-1)$$

$$x^2 + 6x + 5 = x - 1$$

$$x^2 + 5x + 6 = 0$$

$$(x+2)(x+3) = 0$$

$$x + 2 = 0 \Rightarrow x = -2 \quad \text{or} \quad x + 3 = 0 \Rightarrow x = -3$$

Now check each possible solution in the original equation.

$$\ln(-2+5) + \ln(-2+1) \overset{?}{=} \ln(-2-1)$$

$$\ln 3 + \ln(-1) = \ln(-3)$$

This is not possible because logarithms are not defined for negative values.

$$\ln(-3+5) + \ln(-3+1) \overset{?}{=} \ln(-3-1)$$

$$\ln 2 + \ln(-2) = \ln(-4)$$

This is not possible because logarithms are not defined for negative values.
The solution set is $\varnothing$.

9. The year 1987 represents $t = 0$, so the year 2005 represents $t = 18$. Using equation (19) in example 9 in the text, we have

$$6.5 = \frac{35}{1+6e^{-18k}} \Rightarrow 6.5 + 39e^{-18k} = 35 \Rightarrow$$

$$e^{-18k} = \frac{28.5}{39} \Rightarrow -18k = \ln\left(\frac{28.5}{39}\right) \Rightarrow$$

$$k = -\frac{\ln(28.5/39)}{18} \approx .0174 = 1.74\%$$

The growth rate was about 1.74%.

10. $3(0.5)^x + 7 > 19 \Rightarrow 3(0.5)^x > 12 \Rightarrow$

$$(0.5)^x > 4 \Rightarrow x\ln 0.5 > \ln 4 \Rightarrow x < \frac{\ln 4}{\ln 0.5} \Rightarrow$$

$$x < -2$$

11. Since $1 - 3x$ must be positive, $x < \dfrac{1}{3}$.

$$\ln(1-3x) > 2 \Rightarrow 1 - 3x > e^2 \Rightarrow$$

$$-3x > e^2 - 1 \Rightarrow x < \frac{e^2 - 1}{-3} \Rightarrow$$

$$x < \frac{1-e^2}{3} \text{ or } x < -2.13$$

4.4 Basic Concepts and Skills

1. An equation that contains terms of the form a^x is called a(n) <u>exponential</u> equation.

3. The equation $y = \dfrac{m}{1+ae^{-bx}}$ represents a(n) <u>logistic</u> model.

5. False. A logarithmic equation can have a negative solution as long as the solution does not make the parameter of the logarithm negative.

7. $2^x = 16 \Rightarrow 2^x = 2^4 \Rightarrow x = 4$

9. $8^x = 32 \Rightarrow 2^{3x} = 2^5 \Rightarrow 3x = 5 \Rightarrow x = \dfrac{5}{3}$

11. $4^{|x|} = 128 \Rightarrow 2^{2|x|} = 2^7 \Rightarrow 2|x| = 7 \Rightarrow x = \pm\dfrac{7}{2}$

13. $5^{-|x|} = 625 \Rightarrow 5^{-|x|} = 5^4 \Rightarrow -|x| = 4 \Rightarrow$ there is no solution.

15. $\ln x = 0 \Rightarrow x = 1$

17. $\log_2 x = -1 \Rightarrow 2^{-1} = x \Rightarrow x = \dfrac{1}{2}$

19. $\log_3 |x| = 2 \Rightarrow 3^2 = |x| \Rightarrow x = \pm 9$

21. $\dfrac{1}{2}\log x - 2 = 0 \Rightarrow \dfrac{1}{2}\log x = 2 \Rightarrow \log x = 4 \Rightarrow 10^4 = 10,000 = x$

In exercises 23–52, the equations can be solved using either the common logarithm or the natural logarithm.

23. $2^x = 3 \Rightarrow x\ln 2 = \ln 3 \Rightarrow x = \dfrac{\ln 3}{\ln 2} \approx 1.585$

25. $2^{2x+3} = 15 \Rightarrow (2x+3)\ln 2 = \ln 15 \Rightarrow$
$2x + 3 = \dfrac{\ln 15}{\ln 2} \Rightarrow$
$x = \dfrac{\frac{\ln 15}{\ln 2} - 3}{2} = \dfrac{\ln 15 - 3\ln 2}{2\ln 2} \approx 0.453$

27. $5 \cdot 2^x - 7 = 10 \Rightarrow 2^x = \dfrac{17}{5} \Rightarrow$
$x\ln 2 = \ln\left(\dfrac{17}{5}\right) \Rightarrow x = \dfrac{\ln 17 - \ln 5}{\ln 2} \approx 1.766$

29. $3 \cdot 4^{2x-1} + 4 = 14 \Rightarrow 4^{2x-1} = \dfrac{10}{3} \Rightarrow$
$(2x-1)\ln 4 = \ln\left(\dfrac{10}{3}\right) \Rightarrow$
$x = \dfrac{\frac{\ln 10 - \ln 3}{\ln 4} + 1}{2} = \dfrac{\ln 10 - \ln 3 + \ln 4}{2\ln 4} \approx 0.934$

31. $5^{1-x} = 2^x \Rightarrow (1-x)\ln 5 = x\ln 2 \Rightarrow$
$\ln 5 - x\ln 5 = x\ln 2 \Rightarrow \ln 5 = x(\ln 5 + \ln 2) \Rightarrow$
$x = \dfrac{\ln 5}{\ln 5 + \ln 2} \approx 0.699$

33. $2^{1-x} = 3^{4x+6} \Rightarrow (1-x)\ln 2 = (4x+6)\ln 3 \Rightarrow$
$\ln 2 - x\ln 2 = 4x\ln 3 + 6\ln 3 \Rightarrow$
$\ln 2 - 6\ln 3 = 4x\ln 3 + x\ln 2 \Rightarrow$
$\ln 2 - 6\ln 3 = x(4\ln 3 + \ln 2) \Rightarrow$
$\dfrac{\ln 2 - 6\ln 3}{4\ln 3 + \ln 2} = x \Rightarrow x \approx -1.159$

35. $2 \cdot 3^{x-1} = 5^{x+1}$
$\ln\left(2 \cdot 3^{x-1}\right) = \ln\left(5^{x+1}\right)$
$\ln 2 + (x-1)\ln 3 = (x+1)\ln 5$
$\ln 2 + x\ln 3 - \ln 3 = x\ln 5 + \ln 5$
$x\ln 3 - x\ln 5 = \ln 5 - \ln 2 + \ln 3$
$x(\ln 3 - \ln 5) = \ln 5 - \ln 2 + \ln 3$
$x = \dfrac{\ln 5 - \ln 2 + \ln 3}{\ln 3 - \ln 5}$
$= \dfrac{\ln 2 - \ln 3 - \ln 5}{\ln 5 - \ln 3} \approx -3.944$

37. $1.065^t = 2 \Rightarrow t\ln 1.065 = \ln 2 \Rightarrow$
$t = \dfrac{\ln 2}{\ln 1.065} \approx 11.007$

39. Let $y = 2^x$. Then $2^{2x} - 4 \cdot 2^x = 21 \Rightarrow$
$y^2 - 4y - 21 = 0 \Rightarrow (y-7)(y+3) = 0 \Rightarrow$
$y = 7$ or $y = -3$. Reject the negative solution.
$2^x = 7 \Rightarrow x\ln 2 = \ln 7 \Rightarrow x = \dfrac{\ln 7}{\ln 2} \approx 2.807$

41. $9^x - 6 \cdot 3^x + 8 = 0 \Rightarrow 3^{2x} - 6 \cdot 3^x + 8 = 0$.
Let $y = 3^x$. Then $3^{2x} - 6 \cdot 3^x + 8 = 0 \Rightarrow$
$y^2 - 6y + 8 = 0 \Rightarrow (y-4)(y-2) = 0 \Rightarrow y = 2$
or $y = 4$. Substituting, we have $3^x = 2 \Rightarrow$
$x\ln 3 = \ln 2 \Rightarrow x = \dfrac{\ln 2}{\ln 3} \approx 0.631$ or $3^x = 4 \Rightarrow$
$x\ln 3 = \ln 4 \Rightarrow x = \dfrac{\ln 4}{\ln 3} \approx 1.262$.
The solution set is $\{0.631, 1.262\}$.

43. $3^{3x} - 4 \cdot 3^{2x} + 2 \cdot 3^x = 8$
$3^{3x} - 4 \cdot 3^{2x} + 2 \cdot 3^x - 8 = 0$
$3^{2x}\left(3^x - 4\right) + 2\left(3^x - 4\right) = 0$
$\left(3^{2x} + 2\right)\left(3^x - 4\right) = 0$
$3^{2x} + 2 = 0 \Rightarrow 3^{2x} = -2 \Rightarrow$ there is no solution.

(continued on next page)

(*continued*)

$$3^x - 4 = 0 \Rightarrow 3^x = 4 \Rightarrow x \ln 3 = \ln 4 \Rightarrow$$

$$x = \frac{\ln 4}{\ln 3} \approx 1.262$$

Solution set: $\left\{ \dfrac{\ln 4}{\ln 3} \approx 1.262 \right\}$

45. $\dfrac{3^x - 3^{-x}}{3^x + 3^{-x}} = \dfrac{1}{4} \Rightarrow 4 \cdot 3^x - 4 \cdot 3^{-x} = 3^x + 3^{-x} \Rightarrow$

$3 \cdot 3^x - 5 \cdot 3^{-x} = 0 \Rightarrow 3^x \left(3 \cdot 3^x - 5 \cdot 3^{-x} = 0 \right) \Rightarrow$

$3 \cdot 3^{2x} - 5 \cdot 3^0 = 0 \Rightarrow 3 \cdot 3^{2x} = 5 \Rightarrow 3^{2x} = \dfrac{5}{3} \Rightarrow$

$2x \ln 3 = \ln \left(\dfrac{5}{3} \right) \Rightarrow$

$x = \dfrac{\ln (5/3)}{2 \ln 3} = \dfrac{\ln 5 - \ln 3}{2 \ln 3} \approx 0.232$

47. $\dfrac{4}{2 + 3^x} = 1 \Rightarrow 4 = 2 + 3^x \Rightarrow 2 = 3^x \Rightarrow$

$\ln 2 = x \ln 3 \Rightarrow x = \dfrac{\ln 2}{\ln 3} \approx 0.631$

49. $\dfrac{17}{5 - 3^x} = 7 \Rightarrow 17 = 35 - 7 \cdot 3^x \Rightarrow$

$7 \cdot 3^x = 18 \Rightarrow \ln 7 + x \ln 3 = \ln 18 \Rightarrow$

$x = \dfrac{\ln 18 - \ln 7}{\ln 3} \approx 0.860$

51. $\dfrac{5}{2 + 3^x} = 4 \Rightarrow 5 = 4(2 + 3^x) \Rightarrow$

$5 = 8 + 4 \cdot 3^x \Rightarrow -\dfrac{3}{4} = 3^x \Rightarrow$ there is no solution.

53. $3 + \log(2x + 5) = 2 \Rightarrow \log(2x + 5) = -1 \Rightarrow$

$10^{-1} = 2x + 5 \Rightarrow -\dfrac{49}{10} = 2x \Rightarrow x = -\dfrac{49}{20}$

55. $\log \left(x^2 - x - 5 \right) = 0 \Rightarrow x^2 - x - 5 = 10^0 \Rightarrow$

$x^2 - x - 6 = 0 \Rightarrow (x - 3)(x + 2) = 0 \Rightarrow x = -2$
or $x = 3$

57. $\log_4 \left(x^2 - 7x + 14 \right) = 1 \Rightarrow x^2 - 7x + 14 = 4^1 \Rightarrow$

$x^2 - 7x + 10 = (x - 2)(x - 5) \Rightarrow x = 2$ or $x = 5$

59. $\ln(2x - 3) - \ln(x + 5) = 0 \Rightarrow$

$\ln(2x - 3) = \ln(x + 5) \Rightarrow 2x - 3 = x + 5 \Rightarrow x = 8$

61. $\log x + \log(x + 9) = 1 \Rightarrow \log \left(x(x + 9) \right) = 1 \Rightarrow$

$x^2 + 9x = 10 \Rightarrow x^2 + 9x - 10 = 0 \Rightarrow$

$(x + 10)(x - 1) = 0 \Rightarrow x = -10$ or $x = 1$.

Reject the negative solution because the logarithm of a negative number is undefined. The solution is $\{1\}$.

63. $\log_a (5x - 2) - \log_a (3x + 4) = 0 \Rightarrow$

$\log_a \dfrac{5x - 2}{3x + 4} = 0 \Rightarrow \dfrac{5x - 2}{3x + 4} = a^0 = 1 \Rightarrow$

$5x - 2 = 3x + 4 \Rightarrow x = 3$

65. $\log_6(x + 2) + \log_6(x - 3) = 1 \Rightarrow$

$\log_6 \left((x + 2)(x - 3) \right) = 1 \Rightarrow x^2 - x - 6 = 6 \Rightarrow$

$x^2 - x - 12 = 0 \Rightarrow (x - 4)(x + 3) = 0 \Rightarrow$

$x = 4$ or $x = -3$.

Reject the negative solution because the logarithm of a negative number is undefined. The solution is $\{4\}$.

67. $\log_3(2x - 7) - \log_3(4x - 1) = 2 \Rightarrow$

$\log_3 \dfrac{2x - 7}{4x - 1} = 2 \Rightarrow \dfrac{2x - 7}{4x - 1} = 3^2 = 9 \Rightarrow$

$2x - 7 = 36x - 9 \Rightarrow 2 = 34x \Rightarrow x = \dfrac{1}{17}$.

$2 \left(\dfrac{1}{17} \right) - 7 = -\dfrac{117}{17}$, so there is no solution.

69. $\log_7 3x + \log_7(2x - 1) = \log_7(16x - 10) \Rightarrow$

$\log_7 \left(3x(2x - 1) \right) = \log_7(16x - 10) \Rightarrow$

$6x^2 - 3x = 16x - 10 \Rightarrow 6x^2 - 19x + 10 = 0 \Rightarrow$

$(2x - 5)(3x - 2) = 0 \Rightarrow x = \dfrac{5}{2}$ or $x = \dfrac{2}{3}$

71. $f(x) = 20 + a \cdot 2^{kx}; \; f(0) = 50; \; f(1) = 140$

$50 = 20 + a \cdot 2^{k \cdot 0} \Rightarrow 50 = 20 + a \Rightarrow a = 30$

$140 = 20 + 30 \cdot 2^{k \cdot 1} \Rightarrow 4 = 2^k \Rightarrow k = 2$

$f(2) = 20 + 30 \cdot 2^{2 \cdot 2} = 20 + 480 = 500$

73. $f(x) = 16 + a \cdot 3^{kx}; \; f(0) = 21; \; f(4) = 61$

$21 = 16 + a \cdot 3^{k \cdot 0} \Rightarrow a = 5$

$61 = 16 + 5 \cdot 3^{k \cdot 4} \Rightarrow 9 = 3^{4k} \Rightarrow 3^2 = 3^{4k} \Rightarrow$

$2 = 4k \Rightarrow k = \dfrac{1}{2}$

$f(2) = 16 + 5 \cdot 3^{\frac{1}{2} \cdot 2} = 16 + 15 = 31$

75. $f(x) = \dfrac{10}{3 + ae^{kx}}; \; f(0) = 2; \; f(1) = \dfrac{1}{2}$

$2 = \dfrac{10}{3 + ae^{k \cdot 0}} \Rightarrow 2 = \dfrac{10}{3 + a} \Rightarrow 6 + 2a = 10 \Rightarrow$

$2a = 4 \Rightarrow a = 2$

(*continued on next page*)

(*continued*)

$$\frac{1}{2} = \frac{10}{3 + 2e^{k \cdot 1}} \Rightarrow \frac{1}{2} = \frac{10}{3 + 2e^k} \Rightarrow$$

$$3 + 2e^k = 20 \Rightarrow e^k = \frac{17}{2} \Rightarrow k = \ln\frac{17}{2}$$

$$f(2) = \frac{10}{3 + 2e^{2\left(\ln\frac{17}{2}\right)}} \approx 0.068 = \frac{4}{59}$$

77. $f(x) = \dfrac{4}{a + 4e^{kx}}$; $f(0) = 2$; $f(1) = 9$

$$2 = \frac{4}{a + 4e^{k \cdot 0}} = \frac{4}{a + 4} \Rightarrow 2a + 8 = 4 \Rightarrow a = -2$$

$$9 = \frac{4}{-2 + 4e^{k \cdot 1}} = \frac{4}{-2 + 4e^k} \Rightarrow$$

$$-18 + 36e^k = 4 \Rightarrow e^k = \frac{22}{36} = \frac{11}{18} \Rightarrow k = \ln\frac{11}{18}$$

$$f(2) = \frac{4}{-2 + 4e^{2\ln(11/18)}} \approx -7.902$$

79. $5(0.3)^x + 1 \le 11 \Rightarrow (0.3)^x \le 2 \Rightarrow$

$$x\ln 0.3 \le \ln 2 \Rightarrow x \ge \frac{\ln 2}{\ln 0.3} \quad \text{(Note that } \ln 0.3 < 0\text{)}$$

81. $-3(1.2)^x + 11 \ge 8 \Rightarrow -3(1.2)^x \ge -3 \Rightarrow$

$$1.2^x \le 1 \Rightarrow x\ln 1.2 \le \ln 1 \Rightarrow x \le 0$$

83. Note that the domain of $\log(5x + 15)$ is $(-3, \infty)$ since $5x + 15$ must be greater than 0.

$$\log(5x + 15) < 2 \Rightarrow 5x + 15 < 10^2 \Rightarrow$$
$$5x < 85 \Rightarrow x < 17$$

Solution set: $(-3, 17)$

85. Note that the domain of $\ln(x - 5)$ is $(5, \infty)$ since $x - 5$ must be greater than 0.

$$\ln(x - 5) \ge 1 \Rightarrow x - 5 \ge e \Rightarrow x \ge e + 5$$

Solution set: $[e + 5, \infty)$

87. Note that the domain of $\log_2(3x - 7)$ is $\left(\frac{7}{3}, \infty\right)$ since $3x - 7$ must be greater than 0.

$$\log_2(3x - 7) < 3 \Rightarrow 3x - 7 < 2^3 \Rightarrow 3x < 15 \Rightarrow$$
$$x < 5$$

Solution set: $\left(\frac{7}{3}, 5\right)$

4.4 Applying the Concepts

89. a. $18,000 = 10,000\left(1 + \dfrac{0.06}{1}\right)^{(1)t} \Rightarrow$

$$1.8 = 1.06^t \Rightarrow t = \frac{\ln 1.8}{\ln 1.06} \approx 10.087 \text{ years}$$

b. $18,000 = 10,000\left(1 + \dfrac{0.06}{4}\right)^{4t} \Rightarrow$

$$1.8 = 1.015^{4t} \Rightarrow 4t = \frac{\ln 1.8}{\ln 1.015} \Rightarrow t \approx 9.870 \text{ yr}$$

c. $18,000 = 10,000\left(1 + \dfrac{0.06}{12}\right)^{12t} \Rightarrow$

$$1.8 = 1.005^{12t} \Rightarrow 12t = \frac{\ln 1.8}{\ln 1.005} \Rightarrow$$
$$t \approx 9.821 \text{ years}$$

d. $18,000 = 10,000\left(1 + \dfrac{0.06}{365}\right)^{365t} \Rightarrow$

$$1.8 = \left(1 + \frac{0.06}{365}\right)^{365t} \Rightarrow$$

$$365t = \frac{\ln 1.8}{\ln\left(1 + \dfrac{0.06}{365}\right)} \Rightarrow t \approx 9.797 \text{ years}$$

e. $18,000 = 10,000e^{0.06t} \Rightarrow 1.8 = e^{0.06t} \Rightarrow$
$$\ln 1.8 = 0.06t \Rightarrow t \approx 9.796 \text{ years}$$

91. $40,000 = 20,000e^{8r} \Rightarrow 2 = e^{8r} \Rightarrow$

$$\ln 2 = 8r \Rightarrow r = \frac{\ln 2}{8} \approx 0.0866 = 8.66\%$$

$$c_5 = 20,000e^{5(0.0866)} \approx \$30,837.52$$

93. a. $\log\left(\dfrac{I}{12}\right) = -0.025(30) \Rightarrow$

$$\frac{I}{12} = 10^{-0.025(30)} \Rightarrow I \approx 2.134 \text{ lumens}$$

b. $\log\left(\dfrac{4}{12}\right) = -0.025x \Rightarrow x \approx 19.08 \text{ feet}$

95. a. First, find a by substituting $t = 0$ and

$$f(0) = 1000: 1000 = \frac{20,000}{1 + ae^{-k(0)}} \Rightarrow$$

$$1000(1 + a) = 20,000 \Rightarrow 1 + a = 20 \Rightarrow a = 19.$$

Now find k using $t = 4$ and $f(t) = 8999$.

$$8999 = \frac{20,000}{1 + 19e^{-k(4)}} \Rightarrow$$

$$8999 + 170,981e^{-4k} = 20,000 \Rightarrow$$

$$e^{-4k} = \frac{11,001}{170,981} \Rightarrow -4k = \ln\left(\frac{11,001}{170,981}\right) \Rightarrow$$

$$k \approx 0.68589$$

Use this value of k to find the number of people infected:

$$f(8) = \frac{20,000}{1 + 19e^{-0.68589(8)}} \approx 18,542$$

b. $12,400 = \dfrac{20,000}{1+19e^{-0.68589t}} \Rightarrow$

$12,400 + 12,400 \cdot 19e^{-0.68589t} = 20,000 \Rightarrow$

$12,400 \cdot 19e^{-0.68589t} = 7600 \Rightarrow$

$e^{-0.68589t} = \dfrac{7600}{12,400 \cdot 19} \Rightarrow$

$-0.68589t = \ln\left(\dfrac{7600}{12,400 \cdot 19}\right) \Rightarrow$

$t = \dfrac{\ln\left(\dfrac{7600}{12,400 \cdot 19}\right)}{-0.68589} \approx 5$

12,400 people will be infected after 5 weeks.

97. a. $\dfrac{4490}{1+e^{5.4094-1.0255(0)}} \approx 20$

b. The carrying capacity is the numerator, 4490.

c.

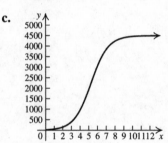

4.4 Beyond the Basics

99. $P = \dfrac{M}{1+e^{-kt}} \Rightarrow P\left(1+e^{-kt}\right) = M \Rightarrow$

$P + Pe^{-kt} = M \Rightarrow \dfrac{M-P}{P} = e^{-kt} \Rightarrow$

$\dfrac{M-P}{P} = \dfrac{1}{e^{kt}} \Rightarrow \dfrac{P}{M-P} = e^{kt} \Rightarrow$

$\ln\left(\dfrac{P}{M-P}\right) = kt \Rightarrow t = \dfrac{1}{k}\ln\left(\dfrac{P}{M-P}\right)$

101. a. $\dfrac{\log x}{2} = \dfrac{\log y}{3} = \dfrac{\log z}{5} = k \Rightarrow$

$\log x = 2k \Rightarrow x = 10^{2k}.$

$\log y = 3k \Rightarrow y = 10^{3k}.$

$\log z = 5k \Rightarrow z = 10^{5k}.$

So, $xy = 10^{2k} \cdot 10^{3k} = 10^{5k} = z.$

b. $\dfrac{\log x}{2} = \dfrac{\log y}{3} = \dfrac{\log z}{5} \Rightarrow$

$3\log x = 2\log y \Rightarrow \log x^3 = \log y^2 \Rightarrow$

$x^3 = y^2.$

$5\log x = 2\log z \Rightarrow \log x^5 = \log z^2 \Rightarrow$

$x^5 = z^2.$

So, $y^2 z^2 = x^3 x^5 = x^8$

103. $(\log x)^2 = \log x \Rightarrow (\log x)^2 - \log x = 0 \Rightarrow$

$\log x(\log x - 1) = 0 \Rightarrow \log x = 0 \Rightarrow x = 1$ or

$\log x - 1 = 0 \Rightarrow \log x = 1 \Rightarrow x = 10$

105. $\left(\log_3 x\right)\left(\log_3 3x\right) = 2 \Rightarrow$

$\left(\log_3 x\right)\left(\log_3 3 + \log_3 x\right) = 2 \Rightarrow$

$\left(\log_3 x\right)\left(1 + \log_3 x\right) = 2$

Let $u = \log_3 x.$ Then we have

$u\left(1+u\right) = 2 \Rightarrow u^2 + u - 2 = 0 \Rightarrow$

$\left(u+2\right)\left(u-1\right) = 0 \Rightarrow u = -2,\ 1 \Rightarrow$

$\log_3 x = -2 \Rightarrow x = 3^{-2} = \dfrac{1}{9}$ or

$\log_3 x = 1 \Rightarrow x = 3^1 = 3$

107. $\log_4 x^2(x-1)^2 - \log_2(x-1) = 1 \Rightarrow$

$\dfrac{\log_2 x^2(x-1)^2}{\log_2 4} - \log_2(x-1) = 1 \Rightarrow$

$\dfrac{\log_2\left[x(x-1)\right]^2}{2} - \log_2(x-1) = 1 \Rightarrow$

$\dfrac{2\log_2\left[x(x-1)\right]}{2} - \log_2(x-1) = 1 \Rightarrow$

$\log_2\left[x(x-1)\right] - \log_2(x-1) = 1 \Rightarrow$

$\log_2\dfrac{x(x-1)}{x-1} = 1 \Rightarrow \log_2 x = 1 \Rightarrow x = 2.$

109. $\dfrac{\log(3x-5)}{2} = \log x \Rightarrow \log(3x-5) = 2\log x \Rightarrow$

$\log(3x-5) = \log x^2 \Rightarrow 3x-5 = x^2 \Rightarrow$

$x^2 - 3x + 5 = 0$

$x = \dfrac{-(-3) \pm \sqrt{(-3)^2 - 4(1)(5)}}{2(1)}$

$= \dfrac{3 \pm \sqrt{9-20}}{2} = \dfrac{3 \pm \sqrt{-11}}{2}$

Since logarithms are not defined for complex numbers, there is no solution. Solution set: $\varnothing$

111. $f(x) = y = 3^x + 5$

Interchange x and y, then solve for y.

$x = 3^y + 5 \Rightarrow x - 5 = 3^y \Rightarrow$

$\log_3(x-5) = y = f^{-1}(x)$

113. $f(x) = y = 3 \cdot 4^x + 7$

Interchange x and y, then solve for y.

$$x = 3 \cdot 4^y + 7 \Rightarrow \frac{x-7}{3} = 4^y \Rightarrow \log_4\left(\frac{x-7}{3}\right) = y = f^{-1}(x)$$

115. $f(x) = y = 1 + \log_2(x-1)$

Interchange x and y, then solve for y.

$$x = 1 + \log_2(y-1) \Rightarrow x - 1 = \log_2(y-1) \Rightarrow 2^{x-1} = y - 1 \Rightarrow 2^{x-1} + 1 = y = f^{-1}(x)$$

117. $f(x) = y = \dfrac{1}{2}\ln\left(\dfrac{x-1}{x+1}\right)$

Interchange x and y, then solve for y.

$$x = \frac{1}{2}\ln\left(\frac{y-1}{y+1}\right) \Rightarrow 2x = \ln\left(\frac{y-1}{y+1}\right) \Rightarrow e^{2x} = \frac{y-1}{y+1} \Rightarrow (y+1)e^{2x} = y - 1 \Rightarrow$$

$$ye^{2x} + e^{2x} = y - 1 \Rightarrow e^{2x} + 1 = y - ye^{2x} \Rightarrow e^{2x} + 1 = y(1 - e^{2x}) \Rightarrow \frac{1+e^{2x}}{1-e^{2x}} = y = f^{-1}(x)$$

119. $7^n > 43^{67} \Rightarrow n\ln 7 > 67\ln 43 \Rightarrow n > \dfrac{67\ln 43}{\ln 7} \Rightarrow n > 129.5$

Thus, the smallest integer for which $7^n > 43^{67}$ is 130.

121. $8^{1/n} < 1.01 \Rightarrow \dfrac{1}{n}\ln 8 < \ln 1.01 \Rightarrow \dfrac{\ln 8}{\ln 1.01} < n \Rightarrow 208.98 < n$

Thus, the smallest integer for which $8^{1/n} < 1.01$ is 209.

123. **If**   **has 567 digits, then**   **(See Section 4.3, Example 4.)**

$$566 \le \log 31^n < 567 \Rightarrow 566 \le n\log 31 < 567 \Rightarrow \frac{566}{\log 31} \le n < \frac{567}{\log 31} \Rightarrow 379.5 \le n < 380.2$$

Thus, $n = 380$.

125. $\dfrac{x^{\log y}}{x^{\log z}} \cdot \dfrac{y^{\log z}}{y^{\log x}} \cdot \dfrac{z^{\log x}}{z^{\log y}} = \dfrac{\left(10^{\log x}\right)^{\log y}}{\left(10^{\log x}\right)^{\log z}} \cdot \dfrac{\left(10^{\log y}\right)^{\log z}}{\left(10^{\log y}\right)^{\log x}} \cdot \dfrac{\left(10^{\log z}\right)^{\log x}}{\left(10^{\log z}\right)^{\log y}} = \dfrac{10^{\log x \log y}}{10^{\log x \log z}} \cdot \dfrac{10^{\log y \log z}}{10^{\log x \log y}} \cdot \dfrac{10^{\log x \log z}}{10^{\log y \log z}}$

$$= \frac{10^{\log x \log y + \log y \log z + \log x \log z}}{10^{\log x \log z + \log x \log y + \log y \log z}} = 1$$

4.4 Critical Thinking/Discussion/Writing

127. $\dfrac{P}{2} = \dfrac{P}{1+ae^{-kt}} \Rightarrow \dfrac{1}{2} = \dfrac{1}{1+ae^{-kt}} \Rightarrow 1 + ae^{-kt} = 2 \Rightarrow ae^{-kt} = 1 \Rightarrow e^{-kt} = \dfrac{1}{a} = a^{-1} \Rightarrow$

$$-kt = \ln\left(a^{-1}\right) = -\ln a \Rightarrow t = \frac{\ln a}{k}$$

128. a. $\log_4(x-1)^2 = 3 \Rightarrow (x-1)^2 = 4^3 = 64 \Rightarrow x^2 - 2x - 63 = 0 \Rightarrow (x+7)(x-9) = 0 \Rightarrow x = -7, 9$

   b. $2\log_4(x-1) = 3 \Rightarrow \log_4(x-1) = \dfrac{3}{2} \Rightarrow x - 1 = 4^{3/2} = 8 \Rightarrow x = 9$

c. $2\log_4 |x-1| = 3 \Rightarrow \log_4 |x-1| = \dfrac{3}{2} \Rightarrow$

$|x-1| = 4^{3/2} = 8 \Rightarrow x-1 = 8 \Rightarrow x = 9$ or

$x-1 = -8 \Rightarrow x = -7$

The three equations do not have identical solutions because the equation in (a) has a quadratic term, the equation in (b) has a linear term, and the equation in (c) has an absolute value.

4.4 Maintaining Skills

129. $9^2 = 81 \Rightarrow \log_9 81 = 2$

131. $2 \cdot 10^x + 1 = 7 \Rightarrow 2 \cdot 10^x = 6 \Rightarrow 10^x = 3 \Rightarrow$
$\log 3 = x$

133. $\log_2 64 = 6 \Rightarrow 2^6 = 64$

135. $\log\left(\dfrac{A}{2}\right) = 3 \Rightarrow 10^3 = \dfrac{A}{2} \Rightarrow A = 2 \cdot 10^3$

137. $\log_5 x = -3 \Rightarrow x = 5^{-3} = \dfrac{1}{125}$

139. $\log_x 1000 = 3 \Rightarrow 1000 = x^3 \Rightarrow 10^3 = x^3 \Rightarrow$
$x = 10$

141. $2^{x+1} = 32 \Rightarrow 2^{x+1} = 2^5 \Rightarrow x+1 = 5 \Rightarrow x = 4$

143.
$$3^{x+1} = 5^{2x-3}$$
$$(x+1)\ln 3 = (2x-3)\ln 5$$
$$x\ln 3 + \ln 3 = 2x\ln 5 - 3\ln 5$$
$$\ln 3 + 3\ln 5 = 2x\ln 5 - x\ln 3$$
$$\ln 3 + 3\ln 5 = x(2\ln 5 - \ln 3)$$
$$x = \frac{\ln 3 + 3\ln 5}{2\ln 5 - \ln 3}$$

4.5 Logarithmic Scales

4.5 Practice Problems

1. a. $\text{pH} = -\log\left[H^+\right] = -\log\left(2.68 \times 10^{-6}\right)$

$= -\left(\log 2.68 + \log 10^{-6}\right)$

$= -\log 2.68 + 6 \approx 5.57$

b. $8.47 = -\log\left[H^+\right]$

$-8.47 = \log\left[H^+\right]$

$\left[H^+\right] = 10^{-8.47} = 10^{0.53} \times 10^{-9} \approx 3.39 \times 10^{-9}$

2. $2.8 = \text{pH}_{\text{acid rain}} = -\log\left[H^+\right]_{\text{acid rain}} \Rightarrow$

$\left[H^+\right]_{\text{acid rain}} = 10^{-2.8}$

$6.2 = \text{pH}_{\text{ordinary rain}} = -\log\left[H^+\right]_{\text{ordinary rain}} \Rightarrow$

$\left[H^+\right]_{\text{ordinary rain}} = 10^{-6.2}$

$\dfrac{\left[H^+\right]_{\text{acid rain}}}{\left[H^+\right]_{\text{ordinary rain}}} = \dfrac{10^{-2.8}}{10^{-6.2}} = 10^{-2.8-(-6.2)}$

$= 10^{3.4} \approx 2512$

This acid rain is about 2512 times more acidic than the ordinary rain.

3. $M = \log\left(\dfrac{I}{I_0}\right)$

$6.5 = \log\left(\dfrac{I}{I_0}\right) \Rightarrow \left(\dfrac{I}{I_0}\right) = 10^{6.5} \Rightarrow$

$I = 10^{6.5} I_0 \approx 3{,}162{,}278 I_0$

4. Let I_M denote the intensity of the Mozambique earthquake and let I_C denote the intensity of the southern California earthquake. Then we have

$7.0 = \log\left(\dfrac{I_M}{I_0}\right) \Rightarrow \dfrac{I_M}{I_0} = 10^{7.0} \Rightarrow I_M = 10^{7.0} I_0$

$5.2 = \log\left(\dfrac{I_C}{I_0}\right) \Rightarrow \dfrac{I_C}{I_0} = 10^{5.2} \Rightarrow I_C = 10^{5.2} I_0$

$\dfrac{I_M}{I_C} = \dfrac{10^{7.0} I_0}{10^{5.2} I_0} = \dfrac{10^{7.0}}{10^{5.2}} = 10^{7.0-5.2} = 10^{1.8} \approx 63.1$

The intensity of the Mozambique earthquake was about 63 times that of the southern California earthquake.

5. Let M_{2011} and E_{2011} represent the magnitude and energy of the 2011 Japan earthquake. Let M_{1997} and E_{1997} represent the magnitude and energy of the 1977 Iran earthquake. From Table 4.7, we have $M_{2011} = 9.0$ and $M_{1997} = 7.5$.

$\dfrac{E_{2011}}{E_{1997}} = 10^{1.5(M_{2011} - M_{1997})} = 10^{1.5(9.0-7.5)} = 10^{2.25}$

≈ 177.8

The energy released by the 2011 Japan earthquake was about 178 times that released by the 1997 Iran earthquake.

6. $I = 200 \times 10^{-7} \text{ W/m}^2$ and
$I_0 = 10^{-12} \text{ W/m}^2$.

$$L = 10\log\left(\frac{I}{I_0}\right) = 10\log\left(\frac{200 \times 10^{-7}}{10^{-12}}\right)$$
$$= 10\log\left(200 \times 10^5\right) = 10\left[\log 200 + \log 10^5\right]$$
$$= 10\left[\log 200 + 5\log 10\right] = 10\left[\log 200 + 5\right]$$
$$= 10\log 200 + 50 \approx 73.01$$

The decibel level is approximately 73 dB.

7. $I = I_0 \times 10^{L/10}$
$I_{75} = I_0 \times 10^{75/10}$
$I_{55} = I_0 \times 10^{55/10}$
$$\frac{I_{75}}{I_{55}} = \frac{I_0 \times 10^{75/10}}{I_0 \times 10^{55/10}} = 10^{20/10} = 10^2 = 100$$

A 75 dB sound is 100 times more intense than a 55 dB sound.

8. $I = I_0 \times 10^{L/10}$
$$I_{48} = 10^{-12} \times 10^{48/10} = 10^{-12} \times 10^{4.8} = 10^{-7.2}$$
$$= 10^{0.8} \times 10^{-8} \approx 6.3 \times 10^{-8} \text{ W/m}^2$$

The intensity of a 48 dB sound is about $6.3 \times 10^{-8} \text{ W/m}^2$.

9. B is two semitones above A, so $P(f) = 200$.
$f_0 = 440$ Hz
$$P(f) = 1200\log_2 \frac{f}{f_0}$$
$$200 = 1200\log_2 \frac{f}{440} \Rightarrow \frac{1}{6} = \log_2 \frac{f}{440} \Rightarrow$$
$$2^{1/6} = \frac{f}{440} \Rightarrow f = 2^{1/6} \cdot 440 \approx 493.9$$

The frequency of B is about 494 Hz.

10. If the reference frequency is f_0 and it increases by 25%, then
$f = f_0 + 0.25f_0 = 1.25f_0$. Then,
$$P(f) = 1200\log_2 \frac{1.25f_0}{f_0}$$
$$= 1200\log_2 1.25 = 1200\frac{\log 1.25}{\log 2}$$
$$\approx 386.3$$
Since an equal tempered major third is 400 cents, the difference is about $400 - 386.3 = 13.7$ cents, or a little less than 14 cents.

11. a. $m_2 - m_1 = 2.5\log\left(\frac{b_1}{b_2}\right)$

$$2 - 0 = \frac{5}{2}\log\left(\frac{b_1}{b_2}\right) \Rightarrow \frac{4}{5} = \log\left(\frac{b_1}{b_2}\right) \Rightarrow$$
$$\frac{b_1}{b_2} = 10^{4/5} \approx 6.3$$

A magnitude 0 star is approximately 6.3 times brighter than a magnitude 2 star.

b. Let $m_1 = 4.6$ and $b_2 = 150b_1$.

$$m_2 - m_1 = 2.5\log\left(\frac{b_1}{b_2}\right)$$
$$m_2 - 4.6 = 2.5\log\left(\frac{b_1}{1.50b_1}\right) = 2.5\log\left(\frac{1}{1.50}\right)$$
$$= 2.5\log 1.50^{-1} = -2.5\log 1.50$$
$$m_2 = -2.5\log 1.50 + 4.6 \approx 4.1598$$

The magnitude of the star that is 50% brighter than a star of magnitude 4.6 is about 4.1598.

4.5 Basic Concepts and Skills

1. If the pH value of a solution is less than 7, the solution is <u>acidic</u>.

3. On the Richter scale, the magnitude of an earthquake $M = \log\dfrac{I}{I_0}$.

5. The loudness L of a sound of intensity I is given by $L = 10\log\left(\dfrac{I}{I_0}\right)$.

7. If two stars have magnitudes m_1 and m_2 with apparent brightness b_1 and b_2, respectively, then $m_2 - m_1 = 2.5\log\dfrac{b_1}{b_2}$.

9. True

11. $\text{pH} = -\log\left[H^+\right] = -\log\left(10^{-8}\right) = 8\log 10 = 8$
The substance is a base.

13. $\text{pH} = -\log\left[H^+\right] = -\log\left(2.3 \times 10^{-5}\right)$
$$= -\left(\log 2.3 + \log 10^{-5}\right)$$
$$= -\left(\log 2.3 - 5\log 10\right) = -\left(\log 2.3 - 5\right)$$
$$= -\log 2.3 + 5 \approx 4.64$$
The substance is an acid.

15. $\text{pH} = 6 = -\log\left[\text{H}^+\right] \Rightarrow \left[\text{H}^+\right] = 10^{-6}$

$\left[\text{H}^+\right]\left[\text{OH}^-\right] = 10^{-14}$

$10^{-6}\left[\text{OH}^-\right] = 10^{-14}$

$\left[\text{OH}^-\right] = \dfrac{10^{-14}}{10^{-6}} = 10^{-8}$

17. $\text{pH} = 9.5 = -\log\left[\text{H}^+\right] \Rightarrow \left[\text{H}^+\right] = 10^{-9.5} \Rightarrow$

$\left[\text{H}^+\right] = 10^{0.5} \times 10^{-10} \approx 3.16 \times 10^{-10}$

$\left[\text{H}^+\right]\left[\text{OH}^-\right] = 10^{-14}$

$\left(3.16 \times 10^{-10}\right)\left[\text{OH}^-\right] = 10^{-14}$

$\left[\text{OH}^-\right] = \dfrac{10^{-14}}{3.16 \times 10^{-10}} = \dfrac{1}{3.16} \times 10^{-4}$

$\approx 0.316 \times 10^{-4} = 3.16 \times 10^{-5}$

19. a. $M = \log\left(\dfrac{I}{I_0}\right)$

$5 = \log\left(\dfrac{I}{I_0}\right) \Rightarrow \left(\dfrac{I}{I_0}\right) = 10^5 \Rightarrow I = 10^5 I_0$

b. $\log E = 4.4 + 1.5M$

$\log E = 4.4 + 1.5(5) = 11.9 \Rightarrow$

$E = 10^{11.9} = 10^{0.9} \times 10^{11}$

$\approx 7.94 \times 10^{11}$ joules

21. a. $M = \log\left(\dfrac{I}{I_0}\right)$

$7.8 = \log\left(\dfrac{I}{I_0}\right) \Rightarrow \left(\dfrac{I}{I_0}\right) = 10^{7.8} \Rightarrow$

$I = 10^{0.8} \times 10^7 I_0 \approx 6.3 \times 10^7 I_0$

b. $\log E = 4.4 + 1.5M$

$\log E = 4.4 + 1.5(7.8) = 16.1 \Rightarrow$

$E = 10^{16.1} = 10^{0.1} \times 10^{16}$

$\approx 1.26 \times 10^{16}$ joules

23. a. $\log E = 4.4 + 1.5M$

$\log 10^{13.4} = 4.4 + 1.5M \Rightarrow 13.4 = 4.4 + 1.5M$

$M = \dfrac{13.4 - 4.4}{1.5} = \dfrac{9}{1.5} = 6$

b. $M = \log\left(\dfrac{I}{I_0}\right)$

$6 = \log\left(\dfrac{I}{I_0}\right) \Rightarrow \left(\dfrac{I}{I_0}\right) = 10^6 \Rightarrow I = 10^6 I_0$

25. a. $\log E = 4.4 + 1.5M$

$\log 10^{12} = 4.4 + 1.5M \Rightarrow 12 = 4.4 + 1.5M$

$M = \dfrac{12 - 4.4}{1.5} = \dfrac{7.6}{1.5} \approx 5.1$

b. $M = \log\left(\dfrac{I}{I_0}\right)$

$5.1 = \log\left(\dfrac{I}{I_0}\right) \Rightarrow \left(\dfrac{I}{I_0}\right) = 10^{5.1}$

$I = 10^{5.1} I_0 = 10^{0.1} \times 10^5 I_0 \approx 1.26 \times 10^5 I_0$

27. $L = 10\log\left(\dfrac{I}{I_0}\right) = 10\log\left(\dfrac{10^{-8}}{10^{-12}}\right) = 10\log 10^4$

$= 40$

29. $L = 10\log\left(\dfrac{I}{I_0}\right) = 10\log\left(\dfrac{3.5 \times 10^{-7}}{10^{-12}}\right)$

$= 10\log\left(3.5 \times 10^5\right) = 10\left(\log 3.5 + \log 10^5\right)$

$= 10\log 3.5 + 50 \approx 55.4$

31. $I = I_0 \times 10^{L/10} = 10^{-12} \times 10^{80/10}$

$= 10^{-12} \times 10^8 = 10^{-4}$

33. $I = I_0 \times 10^{L/10} = 10^{-12} \times 10^{64.7/10}$

$= 10^{-12} \times 10^{6.47} = 10^{-12} \times 10^6 \times 10^{0.47}$

$\approx 2.95 \times 10^{-6}$

35. A# is one semitone above A, so $P(f) = 100$.

$f_0 = 440$ Hz

$P(f) = 1200\log_2 \dfrac{f}{f_0}$

$100 = 1200\log_2 \dfrac{f}{440} \Rightarrow \dfrac{1}{12} = \log_2 \dfrac{f}{440} \Rightarrow$

$2^{1/12} = \dfrac{f}{440} \Rightarrow f = 2^{1/12} \cdot 440 \approx 466$ Hz

The frequency of A# is about 466 Hz.

C is three semitones above A, so $P(f) = 300$.

$f_0 = 440$ Hz

$P(f) = 1200\log_2 \dfrac{f}{f_0}$

$300 = 1200\log_2 \dfrac{f}{440} \Rightarrow \dfrac{1}{4} = \log_2 \dfrac{f}{440} \Rightarrow$

$2^{1/4} = \dfrac{f}{440} \Rightarrow f = 2^{1/4} \cdot 440 \approx 523$ Hz

The frequency of C is about 523 Hz.

37. $P(f) = 1200 \log_2 \dfrac{f}{f_0} = 1200 \log_2 \left(\dfrac{10}{9} \right)$

$\qquad = 1200 \left(\dfrac{\log \left(\frac{10}{9} \right)}{\log 2} \right) \approx 182$

The difference, $200 - 182 = 18$ cents, is noticeable.

39. Let the magnitude and brightness of star A be m_1 and b_1, and let the magnitude of star B be m_2 and b_2.

$m_2 - m_1 = 2.5 \log \left(\dfrac{b_1}{b_2} \right)$

$20 - 4 = 2.5 \log \left(\dfrac{b_1}{b_2} \right) \Rightarrow 16 = 2.5 \log \left(\dfrac{b_1}{b_2} \right) \Rightarrow$

$6.4 = \log \left(\dfrac{b_1}{b_2} \right) \Rightarrow \dfrac{b_1}{b_2} = 10^{6.4} \Rightarrow$

$b_1 = 10^{0.4} \times 10^6 b_2 \approx 2.5 \times 10^6 b_2$

Star A is 2.5×10^6 times as bright as star B.

41. Let the magnitude and brightness of star A be m_1 and b_1, and let the magnitude of star B be $m_2 = m_1 - 2$ and b_2.

$m_2 - m_1 = 2.5 \log \left(\dfrac{b_1}{b_2} \right)$

$(m_1 - 2) - m_1 = 2.5 \log \left(\dfrac{b_1}{b_2} \right) \Rightarrow$

$-2 = 2.5 \log \left(\dfrac{b_1}{b_2} \right) \Rightarrow -0.8 = \log \left(\dfrac{b_1}{b_2} \right) \Rightarrow$

$\dfrac{b_1}{b_2} = 10^{-0.8} \approx \dfrac{1}{6.3} \Rightarrow 6.3 b_1 = b_2$

Star B is about 6.3 times brighter than star A.

4.5 Applying the Concepts

43. $pH = -\log \left[H^+ \right] = -\log \left(3.98 \times 10^{-8} \right)$

$\qquad = -\left(\log 3.98 + \log 10^{-8} \right) = -\left(\log 3.98 - 8 \right)$

$\qquad = -\log 3.98 + 8 \approx 7.4$

The pH of human blood is about 7.4; it is basic.

45. a. $pH = 3.15 = -\log \left[H^+ \right] \Rightarrow \left[H^+ \right] = 10^{-3.15} \Rightarrow$

$\qquad \left[H^+ \right] = 10^{0.85} \times 10^{-4} \approx 7.1 \times 10^{-4}$

b. $pH = 7.2 = -\log \left[H^+ \right] \Rightarrow \left[H^+ \right] = 10^{-7.2} \Rightarrow$

$\qquad \left[H^+ \right] = 10^{0.8} \times 10^{-8} \approx 6.3 \times 10^{-8}$

c. $pH = 7.78 = -\log \left[H^+ \right] \Rightarrow \left[H^+ \right] = 10^{-7.78} \Rightarrow$

$\qquad \left[H^+ \right] = 10^{0.22} \times 10^{-8} \approx 1.7 \times 10^{-8}$

d. $pH = 3 = -\log \left[H^+ \right] \Rightarrow \left[H^+ \right] = 10^{-3} \Rightarrow$

$\qquad \left[H^+ \right] = 10^{-3}$

47. $pH_{\text{acid rain}} = 3.8 = -\log \left[H^+ \right] \Rightarrow$

$\qquad \left[H^+ \right] = 10^{-3.8} \Rightarrow$

$\qquad \left[H^+ \right] = 10^{0.2} \times 10^{-4} \approx 1.6 \times 10^{-4}$

The average concentration of hydrogen ions in the acid rain is 1.6×10^{-4} moles per liter.

$pH_{\text{ordinary rain}} = 6 = -\log \left[H^+ \right] \Rightarrow$

$\qquad \left[H^+ \right] = 10^{-6}$

$\dfrac{\left[H^+ \right]_{\text{acid rain}}}{\left[H^+ \right]_{\text{ordinary rain}}} = \dfrac{1.6 \times 10^{-4}}{10^{-6}} = 1.6 \times 10^2 = 160$

The acid rain is 160 times more acidic than ordinary rain.

49. $\left[H^+_A \right] = 100 \left[H^+_B \right]$

$pH_A = -\log \left[H^+_A \right] = -\log \left(100 \left[H^+_B \right] \right)$

$\qquad = -\left(\log 100 + \log \left[H^+_B \right] \right)$

$\qquad = -\left(2 + \log \left[H^+_B \right] \right) = -\log \left[H^+_B \right] - 2$

$\qquad = pH_B - 2$

51. $\left[H^+_{\text{new}} \right] = 50 \left[H^+_{\text{original}} \right]$

$pH_{\text{new}} = -\log \left(50 \left[H^+_{\text{original}} \right] \right)$

$\qquad = -\log 50 - \log \left[H^+_{\text{original}} \right]$

$\qquad = -\log 50 - pH_{\text{original}}$

The pH decreases by $\log 50 \approx 1.7$. The solution becomes more acidic.

53. a. $M = \log \left(\dfrac{I}{I_0} \right)$

$7.8 = \log \left(\dfrac{I}{I_0} \right) \Rightarrow \left(\dfrac{I}{I_0} \right) = 10^{7.8} \Rightarrow$

$I = 10^{7.8} I_0 = 10^{0.8} \times 10^7 I_0 \approx 6.31 \times 10^7 I_0$

b. $\log E = 4.4 + 1.5(7.8) = 16.1 \Rightarrow$

$E = 10^{16.1} = 10^{0.1} \times 10^{16}$

$\qquad \approx 1.259 \times 10^{16}$ joules

55. a. $M_A = M_B + 1$. Then we have

$$M_A = M_B + 1 = \log\left(\frac{I_A}{I_0}\right) \Rightarrow$$

$$10^{(M_B+1)} = \frac{I_A}{I_0} \Rightarrow 10^{M_B} \times 10 = \frac{I_A}{I_0} \Rightarrow$$

$$10^{M_B} \times 10 I_0 = I_A \quad (1)$$

$$M_B = \log\left(\frac{I_B}{I_0}\right) \Rightarrow 10^{M_B} = \frac{I_B}{I_0} \Rightarrow$$

$$10^{M_B} \times I_0 = I_B$$

Using equation (1), we have

$$10^{M_B} \times 10 I_0 = I_A \Rightarrow 10^{M_B} \times I_0 \times 10 = I_A \Rightarrow$$
$$I_B \times 10 = I_A$$

The intensity of the earthquake A was 10 times that of earthquake B.

b. $\dfrac{E_A}{E_B} = 10^{1.5(M_A - M_B)} = 10^{1.5(M_B + 1 - M_B)}$

$$= 10^{1.5} \approx 31.6 \Rightarrow E_A \approx 31.6 E_B$$

The energy released by the earthquake A was about 31.6 times that released by earthquake B.

57. $M = \log\left(\dfrac{I}{I_0}\right)$

$$I_A = 150 I_B \Rightarrow$$

$$\log\left(\frac{I_A}{I_0}\right) = \log\left(\frac{150 I_B}{I_0}\right) = \log\left(150 \cdot \frac{I_B}{I_0}\right)$$

$$= \log 150 + \log\left(\frac{I_B}{I_0}\right) = \log 150 + M_B$$

The difference in Richter scale readings is $\log 150 \approx 2.18$.

59. $L = 10\log\left(\dfrac{I}{I_0}\right) = 10\log\left(\dfrac{5.2 \times 10^{-5}}{10^{-12}}\right)$

$$= 10\log\left(5.2 \times 10^7\right) = 10\left[\log 5.2 + \log 10^7\right]$$

$$= 10\left[\log 5.2 + 7\right] = 10\log 5.2 + 70 \approx 77.2 \text{ dB}$$

61. $I = I_0 \times 10^{L/10}$

$$I_{130} = I_0 \times 10^{130/10}$$

$$I_{65} = I_0 \times 10^{65/10}$$

$$\frac{I_{130}}{I_{65}} = \frac{I_0 \times 10^{130/10}}{I_0 \times 10^{65/10}} = 10^{65/10} = 10^{6.5}$$

$$= 10^{0.5} \times 10^6 \approx 3.16 \times 10^6$$

The 130 dB sound is about 3.16×10^6 times as intense as the 65 dB sound.

63. A sound at the threshold of pain has intensity $10\,\text{W/m}^2$. A sound that is 1000 times as intense has intensity $10^4\,\text{W/m}^2$.

$$L = 10\log\left(\frac{I}{I_0}\right) = 10\log\left(\frac{10^4}{10^{-12}}\right) = 10\log 10^{16}$$

$$= 160 \text{ dB}$$

65. $L_2 = L_1 + 1$

$$I_{L_2} = I_0 \times 10^{L_2/10} = I_0 \times 10^{0.1(L_1+1)}$$

$$I_{L_1} = I_0 \times 10^{L_1/10} = I_0 \times 10^{0.1 L_1}$$

$$\frac{I_{L_2}}{I_{L_1}} = \frac{I_0 \times 10^{0.1(L_1+1)}}{I_0 \times 10^{0.1 L_1}} = 10^{0.1(L_1 + 1 - L_1)}$$

$$= 10^{0.1} \approx 1.26 \Rightarrow I_{L_2} \approx 1.26 \times I_{L_1}$$

The intensity of the louder sound is about 1.26 times the intensity of the softer sound.

67. $P(f) = 1200 \log_2 \dfrac{f}{f_0}$

$$P(441) = 1200 \log_2\left(\frac{441}{440}\right) = 1200\left(\frac{\log\left(\frac{441}{440}\right)}{\log 2}\right)$$

$$\approx 3.93 \approx 4$$

The difference in pitch is about 4 cents higher.

69. Let m_1 represent the magnitude of the sun and let m_2 represent the magnitude of the full Moon.

$$m_2 - m_1 = 2.5\log\left(\frac{b_1}{b_2}\right)$$

$$-13 - (-27) = 2.5\log\left(\frac{b_1}{b_2}\right) \Rightarrow$$

$$\frac{14}{2.5} = 5.6 = \log\left(\frac{b_1}{b_2}\right) \Rightarrow$$

$$\frac{b_1}{b_2} = 10^{5.6} = 10^{0.6} \times 10^5 \approx 3.98 \times 10^5 \Rightarrow$$

$$b_1 = 3.98 \times 10^5 b_2$$

The sun is about 1.63×10^5 times brighter than the full Moon.

71. Let m_1 represent the magnitude of the Sun and let m_2 represent the magnitude of Venus.

$$m_2 - m_1 = 2.5 \log\left(\frac{b_1}{b_2}\right)$$

$$-4 - (-27) = 2.5 \log\left(\frac{b_1}{b_2}\right) \Rightarrow$$

$$\frac{23}{2.5} = 9.2 = \log\left(\frac{b_1}{b_2}\right) \Rightarrow$$

$$\frac{b_1}{b_2} = 10^{9.2} = 10^{0.2} \times 10^9 \approx 1.58 \times 10^9 \Rightarrow$$

$$b_1 = 1.58 \times 10^9 b_2$$

The Sun is about 1.58×10^9 times brighter than Venus.

73. Let m_1 and b_1 represent the magnitude and brightness of the unknown star and let m_2 and b_2 represent the magnitude and brightness of Saturn.

$$m_2 - m_1 = 2.5 \log\left(\frac{b_1}{b_2}\right); \quad b_1 = 560 b_2$$

$$1.47 - m_1 = 2.5 \log\left(\frac{560 b_2}{b_2}\right) \Rightarrow$$

$$1.47 - m_1 = 2.5 \log 560 \Rightarrow$$
$$1.47 - 2.5 \log 560 = m_1 \Rightarrow -5.4 \approx m_1$$

The magnitude of the unknown star is about -5.4.

4.5 Beyond the Basics

75. $L = 10 \log\left(\frac{4 \times 10^4}{r^2}\right)$

a. $r = 10$ ft

$$L = 10 \log\left(\frac{4 \times 10^4}{10^2}\right) \approx 26 \text{ dB}$$

$r = 25$ ft

$$L = 10 \log\left(\frac{4 \times 10^4}{25^2}\right) \approx 18 \text{ dB}$$

$r = 50$ ft

$$L = 10 \log\left(\frac{4 \times 10^4}{50^2}\right) \approx 12 \text{ dB}$$

$r = 100$ ft

$$L = 10 \log\left(\frac{4 \times 10^4}{100^2}\right) \approx 6 \text{ dB}$$

b. $0 = 10 \log\left(\frac{4 \times 10^4}{r^2}\right) \Rightarrow 0 = \log\left(\frac{4 \times 10^4}{r^2}\right) \Rightarrow$

$$10^0 = 1 = \frac{4 \times 10^4}{r^2} \Rightarrow r^2 = 40{,}000 \Rightarrow$$

$r = 200$ ft

The decibel level drops to 0 when a listener is 200 ft away.

c. $L = 10 \log\left(\frac{4 \times 10^4}{r^2}\right)$

$$= 10\left(\log 4 + \log\left(10^4\right) - \log\left(r^2\right)\right)$$
$$= 10\left((4 + \log 4) - 2 \log r\right)$$
$$= 40 + 10 \log 4 - 20 \log r$$
$$a = 40 + 10 \log 4 = 10(4 + \log 4)$$
$$b = -20$$

d. $L = 10 \log\left(\frac{4 \times 10^4}{r^2}\right) = 10 \log\left(\frac{2 \times 10^2}{r}\right)^2 \Rightarrow$

$$L = 20 \log\left(\frac{200}{r}\right) \Rightarrow \frac{L}{20} = \log\left(\frac{200}{r}\right) \Rightarrow$$

$$10^{L/20} = \frac{200}{r} \Rightarrow r = \frac{200}{10^{L/20}}$$

77. The energy released by the earthquake is given by

$$E = \left(2.5 \times 10^4\right) \times 10^{1.5(8)} = 2.5 \times 10^{16} \text{ joules.}$$

$$\frac{E_{\text{earthquake}}}{E_{\text{bomb}}} = \frac{2.5 \times 10^{16}}{5 \times 10^{15}} = .5 \times 10 = 5$$

The earthquake released five times as much energy as the nuclear bomb did.

79. $10 = 1200 \log_2 \frac{f}{880} \Rightarrow \frac{1}{120} = \log_2 \frac{f}{880} \Rightarrow$

$$2^{1/120} = \frac{f}{880} \Rightarrow f = 880 \times 2^{1/120} \approx 885$$

$$-10 = 1200 \log_2 \frac{f}{880} \Rightarrow -\frac{1}{120} = \log_2 \frac{f}{880} \Rightarrow$$

$$2^{-1/120} = \frac{f}{880} \Rightarrow f = 880 \times 2^{-1/120} \approx 875$$

The frequency range is $[875, 885]$.

81. Let m_1 and b_1 represent the magnitude and brightness of the unknown star and let m_2 and b_2 represent the magnitude and brightness of the given star.

$$m_2 - m_1 = 2.5\log\left(\frac{b_1}{b_2}\right); \quad b_2 = 176b_1$$

$$m_2 - 3.42 = 2.5\log\left(\frac{b_1}{176b_1}\right) \Rightarrow$$

$$m_2 - 3.42 = 2.5\log\left(\frac{1}{176}\right) \Rightarrow$$

$$m_2 = 2.5\log\left(\frac{1}{176}\right) + 3.42 \approx -2.2$$

The magnitude of the unknown star is about -2.2.

4.5 Maintaining Skills

83. $2(x+5)+x=1 \Rightarrow 2x+10+x=1 \Rightarrow$
$3x+10=1 \Rightarrow 3x=-9 \Rightarrow x=-3$
Solution set: $\{-3\}$

85. $6y-y=3(y-6) \Rightarrow 5y=3y-18 \Rightarrow$
$2y=-18 \Rightarrow y=-9$
Solution set: $\{-9\}$

87. $x-(9-5x)=-3(6-2x)$
$x-9+5x=-18+6x$
$6x-9=-18+6x \Rightarrow 0=-9$
Solution set: $\varnothing$

89. $9x+3(2x+1)=5(3x+1)+2$
$9x+6x+3=15x+5+2$
$15x+3=15x+7 \Rightarrow 3=7$
Solution set: $\varnothing$

91. $2x+3y=18 \Rightarrow 3y=-2x+18 \Rightarrow$
$y=-\dfrac{2}{3}x+6$

93. $x=-3$

95. $9x-3y=15 \Rightarrow -3y=-9x+15 \Rightarrow$
$y=3x-5$
The slope of the line we are seeking is 3.
Substitute $m=3$, $x=-3$ and $y=0$ into
$y=mx+b$, then solve for b.
$0=3(-3)+b \Rightarrow b=9$
The equation of the line is $y=3x+9$.

Chapter 4 Review Exercises

Basic Concepts and Skills

1. False. $f(x)=a^x$ is an exponential function if $a>0$ and $a\neq 1$.

3. False. The domain of $f(x)=\log(2-x)$ is $(-\infty, 2)$.

5. True

7. False. $\ln M + \ln N = \ln(MN)$

9. True

11. h   **13.** f   **15.** d   **17.** a

19. Domain: $(-\infty, \infty)$
range: $(0, \infty)$
asymptote: $y=0$

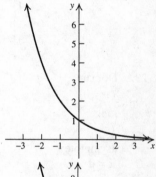

21. Domain: $(-\infty, \infty)$
range: $(3, \infty)$
asymptote: $y=3$

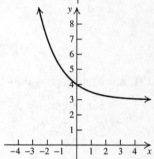

23. Domain: $(-\infty, \infty)$
range: $(0, 1]$
asymptote: $y=0$

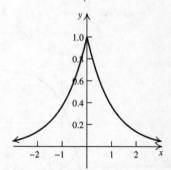

25. Domain:
$(-\infty, 0)$
range:
$(-\infty, \infty)$
asymptote:
$x = 0$

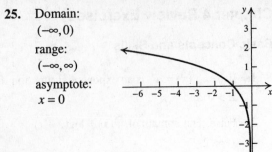

27. Domain:
$(1, \infty)$
range:
$(-\infty, \infty)$
asymptote:
$x = 1$

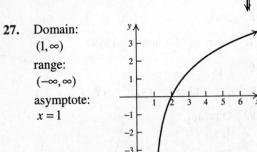

29. Domain:
$(-\infty, 0)$
range:
$(-\infty, \infty)$
asymptote:
$x = 0$

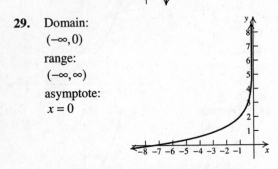

31. a. *y*-intercept: $y = 3 - 2e^0 = 1$

x-intercept: $0 = 3 - 2e^{-x} \Rightarrow 2e^{-x} = 3 \Rightarrow$

$e^{-x} = \dfrac{3}{2} \Rightarrow e^x = \dfrac{2}{3} \Rightarrow x = \ln\left(\dfrac{2}{3}\right)$

b. As $x \to \infty$, $f(x) \to 3$.
As $x \to -\infty$, $f(x) \to -\infty$.

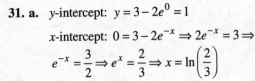

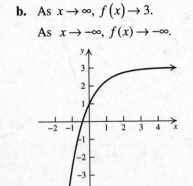

33. a. *y*-intercept: $y = e^{0^2} = 1$

x-intercept: $0 = e^{-x^2} \Rightarrow \ln 0 = -x^2 \Rightarrow$
there is no *x*-intercept.

b. As $x \to \infty$, $f(x) \to 0$.
As $x \to -\infty$, $f(x) \to 0$.

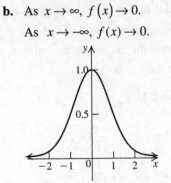

35. $10 = a(2^{0k}) \Rightarrow a = 10$
$640 = 10(2^{3k}) \Rightarrow 64 = 8^k \Rightarrow k = 2$
$f(2) = 10(2^{2(2)}) = 160$

37. $1 = \dfrac{4}{1 + ae^{-k(0)}} = \dfrac{4}{1 + a} \Rightarrow 1 + a = 4 \Rightarrow a = 3$

$\dfrac{1}{2} = \dfrac{4}{1 + 3e^{-3k}} \Rightarrow 1 + 3e^{-3k} = 8 \Rightarrow e^{-3k} = \dfrac{7}{3} \Rightarrow$

$-3k = \ln\left(\dfrac{7}{3}\right) \Rightarrow k \approx -0.2824$

$f(4) = \dfrac{4}{1 + 3e^{-(-0.2824)(4)}} \approx 0.38898$

39. The graph passes through (0, 3), so we have
$3 = ca^0 \Rightarrow c = 3$. The graph passes through
(2, 12), so we have $12 = 3a^2 \Rightarrow 4 = a^2 \Rightarrow$
$a = 2$. The equation is $f(x) = 3 \cdot 2^x$.

41. The graph has been shifted right one unit, so
the equation is of the form $y = \log_a(x - 1)$.
The graph passes through the point (6, 1), so
we have
$1 = \log_a(6 - 1) \Rightarrow 1 = \log_a 5 \Rightarrow a = 5$.
Thus, the equation is $y = \log_5(x - 1)$.

43. a. $y = -\left(2^{x-1} + 3\right) = -2^{x-1} - 3$

b. $y = -2^{x-1} + 3$

45. $\ln(xy^2 z^3) = \ln x + 2\ln y + 3\ln z$

47. $\ln\left[\dfrac{x\sqrt{x^2 + 1}}{(x^2 + 3)^2}\right] = \ln x + \dfrac{1}{2}\ln(x^2 + 1) - 2\ln(x^2 + 3)$

49. $\ln y = \ln x + \ln 3 = \ln(3x) \Rightarrow y = 3x$

51. $\ln y = \ln(x-3) - \ln(y+2) \Rightarrow$
$\ln y + \ln(y+2) = \ln(x-3) \Rightarrow$
$\ln(y(y+2)) = \ln(x-3) \Rightarrow$
$\ln(y^2 + 2y) = \ln(x-3) \Rightarrow y^2 + 2y = x - 3 \Rightarrow$
$y^2 + 2y + 1 = x - 3 + 1 \Rightarrow (y+1)^2 = x - 2 \Rightarrow$
$y + 1 = \pm\sqrt{x-2} \Rightarrow y = -1 \pm \sqrt{x-2}$
We disregard the negative solution, so
$y = -1 + \sqrt{x-2}.$

53. $\ln y = \dfrac{1}{2}\ln(x-1) + \dfrac{1}{2}\ln(x+1) - \ln(x^2+1) \Rightarrow$
$\ln y = \ln\left(\dfrac{\sqrt{x-1}\sqrt{x+1}}{x^2+1}\right) = \ln\left(\dfrac{\sqrt{x^2-1}}{x^2+1}\right) \Rightarrow$
$y = \dfrac{\sqrt{x^2-1}}{x^2+1}$

55. $3^x = 81 \Rightarrow 3^x = 3^4 \Rightarrow x = 4$

57. $2^{x^2+2x} = 16 \Rightarrow 2^{x^2+2x} = 2^4 \Rightarrow x^2 + 2x = 4 \Rightarrow$
$x^2 + 2x - 4 = 0 \Rightarrow x = \dfrac{-2 \pm \sqrt{4+16}}{2} = -1 \pm \sqrt{5}$

59. $3^x = 23 \Rightarrow x\ln 3 = \ln 23 \Rightarrow x = \dfrac{\ln 23}{\ln 3} \approx 2.854$

61. $273^x = 19 \Rightarrow x\ln 273 = \ln 19 \Rightarrow$
$x = \dfrac{\ln 19}{\ln 273} \approx 0.525$

63. $3^{2x} = 7^x \Rightarrow 2x\ln 3 = x\ln 7 \Rightarrow$
$2x\ln 3 - x\ln 7 = 0 \Rightarrow x(2\ln 3 - \ln 7) = 0 \Rightarrow$
$x = 0$

65. $1.7^{3x} = 3^{2x-1} \Rightarrow 3x\ln 1.7 = (2x-1)\ln 3 \Rightarrow$
$3x\ln 1.7 = 2x\ln 3 - \ln 3 \Rightarrow$
$3x\ln 1.7 - 2x\ln 3 = -\ln 3 \Rightarrow$
$x(3\ln 1.7 - 2\ln 3) = -\ln 3 \Rightarrow$
$x = \dfrac{-\ln 3}{3\ln 1.7 - 2\ln 3} \approx 1.815$

67. $\log_3(x+2) - \log_3(x-1) = 1 \Rightarrow$
$\log_3\left(\dfrac{x+2}{x-1}\right) = 1 \Rightarrow \dfrac{x+2}{x-1} = 3 \Rightarrow$
$x + 2 = 3x - 3 \Rightarrow 2x = 5 \Rightarrow x = \dfrac{5}{2}$

69. $\log_2(x+2) + \log_2(x+4) = 3 \Rightarrow$
$\log_2\left[(x+2)(x+4)\right] = 3 \Rightarrow$
$x^2 + 6x + 8 = 2^3 \Rightarrow x^2 + 6x = 0 \Rightarrow$
$x(x+6) = 0 \Rightarrow x = -6, \, 0$
Reject -6 since $\log_2(-6+2) = \log_2(-4)$ is
not defined. Solution set: $\{0\}$

71. $\log_5\left(x^2 - 5x + 6\right) - \log_5(x-2) = 1 \Rightarrow$
$\log_5\left(\dfrac{x^2 - 5x + 6}{x-2}\right) = 1 \Rightarrow \dfrac{x^2 - 5x + 6}{x-2} = 5 \Rightarrow$
$x^2 - 5x + 6 = 5x - 10 \Rightarrow x^2 - 10x + 16 = 0 \Rightarrow$
$(x-2)(x-8) = 0 \Rightarrow x = 2, \, 8$
If $x = 2$, then $\log_5(x-2) = \log_5 0$, which is
undefined, so reject $x = 2$. Solution set: $\{8\}$

73. $\log_6(x-2) + \log_6(x+1)$
$\qquad = \log_6(x+4) + \log_6(x-3) \Rightarrow$
$\log_6\left[(x-2)(x+1)\right] = \log_6\left[(x+4)(x-3)\right] \Rightarrow$
$x^2 - x - 2 = x^2 + x - 12 \Rightarrow -2x = -10 \Rightarrow x = 5$

75. $2\ln 3x = 3\ln x \Rightarrow \ln(3x)^2 = \ln x^3 \Rightarrow$
$9x^2 = x^3 \Rightarrow x = 9$

77. Multipling both sides of $2^x - 8\cdot 2^{-x} - 7 = 0$ by
2^x we have $2^{2x} - 8\cdot 2^{-x+x} - 7\cdot 2^x = 0 \Rightarrow$
$2^{2x} - 7\cdot 2^x - 8 = 0.$ Letting $u = 2^x$, we have
$u^2 - 7u - 8 = 0 \Rightarrow (u-8)(u+1) = 0 \Rightarrow u = 8$
or $u = -1.$ (Reject this solution) $2^x = 8 \Rightarrow x = 3$

79. $3\left(0.2^x\right) + 5 \le 20 \Rightarrow 3\left(0.2^x\right) \le 15 \Rightarrow$
$\left(0.2^x\right) \le 5 \Rightarrow x\ln 0.2 \le \ln 5 \Rightarrow x \ge \dfrac{\ln 5}{\ln 0.2} \Rightarrow$
$x \ge -1$

81. Note that the domain of $\log(2x+7)$ is
$\left(-\dfrac{7}{2}, \infty\right)$ since $2x + 7$ must be greater than 0.
$\log(2x+7) < 2 \Rightarrow 2x + 7 < 10^2 \Rightarrow 2x < 93 \Rightarrow$
$x < \dfrac{93}{2}$
Solution set: $\left(-\dfrac{7}{2}, \dfrac{93}{2}\right)$

Applying the Concepts

83. $2P = P\left(1 + 0.0625\right)^t \Rightarrow 2 = 1.0625^t \Rightarrow$
$\ln 2 = t\ln 1.0625 \Rightarrow t = \dfrac{\ln 2}{\ln 1.0625} \approx 11.4$
It will take about 11.4 years to double the
investment.

85. Find k: $\dfrac{1}{2} = e^{20k} \Rightarrow \ln\left(\dfrac{1}{2}\right) = 20k \Rightarrow$

$$k = \dfrac{\ln(1/2)}{20}$$

$$0.25 = e^{\frac{\ln(1/2)}{20}t} \Rightarrow \ln 0.25 = \dfrac{\ln(1/2)}{20}t \Rightarrow$$

$$t = \dfrac{\ln 0.25}{\dfrac{\ln(1/2)}{20}} = 40$$

It will take 40 hours to decay to 25% of its original value.

87. At 5% compounded yearly,

$$A = 7000\left(1 + \dfrac{0.05}{1}\right)^{7(1)} = 9849.70.$$

At 4.75% compounded monthly,

$$A = 7000\left(1 + \dfrac{0.0475}{12}\right)^{7(12)} = 9754.74.$$

The 5% investment provides the greater return.

89. a. $t = 10; P(10) = 33e^{0.003(10)} \approx 34$ million

b. $60 = 33e^{0.003t} \Rightarrow \dfrac{60}{33} = e^{0.003t} \Rightarrow$

$$\ln\left(\dfrac{60}{33}\right) = 0.003t \Rightarrow t = \dfrac{\ln\left(\dfrac{60}{33}\right)}{0.003} \approx 199.3 \text{ yr}$$

The population will be 60 million sometime during the year 2206 (after 199.3 years).

91. a.

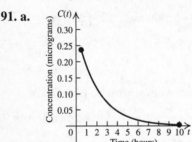

b. $0.029 = 0.3e^{-0.47t} \Rightarrow \dfrac{0.029}{0.3} = e^{-0.47t} \Rightarrow$

$$\ln\left(\dfrac{0.029}{0.3}\right) = -0.47t \Rightarrow t = \dfrac{\ln\left(\dfrac{0.029}{0.3}\right)}{-0.47} \Rightarrow$$

$t \approx 4.97 \approx 5$ hours

93. First find a and k:

$$f(0) = 212 = 75 + ae^{-k(0)} \Rightarrow 137 = a$$

$$f(1) = 200 = 75 + 137e^{-k(1)} \Rightarrow \dfrac{125}{137} = e^{-k} \Rightarrow$$

$$\ln(125/137) = -k \Rightarrow k \approx 0.0917$$

$$150 = 75 + 137e^{-0.0917t} \Rightarrow \dfrac{75}{137} = e^{-0.0917t} \Rightarrow$$

$$t = \dfrac{\ln(75/137)}{-0.0917} \approx 6.57 \text{ minutes}$$

95. a. $D(0) = 5e^{0.08(0)} \approx 5$ thousand $= 5000$ people per square mile

b. $D(5) = 5e^{0.08(5)} \approx 7.459$ thousand $= 7459$ people per square mile

c. $15 = 5e^{0.08t} \Rightarrow 3 = e^{0.08t} \Rightarrow$

$$t = \dfrac{\ln 3}{0.08} \approx 13.73 \text{ miles}$$

97. $I = I_0 e^{-0.73(2)} \approx 0.2322 I_0$

99. a. $s(470,000) = 0.04 + 0.86\ln(470,000)$
$$\approx 11.27 \text{ feet per second}$$

b. $s(450) = 0.04 + 0.86\ln(450)$
$$\approx 5.29 \text{ feet per second}$$

c. $4.6 = 0.04 + 0.86\ln p \Rightarrow \dfrac{4.56}{0.86} = \ln p \Rightarrow$

$$p = e^{4.56/0.86} \approx 200.8 \approx 201 \text{ people}$$

101. a. $m(0) = \dfrac{6}{1 + 5e^{-0.7(0)}} = \dfrac{6}{1+5} = 1$ gram

b. The mass approaches 6 grams

c. $5 = \dfrac{6}{1 + 5e^{-0.7t}} \Rightarrow 5 + 25e^{-0.7t} = 6 \Rightarrow$

$$e^{-0.7t} = \dfrac{1}{25} \Rightarrow t = \dfrac{\ln\left(\dfrac{1}{25}\right)}{-0.7} \approx 4.6 \text{ days}$$

103. $P(f) = 1200\log_2 \dfrac{f}{f_0}$

$$P(f) = 1200\log_2\left(\dfrac{1.2f_0}{f_0}\right) = 1200\log_2(1.2)$$

$$= 1200\left(\dfrac{\log 1.2}{\log 2}\right) \approx 315.6 \approx 316$$

The difference, $316 - 300 = 16$ cents, is noticeable.

105. $\text{pH}_{\text{very acid rain}} = 2.4 = -\log\left[\text{H}^+\right] \Rightarrow$

$\left[\text{H}^+\right] = 10^{-2.4} \Rightarrow$

$\left[\text{H}^+\right] = 10^{0.6} \times 10^{-3} \approx 4.0 \times 10^{-3}$

The average concentration of hydrogen ions in the very acid rain is 4.0×10^{-3} moles per liter.

$\text{pH}_{\text{acid rain}} = 5.6 = -\log\left[\text{H}^+\right] \Rightarrow$

$\left[\text{H}^+\right] = 10^{-5.6} = 10^{0.4} \times 10^{-6} \approx 2.5 \times 10^{-6}$

$\dfrac{\left[\text{H}^+\right]_{\text{very acid rain}}}{\left[\text{H}^+\right]_{\text{acid rain}}} = \dfrac{10^{-2.4}}{10^{-5.6}} = 10^{3.2} = 10^{0.2} \times 10^3$

$\approx 1.585 \times 10^3 = 1585$

The very acid rain is 1585 times more acidic than acid rain.

107. $I = I_0 \times 10^{L/10}$

$I_{115} = I_0 \times 10^{115/10}$

$I_{95} = I_0 \times 10^{95/10}$

$\dfrac{I_{115}}{I_{95}} = \dfrac{I_0 \times 10^{115/10}}{I_0 \times 10^{95/10}} = 10^{20/10} = 10^2 = 100$

The 115 dB sound is 100 times as intense as the 95 dB sound.

109. Let m_1 represent the magnitude of Sirius and let m_2 represent the magnitude of Saturn.

$m_2 - m_1 = 2.5 \log\left(\dfrac{b_1}{b_2}\right)$

$1.47 - (-1) = 2.5 \log\left(\dfrac{b_1}{b_2}\right) \Rightarrow$

$\dfrac{2.47}{2.5} = 0.988 = \log\left(\dfrac{b_1}{b_2}\right) \Rightarrow$

$\dfrac{b_1}{b_2} = 10^{0.988} \approx 9.7 \Rightarrow b_1 = 9.7 b_2$

Sirius is about 9.7 times brighter than Saturn.

Chapter 4 Practice Test A

1. $5^{-x} = 125 \Rightarrow 5^{-x} = 5^3 \Rightarrow x = -3$

2. $\log_2 x = 5 \Rightarrow x = 2^5 = 32$

3. Range: $(-\infty, 1)$; asymptote: $y = 1$

4. $\log_2 \dfrac{1}{8} \Rightarrow 2^x = \dfrac{1}{8} = 2^{-3} \Rightarrow x = -3$

5. $\left(\dfrac{1}{4}\right)^{2-x} = 4 \Rightarrow 4^{x-2} = 4 \Rightarrow x - 2 = 1 \Rightarrow x = 3$

6. $\log 0.001 \Rightarrow 10^x = 0.001 = 10^{-3} \Rightarrow x = -3$

7. $\ln 3 + 5 \ln x = \ln 3 + \ln x^5 = \ln\left(3x^5\right)$

8. $2^{x+1} = 5 \Rightarrow (x+1)\ln 2 = \ln 5 \Rightarrow$

$x \ln 2 + \ln 2 = \ln 5 \Rightarrow x \ln 2 = \ln 5 - \ln 2 \Rightarrow$

$x \ln 2 = \ln\left(\dfrac{5}{2}\right) \Rightarrow x = \dfrac{\ln(5/2)}{\ln 2}$

9. $e^{2x} + e^x - 6 = 0 \Rightarrow (e^x + 3)(e^x - 2) = 0 \Rightarrow$

$e^x = -3 \text{ (reject this) or } e^x - 2 = 0 \Rightarrow$

$e^x = 2 \Rightarrow x = \ln 2$

10. $\ln \dfrac{2x^3}{(x+1)^5} = \ln(2x^3) - \ln(x+1)^5$

$= \ln 2 + 3 \ln x - 5 \ln(x+1)$

11. $\ln e^{-5} = -5$

12. $y = \ln(x-1) + 3$

13.
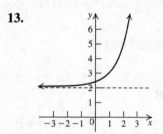

14. Domain: $(-\infty, 0)$

15. $3 \ln x + \ln(x^3 + 2) - \dfrac{1}{2}\ln(3x^2 + 2)$

$= \ln x^3 + \ln(x^3 + 2) - \ln\sqrt{3x^2 + 2}$

$= \ln \dfrac{x^3(x^3 + 2)}{\sqrt{3x^2 + 2}}$

16. $\log x = \log 6 - \log(x-1) \Rightarrow \log x = \log \dfrac{6}{x-1} \Rightarrow$

$x = \dfrac{6}{x-1} \Rightarrow x^2 - x = 6 \Rightarrow x^2 - x - 6 = 0 \Rightarrow$

$(x-3)(x+2) = 0 \Rightarrow x = -2 \text{ (reject this) or } x = 3$

17. $\log_x 9 = 2 \Rightarrow x^2 = 9 \Rightarrow x = 3$

18. $A = 15{,}000\left(1 + \dfrac{0.07}{4}\right)^{4t} = 15{,}000(1.0175)^{4t}$

19. $1,500,000 = 15,000e^{0.2t} \Rightarrow 100 = e^{0.2t} \Rightarrow$

$\ln 100 = 0.2t \Rightarrow t = \dfrac{\ln 100}{0.2} \approx 23.03$

The Hispanic population reached 1.5 million in 1983, 23.03 years after 1960.

20. Let I_{1960} denote the intensity of the 1960 Morocco earthquake and let I_{1972} denote the intensity of the 1972 Nicaragua earthquake. Then,

$5.8 = \log\left(\dfrac{I_{1960}}{I_0}\right) \Rightarrow \dfrac{I_{1960}}{I_0} = 10^{5.8} \Rightarrow$

$I_{1960} = 10^{5.8} I_0$

$6.2 = \log\left(\dfrac{I_{1972}}{I_0}\right) \Rightarrow \dfrac{I_{1972}}{I_0} = 10^{6.2} \Rightarrow$

$I_{1972} = 10^{6.2} I_0$

$\dfrac{I_{1972}}{I_{1960}} = \dfrac{10^{6.2} I_0}{10^{5.8} I_0} = 10^{0.4} \approx 2.5 \Rightarrow I_{1972} \approx 2.5 I_{1960}$

The intensity of the 1972 Nicaragua earthquake was about 2.5 times that of the 1960 Morocco earthquake.

Chapter 4 Practice Test B

1. $3^{-x} = 9 \Rightarrow 3^{-x} = 3^2 \Rightarrow x = -2$.
The answer is B.

2. $\log_5 x = 2 \Rightarrow 5^2 = 25 = x$. The answer is B.

3. The answer is B.

4. $\log_4 64 \Rightarrow 4^x = 64 = 4^3 \Rightarrow x = 3$.
The answer is D.

5. $\left(\dfrac{1}{3}\right)^{1-x} = 3 \Rightarrow 3^{x-1} = 3 \Rightarrow x - 1 = 1 \Rightarrow x = 2$.
The answer is D.

6. $\log 0.01 \Rightarrow 10^x = 10^{-2} \Rightarrow x = -2$.
The answer is B.

7. $\ln 7 + 2\ln x = \ln(7x^2)$. The answer is B.

8. $2^{x^2} = 3^x \Rightarrow x^2 \ln 2 = x \ln 3 \Rightarrow$

$x^2 \ln 2 - x \ln 3 = 0 \Rightarrow x(x \ln 2 - \ln 3) = 0 \Rightarrow$

$x = 0$ or $x \ln 2 - \ln 3 = 0 \Rightarrow x = \dfrac{\ln 3}{\ln 2}$

The answer is C.

9. $e^{2x} - e^x - 6 = 0 \Rightarrow (e^x - 3)(e^x + 2) = 0 \Rightarrow$

$e^x = -2$ (reject this) or $e^x - 3 = 0 \Rightarrow$

$e^x = 3 \Rightarrow x = \ln 3$.
The answer is D.

10. $\ln \dfrac{3x^2}{(x+1)^{10}} = \ln(3x^2) - \ln(x+1)^{10}$

$= \ln 3 + 2\ln x - 10\ln(x+1)$.
The answer is D.

11. $\ln e^{3x} = 3x$. The answer is B.

12. The answer is D.

13. The answer is B.

14. The answer is A.

15. $\ln x - 2\ln(x^2+1) + \dfrac{1}{2}\ln(x^4+1)$

$= \ln x - \ln(x^2+1)^2 + \ln \sqrt{x^4+1}$

$= \ln \dfrac{x\sqrt{x^4+1}}{(x^2+1)^2}$.

The answer is A.

16. $\log x = \log 12 - \log(x+1) \Rightarrow \log x = \log \dfrac{12}{x+1} \Rightarrow$

$x = \dfrac{12}{x+1} \Rightarrow x^2 + x = 12 \Rightarrow x^2 + x - 12 = 0 \Rightarrow$

$(x+4)(x-3) = 0 \Rightarrow x = -4$ (reject this) or
$x = 3$
The answer is D.

17. $\log_x 16 = 4 \Rightarrow x^4 = 16 = 2^4 \Rightarrow x = 2$
The answer is B.

18. $A = 12,000\left(1 + \dfrac{0.105}{12}\right)^{12t} = 12,000(1.00875)^{12t}$
The answer is D.

19. $t = 2020 - 2000 = 20$

$P(20) = 10,000\log_5(20+5) = 10,000\log_5 25$

$= 20,000$. The answer is B.

20. Let I_{1994} denote the intensity of the 1994 Northridge, CA earthquake and let I_{1988} denote the intensity of the 1988 Armenia earthquake. Then,

$6.7 = \log\left(\dfrac{I_{1994}}{I_0}\right) \Rightarrow \dfrac{I_{1994}}{I_0} = 10^{6.7} \Rightarrow$

$I_{1994} = 10^{6.7} I_0$

(continued on next page)

(continued)

$$7.0 = \log\left(\frac{I_{1988}}{I_0}\right) \Rightarrow \frac{I_{1988}}{I_0} = 10^{7.0} \Rightarrow$$

$$I_{1988} = 10^{7.0} I_0$$

$$\frac{I_{1988}}{I_{1994}} = \frac{10^{7.0} I_0}{10^{6.7} I_0} = 10^{0.3} \approx 2.0 \Rightarrow$$

$$I_{1988} \approx 2.0 I_{1994}$$

The intensity of the 1988 Armenia earthquake was about twice that of the 1994 Northridge, CA earthquake. The answer is B.

Cumulative Review Exercises (Chapters P–4)

1. $x^2 + (0-1)^2 = 1 \Rightarrow x^2 = 0 \Rightarrow x = 0.$
$0^2 + (y-1)^2 = 1 \Rightarrow y^2 - 2y + 1 = 1 \Rightarrow$
$y^2 - 2y = 0 \Rightarrow y(y-2) = 0 \Rightarrow y = 0$ or $y = 2.$
The x-intercept is 0.
The y-intercepts are 0 and 2.
$x^2 + (y-1)^2 = 1 \Rightarrow (y-1)^2 = 1 - x^2 \Rightarrow$
$y - 1 = \pm\sqrt{1-x^2} \Rightarrow y = 1 \pm \sqrt{1-x^2} = f(x)$
$f(-x) = 1 \pm \sqrt{1-(-x)^2} = 1 \pm \sqrt{1-x^2} = f(x) \Rightarrow$
$f(x)$ is even $\Rightarrow f$ is symmetric about the y-axis.

2.

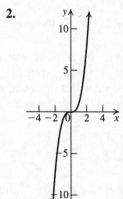

3. $x^2 + 2x + y^2 - 4y - 20 = 0$
Group the terms and complete the squares.
$\left(x^2 + 2x + 1\right) + \left(y^2 - 4y + 4\right) = 20 + 1 + 4$
$$(x+1)^2 + (y-2)^2 = 25$$
The center of the circle is $(-1, 2)$ and its radius is 5.

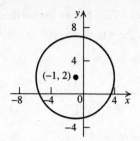

4. $2x + 3y = 17 \Rightarrow y = -\frac{2}{3}x + \frac{17}{3} \Rightarrow$ the slope of the perpendicular line is
$$\frac{3}{2} \cdot 4 = \frac{3}{2}(-2) = b \Rightarrow 7 = b.$$
The equation is $y = \frac{3}{2}x + 7.$

5. Logarithms are ≥ 0, so the domain is
$(-\infty, -3) \cup (2, \infty).$

6. a. $f(-2) = 2(-2) - 3 = -7$

   b. $f(0) = 2(0) - 3 = -3$

   c. $f(3) = 2(3^2) + 1 = 19$

7. Shift the graph of $f(x) = \sqrt{x}$ one unit left, stretch vertically by a factor of 3, then shift the graph two units down.

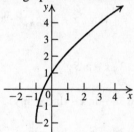

8. $(f \circ g)(x) = f\left(-\frac{1}{x}\right) = 3\sqrt{-\frac{1}{x}} \Rightarrow$ the domain is $(-\infty, 0).$

9. a. The graph of $f(x)$ passes the horizontal line test, so there is an inverse. $y = 2x^3 + 1$
becomes $x = 2y^3 + 1 \Rightarrow \frac{x-1}{2} = y^3 \Rightarrow$
$$y = \sqrt[3]{\frac{x-1}{2}} = f^{-1}(x).$$

   b. The graph of $f(x)$ does not pass the horizontal line test, so there is no inverse.

c. The graph of $f(x)$ passes the horizontal line test, so there is an inverse. $y = \ln x$ becomes $x = \ln y \Rightarrow e^x = y = f^{-1}(x)$.

10. a.
$$x^2 + 1 \overline{)2x^3 + 0x^2 + 3x + 1}$$
$$\underline{2x^3 + 0x^2 + 2x}$$
$$x + 1$$

The quotient is $2x + \dfrac{x+1}{x^2+1}$.

b.
$$\begin{array}{r|rrrrr} -3 & 2 & 7 & 0 & -2 & 3 \\ & & -6 & -3 & 9 & -21 \\ \hline & 2 & 1 & -3 & 7 & -18 \end{array}$$

$$\frac{2x^4 + 7x^3 - 2x + 3}{x+3}$$
$$= 2x^3 + x^2 - 3x + 7 - \frac{18}{x+3}$$

11. a. $(x-1)(x-2)(x+3) = x^3 - 7x + 6$

b. $(x-1)(x+1)(x-i)(x+i) = x^4 - 1$

12. a. $\log_2 5x^3 = \log_2 5 + 3\log_2 x$

b. $\log_a \sqrt[3]{\dfrac{xy^2}{z}} = \dfrac{1}{3}\log_a \dfrac{xy^2}{z}$
$$= \frac{1}{3}(\log_a x + 2\log_a y - \log_a z)$$

c. $\ln \dfrac{3\sqrt{x}}{5y} = \ln 3\sqrt{x} - \ln 5y$
$$= \ln 3 + \frac{1}{2}\ln x - (\ln 5 + \ln y)$$
$$= \ln 3 + \frac{1}{2}\ln x - \ln 5 - \ln y$$

13. a. $\dfrac{1}{2}(\log x + \log y) = \dfrac{1}{2}\log(xy) = \log\sqrt{xy}$

b. $3\ln x - 2\ln y = \ln x^3 - \ln y^2 = \ln\dfrac{x^3}{y^2}$

14. a. $\log_9 17 = \dfrac{\log 17}{\log 9} \approx 1.289$

b. $\log_3 25 = \dfrac{\log 25}{\log 3} \approx 2.930$

c. $\log_{1/2} 0.3 = \dfrac{\log 0.3}{\log(1/2)} \approx 1.737$

15. a. $\log_6(x+3) + \log_6(x-2) = 1 \Rightarrow$
$\log_6\big((x+3)(x-2)\big) = 1 \Rightarrow$
$(x+3)(x-2) = 6 \Rightarrow x^2 + x - 6 = 6 \Rightarrow$
$x^2 + x - 12 = 0 \Rightarrow (x+4)(x-3) = 0 \Rightarrow$
$x = -4$ (reject this) or $x = 3$

b. $5^{x^2-4x+5} = 25 = 5^2 \Rightarrow x^2 - 4x + 5 = 2 \Rightarrow$
$x^2 - 4x + 3 = 0 \Rightarrow (x-3)(x-1) = 0 \Rightarrow$
$x = 1$ or $x = 3$

c. $3.1^{x-1} = 23 \Rightarrow (x-1)\ln 3.1 = \ln 23 \Rightarrow$
$x - 1 = \dfrac{\ln 23}{\ln 3.1} \Rightarrow x = 1 + \dfrac{\ln 23}{\ln 3.1} \approx 3.771$

16. The possible rational zeros are $\{\pm 1, \pm 2, \pm 4\}$. Using synthetic division, we find that one zero is 1:

$$\begin{array}{r|rrrrr} 1 & 1 & -1 & -2 & -2 & 4 \\ & & 1 & 0 & -2 & -4 \\ \hline & 1 & 0 & -2 & -4 & 0 \end{array}$$

Using synthetic division again, we find that 2 is a root of the depressed equation $f(x) = x^3 - 2x - 4$:

$$\begin{array}{r|rrrr} 2 & 1 & 0 & -2 & -4 \\ & & 2 & 4 & 4 \\ \hline & 1 & 2 & 2 & 0 \end{array}$$

There are no rational zeros of the depressed equation $f(x) = x^2 + 2x + 2$. So the only rational zeros are $\{1, 2\}$.

Find the upper bound for the zeros by testing each possible rational zero using synthetic division. The smallest value that makes all the terms in the bottom row nonnegative is the upper bound. The upper bound is 3.

$$\begin{array}{r|rrrrr} 3 & 1 & -1 & -2 & -2 & 4 \\ & & 3 & 6 & 12 & 30 \\ \hline & 1 & 2 & 4 & 10 & 34 \end{array}$$

Find the lower bound for the zeros by testing each possible rational zero using synthetic division. The largest value that makes the terms in the bottom row alternate signs is the lower bound. The lower bound is -1.

$$\begin{array}{r|rrrrr} -1 & 1 & -1 & -2 & -2 & 4 \\ & & -1 & 2 & 0 & 2 \\ \hline & 1 & -2 & 0 & -2 & 6 \end{array}$$

17. a. The zeros are 1 (multiplicity 3), –2 (multiplicity 2), and 3 (multiplicity 1).

b. At $x = 1$ and $x = 3$, the graph crosses the x-axis. At $x = -2$, the graph touches, but does not cross, the x-axis.

c. As $x \to -\infty, f(x) \to \infty$.
As $x \to \infty, f(x) \to \infty$.

d.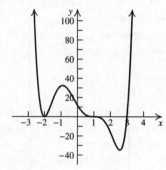

18. a. The vertical asymptotes are $x = 3$ and $x = -4$.

b. $f(x) = \dfrac{(x-1)(x+2)}{(x-3)(x+4)} = \dfrac{x^2 + x - 2}{x^2 + x - 12} \Rightarrow$ the horizontal asymptote is $y = 1$.

c. $f(x)$ lies above the horizontal asymptote on $(-\infty, -4) \cup (3, \infty)$. $f(x)$ lies below the horizontal asymptote on $(-4, 3)$.

d.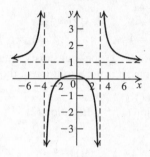

Chapter 5 Systems of Equations and Inequalities

5.1 Systems of Linear Equations in Two Variables

5.1 Practice Problems

1. a. Check $(2, 2)$.

Equation (1) Equation (2)

$x + y = 4$ $3x - y = 0$

$2 + 2 \overset{?}{=} 4$ $3(2) - 2 \overset{?}{=} 0$

$4 = 4 \checkmark$ $4 = 0 \times$

Because $(2, 2)$ does not satisfy both equations, it is not a solution of the system.

b. Check $(1, 3)$.

Equation (1) Equation (2)

$x + y = 4$ $3x - y = 0$

$1 + 3 \overset{?}{=} 4$ $3(1) - 3 \overset{?}{=} 0$

$4 = 4 \checkmark$ $0 = 0 \checkmark$

Because $(1, 3)$ satisfies both equations, it is a solution of the system.

2. The solution is $\{(-1, 3)\}$.

$\begin{cases} x + y = 2 \\ 4x + y = -1 \end{cases}$

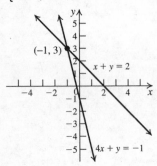

3. $\begin{cases} x - y = 5 & (1) \\ 2x + y = 7 & (2) \end{cases}$

Solve the equation (1) for x, then substitute that expression into equation (2) and solve for y.

$x - y = 5 \Rightarrow x = y + 5$

$2x + y = 7 \Rightarrow 2(y + 5) + y = 7 \Rightarrow$

$3y + 10 = 7 \Rightarrow 3y = -3 \Rightarrow y = -1$

Now substitute the value for y into equation (1) and solve for x.

$x - y = 5 \Rightarrow x - (-1) = 5 \Rightarrow x = 4$

The solution is $\{(4, -1)\}$.

4. $\begin{cases} x - 3y = 1 & (1) \\ -2x + 6y = 3 & (2) \end{cases}$

Solve the equation (1) for x, then substitute that expression into the equation (2) and solve for y.

$x - 3y = 1 \Rightarrow x = 1 + 3y$

$-2x + 6y = 3 \Rightarrow -2(1 + 3y) + 6y = 3 \Rightarrow$

$-2 - 6y + 6y = 3 \Rightarrow -2 = 3$

Since the equation $-2 = 3$ is false, the system is inconsistent. The solution set is $\varnothing$.

5. $\begin{cases} -2x + y = -3 & (1) \\ 4x - 2y = 6 & (2) \end{cases}$

Solve the equation (1) for y, then substitute that expression into equation (2) and solve for x.

$-2x + y = -3 \Rightarrow y = 2x - 3$

$4x - 2y = 6 \Rightarrow 4x - 2(2x - 3) = 6 \Rightarrow$

$4x - 4x + 6 = 6 \Rightarrow 6 = 6$

The equation $6 = 6$ is true for every value of x. Thus, any value of x can be used in the equation $y = 2x - 3$. The solutions of the system are of the form $\{(x, 2x - 3)\}$.

6. $\begin{cases} 3x + 2y = 3 & (1) \\ 9x - 4y = 4 & (2) \end{cases}$

Multiply the equation (1) by 2, then add the two equations.

$\begin{cases} 3x + 2y = 3 \\ 9x - 4y = 4 \end{cases} \Rightarrow \begin{cases} 6x + 4y = 6 \\ 9x - 4y = 4 \end{cases} \Rightarrow 15x = 10 \Rightarrow$

$x = \dfrac{2}{3}$

Now substitute $x = \dfrac{2}{3}$ into the equation (2) and solve for y:

$9x - 4y = 4 \Rightarrow 9\left(\dfrac{2}{3}\right) - 4y = 4 \Rightarrow 6 - 4y = 4 \Rightarrow$

$-4y = -2 \Rightarrow y = \dfrac{1}{2}$

The solution is $\left\{\left(\dfrac{2}{3}, \dfrac{1}{2}\right)\right\}$.

7. $\begin{cases} p = 20 + 0.002x & (1) \\ p = 77 - 0.008x & (2) \end{cases}$

Solve by substitution.

$20 + 0.002x = 77 - 0.008x \Rightarrow 0.010x = 57 \Rightarrow$

$x = 5700$

Substitute this value into equation (1) and solve for p.

(continued on next page)

(*continued*)

$$p = 20 + 0.002x \Rightarrow p = 20 + 0.002(5700) \Rightarrow$$
$$p = 31.4$$

The equilibrium point is (5700, 31.4).

8. Let x = the amount invested at 12%.
   Let y = the amount invested at 8%.
   Then $0.12x$ = the income from the 12% investment and $0.08y$ = the income from the 8% investment. The system of equations is
   $$\begin{cases} x + y = 150,000 & (1) \\ 0.12x + 0.08y = 15,400 & (2) \end{cases}$$
   We will use the elimination method to solve the system.
   $$\begin{array}{ll} -12x - 12y = -1,800,000 & \text{Multiply by } -12. \\ \underline{12x + 8y = 1,540,000} & \text{Multiply by 100.} \\ -4y = -260,000 & \text{Add.} \end{array}$$
   $$y = \frac{-260,000}{-4} = 65,000$$
   Solve for y.

   Back-substitute $y = 65,000$ into equation (1) and solve for x.
   $$x + 65,000 = 150,000$$
   $$x = 85,000$$
   Check:
   12% of 85,000 = 10,200
   8% of 65,000 = 5200
   85,000 + 65,000 = 150,000 and
   10,200 + 5200 = 15,400, as given.
   Solution: $85,000 was invested at 12% and $65,000 was invested at 8%.

5.1 Basic Concepts and Skills

1. The ordered pair (a, b) is a <u>solution</u> of a system of equations in x and y provided that when x is replaced with a and y is replaced with b, the resulting equations are true.

3. If, in the process of solving a system of equations, you get an equation of the form $0 = k$, where k is not zero, then the system is <u>inconsistent</u>.

5. False. A system consisting of two identical equations has an infinite number of solutions.

7. Substituting each ordered pair into the system
   $$\begin{cases} 2x + 3y = 3 \\ 3x - 4y = 13 \end{cases}, \text{ we find that } (3, -1) \text{ is a}$$
   solution.
   $$\begin{cases} 2(3) + 3(-1) = 6 - 3 = 3 \\ 3(3) - 4(-1) = 9 + 4 = 13 \end{cases}$$

9. Substituting each ordered pair into the system
   $$\begin{cases} 5x - 2y = 7 \\ -10x + 4y = 11 \end{cases}, \text{ we find that none of the}$$
   ordered pairs are solutions.

11. Substituting each ordered pair into the system
    $$\begin{cases} x + y = 1 \\ \dfrac{1}{2}x + \dfrac{1}{3}y = 2 \end{cases}, \text{ we find that } (10, -9) \text{ is a}$$
    solution.
    Check: $\begin{cases} 10 - 9 = 1 \\ \dfrac{1}{2}(10) + \dfrac{1}{3}(-9) = 5 - 3 = 2 \end{cases}$

13. The solution is $\{(2, 1)\}$.
    $$\begin{cases} 2 + 1 = 3 \\ 2 - 1 = 1 \end{cases}$$

    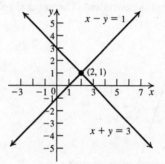

15. The solution is $\{(2, 2)\}$.
    $$\begin{cases} 2 + 2(2) = 2 + 4 = 6 \\ 2(2) + 2 = 4 + 2 = 6 \end{cases}$$

    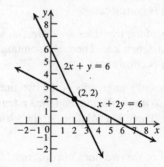

17. The system is inconsistent.

    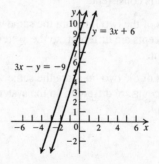

19. The solution is $\left\{\left(\dfrac{7}{3},\dfrac{14}{3}\right)\right\}$.

$$\begin{cases} \dfrac{7}{3}+\dfrac{14}{3}=\dfrac{21}{3}=7 \\ \dfrac{14}{3}=2\left(\dfrac{7}{3}\right) \end{cases}$$

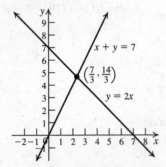

21. The system is dependent. The general solution is $\{(x, 12-3x)\}$.

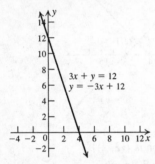

23. The slopes of the two lines are different, so the system is independent. There is a solution, so the system is consistent.

25. The slopes of the two lines are different, so the system is independent. There is a solution, so the system is consistent.

27. The slopes and y-intercepts of the two lines are the same, so they coincide, and the system is dependent. There is a solution, so the system is consistent.

29. The slopes of the two lines are different, so the system is independent. There is a solution, so the system is consistent.

31. The slopes of the two lines are the same while the y-intercepts are different, so the system is inconsistent.

33. The slopes of the two lines are the same while the y-intercepts are different, so the system is inconsistent.

35. The slopes of the two lines are different, so the system is independent. There is a solution, so the system is consistent.

In exercises 37–46, check your answers by substituting the values of x and y into both of the original equations.

37. Substitute $y=2x+1$ into
$5x+2y=9 \Rightarrow 5x+2(2x+1)=9 \Rightarrow$
$5x+4x+2=9 \Rightarrow 9x=7 \Rightarrow x=7/9$

Use this value of x to find y: $y=2\left(\dfrac{7}{9}\right)+1 \Rightarrow$

$y=\dfrac{23}{9}$. The solution is $\left\{\left(\dfrac{7}{9},\dfrac{23}{9}\right)\right\}$.

39. Solve the second equation for y, and then substitute $y=7-x$ into the first equation:
$3x-(7-x)=5 \Rightarrow 3x-7+x=5 \Rightarrow 4x=12 \Rightarrow$
$x=3$. Substitute this value into the second equation to find y : $3+y=7 \Rightarrow y=4$
The solution is $\{(3,4)\}$.

41. Solve the first equation for y, and then substitute $y=2x-5$ into the second equation:
$-4x+2(2x-5)=7 \Rightarrow -4x+4x-10=7 \Rightarrow$
$-10 \neq 7 \Rightarrow$ there is no solution.
Solution set: $\varnothing$

43. Solve the first equation for y, and then substitute $y=3-\dfrac{2}{3}x$ into the second equation:

$3x+2\left(3-\dfrac{2}{3}x\right)=1 \Rightarrow 3x+6-\dfrac{4}{3}x=1 \Rightarrow$

$\dfrac{5}{3}x=-5 \Rightarrow x=-3$. Substitute this value into the first equation to find y:

$\dfrac{2}{3}(-3)+y=3 \Rightarrow y=5$.

The solution is $\{(-3, 5)\}$.

45. Solve the first equation for y, and then substitute $y = \frac{1}{2}(x-5)$ into the second equation:

$-3x + 6\left(\frac{1}{2}(x-5)\right) = -15 \Rightarrow -3x + 3x - 15 = -15 \Rightarrow$

$-15 = -15 \Rightarrow$ the system is dependent. The general form of the solution is

$\left\{\left(x, \frac{1}{2}(x-5)\right)\right\}.$

47. Add the two equations and solve for x:

$\begin{cases} x - y = 1 \\ x + y = 5 \end{cases} \Rightarrow 2x = 6 \Rightarrow x = 3$. Now substitute

$x = 3$ into the second equation and solve for y:
$3 + y = 5 \Rightarrow y = 2$. The solution is $\{(3, 2)\}$.

49. Multiply the first equation by -2:

$\begin{cases} x + y = 0 \\ 2x + 3y = 3 \end{cases} \Rightarrow \begin{cases} -2x - 2y = 0 \\ 2x + 3y = 3 \end{cases}$

Add the equations and solve for y:

$\begin{cases} -2x - 2y = 0 \\ 2x + 3y = 3 \end{cases} \Rightarrow y = 3.$

Substitute this value into the first equation and solve for x: $x + 3 = 0 \Rightarrow x = -3$.
The solution is $\{(-3, 3)\}$.

51. Multiply the first equation by 2:

$\begin{cases} 5x - y = 5 \\ 3x + 2y = -10 \end{cases} \Rightarrow \begin{cases} 10x - 2y = 10 \\ 3x + 2y = -10 \end{cases}$

Add the equations and solve for x:

$\begin{cases} 10x - 2y = 10 \\ 3x + 2y = -10 \end{cases} \Rightarrow 13x = 0 \Rightarrow x = 0.$

Substitute this value into the first equation and solve for y: $5(0) - y = 5 \Rightarrow y = -5$. The solution is $\{(0, -5)\}$.

53. Multiply the first equation by 2:

$\begin{cases} x - y = 2 \\ -2x + 2y = 5 \end{cases} \Rightarrow \begin{cases} 2x - 2y = 4 \\ -2x + 2y = 5 \end{cases}$. Adding the

equations gives $0 = 9 \Rightarrow$ the equations are inconsistent, and there is no solution.
Solution set: $\varnothing$

55. Multiply the second equation by -2:

$\begin{cases} 4x + 6y = 12 \\ 2x + 3y = 6 \end{cases} \Rightarrow \begin{cases} 4x + 6y = 12 \\ -4x - 6y = -12 \end{cases}$. Adding the

equations gives $0 = 0 \Rightarrow$ the equations are dependent. Solving the second equation for y,

we have $2x + 3y = 6 \Rightarrow y = -\frac{2x}{3} + 2$.

The general form of the solution is

$\left\{\left(x, -\frac{2x}{3} + 2\right)\right\}.$

57. Solve the first equation for y, and then substitute this value into the second equation to solve for x: $2x + y = 9 \Rightarrow y = -2x + 9$.

$2x - 3(-2x + 9) = 5 \Rightarrow 8x - 27 = 5 \Rightarrow x = 4.$

Substitute this value into the first equation and solve for y: $2(4) + y = 9 \Rightarrow y = 1$.
The solution is $\{(4, 1)\}$.

59. Multiply the second equation by -2 and then add the equations:

$\begin{cases} 2x + 5y = 2 \\ x + 3y = 2 \end{cases} \Rightarrow \begin{cases} 2x + 5y = 2 \\ -2x - 6y = -4 \end{cases} \Rightarrow -y = -2 \Rightarrow$

$y = 2$. Substitute this value into the second

equation and solve for x: $x + 3(2) = 2 \Rightarrow x = -4$.
The solution is $\{(-4, 2)\}$.

61. Multiply the second equation by -3 and then add the equations:

$\begin{cases} 2x + 3y = 7 \\ 3x + y = 7 \end{cases} \Rightarrow \begin{cases} 2x + 3y = 7 \\ -9x - 3y = -21 \end{cases} \Rightarrow$

$-7x = -14 \Rightarrow x = 2$. Substitute this value into

the second equation and solve for y:
$3(2) + y = 7 \Rightarrow y = 1$. The solution is $\{(2, 1)\}$.

63. Multiply the first equation by -2 and the second equation by 3, then add the equations to solve for x:

$\begin{cases} 2x + 3y = 9 \\ 3x + 2y = 11 \end{cases} \Rightarrow \begin{cases} -4x - 6y = -18 \\ 9x + 6y = 33 \end{cases} \Rightarrow$

$5x = 15 \Rightarrow x = 3$. Substitute this value into the

first equation to solve for y:
$2(3) + 3y = 9 \Rightarrow 3y = 3 \Rightarrow y = 1$.
The solution is $\{(3, 1)\}$.

65. First, simplify both equations:

$\begin{cases} \dfrac{x}{4} + \dfrac{y}{6} = 1 \\ x + 2(x - y) = 7 \end{cases} \Rightarrow \begin{cases} 6x + 4y = 24 \\ 3x - 2y = 7 \end{cases}$. Now

multiply the second equation by 2, and then add the equations:

$\begin{cases} 6x + 4y = 24 \\ 3x - 2y = 7 \end{cases} \Rightarrow \begin{cases} 6x + 4y = 24 \\ 6x - 4y = 14 \end{cases} \Rightarrow$

$12x = 38 \Rightarrow x = \dfrac{19}{6}.$

(continued on next page)

(*continued*)

Substitute this value into the second equation and solve for *y*:

$$\frac{19}{6}+2\left(\frac{19}{6}-y\right)=7 \Rightarrow \frac{57}{6}-2y=7 \Rightarrow$$

$$-2y=-\frac{5}{2} \Rightarrow y=\frac{5}{4}.$$

The solution is $\left\{\left(\frac{19}{6},\frac{5}{4}\right)\right\}$.

67. Simplify the first equation:
$3x=2(x+y) \Rightarrow 3x=2x+2y \Rightarrow x=2y.$
Substitute this expression for *x* into the second equation: $3(2y)-5y=2 \Rightarrow y=2.$ Substitute this value into the first equation to solve for *x*:
$3x=2(x+2) \Rightarrow 3x=2x+4 \Rightarrow x=4.$
The solution is $\{(4, 2)\}$.

69. Multiply the first equation by –2, then add the equations:

$$\begin{cases}0.2x+0.7y=1.5\\0.4x-0.3y=1.3\end{cases} \Rightarrow \begin{cases}-0.4x-1.4y=-3\\0.4x-0.3y=1.3\end{cases} \Rightarrow$$

$-1.7y=-1.7 \Rightarrow y=1$. Substitute this value into the first equation to solve for *x*:
$0.2x+0.7(1)=1.5 \Rightarrow 0.2x=0.8 \Rightarrow x=4.$
The solution is $\{(4, 1)\}$.

71. $\begin{cases}\frac{x}{2}+\frac{y}{3}=1 & (1)\\ \frac{3x}{4}+\frac{y}{2}=1 & (2)\end{cases}$

Clear the fractions by multiplying equation (1) by 6 and equation (2) by 8.

$$\begin{cases}3x+2y=6 & (1)\\6x+4y=8 & (2)\end{cases}$$

Now multiply equation (1) by –2, then add the two equations.

$$\begin{aligned}-6x-4y&=-24\\6x+4y&=\ \ \ 8\\ \hline 0&=-16 \ \text{False}\end{aligned}$$

The system is inconsistent.
The solution set is $\varnothing$.

73. $\begin{cases}\frac{x}{3}-\frac{y}{2}=1 & (1)\\ \frac{3y}{8}-\frac{x}{4}=-\frac{3}{4} & (2)\end{cases}$

Clear the fractions by multiplying equation (1) by 6 and equation (2) by 8.

$$\begin{cases}2x-3y=6 & (1)\\3y-2x=-6 & (2)\end{cases}$$

Now add the two equations.

$$\begin{aligned}2x-3y&=6\\-2x+3y&=-6\\ \hline 0&=0 \ \checkmark\end{aligned}$$

The system is dependent.
Solve equation (1) for *y* in terms of *x*.

$$\frac{x}{3}-\frac{y}{2}=1 \Rightarrow 2x-3y=6 \Rightarrow -3y=-2x+6 \Rightarrow$$

$$3y=2(x-3) \Rightarrow y=\frac{2}{3}(x-3)$$

The solution set is $\left\{\left(x,\frac{2}{3}(x-3)\right)\right\}$.

75. $\begin{cases}\frac{2}{x}+\frac{5}{y}=-5 & (1)\\ \frac{3}{x}-\frac{2}{y}=-17 & (2)\end{cases}$

Let $u=\frac{1}{x}$ and $v=\frac{1}{y}$. Then the system

becomes $\begin{cases}2u+5v=-5\\3u-2v=-17\end{cases}$. Multiply the first

equation by 2 and the second equation by 5,

then add the equations: $\begin{cases}2u+5v=-5\\3u-2v=-17\end{cases} \Rightarrow$

$\begin{cases}4u+10v=-10\\15u-10v=-85\end{cases} \Rightarrow 19u=-95 \Rightarrow u=-5$.

Substitute this value into the first equation to solve for *v*:
$2(-5)+5v=-5 \Rightarrow 5v=5 \Rightarrow v=1.$

$$u=\frac{1}{x} \Rightarrow -5=\frac{1}{x} \Rightarrow x=-\frac{1}{5}.$$

$$v=\frac{1}{y} \Rightarrow 1=\frac{1}{y} \Rightarrow y=1.$$

The solution is $\left\{\left(-\frac{1}{5},1\right)\right\}$.

77. $\begin{cases}\frac{3}{x}+\frac{1}{y}=4 & (1)\\ \frac{6}{x}-\frac{1}{y}=2 & (2)\end{cases}$

Let $u=\frac{1}{x}$ and $v=\frac{1}{y}$. Then the system

becomes $\begin{cases}3u+v=4\\6u-v=2\end{cases}$. Add the equations and

solve for *u*: $9u=6 \Rightarrow u=\frac{2}{3}$. Substitute this

value into the first equation and solve for *v*.

(*continued on next page*)

(*continued*)

$$3\left(\frac{2}{3}\right) + v = 4 \Rightarrow v = 2.$$

$$u = \frac{1}{x} \Rightarrow \frac{2}{3} = \frac{1}{x} \Rightarrow x = \frac{3}{2}.$$

$$v = \frac{1}{y} \Rightarrow 2 = \frac{1}{y} \Rightarrow y = \frac{1}{2}.$$

The solution is $\left\{\left(\frac{3}{2}, \frac{1}{2}\right)\right\}$.

79. $\begin{cases} \dfrac{5}{x} + \dfrac{10}{y} = 3 & (1) \\ \dfrac{2}{x} - \dfrac{12}{y} = -2 & (2) \end{cases}$

Let $u = \dfrac{1}{x}$ and $v = \dfrac{1}{y}$. Then the system

becomes $\begin{cases} 5u + 10v = 3 \\ 2u - 12v = -2 \end{cases}$. Multiply the first

equation by 2 and the second equation by –5, then add the equation to solve for v:

$$\begin{cases} 5u + 10v = 3 \\ 2u - 12v = -2 \end{cases} \Rightarrow \begin{cases} 10u + 20v = 6 \\ -10u + 60v = 10 \end{cases} \Rightarrow$$

$80v = 16 \Rightarrow v = \dfrac{1}{5}.$ Substitute this value into

the first equation to solve for u:

$$5u + 10\left(\frac{1}{5}\right) = 3 \Rightarrow 5u = 1 \Rightarrow u = \frac{1}{5}.$$

$$u = \frac{1}{x} \Rightarrow \frac{1}{5} = \frac{1}{x} \Rightarrow x = 5.$$

$$v = \frac{1}{y} \Rightarrow \frac{1}{5} = \frac{1}{y} \Rightarrow y = 5.$$

The solution is $\{(5, 5)\}$.

81. $\begin{cases} \dfrac{2}{x} + \dfrac{1}{y} = 4 & (1) \\ x + 2y = 6xy & (2) \end{cases}$

Divide equation (2) by $6xy$.

$$\begin{cases} \dfrac{2}{x} + \dfrac{1}{y} = 4 \\ \dfrac{1}{6y} + \dfrac{1}{3x} = 1 \end{cases} \Rightarrow \begin{cases} \dfrac{2}{x} + \dfrac{1}{y} = 4 & (1) \\ \dfrac{1}{3x} + \dfrac{1}{6y} = 1 & (2) \end{cases}$$

Let $u = \dfrac{1}{x}$ and $v = \dfrac{1}{y}$. Then the system

becomes $\begin{cases} 2u + v = 4 \\ \dfrac{1}{3}u + \dfrac{1}{6}v = 1 \end{cases}$.

Clear the fractions in equation (2) by multiplying by 6.

$$\begin{cases} 2u + v = 4 \\ \dfrac{1}{3}u + \dfrac{1}{6}v = 1 \end{cases} \Rightarrow \begin{cases} 2u + v = 4 \\ 2u + v = 6 \end{cases}$$

Multiply equation (2) by –1, then add the equations.

$$\begin{aligned} 2u + v &= 4 \\ \underline{-2u - v} &= \underline{-6} \\ 0 &= -2 \quad \text{False} \end{aligned}$$

The system is inconsistent.
The solution set is $\varnothing$.

5.1 Applying the Concepts

In exercises 83–86, the equilibrium point is the point that satisfies both the demand equation and the supply equation.

83. Add the equations to solve for p:

$$\begin{cases} 2p + x = 140 \\ 12p - x = 280 \end{cases} \Rightarrow 14p = 420 \Rightarrow p = 30.$$

Substitute this value into the first equation to solve for x: $2(30) + x = 140 \Rightarrow x = 80$.
The equilibrium point is (80, 30).

85. Multiply the second equation by –1, then add the equations to solve for x:

$$\begin{cases} 2p + x = 25 \\ x - p = 13 \end{cases} \Rightarrow \begin{cases} 2p + x = 25 \\ p - x = -13 \end{cases} \Rightarrow 3p = 12 \Rightarrow$$

$p = 4.$ Substitute this value into the second equation to solve for x: $x - 4 = 13 \Rightarrow x = 17$.
The equilibrium point is (17, 4).

87. Let x = the diameter of the largest pizza, and let y = the diameter of the smallest pizza.

Then $\begin{cases} x + y = 29 \\ x - y = 13 \end{cases} \Rightarrow 2x = 42 \Rightarrow x = 21.$

$21 + y = 29 \Rightarrow y = 8.$

The largest pizza has diameter 21 inches, and the smallest pizza has diameter 8 inches.

89. Let x = the percentage of paper trash, and let y = the percentage of plastic trash. If the total amount of trash is t, then

$$\begin{cases} x + y = 48 \\ x = 5y \end{cases}.$$

Substitute the expression for x from the second equation into the first equation to solve for y: $5y + y = 48 \Rightarrow y = 8$. Substitute this value into the first equation to solve for x: $x + 8 = 48 \Rightarrow x = 40$. So, 8% of the trash is plastic and 40% is paper.

91. Let x = the number of beads, and let y = the number of doubloons . Then

$$\begin{cases} 0.4x + 0.3y = 265 \\ x + \quad y = 770 \end{cases}.$$ Multiply the second

equation by -0.3 and then add the equations to

solve for x: $\begin{cases} 0.4x + 0.3y = 265 \\ x + \quad y = 770 \end{cases} \Rightarrow$

$\begin{cases} 0.4x + 0.3y = 265 \\ -0.3x - 0.3y = -231 \end{cases} \Rightarrow 0.1x = 34 \Rightarrow x = 340.$

Substitute this value into the second equation to solve for y: $340 + y = 770 \Rightarrow y = 430.$ Levon bought 340 beads and 430 doubloons.

93. Let x = the number of Egg McMuffins, and let y = the number of Breakfast Burritos. Then

$$\begin{cases} 27x + 21y = 123 \\ 17x + 13y = 77 \end{cases}.$$ Multiply the first equation

by 13 and the second equation by -21, then add the equations to solve for x:

$\begin{cases} 27x + 21y = 123 \\ 17x + 13y = 77 \end{cases} \Rightarrow \begin{cases} 351x + 273y = 1599 \\ -357x - 273y = -1617 \end{cases} \Rightarrow$

$-6x = -18 \Rightarrow x = 3.$ Substitute this value into the first equation to solve for y:

$27(3) + 21y = 123 \Rightarrow 21y = 42 \Rightarrow y = 2.$ You will need to eat three Egg McMuffins and two Breakfast Burritos.

95. Let x = the amount invested at 7.5%, and let y = the amount invested at 12%. Then

$$\begin{cases} x + \quad y = 50,000 \\ 0.075x + 0.12y = 5190 \end{cases}.$$ Solve the first

equation for x and substitute this value into the second equation to solve for y:
$x + y = 50,000 \Rightarrow x = 50,000 - y.$

$0.075(50,000 - y) + 0.12y = 5190 \Rightarrow$
$3750 - 0.075y + 0.12y = 5190 \Rightarrow$
$0.045y = 1440 \Rightarrow y = \$32,000.$

Substitute this value into the first equation to solve for x:
$x + 32,000 = 50,000 \Rightarrow x = 18,000.$
Mrs. García invested $18,000 at 7.5% and $32,000 at 12%.

97. Let x = the amount earned tutoring, and let y = the amount earned working at McDougal's.

Then $\begin{cases} x = 2y \\ \dfrac{x + y}{2} = 11.25 \end{cases}.$

Substitute the expression for x from the first equation into the second equation and solve for y.

$\dfrac{2y + y}{2} = 11.25 \Rightarrow 3y = 22.50 \Rightarrow y = 7.50.$

Substitute this value into the first equation to solve for x: $x = 2(7.50) = 15.$ She earned $15 tutoring and $7.50 working at McDougal's.

99. Let x = the number of pounds of the herb, and let y = the number of pounds of tea. Then

$\begin{cases} x + \quad y = 100 \\ 5.50x + 3.20y = 3.66(100) \end{cases}.$ Solve the first

equation for x, and substitute this expression into the second equation to solve for y:
$x + y = 100 \Rightarrow x = 100 - y.$

$5.5(100 - y) + 3.20y = 3.66(100) \Rightarrow$
$550 - 5.5y + 3.2y = 366 \Rightarrow -2.3y = -184 \Rightarrow$
$y = 80.$ Substitute this value into the first equation to solve for x: $x + 80 = 100 \Rightarrow x = 20.$ There are 20 pounds of the herb and 80 pounds of tea in the mixture.

101. Let x = the speed of the plane in still air, and let y = the wind speed. So, the speed of the plane with the wind is $x + y$, and the speed of the plane against the wind is $x - y$. Then, using the fact that rate × time = distance, we have

$\begin{cases} 5(x + y) = 3000 \\ 6(x - y) = 3000 \end{cases} \Rightarrow \begin{cases} x + y = 600 \\ x - y = 500 \end{cases}.$ Add the

equations to solve for x:
$2x = 1100 \Rightarrow x = 550.$ Substitute this value into the first equation to solve for y:
$5(550 + y) = 3000 \Rightarrow$
$550 + y = 600 \Rightarrow y = 50.$ The speed of the plane is 550 kph, and the wind speed is 50 kph.

103. a. $C(x) = y = 2x + 30,000;\ R(x) = y = 3.50x$

b.

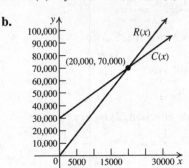

c. $\begin{cases} y = 2x + 30,000 \\ y = 3.5x \end{cases}.$

Substitute the expression for y from the second equation into the first equation to solve for x:
$2x + 30,000 = 3.5x \Rightarrow 1.5x = 30,000 \Rightarrow$
$x = 20,000$ magazines

105. Let x = Shanaysha's weekly sales. Then her salary at store B is $150 + 0.04x$.
$150 + 0.04x > 400 \Rightarrow 0.04x > 250 \Rightarrow x > 6250$.
Her weekly sales should be more than $6250.

5.1 Beyond the Basics

107. $\begin{cases} 2\log_3 x + 3\log_3 y = 8 & (1) \\ 3\log_3 x - \log_3 y = 1 & (2) \end{cases}$

Multiply equation (2) by 3. Then add the two equations.

$2\log_3 x + 3\log_3 y = 8$
$9\log_3 x - 3\log_3 y = 3$
———————————————
$11\log_3 x = 11$

Solve for x.

$11\log_3 x = 11 \Rightarrow \log_3 x = 1 \Rightarrow x = 3$

Substitute $x = 3$ into equation (2) and solve for y:

$3\log_3 3 - \log_3 y = 1 \Rightarrow 3(1) - \log_3 y = 1 \Rightarrow$

$-\log_3 y = -2 \Rightarrow \log_3 y = 2 \Rightarrow y = 3^2 = 9$

Be sure to check your answer in the original system of equations. The solution is $\{(3, 9)\}$.

109. $\begin{cases} 3e^x - 4e^y = 4 & (1) \\ 2e^x + 5e^y = 18 & (2) \end{cases}$

Multiply equation (1) by 5 and equation (2) by 4. Then add the equations.

$15e^x - 20e^y = 20$
$8e^x + 20e^y = 72$
———————————————
$23e^x = 92$

Now solve for x.

$23e^x = 92 \Rightarrow e^x = \dfrac{92}{23} \Rightarrow x = \ln 4$.

Substitute this value in equation (1) and solve for y.

$3e^{\ln 4} - 4e^y = 4 \Rightarrow 3(4) - 4e^y = 4 \Rightarrow$

$-4e^y = -8 \Rightarrow e^y = 2 \Rightarrow y = \ln 2$

111. First solve the system $\begin{cases} x + 2y = 7 \\ 3x + 5y = 11 \end{cases}$.

Multiply the first equation by -3, then add the equations to solve for y:

$\begin{cases} x + 2y = 7 \\ 3x + 5y = 11 \end{cases} \Rightarrow \begin{cases} -3x - 6y = -21 \\ 3x + 5y = 11 \end{cases} \Rightarrow -y = -10 \Rightarrow$

$y = 10$. Substitute this value into the first equation to solve for x: $x + 2(10) = 7 \Rightarrow$

$x = -13$. Now substitute $(-13, 10)$ into the third equation to solve for c:

$-13c + 3(10) = 4 \Rightarrow -13c = -26 \Rightarrow c = 2$.

113. If $(x - 2)$ is a factor of both $f(x)$ and $g(x)$, then the remainder when each function is divided by $(x - 2)$ is zero. Use synthetic division to find the remaining factors.

$$
\begin{array}{r|cccc}
2] & 1 & -4 & a & b \\
 & & 2 & -4 & 2a-8 \\
\hline
 & 1 & -2 & a-4 & 2a+b-8 \\
\end{array}
$$

$$
\begin{array}{r|cccc}
2] & 1 & -a & b & 8 \\
 & & 2 & -2a+4 & -4a+2b+8 \\
\hline
 & 1 & 2-a & -2a+b+4 & -4a+2b+16 \\
\end{array}
$$

This gives us the system

$\begin{cases} 2a + b - 8 = 0 \\ -4a + 2b + 16 = 0 \end{cases} \Rightarrow \begin{cases} 4a + 2b - 16 = 0 \\ -4a + 2b + 16 = 0 \end{cases} \Rightarrow$

$4b = 0 \Rightarrow b = 0$

Substituting $b = 0$ into the first equation gives

$2a + 0 - 8 = 0 \Rightarrow a = 4$

The solution is $a = 4$ and $b = 0$.

115. To find the intersection of the two lines, solve the system $\begin{cases} 2x + y = 3 \\ x - 3y = 12 \end{cases}$. Multiply the first equation by 3, then add the equations to solve for x: $\begin{cases} 2x + y = 3 \\ x - 3y = 12 \end{cases} \Rightarrow \begin{cases} 6x + 3y = 9 \\ x - 3y = 12 \end{cases} \Rightarrow$

$7x = 21 \Rightarrow x = 3$. Substitute this value into the first equation to solve for y:

$2(3) + y = 3 \Rightarrow y = -3$. So the point of intersection is $(3, -3)$.

Solve the equation $3x + 2y = 8$ for y to find the slope: $y = -\dfrac{3}{2}x + 4 \Rightarrow$ the slope is $-\dfrac{3}{2}$.

The equation of the line is

$y + 3 = -\dfrac{3}{2}(x - 3) \Rightarrow y = -\dfrac{3}{2}x + \dfrac{3}{2}$.

Verify graphically:

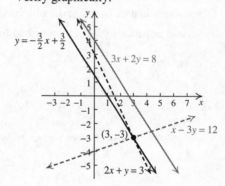

117. To find the intersection of the two lines, solve

the system $\begin{cases} 2x + 5y + 7 = 0 \\ 13x - 10y + 3 = 0 \end{cases}$. Multiply the

first equation by 2, then add the equations to solve for x:

$\begin{cases} 2x + 5y + 7 = 0 \\ 13x - 10y + 3 = 0 \end{cases} \Rightarrow \begin{cases} 4x + 10y + 14 = 0 \\ 13x - 10y + 3 = 0 \end{cases} \Rightarrow$

$17x + 17 = 0 \Rightarrow x = -1$. Substitute this value into the first equation to solve for y:

$2(-1) + 5y + 7 = 0 \Rightarrow 5 + 5y = 0 \Rightarrow y = -1$.

The slope of the line $7x + 13y = 8$ is $-\dfrac{7}{13}$, so

the slope of the perpendicular line is $\dfrac{13}{7}$. The

equation of the line is $y + 1 = \dfrac{13}{7}(x + 1) \Rightarrow$

$y = \dfrac{13}{7}x + \dfrac{6}{7}$.

Verify graphically:

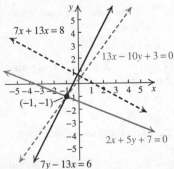

5.1 Critical Thinking/Discussion/Writing

118. a. There is only one solution if the equations are independent. Solve for x by multiplying the first equation by b_2 and multiplying the second equation by $-b_1$, then adding the

equations: $\begin{cases} a_1 x + b_1 y = c_1 \\ a_2 x + b_2 y = c_2 \end{cases} \Rightarrow$

$\begin{cases} b_2 a_1 x + b_2 b_1 y = b_2 c_1 \\ -b_1 a_2 x - b_1 b_2 y = -b_1 c_2 \end{cases} \Rightarrow$

$b_2 a_1 x - b_1 a_2 x = b_2 c_1 - b_1 c_2 \Rightarrow$

$x(b_2 a_1 - b_1 a_2) = b_2 c_1 - b_1 c_2 \Rightarrow$

$x = \dfrac{b_2 c_1 - b_1 c_2}{a_1 b_2 - a_2 b_1}$

Solve for y by multiplying the first equation by a_2 and multiplying the second equation by $-a_1$, then adding the equations:

$\begin{cases} a_1 x + b_1 y = c_1 \\ a_2 x + b_2 y = c_2 \end{cases} \Rightarrow$

$\begin{cases} a_1 a_2 x + a_2 b_1 y = a_2 c_1 \\ -a_1 a_2 x - a_1 b_2 y = -a_1 c_2 \end{cases} \Rightarrow$

$a_2 b_1 y - a_1 b_2 y = a_2 c_1 - a_1 c_2 \Rightarrow$

$y(a_2 b_1 - a_1 b_2) = a_2 c_1 - a_1 c_2 \Rightarrow$

$y = \dfrac{a_2 c_1 - a_1 c_2}{a_2 b_1 - a_1 b_2} = \dfrac{a_1 c_2 - a_2 c_1}{a_1 b_2 - a_2 b_1}$.

Note that the denominator of x and y, $a_1 b_2 - a_2 b_1$, cannot equal zero.

b. There is no solution if the equations are inconsistent. They are inconsistent if they are parallel. The slopes of the lines are

equal and $\dfrac{b_1}{a_1} = \dfrac{b_2}{a_2} \Rightarrow \dfrac{a_1}{a_2} = \dfrac{b_1}{b_2}$. The

intercepts are different. If $c_1 = c_2$, then

$\dfrac{c_1}{a_1} \neq \dfrac{c_2}{a_2}$.

c. There are infinitely many solutions if the equations are dependent. From part (b), we

have $\dfrac{a_1}{a_2} = \dfrac{b_1}{b_2}$. The intercepts of the lines

must be the same, so $\dfrac{a_1}{a_2} = \dfrac{b_1}{b_2} = \dfrac{c_1}{c_2}$.

119. a. The slope of ℓ_1 is $-\dfrac{3}{4}$, so the slope of the

perpendicular is $\dfrac{4}{3}$. The equation of ℓ_2 is

$y - 8 = \dfrac{4}{3}(x - 5) \Rightarrow 3y - 24 = 4x - 20 \Rightarrow$

$3y - 4x = 4$.

b. To find Q, solve the system $\begin{cases} 3x + 4y = 12 \\ -4x + 3y = 4 \end{cases}$.

Multiply the first equation by 4 and the second equation by 3, then add the equations

to solve for x: $\begin{cases} 3x + 4y = 12 \\ -4x + 3y = 4 \end{cases} \Rightarrow$

$\begin{cases} 12x + 16y = 48 \\ -12x + 9y = 12 \end{cases} \Rightarrow 25y = 60 \Rightarrow y = \dfrac{12}{5}$.

Substitute this value into the first equation to

solve for x: $3x + 4\left(\dfrac{12}{5}\right) = 12 \Rightarrow 3x = \dfrac{12}{5} \Rightarrow$

$x = \dfrac{4}{5}$. The coordinates of Q are $\left(\dfrac{4}{5}, \dfrac{12}{5}\right)$.

c. $d(P,Q) = \sqrt{\left(5 - \dfrac{4}{5}\right)^2 + \left(8 - \dfrac{12}{5}\right)^2} = \sqrt{49} = 7$

120. a. The slope of ℓ_1 is $-\dfrac{a}{b}$, so the slope of the

perpendicular is $\dfrac{b}{a}$. The equation of ℓ_2 is

$y - y_1 = \dfrac{b}{a}(x - x_1) \Rightarrow$

$ay - ay_1 = bx - bx_1 \Rightarrow ay - bx = ay_1 - bx_1$ or

$b(x - x_1) - a(y - y_1) = 0.$

b. The intersection of ℓ_1 and ℓ_2 is the solution of the system

$\begin{cases} ax + by = -c \\ -bx + ay = ay_1 - bx_1 \end{cases} \Rightarrow$

$\begin{cases} abx + b^2 y = -bc \\ -abx + a^2 y = a^2 y_1 - abx_1 \end{cases} \Rightarrow$

$y(a^2 + b^2) = -bc + a^2 y_1 - abx_1 \Rightarrow$

$y = \dfrac{-bc + a^2 y_1 - abx_1}{a^2 + b^2}$

$\begin{cases} ax + by = -c \\ -bx + ay = ay_1 - bx_1 \end{cases} \Rightarrow$

$\begin{cases} -a^2 x - aby = ac \\ -b^2 x + aby = aby_1 - b^2 x_1 \end{cases} \Rightarrow$

$-x(a^2 + b^2) = ac + aby_1 - b^2 x_1 \Rightarrow$

$x = \dfrac{-ac - aby_1 + b^2 x_1}{a^2 + b^2}$

$= -\dfrac{ac + aby_1 - b^2 x_1}{a^2 + b^2}$

The intersection of the two lines is

$\left(-\dfrac{ac + aby_1 - b^2 x_1}{a^2 + b^2}, \dfrac{-abx_1 + a^2 y_1 - bc}{a^2 + b^2} \right).$

c. Now find the distance between (x_1, y_1) and

$\left(-\dfrac{ac + aby_1 - b^2 x_1}{a^2 + b^2}, \dfrac{-bc + a^2 y_1 - abx_1}{a^2 + b^2} \right):$

$\sqrt{\left(x_1 + \dfrac{ac + aby_1 - b^2 x_1}{a^2 + b^2} \right)^2 + \left(y_1 - \dfrac{-bc + a^2 y_1 - abx_1}{a^2 + b^2} \right)^2}$

$= \sqrt{\left(\dfrac{a(ax_1 + by_1 + c)}{a^2 + b^2} \right)^2 + \left(\dfrac{b(ax_1 + by_1 + c)}{a^2 + b^2} \right)^2}$

$= \sqrt{\dfrac{a^2(ax_1 + by_1 + c)^2}{\left(a^2 + b^2\right)^2} + \dfrac{b^2(ax_1 + by_1 + c)^2}{\left(a^2 + b^2\right)^2}}$

$= \sqrt{\dfrac{(a^2 + b^2)(ax_1 + by_1 + c)^2}{\left(a^2 + b^2\right)^2}}$

$= \sqrt{\dfrac{(ax_1 + by_1 + c)^2}{a^2 + b^2}} = \dfrac{|ax_1 + by_1 + c|}{\sqrt{a^2 + b^2}}.$

121. a. $a = 1, b = 1, c = -7, x_1 = 2, y_1 = 3$

$\dfrac{|ax_1 + by_1 + c|}{\sqrt{a^2 + b^2}} = \dfrac{|1(2) + 1(3) - 7|}{\sqrt{1^2 + 1^2}} = \dfrac{2}{\sqrt{2}} = \sqrt{2}$

b. $a = 2, b = -1, c = 3, x_1 = -2, y_1 = 5$

$\dfrac{|ax_1 + by_1 + c|}{\sqrt{a^2 + b^2}} = \dfrac{|2(-2) - 1(5) + 3|}{\sqrt{2^2 + (-1)^2}}$

$= \dfrac{6}{\sqrt{5}} = \dfrac{6\sqrt{5}}{5}$

c. $a = 5, b = -2, c = -7, x_1 = 3, y_1 = 4$

$\dfrac{|ax_1 + by_1 + c|}{\sqrt{a^2 + b^2}} = \dfrac{|5(3) - 2(4) - 7|}{\sqrt{3^2 + 4^2}} = 0$

d. $a = a, b = b, c = c, x_1 = 0, y_1 = 0$

$\dfrac{|ax_1 + by_1 + c|}{\sqrt{a^2 + b^2}} = \dfrac{|a(0) + b(0) + c|}{\sqrt{a^2 + b^2}}$

$= \dfrac{|c|}{\sqrt{a^2 + b^2}}$

122.

Let $Q(a, b)$ be the reflection point. Using the hint, we have two equations, one developed from the fact that the midpoint of $\overline{PQ}$ lies on line ℓ, and the other developed from the fact that ℓ is perpendicular to the line containing the segment $\overline{PQ}$.

(*continued on next page*)

(continued)

Equation (1): The midpoint of $\overline{PQ}$ is

$M\left(\dfrac{a+2}{2}, \dfrac{b+1}{2}\right)$. Since M lies on line ℓ, we

have $2\left(\dfrac{a+2}{2}\right) + 3\left(\dfrac{b+1}{2}\right) = 20$. We can

simplify this to

$(a+2) + \dfrac{3}{2}b + \dfrac{3}{2} = 20 \Rightarrow a + \dfrac{3}{2}b = \dfrac{33}{2} \Rightarrow$
$2a + 3b = 33$

Equation (2): The slope of ℓ is $-\dfrac{2}{3}$, so the

slope of the line perpendicular to line ℓ is $\dfrac{3}{2}$.

Using the formula for the slope of $\overline{PQ}$, we
have

$\dfrac{3}{2} = \dfrac{b-1}{a-2} \Rightarrow 3a - 6 = 2b - 2 \Rightarrow 3a - 2b = 4.$

So, the system is

$\begin{cases} 2a + 3b = 33 & (1) \\ 3a - 2b = 4 & (2) \end{cases}$

Multiply the first equation by 2 and the second
equation by 3. Then add the two equations,
then solve for a.

$4a + 6b = 66$
$\underline{9a - 6b = 12}$
$\quad\quad 13a = 78 \Rightarrow a = 6$

Substitute $a = 6$ into equation (2) and solve for
b.

$3(6) - 2b = 4 \Rightarrow -2b = -14 \Rightarrow b = 7$

Be sure to check your answer.
The reflection point is $Q(6, 7)$.

123.

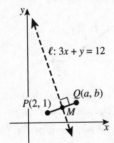

Let $Q(a, b)$ be the reflection point. Using the
hint from exercise 122, we have two equations,
one developed from the fact that the midpoint

of $\overline{PQ}$ lies on line ℓ, and the other developed
from the fact that ℓ is perpendicular to the line
containing the segment $\overline{PQ}$.

Equation (1): The midpoint of $\overline{PQ}$ is

$M\left(\dfrac{a+2}{2}, \dfrac{b+1}{2}\right)$. Since M lies on line ℓ, we

have $3\left(\dfrac{a+2}{2}\right) + \left(\dfrac{b+1}{2}\right) = 12$. We can

simplify this to

$3(a+2) + (b+1) = 24 \Rightarrow$
$3a + 6 + b + 1 = 24 \Rightarrow 3a + b = 17$

Equation (2): The slope of ℓ is -3, so the

slope of the line perpendicular to line ℓ is $\dfrac{1}{3}$.

Using the formula for the slope of $\overline{PQ}$, we
have

$\dfrac{1}{3} = \dfrac{b-1}{a-2} \Rightarrow 3b - 3 = a - 2 \Rightarrow a - 3b = -1.$

So, the system is

$\begin{cases} 3a + b = 17 & (1) \\ a - 3b = -1 & (2) \end{cases}$

Multiply the first equation by 3. Then add the
two equations, then solve for a.

$9a + 3b = 51$
$\underline{a - 3b = -1}$
$\quad 10a = 50 \Rightarrow a = 5$

Substitute $a = 5$ into equation (1) and solve for
b.

$3(5) + b = 17 \Rightarrow b = 2$

Be sure to check your answer.
The reflection point is $Q(5, 2)$.

5.1 Maintaining Skills

125. $y + 2z = 7;\ z = 2$
$y + 2(2) = 7 \Rightarrow y + 4 = 7 \Rightarrow y = 3$

127. $2x + 3y + 5z = 21;\ y = 1;\ z = 2$
$2x + 3(1) + 5(2) = 21 \Rightarrow 2x + 13 = 21 \Rightarrow$
$2x = 8 \Rightarrow x = 4$

129. $2x + 3y - 2z = 5;\ y - z = 4,\ z = -3$
$y - (-3) = 4 \Rightarrow y + 3 = 4 \Rightarrow y = 1$
$2x + 3(1) - 2(-3) = 5 \Rightarrow 2x + 9 = 5 \Rightarrow$
$2x = -4 \Rightarrow x = -2$

5.2 Systems of Linear Equations in Three Variables

5.2 Practice Problems

1. Substituting the ordered triple into the system
$$\begin{cases} x+\ y+\ z = \ 1 \\ 3x+4y+\ z = -4, \\ 2x+\ y+2z = \ 5 \end{cases}$$
we find that $(2, -3, 2)$ is a solution.
$$\begin{cases} 2+(-3)+2 = 1 \\ 3(2)+4(-3)+2 = -4 \\ 2(2)+(-3)+2(2) = 5 \end{cases}$$

2. $\begin{cases} 2x+5y\ \ \ \ \ = 1 & (1) \\ x-3y+2z = 1 & (2) \\ -x+2y+\ z = 7 & (3) \end{cases}$

   Interchange equations (1) and (2).
   $$\begin{cases} x-3y+2z = 1 & (2) \\ 2x+5y\ \ \ \ \ = 1 & (1) \\ -x+2y+\ z = 7 & (3) \end{cases}$$
   Multiply equation (2) by -2, then add the resulting equation and equation (1) to eliminate x, then replace equation (1) with the resulting equation.
   $$\begin{array}{r} -2x+6y-4z = -2 \quad (2) \\ 2x+5y\ \ \ \ \ \ = \ 1 \quad (1) \\ \hline 11y-4z = -1 \quad (4) \end{array}$$
   $$\begin{cases} x-3y+2z = \ 1 & (2) \\ 11y-4z = -1 & (4) \\ -x+2y+\ z = \ 7 & (3) \end{cases}$$
   Now add equations (2) and (3), then replace equation (3) with the resulting equation.
   $$\begin{array}{r} x-3y+2z = 1 \quad (2) \\ -x+2y+\ z = 7 \quad (3) \\ \hline -y+3z = 8 \quad (5) \end{array}$$
   $$\begin{cases} x-3y+2z = \ 1 & (2) \\ 11y-4z = -1 & (4) \\ -y+3z = \ 8 & (5) \end{cases}$$
   Now multiply equation (5) by 11, add the resulting equation to equation (4), then solve for z.
   $$\begin{array}{r} 11y-\ 4z = -1 \quad (4) \\ -11y+33z = 88 \quad (5) \\ \hline 29z = 87 \\ z = \ 3 \quad (6) \end{array}$$
   The system becomes
   $$\begin{cases} x-3y+2z = \ 1 & (2) \\ 11y-4z = -1 & (4). \\ z = \ 3 & (6) \end{cases}$$
   Now substitute $z = 3$ into equation (4) and

solve for y.
$$11y-4(3) = -1 \Rightarrow 11y = 11 \Rightarrow y = 1$$
Substitute the values of y and z into equation (2) and solve for x.
$$x-3(1)+2(3) = 1 \Rightarrow x+3 = 1 \Rightarrow x = -2$$
So, we have $x = -2$, $y = 1$, and $z = 3$. Now check in the original system.
$$\begin{cases} 2(-2)+5(1)\ \ \ \ \ \ \ = 1 & \checkmark \\ -2-3(1)+2(3) = 1 & \checkmark \\ -(-2)+2(1)+\ \ 3 = 7 & \checkmark \end{cases}$$
The solution is $\{(-2, 1, 3)\}$.

3. $\begin{cases} 2x+2y+\ 2z = 12 & (1) \\ -3x+\ y-11z = -6 & (2) \\ 2x+\ y+\ 4z = -8 & (3) \end{cases}$

   Divide equation (1) by 2 and replace equation (1) with the resulting equation.
   $$\begin{cases} x+\ y+\ z = \ 6 & (4) \\ -3x+\ y-11z = -6 & (2) \\ 2x+\ y+\ 4z = -8 & (3) \end{cases}$$
   Multiply equation (4) by 3, add the resulting equation to equation (2), and replace equation (2) with the result. Similarly, multiply equation (4) by -2, add the resulting equation to equation (3), and replace equation (3) with the result.
   $$\begin{cases} x+\ y+\ z = \ \ 6 & (4) \\ 4y-8z = \ 12 & (5) \\ -y+2z = -20 & (6) \end{cases}$$
   Multiply equation (6) by 4, add the resulting equation to equation (5), and replace equation (6) with the result.
   $$\begin{array}{r} 4y-8z = \ 12 \\ -4y+8z = -80 \\ \hline 0 = -68 \end{array}$$
   $$\begin{cases} x+\ y+\ z = \ \ 6 & (4) \\ 4y-8z = \ 12 & (5) \\ 0 = -68 & (7) \end{cases}$$
   Since equation (7) is false, the solution set of the system is $\varnothing$.

4. $\begin{cases} x+\ y+\ z = \ \ 5 & (1) \\ -4x-\ y-8z = -29 & (2) \\ 2x+5y-2z = \ \ 1 & (3) \end{cases}$

   Multiply equation (1) by 4, add the resulting equation to equation (2), and replace equation (2) with the result. Similarly, multiply equation (1) by -2, add the resulting equation to equation (3), and replace equation (3) with the resulting equation.

(continued on next page)

(continued)

$$\begin{cases} x + y + z = 5 & (1) \\ 3y - 4z = -9 & (4) \\ 3y - 4z = -9 & (5) \end{cases}$$

Now multiply equation (4) by -1, add the result to equation (5), and replace equation (5) with the resulting equation.

$$\begin{cases} x + y + z = 5 & (1) \\ 3y - 4z = -9 & (4) \\ 0 = 0 & (6) \end{cases}$$

Solve equation (4) for y.

$$3y - 4z = -9 \Rightarrow y = \frac{-9 + 4z}{3} = -3 + \frac{4}{3}z$$

Substitute this expression into equation (1) and solve for x.

$$x + y + z = 5 \Rightarrow x + \left(-3 + \frac{4}{3}z\right) + z = 5 \Rightarrow$$

$$x + \frac{7}{3}z - 3 = 5 \Rightarrow x = -\frac{7}{3}z + 8$$

The solution set is $\left\{\left(8 - \frac{7}{3}z, -3 + \frac{4}{3}z, z\right)\right\}$.

5. $\begin{cases} x + 3y + 2z = 4 & (1) \\ 2x + 7y - z = 5 & (2) \end{cases}$

Eliminate x.

$$\begin{array}{r} -2x - 6y - 4z = -8 \\ \underline{2x + 7y - z = 5} \\ y - 5z = -3 \end{array}$$

$$\begin{cases} x + 3y + 2z = 4 & (1) \\ y - 5z = -3 & (3) \end{cases}$$

Solve equation (3) for y in terms of z.

$$y - 5z = -3 \Rightarrow y = 5z - 3$$

Substitute this expression into equation (1) and then solve for x in terms of z.

$$x + 3(5z - 3) + 2z = 4 \Rightarrow$$

$$x + 15z - 9 + 2z = 4 \Rightarrow x = -17z + 13$$

The solution set is $\{(-17z + 13, 5z - 3, z)\}$.

6. The system is

$$\begin{cases} x + y = 0.65 & (\text{beam } 1) \\ x + z = 0.70 & (\text{beam } 2) \\ y + z = 0.55 & (\text{beam } 3) \end{cases}.$$

Multiply the first equation by -1, add the result to the second equation, and replace the second equation with the new equation:

$$-1(x + y = 0.65) \Rightarrow -x - y = -0.65$$

$$\begin{cases} -x - y = -0.65 \\ x + z = 0.70 \end{cases} \Rightarrow -y + z = 0.05$$

$$\begin{cases} x + y = 0.65 \\ x + z = 0.70 \\ y + z = 0.55 \end{cases} \Rightarrow \begin{cases} x + y = 0.65 \\ -y + z = 0.05 \\ y + z = 0.55 \end{cases}$$

Add the second and third equations and solve for z: $2z = 0.60 \Rightarrow z = 0.30$. Substituting this value into the original second equation, we have $x + 0.30 = 0.70 \Rightarrow x = 0.40$. Substituting this value into the original first equation, we have $0.40 + y = 0.65 \Rightarrow y = 0.25$. Referring to table 5.1, we see that cell A contains bone (since $x = 0.40$), cell B contains healthy tissue (since $y = 0.25$), and cell C contains tumorous tissue (since $z = 0.30$).

5.2 Basic Concepts and Skills

1. Systems of equations that have the same solution set are called equivalent systems.

3. If any of the equations in a system has no solution, then the system is inconsistent.

5. True

7. Substituting the ordered triple into the system
$$\begin{cases} 2x - 2y - 3z = 1 \\ 3y + 2z = -1, \\ y + z = 0 \end{cases}$$
we find that $(1, -1, 1)$ is a solution.
$$\begin{cases} 2(1) - 2(-1) - 3(1) = 1 \\ 3(-1) + 2(1) = -1 \\ -1 + 1 = 0 \end{cases}$$

9. Substituting the ordered triple into the system
$$\begin{cases} x + 3y - 2z = 0 \\ 2x - y + 4z = 5, \\ x - 11y + 14z = 0 \end{cases}$$ we find that $(-10, 8, 7)$ is not a solution.
$$\begin{cases} -10 + 3(8) - 2(7) = 0 \\ 2(-10) - 8 + 4(7) = 0 \neq 5 \\ -10 - 11(8) + 14(7) = 0 \end{cases}$$

11. To solve the system $\begin{cases} x + y + z = 4 \\ y - 2z = 4, \\ z = -1 \end{cases}$

substitute $z = -1$ into the second equation, and solve for y: $y - 2(-1) = 4 \Rightarrow y = 2$. Now substitute the values for y and z into the first equation, and solve for x:
$$x + 2 - 1 = 4 \Rightarrow x = 3.$$
The solution is $\{(3, 2, -1)\}$.

13. To solve the system $\begin{cases} x - 5y + 3z = -1 \\ \quad y - 2z = -6 \\ \qquad\qquad z = 4 \end{cases}$,

substitute $z = 4$ into the second equation, and solve for y: $y - 2(4) = -6 \Rightarrow y = 2$. Now substitute the values for y and z into the first equation, and solve for x:

$x - 5(2) + 3(4) = -1 \Rightarrow x = -3$.

The solution is $\{(-3, 2, 4)\}$.

15. $\begin{cases} 2x - 2y - 3z = 1 \\ \quad 3y + 2z = -1 \\ \quad\; y + z = 0 \end{cases} \Rightarrow \begin{cases} 2x - 2y - 3z = 1 \\ \quad y + z = 0 \\ \quad 3y + 2z = -1 \end{cases}$.

Multiply the first equation by $1/2$:

$\dfrac{1}{2}(2x - 2y - 3z = 1) \Rightarrow x - y - \dfrac{3}{2}z = \dfrac{1}{2}$.

Now multiply the second equation by -3, add the result to the third equation, and replace the third equation with the new equation:

$-3(y + z = 0) \Rightarrow -3y - 3z = 0$.

$\begin{cases} -3y - 3z = 0 \\ \;\; 3y + 2z = -1 \end{cases} \Rightarrow -z = -1 \Rightarrow z = 1$.

So the system becomes

$\begin{cases} x - y - \dfrac{3}{2}z = \dfrac{1}{2} \\ \quad y + z = 0 \\ \qquad\quad z = 1 \end{cases}$.

17. Multiply the first equation by -2, add the result to the second equation, and replace the second equation with the new equation:

$-2(x + 3y - 2z = 0) \Rightarrow -2x - 6y + 4z = 0$

$\begin{cases} -2x - 6y + 4z = 0 \\ \;\; 2x - \; y + 4z = 5 \end{cases} \Rightarrow -7y + 8z = 5$.

The system becomes:

$\begin{cases} x + 3y - 2z = 0 \\ 2x - \; y + 4z = 5 \\ x - 11y + 14z = 0 \end{cases} \Rightarrow \begin{cases} x + 3y - 2z = 0 \\ \quad -7y + 8z = 5 \\ x - 11y + 14z = 0 \end{cases}$

Multiply the first equation by -1, add the result to the third equation, and replace the third equation with the new equation:

$-1(x + 3y - 2z = 0) \Rightarrow -x - 3y + 2z = 0$

$\begin{cases} -x - 3y + 2z = 0 \\ \;\; x - 11y + 14z = 5 \end{cases} \Rightarrow -14y + 16z = 5$.

The system becomes:

$\begin{cases} x + 3y - 2z = 0 \\ \quad -7y + 8z = 5 \\ x - 11y + 14z = 0 \end{cases} \Rightarrow \begin{cases} x + 3y - 2z = 0 \\ \quad -7y + 8z = 5 \\ -14y + 16z = 5 \end{cases}$

Multiply the second equation by -2, add the result to the third equation, and replace the

third equation with the new equation:

$-2(-7y + 8z = 5) \Rightarrow 14y - 16z = -10$

$\begin{cases} \;\; 14y - 16z = -10 \\ -14y + 16z = 5 \end{cases} \Rightarrow 0 = -5$.

The system becomes:

$\begin{cases} x + 3y - 2z = 0 \\ \quad -7y + 8z = 5 \\ x - 11y + 14z = 0 \end{cases} \Rightarrow \begin{cases} x + 3y - 2z = 0 \\ \quad y - \dfrac{8}{7}z = -\dfrac{5}{7} \\ \qquad\qquad 0 = -5 \end{cases}$

19. $\begin{cases} x - \; y + z = 6 \\ \quad 2y + 3z = 5 \\ \qquad\; 2z = 6 \end{cases} \Rightarrow \begin{cases} x - \; y + z = 6 \\ \quad 2y + 3z = 5 \\ \qquad\quad z = 3 \end{cases}$.

Substitute $z = 3$ into the second equation to solve for y: $2y + 3(3) = 5 \Rightarrow y = -2$.

Now substitute the values for y and z into the first equation to solve for x:

$x - (-2) + 3 = 6 \Rightarrow x = 1$.

The solution is $\{(1, -2, 3)\}$.

21. Multiply the first equation by $-3/4$ add the result to the second equation, then replace the second equation with the new equation:

$-\dfrac{3}{4}(4x + 4y + 4z = 7) \Rightarrow$

$-3x - 3y - 3z = -\dfrac{21}{4}$

$\begin{cases} -3x - 3y - 3z = -\dfrac{21}{4} \\ \;\; 3x - 8y \qquad\quad = 14 \end{cases} \Rightarrow -11y - 3z = \dfrac{35}{4} \Rightarrow$

$y + \dfrac{3}{11}z = -\dfrac{35}{44}$

$\begin{cases} 4x + 4y + 4z = 7 \\ 3x - 8y \qquad = 14 \\ \qquad\qquad 4z = -1 \end{cases} \Rightarrow \begin{cases} x + y + z = \dfrac{7}{4} \\ \quad y + \dfrac{3}{11}z = -\dfrac{35}{44} \\ \qquad\qquad z = -\dfrac{1}{4} \end{cases}$

Substitute $z = -1/4$ into the second equation to solve for y:

$y + \dfrac{3}{11}\left(-\dfrac{1}{4}\right) = -\dfrac{35}{44} \Rightarrow y = -\dfrac{8}{11}$.

Now substitute the values for y and z into the first equation to solve for x:

$4x + 4\left(-\dfrac{8}{11}\right) + 4\left(-\dfrac{1}{4}\right) = 7 \Rightarrow x = \dfrac{30}{11}$.

The solution is $\left\{\left(\dfrac{30}{11}, -\dfrac{8}{11}, -\dfrac{1}{4}\right)\right\}$.

In exercises 23–44, be sure to check the answers by substituting the values into the original system of equations.

23. Multiply the first equation by –1, add the result to the second equation, and replace the second equation with the new equation:

$$-1(x + y + z = 6) \Rightarrow -x - y - z = -6$$

$$\begin{cases} -x - y - z = -6 \\ x - y + z = 2 \end{cases} \Rightarrow -2y = -4$$

$$\begin{cases} x + y + z = 6 \\ x - y + z = 2 \\ 2x + y - z = 1 \end{cases} \Rightarrow \begin{cases} x + y + z = 6 \\ -2y = -4 \\ 2x + y - z = 1 \end{cases}$$

Simplify the new second equation, and then multiply the first equation by –2, add the result to the third equation, and replace the third equation with the new equation:

$$-2(x + y + z = 6) \Rightarrow -2x - 2y - 2z = -12$$

$$\begin{cases} -2x - 2y - 2z = -12 \\ 2x + y - z = 1 \end{cases} \Rightarrow -y - 3z = -11$$

$$\begin{cases} x + y + z = 6 \\ -2y = -4 \\ 2x + y - z = 1 \end{cases} \Rightarrow \begin{cases} x + y + z = 6 \\ y = 2 \\ -y - 3z = -11 \end{cases}$$

Add the second and third equations, replacing the third equation with the new equation:

$$\begin{cases} x + y + z = 6 \\ y = 2 \\ -y - 3z = -11 \end{cases} \Rightarrow \begin{cases} x + y + z = 6 \\ y = 2 \\ -3z = -9 \end{cases} \Rightarrow$$

$$\begin{cases} x + y + z = 6 \\ y = 2 \\ z = 3 \end{cases}$$

Substitute the values for *y* and *z* into the original first equation to solve for *x*:
$$x + 2 + 3 = 6 \Rightarrow x = 1.$$
The solution is {(1, 2, 3)}.

25. Switch the first and second equations.

$$\begin{cases} 2x + 3y + z = 9 \\ x + 2y + 3z = 6 \\ 3x + y + 2z = 8 \end{cases} \Rightarrow \begin{cases} x + 2y + 3z = 6 \\ 2x + 3y + z = 9 \\ 3x + y + 2z = 8 \end{cases}$$

Multiply the first equation by –2, add the result to the second equation, then replace the second equation with the new equation:

$$-2(x + 2y + 3z = 6) = -2x - 4y - 6z = -12$$

$$\begin{cases} -2x - 4y - 6z = -12 \\ 2x + 3y + z = 9 \end{cases} \Rightarrow -y - 5z = -3 \Rightarrow$$

$$y + 5z = 3$$

$$\begin{cases} x + 2y + 3z = 6 \\ 2x + 3y + z = 9 \\ 3x + y + 2z = 8 \end{cases} \Rightarrow \begin{cases} x + 2y + 3z = 6 \\ y + 5z = 3 \\ 3x + y + 2z = 8 \end{cases}$$

Multiply the first equation by –3, add the result to the third equation, then replace the third equation with the new equation:

$$-3(x + 2y + 3z = 6) = -3x - 6y - 9z = -18$$

$$\begin{cases} -3x - 6y - 9z = -18 \\ 3x + y + 2z = 8 \end{cases} \Rightarrow -5y - 7z = -10 \Rightarrow$$

$$5y + 7z = 10$$

$$\begin{cases} x + 2y + 3z = 6 \\ y + 5z = 3 \\ 3x + y + 2z = 8 \end{cases} \Rightarrow \begin{cases} x + 2y + 3z = 6 \\ y + 5z = 3 \\ 5y + 7z = 10 \end{cases}$$

Multiply the second equation by –5, and add the result to the third equation, then replace the third equation with the new equation:

$$-5(y + 5z = 3) \Rightarrow -5y - 25z = -15$$

$$\begin{cases} -5y - 25z = -15 \\ 5y + 7z = 10 \end{cases} \Rightarrow -18z = -5 \Rightarrow z = \frac{5}{18}$$

$$\begin{cases} x + 2y + 3z = 6 \\ y + 5z = 3 \\ 5y + 7z = 10 \end{cases} \Rightarrow \begin{cases} x + 2y + 3z = 6 \\ y + 5z = 3 \\ z = \frac{5}{18} \end{cases}$$

Substitute the value of *z* into the second equation to solve for *y*:

$$y + 5\left(\frac{5}{18}\right) = 3 \Rightarrow y = \frac{29}{18}.$$

Substitute the values for *y* and *z* into the first equation to solve for *x*:

$$x + 2\left(\frac{29}{18}\right) + 3\left(\frac{5}{18}\right) = 6 \Rightarrow x = \frac{35}{18}.$$

The solution is $\left\{\left(\frac{35}{18}, \frac{29}{18}, \frac{5}{18}\right)\right\}$.

27.
$$\begin{cases} x + y - 2z = 5 & (1) \\ 2x - y - z = -4 & (2) \\ x - 2y + z = -2 & (3) \end{cases}$$

Add equations (1) and (2), then replace equation (2) with the new equation (4).

$$\begin{cases} x + y - 2z = 5 & (1) \\ 2x - y - z = -4 & (2) \end{cases} \Rightarrow 3x - 3z = 1 \quad (4)$$

$$\begin{cases} x + y - 2z = 5 \\ 2x - y - z = -4 \\ x - 2y + z = -2 \end{cases} \Rightarrow \begin{cases} x + y - 2z = 5 & (1) \\ 3x - 3z = 1 & (4) \\ x - 2y + z = -2 & (3) \end{cases}$$

Multiply the first equation by 2, add the result to the third equation, then replace the third equation with the new equation (5)

$$2(x + y - 2z = 5) \Rightarrow 2x + 2y - 4z = 10$$

$$\begin{cases} 2x + 2y - 4z = 10 \\ x - 2y + z = -2 \end{cases} \Rightarrow 3x - 3z = 8 \quad (5)$$

$$\begin{cases} x + y - 2z = 5 \\ 3x - 3z = 1 \\ x - 2y + z = -2 \end{cases} \Rightarrow \begin{cases} x + y - 2z = 5 & (1) \\ 3x - 3z = 1 & (4) \\ 3x - 3z = 8 & (5) \end{cases}$$

(continued on next page)

(*continued*)

Multiplying the equation (4) by −1, and adding the result to equation (5) gives:

$-1(3x - 3z = 1) \Rightarrow -3x + 3z = -1$

$\begin{cases} -3x + 3z = -1 \\ 3x - 3z = 8 \end{cases} \Rightarrow 0 = -7 \text{ False}$

Thus, the system is inconsistent and the solution set is $\varnothing$.

29. Multiply the second equation by −2, add the result to the first equation, then replace the second equation with the new equation:

$-2(x + 3y - z = -2) \Rightarrow -2x - 6y + 2z = 4$

$\begin{cases} -2x - 6y + 2z = 4 \\ 2x + 3y + 2z = 7 \end{cases} \Rightarrow -3y + 4z = 11$

$\begin{cases} 2x + 3y + 2z = 7 \\ x + 3y - z = -2 \\ x - y + 2z = 8 \end{cases} \Rightarrow \begin{cases} 2x + 3y + 2z = 7 \\ -3y + 4z = 11 \\ x - y + 2z = 8 \end{cases}$

Multiply the third equation by −2, add the result to the first equation, then replace the third equation with the new equation:

$-2(x - y + 2z = 8) \Rightarrow -2x + 2y - 4z = -16$

$\begin{cases} -2x + 2y - 4z = -16 \\ 2x + 3y + 2z = 7 \end{cases} \Rightarrow 5y - 2z = -9$

$\begin{cases} 2x + 3y + 2z = 7 \\ x + 3y - z = -2 \\ x - y + 2z = 8 \end{cases} \Rightarrow \begin{cases} 2x + 3y + 2z = 7 \\ -3y + 4z = 11 \\ 5y - 2z = -9 \end{cases}$

Multiply the third equation by 2, and add the result to the second equation to solve for y:

$2(5y - 2z = -9) \Rightarrow 10y - 4z = -18$

$\begin{cases} 10y - 4z = -18 \\ -3y + 4z = 11 \end{cases} \Rightarrow 7y = -7 \Rightarrow y = -1$

$\begin{cases} 2x + 3y + 2z = 7 \\ -3y + 4z = 11 \\ 5y - 2z = -9 \end{cases} \Rightarrow \begin{cases} 2x + 3y + 2z = 7 \\ y = -1 \\ 5y - 2z = -9 \end{cases}$

Multiply the second equation by −5, then add the result to the third equation and replace the third equation with the new equation:

$\begin{cases} -5y = 5 \\ 5y - 2z = -9 \end{cases} \Rightarrow -2z = -4 \Rightarrow z = 2$

$\begin{cases} 2x + 3y + 2z = 7 \\ y = -1 \\ 5y - 2z = -9 \end{cases} \Rightarrow \begin{cases} 2x + 3y + 2z = 7 \\ y = -1 \\ z = 2 \end{cases}$

Substitute the values for y and z into the first equation to solve for x:

$2x + 3(-1) + 2(2) = 7 \Rightarrow x = 3.$

The solution is $\{(3, -1, 2)\}$.

31. Switch the first and third equations:

$\begin{cases} 4x - 2y + z = 5 \\ 2x + y - 2z = 4 \\ x + 3y - 2z = 6 \end{cases} \Rightarrow \begin{cases} x + 3y - 2z = 6 \\ 2x + y - 2z = 4 \\ 4x - 2y + z = 5 \end{cases}$

Multiply the first equation by −2, add the result to the second equation, and replace the second equation with the new equation:

$-2(x + 3y - 2z = 6) \Rightarrow -2x - 6y + 4z = -12$

$\begin{cases} -2x - 6y + 4z = -12 \\ 2x + y - 2z = 4 \end{cases} \Rightarrow -5y + 2z = -8$

$\begin{cases} x + 3y - 2z = 6 \\ 2x + y - 2z = 4 \\ 4x - 2y + z = 5 \end{cases} \Rightarrow \begin{cases} x + 3y - 2z = 6 \\ -5y + 2z = -8 \\ 4x - 2y + z = 5 \end{cases}$

Multiply the first equation by −4, add the result to the third equation, and replace the third equation with the new equation:

$-4(x + 3y - 2z = 6) \Rightarrow -4x - 12y + 8z = -24$

$\begin{cases} -4x - 12y + 8z = -24 \\ 4x - 2y + z = 5 \end{cases} \Rightarrow -14y + 9z = -19$

$\begin{cases} x + 3y - 2z = 6 \\ -5y + 2z = -8 \\ 4x - 2y + z = 5 \end{cases} \Rightarrow \begin{cases} x + 3y - 2z = 6 \\ -5y + 2z = -8 \\ -14y + 9z = -19 \end{cases}$

Multiply the second equation by $-\dfrac{14}{5}$ and add the result to the third equation and replace the third equation with the new equation:

$-\dfrac{14}{5}(-5y + 2z = -8) \Rightarrow 14y - \dfrac{28}{5}z = \dfrac{112}{5}$

$\begin{cases} 14y - \dfrac{28}{5}z = \dfrac{112}{5} \\ -14y + 9z = -19 \end{cases} \Rightarrow \dfrac{17}{5}z = \dfrac{17}{5} \Rightarrow z = 1$

$\begin{cases} x + 3y - 2z = 6 \\ -5y + 2z = -8 \\ -14y + 9z = -19 \end{cases} \Rightarrow \begin{cases} x + 3y - 2z = 6 \\ -5y + 2z = -8 \\ z = 1 \end{cases}$

Substituting $z = 1$ into the second equation, we have $-5y + 2(1) = -8 \Rightarrow y = 2$. Substitute the values for y and z into the first equation to solve for x: $x + 3(2) - 2(1) = 6 \Rightarrow x = 2$.

The solution is $\{(2, 2, 1)\}$.

33. Multiply the second equation by −1 and then add the result to the third equation:

$-(x + y - z = 1) \Rightarrow -x - y + z = -1$

$\begin{cases} -x - y + z = -1 \\ x + y + 2z = 4 \end{cases} \Rightarrow 3z = 3 \Rightarrow z = 1$

Multiply the second equation by −2, add the result to the first equation, and replace the second equation with the new equation.

(*continued on next page*)

(*continued*)

$$-2(x+y-z=1) \Rightarrow -2x-2y+2z=-2$$

$$\begin{cases} 2x+y+z=6 \\ -2x-2y+2z=-2 \end{cases} \Rightarrow -y+3z=4$$

$$\begin{cases} 2x+y+z=6 \\ x+y-z=1 \\ x+y+2z=4 \end{cases} \Rightarrow \begin{cases} 2x+y+z=6 \\ -y+3z=4 \\ z=1 \end{cases}$$

Substitute $z = 1$ into the second equation to solve for y: $-y+3(1)=4 \Rightarrow y=-1$.

Substitute the values for y and z into the first equation to solve for x: $2x-1+1=6 \Rightarrow x=3$. The solution is $(3, -1, 1)$.

35. $\begin{cases} x-y-z=3 \\ x+9y+z=3 \\ 2x+3y-z=6 \end{cases}$

Multiply the first equation by -1, add the result to the second equation, and replace the second equation with the new equation:

$$-1(x-y-z=3) \Rightarrow -x+y+z=-3$$

$$\begin{cases} -x+y+z=-3 \\ x+9y+z=3 \end{cases} \Rightarrow 10y+2z=0$$

$$\begin{cases} x-y-z=3 \\ x+9y+z=3 \\ 2x+3y-z=6 \end{cases} \Rightarrow \begin{cases} x-y-z=3 \\ 10y+2z=0 \\ 2x+3y-z=6 \end{cases}$$

Multiply the first equation by -2, add the result to the third equation, and replace the third equation with the new equation:

$$-2(x-y-z=3) \Rightarrow -2x+2y+2z=-6$$

$$\begin{cases} -2x+2y+2z=-6 \\ 2x+3y-z=6 \end{cases} \Rightarrow 5y+z=0$$

$$\begin{cases} x-y-z=3 \\ 10y+2z=0 \\ 2x+3y-z=6 \end{cases} \Rightarrow \begin{cases} x-y-z=3 \\ 10y+2z=0 \\ 5y+z=0 \end{cases}$$

Multiply the third equation by -2, then add the result to the second equation:

$$-2(5y+z=0) \Rightarrow -10y-2z=0$$

$$\begin{cases} 10y+2z=0 \\ -10y-2z=0 \end{cases} \Rightarrow 0=0$$

The equation $0 = 0$ is equivalent to $0z = 0$, which is true for every value of z.

Solving the second equation for y, we have

$$10y+2z=0 \Rightarrow y=-\frac{1}{5}z.$$ Substituting this

into the first equation, we have

$$x-y-z=3 \Rightarrow x-\left(-\frac{1}{5}z\right)-z=3 \Rightarrow$$

$x=3+\dfrac{4}{5}z$. Thus, the solution is

$$\left\{\left(3+\frac{4}{5}z,\ -\frac{1}{5}z,\ z\right)\right\}.$$

37. Rearrange the equations as shown:

$$\begin{cases} x+y=0 \\ y+2z=-4 \\ y+z=4-x \end{cases} \Rightarrow \begin{cases} x+y=0 \\ y+2z=-4 \\ x+y+z=4 \end{cases} \Rightarrow$$

$$\begin{cases} x+y+z=4 \\ y+2z=-4 \\ x+y=0 \end{cases}$$

Subtract the third equation from the first equation, and replace the third equation:

$$\begin{cases} x+y+z=4 \\ y+2z=-4 \\ x+y=0 \end{cases} \Rightarrow \begin{cases} x+y+z=4 \\ y+2z=-4 \\ z=4 \end{cases}$$

Substitute $z = 4$ into the second equation to solve for y: $y+2(4)=-4 \Rightarrow y=-12$.

Substitute the values for x and y into the first equation to solve for x:

$x-12+4=4 \Rightarrow x=12$.
The solution is $\{(12, -12, 4)\}$.

39. $\begin{cases} x+y+z=6 & (1) \\ x+2y+3z=14 & (2) \\ x+4y+7z=30 & (3) \end{cases}$

Multiply equation (1) by -1, add the result to equation (2), and replace equation (2) with the new equation (4).

$$-1(x+y+z=6) \Rightarrow -x-y-z=-6$$

$$\begin{cases} -x-y-z=-6 \\ x+2y+3z=14 \end{cases} \Rightarrow y+2z=8 \quad (4)$$

$$\begin{cases} x+y+z=6 \\ x+2y+3z=3 \\ x+4y+7z=30 \end{cases} \Rightarrow \begin{cases} x+y+z=6 \\ y+2z=8 \quad (4) \\ x+4y+7z=30 \end{cases}$$

Multiply equation (1) by -1, add the result to equation (3), and replace equation (3) with the new equation (5).

$$-1(x+y+z=6) \Rightarrow -x-y-z=-6$$

$$\begin{cases} -x-y-z=-6 \\ x+4y+7z=30 \end{cases} \Rightarrow 3y+6z=24 \quad (5)$$

$$\begin{cases} x+y+z=6 \\ y+2z=8 \quad (4) \\ 3y+6z=24 \quad (5) \end{cases}$$

Multiply equation (4) by -3, add the result to equation (5), then replace equation (5) with the result (6).

(*continued on next page*)

(*continued*)

$$-3(y+2z=8)\Rightarrow -3y-6z=-24$$

$$\begin{cases}-3y-6z=-24\\ 3y+6z=24 \quad (6)\end{cases}\Rightarrow 0=0$$

$$\begin{cases}x+y+z=6 \quad (1)\\ y+2z=8 \quad (4)\\ 0=0 \quad (6)\end{cases}$$

The system is now in triangular form. Solve equation (4) for y: $y+2z=8\Rightarrow y=-2z+8$

Substitute the expression for y into equation (1) and solve for x.

$$x+(-2z+8)+z=6\Rightarrow x-z+8=6\Rightarrow$$
$$x=z-2$$

The solution set for the system is
$$\{(z-2,\,-2z+8,\,z)\}.$$

41. Multiply the second equation by -2, add the result to the third equation, and replace the second equation with the result:
$$-2(2x+y=8)\Rightarrow -4x-2y=-16$$

$$\begin{cases}2y+3z=1\\ -4x-2y \quad =-16\end{cases}\Rightarrow -4x+3z=-15$$

$$\begin{cases}3x-\quad 2z=11\\ 2x+y \quad =8\\ \quad 2y+3z=1\end{cases}\Rightarrow\begin{cases}3x-\quad 2z=11\\ -4x+\quad 3z=-15\\ \quad 2y+3z=1\end{cases}$$

Multiply the first equation by 4 and the second equation by 3, add the two result and replace the second equation:
$$4(3x-2z=11)\Rightarrow 12x-8z=44$$
$$3(-4x+3z=-15)\Rightarrow -12x+9z=-45$$

$$\begin{cases}12x-8z=44\\ -12x+9z=-45\end{cases}\Rightarrow z=-1$$

$$\begin{cases}3x-\quad 2z=11\\ -4x+\quad 3z=-15\\ \quad 2y+3z=1\end{cases}\Rightarrow$$

$$\begin{cases}3x-\quad 2z=11\\ \quad z=-1\\ \quad 2y+3z=1\end{cases}\Rightarrow\begin{cases}3x-\quad 2z=11\\ \quad 2y+3z=1\\ \quad z=-1\end{cases}$$

Substitute $z=-1$ into the first and second equations to solve for x and y:
$$2y+3(-1)=1\Rightarrow y=2.$$

$$3x-2(-1)=11\Rightarrow x=3.$$

The solution is $\{(3, 2, -1)\}$.

43. Multiplying the second equation by -1, and then adding the two equations, we have:
$$\begin{cases}2x+6y+\quad 11=0\\ \quad -6y+18z-1=0\end{cases}\Rightarrow 2x+18z+10=0\Rightarrow$$
$$2x+18z=-10\Rightarrow x=-5-9z$$

$$6y-18z+1=0\Rightarrow y=-\frac{1}{6}+3z.$$

The solution is $\left\{\left(-5-9z,\,-\dfrac{1}{6}+3z,z\right)\right\}$.

The system is dependent.

5.2 Applying the Concepts

45. Let $x=$ the amount invested at 4%, $y=$ the amount invested at 5%, and $z=$ the amount invested at 6%. Then, we have the system:

$$\begin{cases}x+y+z=20,000\\ 0.04x+0.05y+0.06z=1060\\ 0.06z=2(0.05y)\end{cases}\Rightarrow$$

$$\begin{cases}x+\quad y+\quad z=20,000\\ 4x+5y+6z=106,000\\ \quad 6z=10y\end{cases}\Rightarrow$$

$$\begin{cases}x+\quad y+\quad z=20,000\\ 4x+\quad 5y+6z=106,000\\ \quad -10y+6z=0\end{cases}$$

Multiply the first equation by -4, add the result to the second equation, and replace the second equation with the new equation:
$$-4(x+y+z=20,000)\Rightarrow$$
$$-4x-4y-4z=-80,000$$

$$\begin{cases}-4x-4y-4z=-80,000\\ 4x+5y+6z=106,000\end{cases}\Rightarrow y+2z=26,000$$

$$\begin{cases}x+\quad y+\quad z=20,000\\ \quad y+2z=26,000\\ \quad -10y+6z=0\end{cases}$$

Multiply the second equation by 10, add the result to the third equation, and solve for z:
$$10(y+2z=26,000)\Rightarrow 10y+20z=260,000$$

$$\begin{cases}10y+20z=260,000\\ -10y+6z=0\end{cases}\Rightarrow 26z=260,000\Rightarrow$$
$$z=10,000$$

$$\begin{cases}x+\quad y+\quad z=20,000\\ \quad y+2z=26,000\\ \quad -10y+6z=0\end{cases}\Rightarrow$$

$$\begin{cases}x+\quad y+\quad z=20,000\\ \quad y+2z=26,000\\ \quad z=10,000\end{cases}$$

Substitute $z=10,000$ into the second equation to solve for y:
$$y+2(10,000)=26,000\Rightarrow y=6000.$$ Substitute the values for y and z into the first equation to solve for x:
$$x+6000+10,000=20,000\Rightarrow x=4000.$$

Miguel invested \$4000 at 4%, \$6000 at 5%, and \$10,000 at 6%.

47. Let a = the number of hours Alex worked, b = the number of hours Becky worked, and c = the number of hours Courtney worked. Then, we have the system:

$$\begin{cases} a + b + c = 6 \\ 124a + 118b + 132c = 741 \\ b + c = 2a \end{cases} \Rightarrow$$

$$\begin{cases} a + b + c = 6 \\ 124a + 118b + 132c = 741 \\ -2a + b + c = 0 \end{cases}$$

Subtract the third equation from the first, and replace the first equation with the result:

$$\begin{cases} a + b + c = 6 \\ 124a + 118b + 132c = 741 \Rightarrow \\ -2a + b + c = 0 \end{cases}$$

$$\begin{cases} 3a = 6 \\ 124a + 118b + 132c = 741 \Rightarrow \\ -2a + b + c = 0 \end{cases}$$

$$\begin{cases} a = 2 \\ 124a + 118b + 132c = 741 \Rightarrow \\ -2a + b + c = 0 \end{cases}$$

Substitute $a = 2$ into the second and third equations and simplify:

$$\begin{cases} a = 2 \\ 124(2) + 118b + 132c = 741 \Rightarrow \\ -2(2) + b + c = 0 \end{cases}$$

$$\begin{cases} a = 2 \\ 118b + 132c = 493 \\ b + c = 4 \end{cases}$$

Multiply the third equation by -118, and add the result to the second equation to solve for c.

$$-118(b + c = 4) \Rightarrow -118b - 118c = -472$$

$$\begin{cases} 118b + 132c = 493 \\ -118b - 118c = -472 \end{cases} \Rightarrow 14c = 21 \Rightarrow c = 1.5$$

Substitute $c = 1.5$ into the third equation to solve for b: $b + 1.5 = 4 \Rightarrow b = 2.5$.

Alex worked 2 hours, Becky worked 2.5 hours, and Courtney worked 1.5 hours.

49. Let n = the number of nickels, d = the number of dimes, and q = the number of quarters. Then, we have the system

$$\begin{cases} n + d + q = 300 \\ d = 3(n + q) \\ 0.05n + 0.1d + 0.25q = 30.05 \end{cases} \Rightarrow$$

$$\begin{cases} n + d + q = 300 \\ -3n + d - 3q = 0 \\ 5n + 10d + 25q = 3005 \end{cases}$$

Multiply the first equation by 3, add the result to the second equation, and replace the second

equation with the new equation:

$$3(n + d + q = 300) \Rightarrow 3n + 3d + 3q = 900$$

$$\begin{cases} 3n + 3d + 3q = 900 \\ -3n + d - 3q = 0 \end{cases} \Rightarrow 4d = 900 \Rightarrow d = 225$$

$$\begin{cases} n + d + q = 300 \\ d = 225 \\ 5n + 10d + 25q = 3005 \end{cases}$$

Multiply the first equation by -5, add the result to the third equation, and replace the third equation with the new equation:

$$-5(n + d + q = 300) \Rightarrow -5n - 5d - 5q = -1500$$

$$\begin{cases} -5n - 5d - 5q = -1500 \\ 5n + 10d + 25q = 3005 \end{cases} \Rightarrow 5d + 20q = 1505$$

$$\begin{cases} n + d + q = 300 \\ d = 225 \\ 5n + 10d + 25q = 3005 \end{cases} \Rightarrow$$

$$\begin{cases} n + d + q = 300 \\ d = 225 \\ 5d + 20q = 1505 \end{cases}$$

Substitute $d = 225$ into the third equation to solve for q, then substitute the values for d and q into the first equation to solve for n:

$$5(225) + 20q = 1505 \Rightarrow q = 19.$$

$$n + 225 + 19 = 300 \Rightarrow n = 56.$$

There are 56 nickels, 225 dimes, and 19 quarters.

51. Let x = the number of daytime hours Amy worked, y = the number of night hours Amy worked, and z = the number of holiday hours Amy worked. Then, we have the system

$$\begin{cases} x + y + z = 53 \\ 7.40x + 9.20y + 11.75z = 452.20 \Rightarrow \\ x = y + z + 9 \end{cases}$$

$$\begin{cases} x + y + z = 53 \\ 7.40x + 9.20y + 11.75z = 452.20 \\ x - y - z = 9 \end{cases}$$

Add the first and third equations, and replace the first equation with the result:

$$\begin{cases} x + y + z = 53 \\ 7.40x + 9.20y + 11.75z = 452.20 \Rightarrow \\ x - y - z = 9 \end{cases}$$

$$\begin{cases} 2x = 62 \\ 7.40x + 9.20y + 11.75z = 452.20 \Rightarrow \\ x - y - z = 9 \end{cases}$$

$$\begin{cases} x = 31 \\ 7.40x + 9.20y + 11.75z = 452.20 \\ x - y - z = 9 \end{cases}$$

(continued on next page)

(continued)

Substitute $x = 31$ into the second and third equations and simplify:

$$\begin{cases} x & = 31 \\ 7.40x + 9.20y + 11.75z = 452.20 \Rightarrow \\ x - y - z = 9 \end{cases}$$

$$\begin{cases} x & = 31 \\ 7.40(31) + 9.20y + 11.75z = 452.20 \Rightarrow \\ 31 - y - z = 9 \end{cases}$$

$$\begin{cases} x & = 31 \\ 9.20y + 11.75z = 222.80 \\ -y - z = -22 \end{cases}$$

Multiply the third equation by 9.2, add the result to the second equation, and solve for z:

$$9.2(-y - z = -22) \Rightarrow -9.2y - 9.2z = 202.4$$

$$\begin{cases} 9.20y + 11.75z = 222.80 \\ -9.20y - 9.20z = -202.40 \end{cases} \Rightarrow$$

$$2.55z = 20.40 \Rightarrow z = 8$$

$$\begin{cases} x & = 31 \\ 9.20y + 11.75z = 222.80 \Rightarrow \\ -y - z = -22 \end{cases}$$

$$\begin{cases} x & = 31 \\ 9.20y + 11.75z = 222.80 \\ z = 8 \end{cases}$$

$$9.20y + 11.75(8) = 222.80 \Rightarrow y = 14$$

Amy worked 31 daytime hours, 14 night hours, and 8 holiday hours.

53. The system is

$$\begin{cases} x + y & = 0.54 \text{ (beam 1)} \\ x + z = 0.40 \text{ (beam 2)} \\ y + z = 0.52 \text{ (beam 3)} \end{cases}$$

Multiply the first equation by -1, add the result to the second equation, and replace the second equation with the new equation:

$$-1(x + y = 0.54) \Rightarrow -x - y = -0.54$$

$$\begin{cases} -x - y & = -0.54 \\ x + z = 0.40 \end{cases} \Rightarrow -y + z = -0.14$$

$$\begin{cases} x + y & = 0.54 \\ x + z = 0.40 \Rightarrow \\ y + z = 0.52 \end{cases} \begin{cases} x + y & = 0.54 \\ -y + z = -0.14 \\ y + z = 0.52 \end{cases}$$

Add the second and third equations and solve for z: $2z = 0.38 \Rightarrow z = 0.19$. Substituting this value into the original second equation, we have $x + 0.19 = 0.40 \Rightarrow x = 0.21$. Substituting this value into the original first equation, we have $0.21 + y = 0.54 \Rightarrow y = 0.33$.

Referring to table 5.1 in the text, we see that cell A contains healthy tissue (since $x = 0.21$), cell B contains tumorous tissue (since $y = 0.33$), and cell C contains healthy tissue (since $z = 0.19$).

55. The system is

$$\begin{cases} x + y & = 0.51 \text{ (beam 1)} \\ x + z = 0.49 \text{ (beam 2)} \\ y + z = 0.44 \text{ (beam 3)} \end{cases}$$

Multiply the first equation by -1, add the result to the second equation, and replace the second equation with the new equation:

$$-1(x + y = 0.51) \Rightarrow -x - y = -0.51$$

$$\begin{cases} -x - y & = -0.51 \\ x + z = 0.49 \end{cases} \Rightarrow -y + z = -0.02$$

$$\begin{cases} x + y & = 0.51 \\ x + z = 0.49 \Rightarrow \\ y + z = 0.44 \end{cases} \begin{cases} x + y & = 0.51 \\ -y + z = -0.02 \\ y + z = 0.44 \end{cases}$$

Add the second and third equations and solve for z: $2z = 0.42 \Rightarrow z = 0.21$. Substituting this value into the original second equation, we have $x + 0.21 = 0.49 \Rightarrow x = 0.28$. Substituting this value into the original first equation, we have $0.28 + y = 0.51 \Rightarrow y = 0.23$. Referring to table 5.1 in the text, we see that cell A contains healthy tissue (since $x = 0.28$), cell B contains healthy tissue (since $y = 0.23$), and cell C contains healthy tissue (since $z = 0.21$).

5.2 Beyond the Basics

57. Because each of the ordered triples satisfies the given equation, we can find the values of the coefficients by solving the system

$$\begin{cases} 1 + 0b + 0c = d \\ 0 + 1b + 0c = d \Rightarrow \\ 0 + 0b + 1c = d \end{cases} \begin{cases} 1 = d \\ b = d \Rightarrow b = c = d = 1. \\ c = d \end{cases}$$

The equation is $x + y + z = 1$.

59. Because each of the ordered triples satisfies the given equation, we can find the values of the coefficients by solving the system

$$\begin{cases} 3 - 4b + 0c = d \\ 0 + \dfrac{1}{4}b + \dfrac{1}{2}c = d \Rightarrow \\ 1 + 1b - 4c = d \end{cases} \begin{cases} -4b - d = -3 \\ b + 2c - 4d = 0 \\ b - 4c - d = -1 \end{cases}$$

Multiplying the second equation by 2, adding it to the third equation, and replacing the second equation with the new equation, we have

$$2(b + 2c - 4d = 0) \Rightarrow 2b + 4c - 8d = 0$$

$$\begin{cases} 2b + 4c - 8d = 0 \\ b - 4c - d = -1 \end{cases} \Rightarrow 3b - 9d = -1$$

$$\begin{cases} -4b - d = -3 \\ b + 2c - 4d = 0 \Rightarrow \\ b - 4c - d = -1 \end{cases} \begin{cases} -4b - d = -3 \\ 3b - 9d = -1 \\ b - 4c - d = 1 \end{cases}$$

(continued on next page)

(continued)

From the first equation, we have $d = 3 - 4b$. Substituting this into the second equation, we have

$$3b - 9(3 - 4b) = -1 \Rightarrow 39b = 26 \Rightarrow b = \frac{2}{3}.$$

Then $d = 3 - 4\left(\frac{2}{3}\right) - 3 = \frac{1}{3}$.

$$\frac{1}{4}\left(\frac{2}{3}\right) + \frac{1}{2}c = \frac{1}{3} \Rightarrow \frac{1}{2}c = \frac{1}{6} \Rightarrow c = \frac{1}{3}.$$

The equation is $x + \frac{2}{3}y + \frac{1}{3}z = \frac{1}{3}$.

61. Because each of the ordered triples satisfies the given equation, we can find the values of the coefficients by solving the system

$$\begin{cases} a(0)^2 + b(0) + c = 1 \\ a(-1)^2 + b(-1) + c = 0 \Rightarrow \\ a(1)^2 + b(1) + c = 4 \end{cases} \begin{cases} c = 1 \\ a - b + c = 0 \\ a + b + c = 4 \end{cases}$$

Substituting $c = 1$ into the second and third equations and then solving for b and c, we have

$$\begin{cases} a - b + 1 = 0 \\ a + b + 1 = 4 \end{cases} \Rightarrow 2a = 2 \Rightarrow a = 1$$

$$1 - b + 1 = 0 \Rightarrow b = 2$$

The equation is $y = x^2 + 2x + 1$.

63. Because each of the ordered triples satisfies the given equation, we can find the values of the coefficients by solving the system

$$\begin{cases} a(1)^2 + b(1) + c = 2 \\ a(-1)^2 + b(-1) + c = 4 \Rightarrow \\ a(2)^2 + b(2) + c = 4 \end{cases} \begin{cases} c = 2 \\ a - b + c = 4 \\ 4a + 2b + c = 4 \end{cases}$$

Substituting $c = 2$ into the second and third equations and then solving for b and c, we have

$$\begin{cases} a - b + 2 = 4 \\ 4a + 2b + 2 = 4 \end{cases} \Rightarrow \begin{cases} 2a - 2b + 4 = 8 \\ 4a + 2b + 2 = 4 \end{cases} \Rightarrow$$

$$6a = 6 \Rightarrow a = 1$$

$$1 - b + 2 = 4 \Rightarrow b = -1$$

The equation is $y = x^2 - x + 2$.

65. Because each of the ordered triples satisfies the given equation, we can find the values of the coefficients by solving the system

$$\begin{cases} 0^2 + 4^2 + a(0) + b(4) + c = 0 \\ \left(2\sqrt{2}\right)^2 + \left(2\sqrt{2}\right)^2 + a\left(2\sqrt{2}\right) + b\left(2\sqrt{2}\right) + c = 0 \Rightarrow \\ (-4)^2 + 0^2 + a(-4) + b(0) + c = 0 \end{cases}$$

$$\begin{cases} 4b + c = -16 \\ \left(2\sqrt{2}\right)a + \left(2\sqrt{2}\right)b + c = -16 \\ -4a + c = -16 \end{cases}$$

From the third equation, we have $c = 4a - 16$. Substituting this expression into the first and second equations, we have

$$\begin{cases} 4b + 4a - 16 = -16 \\ \left(2\sqrt{2}\right)a + \left(2\sqrt{2}\right)b + 4a - 16 = -16 \end{cases} \Rightarrow$$

$$\begin{cases} 4a + 4b = 0 \\ \left(2\sqrt{2} + 4\right)a + \left(2\sqrt{2}\right)b = 0 \end{cases} \Rightarrow$$

$$\begin{cases} a + b = 0 \\ \left(2\sqrt{2} + 4\right)a + \left(2\sqrt{2}\right)b = 0 \end{cases} \Rightarrow$$

$$\begin{cases} a = -b \\ \left(2\sqrt{2} + 4\right)a + \left(2\sqrt{2}\right)b = 0 \end{cases} \Rightarrow$$

$$\left(2\sqrt{2} + 4\right)(-b) + \left(2\sqrt{2}\right)b = 0 \Rightarrow$$

$-4b = 0 \Rightarrow b = 0$. Substituting this value into the first equation, we have $4(0) + c = -16 \Rightarrow c = -16$. Substituting into the third equation, we have $-4a - 16 = -16 \Rightarrow a = 0$. The equation is $x^2 + y^2 - 16 = 0$.

67. Because each of the ordered triples satisfies the given equation, we can find the values of the coefficients by solving the system

$$\begin{cases} 1^2 + 2^2 + a(1) + b(2) + c = 0 \\ 6^2 + (-3)^2 + a(6) + b(-3) + c = 0 \Rightarrow \\ (4)^2 + 1^2 + a(4) + b(1) + c = 0 \end{cases}$$

$$\begin{cases} a + 2b + c = -5 \\ 6a - 3b + c = -45 \\ 4a + b + c = -17 \end{cases}$$

Multiply the first equation by –6, add the result to the second equation, then replace the second equation with the new equation. Similarly, multiply the first equation by –4, add that result to the third equation, then replace the third equation with the new equation:

$$-6(a + 2b + c = -5) \Rightarrow -6a - 12b - 6c = 30$$

$$\begin{cases} -6a - 12b - 6c = 30 \\ 6a - 3b + c = -45 \end{cases} \Rightarrow -15b - 5c = -15 \Rightarrow$$

$$3b + c = 3$$

$$-4(a + 2b + c = -5) \Rightarrow -4a - 8b - 4c = 20$$

$$\begin{cases} -4a - 8b - 4c = 20 \\ 4a + b + c = -17 \end{cases} \Rightarrow -7b - 3c = 3$$

$$\begin{cases} a + 2b + c = -5 \\ 6a - 3b + c = -45 \Rightarrow \\ 4a + b + c = -17 \end{cases} \begin{cases} a + 2b + c = -5 \\ 3b + c = 3 \\ -7b - 3c = 3 \end{cases}$$

(continued on next page)

(*continued*)

From the second equation, we have
$c = -3b + 3$. Substitute this expression into the third equation, and solve for b.

$-7b - 3(-3b + 3) = 3 \Rightarrow 2b = 12 \Rightarrow b = 6$.

So, $c = -3(6) + 3 \Rightarrow c = -15$. Substituting into the original first equation, we have
$a + 2(6) - 15 = -5 \Rightarrow a = -2$.

The equation is $x^2 + y^2 - 2x + 6y - 15 = 0$.

69. Letting $u = 1/x, v = 1/y, w = 1/z$, we have

$$\begin{cases} \dfrac{1}{x} + \dfrac{3}{y} - \dfrac{1}{z} = 5 \\ \dfrac{2}{x} + \dfrac{4}{y} + \dfrac{6}{z} = 4 \\ \dfrac{2}{x} + \dfrac{3}{y} + \dfrac{1}{z} = 3 \end{cases} \Rightarrow \begin{cases} u + 3v - w = 5 \\ 2u + 4v + 6w = 4 \\ 2u + 3v + w = 3 \end{cases}$$

Multiplying the first equation by -2, adding the result to the second and third equations, and replacing those equations with the new equations, we have
$-2(u + 3v - w = 5) \Rightarrow -2u - 6v + 2w = -10$

$$\begin{cases} -2u - 6v + 2w = -10 \\ 2u + 4v + 6w = 4 \end{cases} \Rightarrow -2v + 8w = -6 \Rightarrow$$
$v - 4w = 3$

$$\begin{cases} -2u - 6v + 2w = -10 \\ 2u + 3v + w = 3 \end{cases} \Rightarrow -3v + 3w = -7$$

$$\begin{cases} u + 3v - w = 5 \\ 2u + 4v + 6w = 4 \\ 2u + 3v + w = 3 \end{cases} \Rightarrow \begin{cases} u + 3v - w = 5 \\ v - 4w = 3 \\ -3v + 3w = -7 \end{cases}$$

Multiplying the second equation by 3 and adding the result to the third equation, we have
$3(v - 4w = 3) \Rightarrow 3v - 12w = 9$

$$\begin{cases} 3v - 12w = 9 \\ -3v + 3w = -7 \end{cases} \Rightarrow -9w = 2 \Rightarrow w = -\dfrac{2}{9}.$$

Substituting, we have

$v - 4\left(-\dfrac{2}{9}\right) = 3 \Rightarrow v = \dfrac{19}{9}$ and

$u + 3\left(\dfrac{19}{9}\right) - \left(-\dfrac{2}{9}\right) = 5 \Rightarrow u = -\dfrac{14}{9}.$

So, $x = -\dfrac{9}{14}, y = \dfrac{9}{19}, z = -\dfrac{9}{2}.$

The solution set is $\left\{\left(-\dfrac{9}{14}, \dfrac{9}{19}, -\dfrac{9}{2}\right)\right\}.$

71. Solve the first three equations in the system for x, y, and z:
$-3(x + 2y - 5z = 9) \Rightarrow -3x - 6y + 15z = -27$

$$\begin{cases} -3x - 6y + 15z = -27 \\ 3x - y + 2z = 14 \end{cases} \Rightarrow -7y + 17z = -13$$

$-2(x + 2y - 5z = 9) \Rightarrow -2x - 4y + 10z = -18$

$$\begin{cases} -2x - 4y + 10z = -18 \\ 2x + 3y - z = 3 \end{cases} \Rightarrow -y + 9z = -15$$

$$\begin{cases} x + 2y - 5z = 9 \\ 3x - y + 2z = 14 \\ 2x + 3y - z = 3 \end{cases} \Rightarrow \begin{cases} x + 2y - 5z = 9 \\ -7y + 17z = -13 \\ -y + 9z = -15 \end{cases}$$

$-7(-y + 9z = -15) \Rightarrow 7y - 63z = 105$

$$\begin{cases} -7y + 17z = -13 \\ 7y - 63z = 105 \end{cases} \Rightarrow -46z = 92 \Rightarrow z = -2$$

$-7y + 17(-2) = -13 \Rightarrow -7y = 21 \Rightarrow y = -3$
$x + 2(-3) - 5(-2) = 9 \Rightarrow x = 5.$

Substituting $x = 5$, $y = -3$, and $z = -2$ into the fourth equation of the original system, we have
$5c - 5(-3) + (-2) + 3 = 0 \Rightarrow$

$5c = -16 \Rightarrow c = -\dfrac{16}{5}.$

73. Each of the ordered triples satisfies the given equation, so find the values of the coefficients by solving the system

$$\begin{cases} a(-1)^2 + b(-1) + c = -1 \\ a(0)^2 + b(0) + c = 5 \\ a(2)^2 + b(2) + c = 5 \end{cases} \Rightarrow \begin{cases} a - b + c = -1 \\ c = 5 \\ 4a + 2b + c = 5 \end{cases}$$

Substitute $c = 5$ into the first and third equations and then solve those equations for a and b

$$\begin{cases} a - b + 5 = -1 \\ 4a + 2b + 5 = 5 \end{cases} \Rightarrow \begin{cases} a - b = -6 \\ 2a + b = 0 \end{cases} \Rightarrow$$
$3a = -6 \Rightarrow a = -2$
$-2 - b + 5 = -1 \Rightarrow b = -4.$

The equation is $y = -2x^2 + 4x + 5.$

5.2 Critical Thinking/Discussion/Writing

Answers may vary for exercises 75 and 76.

75. $\begin{cases} 1 + (-1) - 2 = -2 \\ 2(1) - (-1) + 3(2) = 9 \\ 1 + (-1) + 2 = 2 \end{cases} \Rightarrow \begin{cases} x + y - z = -2 \\ 2x - y + 3z = 9 \\ x + y + z = 2 \end{cases}$

76. a. $\begin{cases} x + 3y + 3z = 15 \\ x + 2y + z = 1 \\ 2y + 4z = 11 \end{cases}$ **b.** $\begin{cases} x + 2y - z = 4 \\ x + 3y + 2z = 5 \\ y + 3z = 1 \end{cases}$

5.2 Maintaining Skills

77. $\dfrac{1}{2\cdot 3} = \dfrac{1}{2} + \dfrac{B}{3} \Rightarrow 1 = 3 + 2B \Rightarrow -2 = 2B \Rightarrow$
$B = -1$

79. $\dfrac{1}{n\cdot(n+1)} = \dfrac{A}{n} - \dfrac{1}{n+1} \Rightarrow 1 = A(n+1) - n \Rightarrow$
$1 + n = A(n+1) \Rightarrow 1 = A$

81. $2x + 3 = A(x+3) + B(x-1)$
$2x + 3 = Ax + 3A + Bx - B = (A+B)x + 3A - B$
Equating the coefficients gives the system
$\begin{cases} A+B=2 \\ 3A-B=3 \end{cases} \Rightarrow 4A = 5 \Rightarrow A = \dfrac{5}{4}$
$\dfrac{5}{4} + B = 2 \Rightarrow B = \dfrac{3}{4}$

83. $x^2 + 5x + 6 = (x+2)(x+3)$

85. $2x^2 + 5x - 3 = (2x-1)(x+3)$

87. $4x^2 - 9 = (2x-3)(2x+3)$

5.3 Partial-Fraction Decomposition

5.3 Practice Problems

1. $\dfrac{2x-7}{(x+1)(x-2)} = \dfrac{A}{x+1} + \dfrac{B}{x-2} \Rightarrow$
$2x - 7 = A(x-2) + B(x+1) \Rightarrow$
$2x - 7 = (A+B)x + (-2A+B) \Rightarrow$
$\begin{cases} A+\ B=\ 2 \\ -2A+\ B=-7 \end{cases} \Rightarrow \begin{cases} -A-B=-2 \\ -2A+B=-7 \end{cases} \Rightarrow$
$-3A = -9 \Rightarrow A = 3$
$3 + B = 2 \Rightarrow B = -1$
$\dfrac{2x-7}{(x+1)(x-2)} = \dfrac{3}{x+1} - \dfrac{1}{x-2}$

2. $\dfrac{3x^2+4x+3}{x^3-x} = \dfrac{3x^2+4x+3}{x(x-1)(x+1)}$
$\qquad = \dfrac{A}{x} + \dfrac{B}{x-1} + \dfrac{C}{x+1} \Rightarrow$
$3x^2 + 4x + 3$
$\quad = A(x+1)(x-1) + Bx(x+1) + Cx(x-1)$
$\quad = Ax^2 - A + Bx^2 + Bx + Cx^2 - Cx$
$\quad = (A+B+C)x^2 + (B-C)x - A$
$\begin{cases} A+B+C=3 \\ \quad\ B-C=4 \Rightarrow A = -3 \\ -A \qquad\quad =3 \end{cases}$

$\begin{cases} -3+B+C=3 \\ \qquad B-C=4 \end{cases} \Rightarrow -3+2B=7 \Rightarrow B=5$
$5 - C = 4 \Rightarrow C = 1$
$\dfrac{3x^2+4x+3}{x^3-x} = -\dfrac{3}{x} + \dfrac{5}{x-1} + \dfrac{1}{x+1}$

3. $\dfrac{x+5}{x(x-1)^2} = \dfrac{A}{x} + \dfrac{B}{(x-1)} + \dfrac{C}{(x-1)^2} \Rightarrow$
$x + 5 = A(x-1)^2 + Bx(x-1) + Cx \Rightarrow$
$x + 5 = (A+B)x^2 + (-2A-B+C)x + A$
$\begin{cases} A+B \quad\quad = 0 \\ -2A-B+C=1 \Rightarrow A = 5,\ B = -5 \\ A \qquad\quad = 5 \end{cases}$
$-2A - B + C = 1 \Rightarrow -2(5) + 5 + C = 1 \Rightarrow C = 6$
$\dfrac{x+5}{x(x-1)^2} = \dfrac{5}{x} - \dfrac{5}{(x-1)} + \dfrac{6}{(x-1)^2}$

4. $\dfrac{3x^2+5x-2}{x(x^2+2)} = \dfrac{A}{x} + \dfrac{Bx+C}{x^2+2} \Rightarrow$
$3x^2 + 5x - 2 = A(x^2+2) + (Bx+C)x \Rightarrow$
$3x^2 + 5x - 2 = (A+B)x^2 + Cx + 2A \Rightarrow$
$\begin{cases} A+B \quad\quad = 3 \\ \qquad C = 5 \Rightarrow A = -1,\ C = 5 \\ 2A \qquad\quad = -2 \end{cases}$
$A + B = 3 \Rightarrow -1 + B = 3 \Rightarrow B = 4$
$\dfrac{3x^2+5x-2}{x(x^2+2)} = -\dfrac{1}{x} + \dfrac{4x+5}{x^2+2}$

5. $\dfrac{x^2+3x+1}{(x^2+1)^2} = \dfrac{Ax+B}{x^2+1} + \dfrac{Cx+D}{(x^2+1)^2} \Rightarrow$
$x^2 + 3x + 1 = (Ax+B)(x^2+1) + Cx + D \Rightarrow$
$x^2 + 3x + 1 = Ax^3 + Bx^2 + (A+C)x + (B+D) \Rightarrow$
$\begin{cases} A \qquad\qquad = 0 \\ \quad\ B \qquad\quad = 1 \\ A+\quad C \quad\ = 3 \\ \quad\ B+\ D = 1 \end{cases} \Rightarrow A=0, B=1, C=3, D=0$
$\dfrac{x^2+3x+1}{(x^2+1)^2} = \dfrac{1}{x^2+1} + \dfrac{3x}{(x^2+1)^2}$

6. $\dfrac{1}{4\cdot 5} + \dfrac{1}{5\cdot 6} + \dfrac{1}{6\cdot 7} + \cdots + \dfrac{1}{3111\cdot 3112}$
$= \left(\dfrac{1}{4} - \dfrac{1}{5}\right) + \left(\dfrac{1}{5} - \dfrac{1}{6}\right) + \left(\dfrac{1}{6} - \dfrac{1}{7}\right) + \cdots$
$\qquad\qquad + \left(\dfrac{1}{3111} - \dfrac{1}{3112}\right)$
$= \dfrac{1}{4} - \dfrac{1}{3112} = \dfrac{778-1}{3112} = \dfrac{777}{3112}$

7. $R = \dfrac{(x+1)(x+2)}{4x+7} \Rightarrow \dfrac{1}{R} = \dfrac{4x+7}{(x+1)(x+2)}$

$\dfrac{4x+7}{(x+1)(x+2)} = \dfrac{A}{x+1} + \dfrac{B}{x+2} \Rightarrow$

$4x+7 = A(x+2) + B(x+1) \Rightarrow$

$4x+7 = (A+B)x + (2A+B) \Rightarrow$

$\begin{cases} A+B = 4 \\ 2A+B = 7 \end{cases} \Rightarrow A = 3, B = 1$

$\dfrac{4x+7}{(x+1)(x+2)} = \dfrac{3}{x+1} + \dfrac{1}{x+2}$

$\qquad\qquad = \dfrac{\dfrac{1}{x+1}}{3} + \dfrac{1}{x+2} = \dfrac{1}{R}$

This means that if two resistances $R_1 = \dfrac{x+1}{3}$

and $R_2 = x+2$ are connected in parallel, they will produce a total resistance given by
$\dfrac{(x+1)(x+2)}{4x+7}$.

5.3 Basic Concepts and Skills

1. In a rational expression, if the degree of the numerator, $P(x)$, is less than the degree of the denominator, $Q(x)$, then the expression is <u>proper</u>.

3. In a rational expression, if $(x - 8)$ is a linear factor that is repeated three times in the denominator, then the portion of the expression's partial-fraction decomposition that corresponds to $(x - 8)^3$ has <u>three</u> terms.

5. False. The factors are repeated linear factors.

7. $\dfrac{1}{(x-1)(x+2)} = \dfrac{A}{x-1} + \dfrac{B}{x+2}$

9. $\dfrac{1}{x^2+7x+6} = \dfrac{1}{(x+6)(x+1)} = \dfrac{A}{x+6} + \dfrac{B}{x+1}$

11. $\dfrac{2}{x^3-x^2} = \dfrac{2}{x^2(x-1)} = \dfrac{A}{x^2} + \dfrac{B}{x} + \dfrac{C}{x-1}$

13. $\dfrac{x^2-3x+3}{(x+1)(x^2-x+1)} = \dfrac{A}{x+1} + \dfrac{Bx+C}{x^2-x+1}$

15. $\dfrac{3x-4}{(x^2+1)^2} = \dfrac{Ax+B}{x^2+1} + \dfrac{Cx+D}{(x^2+1)^2}$

17. $\dfrac{2x+1}{(x+1)(x+2)} = \dfrac{A}{x+1} + \dfrac{B}{x+2} \Rightarrow$

$2x+1 = A(x+2) + B(x+1) \Rightarrow$

$2x+1 = (A+B)x + (2A+B) \Rightarrow$

$\begin{cases} A+B = 2 \\ 2A+B = 1 \end{cases} \Rightarrow \begin{cases} -A-B = -2 \\ 2A+B = 1 \end{cases} \Rightarrow A = -1, B = 3$

$\dfrac{2x+1}{(x+1)(x+2)} = -\dfrac{1}{x+1} + \dfrac{3}{x+2} = \dfrac{3}{x+2} - \dfrac{1}{x+1}$

19. $\dfrac{1}{x^2+4x+3} = \dfrac{1}{(x+3)(x+1)} = \dfrac{A}{x+3} + \dfrac{B}{x+1} \Rightarrow$

$1 = A(x+1) + B(x+3) = (A+B)x + (A+3B) \Rightarrow$

$\begin{cases} A+\ B = 0 \\ A+3B = 1 \end{cases} \Rightarrow \begin{cases} -A-\ B = 0 \\ A+3B = 1 \end{cases} \Rightarrow 2B = 1 \Rightarrow$

$B = \dfrac{1}{2}, A = -\dfrac{1}{2}$

$\dfrac{1}{x^2+4x+3} = -\dfrac{1}{2(x+3)} + \dfrac{1}{2(x+1)}$

21. $\dfrac{2}{x^2+2x} = \dfrac{2}{x(x+2)} = \dfrac{A}{x} + \dfrac{B}{x+2} \Rightarrow$

$2 = A(x+2) + Bx = (A+B)x + 2A \Rightarrow$

$\begin{cases} A+B = 0 \\ 2A = 2 \end{cases} \Rightarrow A = 1, B = -1 \Rightarrow$

$\dfrac{2}{x^2+2x} = \dfrac{1}{x} - \dfrac{1}{x+2}$

23. $\dfrac{x}{(x+1)(x+2)(x+3)} = \dfrac{A}{x+1} + \dfrac{B}{x+2} + \dfrac{C}{x+3} \Rightarrow$

$x = A(x+2)(x+3) + B(x+1)(x+3)$
$\qquad\qquad + C(x+1)(x+2) \Rightarrow$

$x = A(x^2+5x+6) + B(x^2+4x+3)$
$\qquad\qquad + C(x^2+3x+2) \Rightarrow$

$x = (A+B+C)x^2 + (5A+4B+3C)x$
$\qquad\qquad + (6A+3B+2C) \Rightarrow$

$\begin{cases} A+\ B+\ C = 0 \\ 5A+4B+3C = 1 \\ 6A+3B+2C = 0 \end{cases}$

$-5(A+\ B+\ C = 0) = -5A - 5B - 5C = 0$

$\begin{cases} -5A-5B-5C = 0 \\ \ \ 5A+4B+3C = 1 \end{cases} \Rightarrow -B - 2C = 1$

$-6(A+B+C = 0) = -6A - 6B - 6C = 0$

$\begin{cases} -6A-6B-6C = 0 \\ \ \ 6A+3B+2C = 0 \end{cases} \Rightarrow -3B - 4C = 0$

$\begin{cases} A+\ B+\ C = 0 \\ 5A+3B+3C = 1 \\ 6A+3B+2C = 0 \end{cases} \Rightarrow \begin{cases} A+\ B+\ C = 0 \\ \quad\ -B - 2C = 1 \\ \quad\ -3B - 4C = 0 \end{cases}$

(continued on next page)

(*continued*)

$$-3(-B - 2C = 1) = 3B + 6C = -3$$

$$\begin{cases} 3B + 6C = -3 \\ -3B - 4C = 0 \end{cases} \Rightarrow 2C = -3 \Rightarrow C = -\frac{3}{2}$$

$$-B - 2\left(-\frac{3}{2}\right) = 1 \Rightarrow -B = -4 \Rightarrow B = 2$$

$$A + 2 - \frac{3}{2} = 0 \Rightarrow A = -\frac{1}{2}$$

$$\frac{x}{(x+1)(x+2)(x+3)} = -\frac{1}{2(x+1)} + \frac{2}{x+2} - \frac{3}{2(x+3)}$$

25. $\dfrac{x-1}{(x+1)^2} = \dfrac{A}{(x+1)^2} + \dfrac{B}{(x+1)} \Rightarrow x - 1 = A + B(x+1) = Bx + (A+B) \Rightarrow \begin{cases} B = 1 \\ A + B = -1 \end{cases} \Rightarrow B = 1, A = -2$

$$\frac{x-1}{(x+1)^2} = -\frac{2}{(x+1)^2} + \frac{1}{(x+1)}$$

27. $\dfrac{2x^2 + x}{(x+1)^3} = \dfrac{A}{(x+1)^3} + \dfrac{B}{(x+1)^2} + \dfrac{C}{x+1} \Rightarrow$

$$2x^2 + x = A + B(x+1) + C(x+1)^2 = A + Bx + B + C(x^2 + 2x + 1) = Cx^2 + (B + 2C)x + (A + B + C) \Rightarrow$$

$$\begin{cases} A + B + C = 0 \\ B + 2C = 1 \Rightarrow C = 2, B = -3, A = 1 \\ C = 2 \end{cases}$$

$$\frac{2x^2 + x}{(x+1)^3} = \frac{2}{x+1} - \frac{3}{(x+1)^2} + \frac{1}{(x+1)^3}$$

29. $\dfrac{-x^2 + 3x + 1}{x^3 + 2x^2 + x} = \dfrac{-x^2 + 3x + 1}{x(x+1)^2} = \dfrac{A}{(x+1)^2} + \dfrac{B}{x+1} + \dfrac{C}{x} \Rightarrow$

$$-x^2 + 3x + 1 = Ax + Bx(x+1) + C(x+1)^2 = Ax + Bx^2 + Bx + Cx^2 + 2Cx + C = (B+C)x^2 + (A+B+2C)x + C$$

$$\begin{cases} B + C = -1 \Rightarrow C = 1, B = -2 \\ A + B + 2C = 3 \\ C = 1 \end{cases}$$

$$A + 2(-2) + 3(1) = 2 \Rightarrow A = 3$$

$$\frac{-x^2 + 3x + 1}{x^3 + 2x^2 + x} = \frac{1}{x} - \frac{2}{x+1} + \frac{3}{(x+1)^2}$$

31. $\dfrac{1}{x^2(x+1)^2} = \dfrac{A}{x} + \dfrac{B}{x^2} + \dfrac{C}{x+1} + \dfrac{D}{(x+1)^2} \Rightarrow 1 = Ax(x+1)^2 + B(x+1)^2 + Cx^2(x+1) + Dx^2$

Letting $x = -1$, we have $D = 1$. Letting $x = 0$, we have $B = 1$. Substitute the values for B and D, expand the equation, and simplify:

$$1 = Ax(x+1)^2 + 1(x+1)^2 + Cx^2(x+1) + 1x^2 = (A+C)x^3 + (2 + 2A + C)x^2 + (2 + A)x + 1 \Rightarrow$$

$$\begin{cases} A + C = 0 \\ 2A + C = -2 \Rightarrow A = -2, C = 2 \\ A = -2 \end{cases}$$

$$\frac{1}{x^2(x+1)^2} = -\frac{2}{x} + \frac{1}{x^2} + \frac{2}{x+1} + \frac{1}{(x+1)^2}$$

33. $\dfrac{1}{(x^2-1)^2} = \dfrac{1}{((x-1)(x+1))^2} = \dfrac{A}{x-1} + \dfrac{B}{(x-1)^2} + \dfrac{C}{x+1} + \dfrac{D}{(x+1)^2} \Rightarrow$

$\qquad 1 = A(x-1)(x+1)^2 + B(x+1)^2 + C(x-1)^2(x+1) + D(x-1)^2$

Letting $x = 1$, we have $1 = B(2)^2 \Rightarrow B = 1/4$. Letting $x = -1$, we have $1 = D(-2)^2 \Rightarrow D = 1/4$.

Substitute the values for B and D, expand the equation, and simplify:

$1 = A(x-1)(x+1)^2 + \dfrac{1}{4}(x+1)^2 + C(x-1)^2(x+1) + \dfrac{1}{4}(x-1)^2 \Rightarrow$

$1 = (A+C)x^3 + \left(\dfrac{1}{2} + A - C\right)x^2 + (-A-C)x + \left(\dfrac{1}{2} - A + C\right) \Rightarrow$

$\begin{cases} A+C = 0 \\ A-C = -\dfrac{1}{2} \\ -A-C = 0 \\ -A+C = \dfrac{1}{2} \end{cases} \Rightarrow A = -\dfrac{1}{4}, C = \dfrac{1}{4}$

$\dfrac{1}{(x^2-1)^2} = -\dfrac{1}{4(x-1)} + \dfrac{1}{4(x-1)^2} + \dfrac{1}{4(x+1)} + \dfrac{1}{4(x+1)^2}$

35. $\dfrac{x-1}{(2x-3)^2} = \dfrac{A}{2x-3} + \dfrac{B}{(2x-3)^2} \Rightarrow x-1 = A(2x-3) + B \Rightarrow x-1 = 2Ax + (-3A+B)$

$\begin{cases} 2A = 1 \\ -3A+B = -1 \end{cases} \Rightarrow A = \dfrac{1}{2}$

$-3\left(\dfrac{1}{2}\right) + B = -1 \Rightarrow B = \dfrac{1}{2}$

$\dfrac{x-1}{(2x-3)^2} = \dfrac{1}{2(2x-3)} + \dfrac{1}{2(2x-3)^2}$

37. $\dfrac{6x+7}{4x^2+12x+9} = \dfrac{6x+7}{(2x+3)^2} = \dfrac{A}{2x+3} + \dfrac{B}{(2x+3)^2} \Rightarrow 6x+7 = A(2x+3) + B \Rightarrow 6x+7 = 2Ax + (3A+B) \Rightarrow$

$\begin{cases} 2A = 6 \\ 3A+B = 7 \end{cases} \Rightarrow A = 3, B = -2$

$\dfrac{6x+7}{4x^2+12x+9} = \dfrac{3}{2x+3} - \dfrac{2}{(2x+3)^2}$

39. $\dfrac{x-3}{x^3+x^2} = \dfrac{x-3}{x^2(x+1)} = \dfrac{A}{x} + \dfrac{B}{x^2} + \dfrac{C}{x+1} \Rightarrow x-3 = Ax(x+1) + B(x+1) + Cx^2 \Rightarrow$

$x-3 = (A+C)x^2 + (A+B)x + B \Rightarrow$

$\begin{cases} A+C = 0 \\ A+B = 1 \\ B = -3 \end{cases} \Rightarrow B = -3, A = 4, C = -4$

$\dfrac{x-3}{x^3+x^2} = \dfrac{4}{x} - \dfrac{3}{x^2} - \dfrac{4}{x+1}$

41. $\dfrac{x^2+2x+4}{x^3+x^2} = \dfrac{x^2+2x+4}{x^2(x+1)} = \dfrac{A}{x} + \dfrac{B}{x^2} + \dfrac{C}{x+1} \Rightarrow x^2+2x+4 = Ax(x+1) + B(x+1) + Cx^2 \Rightarrow$

$x^2+2x+4 = (A+C)x^2 + (A+B)x + B \Rightarrow$

(*continued on next page*)

(continued)

$$\begin{cases} A+\ \ C=1 \\ A+B\ \ \ =2 \\ \ \ \ B\ \ =4 \end{cases} \Rightarrow B=4, A=-2, C=3$$

$$\frac{x^2+2x+4}{x^3+x^2}=\frac{3}{x+1}-\frac{2}{x}+\frac{4}{x^2}$$

43. $\dfrac{x}{x^4-1}=\dfrac{x}{(x-1)(x+1)(x^2+1)}=\dfrac{A}{x-1}+\dfrac{B}{x+1}+\dfrac{Cx+D}{x^2+1}\Rightarrow$

$x=A(x+1)(x^2+1)+B(x-1)(x^2+1)+(Cx+D)(x^2-1)\Rightarrow$

$x=(A+B+C)x^3+(A-B+D)x^2+(A+B-C)x+(A-B-D)\Rightarrow$

$$\begin{cases} A+B+C\ \ \ =0 \\ A-B+\ \ D=0 \\ A+B-C\ \ \ =1 \\ A-B-\ \ D=0 \end{cases} \Rightarrow A=\frac{1}{4}, B=\frac{1}{4}, C=-\frac{1}{2}, D=0$$

$$\frac{x}{x^4-1}=\frac{1}{4(x-1)}+\frac{1}{4(x+1)}-\frac{x}{2(x^2+1)}$$

45. $\dfrac{1}{x(x^2+1)^2}=\dfrac{A}{x}+\dfrac{Bx+C}{x^2+1}+\dfrac{Dx+E}{(x^2+1)^2}\Rightarrow$

$1=A(x^2+1)^2+(Bx+C)x(x^2+1)+(Dx+E)x\Rightarrow 1=(A+B)x^4+Cx^3+(2A+B+D)x^2+(C+E)x+A\Rightarrow$

$$\begin{cases} A+B\ \ \ \ \ =0 \\ \ \ \ C\ \ \ \ =0 \\ 2A+B+\ \ D\ \ =0 \\ \ \ \ C+\ E=0 \\ A\ \ \ \ \ \ =1 \end{cases} \Rightarrow A=1, B=-1, C=0, D=-1, E=0$$

$$\frac{1}{x(x^2+1)^2}=\frac{1}{x}-\frac{x}{x^2+1}-\frac{x}{(x^2+1)^2}$$

47. $\dfrac{2x^2+3x}{(x^2+1)(x^2+2)}=\dfrac{Ax+B}{x^2+1}+\dfrac{Cx+D}{x^2+2}\Rightarrow 2x^2+3x=(Ax+B)(x^2+2)+(Cx+D)(x^2+1)\Rightarrow$

$2x^2+3x=(A+C)x^3+(B+D)x^2+(2A+C)x+(2B+D)\Rightarrow$

$$\begin{cases} A+\ \ \ C\ \ =0 \\ \ \ \ B+\ \ D=2 \\ 2A+\ \ C\ \ =3 \\ \ \ \ 2B+\ D=0 \end{cases} \Rightarrow A=3, B=-2, C=-3, D=4$$

$$\frac{2x^2+3x}{(x^2+1)(x^2+2)}=\frac{3x-2}{x^2+1}+\frac{-3x+4}{x^2+2}$$

5.3 Applying the Concepts

49. $\dfrac{1}{1\cdot2}+\dfrac{1}{2\cdot3}+\dfrac{1}{3\cdot4}+\cdots+\dfrac{1}{n(n+1)}=\left(1-\dfrac{1}{2}\right)+\left(\dfrac{1}{2}-\dfrac{1}{3}\right)+\left(\dfrac{1}{3}-\dfrac{1}{4}\right)+\cdots+\left(\dfrac{1}{n}-\dfrac{1}{n+1}\right)=1-\dfrac{1}{n+1}=\dfrac{n}{n+1}$

51. Each term is of the form $\dfrac{2}{(2n-1)(2n+1)}$. Decomposing $\dfrac{2}{(2n-1)(2n+1)}$ we have

$$\frac{2}{(2n-1)(2n+1)} = \frac{A}{2n-1} + \frac{B}{2n+1} \Rightarrow 2 = A(2n+1) + B(2n-1) \Rightarrow 2 = (2A+2B)n + (A-B) \Rightarrow$$

$$\begin{cases} 2A+2B=0 \\ A-\ B=2 \end{cases} \Rightarrow A=1, B=-1$$

$$\frac{2}{(2n-1)(2n+1)} = \frac{1}{2n-1} - \frac{1}{2n+1}.$$

Thus,

$$\frac{2}{1\cdot 3} + \frac{2}{3\cdot 5} + \frac{2}{5\cdot 7} + \cdots + \frac{2}{(2n-1)(2n+1)} = \left(\frac{1}{1} - \frac{1}{3}\right) + \left(\frac{1}{3} - \frac{1}{5}\right) + \left(\frac{1}{5} - \frac{1}{7}\right) + \cdots + \frac{1}{2n-1} - \frac{1}{2n+1}$$

$$= 1 - \frac{1}{2n+1} = \frac{2n}{2n+1}$$

53. $\dfrac{1}{R} = \dfrac{2x+4}{(x+1)(x+3)}$

$$\frac{2x+4}{(x+1)(x+3)} = \frac{A}{x+1} + \frac{B}{x+3} \Rightarrow 2x+4 = A(x+3) + B(x+1) \Rightarrow 2x+4 = (A+B)x + (3A+B) \Rightarrow$$

$$\begin{cases} A+B=2 \\ 3A+B=4 \end{cases} \Rightarrow A=1, B=1$$

$$\frac{2x+4}{(x+1)(x+3)} = \frac{1}{x+1} + \frac{1}{x+3} = \frac{1}{R}.$$

This means that if two resistances $R_1 = x+1$ and $R_2 = x+3$ are connected in parallel, they will produce a total resistance given by $\dfrac{(x+1)(x+3)}{2x+4}$.

55. $\dfrac{1}{R} = \dfrac{R_1R_2 + R_2R_3 + R_3R_1}{R_1R_2R_3} = \dfrac{A}{R_1} + \dfrac{B}{R_2} + \dfrac{C}{R_3} \Rightarrow R_1R_2 + R_2R_3 + R_3R_1 = AR_2R_3 + BR_1R_3 + CR_1R_2 \Rightarrow \begin{cases} A=1 \\ B=1 \\ C=1 \end{cases}$

$$\frac{R_1R_2 + R_2R_3 + R_3R_1}{R_1R_2R_3} = \frac{1}{R_1} + \frac{1}{R_2} + \frac{1}{R_3} = \frac{1}{R}$$

This means that if three resistances $R_1, R_2,$ and R_3 are connected in parallel, they will produce a total resistance given by $\dfrac{R_1R_2R_3}{R_1R_2 + R_2R_3 + R_3R_1}$.

5.3 Beyond the Basics

57. $\begin{cases} A+B+\ C=0 \\ \quad\ 2B-2C=10 \\ -4A\qquad\quad = -4 \end{cases} \Rightarrow \begin{cases} A+B+C=0 \\ \quad\ B-C=5 \\ A\qquad\quad =1 \end{cases}$

Subtract the third equation from the first equation, and replace the first equation with the new equation:

$\begin{cases} A+B+C=0 \\ \quad\ B-C=5 \\ A\qquad\ =1 \end{cases} \Rightarrow \begin{cases} \quad\ B+C=-1 \\ \quad\ B-C=5 \\ A\qquad\quad =1 \end{cases}$

Add the first and second equations to solve for B:

$\begin{cases} \quad\ B+C=-1 \\ \quad\ B-C=5 \\ A\qquad\quad =1 \end{cases} \Rightarrow \begin{cases} \quad\ B+C=-1 \\ \quad\ 2B\qquad =4 \\ A\qquad\quad =1 \end{cases} \Rightarrow \begin{cases} \quad\ B+C=-1 \\ \quad\ B\qquad =2 \\ A\qquad\quad =1 \end{cases}$

(*continued on next page*)

(continued)

Substitute the value for B into the first equation to solve for C:

$$\begin{cases} B+C=-1 \\ B\quad =2 \\ A\qquad =1 \end{cases} \Rightarrow \begin{cases} 2+C=-1 \\ B\quad =2 \\ A\qquad =1 \end{cases} \Rightarrow \begin{cases} C=-3 \\ B=2 \\ A=1 \end{cases}$$

59. $\dfrac{4x}{(x^2-1)^2} = \dfrac{4x}{(x-1)^2(x+1)^2} = \dfrac{A}{x-1} + \dfrac{B}{(x-1)^2} + \dfrac{C}{x+1} + \dfrac{D}{(x+1)^2} \Rightarrow$

$4x = A(x-1)(x+1)^2 + B(x+1)^2 + C(x-1)^2(x+1) + D(x-1)^2 \Rightarrow$

$4x = (A+C)x^3 + (A+B-C+D)x^2 + (-A+2B-C-2D)x + (-A+B+C+D) \Rightarrow$

$$\begin{cases} A+\quad C\qquad =0 \quad (1) \\ A+\ B-C+\ D=0 \quad (2) \\ -A+2B-C-2D=4 \quad (3) \\ -A+\ B+C+\ D=0 \quad (4) \end{cases}$$

Add equations (1) and (3), and (2) and (4), and replace equations (3) and (4):

$$\begin{cases} A+\quad C\qquad =0 \\ A+\ B-C+\ D=0 \\ -A+2B-C-2D=4 \\ -A+\ B+C+\ D=0 \end{cases} \Rightarrow \begin{cases} A+\quad C\qquad =0 \quad (1) \\ A+\ B-C+\ D=0 \quad (2) \\ \quad 2B-2D\quad =4 \quad (3) \\ \quad 2B+2D\quad =0 \quad (4) \end{cases} \Rightarrow \begin{cases} A+\quad C\qquad =0 \\ A+\ B-C+\ D=0 \\ \quad 4B\qquad =4 \\ \quad 2B+2D\quad =0 \end{cases} \Rightarrow B=1, D=-1, A=0, C=0$$

$$\frac{4x}{(x^2-1)^2} = \frac{1}{(x-1)^2} - \frac{1}{(x+1)^2}$$

61. $\dfrac{x+1}{(x^2+1)(x-1)^2} = \dfrac{A}{x-1} + \dfrac{B}{(x-1)^2} + \dfrac{Cx+D}{x^2+1} \Rightarrow x+1 = A(x-1)(x^2+1) + B(x^2+1) + (Cx+D)(x-1)^2 \Rightarrow$

$x+1 = (A+C)x^3 + (-A+B-2C+D)x^2 + (A+B+C-2D)x + (-A+B+D) \Rightarrow$

$$\begin{cases} A+\quad C\qquad =0 \\ -A+B-2C+\ D=0 \\ A+B+\ C-2D=1 \\ -A+B+\quad D=1 \end{cases}$$

From the first equation, we have $C=-A$. Substitute this into the last three equations and simplify.

$$\begin{cases} A+\quad C\qquad =0 \\ -A+B-2C+\ D=0 \\ A+\quad C-2D=1 \\ -A+B+\quad D=1 \end{cases} \Rightarrow \begin{cases} C=-A \\ -A+B-2(-A)+D=0 \\ A+\ (-A)-2D=1 \\ -A+B+\quad D=1 \end{cases} \Rightarrow \begin{cases} C=-A \\ A+B+\ D=0 \\ \quad -2D=1 \\ -A+B+\ D=1 \end{cases} \Rightarrow \begin{cases} C=-A \\ A+B+\ D=0 \\ D=-\dfrac{1}{2} \\ -A+B+\ D=1 \end{cases}$$

Substitute the value for D into the second and fourth equations, and simplify:

$$\begin{cases} C=-A \\ A+B=\dfrac{1}{2} \\ D=-\dfrac{1}{2} \\ -A+B=\dfrac{3}{2} \end{cases} \Rightarrow B=1, A=-\dfrac{1}{2}, C=\dfrac{1}{2}$$

$$\frac{x+1}{(x^2+1)(x-1)^2} = -\frac{1}{2(x-1)} + \frac{1}{(x-1)^2} + \frac{x-1}{2(x^2+1)}$$

63. $\dfrac{x^3}{(x+1)^2(x+2)^2} = \dfrac{A}{x+1} + \dfrac{B}{(x+1)^2} + \dfrac{C}{x+2} + \dfrac{D}{(x+2)^2} \Rightarrow$

$x^3 = A(x+1)(x+2)^2 + B(x+2)^2 + C(x+2)(x+1)^2 + D(x+1)^2$

Letting $x = -2$, we have $(-2)^3 = D(-1)^2 \Rightarrow D = -8$. Letting $x = -1$, we have $(-1)^3 = B(1)^2 \Rightarrow B = -1$.

Expand the equation and substitute the values for B and D:

$x^3 = (A+C)x^3 + (-9+5A+4C)x^2 + (-20+8A+5C)x + (-12+4A+2C) \Rightarrow$

$\begin{cases} A+\ C = 1 \\ 5A+4C = 9 \\ 8A+5C = 20 \\ 4A+2C = 12 \end{cases} \Rightarrow A = 5, C = -4$

$\dfrac{x^3}{(x+1)^2(x+2)^2} = \dfrac{5}{x+1} - \dfrac{1}{(x+1)^2} - \dfrac{4}{x+2} - \dfrac{8}{(x+2)^2}$

65. $\dfrac{15x}{x^3-27} = \dfrac{15x}{(x-3)(x^2+3x+9)} = \dfrac{A}{x-3} + \dfrac{Bx+C}{(x^2+3x+9)} \Rightarrow 15x = A(x^2+3x+9) + (Bx+C)(x-3)$

Letting $x = 3$, we have $15(3) = A(3^2+3(3)+9) \Rightarrow A = 5/3$.

Expand the equation and substitute the value for A:

$15x = \left(\dfrac{5}{3}+B\right)x^2 + (5-3B+C)x + (15-3C) \Rightarrow \begin{cases} B\ = -5/3 \\ -3B+C = 10 \\ -3C = -15 \end{cases} \Rightarrow B = -\dfrac{5}{3}, C = 5$

$\dfrac{15x}{x^3-27} = \dfrac{5}{3(x-3)} + \dfrac{-5x+15}{3(x^2+3x+9)}$

67. Using the hint provided, we have

$\dfrac{4x}{x^4+4} = \dfrac{4x}{(x^2-2x+2)(x^2+2x+2)} = \dfrac{Ax+B}{x^2-2x+2} + \dfrac{Cx+D}{x^2+2x+2} \Rightarrow$

$4x = (Ax+B)(x^2+2x+2) + (Cx+D)(x^2-2x+2) \Rightarrow$

$4x = (A+C)x^3 + (2A+B-2C+D)x^2 + (2A+2B+2C-2D)x + (2B+2D) \Rightarrow$

$\begin{cases} A+\ \ \ \ \ \ C\ \ \ \ \ \ = 0 \\ 2A+\ B-2C+\ D = 0 \\ 2A+2B+2C-2D = 4 \\ \ \ \ \ 2B+\ \ \ \ \ \ 2D = 0 \end{cases} \Rightarrow A = 0, B = 1, C = 0, D = -1$

$\dfrac{4x}{x^4+4} = \dfrac{1}{x^2-2x+2} - \dfrac{1}{x^2+2x+2}$

69. $x^2+3x+2 \overline{)x^2+4x+5}$ with quotient 1

$\underline{x^2+3x+2}$

$x+3$

$\dfrac{x^2+4x+5}{x^2+3x+2} = 1 + \dfrac{x+3}{x^2+3x+2}$

Decompose $\dfrac{x+3}{x^2+3x+2}$:

$\dfrac{x+3}{x^2+3x+2} = \dfrac{x+3}{(x+1)(x+2)} = \dfrac{A}{x+1} + \dfrac{B}{x+2} \Rightarrow x+3 = A(x+2) + B(x+1) \Rightarrow$

$x+3 = (A+B)x + (2A+B) \Rightarrow \begin{cases} A+B = 1 \\ 2A+B = 3 \end{cases} \Rightarrow A = 2, B = -1$

$\dfrac{x^2+4x+5}{x^2+3x+2} = 1 + \dfrac{2}{x+1} - \dfrac{1}{x+2}$

71. $\dfrac{2x^4+x^3+2x^2-2x-1}{(x-1)(x^2+1)}=\dfrac{2x^4+x^3+2x^2-2x-1}{x^3-x^2+x-1}$

$$x^3-x^2+x-1\,\overline{\big)\,2x^4+x^3+2x^2-2x-1}$$

$$
\begin{array}{r}
2x+3\\
\underline{2x^4-2x^3+2x^2-2x}\\
3x^3-1\\
\underline{3x^3-3x^2+3x-3}\\
3x^2-3x+2
\end{array}
$$

$\dfrac{2x^4+x^3+2x^2-2x-1}{(x-1)(x^2+1)}=2x+3+\dfrac{3x^2-3x+2}{(x-1)(x^2+1)}$

Decompose $\dfrac{3x^2-3x+2}{(x-1)(x^2+1)}$:

$\dfrac{3x^2-3x+2}{(x-1)(x^2+1)}=\dfrac{A}{x-1}+\dfrac{Bx+C}{x^2+1}\Rightarrow 3x^2-3x+2=A(x^2+1)+(Bx+C)(x-1)\Rightarrow$

$3x^2-3x+2=(A+B)x^2+(-B+C)x+(A-C)\Rightarrow \begin{cases}A+B=3\\ -B+C=-3\\ A-C=2\end{cases}\Rightarrow A=1, B=2, C=-1$

$\dfrac{3x^2-3x+2}{(x-1)(x^2+1)}=\dfrac{1}{x-1}+\dfrac{2x-1}{x^2+1}$

Thus, $\dfrac{2x^4+x^3+2x^2-2x-1}{(x-1)(x^2+1)}=2x+3+\dfrac{1}{x-1}+\dfrac{2x-1}{x^2+1}$.

5.3 Critical Thinking/Discussion/Writing

73. Two equal polynomials have equal corresponding coefficients. See page 755 in the text.

74. $f(x)=\dfrac{x+8}{x^2+x-2}=\dfrac{x+8}{(x+2)(x-1)}$

$=\dfrac{A}{x+2}+\dfrac{B}{x-1}\Rightarrow$

$x+8=A(x-1)+B(x+2)$

$=(A+B)x+(-A+2B)$

$\begin{cases}A+B=1\\ -A+2B=8\end{cases}\Rightarrow 3B=9\Rightarrow B=3,\ A=-2$

$\dfrac{x+8}{x^2+x-2}=-\dfrac{2}{x+2}+\dfrac{3}{x-1}$

$g(x)=-\dfrac{2}{x+2},\ \ h(x)=\dfrac{3}{x-1}$

a.

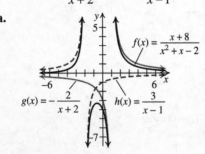

b. The graphs of f and g have the same vertical asymptote, $x=-2$, and the same horizontal asymptote, $y=0$, or the x-axis. Also g is an asymptote of f as x approaches 1 from the right or from the left.

c. The graphs of f and h have the same vertical asymptote, $x=1$, and the same horizontal asymptote, $y=0$, or the x-axis. Also h is an asymptote of f as x approaches -2 from the right or from the left.

5.3 Maintaining Skills

75. $x^2+10x+21=0\Rightarrow(x+7)(x+3)=0\Rightarrow$
$x=-7,\ x=-3$

77. $2x^2+x-10=0\Rightarrow(2x+5)(x-2)=0\Rightarrow$
$x=-\dfrac{5}{2},\ x=2$

In exercises 79–86, be sure to check the solution in both equations.

79. $\begin{cases}x+y=3 & (1)\\ 3x+2y=7 & (2)\end{cases}$

From equation (1), we have $y=3-x$.
Substituting in equation (2), we have
$3x+2(3-x)=7\Rightarrow x+6=7\Rightarrow x=1$

Substituting $x=1$ in equation (1) gives
$1+y=3\Rightarrow y=2$.

Solution set: $\{(1,2)\}$

81. $\begin{cases} x - 2y = 1 & (1) \\ 2x + 3y = 16 & (2) \end{cases}$

From equation (1), we have $x = 2y + 1$.
Substituting in equation (2), we have
$2(2y + 1) + 3y = 16 \Rightarrow 7y + 2 = 16 \Rightarrow$
$7y = 14 \Rightarrow y = 2$
Substituting $y = 2$ in equation (1) gives
$x - 2(2) = 1 \Rightarrow x - 4 = 1 \Rightarrow x = 5$
Solution set: $\{(5, 2)\}$

83. $\begin{cases} 2x - y = 4 & (1) \\ 3x + 2y = 13 & (2) \end{cases}$

Multiply equation (1) by 2, then add the two equations and solve for x.
$4x - 2y = 8$
$\underline{3x + 2y = 13}$
$7x = 21 \Rightarrow x = 3$
Substitute $x = 3$ in equation (1), then solve for y.
$2(3) - y = 4 \Rightarrow 6 - y = 4 \Rightarrow y = 2$
Solution set: $\{(3, 2)\}$

85. $\begin{cases} 2x + 5y = 1 & (1) \\ 3x - 2y = -8 & (2) \end{cases}$

Multiply equation (1) by 2 and equation (2) by 5 to eliminate y. Add the resulting equations and solve for x.
$4x + 10y = 2$
$\underline{15x - 10y = -40}$
$19x = -38 \Rightarrow x = -2$
Substitute $x = -2$ in equation (1), then solve for y.
$2(-2) + 5y = 1 \Rightarrow -4 + 5y = 1 \Rightarrow 5y = 5 \Rightarrow$
$y = 1$
Solution set: $\{(-2, 1)\}$

5.4 Systems of Nonlinear Equations

5.4 Practice Problems

1. $\begin{cases} x^2 + y = 2 \\ 2x + y = 3 \end{cases} \Rightarrow \begin{cases} y = 2 - x^2 \\ 2x + y = 3 \end{cases} \Rightarrow$
$2x + (2 - x^2) = 3 \Rightarrow -x^2 + 2x - 1 = 0 \Rightarrow$
$x^2 - 2x + 1 = 0 \Rightarrow (x - 1)^2 = 0 \Rightarrow x = 1$
$2(1) + y = 3 \Rightarrow y = 1$
The solution is $\{(1, 1)\}$.

2. $\begin{cases} x^2 + 2y^2 = 34 \\ x^2 - y^2 = 7 \end{cases} \Rightarrow 3y^2 = 27 \Rightarrow y^2 = 9 \Rightarrow$
$y = \pm 3$

$x^2 - (-3)^2 = 7 \Rightarrow x^2 = 16 \Rightarrow x = \pm 4$
$x^2 - (3)^2 = 7 \Rightarrow x^2 = 16 \Rightarrow x = \pm 4$
The solution is $\{(-4, -3), (-4, 3), (4, -3), (4, 3)\}$.

3. Let x = the number of shares received as dividends, and let p = the selling price per share. Then $xp = 1950$. The number of shares she sold at p dollars per share is $240 + x$. Thus, revenue = $(240 + x)p$ and the cost of her stock was $(240)(40) + 100 = 9700$.
Since revenue − cost = profit, we have
$(240 + x)p - 9700 = 7850 \Rightarrow$
$240p + xp = 17,550$
Thus, the system of equations is
$\begin{cases} xp = 1950 \\ 240p + xp = 17,550 \end{cases}.$

$\begin{cases} xp = 1950 \\ 240p + xp = 17,550 \end{cases} \Rightarrow$
$240p + 1950 = 17,550 \Rightarrow 240p = 15,600 \Rightarrow$
$p = 65$
$65x = 1950 \Rightarrow x = 30$
Danielle received 30 shares as dividends and sold her stock at $65 per share.

5.4 Basic Concepts and Skills

1. In a system of nonlinear equations, <u>at least one</u> equation must be nonlinear.

3. The solutions of the system
$\begin{cases} ax + by = c & (1) \\ (x - h)^2 + (y - k)^2 = r^2 & (2) \end{cases}$
represent the points of <u>intersection</u> of the graphs of equations (1) and (2).

5. True

7. Substituting each ordered pair into the system
$\begin{cases} 2x + 3y = 3 \\ x - y^2 = 2 \end{cases}$
we find that $(3, -1)$ is a solution.
$\begin{cases} 2(3) + 3(-1) = 6 - 3 = 3 \\ 3 - (-1)^2 = 3 - 1 = 2 \end{cases}$

9. Substituting each ordered pair into the system
$\begin{cases} 5x - 2y = 7 \\ x^2 + y^2 = 2 \end{cases},$
we find that $(1, -1)$ is a solution:
$\begin{cases} 5(1) - 2(-1) = 5 + 2 = 7 \\ (1)^2 + (-1)^2 = 1 + 1 = 2 \end{cases}$

11. Substituting each ordered pair into the system
$$\begin{cases} 4x^2 + 5y^2 = 180 \\ x^2 - y^2 = 9 \end{cases}$$
we find that (5, 4), (−5, 4), and (−5, −4) are solutions.
$$\begin{cases} 4(5)^2 + 5(4)^2 = 100 + 80 = 180 \\ 5^2 - 4^2 = 25 - 16 = 9 \end{cases}$$
$$\begin{cases} 4(-5)^2 + 5(4)^2 = 100 + 80 = 180 \\ (-5)^2 - (4)^2 = 25 - 16 = 9 \end{cases}$$
$$\begin{cases} 4(-5)^2 + 5(-4)^2 = 100 + 80 = 180 \\ (-5)^2 - (-4)^2 = 25 - 16 = 9 \end{cases}$$

13. Substituting each ordered pair into the system
$$\begin{cases} y = e^{x-1} \\ y = 2x - 1 \end{cases}$$
we find that (1, 1) is a solution:
$$\begin{cases} y = e^{1-1} = e^0 = 1 \\ y = 2(1) - 1 = 1 \end{cases}$$

15. $\begin{cases} y = x^2 \\ y = x + 2 \end{cases} \Rightarrow x^2 = x + 2 \Rightarrow x^2 - x - 2 = 0 \Rightarrow$
$(x - 2)(x + 1) = 0 \Rightarrow x = 2 \text{ or } x = -1$
$y = 2^2 = 4 \text{ or } y = (-1)^2 = 1$
The solution is {(2, 4), (−1, 1)}.

17. $\begin{cases} x^2 - y = 6 \\ x - y = 0 \end{cases} \Rightarrow \begin{cases} x^2 - y = 6 \\ x = y \end{cases} \Rightarrow x^2 - x = 6 \Rightarrow$
$x^2 - x - 6 = 0 \Rightarrow (x - 3)(x + 2) = 0 \Rightarrow$
$x = 3 \text{ or } x = -2$
$3 - y = 0 \Rightarrow y = 3 \text{ or } -2 - y = 0 \Rightarrow y = -2$
The solution is {(3, 3), (−2, −2)}.

19. $\begin{cases} x^2 + y^2 = 9 \\ x = 3 \end{cases} \Rightarrow 3^2 + y^2 = 9 \Rightarrow y = 0$
The solution is {(3, 0)}.

21. $\begin{cases} x^2 + y^2 = 5 \\ x - y = -3 \end{cases} \Rightarrow \begin{cases} x^2 + y^2 = 5 \\ x = y - 3 \end{cases} \Rightarrow$
$(y - 3)^2 + y^2 = 5 \Rightarrow 2y^2 - 6y + 4 = 0 \Rightarrow$
$y^2 - 3y + 2 = 0 \Rightarrow (y - 2)(y - 1) = 0 \Rightarrow$
$y = 2 \text{ or } y = 1$
$x - 2 = -3 \Rightarrow x = -1$
$x - 1 = -3 \Rightarrow x = -2$
The solution is {(−2, 1), (−1, 2)}.

23. $\begin{cases} x^2 - 4x + y^2 = -2 \\ x - y = 2 \end{cases} \Rightarrow \begin{cases} x^2 - 4x + y^2 = -2 \\ x = y + 2 \end{cases} \Rightarrow$
$(y + 2)^2 - 4(y + 2) + y^2 = -2 \Rightarrow 2y^2 = 2 \Rightarrow$
$y = \pm 1$
$x - (-1) = 2 \Rightarrow x = 1$
$x - 1 = 2 \Rightarrow x = 3$
The solution is {(1, −1), (3, 1)}.

25. $\begin{cases} x - y = -2 \\ xy = 3 \end{cases} \Rightarrow \begin{cases} x = y - 2 \\ xy = 3 \end{cases} \Rightarrow y(y - 2) = 3 \Rightarrow$
$y^2 - 2y - 3 = 0 \Rightarrow (y - 3)(y + 1) = 0 \Rightarrow$
$y = 3 \text{ or } y = -1$
$x - 3 = -2 \Rightarrow x = 1$
$x - (-1) = -2 \Rightarrow x = -3$
The solution is {(1, 3), (−3, −1)}.

27. $\begin{cases} 4x^2 + y^2 = 25 \\ x + y = 5 \end{cases} \Rightarrow \begin{cases} 4x^2 + y^2 = 25 \\ x = 5 - y \end{cases} \Rightarrow$
$4(5 - y)^2 + y^2 = 25 \Rightarrow 5y^2 - 40y + 75 = 0 \Rightarrow$
$y^2 - 8y + 15 = 0 \Rightarrow (y - 5)(y - 3) = 0 \Rightarrow$
$y = 5 \text{ or } y = 3$
$x + 5 = 5 \Rightarrow x = 0$
$x + 3 = 5 \Rightarrow x = 2$
The solution is {(0, 5), (2, 3)}.

29. $\begin{cases} x^2 - y^2 = 24 \\ 5x - 7y = 0 \end{cases} \Rightarrow \begin{cases} x^2 - y^2 = 24 \\ x = \dfrac{7}{5}y \end{cases} \Rightarrow$
$\left(\dfrac{7}{5}y\right)^2 - y^2 = 24 \Rightarrow \dfrac{24}{25}y^2 = 24 \Rightarrow y^2 = 25 \Rightarrow$
$y = \pm 5$
$5x - 7(-5) = 0 \Rightarrow x = -7$
$5x - 7(5) = 0 \Rightarrow x = 7$
The solution is {(7, 5), (−7, −5)}.

31. $\begin{cases} x^2 + y^2 = 20 \\ x^2 - y^2 = 12 \end{cases} \Rightarrow 2x^2 = 32 \Rightarrow x = \pm 4$
$(-4)^2 + y^2 = 20 \Rightarrow y^2 = 4 \Rightarrow y = \pm 2$
$(4)^2 + y^2 = 20 \Rightarrow y^2 = 4 \Rightarrow y = \pm 2$
The solution is {(−4, −2), (−4, 2), (4, −2), (4, 2)}.

33. $\begin{cases} x^2 + 2y^2 = 12 \\ 7y^2 - 5x^2 = 8 \end{cases} \Rightarrow \begin{cases} 5x^2 + 10y^2 = 60 \\ 7y^2 - 5x^2 = 8 \end{cases} \Rightarrow$
$17y^2 = 68 \Rightarrow y^2 = 4 \Rightarrow y = \pm 2$
$x^2 + 2(-2)^2 = 12 \Rightarrow x^2 = 4 \Rightarrow x = \pm 2$
$x^2 + 2(2)^2 = 12 \Rightarrow x^2 = 4 \Rightarrow x = \pm 2$
The solution is {(−2, −2), (−2, 2), (2, −2), (2, 2)}.

35. $\begin{cases} x^2 - y = 2 \\ 2x - y = 4 \end{cases} \Rightarrow \begin{cases} x^2 - y = 2 \\ -2x + y = -4 \end{cases} \Rightarrow$

$x^2 - 2x = -2 \Rightarrow x^2 - 2x + 2 = 0 \Rightarrow$

$x = \dfrac{2 \pm \sqrt{4-8}}{2} = 1 \pm i \Rightarrow$ there are no real

solutions. Solution set: $\varnothing$

37. $\begin{cases} x^2 + y^2 = 5 \\ 3x^2 - 2y^2 = -5 \end{cases} \Rightarrow \begin{cases} 2x^2 + 2y^2 = 10 \\ 3x^2 - 2y^2 = -5 \end{cases} \Rightarrow$

$5x^2 = 5 \Rightarrow x = \pm 1$

$(1)^2 + y^2 = 5 \Rightarrow y = \pm 2$

$(-1)^2 + y^2 = 5 \Rightarrow y = \pm 2$

The solution is $\{(-1, -2), (-1, 2), (1, -2), (1, 2)\}$.

39. $\begin{cases} x^2 + y^2 + 2x = 9 \\ x^2 + 4y^2 + 3x = 14 \end{cases} \Rightarrow$

$\begin{cases} 4x^2 + 4y^2 + 8x = 36 \\ x^2 + 4y^2 + 3x = 14 \end{cases} \Rightarrow 3x^2 + 5x - 22 = 0 \Rightarrow$

$(x - 2)(3x + 11) = 0 \Rightarrow x = 2$ or $x = -\dfrac{11}{3}$

$(2)^2 + y^2 + 2(2) = 9 \Rightarrow y = \pm 1$

$\left(-\dfrac{11}{3}\right)^2 + y^2 + 2\left(-\dfrac{11}{3}\right) = 9 \Rightarrow y^2 + \dfrac{55}{9} = 9 \Rightarrow$

$y = \pm \dfrac{\sqrt{26}}{3}$

The solution is $\left\{ (2, -1), (2, 1), \left(-\dfrac{11}{3}, -\dfrac{\sqrt{26}}{3}\right), \right.$

$\left. \left(-\dfrac{11}{3}, \dfrac{\sqrt{26}}{3}\right) \right\}$.

41. Using substitution, we have

$\begin{cases} x + y = 8 \\ xy = 15 \end{cases} \Rightarrow \begin{cases} x = 8 - y \\ xy = 15 \end{cases} \Rightarrow$

$y(8 - y) = 15 \Rightarrow -y^2 + 8y - 15 = 0 \Rightarrow$

$y^2 - 8y + 15 = 0 \Rightarrow (y - 3)(y - 5) = 0 \Rightarrow$

$y = 3$ or $y = 5$

$x + 3 = 8 \Rightarrow x = 5$

$x + 5 = 8 \Rightarrow x = 3$

The solution is $\{(3, 5), (5, 3)\}$.

43. Using elimination, we have

$\begin{cases} x^2 + y^2 = 2 \\ 3x^2 + 3y^2 = 9 \end{cases} \Rightarrow \begin{cases} -3x^2 - 3y^2 = -6 \\ 3x^2 + 3y^2 = 9 \end{cases} \Rightarrow$

$0 = 3 \Rightarrow$ there is no solution.

Solution set: $\varnothing$

45. Using substitution, we have

$\begin{cases} y^2 = 4x + 4 \\ y = 2x - 2 \end{cases} \Rightarrow (2x - 2)^2 = 4x + 4 \Rightarrow$

$4x^2 - 12x = 0 \Rightarrow 4x(x - 3) = 0 \Rightarrow x = 0$ or $x = 3$

$y = 2(0) - 2 = -2$

$y = 2(3) - 2 = 4$

The solution is $\{(0, -2), (3, 4)\}$.

47. Using substitution, we have

$\begin{cases} x^2 + 4y^2 = 25 \\ x - 2y + 1 = 0 \end{cases} \Rightarrow \begin{cases} x^2 + 4y^2 = 25 \\ x = 2y - 1 \end{cases} \Rightarrow$

$(2y - 1)^2 + 4y^2 = 25 \Rightarrow 8y^2 - 4y - 24 = 0 \Rightarrow$

$4(y - 2)(2y + 3) = 0 \Rightarrow y = 2$ or $y = -\dfrac{3}{2}$

$x - 2(2) + 1 = 0 \Rightarrow x = 3$

$x - 2\left(-\dfrac{3}{2}\right) + 1 = 0 \Rightarrow x = -4$

The solution is $\left\{ (3, 2), \left(-4, -\dfrac{3}{2}\right) \right\}$.

49. Using elimination, we have

$\begin{cases} x^2 - 3y^2 = 1 \\ x^2 + 4y^2 = 8 \end{cases} \Rightarrow -7y^2 = -7 \Rightarrow y = \pm 1$

$x^2 - 3(-1)^2 = 1 \Rightarrow x = \pm 2$

$x^2 - 3(1)^2 = 1 \Rightarrow x = \pm 2$

The solution is $\{(-2, -1), (-2, 1), (2, -1), (2, 1)\}$.

51. Using elimination, we have

$\begin{cases} x^2 - xy + 5x = 4 \\ 2x^2 - 3xy + 10x = -2 \end{cases} \Rightarrow$

$\begin{cases} -3x^2 + 3xy - 15x = -12 \\ 2x^2 - 3xy + 10x = -2 \end{cases} \Rightarrow -x^2 - 5x = -14 \Rightarrow$

$x^2 + 5x - 14 = 0 \Rightarrow (x + 7)(x - 2) = 0 \Rightarrow$

$x = -7$ or $x = 2$

$(-7)^2 - (-7)y + 5(-7) = 4 \Rightarrow y = -\dfrac{10}{7}$

$(2)^2 - 2y + 5(2) = 4 \Rightarrow y = 5$

The solution is $\left\{ \left(-7, -\dfrac{10}{7}\right), (2, 5) \right\}$.

53. Using elimination, we have
$$\begin{cases} x^2 + y^2 - 8x = -8 \\ x^2 - 4y^2 + 6x = 0 \end{cases} \Rightarrow$$
$$\begin{cases} 4x^2 + 4y^2 - 32x = -32 \\ x^2 - 4y^2 + 6x = 0 \end{cases} \Rightarrow 5x^2 - 26x = -32 \Rightarrow$$
$$5x^2 - 26x + 32 = 0 \Rightarrow (x-2)(5x-16) = 0 \Rightarrow$$
$$x = 2 \text{ or } x = \frac{16}{5}$$
$$2^2 + y^2 - 8(2) = -8 \Rightarrow y = \pm 2$$
$$\left(\frac{16}{5}\right)^2 + y^2 - 8\left(\frac{16}{5}\right) = -8 \Rightarrow y^2 = \frac{184}{25} \Rightarrow$$
$$y = \pm \frac{2\sqrt{46}}{5}$$
The solution is
$$\left\{ (2,-2), (2,2), \left(\frac{16}{5}, -\frac{2\sqrt{46}}{5}\right), \left(\frac{16}{5}, \frac{2\sqrt{46}}{5}\right) \right\}.$$

5.4 Applying the Concepts

55. Using elimination, we have
$$\begin{cases} p + 2x^2 = 96 \\ p - 13x = 39 \end{cases} \Rightarrow 2x^2 + 13x = 57 \Rightarrow$$
$$2x^2 + 13x - 57 = 0 \Rightarrow (x-3)(2x+19) = 0 \Rightarrow$$
$$x = 3 \text{ or } x = -\frac{19}{2} \text{ (reject this)}$$
$$p - 13(3) = 39 \Rightarrow p = 78$$
Market equilibrium occurs when 3 (hundred) units are sold and the price is $78/unit.

57. Let x = the first positive number and let y = the second positive number. Then
$$\begin{cases} x + y = 24 \\ xy = 143 \end{cases} \Rightarrow \begin{cases} y = 24 - x \\ xy = 143 \end{cases} \Rightarrow$$
$$x(24 - x) = 143 \Rightarrow -x^2 + 24x - 143 = 0 \Rightarrow$$
$$-(x-13)(x-11) = 0 \Rightarrow x = 11 \text{ or } x = 13$$
$$11 + y = 24 \Rightarrow y = 13$$
$$13 + y = 24 \Rightarrow y = 11$$
The numbers are 11 and 13.

59. a. Let x = the length of the two equal sides and let y = the length of the third side.

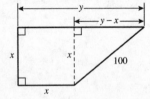

Using the Pythagorean theorem, we have $x^2 + (y-x)^2 = 100^2$. So,
$$\begin{cases} 2x + y + 100 = 360 \\ x^2 + (y-x)^2 = 100^2 \end{cases} \Rightarrow$$
$$\begin{cases} y = 260 - 2x \\ x^2 + (y-x)^2 = 100^2 \end{cases} \Rightarrow$$
$$x^2 + ((260 - 2x) - x)^2 = 10,000 \Rightarrow$$
$$10x^2 - 1560x + 57,600 = 0 \Rightarrow$$
$$10(x-60)(x-96) = 0 \Rightarrow x = 60 \text{ or } x = 96$$
$$2(60) + y + 100 = 360 \Rightarrow y = 140$$
$$2(96) + y + 100 = 360 \Rightarrow y = 68$$
Because $y > x$ (from the diagram), we reject $x = 96$ and $y = 68$.
So, $x = 60$ m and $y = 140$ m

b. The area is $60^2 + \frac{1}{2}(60)(80) = 6000$ m^2.

61. Let x = the original number of students in the group and let y = the original cost per student. Then
$$\begin{cases} xy = 960 \\ (x+8)(y-6) = 960 \end{cases} \Rightarrow$$
$$\begin{cases} y = \frac{960}{x} \\ (x+8)(y-6) = 960 \end{cases} \Rightarrow$$
$$(x+8)\left(\frac{960}{x} - 6\right) = 960 \Rightarrow$$
$$-6x + \frac{7680}{x} + 912 = 960 \Rightarrow$$
$$-6x + \frac{7680}{x} - 48 = 0 \Rightarrow$$
$$-6x^2 - 48x + 7680 = 0 \Rightarrow$$
$$-6(x-32)(x+40) = 0 \Rightarrow x = 32 \text{ or }$$
$$x = -40 \text{ (reject this)}$$
$$32y = 960 \Rightarrow y = 30$$
There were originally 32 students at a cost of $30 each.

63. Let x = the number of shares of stock she bought and let y = the original price per share. Then
$$\begin{cases} xy + 100 = 10,000 \\ (x+30)(y+3) = 11,900 + 100 \end{cases} \Rightarrow$$
$$\begin{cases} y = \frac{9900}{x} \\ (x+30)(y+3) = 12,000 \end{cases} \Rightarrow$$
$$(x+30)\left(\frac{9900}{x} + 3\right) = 12,000 \Rightarrow$$

(continued on next page)

(continued)

$$3x + \frac{297,000}{x} - 2010 = 0 \Rightarrow$$

$$3x^2 - 2010x + 297,000 = 0 \Rightarrow$$

$$3(x - 450)(x - 220) = 0 \Rightarrow x = 450 \text{ or } x = 220$$

The problem says that she bought more than 400 shares, so reject $x = 220$.

$$450y + 100 = 10,000 \Rightarrow y = 22$$

She bought 450 shares at $22 per share.

5.4 Beyond the Basics

65. $\begin{cases} x^2 + y^2 - 7x + 5y + 6 = 0 \\ y = 0 \end{cases} \Rightarrow$

$$x^2 - 7x + 6 = 0 \Rightarrow (x - 6)(x - 1) = 0 \Rightarrow$$
$$x = 6 \text{ or } x = 1$$

The circle intersects the x-axis at $A(1, 0)$ and $B(6, 0)$. $d(A, B) = \sqrt{(6-1)^2 + (0-0)^2} = 5$.

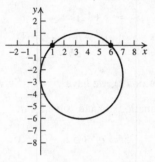

67. $\begin{cases} x^2 + y^2 + 2x - 4y - 5 = 0 \\ x - y + 1 = 0 \end{cases} \Rightarrow$

$$\begin{cases} x^2 + y^2 + 2x - 4y - 5 = 0 \\ x = y - 1 \end{cases} \Rightarrow$$

$$(y - 1)^2 + y^2 + 2(y - 1) - 4y - 5 = 0 \Rightarrow$$
$$2y^2 - 4y - 6 = 0 \Rightarrow 2(y - 3)(y + 1) = 0 \Rightarrow$$
$$y = 3 \text{ or } y = -1$$
$$x - 3 + 1 = 0 \Rightarrow x = 2$$
$$x - (-1) + 1 \Rightarrow x = -2$$

The circle and the line intersect at $A(-2, -1)$ and $B(2, 3)$.

$$d(A, B) = \sqrt{(-2 - 2)^2 + (-1 - 3)^2} =$$
$$\sqrt{16 + 16} = \sqrt{32} = 4\sqrt{2}.$$

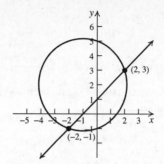

69. $\begin{cases} x^2 + y^2 - 2x + 2y - 3 = 0 \\ x + 2y + 6 = 0 \end{cases} \Rightarrow$

$$\begin{cases} x^2 + y^2 - 2x + 2y - 3 = 0 \\ x = -2y - 6 \end{cases} \Rightarrow$$

$$(-2y - 6)^2 + y^2 - 2(-2y - 6) + 2y - 3 = 0 \Rightarrow$$
$$5y^2 + 30y + 45 = 0 \Rightarrow 5(y + 3)^2 = 0 \Rightarrow y = -3$$
$$x + 2(-3) + 6 = 0 \Rightarrow x = 0$$

The line and the circle intersect at only one point $(0, -3)$, so the line is tangent to the circle.

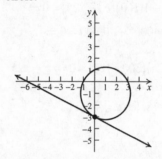

71. $\begin{cases} x + y = 2 & (1) \\ x^2 + y^2 = c^2 & (2) \end{cases}$

From equation (1), we have $y = 2 - x$. Substituting in equation (2) and then solving for x, we have

$$x^2 + (2 - x)^2 = c^2 \Rightarrow x^2 + 4 - 4x + x^2 = c^2 \Rightarrow$$
$$2x^2 - 4x + 4 - c^2 = 0 \Rightarrow$$

$$x = \frac{-(-4) \pm \sqrt{(-4)^2 - 4(2)(4 - c^2)}}{2(2)}$$

Since the equations are tangent, there is only one solution. Recall that the value of the discriminant is 0 when there is exactly one real solution, so solve

$$(-4)^2 - 4(2)(4 - c^2) = 0 \Rightarrow$$
$$16 - 32 + 8c^2 = 0 \Rightarrow c^2 = 2 \Rightarrow c = \pm\sqrt{2}$$

73. $\begin{cases} 4x^2 + y^2 = 25 & (1) \\ 8x + 3y = c & (2) \end{cases}$

From equation (2), we have $y = \dfrac{c - 8x}{3}$.

Substituting in equation (1) and then solving for x, we have

$4x^2 + \left(\dfrac{c - 8x}{3}\right)^2 = 25 \Rightarrow$

$4x^2 + \dfrac{c^2 - 16cx + 64x^2}{9} = 25 \Rightarrow$

$36x^2 + c^2 - 16cx + 64x^2 = 225 \Rightarrow$

$100x^2 - 16cx + c^2 - 225 = 0$

$x = \dfrac{-(-16c) \pm \sqrt{(-16c)^2 - 4(100)(c^2 - 225)}}{2(100)}$

Since the equations are tangent, there is only one solution. Recall that the value of the discriminant is 0 when there is exactly one real solution, so solve

$(-16c)^2 - 4(100)(c^2 - 225) = 0 \Rightarrow$

$256c^2 - 400c^2 + 90,000 = 0 \Rightarrow$

$90,000 - 144c^2 = 0 \Rightarrow$

$(300 - 12c)(300 + 12c) = 0 \Rightarrow c = \pm 25$

In exercises 75–78, let $u = \dfrac{1}{x}$ and $v = \dfrac{1}{y}$.

75. $\begin{cases} \dfrac{1}{x} - \dfrac{1}{y} = 5 \\ \dfrac{1}{xy} - 24 = 0 \end{cases} \Rightarrow \begin{cases} u - v = 5 & (1) \\ uv - 24 = 0 & (2) \end{cases}$

From equation (1), we have $u = v + 5$. Substitute this into equation (2) and solve for v.

$v(v + 5) - 24 = 0 \Rightarrow v^2 + 5v - 24 = 0 \Rightarrow$

$(v + 8)(v - 3) = 0 \Rightarrow v = -8, 3$

$v = -8 \Rightarrow u + 8 = 5 \Rightarrow u = -3 \Rightarrow$

$x = -\dfrac{1}{3}, \ y = -\dfrac{1}{8}$

$v = 3 \Rightarrow u - 3 = 5 \Rightarrow u = 8 \Rightarrow x = \dfrac{1}{8}, \ y = \dfrac{1}{3}$

Solution set: $\left\{ \left(-\dfrac{1}{3}, -\dfrac{1}{8}\right), \left(\dfrac{1}{8}, \dfrac{1}{3}\right) \right\}$

77. $\begin{cases} \dfrac{5}{x^2} - \dfrac{3}{y^2} = 2 \\ \dfrac{6}{x^2} + \dfrac{1}{y^2} = 7 \end{cases} \Rightarrow \begin{cases} 5u^2 - 3v^2 = 2 & (1) \\ 6u^2 + v^2 = 7 & (2) \end{cases}$

Solve equation (2) for v^2, then substitute this expression in equation (1) and solve for u.

$v^2 = 7 - 6u^2$

$5u^2 - 3\left(7 - 6u^2\right) = 2 \Rightarrow 23u^2 - 21 = 2 \Rightarrow$

$u^2 = 1 \Rightarrow u = \pm 1 \Rightarrow x = \pm 1$

If $u = 1$, then

$6\left(1^2\right) + v^2 = 7 \Rightarrow v^2 = 1 \Rightarrow v = \pm 1 \Rightarrow y = \pm 1$

If $u = -1$, then

$6\left(-1^2\right) + v^2 = 7 \Rightarrow v^2 = 1 \Rightarrow v = \pm 1 \Rightarrow y = \pm 1$

Solution set: $\{(-1, -1), (-1, 1), (1, -1), (1, 1)\}$

79. $(a + bi)^2 = -5 + 12i \Rightarrow$

$a^2 + 2abi - b^2 = -5 + 12i \Rightarrow$

$\begin{cases} a^2 - b^2 = -5 \\ 2ab = 12 \end{cases} \Rightarrow \begin{cases} a^2 - b^2 = -5 & (1) \\ ab = 6 & (2) \end{cases}$

From equation (2), we have $b = \dfrac{6}{a}$. Substitute this into equation (1) and solve for a.

$a^2 - \left(\dfrac{6}{a}\right)^2 = -5 \Rightarrow a^4 + 5a^2 - 36 = 0 \Rightarrow$

$\left(a^2 + 9\right)\left(a^2 - 4\right) = 0 \Rightarrow$

$\left(a^2 + 9\right)(a - 2)(a + 2) = 0 \Rightarrow a = \pm 2, \ \pm 3i$

Since a is real, reject $\pm 3i$. Using equation (2), if $a = -2$, then $b = -3$. If $a = 2$, then $b = 3$. Thus, the square roots of $w = -5 + 12i$ are $-2 - 3i$ and $2 + 3i$.

81. $(a + bi)^2 = 7 - 24i \Rightarrow$

$a^2 + 2abi - b^2 = 7 - 24i \Rightarrow$

$\begin{cases} a^2 - b^2 = 7 \\ 2ab = -24 \end{cases} \Rightarrow \begin{cases} a^2 - b^2 = 7 & (1) \\ ab = -12 & (2) \end{cases}$

From equation (2), we have $b = -\dfrac{12}{a}$.

Substitute this into equation (1) and solve for a.

$a^2 - \left(-\dfrac{12}{a}\right)^2 = 7 \Rightarrow a^4 - 7a^2 - 144 = 0 \Rightarrow$

$\left(a^2 - 16\right)\left(a^2 + 9\right) = 0 \Rightarrow$

$\left(a^2 + 9\right)(a - 4)(a + 4) = 0 \Rightarrow a = \pm 4, \ \pm 3i$

Since a is real, reject $\pm 3i$.

(*continued on next page*)

(*continued*)

Using equation (2), if $a = -4$, then $b = 3$.
If $a = 4$, then $b = -3$. Thus, the square roots of
$w = 7 - 24i$ are $-4 + 3i$ and $4 - 3i$.

83. $(a + bi)^2 = 13 + 8\sqrt{3}i \Rightarrow$

$a^2 + 2abi - b^2 = 13 + 8\sqrt{3}i \Rightarrow$

$\begin{cases} a^2 - b^2 = 13 \\ 2ab = 8\sqrt{3} \end{cases} \Rightarrow \begin{cases} a^2 - b^2 = 13 & (1) \\ ab = 4\sqrt{3} & (2) \end{cases}$

From equation (2), we have $b = \dfrac{4\sqrt{3}}{a}$.

Substitute this into equation (1) and solve for a.

$a^2 - \left(\dfrac{4\sqrt{3}}{a}\right)^2 = 13 \Rightarrow a^4 - 13a^2 - 48 = 0 \Rightarrow$

$(a^2 - 16)(a^2 + 3) = 0 \Rightarrow$

$(a^2 + 3)(a - 4)(a + 4) = 0 \Rightarrow a = \pm 4, \ \pm\sqrt{3}i$

Since a is real, reject $\pm\sqrt{3}i$. Using equation
(2), if $a = -4$, then $b = -\sqrt{3}$. If $a = 4$, then
$b = \sqrt{3}$. Thus, the square roots of
$w = 13 + 8\sqrt{3}i$ are $-4 - \sqrt{3}i$ and $4 + \sqrt{3}i$.

85. Using substitution, we have

$\begin{cases} y = 3^x + 4 \\ y = 3^{2x} - 2 \end{cases} \Rightarrow 3^{2x} - 2 = 3^x + 4 \Rightarrow$

$3^{2x} - 3^x - 6 = 0$

Let $u = 3^x$. Then $u^2 - u - 6 = 0 \Rightarrow$
$(u - 3)(u + 2) = 0 \Rightarrow u = 3$ or $u = -2$
(Reject the negative solution.)

$3 = 3^x \Rightarrow x = 1$

$y = 3^1 + 4 = 7$

The solution is $\{(1, 7)\}$.

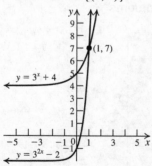

87. Using substitution, we have

$\begin{cases} y = 2^x + 3 \\ y = 2^{2x} + 1 \end{cases} \Rightarrow 2^x + 3 = 2^{2x} + 1 \Rightarrow$

$2^{2x} - 2^x - 2 = 0.$

Let $u = 2^x$. Then $u^2 - u - 2 = 0 \Rightarrow$
$(u - 2)(u + 1) = 0 \Rightarrow u = 2$ or $u = -1$
(reject the negative solution.)

$2 = 2^x \Rightarrow x = 1; \quad y = 2^1 + 3 = 5$

The solution is $\{(1, 5)\}$.

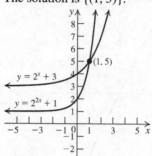

89. $\begin{cases} x = 3^y \\ 3^{2y} = 3x - 2 \end{cases} \Rightarrow 3^{2y} = 3 \cdot 3^y - 2 \Rightarrow$

$3^{2y} - 3 \cdot 3^y + 2 = 0.$ Let $u = 3^y$. Then

$u^2 - 3u + 2 = 0 \Rightarrow (u - 2)(u - 1) = 0 \Rightarrow$
$u = 2$ or $u = 1$

$2 = 3^y \Rightarrow \ln 2 = y \ln 3 \Rightarrow \dfrac{\ln 2}{\ln 3} = y;$

$x = 3^{\ln 2 / \ln 3} = 2$ or $1 = 3^y \Rightarrow y = 0$

$x = 3^0 = 1$

The solution is $\left\{(1, 0), \left(2, \dfrac{\ln 2}{\ln 3}\right)\right\}$.

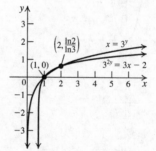

5.4 Critical Thinking/Discussion/Writing

91. a. Not possible

   b. Possible

   c. Possible

   d. Possible

   e. Not possible

   f. Not possible.

92. a. Possible

   b. Possible

   c. Possible

d. Possible

e. Possible

f. Not possible.

5.4 Maintaining Skills

93.

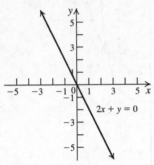

95.

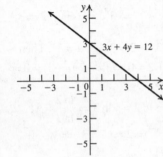

97.

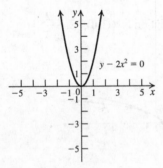

99.

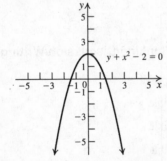

101.

Point	$5x + 3y = 0$?
$(0, 0)$	$5(0)+3(0)\stackrel{?}{=}0$ $0 = 0$ ✓
$(-3, 5)$	$5(-3)+3(5)\stackrel{?}{=}0$ $0 = 0$ ✓
$(1, -1)$	$5(-1)+3(1)\stackrel{?}{=}0$ $-2 = 0$ ✗
$(-1, 1)$	$5(1)+3(-1)\stackrel{?}{=}0$ $2 = 0$ ✗

$(0, 0)$ and $(-3, 5)$ lie on the graph of the equation $5x + 3y = 0$.

103.

Point	$y = x^2 + 3$?
$(0, 0)$	$0 \stackrel{?}{=} 0^2 + 3$ $0 = 3$ ✗
$(1, 4)$	$4 \stackrel{?}{=} 1^2 + 3$ $3 = 3$ ✓
$(-1, 4)$	$4 \stackrel{?}{=} (-1)^2 + 3$ $4 = 4$ ✓
$(2, 5)$	$5 \stackrel{?}{=} 2^2 + 3$ $5 = 7$ ✗

$(1, 4)$ and $(-1, 4)$ lie on the graph of the equation $y = x^2 + 3$.

5.5 Systems of Inequalities

5.5 Practice Problems

1.

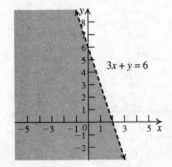

2.

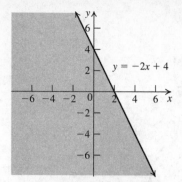

$y = -2x + 4$

3.

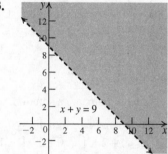

$x + y = 9$

4.

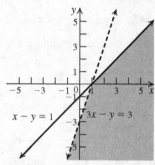

$x - y = 1$ $3x - y = 3$

5. Solve the systems $\begin{cases} 2x + 3y = 16 \\ 4x + 2y = 16 \end{cases}$,

$\begin{cases} 4x + 2y = 16 \\ 2x - \ y = 8 \end{cases}$, and $\begin{cases} 2x + 3y = 16 \\ 2x - \ y = 8 \end{cases}$ to find the

corner points: $(4, 0)$, $(5, 2)$, and $(2, 4)$.

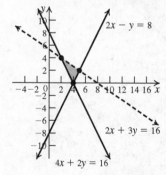

$2x - y = 8$

$2x + 3y = 16$

$4x + 2y = 16$

6.

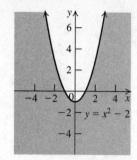

$y = x^2 - 2$

7.

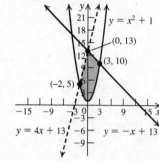

$y = x^2 + 1$

$(0, 13)$

$(3, 10)$

$(-2, 5)$

$y = 4x + 13$ $y = -x + 13$

5.5 Basic Concepts and Skills

1. In the graph of $x - 3y > 1$, the corresponding
equation $x - 3y = 1$ is graphed as a <u>dashed</u>
line.

3. In a system of inequalities containing both
$2x + y > 5$ and $2x - y \le 3$, the point of
intersection of the lines $2x + y = 5$ and
$2x - y = 3$ is <u>not a solution of the system</u>.

5. True

7.

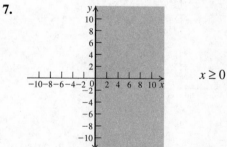

$x \ge 0$

9.

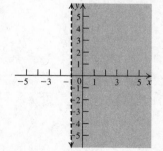

$x > -1$

11.

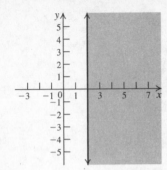

$x \geq 2$

13.

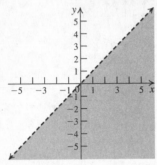

$y - x < 0$

15.

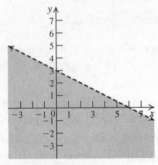

$x + 2y < 6$

17.

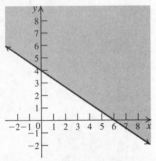

$2x + 3y \geq 12$

19.

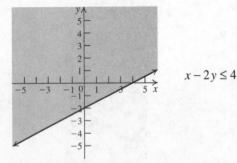

$x - 2y \leq 4$

21.

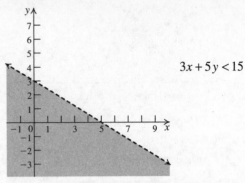

$3x + 5y < 15$

23. Substitute each ordered pair into the system:

$\begin{cases} 0 + 0 < 2 \\ 2(0) + 0 \geq 6 \end{cases}$ $\begin{cases} -4 - 1 < 2 \\ 2(-4) - 1 \geq 6 \end{cases}$

$\begin{cases} 3 + 0 < 2 \\ 2(3) + 0 \geq 6 \end{cases}$ $\begin{cases} 0 + 3 < 2 \\ 2(0) + 3 \geq 6 \end{cases}$

None of the ordered pairs are a solution.

25. Substitute each ordered pair into the system:

$\begin{cases} 0 < 0 + 2 \\ 0 + 0 \leq 4 \end{cases}$ $\begin{cases} 0 < 1 + 2 \\ 1 + 0 \leq 4 \end{cases}$ $\begin{cases} 1 < 0 + 2 \\ 0 + 1 \leq 4 \end{cases}$ $\begin{cases} 1 < 1 + 2 \\ 1 + 1 \leq 4 \end{cases}$

All of the ordered pairs are solutions.

27. Substitute each ordered pair into the system:

$\begin{cases} 3(0) - 4(0) \leq 12 \\ 0 + 0 \leq 4 \\ 5(0) - 2(0) \geq 6 \end{cases}$ $\begin{cases} 3(2) - 4(0) \leq 12 \\ 2 + 0 \leq 4 \\ 5(2) - 2(0) \geq 6 \end{cases}$

$\begin{cases} 3(3) - 4(1) \leq 12 \\ 3 + 1 \leq 4 \\ 5(3) - 2(1) \geq 6 \end{cases}$ $\begin{cases} 3(2) - 4(2) \leq 12 \\ 2 + 2 \leq 4 \\ 5(2) - 2(2) \geq 6 \end{cases}$

The solutions are (2, 0), (3, 1) and (2, 2).

29.

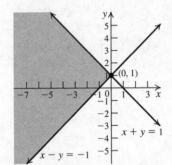

31.

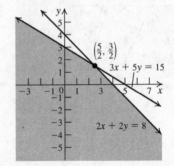

33.

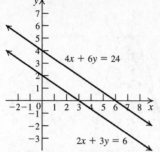

4x + 6y = 24

2x + 3y = 6

The system is inconsistent. There are no vertices of the solution set.

35.

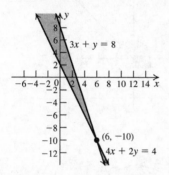

3x + y = 8

(6, −10)

4x + 2y = 4

37.

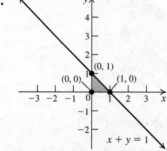

(0, 1)

(0, 0)   (1, 0)

x + y = 1

39.

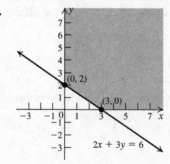

(0, 2)

(3, 0)

2x + 3y = 6

41.

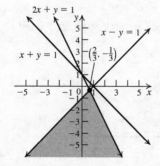

2x + y = 1

x − y = 1

x + y = 1 $\left(\frac{2}{3}, -\frac{1}{3}\right)$

43.

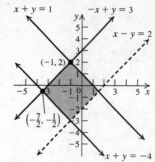

x + y = 1   −x + y = 3

x − y = 2

(−1, 2)

$\left(-\frac{7}{2}, -\frac{1}{2}\right)$

x + y = −4

45. A

47. B

49. E

51.

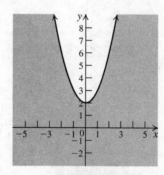

53.

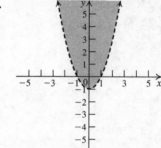

55.

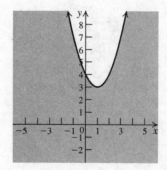

57.

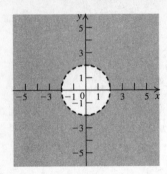

59.

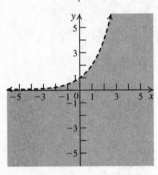

61.

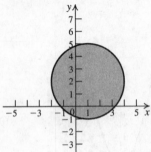

63.

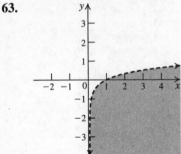

65. D, K   **67.** E, L   **69.** E, B

71.

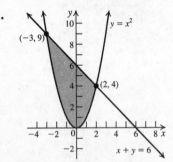

73.

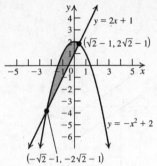

75.

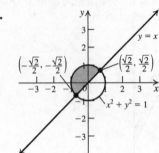

77.

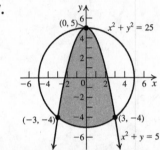

5.5 Applying the Concepts

79. Let x = the number of cases of Coke and y = the number of cases of Sprite. Then we have

$$\begin{cases} x + y \le 40 & (1) \\ x \ge 10 & (2) \\ y \ge 5 & (3) \end{cases}$$

To find the vertices, solve the systems

$$\begin{cases} x + y = 40 & (1) \\ x = 10 & (2) \end{cases}, \begin{cases} x + y = 40 & (1) \\ y = 5 & (3) \end{cases}, \text{ and}$$

$$\begin{cases} x = 10 & (2) \\ y = 5 & (3) \end{cases}.$$

The vertices are (10, 30), (35, 5), and (10, 5).

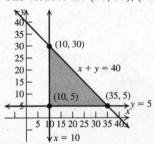

81. Let x = the number of two story houses and y = the number of one story houses. Then we have

$$\begin{cases} 7x + 5y \le 43 \\ 4x + 3y \le 25 \\ \quad\;\; x \ge 0 \\ \quad\;\; y \ge 0 \end{cases}$$

To find the vertices, solve the systems

$$\begin{cases} 7x + 5y = 43 \\ 4x + 3y = 25 \end{cases}, \begin{cases} 7x + 5y = 43 \\ \quad\;\; y = 0 \end{cases}, \begin{cases} 4x + 3y = 25 \\ \quad\;\; x = 0 \end{cases},$$

and $\begin{cases} x = 0 \\ y = 0 \end{cases}$.

$$\begin{cases} 7x + 5y = 43 \\ 4x + 3y = 25 \end{cases} \Rightarrow \begin{cases} 28x + 20y = 172 \\ -28x - 21y = -175 \end{cases} \Rightarrow$$
$$-y = -3 \Rightarrow y = 3$$
$$7x + 15 = 43 \Rightarrow x = 4$$

$$\begin{cases} 7x + 5y = 43 \\ \quad\;\; y \ge 0 \end{cases} \Rightarrow x = \frac{43}{7}$$

$$\begin{cases} 4x + 3y = 25 \\ \quad\;\; x = 0 \end{cases} \Rightarrow y = \frac{25}{3}$$

The vertices are $(0, 0)$, $\left(0, \frac{25}{3} \right)$, $\left(\frac{43}{7}, 0 \right)$, and $(4, 3)$.

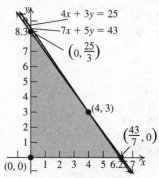

83. Let x = number of 3 hp engines and y = the number of 5 hp engines. Then we have

$$\begin{cases} \quad 3x + \;\; 4.5y \le 360 \quad (1) \\ \quad 2x + \quad\;\; y \le 200 \quad (2) \\ 0.5x + 0.75y \le 60 \quad\;\; (3) \\ \qquad\qquad\quad x \ge 0 \\ \qquad\qquad\quad y \ge 0 \end{cases}$$

Equations (1) and (3) coincide. To find the vertices, solve the systems

$$\begin{cases} 3x + 4.5y = 360 \\ 2x + \quad\; y = 200 \end{cases}, \begin{cases} 3x + 4.5y = 360 \\ \qquad\qquad x = 0 \end{cases}, \text{ and}$$

$$\begin{cases} 2x + y = 200. \\ \qquad\; y = 0 \end{cases}$$

$$\begin{cases} 3x + 4.5y = 360 \\ 2x + \quad\; y = 200 \end{cases} \Rightarrow$$
$$3x + 4.5(200 - 2x) = 360 \Rightarrow 900 - 6x = 360 \Rightarrow$$
$$x = 90$$
$$2(90) + y = 200 \Rightarrow y = 20$$

$$\begin{cases} 3x + 4.5y = 360 \\ \qquad\qquad x = 0 \end{cases} \Rightarrow 4.5y = 360 \Rightarrow y = 80$$

$$\begin{cases} 2x + y = 200 \\ \qquad\; y = 0 \end{cases} \Rightarrow 2x = 200 \Rightarrow x = 100$$

The vertices are $(0, 0)$ $(100, 0)$, $(0, 80)$, and $(90, 20)$.

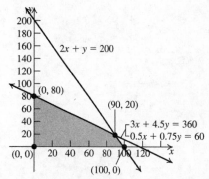

5.5 Beyond the Basics

85. The equation of the line connecting $(0, 2)$ and $(3, 0)$ is $y = -\frac{2}{3}x + 2$. The system is

$$\begin{cases} y \le -\frac{2}{3}x + 2 \\ x \ge 0 \\ y \ge 0 \end{cases}.$$

87. $\begin{cases} x \ge -1 \\ x \le 3 \end{cases}$

89. The equation of the line connecting $(0, 4)$ and $(2, 3)$ is $y = -\frac{1}{2}x + 4$. The equation of the line connecting $(2, 3)$ and $(4, 0)$ is $y = -\frac{3}{2}x + 6$. The system is

$$\begin{cases} x \ge 0 \\ y \ge 0 \\ y \le -\frac{1}{2}x + 4 \\ y \le -\frac{3}{2}x + 6 \end{cases}.$$

91. The equation of the line connecting (0, 16) and (10, 6) is $y = -x + 16$. The equation of the line connecting (10, 6) and (5, 1) is $y = x - 4$. The equation of the line connecting (5, 1) and (0, 6) is $y = -x + 6$. The system is

$$\begin{cases} x \geq 0 \\ y \leq -x + 16 \\ y \geq x - 4 \\ y \geq -x + 6 \end{cases}.$$

93.

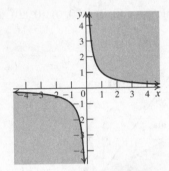

95.

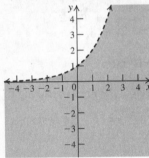

97.

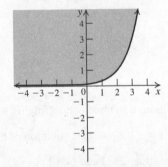

99.

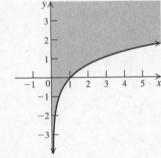

101.

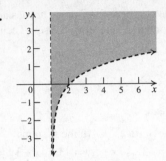

103.

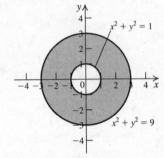

5.5 Critical Thinking/Discussion/Writing

105.

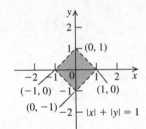

106.

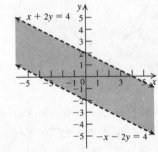

5.5 Maintaining Skills

107. To find the point of intersection, solve the system

$$\begin{cases} x + 2y = 40 & (1) \\ 3x + y = 30 & (2) \end{cases}$$

Solve equation (2) for y in terms of x, then substitute the expression in equation (1) and solve for x.

$3x + y = 30 \Rightarrow y = -3x + 30$

$x + 2(-3x + 30) = 40 \Rightarrow -5x + 60 = 40 \Rightarrow$

$-5x = -20 \Rightarrow x = 4$

(continued on next page)

(*continued*)

Substitute $x = 4$ into equation (1) and solve for y.
$$4 + 2y = 40 \Rightarrow 2y = 36 \Rightarrow y = 18$$
The point of intersection is (4, 18).

109. To find the point of intersection, solve the system
$$\begin{cases} x - 2y = 2 & (1) \\ 3x + 2y = 12 & (2) \end{cases}$$
Add the two equations, then solve for x.
$$\begin{array}{l} x - 2y = 2 \\ \underline{3x + 2y = 12} \\ \quad 4x = 14 \Rightarrow x = \dfrac{14}{4} = \dfrac{7}{2} \end{array}$$

Substitute $x = \dfrac{7}{2}$ into equation (1) and solve for y.
$$\dfrac{7}{2} - 2y = 2 \Rightarrow -2y = -\dfrac{3}{2} \Rightarrow y = \dfrac{3}{4}$$
The point of intersection is $\left(\dfrac{7}{2}, \dfrac{3}{4} \right)$.

111.

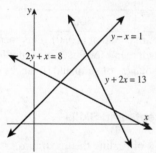

Solve each pair of equations to find the points of intersection.

I: $\begin{cases} 2y + x = 8 \\ y - x = 1 \end{cases}$

II: $\begin{cases} y - x = 1 \\ y + 2x = 13 \end{cases}$

III: $\begin{cases} y + 2x = 13 \\ 2y + x = 8 \end{cases}$

For system I, we will use substitution.
$$y = x + 1$$
$$2(x + 1) + x = 8 \Rightarrow 3x + 2 = 8 \Rightarrow 3x = 6 \Rightarrow$$
$$x = 2$$
$$y - 2 = 1 \Rightarrow y = 3$$
The point of intersection is (2, 3).
For system II, we will use substitution.
$$y = x + 1$$
$$(x + 1) + 2x = 13 \Rightarrow 3x + 1 = 13 \Rightarrow 3x = 12 \Rightarrow$$
$$x = 4; \quad y - 4 = 1 \Rightarrow y = 5$$
The point of intersection is (4, 5).

For system III, we will use substitution.
$$y = -2x + 13$$
$$2(-2x + 13) + x = 8 \Rightarrow -3x + 26 = 8 \Rightarrow$$
$$-3x = -18 \Rightarrow x = 6$$
$$y + 2(6) = 13 \Rightarrow y + 12 = 13 \Rightarrow y = 1$$
The point of intersection is (6, 1).
The vertices of the triangle are (2, 3), (4, 5), and (6, 1).

113. a.

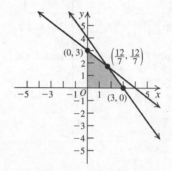

b. Solve the systems
$$\begin{cases} x = 0 \\ y = 0 \end{cases}, \begin{cases} 3x + 4y = 12 \\ y = 0 \end{cases}, \begin{cases} 3x + 4y = 12 \\ 4x + 3y = 12 \end{cases},$$
and $\begin{cases} 4x + 3y = 12 \\ x = 0 \end{cases}$

to find the vertices (0, 0), (0, 3), $\left(\dfrac{12}{7}, \dfrac{12}{7} \right)$, and (3, 0).

5.6 Linear Programming

5.6 Practice Problems

1. Maximize $f = 4x + 5y$ subject to the constraints
$$\begin{cases} 5x + 7y \le 35 \\ x \ge 0 \\ y \ge 0 \end{cases}.$$
First, graph the solution set of the constraints.

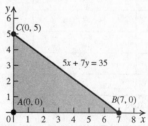

Solve the systems $\begin{cases} x = 0 \\ y = 0 \end{cases}, \begin{cases} 5x + 7y = 15 \\ y = 0 \end{cases}$, and

$\begin{cases} 5x + 7y = 15 \\ x = 0 \end{cases}$ to find the vertices: (0, 0),

(7, 0) and (0, 5).

(*continued on next page*)

(*continued*)

Now find the values of the objective function at each vertex:

Ordered pair	$f = 4x + 5y$
(0,0)	0
(7, 0)	28
(0, 5)	25

The maximum is 28 at (7, 0).

2. Minimize $f = \dfrac{1}{2}x + y$ subject to the constraints

$$\begin{cases} x + y \le 8 \\ x + 2y \ge 4 \\ 3x + 2y \ge 6. \\ x \ge 0 \\ y \ge 0 \end{cases}$$

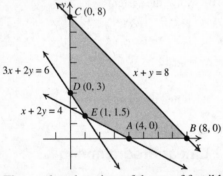

The graph and vertices of the set of feasible solutions are as given in Example 2. The vertices are: (0, 8), (0, 3), (1, 1.5), (4, 0), and (8, 0). Now find the values of the objective function at each vertex:

Ordered pair	$f = \dfrac{1}{2}x + y$
(0, 8)	8
(0, 3)	3
(1, 1.5)	2
(4, 0)	2
(8, 0)	4

The minimum is 2 at (1, 1.5) and (4, 0).

3. Let $x =$ the number of ounces of soup and let $y =$ the number of ounces of salad. The number of calories in the two items is $f = 30x + 60y$. (This is the objective function.) The constraints are as given in Example 3.

$$\begin{cases} x + y \ge 10 \\ 3x + 2y \ge 24 \\ x \ge 0 \\ y \ge 0 \end{cases}.$$

The graph and vertices of the set of feasible solutions are as given in Example 3. The vertices are (0, 12), (4, 6), and (10, 0).

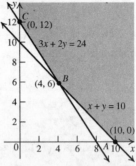

Now find the values of the objective function at each vertex:

Ordered pair	$f = 30x + 60y$
(0,12)	720
(4, 6)	480
(10, 0)	300

The minimum is 300 at (10, 0). This means that the lunch menu for Fat Albert should contain 10 ounces of soup and 0 ounces of salad.

5.6 Basic Concepts and Skills

1. The process of finding the maximum or minimum value of a quantity is called <u>optimization</u>.

3. The inequalities that determine the region S in a linear programming problem are called <u>constraints</u>, and S is called the set of <u>feasible</u> solutions.

5. True

7.

Ordered pair	$f = x + y$
(10, 0)	10
(3, 2)	5
(1, 4)	5
(0, 8)	8
(10, 8)	18

Maximum: 18; minimum: 5

9.

Ordered pair	$f = x + 2y$
(10, 0)	10
(3, 2)	7
(1, 4)	9
(0, 8)	16
(10, 8)	26

Maximum: 26; minimum: 7

11.

Ordered pair	$f = 5x + 2y$
(10, 0)	50
(3, 2)	19
(1, 4)	13
(0, 8)	16
(10, 8)	66

Maximum: 66; minimum: 13

13.

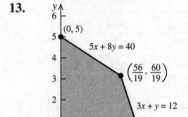

Solve the systems

$$\begin{cases} x = 0 \\ 5x + 8y = 40 \end{cases} \begin{cases} 5x + 8y = 40 \\ 3x + y = 12 \end{cases} \begin{cases} 3x + y = 12 \\ y = 0 \end{cases},$$

and $\begin{cases} x = 0 \\ y = 0 \end{cases}$ to find the vertices: $(0, 5)$,

$\left(\dfrac{56}{19}, \dfrac{60}{19}\right)$, $(4, 0,$ and $(0, 0)$. Now find the

values of the objective function at each vertex:

Ordered pair	$f = 9x + 13y$
(0, 5)	65
$\left(\dfrac{56}{19}, \dfrac{60}{19}\right)$	$\dfrac{1284}{19}$
(4, 0)	36
(0, 0)	0

The maximum is $\dfrac{1284}{19}$ at $\left(\dfrac{56}{19}, \dfrac{60}{19}\right)$.

15.

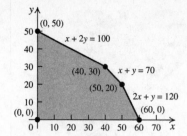

Solve the systems

$$\begin{cases} x = 0 \\ x + 2y = 100 \end{cases} \begin{cases} x + 2y = 100 \\ x + y = 70 \end{cases}, \begin{cases} x + y = 70 \\ 2x + y = 120 \end{cases},$$

$\begin{cases} 2x + y = 120 \\ y = 0 \end{cases}$, and $\begin{cases} x = 0 \\ y = 0 \end{cases}$ to find the

vertices: $(0, 50)$, $(40, 30)$, $(50, 20)$, $(60, 0)$,
and $(0, 0)$. Now find the values of the
objective function at each vertex:

Ordered pair	$f = 5x + 7y$
(0, 50)	350
(40, 30)	410
(50, 20)	390
(60, 0)	300
(0, 0)	0

The maximum is 410 at (40, 30).

17.

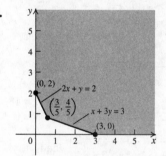

Solve the systems

$$\begin{cases} x = 0 \\ 2x + y = 2 \end{cases} \begin{cases} 2x + y = 2 \\ x + 3y = 3 \end{cases}, \text{and} \begin{cases} x + 3y = 3 \\ y = 0 \end{cases} \text{to}$$

find the vertices: $(0, 2)$, $\left(\dfrac{3}{5}, \dfrac{4}{5}\right)$, and $(3, 0)$.

Now find the values of the objective function
at each vertex:

Ordered pair	$f = x + 4y$
(0, 2)	8
$\left(\dfrac{3}{5}, \dfrac{4}{5}\right)$	$\dfrac{19}{5}$
(3, 0)	3

The minimum is 3 at (3, 0).

19.

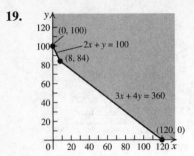

Solve the systems

$$\begin{cases} x = 0 \\ 2x + y = 100 \end{cases}, \begin{cases} 2x + y = 100 \\ 3x + 4y = 360 \end{cases}, \text{ and}$$

$$\begin{cases} 3x + 4y = 360 \\ y = 0 \end{cases}$$ to find the vertices: (0, 100),

(8, 84), and (120, 0). Now find the values of the objective function at each vertex.

Ordered pair	$f = 13x + 15y$
(0,100)	1500
(8, 84)	1364
(120, 0)	1560

The minimum is 1364 at (8, 84).

21.

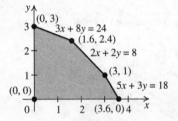

Solve the systems $\begin{cases} x = 0 \\ y = 0 \end{cases}, \begin{cases} 5x + 3y = 18 \\ y = 0 \end{cases}$,

$$\begin{cases} 2x + 2y = 8 \\ 5x + 3y = 18 \end{cases}, \begin{cases} 3x + 8y = 24 \\ 2x + 2y = 8 \end{cases}, \text{ and}$$

$$\begin{cases} 3x + 8y = 24 \\ x = 0 \end{cases}$$ to find the vertices: (0, 0),

(3.6, 0), (3, 1), (1.6, 2.4), and (0, 3). Now find the values of the objective function at each vertex:

Ordered pair	$f = 15x + 7y$
(0, 0)	0
(3.6, 0)	54
(3, 1)	52
(1.6, 2.4)	40.8
(0, 3)	21

The maximum is 54 at (3.6, 0).

23.

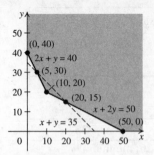

Solve the systems

$$\begin{cases} y = 0 \\ x + 2y = 50 \end{cases}, \begin{cases} x + 2y = 50 \\ x + y = 35 \end{cases}, \begin{cases} x + y = 35 \\ 2x + y = 40 \end{cases},$$

and $\begin{cases} 2x + y = 40 \\ x = 0 \end{cases}$ to find the vertices: (50, 0),

(20, 15), (30, 5), and (0, 40). Now find the values of the objective function at each vertex:

Ordered pair	$f = 8x + 16y$
(50, 0)	400
(20, 15)	400
(10, 20)	400
(5, 30)	520
(0, 40)	640

The minimum is 400 for all (x, y) on the line segments between (10, 20) and (20, 15), and (20, 15) and (50, 0).

5.6 Applying the Concepts

25. Let $x =$ the number of corn acres, and let $y =$ the number of soybean acres. Then, the profit $p = 50x + 40y$. The constraints are $x \geq 0$, $y \geq 0$, $x + y \leq 240$ (number of acres) and $2x + y \leq 320$ (number of labor hours). Solve the systems

$$\begin{cases} x = 0 \\ x + y = 240 \end{cases}, \begin{cases} x + y = 240 \\ 2x + y = 320 \end{cases}, \text{ and}$$

$$\begin{cases} 2x + y = 320 \\ y = 0 \end{cases}$$ to find the vertices: (0, 240),

(80, 160), and (160, 0). Note that (0, 0) is also a vertex.

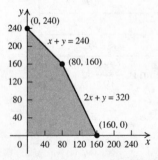

(continued on next page)

(*continued*)

Now find the values of the profit function at each vertex:

Ordered pair	$p = 50x + 40y$
(0, 240)	9600
(80, 160)	10,400
(160, 0)	8000

The maximum profit is $10,400 when 80 acres of corn and 160 acres of soybeans are planted.

27. Let x = the number of hours machine I operates, and let y = the number hours machine II operates. Then, the cost is $c = 50x + 80y$. The constraints are $x \geq 0, y \geq 0, 20x + 30y \geq 1400$ (number of units of Grade A plywood) and $10x + 40y \geq 1200$ (number of units of Grade B plywood.) Solve the systems

$$\begin{cases} x = 0 \\ 20x + 30y = 1400 \end{cases} \begin{cases} 20x + 30y = 1400 \\ 10x + 40y = 1200 \end{cases}, \text{ and}$$

$$\begin{cases} 10x + 40y = 1200 \\ y = 0 \end{cases} \text{ to find the vertices}$$

$\left(0, \dfrac{140}{3}\right)$, (40, 20), and (120, 0).

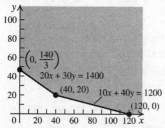

Now find the values of the cost function at each vertex:

Ordered pair	$c = 50x + 80y$
$\left(0, \dfrac{140}{3}\right)$	≈ 3733.33
(40, 20)	3600
(120, 0)	6000

The minimum cost is $3600 when machine I operates for 40 hours and machine II operates for 20 hours.

29. Let x = the number of minutes of television time, and let y = the number of pages of newspaper advertising. Then, the exposure is $f = 60,000x + 20,000y$. The constraints are $x \geq 1, y \geq 2,$ and $1000x + 500y \leq 6000$ (budget.) Solve the systems

$$\begin{cases} x = 1 \\ 1000x + 500y = 6000 \end{cases}, \begin{cases} 1000x + 500y = 6000 \\ y = 2 \end{cases},$$

and $\begin{cases} x = 1 \\ y = 2 \end{cases}$ to find the vertices (1, 10), (5, 2), and (1, 2).

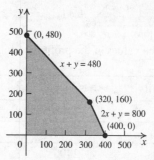

Now find the values of the exposure function at each vertex:

Ordered pair	$f = 60,000x + 20,000y$
(1, 10)	260,000
(5, 2)	340,000
(1, 2)	100,000

There are a maximum of 340,000 viewers if the company buys 5 minutes of television time and 2 pages of newspaper advertising.

31. Let x = the number of orange acres, and let y = the number of grapefruit acres. Then, the profit is $p = 40x + 30y - 3000$. The constraints are $x \geq 0, \ y \geq 0, \ x + y \leq 480$ (number of acres) and $2x + y \leq 800$ (number of labor hours). Solve the systems

$$\begin{cases} x = 0 \\ x + y = 480 \end{cases} \begin{cases} x + y = 480 \\ 2x + y = 800 \end{cases}, \text{ and}$$

$$\begin{cases} 2x + y = 800 \\ y = 0 \end{cases} \text{ to find the vertices: (0, 480),}$$

(320, 160), and (400, 0). Note that (0, 0) is also a vertex.

(*continued on next page*)

(continued)

Now find the values of the profit function at each vertex:

Ordered pair	$40x + 30y - 3000$
(0, 480)	11,400
(320, 160)	14,600
(400, 0)	13,000

The maximum profit is $14,600 when 320 acres of oranges and 160 acres of grapefruits are planted.

33. Let x = the number of rectangular tables, and let y = the number of circular tables. The profit is $p = 3x + 4y$. The number of hours to assemble the rectangular tables is x and the number of hours to assemble the circular tables is y. The number of hours to finish the rectangular tables is x and the number of hours to finish the circular tables is $2y$. The 20 assemblers work a total of $(20)(40) = 800$ hours, and the 30 finishers work a total of $(30)(40) = 1200$ hours. So, the constraints are $x \geq 0$, $y \geq 0$, $x + y \leq 800$, and $x + 2y \leq 1200$. Solve the systems

$$\begin{cases} x = 0 \\ x + 2y = 1200 \end{cases}, \begin{cases} x + 2y = 1200 \\ x + y = 800 \end{cases}, \text{ and}$$

$$\begin{cases} x + y = 800 \\ y = 0 \end{cases}$$ to find the vertices: (0, 600), (400, 400), and (800, 0). Note that (0, 0) is also a vertex.

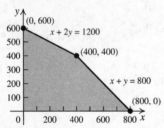

Now find the values of the profit function at each vertex:

Ordered pair	$p = 3x + 4y$.
(0, 600)	2400
(400, 400)	2800
(800, 0)	2400

The maximum profit is $2800 for 400 rectangular tables and 400 circular tables.

35. Let x = the number of terraced houses, and let y = the number of cottages. Then, the revenue is $r = 40,000x + 45,000y$. The number of units of concrete for the two house types is $x + y$; the number of units of wood for the two house types is $2x + y$; and the number of units of glass for the two house types is $2x + 5y$. The constraints are $x \geq 0$, $y \geq 0$, $x + y \leq 100$, $2x + y \leq 160$, and $2x + 5y \leq 400$. Solve the systems

$$\begin{cases} x = 0 \\ 2x + 5y = 400 \end{cases}, \begin{cases} 2x + 5y = 400 \\ x + y = 100 \end{cases},$$

$$\begin{cases} x + y = 100 \\ 2x + y = 160 \end{cases}, \text{ and } \begin{cases} 2x + y = 160 \\ y = 0 \end{cases} \text{ to find the}$$

vertices: (0, 80), $\left(\dfrac{100}{3}, \dfrac{200}{3} \right)$, (60, 40), and (80, 0). Note that (0, 0) is also a vertex.

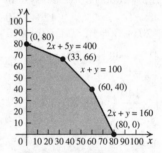

Now find the values of the profit function at each vertex. Note that we cannot use $\left(\dfrac{100}{3}, \dfrac{200}{3} \right)$ because there cannot be a fraction of a house. This value becomes (33, 66).

Ordered pair	$r = 40,000x + 45,000y$
(0, 80)	3,600,000
(33, 66)	4,290,000
(60, 40)	4,200,000
(80, 0)	3,200,000

The maximum profit is $4,290,000 when 33 terraced houses and 66 cottages are built.

37. Let x = the number of female guests, and let y = the number of male guests. Then, the number of guests who eat in the restaurant is $0.5x + 0.3y$, and the number of guests who do not eat in the restaurant is $0.5x + 0.7y$. The profit $p = 15(0.5x + 0.3y) - 2(0.5x + 0.7y)$ $= 6.5x + 3.1y$.

(continued on next page)

(continued)

The constraints are $x \geq 0$, $y \geq 0$, $x \leq 125$, $y \leq 220$, $x + y \geq 100$ and $x + y \leq 300$. Solve the systems $\begin{cases} x = 0, \\ y = 220 \end{cases}$,

$\begin{cases} y = 220 \\ x + y = 300 \end{cases}$, $\begin{cases} x + y = 300 \\ x = 125 \end{cases}$, $\begin{cases} x = 125 \\ y = 0 \end{cases}$,

$\begin{cases} y = 0 \\ x + y = 100 \end{cases}$, and $\begin{cases} x + y = 100 \\ x = 0 \end{cases}$ to find the

vertices: $(0, 220)$, $(80, 220)$, $(125, 175)$, $(125, 0)$, $(100, 0)$, and $(0, 100)$.

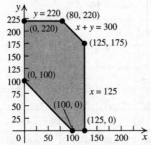

Now find the values of the profit function at each vertex:

Ordered pair	$p = 6.5x + 3.1y$
$(0, 220)$	682
$(80, 220)$	1202
$(125, 175)$	1355
$(125, 0)$	812.50
$(100, 0)$	650
$(0, 100)$	310

The maximum profit is \$1355 when there are 125 female guests and 175 male guests.

39. Let $x =$ the number of pounds of enchiladas, and let $y =$ the number of pounds of vegetable loaf. The cost $c = 3.50x + 2.25y$. The number of units of carbohydrate is $5x + 2y$. The number of units of protein is $3x + 2y$. The number of units of fat is $5x + y$. The constraints are $x \geq 0$, $y \geq 0$, $5x + 2y \geq 60$, $3x + 2y \geq 40$, and $5x + y \geq 35$. Solve the systems

$\begin{cases} x = 0, \\ 5x + y = 35 \end{cases}$, $\begin{cases} 5x + y = 35 \\ 5x + 2y = 60 \end{cases}$ $\begin{cases} 5x + 2y = 60 \\ 3x + 2y = 40 \end{cases}$,

and $\begin{cases} 3x + 2y = 40 \\ x = 0 \end{cases}$ to find the vertices:

$(0, 35)$, $(2, 25)$, $(10, 5)$, and $(40/3, 0)$.

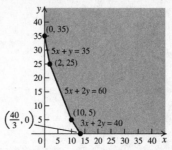

Now find the values of the profit function at each vertex:

Ordered pair	$c = 3.50x + 2.25y$
$(0, 35)$	78.75
$(2, 25)$	63.25
$(10, 5)$	46.25
$(40/3, 0)$.	46.67

The minimum cost is \$46.25 when Elisa buys 10 enchilada meals and 5 vegetable loafs.

5.6 Beyond the Basics

41.

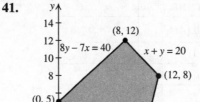

Solve the systems

$\begin{cases} x + 2y = 10 \\ 4x - y = 40 \end{cases}$ $\begin{cases} 4x - y = 40 \\ x + y = 20 \end{cases}$ $\begin{cases} x + y = 20 \\ 8y - 7x = 40 \end{cases}$,

and $\begin{cases} 8y - 7x = 40 \\ x + 2y = 10 \end{cases}$ to find the vertices:

$(10, 0)$, $(12, 8)$, $(8, 12)$, and $(0, 5)$. Now find the values of the objective function at each vertex:

Ordered pair	$f = 8x + 7y$
$(10, 0)$	80
$(12, 8)$	152
$(8, 12)$	148
$(0, 5)$	35

The maximum is 152 at $(12, 8)$.

43.

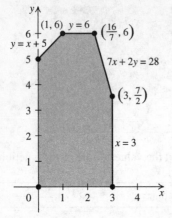

Solve the systems

$$\begin{cases} x=0 \\ y=0 \end{cases} \begin{cases} y=0 \\ x=3 \end{cases} \begin{cases} x=3 \\ 7x+2y=28 \end{cases} \begin{cases} 7x+2y=28 \\ y=6 \end{cases},$$

and $\begin{cases} y=6 \\ y=x+5 \end{cases}$ to find the vertices: $(0, 0)$,

$(3, 0)$, $(3, 7/2)$, $\left(\dfrac{16}{7}, 6\right)$, $(1, 6)$, and $(0, 5)$.

Now find the values of the objective function at each vertex:

Ordered pair	$f = 5x + 6y$
$(0, 0)$	0
$(3, 0)$	15
$\left(3, \dfrac{7}{2}\right)$	36
$\left(\dfrac{16}{7}, 6\right)$	$\dfrac{332}{7}$
$(1, 6)$	41
$(0, 5)$	30

The minimum is 0 at $(0, 0)$.

45. $ax_1 + by_1 = M = ax_2 + by_2 \Rightarrow$

$ax_1 - ax_2 = by_2 - by_1 \Rightarrow$

$a(x_1 - x_2) = b(y_2 - y_1) \Rightarrow$

$\dfrac{a}{b} = \dfrac{y_2 - y_1}{x_1 - x_2} = -\dfrac{y_2 - y_1}{x_2 - x_1} = -m$, where m is the

slope of $\overline{PQ}$.

Since the slope is the same between any two points on a given line segment, we can conclude that f has the same value M at every point of the line segment PQ.

5.6 Critical Thinking/Discussion/Writing

47.

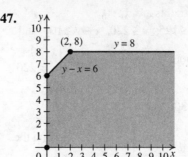

a. S is unbounded because there is no upper limit on the values for x.

b.

Ordered pair	$f = 2x + 3y$
$(0, 0)$	0
$(0, 6)$	18
$(2, 8)$	28
$(x, 8)$	$2x + 24$

As x increases, the value of the objective function $f = 2x + 3y$ increases. Since S is unbounded, there is no solution.

48. Minimize $f = 3.2x + 2.5y + 1.3z$ subject to the constraints

$x + 2y + 5 \geq 1$
$10x + 5y + 4z \geq 50$
$6x + 20y + 7z \geq 10$
$x \geq 0, \ y \geq 0, \ z \geq 0$

5.6 Maintaining Skills

49. $\begin{cases} x - 2y = 4 & (1) \\ -3x + 5y = -7 & (2) \end{cases}$

The first equation already has a leading coefficient of 1. To eliminate x from equation (2), add 3 times equation (1) to equation (2). Then solve for y.

$\begin{array}{ll} 3x - 6y = 12 & 3(x - 2y = 12) \\ \underline{-3x + 5y = -7} & (2) \\ \qquad -y = 5 & \\ \qquad\quad y = -5 & (3) \end{array}$

The equivalent system in triangular form is

$\begin{cases} x - 2y = 4 & (1) \\ \quad\ \ y = -5 & (3) \end{cases}$

Back-substitute the value of y into equation (1) and solve for x.

$x - 2(-5) = 4 \Rightarrow x + 10 = 4 \Rightarrow x = 6$

The solution set is $\{(6, -5)\}$. Be sure to check the solution in each of the original equations.

51. $\begin{cases} x - 4y - z = 11 & (1) \\ 2x - 5y + 2z = 39 & (2) \\ -3x + 2y + z = 1 & (3) \end{cases}$

The first equation already has a leading coefficient of 1. To eliminate x from equation (2), add -2 times equation (1) to equation (2).

$$\begin{array}{rl} -2x + 8y + 2z = -22 & (-2)(x - 4y - z = 11) \\ \underline{2x - 5y + 2z = \;\;39} & (2) \\ 3y + 4z = \;\;17 & (4) \end{array}$$

Multiply equation (4) by $\frac{1}{3}$.

$\begin{cases} x - 4y - z = 11 & (1) \\ y + \dfrac{4}{3}z = \dfrac{17}{4} & (4) \\ -3x + 2y + z = 1 & (3) \end{cases}$

To eliminate x from equation (3), add 3 times equation (1) to equation (3).

$$\begin{array}{rl} 3x - 12y - 3z = 33 & (3)(x - 4y - z = 11) \\ \underline{-3x + \;\;2y + \;\;z = 1} & (3) \\ -10y - 2z = 34 & (5) \end{array}$$

Multiply equation (5) by $-\frac{1}{10}$.

$\begin{cases} x - 4y - z = 11 & (1) \\ y + \dfrac{4}{3}z = \dfrac{17}{3} & (4) \\ y + \dfrac{1}{5}z = -\dfrac{17}{5} & (5) \end{cases}$

Eliminate y in equation (5) by adding -1 times equation (2). Then solve for z.

$$\begin{array}{rl} -y - \dfrac{4}{3}z = -\dfrac{17}{3} & (-1)\left(y + \frac{4}{3}z = \frac{17}{4}\right) \\ \underline{y + \dfrac{1}{5}z = -\dfrac{17}{5}} & (5) \\ -\dfrac{17}{15}z = -\dfrac{136}{15} & \\ z = 8 & (6) \end{array}$$

The equivalent system in triangular form is

$\begin{cases} x - 4y - z = 11 & (1) \\ y + \dfrac{4}{3}z = \dfrac{17}{3} & (4) \\ z = 8 & (6) \end{cases}$

Back-substitute the value of z into equation (4) and solve for y.

$y + \dfrac{4}{3}(8) = \dfrac{17}{3} \Rightarrow y + \dfrac{32}{3} = \dfrac{17}{3} \Rightarrow y = -5$

Back-substitute $z = 8$ and $y = -5$ into equation (1) and solve for x.

$x - 4(-5) - 8 = 11 \Rightarrow x + 12 = 11 \Rightarrow x = -1$

The solution set is $\{(-1, -5, 8)\}$. Be sure to check the solution in each of the original equations.

For exercises 53–56, refer to the following array.

$\begin{bmatrix} 2 & 4 & -6 & 10 \\ 3 & -2 & 4 & 12 \end{bmatrix} \begin{matrix} \leftarrow \text{ row 1} \\ \leftarrow \text{ row 2} \end{matrix}$

53. Multiply each number in row 1 by $\frac{1}{2}$.

$\begin{bmatrix} 1 & 2 & -3 & 5 \\ 3 & -2 & 4 & 12 \end{bmatrix}$

55. Using the rectangular array in exercise 54, multiply each number in row 2 by $-\frac{1}{8}$.

$\begin{bmatrix} 1 & 2 & -3 & 5 \\ 0 & 1 & -\dfrac{13}{8} & \dfrac{3}{8} \end{bmatrix}$

Chapter 5 Review Exercises

Basic Concepts and Skills

1. Using elimination, we have

$\begin{cases} 3x - y = -5 \\ x + 2y = 3 \end{cases} \Rightarrow \begin{cases} 6x - 2y = -10 \\ x + 2y = 3 \end{cases} \Rightarrow 7x = -7 \Rightarrow$

$x = -1$

$3(-1) - y = -5 \Rightarrow y = 2$

The solution is $\{(-1, 2)\}$.

3. Using elimination, we have

$\begin{cases} 2x + 4y = 3 \\ 3x + 6y = 10 \end{cases} \Rightarrow \begin{cases} -6x - 12y = -9 \\ 6x + 12y = 20 \end{cases} \Rightarrow 0 = 11 \Rightarrow$

there is no solution. Solution set: $\varnothing$

5. Using elimination, we have

$\begin{cases} 3x - y = 3 \\ \dfrac{1}{2}x + \dfrac{1}{3}y = 2 \end{cases} \Rightarrow \begin{cases} x - \dfrac{1}{3}y = 1 \\ \dfrac{1}{2}x + \dfrac{1}{3}y = 2 \end{cases} \Rightarrow \dfrac{3}{2}x = 3 \Rightarrow$

$x = 2$

$3(2) - y = 3 \Rightarrow y = 3$

The solution is $\{(2, 3)\}$.

7. Multiply the first equation by -2, add the result to the second equation, and replace the second equation with the new equation:

$\begin{cases} x + 3y + z = 0 \\ 2x - y + z = 5 \\ 3x - 3y + 2z = 10 \end{cases} \Rightarrow \begin{cases} x + 3y + z = 0 \\ -7y - z = 5 \\ 3x - 3y + 2z = 10 \end{cases}$

Multiply the first equation by -3, add the result to the third equation, and replace the third equation with the new equation:

$\begin{cases} x + 3y + z = 0 \\ -7y - z = 5 \\ 3x - 3y + 2z = 10 \end{cases} \Rightarrow \begin{cases} x + 3y + z = 0 \\ -7y - z = 5 \\ -12y - z = 10 \end{cases}$

(*continued on next page*)

(*continued*)

Subtract the third equation from the second equation, and replace the second equation with the result:

$$\begin{cases} x+3y+z=0 \\ -7y-z=5 \\ -12y-z=10 \end{cases} \Rightarrow \begin{cases} x+3y+z=0 \\ 5y=-5 \\ -12y-z=10 \end{cases} \Rightarrow$$

$$\begin{cases} x+3y+z=0 \\ y=-1 \\ -12y-z=10 \end{cases}$$

Substitute $y=-1$ into the third equation to solve for z and then substitute the values for z and y into the first equation to solve for x:

$$\begin{cases} x+3y+z=0 \\ y=-1 \\ -12y-z=10 \end{cases} \Rightarrow \begin{cases} x+3y+z=0 \\ y=-1 \\ -12(-1)-z=10 \end{cases} \Rightarrow$$

$$\begin{cases} x+3y+z=0 \\ y=-1 \\ z=2 \end{cases} \Rightarrow \begin{cases} x+3(-1)+2=0 \\ y=-1 \\ z=2 \end{cases} \Rightarrow$$

$$\begin{cases} x=1 \\ y=-1 \\ z=2 \end{cases}$$

The solution is $\{(1, -1, 2)\}$.

9. Switch the second and third equations, multiply the first equation by -2, add the result to the second equation, and replace the second equation with the result.

$$\begin{cases} x+y=1 \\ 3y+2z=0 \\ 2x-3z=7 \end{cases} \Rightarrow \begin{cases} x+y=1 \\ 2x-3z=7 \\ 3y+2z=0 \end{cases} \Rightarrow$$

$$\begin{cases} x+y=1 \\ -2y-3z=5 \\ 3y+2z=0 \end{cases}$$

Multiply the second equation by 3, multiply the third equation by 2, add the results, replace the third equation with the new equation, and solve for z:

$$\begin{cases} x+y=1 \\ -2y-3z=5 \\ 3y+2z=0 \end{cases} \Rightarrow \begin{cases} x+y=1 \\ -2y-3z=5 \\ -5z=15 \end{cases} \Rightarrow$$

$$\begin{cases} x+y=1 \\ -2y-3z=5 \\ z=-3 \end{cases}$$

Substitute $z=-3$ into the second equation to solve for y, and then substitute the value for y into the first equation to solve for x:

$$\begin{cases} x+y=1 \\ -2y-3z=5 \\ z=-3 \end{cases} \Rightarrow \begin{cases} x+y=1 \\ -2y-3(-3)=5 \\ z=-3 \end{cases} \Rightarrow$$

$$\begin{cases} x+y=1 \\ y=2 \\ z=-3 \end{cases} \Rightarrow \begin{cases} x+2=1 \\ y=2 \\ z=-3 \end{cases} \Rightarrow \begin{cases} x=-1 \\ y=2 \\ z=-3 \end{cases}$$

The solution is $\{(-1, 2, -3)\}$.

11. Multiply the first equation by -5, add the result to the second equation, and replace the second equation with the result, then switch the first and second equations:

$$\begin{cases} x+y+z=1 \\ x+5y+5z=-1 \\ 3x-y-z=4 \end{cases} \Rightarrow \begin{cases} x+y+z=1 \\ -4x=-6 \\ 3x-y-z=4 \end{cases} \Rightarrow$$

$$\begin{cases} x=\dfrac{3}{2} \\ x+y+z=1 \\ 3x-y-z=4 \end{cases}$$

Add the second and third equations, and replace the second equation with the result.

$$\begin{cases} x=\dfrac{3}{2} \\ x+y+z=1 \\ 3x-y-z=4 \end{cases} \Rightarrow \begin{cases} x=\dfrac{3}{2} \\ 4x=5 \\ 3x-y-z=4 \end{cases} \Rightarrow$$

$$\begin{cases} x=3/2 \\ x=5/4 \Rightarrow \text{There is no solution.} \\ 3x-y-z=4 \end{cases}$$

Solution set: $\varnothing$

13. The system is a dependent system because there are two equations in three unknowns. Multiply the first equation by -5, and multiply the second equation by 4, then add the resulting equations and solve for z:

$$\begin{cases} x+4y+3z=1 \\ 2x+5y+4z=4 \end{cases} \Rightarrow \begin{cases} -5x-20y-15z=-5 \\ 8x+20y+16z=16 \end{cases} \Rightarrow$$

$3x+z=11 \Rightarrow z=-3x+11$

Substitute the expression for z into the first equation and solve for y:

$x+4y+3(-3x+11)=1 \Rightarrow -8x+4y+33=1 \Rightarrow$
$y=2x-8$.

The solution is $\{(x, 2x-8, -3x+11)\}$.

15. Add the first and third equations, and replace the first equation with the result:

$$\begin{cases} 3x-y=2 \\ x+2y=9 \\ 3x+y=10 \end{cases} \Rightarrow \begin{cases} 6x=12 \\ x+2y=9 \\ 3x+y=10 \end{cases} \Rightarrow \begin{cases} x=2 \\ x+2y=9 \\ 3x+y=10 \end{cases}$$

Substitute $x=2$ into the second equation and solve for y.

(continued on next page)

(continued)

$$\begin{cases} x = 2 \\ x+2y=9 \\ 3x+y=10 \end{cases} \Rightarrow \begin{cases} x = 2 \\ 2+2y=9 \\ 3x+y=10 \end{cases} \Rightarrow \begin{cases} x = 2 \\ y = \dfrac{7}{2} \\ 3x+y=10 \end{cases}$$

$3(2)+\dfrac{7}{2}=\dfrac{19}{2}\neq 10 \Rightarrow$ there is no solution.

Solution set: $\varnothing$

17. $\begin{cases} x+y+z=3 & (1) \\ 2x-y+z=4 & (2) \\ x+4y+2z=5 & (3) \end{cases}$

Multiply equation (1) by -2, then add the result to equation (2), and replace equation (2) with the new equation:

$$\begin{cases} x+y+z=3 & (1) \\ 2x-y+z=4 & (2) \\ x+4y+2z=5 & (3) \end{cases} \Rightarrow$$

$$\begin{cases} x+y+z=3 & (1) \\ -3y-z=-2 & (4) \\ x+4y+2z=5 & (3) \end{cases}$$

Multiply equation (1) by -1, then add the result to equation (3), and replace equation (3) with the new equation:

$$\begin{cases} x+y+z=3 & (1) \\ 2x-y+z=4 & (2) \\ x+4y+2z=5 & (3) \end{cases} \Rightarrow \begin{cases} x+y+z=3 & (1) \\ -3y-z=-2 & (4) \\ 3y+z=2 & (5) \end{cases}$$

Add equation (4) to equation (5) to solve for z:

$$\begin{cases} x+y+z=3 & (1) \\ -3y-z=-2 & (4) \\ 3y+z=2 & (5) \end{cases} \Rightarrow 0=0$$

The equation $0=0$ is equivalent to $0z=0$, which is true for every value of z. Solving the second equation for y, we have

$-3y-z=-2 \Rightarrow y=\dfrac{2}{3}-\dfrac{1}{3}z$. Substituting this

into the first equation, we have $x+y+z=3 \Rightarrow$

$x+\left(\dfrac{2}{3}-\dfrac{1}{3}z\right)-z=3 \Rightarrow x=\dfrac{7}{3}-\dfrac{2}{3}z$

Thus, the solution is $\left\{\left(\dfrac{7}{3}-\dfrac{2}{3}z,\ \dfrac{2}{3}-\dfrac{1}{3}z,\ z\right)\right\}$.

19.

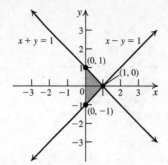

21.

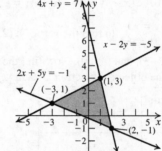

23.

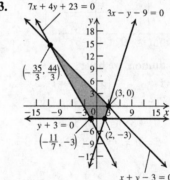

25.

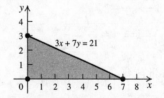

Solve the systems

$\begin{cases} x=0 \\ 3x+7y=21 \end{cases}$, $\begin{cases} 3x+7y=21 \\ y=0 \end{cases}$, and $\begin{cases} y=0 \\ x=0 \end{cases}$ to

find the vertices: $(0, 3)$, $(7, 0)$, and $(0, 0)$. Now find the values of the objective function at each vertex:

Ordered pair	$z=2x+3y$
$(0,3)$	9
$(7, 0)$	14
$(0, 0)$	0

The maximum is 14 at $(7, 0)$.

27.

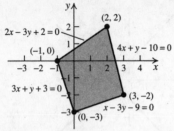

Solve the systems
$$\begin{cases} 2x-3y=-2 \\ 4x+y=10 \end{cases}, \begin{cases} 4x+y=10 \\ x-3y=9 \end{cases}, \begin{cases} x-3y=9 \\ 3x+y=-3 \end{cases},$$

and $\begin{cases} 3x+y=-3 \\ 2x-3y=-2 \end{cases}$ to find the vertices: (2,2),

(3, −2), (0, −3), and (−1, 0). Now find the
values of the objective function at each vertex:

Ordered pair	$z = x + 3y$
(2, 2)	8
(3, −2)	−3
(0, −3)	−9
(−1, 0)	−1

The minimum is −9 at (0, −3).

29. Using substitution, we have
$$\begin{cases} x+3y=1 \\ x^2-3x=7y+3 \end{cases} \Rightarrow \begin{cases} x=1-3y \\ x^2-3x=7y+3 \end{cases} \Rightarrow$$
$$(1-3y)^2-3(1-3y)=7y+3 \Rightarrow$$
$$9y^2-4y-5=0 \Rightarrow (y-1)(9y+5)=0 \Rightarrow$$
$$y=1 \text{ or } y=-\frac{5}{9}$$
$$x+3(1)=1 \Rightarrow x=-2$$
$$x+3\left(-\frac{5}{9}\right)=1 \Rightarrow x=\frac{24}{9}=\frac{8}{3}$$

The solution is $\left\{(-2,1),\left(\frac{8}{3},-\frac{5}{9}\right)\right\}$.

31. Using substitution, we have
$$\begin{cases} x-y=4 \\ 5x^2+y^2=24 \end{cases} \Rightarrow \begin{cases} y=x-4 \\ 5x^2+y^2=24 \end{cases} \Rightarrow$$
$$5x^2+(x-4)^2=24 \Rightarrow 6x^2-8x-8=0 \Rightarrow$$
$$2(x-2)(3x+2)=0 \Rightarrow x=2 \text{ or } x=-\frac{2}{3}$$
$$2-y=4 \Rightarrow y=-2$$
$$-\frac{2}{3}-y=4 \Rightarrow y=-\frac{14}{3}$$

The solution is $\left\{(2,-2),\left(-\frac{2}{3},-\frac{14}{3}\right)\right\}$.

33. Using substitution, we have
$$\begin{cases} xy=2 \\ x^2+2y^2=9 \end{cases} \Rightarrow \begin{cases} y=\dfrac{2}{x} \\ x^2+2y^2=9 \end{cases} \Rightarrow$$
$$x^2+2\left(\frac{2}{x}\right)^2=9 \Rightarrow x^2-9+\frac{8}{x^2}=0 \Rightarrow$$
$$x^4-9x^2+8=0$$
Let $u=x^2$. Then $u^2-9u+8=0 \Rightarrow$
$$(u-8)(u-1)=0 \Rightarrow u=8 \Rightarrow x^2=8 \Rightarrow$$
$$x=\pm2\sqrt{2} \text{ or } u=1 \Rightarrow x^2=1 \Rightarrow x^2=\pm1$$
$$\left(2\sqrt{2}\right)y=2 \Rightarrow y=\frac{\sqrt{2}}{2}$$
$$\left(-2\sqrt{2}\right)y=2 \Rightarrow y=-\frac{\sqrt{2}}{2}$$
$$(1)y=2 \Rightarrow y=2; (-1)y=2 \Rightarrow y=-2$$
None of the values are extraneous, so the

solution is $\left\{(1,2),(-1,-2),\left(2\sqrt{2},\frac{\sqrt{2}}{2}\right),\right.$

$\left.\left(-2\sqrt{2},-\frac{\sqrt{2}}{2}\right)\right\}$.

35.

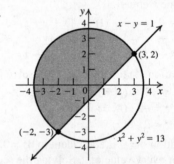

37.

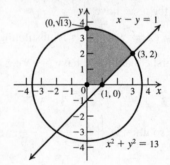

39.

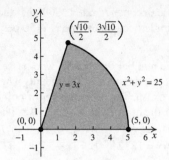

41. $\dfrac{x+4}{x^2+5x+6} = \dfrac{x+4}{(x+2)(x+3)} = \dfrac{A}{x+2} + \dfrac{B}{x+3} \Rightarrow$

$x+4 = A(x+3) + B(x+2) \Rightarrow$

$x+4 = (A+B)x + (3A+2B) \Rightarrow$

$\begin{cases} A+\ B=1 \\ 3A+2B=4 \end{cases} \Rightarrow \begin{cases} -2A-2B=-2 \\ 3A+2B=4 \end{cases} \Rightarrow$

$A=2, B=-1$

$\dfrac{x+4}{x^2+5x+6} = \dfrac{2}{x+2} - \dfrac{1}{x+3}$

43. $\dfrac{3x^2+x+1}{x(x-1)^2} = \dfrac{A}{x} + \dfrac{B}{x-1} + \dfrac{C}{(x-1)^2} \Rightarrow$

$3x^2+x+1 = A(x-1)^2 + Bx(x-1) + Cx \Rightarrow$

$3x^2+x+1$

$\qquad = (A+B)x^2 + (-2A-B+C)x + A \Rightarrow$

$\begin{cases} A+B\quad\ =3 \\ -2A-B+C=1 \\ A\qquad\quad =1 \end{cases} \Rightarrow A=1, B=2, C=5$

$\dfrac{3x^2+x+1}{x(x-1)^2} = \dfrac{1}{x} + \dfrac{2}{x-1} + \dfrac{5}{(x-1)^2}$

45. $\dfrac{x^2+2x+3}{(x^2+4)^2} = \dfrac{Ax+B}{x^2+4} + \dfrac{Cx+D}{(x^2+4)^2} \Rightarrow$

$x^2+2x+3 = (Ax+B)(x^2+4) + Cx+D \Rightarrow$

x^2+2x+3

$\qquad = Ax^3 + Bx^2 + (4A+C)x + (4B+D) \Rightarrow$

$\begin{cases} A=0 \\ B=1 \\ 4A+C=2 \\ 4B+D=3 \end{cases} \Rightarrow A=0, B=1, C=2, D=-1$

$\dfrac{x^2+2x+3}{(x^2+4)^2} = \dfrac{1}{x^2+4} + \dfrac{2x-1}{(x^2+4)^2}$

Applying the Concepts

47. Let x = the amount invested at high-risk. Let y = the amount invested at 4% interest. Then

$\begin{cases} x+\qquad y=15{,}000 \\ 0.12x+0.04y=1300 \end{cases} \Rightarrow$

$\begin{cases} -0.04x-0.04y=-600 \\ \ \ 0.12x+0.04y=1300 \end{cases} \Rightarrow 0.08x=700 \Rightarrow$

$x=8750$

$8750+y=15{,}000 \Rightarrow y=6250$

The speculator invested $8750 at 12% and $6250 at 4%.

49. Let x = the length of the rectangle. Let y = the width of the rectangle. Then

$\begin{cases} xy=63 \\ 2x+2y=33 \end{cases} \Rightarrow \begin{cases} y=\dfrac{63}{x} \\ 2x+2y=33 \end{cases} \Rightarrow$

$2x+2\left(\dfrac{63}{x}\right)=33 \Rightarrow 2x-33+\dfrac{126}{x}=0 \Rightarrow$

$2x^2-33x+126=0 \Rightarrow (x-6)(2x-21)=0 \Rightarrow$

$x=6$ or $x=\dfrac{21}{2}=10.5; \ 6y=63 \Rightarrow y=\dfrac{21}{2}=10.5$

$\dfrac{21}{2}y=63 \Rightarrow y=6$

The rectangle is 10.5 feet by 6 feet.

51. Let x = the length of one leg. Let y = the length of the other leg. Then

$\begin{cases} x^2+y^2=17^2 \\ (x+1)^2+(y+4)^2=20^2 \end{cases} \Rightarrow$

$\begin{cases} y^2=289-x^2 \\ x^2+2x+y^2+8y-383=0 \end{cases} \Rightarrow$

$x^2+2x+289-x^2+8\sqrt{289-x^2}-383=0 \Rightarrow$

$8\sqrt{289-x^2}=-2x+94 \Rightarrow$

$64(289-x^2)=(-2x+94)^2 \Rightarrow$

$18{,}496-64x^2=4x^2-376x+8836 \Rightarrow$

$68x^2-376x-9660=0 \Rightarrow$

$4(x-15)(17x+161)=0 \Rightarrow x=15$ or $x=-\dfrac{161}{7}$

(Reject the negative solution.)

$15^2+y^2=17^2 \Rightarrow y=8$

The legs of the triangle are 15 and 8.

53. Let x = the width of the smaller pastures. Let y = the length of the pastures.

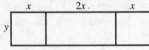

Then,

$$\begin{cases} 4xy = 6400 \\ 4x + 2y = 240 \end{cases} \Rightarrow \begin{cases} y = \dfrac{1600}{x} \\ 4x + 2y = 240 \end{cases} \Rightarrow$$

$$4x + 2\left(\dfrac{1600}{x}\right) = 240 \Rightarrow 4x - 240 + \dfrac{3200}{x} = 0 \Rightarrow$$

$$4x^2 - 240x + 3200 = 0 \Rightarrow$$
$$4(x - 40)(x - 20) = 0 \Rightarrow x = 40 \text{ or } x = 20$$
$$4(40)y = 6400 \Rightarrow y = 40$$
$$4(20)y = 6400 \Rightarrow y = 80$$

The dimensions of the pasture are 40 meters by 160 meters.

55. a. $C(x) = 12x + 60{,}000;\ R(x) = 20x$

b.

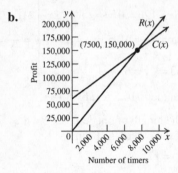

Number of timers

c. $C(x) = R(x) \Rightarrow 12x + 60{,}000 = 20x \Rightarrow$
$x = 7500$
$R(12) = 20(7500) = 150{,}000$
The breakeven point is (7500, 150,000).

d. $P(x) = R(x) - C(x) = 0.15C(x) \Rightarrow$
$20x - (12x + 60{,}000)$
$\qquad\qquad = 0.15(12x + 60{,}000) \Rightarrow$
$8x - 60{,}000 = 1.8x + 9000 \Rightarrow$
$6.2x = 69{,}000 \Rightarrow x \approx 11{,}129$
11,129 timers should be sold.

57. Let x = the double occupancy rate. Let y = the single occupancy rate. Each pays half of the double occupancy rate, so Alisha paid

$$6\left(\dfrac{x}{2}\right) + 3(x - y) = 6x - 3y \text{ for the nine}$$

months, and Sunita paid $6\left(\dfrac{x}{2}\right) + 3y = 3x + 3y$

for the nine months.

$$\begin{cases} 6x - 3y = 2250 \\ 3x + 3y = 3150 \end{cases} \Rightarrow 9x = 5400 \Rightarrow x = 600$$
$$3(600) + 3y = 3150 \Rightarrow y = 450$$

The double occupancy rate is $600 per month, and the single occupancy rate is $450 per month.

59. Let x = the number of students who passed the exam. Let y = the number of students who failed the exam. The total number of points scored for the class is $26 \times 72 = 1872$. Then

$$\begin{cases} x + y = 26 \\ 78x + 26y = 1872 \end{cases} \Rightarrow \begin{cases} -26x - 26y = -676 \\ 78x + 26y = 1872 \end{cases} \Rightarrow$$
$$52x = 1196 \Rightarrow x = 23,\ y = 3$$

Three students failed the exam.

61. Let x = Steve's age now. Let y = Janet's age now. Then $x - y$ = the difference in their ages.

$$\begin{cases} y = 2(y - (x - y)) \\ x + ((x - y) + x) = 119 \end{cases} \Rightarrow \begin{cases} y = 4y - 2x \\ 3x - y = 119 \end{cases} \Rightarrow$$
$$\begin{cases} -2x + 3y = 0 \\ 3x - y = 119 \end{cases} \Rightarrow \begin{cases} -2x + 3y = 0 \\ 9x - 3y = 357 \end{cases} \Rightarrow$$
$$7x = 357 \Rightarrow x = 51$$
$$51 + (51 - y) + 51 = 119 \Rightarrow y = 34$$

So, Steve is 51 years old now and Janet is 34 years old now.

63. Let x = Butch's amount. Let y = Sundance's amount. Let z = Billy's amount. Then

$$\begin{cases} x = 0.75(y + z) \\ y = 0.5(x + z) + 500 \Rightarrow \\ z = 3(x - y) - 1000 \end{cases}$$
$$\begin{cases} x - 0.75y - 0.75z = 0 \\ -0.5x + y - 0.5z = 500 \\ -3x + 3y + z = -1000 \end{cases}$$

Multiply the first equation by 0.5, add the result to the second equation, and replace the second equation with the new equation. Then, multiply the first equation by 3, add the result to the third equation, and replace the third equation with the new equation.

$$\begin{cases} x - 0.75y - 0.75z = 0 \\ -0.5x + y - 0.5z = 500 \\ -3x + 3y + z = -1000 \end{cases} \Rightarrow$$
$$\begin{cases} x - 0.75y - 0.75z = 0 \\ 0.625y - 0.875z = 500 \\ 0.75y - 1.25z = -1000 \end{cases}$$

Multiply the second equation by -1.2, add the result to the third equation, replace the third equation with the new equation, then solve for z.

(continued on next page)

(*continued*)

$$\begin{cases} x - 0.75y - 0.75z = 0 \\ 0.625y - 0.875z = 500 \\ 0.75y - 1.25z = -1000 \end{cases} \Rightarrow$$

$$\begin{cases} x - 0.75y - 0.75z = 0 \\ 0.625y - 0.875z = 500 \\ -0.2z = -1600 \end{cases} \Rightarrow$$

$$\begin{cases} x - 0.75y - 0.75z = 0 \\ 0.625y - 0.875z = 500 \\ z = 8000 \end{cases}$$

Substitute $z = 8000$ into the second equation to solve for y, then substitute the values for y and z into the first equation to solve for x:

$$\begin{cases} x - 0.75y - 0.75z = 0 \\ 0.625y - 0.875z = 500 \\ z = 8000 \end{cases} \Rightarrow$$

$$\begin{cases} x - 0.75y - 0.75z = 0 \\ 0.625y - 0.875(8000) = 500 \\ z = 8000 \end{cases} \Rightarrow$$

$$\begin{cases} x - 0.75y - 0.75z = 0 \\ y = 12,000 \\ z = 8000 \end{cases} \Rightarrow$$

$$\begin{cases} x - 0.75(12,000) - 0.75(8000) = 0 \\ y = 12,000 \\ z = 8000 \end{cases} \Rightarrow$$

$$\begin{cases} x = 15,000 \\ y = 12,000 \\ z = 8000 \end{cases}$$

So, they stole \$35,000. Butch received \$15,000, Sundance received \$12,000, and Billy received \$8000.

65. $2^x = 8^y \Rightarrow 2^x = 2^{3y} \Rightarrow x = 3y$
$9^y = 3^{x-2} \Rightarrow 3^{2y} = 3^{x-2} \Rightarrow 2y = x - 2$
$\begin{cases} x = 3y \\ 2y = x - 2 \end{cases} \Rightarrow 2y = 3y - 2 \Rightarrow y = 2, x = 6$

67. Let x = the number of two-story houses. Let y = the number of one-story houses. The profit $p = 10,000x + 4000y.$ The two-story houses use $7x$ units of material, while the one-story houses use y units of material. The two-story houses use x units of labor, while the one-story houses use $2y$ units of labor. So, the constraints are $x \geq 0, y \geq 0, 7x + y \leq 42$, and $x + 2y \leq 32.$

Solve the systems $\begin{cases} x = 0 \\ x + 2y = 32 \end{cases}, \begin{cases} x + 2y = 32 \\ 7x + y = 42 \end{cases},$

and $\begin{cases} 7x + y = 42 \\ y = 0 \end{cases}$ to find the vertices: (0, 16),

(4, 14), and (6, 0).

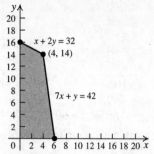

Now find the values of the profit function at each vertex:

Ordered pair	$p = 10,000x + 4000y.$
(0, 16)	64,000
(4, 14)	96,000
(6, 0)	60,000

The builder should build 4 two-story houses and 14 one-story houses for a maximum profit of \$96,000.

69. Let x = the cost of the small lobster, y = the cost of the medium lobster, and z = the cost of the large lobster. Then we have

$$\begin{cases} 4x + 2y + z = 344 \quad (1) \\ 3x + 2z = 255 \quad (2) \\ 5x + 2y + 2z = 449 \quad (3) \end{cases}$$

Interchange equations (2) and (3).

$$\begin{cases} 4x + 2y + z = 344 \quad (1) \\ 5x + 2y + 2z = 449 \quad (3) \\ 3x + 2z = 255 \quad (2) \end{cases}$$

Multiply equation (1) by -1, add the result to equation (3), then replace equation (1) with the result (4).

$$\begin{cases} 4x + 2y + z = 344 \\ 5x + 2y + 2z = 449 \\ 3x + 2z = 255 \end{cases} \Rightarrow$$

$$\begin{cases} -4x - 2y - z = -344 \\ 5x + 2y + 2z = 449 \\ 3x + 2z = 255 \end{cases} \Rightarrow$$

$$\begin{cases} x + z = 105 \quad (4) \\ 5x + 2y + 2z = 449 \quad (3) \\ 3x + 2z = 255 \quad (2) \end{cases}$$

Multiply equation (4) by -3, add the result to equation (2), then replace equation (2) with the result (5).

$$\begin{cases} x + z = 105 \quad (4) \\ 5x + 2y + 2z = 499 \quad (3) \\ 3x + 2z = 255 \quad (2) \end{cases} \Rightarrow$$

(*continued on next page*)

(continued)

$$\begin{cases} -3x - 3z = -315 \\ 5x + 2y + 2z = 449 \\ 3x + 2z = 255 \end{cases} \Rightarrow$$

$$\begin{cases} x + z = 105 & (4) \\ 5x + 2y + 2z = 449 & (3) \\ -z = -60 \Rightarrow z = 60 & (5) \end{cases}$$

Substitute $z = 60$ in equation (4) and solve for x: $x + 60 = 105 \Rightarrow x = 45$

Now substitute $x = 45$ and $z = 60$ into equation (3) and solve for y:

$5(45) + 2y + 2(60) = 449 \Rightarrow 2y = 104 \Rightarrow$
$y = 52$

A small lobster costs \$45, a medium lobster costs \$52, and a large lobster costs \$60.

Chapter 5 Practice Test A

1. Using elimination, we have
$$\begin{cases} 2x - y = 4 \\ 2x + y = 4 \end{cases} \Rightarrow 4x = 8 \Rightarrow x = 2$$
$2(2) + y = 4 \Rightarrow y = 0$
The solution is $\{(2, 0)\}$.

2. Using elimination, we have
$$\begin{cases} x + 2y = 8 \\ 3x + 6y = 24 \end{cases} \Rightarrow \begin{cases} 3x + 6y = 24 \\ 3x + 6y = 24 \end{cases} \Rightarrow \text{ the system}$$

is consistent. The solution is $\left\{ \left(x, 4 - \dfrac{x}{2} \right) \right\}$.

3. Using substitution, we have
$$\begin{cases} -2x + y = 4 \\ 4x - 2y = 4 \end{cases} \Rightarrow \begin{cases} y = 2x + 4 \\ 4x - 2y = 4 \end{cases} \Rightarrow$$
$4x - 2(2x + 4) = 4 \Rightarrow -8 = 4 \Rightarrow$ there is no solution.
Solution set: $\varnothing$

4. $\begin{cases} 3x + 3y = -15 \\ 2x - 2y = -10 \end{cases} \Rightarrow \begin{cases} x + y = -5 \\ x - y = -5 \end{cases} \Rightarrow 2x = -10 \Rightarrow$
$x = -5$
$3(-5) + 3y = -15 \Rightarrow y = 0$
The solution is $\{-5, 0)\}$.

5. Using elimination, we have
$$\begin{cases} \dfrac{5}{3}x + \dfrac{y}{2} = 14 \\ \dfrac{2}{3}x - \dfrac{y}{8} = 3 \end{cases} \Rightarrow \begin{cases} 10x + 3y = 84 \\ 16x - 3y = 72 \end{cases} \Rightarrow$$
$26x = 156 \Rightarrow x = 6$
$\dfrac{5}{3}(6) + \dfrac{y}{2} = 14 \Rightarrow y = 8$
The solution is $\{(6, 8)\}$.

6. Using substitution, we have
$$\begin{cases} y = x^2 \\ 3x - y + 4 = 0 \end{cases} \Rightarrow 3x - x^2 + 4 = 0 \Rightarrow$$
$x^2 - 3x - 4 = 0 \Rightarrow (x - 4)(x + 1) = 0 \Rightarrow$
$x = 4$ or $x = -1$
$y = 4^2 = 16$
$y = (-1)^2 = 1$
The solution is $\{(4, 16), (-1, 1)\}$.

7. Using elimination, we have
$$\begin{cases} x - 3y = -4 \\ 2x^2 + 3x - 3y = 8 \end{cases} \Rightarrow -2x^2 - 2x = -12 \Rightarrow$$
$x^2 + x - 6 = 0 \Rightarrow (x + 3)(x - 2) = 0 \Rightarrow$
$x = -3$ or $x = 2$
$-3 - 3y = -4 \Rightarrow y = \dfrac{1}{3}$
$2 - 3y = -4 \Rightarrow y = 2$
The solution is $\left\{ \left(-3, \dfrac{1}{3} \right), (2, 2) \right\}$.

8. a. Let $x =$ the weight of one bar. Let $y =$ the weight of the other bar. Then
$$\begin{cases} x + y = 485 \\ x - y = 15 \end{cases}$$

   b. $\begin{cases} x + y = 485 \\ x - y = 15 \end{cases} \Rightarrow 2x = 500 \Rightarrow x = 250$
   $250 + y = 485 \Rightarrow y = 235$
   One bar weight 250 pounds and the other weighs 235 pounds.

9. Multiply the first equation by -2, add the result to the second equation, and replace the second equation with the new equation. Then divide the second equation by 4.
$$\begin{cases} 2x + y + 2z = 4 \\ 4x + 6y + z = 15 \\ 2x + 2y + 7z = -1 \end{cases} \Rightarrow \begin{cases} 2x + y + 2z = 4 \\ y - \dfrac{3}{4}z = \dfrac{7}{4} \\ 2x + 2y + 7z = -1 \end{cases}$$

Subtract the third equation from the first equation and replace the third equation with the new equation:
$$\begin{cases} 2x + y + 2z = 4 \\ y - \dfrac{3}{4}z = \dfrac{7}{4} \\ 2x + 2y + 7z = -1 \end{cases} \Rightarrow \begin{cases} 2x + y + 2z = 4 \\ y - \dfrac{3}{4}z = \dfrac{7}{4} \\ -y - 5z = 5 \end{cases}$$

Add the second and third equations, replace the third equation with the new equation.

(continued on next page)

(continued)

$$\begin{cases} 2x+ y+2z=4 \\ \quad y-\dfrac{3}{4}z=\dfrac{7}{4} \\ \quad -y-5z=5 \end{cases} \Rightarrow \begin{cases} 2x+ y+2z=4 \\ \quad y-\dfrac{3}{4}z=\dfrac{7}{4} \\ \quad -\dfrac{23}{4}z=\dfrac{27}{4} \end{cases} \Rightarrow$$

Solve for the third equation for z. Divide the first equation by 2:

$$\begin{cases} x+\dfrac{1}{2}y+ z=2 \\ \quad y-\dfrac{3}{4}z=\dfrac{7}{4} \\ \quad z=-\dfrac{27}{23} \end{cases}$$

10. $\begin{cases} x-6y+3z=-2 \\ 9y-5z=2 \\ 2z=10 \end{cases} \Rightarrow \begin{cases} x-6y+3z=-2 \\ 9y-5z=2 \\ z=5 \end{cases}$

$9y-5(5)=2 \Rightarrow y=3$

$x-6(3)+3(5)=-2 \Rightarrow x=1$

The solution is $\{(1, 3, 5)\}$.

11. Subtract the third equation from the first equation, replace the third equation with the new equation, and solve for z.

$$\begin{cases} x+ y+ z=8 \\ 2x-2y+2z=4 \\ x+ y- z=12 \end{cases} \Rightarrow \begin{cases} x+ y+ z=8 \\ 2x-2y+2z=4 \\ z=-2 \end{cases}$$

Multiply the first equation by -2, add the result to the second equation, and replace the second equation with the new equation. Solve for y:

$$\begin{cases} x+ y+ z=8 \\ 2x-2y+2z=4 \\ z=-2 \end{cases} \Rightarrow \begin{cases} x+y+z=8 \\ y=3 \\ z=-2 \end{cases}$$

Substitute the values for y and z into the first equation to solve for x:

$$\begin{cases} x+3-2=8 \\ y=3 \\ z=-2 \end{cases} \Rightarrow \begin{cases} x=7 \\ y=3 \\ z=-2 \end{cases}$$

The solution is $\{(7, 3, -2)\}$.

12. Add the second and third equations, and replace the third equation with the result:

$$\begin{cases} x+ z=-1 \\ 3y+2z=5 \\ 3x-3y+ z=-8 \end{cases} \Rightarrow \begin{cases} x+ z=-1 \\ 3y+2z=5 \\ 3x +3z=-3 \end{cases} \Rightarrow$$

$$\begin{cases} x+ z=-1 \\ 3y+2z=5 \\ x + z=-1 \end{cases}$$

The first and third equations are the same, so we have two equations in three unknowns.

Solve the first equation for z, then substitute that expression into the second equation to solve for y:

$z=-x-1; \ 3y+2(-x-1)=5 \Rightarrow$

$3y=2x+7 \Rightarrow y=\dfrac{2}{3}x+\dfrac{7}{3}$

The solution is $\left\{\left(x,\dfrac{2}{3}x+\dfrac{7}{3},-x-1\right)\right\}$.

13. Add the first and second equations, replace the first equation with the solution, then solve for

x: $\begin{cases} 2x- y+z=2 \\ x+ y-z=-1 \\ x-5y+5z=7 \end{cases} \Rightarrow \begin{cases} x =\dfrac{1}{3} \\ x+ y-z=-1 \\ x-5y+5z=7 \end{cases}$

Substitute $x=1/3$ into the second and third equations, then solve the system

$$\begin{cases} \dfrac{1}{3}+ y- z=-1 \\ \dfrac{1}{3}-5y+5z=7 \end{cases} \Rightarrow \begin{cases} 5y-5z=-\dfrac{20}{3} \\ -5y+5z=\dfrac{20}{3} \end{cases} \Rightarrow 0=0$$

So, solve the original second equation for y:

$\dfrac{1}{3}+y-z=-1 \Rightarrow z=z-\dfrac{4}{3}$.

The solution is $\left\{\left(\dfrac{1}{3}, -\dfrac{4}{3}+z, z\right)\right\}$.

14. Let $n=$ the number of nickels. Let $d=$ the number of dimes. Let $q=$ the number of quarters. Then

$$\begin{cases} n+d+q=300 \\ d=3(n+q) \\ 0.05n+0.1d+0.25q=30.65 \end{cases} \Rightarrow$$

$$\begin{cases} n+ d+ q=300 \\ -3n+ d- 3q=0 \\ 5n+10d+25q=3065 \end{cases}$$

Multiply the first equation by 3, add the result to the second equation, and replace the second equation with the new equation

$$\begin{cases} n+ d+ q=300 \\ -3n+ d- 3q=0 \\ 5n+10d+25q=3065 \end{cases} \Rightarrow$$

$$\begin{cases} n+ d+ q=300 \\ 4d=900 \\ 5n+10d+25q=3065 \end{cases} \Rightarrow$$

$$\begin{cases} n+ d+ q=300 \\ d=225 \\ 5n+10d+25q=3065 \end{cases}$$

(continued on next page)

(continued)

Multiply the first equation by −5, add the result to the third equation, and replace the third equation with the new equation:

$$\begin{cases} n + d + q = 300 \\ \quad\quad d \quad\quad = 225 \\ 5n + 10d + 25q = 3065 \end{cases} \Rightarrow$$

$$\begin{cases} n + d + q = 300 \\ \quad d \quad = 225 \\ \quad 5d + 20q = 1565 \end{cases}$$

Substitute $d = 225$ into the third equation, solve for q, then substitute the values for d and q into the first equation to solve for n:

$5(225) + 20q = 1565 \Rightarrow q = 22$

$n + 225 + 22 = 300 \Rightarrow n = 53$

There are 53 nickels, 225 dimes, and 22 quarters.

15. $\dfrac{2x}{(x-5)(x+1)} = \dfrac{A}{x-5} + \dfrac{B}{x+1}$

16. $\dfrac{-5x^2 + x - 8}{(x-2)(x^2+1)^2} = \dfrac{A}{x-2} + \dfrac{Bx+C}{x^2+1} + \dfrac{Dx+E}{(x^2+1)^2}$

17. $\dfrac{x+3}{(x+4)^2(x-7)} = \dfrac{A}{x-7} + \dfrac{B}{x+4} + \dfrac{C}{(x+4)^2} \Rightarrow$

$x + 3 = A(x+4)^2 + B(x-7)(x+4) + C(x-7)$

Letting $x = 7$, we have $10 = 121A \Rightarrow A = \dfrac{10}{121}$.

Letting $x = -4$, we have

$-1 = -11C \Rightarrow \dfrac{1}{11} = C$. Substitute the values

for A and C into the equation and expand.

$x + 3 = \left(B + \dfrac{10}{121}\right)x^2 + \left(\dfrac{91}{121} - 3B\right)x$

$\quad\quad\quad + \left(\dfrac{83}{121} - 28B\right) \Rightarrow$

$B + \dfrac{10}{121} = 0 \Rightarrow B = -\dfrac{10}{121}$

$\dfrac{x+3}{(x+4)^2(x-7)}$

$\quad = \dfrac{10}{121(x-7)} - \dfrac{10}{121(x+4)} + \dfrac{1}{11(x+4)^2}$

18.

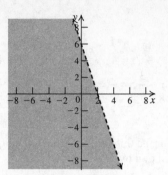

19.

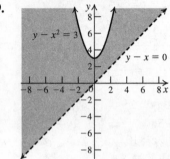

20. Solve the systems $\begin{cases} x = 0 \\ x + 3y = 3 \end{cases}$, $\begin{cases} x + 3y = 3 \\ y = 0 \end{cases}$,

and $\begin{cases} y = 0 \\ x = 0 \end{cases}$ to find the vertices: $(0, 1)$, $(3, 0)$, and $(0, 0)$.

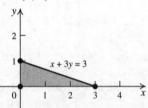

Now find the values of the objective function at each vertex:

Ordered pair	$z = 2x + y$
$(0, 1)$	1
$(3, 0)$	6
$(0, 0)$	0

The maximum value is 6 at $(3, 0)$.

Chapter 5 Practice Test B

1. Substitute each of the ordered pairs into the system.

$$\begin{cases} 2(5) - 3(-2) = 16 \\ 5 - (-2) = 7 \end{cases}$$

The answer is A.

2. Substitute each of the ordered pairs into the system.
$$\begin{cases} 6\left(\dfrac{44}{3}\right) - 9(10) = -2 \\ 3\left(\dfrac{44}{3}\right) - 5(10) = -6 \end{cases}$$
The answer is C.

3. Substitute each of the ordered pairs into the system.
$$\begin{cases} 2\left(\dfrac{5}{2}y + \dfrac{9}{2}\right) - 5y = 9 \\ 4\left(\dfrac{5}{2}y + \dfrac{9}{2}\right) - 10y = 18 \end{cases}$$
The answer is B.

4. Substituting each of the ordered pairs into the system, we find that none are solutions. The answer is C.

5. Substitute each of the ordered pairs into the system.
$$\begin{cases} \dfrac{1}{5}(1) + \dfrac{2}{5}(2) = 1 \\ \dfrac{1}{4}(1) - \dfrac{1}{3}(2) = -\dfrac{5}{12} \end{cases}$$
The answer is B.

6. Substitute each of the ordered pairs into the system.
$$\begin{cases} \left(3y + \dfrac{1}{2}\right) - 3y = \dfrac{1}{2} \\ -2\left(3y + \dfrac{1}{2}\right) + 6y = -1 \end{cases}$$
The answer is D.

7. Substitute each of the ordered pairs into the system.
$$\begin{cases} 3(4) + 4(0) = 12 \\ 3(4)^2 + 16(0)^2 = 48 \end{cases} \quad \begin{cases} 3(2) + 4\left(\dfrac{3}{2}\right) = 12 \\ 3(2)^2 + 16\left(\dfrac{3}{2}\right)^2 = 48 \end{cases}$$
The answer is D.

8. Let x = the number of student tickets and y = the number of nonstudent tickets. Then
$$\begin{cases} x + y = 300 \\ 2x + 5y = 975 \end{cases} \Rightarrow \begin{cases} -2x - 2y = -600 \\ 2x + 5y = 975 \end{cases} \Rightarrow$$
$$3y = 375 \Rightarrow y = 125$$
$$x + 125 = 300 \Rightarrow x = 175$$
The answer is D.

9. Multiply the first equation by 2, subtract the second equation from that result, then replace the second equation with the new equation:
$$\begin{cases} x + 3y + 3z = 4 \\ 2x + 5y + 4x = 5 \\ x + 2y + 2z = 6 \end{cases} \Rightarrow \begin{cases} x + 3y + 3z = 4 \\ y + 2z = 3 \\ x + 2y + 2z = 6 \end{cases}$$
Subtract the third equation from the first equation, then replace the third equation with the new equation:
$$\begin{cases} x + 3y + 3z = 4 \\ y + 2z = 3 \\ x + 2y + 2z = 6 \end{cases} \Rightarrow \begin{cases} x + 3y + 3z = 4 \\ y + 2z = 3 \\ y + z = -2 \end{cases}$$
Subtract the third equation from the second equation and replace the third equation with the new equation:
$$\begin{cases} x + 3y + 3z = 4 \\ y + 2z = 3 \\ y + z = -2 \end{cases} \Rightarrow \begin{cases} x + 3y + 3z = 4 \\ y + 2z = 3 \\ z = 5 \end{cases}$$
The answer is C.

10. From the third equation, $z = 3$.
$$10y - 2(3) = 4 \Rightarrow y = 1; 2x + 1 + 3 = 0 \Rightarrow$$
$$x = -2$$
The answer is A.

11. Substitute each of the ordered pairs into the system.
$$\begin{cases} 2(1) + 13(-1) + 6(2) = 1 \\ 3(1) + 10(-1) + 11(2) = 15 \\ 2(1) + 10(-1) + 8(2) = 8 \end{cases}$$
The answer is B.

12. Substituting each of the ordered pairs into the system, we find that none are solutions. The answer is D.

13. Substitute each of the ordered pairs into the system.
$$\begin{cases} x - 2(5x + 1) + 3(3x + 2) = 4 \\ 2x - (5x + 1) + (3x + 2) = 1 \\ x + (5x + 1) - 2(3x + 2) = -3 \end{cases}$$
The answer is C.

14. Let x = the number of students going to France. Let y = the number of students going to Italy. Let z = the number of students going to Spain. Then

$$\begin{cases} x+y+z=46 \\ x+y=4+z \\ x=z-2 \end{cases} \Rightarrow \begin{cases} x+y+z=46 \\ x+y-z=4 \\ x-z=-2 \end{cases} \text{ Subtract}$$

the second equation from the first equation, replace the second equation with the new equation, and solve for z:

$$\begin{cases} x+y+z=46 \\ x+y-z=4 \\ x-z=-2 \end{cases} \Rightarrow \begin{cases} x+y+z=46 \\ 2z=42 \\ x-z=-2 \end{cases} \Rightarrow$$

$$\begin{cases} x+y+z=46 \\ z=21 \\ x-z=-2 \end{cases}$$

Substitute $z=21$ into the third equation to solve for x. Then substitute the values for x and z into the first equation to solve for y:

$x-21=-2 \Rightarrow x=19; \; 19+y+21=46 \Rightarrow$
$y=6$

The answer is B.

15. C **16.** A

17. $\dfrac{x^2+15x+18}{x^3-9x} = \dfrac{x^2+15x+18}{x(x-3)(x+3)}$

$$= \dfrac{A}{x} + \dfrac{B}{x-3} + \dfrac{C}{x+3} \Rightarrow$$

$x^2+15x+18$
$\quad = A(x^2-9) + Bx(x+3) + Cx(x-3)$

Letting $x=-3$, we have

$(-3)^2+15(-3)+18 = -3C(-3-3) \Rightarrow$
$-18=18C \Rightarrow -1=C$

Letting $x=3$, we have

$(3)^2+15(3)+18 = 3B(3+3) \Rightarrow$
$72=18B \Rightarrow 4=B$

Substituting the values for B and C and expanding the equation, we have

$x^2+15x+18$
$\quad = A(x^2-9) + 4x(x+3) - x(x-3)$
$\quad = (A+3)x^2+15x-9A \Rightarrow$

$$\begin{cases} A+3=1 \\ -9A=18 \end{cases} \Rightarrow A=-2$$

$\dfrac{x^2+15x+18}{x^3-9x} = -\dfrac{2}{x} + \dfrac{4}{x-3} - \dfrac{1}{x+3}$

The answer is C.

18. B **19.** A

20. Solve the systems

$$\begin{cases} x=0 \\ 2x+3y=16 \end{cases}, \begin{cases} 2x+3y=16 \\ 2x+y=8 \end{cases}, \begin{cases} 2x+y=8 \\ y=0 \end{cases} \text{ and}$$

$$\begin{cases} y=0 \\ x=0 \end{cases} \text{ to find the vertices: } \left(0, \dfrac{16}{3}\right),$$

$(2,4)$, $(4,0)$, and $(0,0)$.

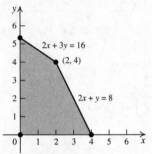

Now find the values of the objective function at each vertex:

Ordered pair	$z=3x+21y$
$\left(0, \dfrac{16}{3}\right)$	112
$(2,4)$	90
$(4,0)$	12
$(0,0)$	0

The maximum value is 112 at $\left(0, \dfrac{16}{3}\right)$.

The answer is B.

Cumulative Review Exercises (Chapters P–5)

1. $\dfrac{1}{x-1} + \dfrac{4}{x-4} = \dfrac{5}{x-5} \Rightarrow$

$(x-4)(x-5) + 4(x-1)(x-5)$
$\qquad\qquad\qquad = 5(x-1)(x-4) \Rightarrow$

$5x^2-33x+40 = 5x^2-25x+20 \Rightarrow$

$-8x=-20 \Rightarrow x=\dfrac{5}{2}$

2. Let $u = x + \dfrac{1}{x}$. Then

$2u^2-7u+5=0 \Rightarrow (u-1)(2u-5)=0 \Rightarrow$
$u=1 \text{ or } u=5/2$

$x+\dfrac{1}{x}=1 \Rightarrow x^2-x+1=0 \Rightarrow$

$x = \dfrac{1\pm\sqrt{1-4}}{2} = \dfrac{1\pm i\sqrt{3}}{2} = \dfrac{1}{2} \pm \dfrac{i\sqrt{3}}{2}$

(continued on next page)

(*continued*)

$$x + \frac{1}{x} = \frac{5}{2} \Rightarrow 2x^2 - 5x + 2 = 0 \Rightarrow$$
$$(x-2)(2x-1) = 0 \Rightarrow x = 2 \text{ or } x = 1/2$$

The solution is $\left\{ \frac{1}{2} \pm \frac{i\sqrt{3}}{2}, 2, \frac{1}{2} \right\}$.

3. $\sqrt{3x-5} = x-3 \Rightarrow 3x-5 = (x-3)^2 \Rightarrow$
   $x^2 - 9x + 14 = 0 \Rightarrow (x-7)(x-2) = 0 \Rightarrow$
   $x = 7 \text{ or } x = 2$.
   Check each answer to see if either is extraneous:
   $$\sqrt{3(7)-5} = \sqrt{16} = 4 = 7-3$$
   $$\sqrt{3(2)-5} = \sqrt{1} = 1 \neq 2-3 \Rightarrow 2 \text{ is extraneous.}$$
   The solution is $\{7\}$.

4. $\frac{x-1}{x+3} = 0 \Rightarrow x = 1$ is the x-intercept. The
   vertical asymptote is $x = -3$. The intervals to
   be tested are $(-\infty, -3)$, $(-3, 1]$, and $[1, \infty)$.

Interval	Test Point	Value of $\frac{x-1}{x+3}$	Result
$(-\infty, -3)$	-4	5	$+$
$(-3, 1]$	0	$-\frac{1}{3}$	$-$
$[1, \infty)$	2	$\frac{1}{5}$	$+$

   The solution is $(-3, 1]$.

5. Solve the associated equation to find the test intervals:
   $$x^2 - 9x + 20 = 0 \Rightarrow (x-4)(x-5) = 0 \Rightarrow$$
   $$x = 4 \text{ or } x = 5.$$
   The intervals to be tested are $(-\infty, 4), (4, 5)$,
   and $(5, \infty)$.

Interval	Test Point	Value of $x^2 - 9x + 20$	Result
$(-\infty, 4)$	0	20	$+$
$(4, 5)$	4.5	-0.25	$-$
$(5, \infty)$	10	30	$+$

   The solution is $(-\infty, 4) \cup (5, \infty)$.

6. $2^{x-1} = 5 \Rightarrow (x-1)\ln 2 = \ln 5 \Rightarrow$
   $$x\ln 2 - \ln 2 = \ln 5 \Rightarrow x = \frac{\ln 5 + \ln 2}{\ln 2}$$

7. $\log_x 16 = 4 \Rightarrow x^4 = 16 \Rightarrow x^4 = 2^4 \Rightarrow x = 2$

8. $\log(x-3) + \log(x-1) = \log(2x-5) \Rightarrow$
   $\log((x-3)(x-1)) = \log(2x-5) \Rightarrow$
   $x^2 - 4x + 3 = 2x - 5 \Rightarrow x^2 - 6x + 8 = 0 \Rightarrow$
   $(x-4)(x-2) = 0 \Rightarrow x = 4 \text{ or } x = 2$
   Reject $x = 2$ because it makes
   $\log(2-3) = \log(-1)$, which is not possible.
   The solution is $\{4\}$.

9. Shift the graph of $y = |x|$ one unit left and two units down.

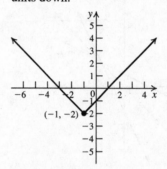

10. Shift the graph of $y = x^2$ two units right, reflect it across the x-axis, then shift it three units up.

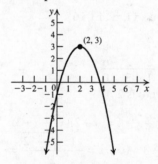

11. Shift the graph of $y = 2^x$ one unit right and three units down.

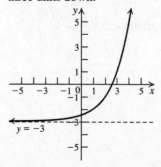

12. Shift the graph of $y = \sqrt{x}$ one unit right and three units up.

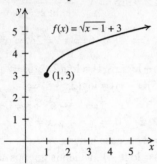

13. a. Switch the variables and solve for y:

$$y = 2x - 2 \Rightarrow x = 2y - 2 \Rightarrow y = f^{-1} = \frac{x+2}{2}$$

b.

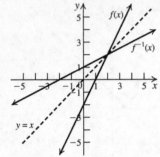

14. a. The factors of the constant term are $\{\pm 1, \pm 2, \pm 3, \pm 6\}$. The factors of the leading coefficient are $\{\pm 1\}$. The possible rational zeros are $\{\pm 1, \pm 2, \pm 3, \pm 6\}$.

b.
$$\begin{array}{r|rrrr} 2 & 1 & -1 & 1 & -6 \\ & & 2 & 2 & 6 \\ \hline & 1 & 1 & 3 & 0 \end{array}$$
A real zero is 2.

c. $x^3 - x^2 + x - 6 = (x - 2)(x^2 + x + 3)$

$$x^2 + x + 3 = 0 \Rightarrow x = \frac{-1 \pm \sqrt{1 - 12}}{2} \Rightarrow$$

$$x = \frac{-1 \pm i\sqrt{11}}{2}$$

The zeros are $\left\{ 2, -\frac{1}{2} \pm \frac{i\sqrt{11}}{2} \right\}$.

15. $\log_3(9x^4) = \log_3 9 + \log_3 x^4 = 2 + 4\log_3 x$

16. a. $0.5 = e^{-0.05t} \Rightarrow \ln 0.5 = -0.05t \Rightarrow$
$$t = -\frac{\ln 0.5}{0.05} \approx 13.86 \text{ years}$$

b. $0.5 = e^{-0.0002t} \Rightarrow \ln 0.5 = -0.0002t \Rightarrow$
$$t = -\frac{\ln 0.5}{0.0002} \approx 3465.74 \text{ years}$$

17. $f^{-1}(9) = 2$

18. $A = Pe^{rt} \Rightarrow 2P = Pe^{0.075t} \Rightarrow 2 = e^{0.075t} \Rightarrow$
$$\ln 2 = 0.075t \Rightarrow t = \frac{\ln 2}{0.075} \approx 9.24$$
It will take about 9.24 years to double the money.

19. $\begin{cases} 5x - 2y + 25 = 0 \\ 4y - 3x - 29 = 0 \end{cases} \Rightarrow \begin{cases} 10x - 4y = -50 \\ -3x + 4y = 29 \end{cases} \Rightarrow$
$$7x = -21 \Rightarrow x = -3$$
$$4y - 3(-3) - 29 = 0 \Rightarrow y = 5$$
The solution is $\{(-3, 5)\}$.

20. Multiply the second equation by -2, add the result to the first equation, and replace the second equation with the new equation:
$$\begin{cases} 2x - y + z = 3 \\ x + 3y - 2z = 11 \\ 3x - 2y + z = 4 \end{cases} \Rightarrow \begin{cases} 2x - y + z = 3 \\ -7y + 5z = -19 \\ 3x - 2y + z = 4 \end{cases}$$
Multiply the first equation by 3, multiply the third equation by -2, add the two and replace the third equation with the new equation:
$$\begin{cases} 2x - y + z = 3 \\ -7y + 5z = -19 \\ 3x - 2y + z = 4 \end{cases} \Rightarrow \begin{cases} 2x - y + z = 3 \\ -7y + 5z = -19 \\ y + z = 1 \end{cases}$$
Multiply the third equation by 7, add the result to the second equation, replace the third equation and solve for z:
$$\begin{cases} 2x - y + z = 3 \\ -7y + 5z = -19 \\ y + z = 1 \end{cases} \Rightarrow \begin{cases} 2x - y + z = 3 \\ -7y + 5z = -19 \\ 12z = -12 \end{cases} \Rightarrow$$
$$z = -1$$
$$-7y + 5(-1) = -19 \Rightarrow y = 2$$
$$2x - 2 - 1 = 3 \Rightarrow x = 3$$
The solution is $\{(3, 2, -1)\}$.

Chapter 6 Matrices and Determinants

6.1 Matrices and Systems of Equations

6.1 Practice Problems

1. a. 3×2 **b.** 1×2

2. $\begin{bmatrix} 0 & 3 & -1 & 8 \\ 1 & 4 & 0 & 14 \\ 0 & -2 & 9 & 0 \end{bmatrix}$

3. $\begin{bmatrix} 3 & 4 & 5 \\ 2 & 4 & 6 \end{bmatrix} \xrightarrow{R_1 \leftrightarrow R_2} \begin{bmatrix} 2 & 4 & 6 \\ 3 & 4 & 5 \end{bmatrix} \xrightarrow{\frac{1}{2}R_1}$

$\begin{bmatrix} 1 & 2 & 3 \\ 3 & 4 & 5 \end{bmatrix} \xrightarrow{-3R_1+R_2 \to R_2} \begin{bmatrix} 1 & 2 & 3 \\ 0 & -2 & -4 \end{bmatrix}$

4. First write the system as an augmented matrix.

$\begin{cases} x - 6y + 3z = -2 \\ 3x + 3y - 2z = -2 \\ 2x - 3y + z = -2 \end{cases} \Rightarrow \begin{bmatrix} 1 & -6 & 3 & -2 \\ 3 & 3 & -2 & -2 \\ 2 & -3 & 1 & -2 \end{bmatrix}$

Now perform the row operations.

$\begin{bmatrix} 1 & -6 & 3 & -2 \\ 3 & 3 & -2 & -2 \\ 2 & -3 & 1 & -2 \end{bmatrix}$

$\xrightarrow[2R_1-R_3 \to R_3]{3R_1-R_2 \to R_2} \begin{bmatrix} 1 & -6 & 3 & -2 \\ 0 & -21 & 11 & -4 \\ 0 & -9 & 5 & -2 \end{bmatrix}$

$\xrightarrow{9R_2-21R_3 \to R_3} \begin{bmatrix} 1 & -6 & 3 & -2 \\ 0 & -21 & 11 & -4 \\ 0 & 0 & -6 & 6 \end{bmatrix}$

$\xrightarrow[-\frac{1}{6}R_3 \to R_3]{-\frac{1}{21}R_2 \to R_2} \begin{bmatrix} 1 & -6 & 3 & -2 \\ 0 & 1 & -\frac{11}{21} & \frac{4}{21} \\ 0 & 0 & 1 & -1 \end{bmatrix}$

Thus, $z = -1$.
Using back-substitution, we have

$y - \frac{11}{21}(-1) = \frac{4}{21} \Rightarrow y = -\frac{7}{21} = -\frac{1}{3}$. Then,

$x - 6\left(-\frac{1}{3}\right) + 3(-1) = -2 \Rightarrow x - 1 = -2 \Rightarrow$
$x = -1$

The solution is $\left\{ \left(-1, -\frac{1}{3}, -1 \right) \right\}$.

5. First write the system as an augmented matrix.

$\begin{cases} x - 2y = 1 \\ 2x + 3y = 16 \end{cases} \Rightarrow \begin{bmatrix} 1 & -2 & 1 \\ 2 & 3 & 16 \end{bmatrix}$

Now perform the row operations.

$\begin{bmatrix} 1 & -2 & 1 \\ 2 & 3 & 16 \end{bmatrix} \xrightarrow{-2R_1+R_2 \to R_2} \begin{bmatrix} 1 & -2 & 1 \\ 0 & 7 & 14 \end{bmatrix}$

$\xrightarrow{\frac{1}{7}R_2 \to R_2} \begin{bmatrix} 1 & -2 & 1 \\ 0 & 1 & 2 \end{bmatrix}$

Thus, $y = 2$. Using back-substitution, we have
$x - 2(2) = 1 \Rightarrow x = 5$. The solution is $\{(5, 2)\}$.

6. First write the system as an augmented matrix.

$\begin{cases} 2x + y - z = 7 \\ x - 3y - 3z = 4 \\ 4x + y + z = 3 \end{cases} \Rightarrow \begin{bmatrix} 2 & 1 & -1 & 7 \\ 1 & -3 & -3 & 4 \\ 4 & 1 & 1 & 3 \end{bmatrix}$

Now perform the row operations.

$\begin{bmatrix} 2 & 1 & -1 & 7 \\ 1 & -3 & -3 & 4 \\ 4 & 1 & 1 & 3 \end{bmatrix}$

$\xrightarrow{R_1 \leftrightarrow R_2} \begin{bmatrix} 1 & -3 & -3 & 4 \\ 2 & 1 & -1 & 7 \\ 4 & 1 & 1 & 3 \end{bmatrix}$

$\xrightarrow[-4R_1+R_3 \to R_3]{-2R_1+R_2 \to R_2} \begin{bmatrix} 1 & -3 & -3 & 4 \\ 0 & 7 & 5 & -1 \\ 0 & 13 & 13 & -13 \end{bmatrix}$

$\xrightarrow[\frac{1}{13}R_3 \to R_3]{\frac{1}{7}R_2 \to R_2} \begin{bmatrix} 1 & -3 & -3 & 4 \\ 0 & 1 & \frac{5}{7} & -\frac{1}{7} \\ 0 & 1 & 1 & -1 \end{bmatrix}$

$\xrightarrow{\frac{7}{2}(-R_2+R_3) \to R_3} \begin{bmatrix} 1 & -3 & -3 & 4 \\ 0 & 1 & \frac{5}{7} & -\frac{1}{7} \\ 0 & 0 & 1 & -3 \end{bmatrix}$

Thus, $z = -3$. Using back-substitution, we
have $y + \frac{5}{7}(-3) = -\frac{1}{7} \Rightarrow y = 2$ and
$x - 3(2) - 3(-3) = 4 \Rightarrow x = 1$.
The solution is $\{(1, 2, -3)\}$.

7. $\begin{cases} 6x + 8y - 14z = 3 \\ 3x + 4y - 7z = 12 \\ 6x + 3y + z = 0 \end{cases} \Rightarrow \begin{bmatrix} 6 & 8 & -14 & | & 3 \\ 3 & 4 & -7 & | & 12 \\ 6 & 3 & 1 & | & 0 \end{bmatrix}$

Now perform the row operations.

$\begin{bmatrix} 6 & 8 & -14 & | & 3 \\ 3 & 4 & -7 & | & 12 \\ 6 & 3 & 1 & | & 0 \end{bmatrix}$

$\xrightarrow{R_1 \leftrightarrow R_2} \begin{bmatrix} 3 & 4 & -7 & | & 12 \\ 6 & 8 & -14 & | & 3 \\ 6 & 3 & 1 & | & 0 \end{bmatrix}$

$\xrightarrow{\frac{1}{3}R_1 \to R_1} \begin{bmatrix} 1 & \frac{4}{3} & -\frac{7}{3} & | & 4 \\ 6 & 8 & -14 & | & 3 \\ 6 & 3 & 1 & | & 0 \end{bmatrix}$

$\xrightarrow{-6R_1 + R_2 \to R_2} \begin{bmatrix} 1 & \frac{4}{3} & -\frac{7}{3} & | & 4 \\ 0 & 0 & 0 & | & -21 \\ 6 & 3 & 1 & | & 0 \end{bmatrix}$

Since the second row is equivalent to $0 = -21$, the system is inconsistent and the solution is $\varnothing$.

8. $\begin{cases} 2x - 3y - 2z = 0 \\ x + y - 2z = 7 \\ 3x - 5y - 5z = 3 \end{cases} \Rightarrow \begin{bmatrix} 2 & -3 & -2 & | & 0 \\ 1 & 1 & -2 & | & 7 \\ 3 & -5 & -5 & | & 3 \end{bmatrix}$

$\begin{bmatrix} 2 & -3 & -2 & | & 0 \\ 1 & 1 & -2 & | & 7 \\ 3 & -5 & -5 & | & 3 \end{bmatrix} \xrightarrow{R_1 \leftrightarrow R_2} \begin{bmatrix} 1 & 1 & -2 & | & 7 \\ 2 & -3 & -2 & | & 0 \\ 3 & -5 & -5 & | & 3 \end{bmatrix}$

$\xrightarrow[-3R_1 + R_3 \to R_3]{-2R_1 + R_2 \to R_2} \begin{bmatrix} 1 & 1 & -2 & | & 7 \\ 0 & -5 & 2 & | & -14 \\ 0 & -8 & 1 & | & -18 \end{bmatrix}$

$\xrightarrow{-\frac{1}{5}R_2 \to R_2} \begin{bmatrix} 1 & 1 & -2 & | & 7 \\ 0 & 1 & -\frac{2}{5} & | & \frac{14}{5} \\ 0 & -8 & 1 & | & -18 \end{bmatrix}$

$\xrightarrow[8R_2 + R_3 \to R_3]{R_1 - R_2 \to R_1} \begin{bmatrix} 1 & 0 & -\frac{8}{5} & | & \frac{21}{5} \\ 0 & 1 & -\frac{2}{5} & | & \frac{14}{5} \\ 0 & 0 & -\frac{11}{5} & | & \frac{22}{5} \end{bmatrix}$

$\xrightarrow{-\frac{5}{11}R_3 \to R_3} \begin{bmatrix} 1 & 0 & -\frac{8}{5} & | & \frac{21}{5} \\ 0 & 1 & -\frac{2}{5} & | & \frac{14}{5} \\ 0 & 0 & 1 & | & -2 \end{bmatrix}$

$\xrightarrow[R_2 + \frac{2}{5}R_3 \to R_2]{R_1 + \frac{8}{5}R_3 \to R_1} \begin{bmatrix} 1 & 0 & 0 & | & 1 \\ 0 & 1 & 0 & | & 2 \\ 0 & 0 & 1 & | & -2 \end{bmatrix}$

The solution is $\{(1, 2, -2)\}$.

9. $\begin{cases} x + z = -1 \\ 3y + 2z = 5 \\ 3x - 3y + z = -8 \end{cases} \Rightarrow \begin{bmatrix} 1 & 0 & 1 & | & -1 \\ 0 & 3 & 2 & | & 5 \\ 3 & -3 & 1 & | & -8 \end{bmatrix}$

$\begin{bmatrix} 1 & 0 & 1 & | & -1 \\ 0 & 3 & 2 & | & 5 \\ 3 & -3 & 1 & | & -8 \end{bmatrix}$

$\xrightarrow{-3R_1 + R_3 \to R_3} \begin{bmatrix} 1 & 0 & 1 & | & -1 \\ 0 & 3 & 2 & | & 5 \\ 0 & -3 & -2 & | & -5 \end{bmatrix}$

$\xrightarrow[R_2 + R_3 \to R_3]{\frac{1}{3}R_2 \to R_2} \begin{bmatrix} 1 & 0 & 1 & | & -1 \\ 0 & 1 & \frac{2}{3} & | & \frac{5}{3} \\ 0 & 0 & 0 & | & 0 \end{bmatrix}$

The matrix is in row echelon form. The equivalent system is $\begin{cases} x + z = -1 \\ y + \frac{2}{3}z = \frac{5}{3} \end{cases}$. Solving for x and y in terms of z, we have $x = -z - 1$ and $y = \frac{5}{3} - \frac{2}{3}z$.

The solution set is $\left\{\left(-z - 1, \frac{5}{3} - \frac{2}{3}z, z\right)\right\}$.

10. a. $\begin{cases} s = 0.01s + 0.1c + 0.1t + 12.46 \\ c = 0.2s + 0.02c + 0.01t + 3 \\ t = 0.25s + 0.3c + 2.7 \end{cases}$ or

$\begin{cases} 0.99s - 0.1c - 0.1t = 12.46 \\ -0.2s + 0.98c - 0.01t = 3 \\ -0.25s - 0.3c + t = 2.7 \end{cases}$

b. $0.99(14) - 0.1(6) - 0.1(8) = 12.46$ ✓
$-0.2(14) + 0.98(6) - 0.01(8) = 3$ ✓
$-0.25(14) - 0.3(6) + (8) = 2.7$ ✓

6.1 Basic Concepts and Skills

1. A matrix is any rectangular array of <u>numbers</u>.

3. Two matrices are row equivalent if one can be obtained form the other by a sequence of <u>row operations</u>.

5. False. A 3×4 matrix has three rows with four entries in each row. In other words, the matrix has three rows and four columns.

7. False. The augmented matrix of the system is
$$\left[\begin{array}{ccc|c} 2 & 3 & 0 & 5 \\ 0 & 3 & -4 & -1 \\ 1 & 0 & 2 & 3 \end{array}\right].$$

9. 1×1 **11.** 2×4 **13.** 2×3

15. $a_{13} = 3, a_{31} = 9, a_{33} = 11, a_{34} = 12$

17. No, the array is not a matrix because it is not a rectangular array of numbers.

19. $\left[\begin{array}{cc|c} 2 & 4 & 2 \\ 1 & -3 & 1 \end{array}\right]$

21. $\left[\begin{array}{cc|c} 5 & -7 & 11 \\ -13 & 17 & 19 \end{array}\right]$

23. $\left[\begin{array}{ccc|c} -1 & 2 & 3 & 8 \\ 2 & -3 & 9 & 16 \\ 4 & -5 & -6 & 32 \end{array}\right]$

25. $\begin{cases} x + 2y - 3z = 4 \\ -2x - 3y + z = 5 \\ 3x - 3y + 2z = 7 \end{cases}$

27. $\begin{cases} x - y + z = 2 \\ 2x + y - 3z = 6 \end{cases}$

29.(i) $\begin{bmatrix} 2 & 3 & 5 \\ 1 & 2 & 3 \end{bmatrix} \xrightarrow{R_1 \leftrightarrow R_2} \begin{bmatrix} 1 & 2 & 3 \\ 2 & 3 & 5 \end{bmatrix}$

(ii) $\begin{bmatrix} 1 & 2 & 3 \\ 2 & 3 & 5 \end{bmatrix} \xrightarrow{-2R_1 + R_2 \rightarrow R_2} \begin{bmatrix} 1 & 2 & 3 \\ 0 & -1 & -1 \end{bmatrix}$

(iii) $\begin{bmatrix} 1 & 2 & 3 \\ 0 & -1 & -1 \end{bmatrix} \xrightarrow{-R_2} \begin{bmatrix} 1 & 2 & 3 \\ 0 & 1 & 1 \end{bmatrix}$

31.(i) $\begin{bmatrix} 1 & 2 & 3 & 4 \\ 0 & 4 & -3 & 11 \\ 0 & 1 & 5 & -3 \end{bmatrix} \xrightarrow{R_2 \leftrightarrow R_3}$

$\begin{bmatrix} 1 & 2 & 3 & 4 \\ 0 & 1 & 5 & -3 \\ 0 & 4 & -3 & 11 \end{bmatrix}$

(ii) $\begin{bmatrix} 1 & 2 & 3 & 4 \\ 0 & 1 & 5 & -3 \\ 0 & 4 & -3 & 11 \end{bmatrix} \xrightarrow{-4R_2 + R_3 \rightarrow R_3}$

$\begin{bmatrix} 1 & 2 & 3 & 4 \\ 0 & 1 & 5 & -3 \\ 0 & 0 & -23 & 23 \end{bmatrix}$

(iii) $\begin{bmatrix} 1 & 2 & 3 & 4 \\ 0 & 1 & 5 & -3 \\ 0 & 0 & -23 & 23 \end{bmatrix} \xrightarrow{-\frac{1}{23}R_3}$

$\begin{bmatrix} 1 & 2 & 3 & 4 \\ 0 & 1 & 5 & -3 \\ 0 & 0 & 1 & -1 \end{bmatrix}$

33. $\begin{bmatrix} 4 & 5 & -7 \\ 5 & 4 & -2 \end{bmatrix} \xrightarrow{\frac{1}{4}R_1} \begin{bmatrix} 1 & \frac{5}{4} & -\frac{7}{4} \\ 5 & 4 & -2 \end{bmatrix}$

$\xrightarrow{-5R_1 + R_2 \rightarrow R_2} \begin{bmatrix} 1 & \frac{5}{4} & -\frac{7}{4} \\ 0 & -\frac{9}{4} & \frac{27}{4} \end{bmatrix}$

$\xrightarrow{-\frac{4}{9}R_2 \rightarrow R_2} \begin{bmatrix} 1 & \frac{5}{4} & -\frac{7}{4} \\ 0 & 1 & -3 \end{bmatrix}$

35. $\begin{bmatrix} 1 & 4 & 3 & 1 \\ 0 & -3 & -2 & 0 \\ 0 & 7 & 5 & -3 \end{bmatrix} \xrightarrow{-\frac{1}{3}R_2} \begin{bmatrix} 1 & 4 & 3 & 1 \\ 0 & 1 & \frac{2}{3} & 0 \\ 0 & 7 & 5 & -3 \end{bmatrix}$

$\xrightarrow{-7R_2 + R_3 \rightarrow R_3} \begin{bmatrix} 1 & 4 & 3 & 1 \\ 0 & 1 & \frac{2}{3} & 0 \\ 0 & 0 & \frac{1}{3} & -3 \end{bmatrix}$

$\xrightarrow{3R_3} \begin{bmatrix} 1 & 4 & 3 & 1 \\ 0 & 1 & \frac{2}{3} & 0 \\ 0 & 0 & 1 & -9 \end{bmatrix}$

37. No. The matrix does not have a step-like pattern that moves down and to the right. (Property 2)

39. Yes. The matrix is in reduced row-echelon form.

41. Yes. The matrix is in reduced row-echelon form.

43. Yes. The matrix is in reduced row-echelon form.

45. $\begin{cases} x + 2y = 1 \\ \quad\quad y = -2 \end{cases} \Rightarrow x + 2(-2) = 1 \Rightarrow x = 5$

The solution is $\{(5, -2)\}$.

47. $\begin{cases} x + 4y + 2z = 2 \\ \quad\quad\quad\quad z = 3 \end{cases} \Rightarrow x + 4y = -4 \Rightarrow x = -4y - 4$

The solution is $\left\{(-4y - 4, y, 3)\right\}$.

49. $\begin{cases} x + 2y + 3z = 2 \\ \quad y - 2z = 4 \\ \quad\quad\quad z = -1 \end{cases} \Rightarrow y - 2(-1) = 4 \Rightarrow y = 2$

$x + 2(2) + 3(-1) = 2 \Rightarrow x = 1$

The solution is $\{(1, 2, -1)\}$.

51. $\begin{cases} x = 2 \\ y = -5 \\ z + 2w = 3 \end{cases} \Rightarrow z = -2w + 3$

The solution is $\{(2, -5, -2w + 3, w)\}$.

53. $\begin{cases} x \quad\quad\quad = -5 \\ y \quad\quad\quad = 4 \\ z + 2w = 3 \\ \quad\quad w = 0 \end{cases} \Rightarrow z + 2(0) = 3 \Rightarrow z = 3$

The solution is $\{(-5, 4, 3, 0)\}$.

55. $\begin{bmatrix} 1 & -2 & | & 11 \\ 2 & -1 & | & 13 \end{bmatrix} \xrightarrow{-2R_1 + R_2 \to R_2} \begin{bmatrix} 1 & -2 & | & 11 \\ 0 & 3 & | & -9 \end{bmatrix}$

$\xrightarrow{\frac{1}{3}R_2 \to R_2} \begin{bmatrix} 1 & -2 & | & 11 \\ 0 & 1 & | & -3 \end{bmatrix} \Rightarrow$

$y = -3; x - 2(-3) = 11 \Rightarrow x = 5$

The solution is $\{(5, -3)\}$.

57. $\begin{bmatrix} 2 & -3 & | & 3 \\ 4 & -1 & | & 11 \end{bmatrix} \xrightarrow{\frac{1}{2}R_1 \to R_1} \begin{bmatrix} 1 & -\frac{3}{2} & | & \frac{3}{2} \\ 4 & -1 & | & 11 \end{bmatrix}$

$\xrightarrow{-4R_1 + R_2 \to R_2} \begin{bmatrix} 1 & -\frac{3}{2} & | & \frac{3}{2} \\ 0 & 5 & | & 5 \end{bmatrix}$

$\xrightarrow{\frac{1}{5}R_2 \to R_2} \begin{bmatrix} 1 & -\frac{3}{2} & | & \frac{3}{2} \\ 0 & 1 & | & 1 \end{bmatrix} \Rightarrow$

$y = 1; x - \frac{3}{2}(1) = \frac{3}{2} \Rightarrow x = 3$

The solution is $\{(3, 1)\}$.

59. $\begin{bmatrix} 3 & -5 & | & 4 \\ 4 & -15 & | & 13 \end{bmatrix} \xrightarrow{\frac{1}{3}R_1 \to R_1} \begin{bmatrix} 1 & -\frac{5}{3} & | & \frac{4}{3} \\ 4 & -15 & | & 13 \end{bmatrix}$

$\xrightarrow{-4R_1 + R_2 \to R_2} \begin{bmatrix} 1 & -\frac{5}{3} & | & \frac{4}{3} \\ 0 & -\frac{25}{3} & | & \frac{23}{3} \end{bmatrix}$

$\xrightarrow{-\frac{3}{25}R_2 \to R_2} \begin{bmatrix} 1 & -\frac{5}{3} & | & \frac{4}{3} \\ 0 & 1 & | & -\frac{23}{25} \end{bmatrix} \Rightarrow$

$y = -\frac{23}{25}; x - \left(\frac{5}{3}\right)\left(-\frac{23}{25}\right) = \frac{4}{3} \Rightarrow x = -\frac{1}{5}$

The solution is $\left\{\left(-\frac{1}{5}, -\frac{23}{25}\right)\right\}$.

61. $\begin{bmatrix} 1 & -1 & | & 1 \\ 2 & 1 & | & 5 \\ 3 & -4 & | & 2 \end{bmatrix} \xrightarrow[-3R_1 + R_3 \to R_3]{-2R_1 + R_2 \to R_2} \begin{bmatrix} 1 & -1 & | & 1 \\ 0 & 3 & | & 3 \\ 0 & -1 & | & -1 \end{bmatrix}$

$\xrightarrow[-R_3 \to R_3]{\frac{1}{3}R_2 \to R_2} \begin{bmatrix} 1 & -1 & | & 1 \\ 0 & 1 & | & 1 \\ 0 & 1 & | & 1 \end{bmatrix} \Rightarrow$

$y = 1; x - 1 = 1 \Rightarrow x = 2$

The solution is $\{(2, 1)\}$.

63. $\begin{bmatrix} 1 & 1 & 1 & | & 6 \\ 1 & -1 & 1 & | & 2 \\ 2 & 1 & -1 & | & 1 \end{bmatrix}$

$\xrightarrow[-2R_1 + R_3 \to R_3]{R_1 - R_2 \to R_2} \begin{bmatrix} 1 & 1 & 1 & | & 6 \\ 0 & 2 & 0 & | & 4 \\ 0 & -1 & -3 & | & -11 \end{bmatrix}$

$\xrightarrow{\frac{1}{2}R_2 \to R_2} \begin{bmatrix} 1 & 1 & 1 & | & 6 \\ 0 & 1 & 0 & | & 2 \\ 0 & -1 & -3 & | & -11 \end{bmatrix}$

$\xrightarrow{R_2 + R_3 \to R_3} \begin{bmatrix} 1 & 1 & 1 & | & 6 \\ 0 & 1 & 0 & | & 2 \\ 0 & 0 & -3 & | & -9 \end{bmatrix}$

$\xrightarrow{-\frac{1}{3}R_3 \to R_3} \begin{bmatrix} 1 & 1 & 1 & | & 6 \\ 0 & 1 & 0 & | & 2 \\ 0 & 0 & 1 & | & 3 \end{bmatrix} \Rightarrow$

$z = 3; y = 2; x + 2 + 3 = 6 \Rightarrow x = 1$

The solution is $\{(1, 2, 3)\}$.

65. $\begin{bmatrix} 2 & 3 & -1 & | & 9 \\ 1 & 1 & 1 & | & 9 \\ 3 & -1 & -1 & | & -1 \end{bmatrix}$

$\xrightarrow{R_1 \leftrightarrow R_2} \begin{bmatrix} 1 & 1 & 1 & | & 9 \\ 2 & 3 & -1 & | & 9 \\ 3 & -1 & -1 & | & -1 \end{bmatrix}$

$\xrightarrow[{-3R_1+R_3 \to R_3}]{-2R_1+R_2 \to R_2} \begin{bmatrix} 1 & 1 & 1 & | & 9 \\ 0 & 1 & -3 & | & -9 \\ 0 & -4 & -4 & | & -28 \end{bmatrix}$

$\xrightarrow{4R_2+R_3 \to R_3} \begin{bmatrix} 1 & 1 & 1 & | & 9 \\ 0 & 1 & -3 & | & -9 \\ 0 & 0 & -16 & | & -64 \end{bmatrix}$

$\xrightarrow{-\frac{1}{16}R_3 \to R_3} \begin{bmatrix} 1 & 1 & 1 & | & 9 \\ 0 & 1 & -3 & | & -9 \\ 0 & 0 & 1 & | & 4 \end{bmatrix} \Rightarrow$

$z = 4; \ y - 3(4) = -9 \Rightarrow y = 3$
$x + 3 + 4 = 9 \Rightarrow x = 2$
The solution is $\{(2, 3, 4)\}$.

67. $\begin{bmatrix} 3 & 2 & 4 & | & 19 \\ 2 & -1 & 1 & | & 3 \\ 6 & 7 & -1 & | & 17 \end{bmatrix}$

$\xrightarrow{R_1 \leftrightarrow \frac{1}{2}R_2} \begin{bmatrix} 1 & -\frac{1}{2} & \frac{1}{2} & | & \frac{3}{2} \\ 3 & 2 & 4 & | & 19 \\ 6 & 7 & -1 & | & 17 \end{bmatrix}$

$\xrightarrow[{-6R_1+R_3 \to R_3}]{-3R_1+R_2 \to R_2} \begin{bmatrix} 1 & -\frac{1}{2} & \frac{1}{2} & | & \frac{3}{2} \\ 0 & \frac{7}{2} & \frac{5}{2} & | & \frac{29}{2} \\ 0 & 10 & -4 & | & 8 \end{bmatrix}$

$\xrightarrow{\frac{2}{7}R_2 \to R_2} \begin{bmatrix} 1 & -\frac{1}{2} & \frac{1}{2} & | & \frac{3}{2} \\ 0 & 1 & \frac{5}{7} & | & \frac{29}{7} \\ 0 & 10 & -4 & | & 8 \end{bmatrix}$

$\xrightarrow{-10R_2+R_3 \to R_3} \begin{bmatrix} 1 & -\frac{1}{2} & \frac{1}{2} & | & \frac{3}{2} \\ 0 & 1 & \frac{5}{7} & | & \frac{29}{7} \\ 0 & 0 & -\frac{78}{7} & | & -\frac{234}{7} \end{bmatrix}$

$\xrightarrow{-\frac{7}{78}R_3 \to R_3} \begin{bmatrix} 1 & -\frac{1}{2} & \frac{1}{2} & | & \frac{3}{2} \\ 0 & 1 & \frac{5}{7} & | & \frac{29}{7} \\ 0 & 0 & 1 & | & 3 \end{bmatrix} \Rightarrow$

$z = 3$
$y + \left(\frac{5}{7}\right)(3) = \frac{29}{7} \Rightarrow y = 2$
$x - \frac{1}{2}(2) + \frac{1}{2}(3) = \frac{3}{2} \Rightarrow x = 1$
The solution is $\{(1, 2, 3)\}$.

69. $\begin{bmatrix} 1 & -1 & 0 & | & 1 \\ 1 & 0 & -1 & | & -1 \\ 2 & 1 & -1 & | & 3 \end{bmatrix}$

$\xrightarrow[{-2R_1+R_3 \to R_3}]{R_1-R_2 \to R_2} \begin{bmatrix} 1 & -1 & 0 & | & 1 \\ 0 & -1 & 1 & | & 2 \\ 0 & 3 & -1 & | & 1 \end{bmatrix}$

$\xrightarrow[{3R_2+R_3 \to R_3}]{-R_2+R_1 \to R_1} \begin{bmatrix} 1 & 0 & -1 & | & -1 \\ 0 & -1 & 1 & | & 2 \\ 0 & 0 & 2 & | & 7 \end{bmatrix}$

$\xrightarrow{\frac{1}{2}R_3 \to R_3} \begin{bmatrix} 1 & 0 & -1 & | & -1 \\ 0 & -1 & 1 & | & 2 \\ 0 & 0 & 1 & | & \frac{7}{2} \end{bmatrix}$

$\xrightarrow[{-R_2+R_3 \to R_2}]{R_3+R_1 \to R_1} \begin{bmatrix} 1 & 0 & 0 & | & \frac{5}{2} \\ 0 & 1 & 0 & | & \frac{3}{2} \\ 0 & 0 & 1 & | & \frac{7}{2} \end{bmatrix} \Rightarrow$

$x = \frac{5}{2}, y = \frac{3}{2}, z = \frac{7}{2}$
The solution is $\left\{ \left(\frac{5}{2}, \frac{3}{2}, \frac{7}{2} \right) \right\}$.

71. $\begin{bmatrix} 1 & 1 & -1 & | & 4 \\ 1 & 3 & 5 & | & 10 \\ 3 & 5 & 3 & | & 18 \end{bmatrix}$

$\xrightarrow[-3R_1+R_3\to R_3]{R_1-R_2\to R_2} \begin{bmatrix} 1 & 1 & -1 & | & 4 \\ 0 & -2 & -6 & | & -6 \\ 0 & 2 & 6 & | & 6 \end{bmatrix}$

$\xrightarrow{R_2+R_3\to R_3} \begin{bmatrix} 1 & 1 & -1 & | & 4 \\ 0 & -2 & -6 & | & -6 \\ 0 & 0 & 0 & | & 0 \end{bmatrix}$

$\xrightarrow{-\frac{1}{2}R_2\to R_2} \begin{bmatrix} 1 & 1 & -1 & | & 4 \\ 0 & 1 & 3 & | & 3 \\ 0 & 0 & 0 & | & 0 \end{bmatrix}$

$\xrightarrow{R_1-R_2\to R_1} \begin{bmatrix} 1 & 0 & -4 & | & 1 \\ 0 & 1 & 3 & | & 3 \\ 0 & 0 & 0 & | & 0 \end{bmatrix} \Rightarrow$

$x - 4z = 1 \Rightarrow x = 1 + 4z$
$y + 3z = 3 \Rightarrow y = 3 - 3z$
The solution is $\{(1+4z, 3-3z, z)\}$

73. $\begin{bmatrix} 1 & 2 & -1 & | & 6 \\ 3 & 1 & 2 & | & 3 \\ 2 & 5 & 3 & | & 9 \end{bmatrix}$

$\xrightarrow[-2R_1+R_3\to R_3]{-3R_1+R_2\to R_2} \begin{bmatrix} 1 & 2 & -1 & | & 6 \\ 0 & -5 & 5 & | & -15 \\ 0 & 1 & 5 & | & -3 \end{bmatrix}$

$\xrightarrow{-\frac{1}{5}R_2\to R_2} \begin{bmatrix} 1 & 2 & -1 & | & 6 \\ 0 & 1 & -1 & | & 3 \\ 0 & 1 & 5 & | & -3 \end{bmatrix}$

$\xrightarrow[R_2-R_3\to R_3]{-2R_2+R_1\to R_1} \begin{bmatrix} 1 & 0 & 1 & | & 0 \\ 0 & 1 & -1 & | & 3 \\ 0 & 0 & -6 & | & 6 \end{bmatrix}$

$\xrightarrow{-\frac{1}{6}R_3\to R_3} \begin{bmatrix} 1 & 0 & 1 & | & 0 \\ 0 & 1 & -1 & | & 3 \\ 0 & 0 & 1 & | & -1 \end{bmatrix}$

$\xrightarrow[R_2+R_3\to R_2]{R_1-R_3\to R_1} \begin{bmatrix} 1 & 0 & 0 & | & 1 \\ 0 & 1 & 0 & | & 2 \\ 0 & 0 & 1 & | & -1 \end{bmatrix} \Rightarrow$

$x = 1, y = 2, z = -1$
The solution is $\{(1, 2, -1)\}$.

75. a. There is one solution: $\{(2, 3)\}$.

  b. There are infinitely many solutions:
  $\{(-2y + 2, y, 2)\}$.

  c. There is no solution.

77. False. Each row of matrix A has seven entries.

6.1 Applying the Concepts

79. a. $\begin{cases} a = 0.1b + 1000 \\ b = 0.2a + 780 \end{cases}$

  b. Rewrite the system as
  $\begin{cases} a - 0.1b = 1000 \\ -0.2a + b = 780 \end{cases} \Rightarrow \begin{bmatrix} 1 & -0.1 & | & 1000 \\ -0.2 & 1 & | & 780 \end{bmatrix}$

  c. $\begin{bmatrix} 1 & -0.1 & | & 1000 \\ -0.2 & 1 & | & 780 \end{bmatrix}$

  $\xrightarrow{0.2R_1+R_2\to R_2} \begin{bmatrix} 1 & -0.1 & | & 1000 \\ 0 & 0.98 & | & 980 \end{bmatrix}$

  $\xrightarrow{\frac{100}{98}R_2\to R_2} \begin{bmatrix} 1 & -0.1 & | & 1000 \\ 0 & 1 & | & 1000 \end{bmatrix}$

  $\xrightarrow{0.1R_2+R_1\to R_1} \begin{bmatrix} 1 & 0 & | & 1100 \\ 0 & 1 & | & 1000 \end{bmatrix} \Rightarrow$

  $a = 1100, b = 1000$

81. a. $\begin{cases} l = 0.4t + 0.2f + 10,000 \\ t = 0.5l + 0.3t + 20,000 \\ f = 0.5l + 0.05t + 0.35f + 10,000 \end{cases}$

  b. Rewrite the system as
  $\begin{cases} l - 0.4t - 0.2f = 10,000 \\ -0.5l + 0.7t - 0f = 20,000 \\ -0.5l - 0.05t + 0.65f = 10,000 \end{cases} \Rightarrow$

  $\begin{bmatrix} 1 & -0.4 & -0.2 & | & 10,000 \\ -0.5 & 0.7 & 0 & | & 20,000 \\ -0.5 & -0.05 & 0.65 & | & 10,000 \end{bmatrix}$

  c. $\begin{bmatrix} 1 & -0.4 & -0.2 & | & 10,000 \\ -0.5 & 0.7 & 0 & | & 20,000 \\ -0.5 & -0.05 & 0.65 & | & 10,000 \end{bmatrix}$

  $\xrightarrow[R_1+2R_3\to R_3]{R_1+2R_2\to R_2} \begin{bmatrix} 1 & -0.4 & -0.2 & | & 10,000 \\ 0 & 1 & -0.2 & | & 50,000 \\ 0 & -0.5 & 1.1 & | & 30,000 \end{bmatrix}$

  $\xrightarrow{R_2+2R_3\to R_3} \begin{bmatrix} 1 & -0.4 & -0.2 & | & 10,000 \\ 0 & 1 & -0.2 & | & 50,000 \\ 0 & 0 & 2 & | & 110,000 \end{bmatrix}$

  $\xrightarrow[-\frac{1}{2}R_3\to R_3]{R_1+0.4R_2\to R_1} \begin{bmatrix} 1 & 0 & -0.28 & | & 30,000 \\ 0 & 1 & -0.2 & | & 50,000 \\ 0 & 0 & 1 & | & 55,000 \end{bmatrix}$

  $\xrightarrow[R_2+0.2R_3\to R_2]{R_1+0.28R_3\to R_1} \begin{bmatrix} 1 & 0 & 0 & | & 45,400 \\ 0 & 1 & 0 & | & 61,000 \\ 0 & 0 & 1 & | & 55,000 \end{bmatrix}$

  $l = \$45,400, t = \$61,000, f = \$55,000$

83. $T_1 = \dfrac{0+0+300+T_2}{4} \Rightarrow 4T_1 - T_2 = 300$

$T_2 = \dfrac{T_1+T_3+190+200}{4} \Rightarrow -T_1 + 4T_2 - T_3 = 390$

$T_3 = \dfrac{0+0+100+T_2}{4} \Rightarrow -T_2 + 4T_3 = 100$

$$\begin{bmatrix} 4 & -1 & 0 & | & 300 \\ -1 & 4 & -1 & | & 390 \\ 0 & -1 & 4 & | & 100 \end{bmatrix}$$

$\xrightarrow{R_1 \leftrightarrow -R_2} \begin{bmatrix} 1 & -4 & 1 & | & -390 \\ 4 & -1 & 0 & | & 300 \\ 0 & -1 & 4 & | & 100 \end{bmatrix}$

$\xrightarrow{-4R_1 + R_2 \to R_2} \begin{bmatrix} 1 & -4 & -1 & | & -390 \\ 0 & 15 & -4 & | & 1860 \\ 0 & -1 & 4 & | & 100 \end{bmatrix}$

$\xrightarrow{R_2 + R_3 \to R_2} \begin{bmatrix} 1 & -4 & 1 & | & -390 \\ 0 & 14 & 0 & | & 1960 \\ 0 & -1 & 4 & | & 100 \end{bmatrix}$

$\xrightarrow{\frac{1}{14}R_2 \to R_2} \begin{bmatrix} 1 & -4 & 1 & | & -390 \\ 0 & 1 & 0 & | & 140 \\ 0 & -1 & 4 & | & 100 \end{bmatrix}$

$\xrightarrow{R_2 + R_3 \to R_3} \begin{bmatrix} 1 & -4 & 1 & | & -390 \\ 0 & 1 & 0 & | & 140 \\ 0 & 0 & 4 & | & 240 \end{bmatrix}$

$\xrightarrow[\frac{1}{4}R_3 \to R_3]{R_1 + 4R_2 \to R_1} \begin{bmatrix} 1 & 0 & 1 & | & 170 \\ 0 & 1 & 0 & | & 140 \\ 0 & 0 & 1 & | & 60 \end{bmatrix}$

$\xrightarrow{R_1 - R_3 \to R_1} \begin{bmatrix} 1 & 0 & 0 & | & 110 \\ 0 & 1 & 0 & | & 140 \\ 0 & 0 & 1 & | & 60 \end{bmatrix} \Rightarrow$

$T_1 = 110,\ T_2 = 140,\ T_3 = 60$

85. a. $\begin{cases} x - y & = 270 - 200 \\ -x + \quad z = 180 - 300 \\ \quad y - z = 40 + 70 - 60 \end{cases}$

$\Rightarrow \begin{cases} x - y & = 70 \\ -x + \quad z = -120 \\ \quad y - z = 50 \end{cases}$

b.
$$\begin{bmatrix} 1 & -1 & 0 & | & 70 \\ -1 & 0 & 1 & | & -120 \\ 0 & 1 & -1 & | & 50 \end{bmatrix}$$

$\xrightarrow{R_1 + R_2 \to R_2} \begin{bmatrix} 1 & -1 & 0 & | & 70 \\ 0 & -1 & 1 & | & -50 \\ 0 & 1 & -1 & | & 50 \end{bmatrix}$

$\xrightarrow[R_2 + R_3 \to R_3]{R_1 - R_2 \to R_1} \begin{bmatrix} 1 & 0 & -1 & | & 120 \\ 0 & -1 & 1 & | & -50 \\ 0 & 0 & 0 & | & 0 \end{bmatrix} \Rightarrow$

$x - z = 120 \Rightarrow x = z + 120$

$-y + z = -50 \Rightarrow y = z + 50$

The solution is $\{(z+120, z+50, z)\}$.

There are $300 + 200 + 60 = 560$ cars entering the system, so

$0 \le (z+120) + (z+50) + z \le 560$

$0 \le 3z + 170 \le 560$

$0 \le 3z \le 390$

$0 \le z \le 130$

Thus, the system has 131 solutions.

87. $9 = a(-1)^2 + b(-1) + c = a - b + c$

$3 = a(1)^2 + b(1) + c = a + b + c$

$6 = a(2)^2 + b(2) + c = 4a + 2b + c$

$$\begin{bmatrix} 1 & -1 & 1 & | & 9 \\ 1 & 1 & 1 & | & 3 \\ 4 & 2 & 1 & | & 6 \end{bmatrix}$$

$\xrightarrow[-4R_1 + R_3 \to R_3]{-R_1 + R_2 \to R_2} \begin{bmatrix} 1 & -1 & 1 & | & 9 \\ 0 & 2 & 0 & | & -6 \\ 0 & 6 & -3 & | & -30 \end{bmatrix}$

$\xrightarrow[-\frac{1}{3}R_3 \to R_3]{\frac{1}{2}R_2 \to R_2} \begin{bmatrix} 1 & -1 & 1 & | & 9 \\ 0 & 1 & 0 & | & -3 \\ 0 & -2 & 1 & | & 10 \end{bmatrix}$

$\xrightarrow{2R_2 + R_3 \to R_3} \begin{bmatrix} 1 & -1 & 1 & | & 9 \\ 0 & 1 & 0 & | & -3 \\ 0 & 0 & 1 & | & 4 \end{bmatrix}$

$\xrightarrow{R_1 + R_2 \to R_1} \begin{bmatrix} 1 & 0 & 1 & | & 6 \\ 0 & 1 & 0 & | & -3 \\ 0 & 0 & 1 & | & 4 \end{bmatrix}$

$\xrightarrow{R_1 - R_3 \to R_1} \begin{bmatrix} 1 & 0 & 0 & | & 2 \\ 0 & 1 & 0 & | & -3 \\ 0 & 0 & 1 & | & 4 \end{bmatrix}$

Thus, $a = 2$, $b = -3$, and $c = 4$.

The equation is $y = 2x^2 - 3x + 4$.

6.1 Beyond the Basics

89. $\begin{bmatrix} 1 & 1 & 1 & 1 & | & 0 \\ 1 & 3 & 2 & 4 & | & 0 \\ 2 & 0 & 1 & -1 & | & 0 \end{bmatrix}$

$\xrightarrow[\;-2R_1+R_3 \to R_3\;]{R_1-R_2 \to R_2}$ $\begin{bmatrix} 1 & 1 & 1 & 1 & | & 0 \\ 0 & -2 & -1 & -3 & | & 0 \\ 0 & -2 & -1 & -3 & | & 0 \end{bmatrix}$

$\xrightarrow[\;R_2-R_3 \to R_3\;]{\frac{1}{2}R_2+R_1 \to R_1}$ $\begin{bmatrix} 1 & 0 & \frac{1}{2} & -\frac{1}{2} & | & 0 \\ 0 & 1 & \frac{1}{2} & \frac{3}{2} & | & 0 \\ 0 & 0 & 0 & 0 & | & 0 \end{bmatrix}$

$\xrightarrow[]{-\frac{1}{2}R_2 \to R_2}$ $\begin{bmatrix} 1 & 0 & \frac{1}{2} & -\frac{1}{2} & | & 0 \\ 0 & 1 & \frac{1}{2} & \frac{3}{2} & | & 0 \\ 0 & 0 & 0 & 0 & | & 0 \end{bmatrix} \Rightarrow$

$y + \frac{1}{2}z + \frac{3}{2}w = 0 \Rightarrow z = -2y - 3w$

$x + \frac{1}{2}z - \frac{1}{2}w = 0 \Rightarrow z = w - 2x$

$-2y - 3w = w - 2x \Rightarrow x = 2w + y$

The solution is $\left\{ \left(y + 2w,\ y,\ -2y - 3w,\ w \right) \right\}$.

91. a. $\begin{bmatrix} 1 & 2 & -3 & 1 \\ 1 & 0 & -3 & 2 \\ 0 & 1 & 1 & 0 \\ 2 & 3 & 0 & -2 \end{bmatrix}$

$\xrightarrow[]{R_1 \leftrightarrow R_4}$ $\begin{bmatrix} 2 & 3 & 0 & -2 \\ 1 & 0 & -3 & 2 \\ 0 & 1 & 1 & 0 \\ 1 & 2 & -3 & 1 \end{bmatrix}$

$\xrightarrow[\;R_1-2R_4 \to R_4\;]{R_2-R_4 \to R_2}$ $\begin{bmatrix} 2 & 3 & 0 & -2 \\ 0 & -2 & 0 & 1 \\ 0 & 1 & 1 & 0 \\ 0 & -1 & 6 & -4 \end{bmatrix}$

$\xrightarrow[\;R_2+2R_3 \to R_3\;]{R_2-2R_4 \to R_4}$ $\begin{bmatrix} 2 & 3 & 0 & -2 \\ 0 & -2 & 0 & 1 \\ 0 & 0 & 2 & 1 \\ 0 & 0 & -12 & 9 \end{bmatrix}$

$\xrightarrow[]{6R_3+R_4 \to R_4}$ $\begin{bmatrix} 2 & 3 & 0 & -2 \\ 0 & -2 & 0 & 1 \\ 0 & 0 & 2 & 1 \\ 0 & 0 & 0 & 15 \end{bmatrix}$

$\xrightarrow[\substack{\frac{1}{2}R_3 \to R_3 \\ \frac{1}{15}R_4 \to R_4}]{\substack{\frac{1}{2}R_1 \to R_1 \\ -\frac{1}{2}R_2 \to R_2}}$ $\begin{bmatrix} 1 & \frac{3}{2} & 0 & -1 \\ 0 & 1 & 0 & -\frac{1}{2} \\ 0 & 0 & 1 & \frac{1}{2} \\ 0 & 0 & 0 & 1 \end{bmatrix} = B$

$\begin{bmatrix} 1 & 2 & -3 & 1 \\ 1 & 0 & -3 & 2 \\ 0 & 1 & 1 & 0 \\ 2 & 3 & 0 & -2 \end{bmatrix}$

$\xrightarrow[\;2R_1-R_4 \to R_4\;]{R_1-R_2 \to R_2}$ $\begin{bmatrix} 1 & 2 & -3 & 1 \\ 0 & 2 & 0 & -1 \\ 0 & 1 & 1 & 0 \\ 0 & 1 & -6 & 4 \end{bmatrix}$

$\xrightarrow[]{R_2-2R_4 \to R_4}$ $\begin{bmatrix} 1 & 2 & -3 & 1 \\ 0 & 2 & 0 & -1 \\ 0 & 1 & 1 & 0 \\ 0 & 0 & 12 & -9 \end{bmatrix}$

$\xrightarrow[\;\frac{1}{12}R_4 \to R_4\;]{\frac{1}{2}R_2 \to R_2}$ $\begin{bmatrix} 1 & 2 & -3 & 1 \\ 0 & 1 & 0 & -\frac{1}{2} \\ 0 & 1 & 1 & 0 \\ 0 & 0 & 1 & -\frac{3}{4} \end{bmatrix}$

$\xrightarrow[]{R_2-R_3 \to R_3}$ $\begin{bmatrix} 1 & 2 & -3 & 1 \\ 0 & 1 & 0 & -\frac{1}{2} \\ 0 & 0 & -1 & -\frac{1}{2} \\ 0 & 0 & 1 & -\frac{3}{4} \end{bmatrix}$

$\xrightarrow[]{R_3+R_4 \to R_4}$ $\begin{bmatrix} 1 & 2 & -3 & 1 \\ 0 & 1 & 0 & -\frac{1}{2} \\ 0 & 0 & -1 & -\frac{1}{2} \\ 0 & 0 & 0 & -\frac{5}{4} \end{bmatrix}$

$\xrightarrow[\;-\frac{4}{5}R_4 \to R_4\;]{-R_3 \to R_3}$ $\begin{bmatrix} 1 & 2 & -3 & 1 \\ 0 & 1 & 0 & -\frac{1}{2} \\ 0 & 0 & 1 & \frac{1}{2} \\ 0 & 0 & 0 & 1 \end{bmatrix} = C$

b. Use a calculator to show that the reduced row-echelon form of all three matrices is

$$\begin{bmatrix} 1 & 0 & 0 & 0 \\ 0 & 1 & 0 & 0 \\ 0 & 0 & 1 & 0 \\ 0 & 0 & 0 & 1 \end{bmatrix}.$$

93. a.
$$\begin{bmatrix} a & b & m \\ c & d & n \end{bmatrix} \xrightarrow[\frac{1}{c}R_2 \to R_2]{\frac{1}{a}R_1 \to R_1} \begin{bmatrix} 1 & \dfrac{b}{a} & \dfrac{m}{a} \\ 1 & \dfrac{d}{c} & \dfrac{n}{c} \end{bmatrix}$$

$$\xrightarrow{R_1 - R_2 \to R_2} \begin{bmatrix} 1 & \dfrac{b}{a} & \dfrac{m}{a} \\ 0 & \dfrac{b}{a} - \dfrac{d}{c} & \dfrac{m}{a} - \dfrac{n}{c} \end{bmatrix}$$

$$= \begin{bmatrix} 1 & \dfrac{b}{a} & \dfrac{m}{a} \\ 0 & \dfrac{bc - ad}{ac} & \dfrac{cm - an}{ac} \end{bmatrix}$$

$$\xrightarrow{\frac{ac}{bc-da}R_2 \to R_2} \begin{bmatrix} 1 & \dfrac{b}{a} & \dfrac{m}{a} \\ 0 & 1 & \dfrac{cm - an}{bc - ad} \end{bmatrix}$$

$$\xrightarrow{-\frac{b}{a}R_2 + R_1 \to R_1} \begin{bmatrix} 1 & 0 & \dfrac{dm - bn}{ad - bc} \\ 0 & 1 & \dfrac{cm - an}{bc - ad} \end{bmatrix}$$

The solution is $\left(\dfrac{dm - bn}{ad - bc}, \dfrac{cm - an}{bc - ad} \right)$.

b. **(i)** There is a unique solution if $bc \neq ad$.

    (ii) There is no solution if $bc = ad$ and
$$\dfrac{m}{b} \neq \dfrac{n}{d}.$$

    (iii) There are infinitely many solutions if
$$bc = ad \text{ and } \dfrac{m}{b} = \dfrac{n}{d}.$$

95. Using the hint suggested, we have
$$\begin{bmatrix} 1 & 1 & 1 & | & 6 \\ 3 & -1 & 3 & | & 10 \\ 5 & 5 & -4 & | & 3 \end{bmatrix}$$

$$\xrightarrow[-5R_1 + R_3 \to R_3]{-3R_1 + R_2 \to R_2} \begin{bmatrix} 1 & 1 & 1 & | & 6 \\ 0 & -4 & 0 & | & -8 \\ 0 & 0 & -9 & | & -27 \end{bmatrix}$$

$$\xrightarrow[-\frac{1}{9}R_3 \to R_3]{-\frac{1}{4}R_2 \to R_2} \begin{bmatrix} 1 & 1 & 1 & | & 6 \\ 0 & 1 & 0 & | & 2 \\ 0 & 0 & 1 & | & 3 \end{bmatrix} \Rightarrow$$

$w = 3, v = 2, u + 2 + 3 = 6 \Rightarrow u = 1$
$\log x = 1 \Rightarrow x = 10;$
$\log y = 2 \Rightarrow y = 100$
$\log z = 3 \Rightarrow y = 1000$
The solution is $\{(10, 100, 1000)\}$.

97. $\begin{cases} a(1)^3 + b(1)^2 + c(1) + d = 5 \\ a(-1)^3 + b(-1)^2 + c(-1) + d = 1 \\ a(2)^3 + b(2)^2 + c(2) + d = 7 \\ a(-2)^3 + b(-2)^2 + c(-2) + d = 11 \end{cases} \Rightarrow$

$\begin{cases} a + b + c + d = 5 \\ -a + b - c + d = 1 \\ 8a + 4b + 2c + d = 7 \\ -8a + 4b - 2c + d = 11 \end{cases}$

Using a graphing calculator, we have

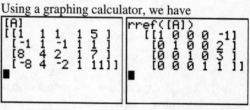

The equation is $y = -x^3 + 2x^2 + 3x + 1$.

6.1 Critical Thinking/Discussion/Writing

99. If $a \neq 0$,

$$\begin{bmatrix} a \\ b \end{bmatrix} \xrightarrow[\frac{1}{b}R_2 \to R_2]{\frac{1}{a}R_1 \to R_1} \begin{bmatrix} 1 \\ 1 \end{bmatrix} \xrightarrow{R_1 - R_2 \to R_2} \begin{bmatrix} 1 \\ 0 \end{bmatrix}.$$

If $a = 0$, $b = 0$, $\begin{bmatrix} a \\ b \end{bmatrix} = \begin{bmatrix} 0 \\ 0 \end{bmatrix}$.

100. If the matrix is $\begin{bmatrix} 0 & 0 \\ 0 & 0 \end{bmatrix}$, then it is in reduced row-echelon form already. If the matrix is $\begin{bmatrix} 1 & k \\ 0 & 0 \end{bmatrix}$ (for k any real number), then it is in reduced row-echelon form already.

If the matrix is $\begin{bmatrix} 0 & 1 \\ 0 & 0 \end{bmatrix}$ then it is in reduced row-echelon form already.

(continued on next page)

(continued)

If the matrix is $\begin{bmatrix} a & b \\ c & d \end{bmatrix}$, then we have

$$\begin{bmatrix} a & b \\ c & d \end{bmatrix} \xrightarrow[\frac{1}{c}R_2 \to R_2]{\frac{1}{a}R_1 \to R_1} \begin{bmatrix} 1 & \dfrac{b}{a} \\ 1 & \dfrac{d}{c} \end{bmatrix}$$

$$\xrightarrow{R_1 - R_2 \to R_2} \begin{bmatrix} 1 & \dfrac{b}{a} \\ 0 & \dfrac{b}{a} - \dfrac{d}{c} \end{bmatrix} = \begin{bmatrix} 1 & \dfrac{b}{a} \\ 0 & \dfrac{bc-ad}{ac} \end{bmatrix}$$

$$\xrightarrow{\frac{ac}{bc-ad}R_2 \to R_2} \begin{bmatrix} 1 & \dfrac{b}{a} \\ 0 & 1 \end{bmatrix}$$

$$\xrightarrow{-\frac{b}{a}R_2 + R_1 \to R_1} \begin{bmatrix} 1 & 0 \\ 0 & 1 \end{bmatrix}$$

101. Yes. Use the inverse of the operation used to transform A to B.

102. a. True. For example,

$$\begin{bmatrix} 1 & 0 & 0 & 6 \\ 0 & 1 & 0 & 3 \\ 0 & 0 & 1 & 2 \end{bmatrix} \to \begin{bmatrix} 1 & 0 & 0 & 6 \\ 0 & 0 & 1 & 2 \end{bmatrix}$$

b. False. For example,

$$\begin{bmatrix} 1 & 0 & 0 & 6 \\ 0 & 1 & 0 & 3 \\ 0 & 0 & 1 & 2 \end{bmatrix} \to \begin{bmatrix} 1 & 0 & 6 \\ 0 & 0 & 3 \\ 0 & 1 & 2 \end{bmatrix}$$

6.1 Maintaining Skills

103. $3(x-1) = 5 - x \Rightarrow 3x - 3 = 5 - x \Rightarrow$
$4x = 8 \Rightarrow x = 2$
Solution set: $\{2\}$

105. $-3(x+4) + 2 = 8 - x \Rightarrow -3x - 12 + 2 = 8 - x \Rightarrow$
$-3x - 10 = 8 - x \Rightarrow -2x = 18 \Rightarrow x = -9$
Solution set: $\{-9\}$

Be sure to check your solutions in the original equations in exercises 107–110.

107. $\begin{cases} 2x - y = 5 & (1) \\ x + 2y = 25 & (2) \end{cases}$

From equation (1) we have $y = 2x - 5$.
Substituting this expression in equation (2) gives
$x + 2(2x - 5) = 25 \Rightarrow x + 4x - 10 = 25 \Rightarrow$
$5x = 35 \Rightarrow x = 7$
Substitute $x = 7$ in equation (1) and solve for y.
$2(7) - y = 5 \Rightarrow 14 - y = 5 \Rightarrow y = 9$
Solution set: $\{(7, 9)\}$

109. $\begin{cases} x + 3y = 6 & (1) \\ 2x + 6y = 8 & (2) \end{cases}$

Multiply equation (1) by -2, then add the two equations.

$$\begin{array}{r} -2x - 6y = -12 \\ 2x + 6y = 8 \\ \hline 0 = -4 \quad \text{False} \end{array}$$

Solution set: $\varnothing$

111. True

113. True. This is an example of the distributive property.

6.2 Matrix Algebra

6.2 Practice Problems

1. $\begin{bmatrix} 1 & 2x - y \\ x + y & 5 \end{bmatrix} = \begin{bmatrix} 1 & 1 \\ 2 & 5 \end{bmatrix} \Rightarrow$
$\begin{cases} 2x - y = 1 \\ x + y = 2 \end{cases} \Rightarrow x = 1, \ y = 1$

2. $\begin{bmatrix} 2 & -1 & 4 \\ 5 & 0 & 9 \end{bmatrix} + \begin{bmatrix} -8 & 2 & 9 \\ 7 & 3 & 6 \end{bmatrix} = \begin{bmatrix} -6 & 1 & 13 \\ 12 & 3 & 15 \end{bmatrix}$

3. $2A - 3B = 2\begin{bmatrix} 7 & -4 \\ 3 & 6 \\ 0 & -2 \end{bmatrix} - 3\begin{bmatrix} 1 & -3 \\ 2 & 2 \\ 5 & 8 \end{bmatrix}$

$$= \begin{bmatrix} 14 & -8 \\ 6 & 12 \\ 0 & -4 \end{bmatrix} - \begin{bmatrix} 3 & -9 \\ 6 & 6 \\ 15 & 24 \end{bmatrix}$$

$$= \begin{bmatrix} 11 & 1 \\ 0 & 6 \\ -15 & -28 \end{bmatrix}$$

4. $5\begin{bmatrix} 1 & -1 \\ 3 & 5 \end{bmatrix} + 3X = 2\begin{bmatrix} 2 & 7 \\ -3 & -5 \end{bmatrix} \Rightarrow \begin{bmatrix} 5 & -5 \\ 15 & 25 \end{bmatrix} + 3X = \begin{bmatrix} 4 & 14 \\ -6 & -10 \end{bmatrix} \Rightarrow$

$3X = \begin{bmatrix} 4 & 14 \\ -6 & -10 \end{bmatrix} - \begin{bmatrix} 5 & -5 \\ 15 & 25 \end{bmatrix} = \begin{bmatrix} -1 & 19 \\ -21 & -35 \end{bmatrix} \Rightarrow X = \begin{bmatrix} -\dfrac{1}{3} & \dfrac{19}{3} \\ -7 & -\dfrac{35}{3} \end{bmatrix}$

5. Yes, the product matrix AB is defined. A is a 2×3 matrix, while B is a 3×1 matrix. The product matrix has order 2×1.

6. $N = \begin{bmatrix} S & C & M \end{bmatrix} = \begin{bmatrix} 10 & 30 & 45 \end{bmatrix}$; $P = \begin{bmatrix} 41 \\ 26 \\ 19 \end{bmatrix}$

$NP = \begin{bmatrix} 10 & 30 & 45 \end{bmatrix}\begin{bmatrix} 41 \\ 26 \\ 19 \end{bmatrix} = 10(41) + 30(26) + 45(19) = \$2045 \text{ thousand}$

7. $\begin{bmatrix} 3 & -1 & 2 & 7 \end{bmatrix}\begin{bmatrix} -2 \\ 0 \\ 1 \\ 5 \end{bmatrix} = \begin{bmatrix} 3(-2) - 1(0) + 2(1) + 7(5) \end{bmatrix} = \begin{bmatrix} 31 \end{bmatrix}$

8. AB is not defined because A is a 2×2 matrix and B is a 3×2 matrix.

$BA = \begin{bmatrix} 8 & 1 \\ -2 & 6 \\ 0 & 4 \end{bmatrix}\begin{bmatrix} 5 & 0 \\ 2 & -1 \end{bmatrix} = \begin{bmatrix} 8(5) + 1(2) & 8(0) + 1(-1) \\ -2(5) + 6(2) & -2(0) + 6(-1) \\ 0(5) + 4(2) & 0(0) + 4(-1) \end{bmatrix} = \begin{bmatrix} 42 & -1 \\ 2 & -6 \\ 8 & -4 \end{bmatrix}$

9. $AB = \begin{bmatrix} 7 & 1 \\ 0 & 3 \end{bmatrix}\begin{bmatrix} 2 & -1 \\ 4 & 4 \end{bmatrix} = \begin{bmatrix} 7(2) + 1(4) & 7(-1) + 1(4) \\ 0(2) + 3(4) & 0(-1) + 3(4) \end{bmatrix} = \begin{bmatrix} 18 & -3 \\ 12 & 12 \end{bmatrix}$

$BA = \begin{bmatrix} 2 & -1 \\ 4 & 4 \end{bmatrix}\begin{bmatrix} 7 & 1 \\ 0 & 3 \end{bmatrix} = \begin{bmatrix} 2(7) - 1(0) & 2(1) - 1(3) \\ 4(7) + 4(0) & 4(1) + 4(3) \end{bmatrix} = \begin{bmatrix} 14 & -1 \\ 28 & 16 \end{bmatrix}$

10. $AD = \begin{bmatrix} 0 & 1 \\ 1 & 0.25 \end{bmatrix}\begin{bmatrix} 0 & 4 & 4 & 1 & 1 & 0 \\ 0 & 0 & 1 & 1 & 6 & 6 \end{bmatrix}$

$= \begin{bmatrix} 0(0) + 1(0) & 0(4) + 1(0) & 0(4) + 1(1) & 0(1) + 1(1) & 0(1) + 1(6) & 0(0) + 1(6) \\ 1(0) + 0.25(0) & 1(4) + 0.25(0) & 1(4) + 0.25(1) & 1(1) + 0.25(1) & 1(1) + 0.25(6) & 1(0) + 0.25(6) \end{bmatrix}$

$= \begin{bmatrix} 0 & 0 & 1 & 1 & 6 & 6 \\ 0 & 4 & 4.25 & 1.25 & 2.5 & 1.5 \end{bmatrix}$

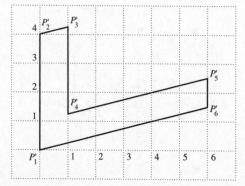

6.2 Basic Concepts and Skills

1. Two $m \times n$ matrices $A = \begin{bmatrix} a_{ij} \end{bmatrix}$ and $B = \begin{bmatrix} b_{ij} \end{bmatrix}$ are equal if $\underline{a_{ij} = b_{ij}}$ for all i and j.

3. The product of a $1 \times n$ matrix A and $n \times 1$ matrix B is $\underline{1 \times 1}$ matrix.

5. False. It is possible that $AB = BA$, but not necessarily true.

7. $\begin{bmatrix} 2 \\ -3 \end{bmatrix} = \begin{bmatrix} x \\ u \end{bmatrix} \Rightarrow x = 2, u = -3$

9. $\begin{bmatrix} 2 & x \\ y & -3 \end{bmatrix} = \begin{bmatrix} 2 & -1 \\ 3 & -3 \end{bmatrix} \Rightarrow x = -1, y = 3$

11. $\begin{bmatrix} 2x-3y & -4 \\ 5 & 3x+y \end{bmatrix} = \begin{bmatrix} 1 & -4 \\ 5 & 7 \end{bmatrix} \Rightarrow$

$\begin{cases} 2x-3y = 1 \\ 3x+y = 7 \end{cases} \Rightarrow x = 2, y = 1$

13. $\begin{bmatrix} x-y & 1 & 2 \\ 4 & 3x-2y & 3 \\ 5 & 6 & 5x-10y \end{bmatrix} = \begin{bmatrix} -1 & 1 & 2 \\ 4 & -1 & 3 \\ 5 & 6 & 6 \end{bmatrix} \Rightarrow$

$\begin{cases} x- y = -1 \\ 3x- 2y = -1 \\ 5x-10y = 6 \end{cases} \Rightarrow \begin{bmatrix} 1 & -1 & | & -1 \\ 3 & -2 & | & -1 \\ 5 & -10 & | & 6 \end{bmatrix}$

$\xrightarrow[-5R_1+R_3 \to R_3]{-3R_1+R_2 \to R_2} \begin{bmatrix} 1 & -1 & | & -1 \\ 0 & 1 & | & 2 \\ 0 & -5 & | & 11 \end{bmatrix}$

$\xrightarrow{5R_2+R_3 \to R_3} \begin{bmatrix} 1 & -1 & | & -1 \\ 0 & 1 & | & 2 \\ 0 & 0 & | & 21 \end{bmatrix} \Rightarrow 0 = 21$

There is no solution.

15. a. $A + B = \begin{bmatrix} 1 & 2 \\ 3 & 4 \end{bmatrix} + \begin{bmatrix} -1 & 0 \\ 2 & -3 \end{bmatrix} = \begin{bmatrix} 0 & 2 \\ 5 & 1 \end{bmatrix}$

b. $A - B = \begin{bmatrix} 1 & 2 \\ 3 & 4 \end{bmatrix} - \begin{bmatrix} -1 & 0 \\ 2 & -3 \end{bmatrix} = \begin{bmatrix} 2 & 2 \\ 1 & 7 \end{bmatrix}$

c. $-3A = -3\begin{bmatrix} 1 & 2 \\ 3 & 4 \end{bmatrix} = \begin{bmatrix} -3 & -6 \\ -9 & -12 \end{bmatrix}$

d. $3A - 2B = 3\begin{bmatrix} 1 & 2 \\ 3 & 4 \end{bmatrix} - 2\begin{bmatrix} -1 & 0 \\ 2 & -3 \end{bmatrix}$

$= \begin{bmatrix} 3 & 6 \\ 9 & 12 \end{bmatrix} - \begin{bmatrix} -2 & 0 \\ 4 & -6 \end{bmatrix} = \begin{bmatrix} 5 & 6 \\ 5 & 18 \end{bmatrix}$

e. $(A+B)^2 = \left(\begin{bmatrix} 1 & 2 \\ 3 & 4 \end{bmatrix} + \begin{bmatrix} -1 & 0 \\ 2 & -3 \end{bmatrix} \right)^2$

$= \left(\begin{bmatrix} 0 & 2 \\ 5 & 1 \end{bmatrix} \right)^2 = \begin{bmatrix} 0 & 2 \\ 5 & 1 \end{bmatrix} \begin{bmatrix} 0 & 2 \\ 5 & 1 \end{bmatrix}$

$= \begin{bmatrix} 0(0)+2(5) & 0(2)+2(1) \\ 5(0)+1(5) & 5(2)+1(1) \end{bmatrix}$

$= \begin{bmatrix} 10 & 2 \\ 5 & 11 \end{bmatrix}$

f. $A^2 = \begin{bmatrix} 1 & 2 \\ 3 & 4 \end{bmatrix} \begin{bmatrix} 1 & 2 \\ 3 & 4 \end{bmatrix}$

$= \begin{bmatrix} 1(1)+2(3) & 1(2)+2(4) \\ 3(1)+4(3) & 3(2)+4(4) \end{bmatrix} = \begin{bmatrix} 7 & 10 \\ 15 & 22 \end{bmatrix}$

$B^2 = \begin{bmatrix} -1 & 0 \\ 2 & -3 \end{bmatrix} \begin{bmatrix} -1 & 0 \\ 2 & -3 \end{bmatrix}$

$= \begin{bmatrix} -1(-1)+0(2) & -1(0)+0(-3) \\ 2(-1)-3(2) & 2(0)-3(-3) \end{bmatrix}$

$= \begin{bmatrix} 1 & 0 \\ -8 & 9 \end{bmatrix}$

$A^2 - B^2 = \begin{bmatrix} 7 & 10 \\ 15 & 22 \end{bmatrix} - \begin{bmatrix} 1 & 0 \\ -8 & 9 \end{bmatrix} = \begin{bmatrix} 6 & 10 \\ 23 & 13 \end{bmatrix}$

17. a. $A + B$ is not defined.

b. $A - B$ is not defined.

c. $-3A = -3\begin{bmatrix} 2 & 3 \\ -4 & 5 \end{bmatrix} = \begin{bmatrix} -6 & -9 \\ 12 & -15 \end{bmatrix}$

d. $3A - 2B$ is not defined

e. $(A+B)^2$ is not defined.

f. $A^2 - B^2$ is not defined.

19. a. $A + B = \begin{bmatrix} 4 & 0 & -1 \\ -2 & 5 & 2 \\ 0 & 0 & 1 \end{bmatrix} + \begin{bmatrix} 3 & 1 & 0 \\ 1 & -4 & 2 \\ 2 & 1 & 3 \end{bmatrix}$

$= \begin{bmatrix} 7 & 1 & -1 \\ -1 & 1 & 4 \\ 2 & 1 & 4 \end{bmatrix}$

b. $A - B = \begin{bmatrix} 4 & 0 & -1 \\ -2 & 5 & 2 \\ 0 & 0 & 1 \end{bmatrix} - \begin{bmatrix} 3 & 1 & 0 \\ 1 & -4 & 2 \\ 2 & 1 & 3 \end{bmatrix}$

$= \begin{bmatrix} 1 & -1 & -1 \\ -3 & 9 & 0 \\ -2 & -1 & -2 \end{bmatrix}$

c. $-3A = -3\begin{bmatrix} 4 & 0 & -1 \\ -2 & 5 & 2 \\ 0 & 0 & 1 \end{bmatrix} = \begin{bmatrix} -12 & 0 & 3 \\ 6 & -15 & -6 \\ 0 & 0 & -3 \end{bmatrix}$

d. $3A - 2B = 3\begin{bmatrix} 4 & 0 & -1 \\ -2 & 5 & 2 \\ 0 & 0 & 1 \end{bmatrix} - 2\begin{bmatrix} 3 & 1 & 0 \\ 1 & -4 & 2 \\ 2 & 1 & 3 \end{bmatrix} = \begin{bmatrix} 12 & 0 & -3 \\ -6 & 15 & 6 \\ 0 & 0 & 3 \end{bmatrix} - \begin{bmatrix} 6 & 2 & 0 \\ 2 & -8 & 4 \\ 4 & 2 & 6 \end{bmatrix} = \begin{bmatrix} 6 & -2 & -3 \\ -8 & 23 & 2 \\ -4 & -2 & -3 \end{bmatrix}$

e. $(A+B)^2 = \begin{bmatrix} 7 & 1 & -1 \\ -1 & 1 & 4 \\ 2 & 1 & 4 \end{bmatrix}\begin{bmatrix} 7 & 1 & -1 \\ -1 & 1 & 4 \\ 2 & 1 & 4 \end{bmatrix} = \begin{bmatrix} 7(7)+1(-1)-1(2) & 7(1)+1(1)-1(1) & 7(-1)+1(4)-1(4) \\ -1(7)+1(-1)+4(2) & -1(1)+1(1)+4(1) & -1(-1)+1(4)+4(4) \\ 2(7)+1(-1)+4(2) & 2(1)+1(1)+4(1) & 2(-1)+1(4)+4(4) \end{bmatrix}$

$= \begin{bmatrix} 46 & 7 & -7 \\ 0 & 4 & 21 \\ 21 & 7 & 18 \end{bmatrix}$

f. $A^2 = \begin{bmatrix} 4 & 0 & -1 \\ -2 & 5 & 2 \\ 0 & 0 & 1 \end{bmatrix}\begin{bmatrix} 4 & 0 & -1 \\ -2 & 5 & 2 \\ 0 & 0 & 1 \end{bmatrix} = \begin{bmatrix} 4(4)+0(-2)-1(0) & 4(0)+0(5)-1(0) & 4(-1)+0(2)-1(1) \\ -2(4)+5(-2)+2(0) & -2(0)+5(5)+2(0) & -2(-1)+5(2)+2(1) \\ 0(4)+0(-2)+1(0) & 0(0)+0(5)+1(0) & 0(-1)+0(2)+1(1) \end{bmatrix}$

$= \begin{bmatrix} 16 & 0 & -5 \\ -18 & 25 & 14 \\ 0 & 0 & 1 \end{bmatrix}$

$B^2 = \begin{bmatrix} 3 & 1 & 0 \\ 1 & -4 & 2 \\ 2 & 1 & 3 \end{bmatrix}\begin{bmatrix} 3 & 1 & 0 \\ 1 & -4 & 2 \\ 2 & 1 & 3 \end{bmatrix} = \begin{bmatrix} 3(3)+1(1)+0(2) & 3(1)+1(-4)+0(1) & 3(0)+1(2)+0(3) \\ 1(3)-4(1)+2(2) & 1(1)-4(-4)+2(1) & 1(0)-4(2)+2(3) \\ 2(3)+1(1)+3(2) & 2(1)+1(-4)+3(1) & 2(0)+1(2)+3(3) \end{bmatrix} = \begin{bmatrix} 10 & -1 & 2 \\ 3 & 19 & -2 \\ 13 & 1 & 11 \end{bmatrix}$

$A^2 - B^2 = \begin{bmatrix} 16 & 0 & -5 \\ -18 & 25 & 14 \\ 0 & 0 & 1 \end{bmatrix} - \begin{bmatrix} 10 & -1 & 2 \\ 3 & 19 & -2 \\ 13 & 1 & 11 \end{bmatrix} = \begin{bmatrix} 6 & 1 & -7 \\ -21 & 6 & 16 \\ -13 & -1 & -10 \end{bmatrix}$

21. a. $A + B = \begin{bmatrix} 1 & 2 & -3 \\ 3 & 4 & 5 \\ 2 & -1 & 0 \end{bmatrix} + \begin{bmatrix} 3 & 1 & 0 \\ 1 & -4 & 2 \\ 2 & 1 & 3 \end{bmatrix} = \begin{bmatrix} 4 & 3 & -3 \\ 4 & 0 & 7 \\ 4 & 0 & 3 \end{bmatrix}$

b. $A - B = \begin{bmatrix} 1 & 2 & -3 \\ 3 & 4 & 5 \\ 2 & -1 & 0 \end{bmatrix} - \begin{bmatrix} 3 & 1 & 0 \\ 1 & -4 & 2 \\ 2 & 1 & 3 \end{bmatrix} = \begin{bmatrix} -2 & 1 & -3 \\ 2 & 8 & 3 \\ 0 & -2 & -3 \end{bmatrix}$

c. $-3A = -3\begin{bmatrix} 1 & 2 & -3 \\ 3 & 4 & 5 \\ 2 & -1 & 0 \end{bmatrix} = \begin{bmatrix} -3 & -6 & 9 \\ -9 & -12 & -15 \\ -6 & 3 & 0 \end{bmatrix}$

d. $3A - 2B = 3\begin{bmatrix} 1 & 2 & -3 \\ 3 & 4 & 5 \\ 2 & -1 & 0 \end{bmatrix} - 2\begin{bmatrix} 3 & 1 & 0 \\ 1 & -4 & 2 \\ 2 & 1 & 3 \end{bmatrix} = \begin{bmatrix} 3 & 6 & -9 \\ 9 & 12 & 15 \\ 6 & -3 & 0 \end{bmatrix} - \begin{bmatrix} 6 & 2 & 0 \\ 2 & -8 & 4 \\ 4 & 2 & 6 \end{bmatrix} = \begin{bmatrix} -3 & 4 & -9 \\ 7 & 20 & 11 \\ 2 & -5 & -6 \end{bmatrix}$

e. $(A+B)^2 = \begin{bmatrix} 4 & 3 & -3 \\ 4 & 0 & 7 \\ 4 & 0 & 3 \end{bmatrix}\begin{bmatrix} 4 & 3 & -3 \\ 4 & 0 & 7 \\ 4 & 0 & 3 \end{bmatrix} = \begin{bmatrix} 4(4)+3(4)-3(4) & 4(3)+3(0)-3(0) & 4(-3)+3(7)-3(3) \\ 4(4)+0(4)+7(4) & 4(3)+0(0)+7(0) & 4(-3)+0(7)+7(3) \\ 4(4)+0(4)+3(4) & 4(3)+0(0)+3(0) & 4(-3)+0(7)+3(3) \end{bmatrix}$

$= \begin{bmatrix} 16 & 12 & 0 \\ 44 & 12 & 9 \\ 28 & 12 & -3 \end{bmatrix}$

f. $A^2 = \begin{bmatrix} 1 & 2 & -3 \\ 3 & 4 & 5 \\ 2 & -1 & 0 \end{bmatrix}\begin{bmatrix} 1 & 2 & -3 \\ 3 & 4 & 5 \\ 2 & -1 & 0 \end{bmatrix} = \begin{bmatrix} 1(1)+2(3)-3(2) & 1(2)+2(4)-3(-1) & 1(-3)+2(5)-3(0) \\ 3(1)+4(3)+5(2) & 3(2)+4(4)+5(-1) & 3(-3)+4(5)+5(0) \\ 2(1)-1(3)+0(2) & 2(2)-1(4)+0(-1) & 2(-3)-1(5)+0(0) \end{bmatrix}$

$= \begin{bmatrix} 1 & 13 & 7 \\ 25 & 17 & 11 \\ -1 & 0 & -11 \end{bmatrix}$

$B^2 = \begin{bmatrix} 3 & 1 & 0 \\ 1 & -4 & 2 \\ 2 & 1 & 3 \end{bmatrix}\begin{bmatrix} 3 & 1 & 0 \\ 1 & -4 & 2 \\ 2 & 1 & 3 \end{bmatrix} = \begin{bmatrix} 3(3)+1(1)+0(2) & 3(1)+1(-4)+0(1) & 3(0)+1(2)+0(3) \\ 1(3)-4(1)+2(2) & 1(1)-4(-4)+2(1) & 1(0)-4(2)+2(3) \\ 2(3)+1(1)+3(2) & 2(1)+1(-4)+3(1) & 2(0)+1(2)+3(3) \end{bmatrix} = \begin{bmatrix} 10 & -1 & 2 \\ 3 & 19 & -2 \\ 13 & 1 & 11 \end{bmatrix}$

$A^2 - B^2 = \begin{bmatrix} 1 & 13 & 7 \\ 25 & 17 & 11 \\ -1 & 0 & -11 \end{bmatrix} - \begin{bmatrix} 10 & -1 & 2 \\ 3 & 19 & -2 \\ 13 & 1 & 11 \end{bmatrix} = \begin{bmatrix} -9 & 14 & 5 \\ 22 & -2 & 13 \\ -14 & -1 & -22 \end{bmatrix}$

23. $\begin{bmatrix} 2 & 3 & -1 \\ 1 & -2 & 4 \end{bmatrix} + X = \begin{bmatrix} -2 & 1 & 0 \\ 2 & 3 & 4 \end{bmatrix} \Rightarrow X = \begin{bmatrix} -2 & 1 & 0 \\ 2 & 3 & 4 \end{bmatrix} - \begin{bmatrix} 2 & 3 & -1 \\ 1 & -2 & 4 \end{bmatrix} = \begin{bmatrix} -4 & -2 & 1 \\ 1 & 5 & 0 \end{bmatrix}$

25. $2X - \begin{bmatrix} 2 & 3 & -1 \\ 1 & -2 & 4 \end{bmatrix} = \begin{bmatrix} -2 & 1 & 0 \\ 2 & 3 & 4 \end{bmatrix} \Rightarrow 2X = \begin{bmatrix} -2 & 1 & 0 \\ 2 & 3 & 4 \end{bmatrix} + \begin{bmatrix} 2 & 3 & -1 \\ 1 & -2 & 4 \end{bmatrix} \Rightarrow$

$2X = \begin{bmatrix} 0 & 4 & -1 \\ 3 & 1 & 8 \end{bmatrix} \Rightarrow X = \begin{bmatrix} 0 & 2 & -\dfrac{1}{2} \\ \dfrac{3}{2} & \dfrac{1}{2} & 4 \end{bmatrix}$

27. $2X + 3\begin{bmatrix} 2 & 3 & -1 \\ 1 & -2 & 4 \end{bmatrix} = \begin{bmatrix} -2 & 1 & 0 \\ 2 & 3 & 4 \end{bmatrix} \Rightarrow 2X + \begin{bmatrix} 6 & 9 & -3 \\ 3 & -6 & 12 \end{bmatrix} = \begin{bmatrix} -2 & 1 & 0 \\ 2 & 3 & 4 \end{bmatrix} \Rightarrow$

$2X = \begin{bmatrix} -2 & 1 & 0 \\ 2 & 3 & 4 \end{bmatrix} - \begin{bmatrix} 6 & 9 & -3 \\ 3 & -6 & 12 \end{bmatrix} \Rightarrow 2X = \begin{bmatrix} -8 & -8 & 3 \\ -1 & 9 & -8 \end{bmatrix} \Rightarrow X = \begin{bmatrix} -4 & -4 & \dfrac{3}{2} \\ -\dfrac{1}{2} & \dfrac{9}{2} & -4 \end{bmatrix}$

29. $2\begin{bmatrix} 2 & 3 & -1 \\ 1 & -2 & 4 \end{bmatrix} + 3\begin{bmatrix} -2 & 1 & 0 \\ 2 & 3 & 4 \end{bmatrix} + 4X = 0 \Rightarrow \begin{bmatrix} 4 & 6 & -2 \\ 2 & -4 & 8 \end{bmatrix} + \begin{bmatrix} -6 & 3 & 0 \\ 6 & 9 & 12 \end{bmatrix} = -4X \Rightarrow$

$\begin{bmatrix} -2 & 9 & -2 \\ 8 & 5 & 20 \end{bmatrix} = -4X \Rightarrow X = \begin{bmatrix} \dfrac{1}{2} & -\dfrac{9}{4} & \dfrac{1}{2} \\ -2 & -\dfrac{5}{4} & -5 \end{bmatrix}$

31. a. $\begin{bmatrix} 1 & 2 \\ 3 & 4 \end{bmatrix}\begin{bmatrix} -2 & 1 \\ 3 & 5 \end{bmatrix} = \begin{bmatrix} 1(-2)+2(3) & 1(1)+2(5) \\ 3(-2)+4(3) & 3(1)+4(5) \end{bmatrix} = \begin{bmatrix} 4 & 11 \\ 6 & 23 \end{bmatrix}$

b. $\begin{bmatrix} -2 & 1 \\ 3 & 5 \end{bmatrix}\begin{bmatrix} 1 & 2 \\ 3 & 4 \end{bmatrix} = \begin{bmatrix} -2(1)+1(3) & -2(2)+1(4) \\ 3(1)+5(3) & 3(2)+5(4) \end{bmatrix}$

$= \begin{bmatrix} 1 & 0 \\ 18 & 26 \end{bmatrix}$

33. a. $\begin{bmatrix} 2 & -1 & 0 \\ -3 & 1 & 2 \end{bmatrix}\begin{bmatrix} 1 & 5 \\ -2 & 3 \\ 4 & 0 \end{bmatrix} = \begin{bmatrix} 2(1)-1(-2)+0(4) & 2(5)-1(3)+0(0) \\ -3(1)+1(-2)+2(4) & -3(5)+1(3)+2(0) \end{bmatrix} = \begin{bmatrix} 4 & 7 \\ 3 & -12 \end{bmatrix}$

b. $\begin{bmatrix} 1 & 5 \\ -2 & 3 \\ 4 & 0 \end{bmatrix}\begin{bmatrix} 2 & -1 & 0 \\ -3 & 1 & 2 \end{bmatrix} = \begin{bmatrix} 1(2)+5(-3) & 1(-1)+5(1) & 1(0)+5(2) \\ -2(2)+3(-3) & -2(-1)+3(1) & -2(0)+3(2) \\ 4(2)+0(-3) & 4(-1)+0(1) & 4(0)+0(2) \end{bmatrix} = \begin{bmatrix} -13 & 4 & 10 \\ -13 & 5 & 6 \\ 8 & -4 & 0 \end{bmatrix}$

35. a. $\begin{bmatrix} 2 & 3 & 5 \end{bmatrix}\begin{bmatrix} 1 \\ -2 \\ 4 \end{bmatrix} = \begin{bmatrix} 2(1)+3(-2)+5(4) \end{bmatrix} = \begin{bmatrix} 16 \end{bmatrix}$

b. $\begin{bmatrix} 1 \\ -2 \\ 4 \end{bmatrix}\begin{bmatrix} 2 & 3 & 5 \end{bmatrix} = \begin{bmatrix} 1(2) & 1(3) & 1(5) \\ -2(2) & -2(3) & -2(5) \\ 4(2) & 4(3) & 4(5) \end{bmatrix} = \begin{bmatrix} 2 & 3 & 5 \\ -4 & -6 & -10 \\ 8 & 12 & 20 \end{bmatrix}$

37. a. $\begin{bmatrix} 1 & 2 & 3 \end{bmatrix}\begin{bmatrix} 1 & 2 & -1 \\ 0 & 3 & 1 \\ 2 & 0 & -3 \end{bmatrix} = \begin{bmatrix} 1(1)+2(0)+3(2) & 1(2)+2(3)+3(0) & 1(-1)+2(1)+3(-3) \end{bmatrix} = \begin{bmatrix} 7 & 8 & -8 \end{bmatrix}$

b. The product *BA* is not defined.

39. a. $\begin{bmatrix} 2 & 0 & 1 \\ 1 & 4 & 2 \\ 3 & -1 & 0 \end{bmatrix}\begin{bmatrix} 3 & 1 & 0 \\ -1 & 2 & 0 \\ 4 & 5 & 2 \end{bmatrix} = \begin{bmatrix} 2(3)+0(-1)+1(4) & 2(1)+0(2)+1(5) & 2(0)+0(0)+1(2) \\ 1(3)+4(-1)+2(4) & 1(1)+4(2)+2(5) & 1(0)+4(0)+2(2) \\ 3(3)-1(-1)+0(4) & 3(1)-1(2)+0(5) & 3(0)-1(0)+0(2) \end{bmatrix} = \begin{bmatrix} 10 & 7 & 2 \\ 7 & 19 & 4 \\ 10 & 1 & 0 \end{bmatrix}$

b. $\begin{bmatrix} 3 & 1 & 0 \\ -1 & 2 & 0 \\ 4 & 5 & 2 \end{bmatrix}\begin{bmatrix} 2 & 0 & 1 \\ 1 & 4 & 2 \\ 3 & -1 & 0 \end{bmatrix} = \begin{bmatrix} 3(2)+1(1)+0(3) & 3(0)+1(4)+0(-1) & 3(1)+1(2)+0(0) \\ -1(2)+2(1)+0(3) & -1(0)+2(4)+0(-1) & -1(1)+2(2)+0(0) \\ 4(2)+5(1)+2(3) & 4(0)+5(4)+2(-1) & 4(1)+5(2)+2(0) \end{bmatrix} = \begin{bmatrix} 7 & 4 & 5 \\ 0 & 8 & 3 \\ 19 & 18 & 14 \end{bmatrix}$

41. $AB = \begin{bmatrix} 3 & 1 & 2 \\ 0 & 4 & 3 \\ 1 & -2 & 2 \end{bmatrix}\begin{bmatrix} 2 & 5 & 0 \\ 1 & 2 & -1 \\ 3 & 0 & 2 \end{bmatrix} = \begin{bmatrix} 3(2)+1(1)+2(3) & 3(5)+1(2)+2(0) & 3(0)+1(-1)+2(2) \\ 0(2)+4(1)+3(3) & 0(5)+4(2)+3(0) & 0(0)+4(-1)+3(2) \\ 1(2)-2(1)+2(3) & 1(5)-2(2)+2(0) & 1(0)-2(-1)+2(2) \end{bmatrix} = \begin{bmatrix} 13 & 17 & 3 \\ 13 & 8 & 2 \\ 6 & 1 & 6 \end{bmatrix}$

$BA = \begin{bmatrix} 2 & 5 & 0 \\ 1 & 2 & -1 \\ 3 & 0 & 2 \end{bmatrix}\begin{bmatrix} 3 & 1 & 2 \\ 0 & 4 & 3 \\ 1 & -2 & 2 \end{bmatrix} = \begin{bmatrix} 2(3)+5(0)+0(1) & 2(1)+5(4)+0(-2) & 2(2)+5(3)+0(2) \\ 1(3)+2(0)-1(1) & 1(1)+2(4)-1(-2) & 1(2)+2(3)-1(2) \\ 3(3)+0(0)+2(1) & 3(1)+0(4)+2(-2) & 3(2)+0(3)+2(2) \end{bmatrix} = \begin{bmatrix} 6 & 22 & 19 \\ 2 & 11 & 6 \\ 11 & -1 & 10 \end{bmatrix}$

So, $AB \neq BA$.

43. $(AB)C = \left(\begin{bmatrix} 1 & 2 \\ 3 & 4 \end{bmatrix} \begin{bmatrix} 2 & -3 \\ 3 & 5 \end{bmatrix} \right) \begin{bmatrix} 0 & 1 \\ 2 & 4 \end{bmatrix} = \begin{bmatrix} 1(2)+2(3) & 1(-3)+2(5) \\ 3(2)+4(3) & 3(-3)+4(5) \end{bmatrix} \begin{bmatrix} 0 & 1 \\ 2 & 4 \end{bmatrix} = \begin{bmatrix} 8 & 7 \\ 18 & 11 \end{bmatrix} \begin{bmatrix} 0 & 1 \\ 2 & 4 \end{bmatrix}$

$= \begin{bmatrix} 8(0)+7(2) & 8(1)+7(4) \\ 18(0)+11(2) & 18(1)+11(4) \end{bmatrix} = \begin{bmatrix} 14 & 36 \\ 22 & 62 \end{bmatrix}$

$A(BC) = \begin{bmatrix} 1 & 2 \\ 3 & 4 \end{bmatrix} \left(\begin{bmatrix} 2 & -3 \\ 3 & 5 \end{bmatrix} \begin{bmatrix} 0 & 1 \\ 2 & 4 \end{bmatrix} \right) = \begin{bmatrix} 1 & 2 \\ 3 & 4 \end{bmatrix} \begin{bmatrix} 2(0)-3(2) & 2(1)-3(4) \\ 3(0)+5(2) & 3(1)+5(4) \end{bmatrix} = \begin{bmatrix} 1 & 2 \\ 3 & 4 \end{bmatrix} \begin{bmatrix} -6 & -10 \\ 10 & 23 \end{bmatrix}$

$= \begin{bmatrix} 1(-6)+2(10) & 1(-10)+2(23) \\ 3(-6)+4(10) & 3(-10)+4(23) \end{bmatrix} = \begin{bmatrix} 14 & 36 \\ 22 & 62 \end{bmatrix} \Rightarrow (AB)C = A(BC)$

45. $(A+B)C = \left(\begin{bmatrix} 1 & 2 \\ 3 & 4 \end{bmatrix} + \begin{bmatrix} 2 & -3 \\ 3 & 5 \end{bmatrix} \right) \begin{bmatrix} 0 & 1 \\ 2 & 4 \end{bmatrix} = \begin{bmatrix} 3 & -1 \\ 6 & 9 \end{bmatrix} \begin{bmatrix} 0 & 1 \\ 2 & 4 \end{bmatrix} = \begin{bmatrix} 3(0)-1(2) & 3(1)-1(4) \\ 6(0)+9(2) & 6(1)+9(4) \end{bmatrix} = \begin{bmatrix} -2 & -1 \\ 18 & 42 \end{bmatrix}$

$AC+BC = \begin{bmatrix} 1 & 2 \\ 3 & 4 \end{bmatrix} \begin{bmatrix} 0 & 1 \\ 2 & 4 \end{bmatrix} + \begin{bmatrix} 2 & -3 \\ 3 & 5 \end{bmatrix} \begin{bmatrix} 0 & 1 \\ 2 & 4 \end{bmatrix} = \begin{bmatrix} 1(0)+2(2) & 1(1)+2(4) \\ 3(0)+4(2) & 3(1)+4(4) \end{bmatrix} + \begin{bmatrix} 2(0)-3(2) & 2(1)-3(4) \\ 3(0)+5(2) & 3(1)+5(4) \end{bmatrix}$

$= \begin{bmatrix} 4 & 9 \\ 8 & 19 \end{bmatrix} + \begin{bmatrix} -6 & -10 \\ 10 & 23 \end{bmatrix} = \begin{bmatrix} -2 & -1 \\ 18 & 42 \end{bmatrix} \Rightarrow (A+B)C = AC+BC$

6.2 Applying the Concepts

47.

$A+B = \begin{bmatrix} 7 & 3 & 18 \\ 4 & 1 & 3 \end{bmatrix} + \begin{bmatrix} 6 & 2 & 20 \\ 3 & 1 & 4 \end{bmatrix} = \begin{bmatrix} 13 & 5 & 38 \\ 7 & 2 & 7 \end{bmatrix}$

Steel Glass Wood; Material / Transportation

49. $\begin{bmatrix} 100 & 300 & 400 \end{bmatrix} \begin{bmatrix} 60 \\ 38 \\ 17 \end{bmatrix} = \begin{bmatrix} 100(60)+300(38)+400(17) \end{bmatrix} = \begin{bmatrix} 24,200 \end{bmatrix}$

The total cost is $24,200.

51. a.

	Chairman	President	Vice President
Salary	2,500,000	1,250,000	100,000
Bonus	1,500,000	750,000	150,000
Stock	50,000	25,000	5000

b. $\begin{bmatrix} 1 \\ 1 \\ 4 \end{bmatrix}$

c. $\begin{bmatrix} 2,500,000 & 1,250,000 & 100,000 \\ 1,500,000 & 750,000 & 150,000 \\ 50,000 & 25,000 & 5000 \end{bmatrix} \begin{bmatrix} 1 \\ 1 \\ 4 \end{bmatrix} = \begin{bmatrix} 2,500,000(1)+1,250,000(1)+100,000(4) \\ 1,500,000(1)+750,000(1)+150,000(4) \\ 50,000(1)+25,000(1)+5000(4) \end{bmatrix} = \begin{bmatrix} 4,150,000 \\ 2,850,000 \\ 95,000 \end{bmatrix}$ Totals: Salary / Bonus / Stock

53.

$AD = \begin{bmatrix} 1 & 0 \\ 0 & -1 \end{bmatrix} \begin{bmatrix} 0 & 4 & 4 & 1 & 1 & 0 \\ 0 & 0 & 1 & 1 & 6 & 6 \end{bmatrix}$

$= \begin{bmatrix} 1(0)+0(0) & 1(4)+0(0) & 1(4)+0(1) & 1(1)+0(1) & 1(1)+0(6) & 1(0)+0(6) \\ 0(0)-1(0) & 0(4)-1(0) & 0(4)-1(1) & 0(1)-1(1) & 0(1)-1(6) & 0(0)-1(6) \end{bmatrix}$

$= \begin{bmatrix} 0 & 4 & 4 & 1 & 1 & 0 \\ 0 & 0 & -1 & -1 & -6 & -6 \end{bmatrix}$

55.
$$AD = \begin{bmatrix} 1 & 0 \\ 0.25 & 1 \end{bmatrix} \begin{bmatrix} 0 & 4 & 4 & 1 & 1 & 0 \\ 0 & 0 & 1 & 1 & 6 & 6 \end{bmatrix}$$

$$= \begin{bmatrix} 1(0)+0(0) & 1(4)+0(0) & 1(4)+0(1) & 1(1)+0(1) & 1(1)+0(6) & 1(0)+0(6) \\ 0.25(0)+1(0) & 0.25(4)+1(0) & 0.25(4)+1(1) & 0.25(1)+1(1) & 0.25(1)+1(6) & 0.25(0)+1(6) \end{bmatrix}$$

$$= \begin{bmatrix} 0 & 4 & 4 & 1 & 1 & 0 \\ 0 & 1 & 2 & 1.25 & 6.25 & 6 \end{bmatrix}$$

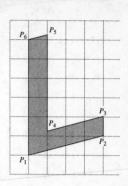

6.2 Beyond the Basics

57.
$$AB = \begin{bmatrix} 0 & 3 \\ 0 & 0 \end{bmatrix} \begin{bmatrix} 2 & 1 \\ 3 & 0 \end{bmatrix} = \begin{bmatrix} 0(2)+3(3) & 0(1)+3(0) \\ 0(2)+0(3) & 0(1)+0(0) \end{bmatrix} = \begin{bmatrix} 9 & 0 \\ 0 & 0 \end{bmatrix}$$

$$AC = \begin{bmatrix} 0 & 3 \\ 0 & 0 \end{bmatrix} \begin{bmatrix} 5 & 4 \\ 3 & 0 \end{bmatrix} = \begin{bmatrix} 0(5)+3(3) & 0(4)+3(0) \\ 0(5)+0(3) & 0(4)+0(0) \end{bmatrix} = \begin{bmatrix} 9 & 0 \\ 0 & 0 \end{bmatrix}$$

So, $AB = AC$ does not imply $B = C$.

59. Answers will vary. Let $A = \begin{bmatrix} 1 & 2 \\ 3 & 4 \end{bmatrix}$ and $B = \begin{bmatrix} -1 & 0 \\ 2 & -3 \end{bmatrix}$. Then $A + B = \begin{bmatrix} 1 & 2 \\ 3 & 4 \end{bmatrix} + \begin{bmatrix} -1 & 0 \\ 2 & -3 \end{bmatrix} = \begin{bmatrix} 0 & 2 \\ 5 & 1 \end{bmatrix}$

$$(A+B)^2 = \begin{bmatrix} 0 & 2 \\ 5 & 1 \end{bmatrix} \begin{bmatrix} 0 & 2 \\ 5 & 1 \end{bmatrix} = \begin{bmatrix} 0(0)+2(5) & 0(2)+2(1) \\ 5(0)+1(5) & 5(2)+1(1) \end{bmatrix} = \begin{bmatrix} 10 & 2 \\ 5 & 11 \end{bmatrix}$$

$$A^2 = \begin{bmatrix} 1 & 2 \\ 3 & 4 \end{bmatrix} \begin{bmatrix} 1 & 2 \\ 3 & 4 \end{bmatrix} = \begin{bmatrix} 1(1)+2(3) & 1(2)+2(4) \\ 3(1)+4(3) & 3(2)+4(4) \end{bmatrix} = \begin{bmatrix} 7 & 10 \\ 15 & 22 \end{bmatrix}$$

$$2AB = \begin{bmatrix} 1 & 2 \\ 3 & 4 \end{bmatrix} \begin{bmatrix} -1 & 0 \\ 2 & -3 \end{bmatrix} = 2\begin{bmatrix} 1(-1)+2(2) & 1(0)+2(-3) \\ 3(-1)+4(2) & 3(0)+4(-3) \end{bmatrix} = 2\begin{bmatrix} 3 & -6 \\ 5 & -12 \end{bmatrix} = \begin{bmatrix} 6 & -12 \\ 10 & -24 \end{bmatrix}$$

$$B^2 = \begin{bmatrix} -1 & 0 \\ 2 & -3 \end{bmatrix} \begin{bmatrix} -1 & 0 \\ 2 & -3 \end{bmatrix} = \begin{bmatrix} -1(-1)+0(2) & -1(0)+0(-3) \\ 2(-1)-3(2) & 2(0)-3(-3) \end{bmatrix} = \begin{bmatrix} 1 & 0 \\ -8 & 9 \end{bmatrix}$$

$$A^2 + 2AB + B^2 = \begin{bmatrix} 7 & 10 \\ 15 & 22 \end{bmatrix} + \begin{bmatrix} 6 & -12 \\ 10 & -24 \end{bmatrix} + \begin{bmatrix} 1 & 0 \\ -8 & 9 \end{bmatrix} = \begin{bmatrix} 14 & -2 \\ 17 & 7 \end{bmatrix}$$

So, $(A+B)^2 \neq A^2 + 2AB + B^2$.

61.

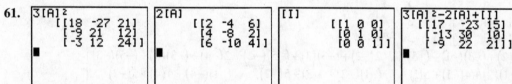

63. Let $B = \begin{bmatrix} x & y \\ z & w \end{bmatrix}$

$\begin{bmatrix} 2 & 3 \\ 1 & 2 \end{bmatrix}\begin{bmatrix} x & y \\ z & w \end{bmatrix} = \begin{bmatrix} 1 & 0 \\ 0 & 1 \end{bmatrix} \Rightarrow \begin{bmatrix} 2x+3z & 2y+3w \\ x+2z & y+2w \end{bmatrix} = \begin{bmatrix} 1 & 0 \\ 0 & 1 \end{bmatrix} \Rightarrow \begin{cases} 2x+3z = 1 \\ 2y+3w = 0 \\ x+2z = 0 \\ y+2w = 1 \end{cases} \Rightarrow$

$\left[\begin{array}{cccc|c} 2 & 0 & 3 & 0 & 1 \\ 0 & 2 & 0 & 3 & 0 \\ 1 & 0 & 2 & 0 & 0 \\ 0 & 1 & 0 & 2 & 1 \end{array}\right] \xrightarrow[-(R_2-2R_4)\to R_4]{-(R_1-2R_3)\to R_3} \left[\begin{array}{cccc|c} 2 & 0 & 3 & 0 & 1 \\ 0 & 2 & 0 & 3 & 0 \\ 0 & 0 & 1 & 0 & -1 \\ 0 & 0 & 0 & 1 & 2 \end{array}\right] \xrightarrow[\frac{1}{2}(R_2-3R_4)\to R_2]{\frac{1}{2}(R_1-3R_3)\to R_1} \left[\begin{array}{cccc|c} 1 & 0 & 0 & 0 & 2 \\ 0 & 1 & 0 & 0 & -3 \\ 0 & 0 & 1 & 0 & -1 \\ 0 & 0 & 0 & 1 & 2 \end{array}\right] \Rightarrow$

$x = 2, y = -3, z = -1, w = 2 \Rightarrow B = \begin{bmatrix} 2 & -3 \\ -1 & 2 \end{bmatrix}$

65. $\left(4\begin{bmatrix} 2 & -1 & 3 \\ 1 & 0 & 2 \end{bmatrix} - \begin{bmatrix} 1 & 2 & -1 \\ 2 & -3 & 4 \end{bmatrix}\right)\begin{bmatrix} 2 \\ -1 \\ 1 \end{bmatrix} = \begin{bmatrix} x \\ y \end{bmatrix} \Rightarrow \left(\begin{bmatrix} 8 & -4 & 12 \\ 4 & 0 & 8 \end{bmatrix} - \begin{bmatrix} 1 & 2 & -1 \\ 2 & -3 & 4 \end{bmatrix}\right)\begin{bmatrix} 2 \\ -1 \\ 1 \end{bmatrix} = \begin{bmatrix} x \\ y \end{bmatrix} \Rightarrow$

$\begin{bmatrix} 7 & -6 & 13 \\ 2 & 3 & 4 \end{bmatrix}\begin{bmatrix} 2 \\ -1 \\ 1 \end{bmatrix} = \begin{bmatrix} x \\ y \end{bmatrix} \Rightarrow \begin{bmatrix} 7(2)-6(-1)+13(1) \\ 2(2)+3(-1)+4(1) \end{bmatrix} = \begin{bmatrix} x \\ y \end{bmatrix} \Rightarrow \begin{bmatrix} 33 \\ 5 \end{bmatrix} = \begin{bmatrix} x \\ y \end{bmatrix} \Rightarrow x = 33, y = 5$

67. $AB = \begin{bmatrix} 0 & 1 \\ 0 & 0 \end{bmatrix}\begin{bmatrix} 1 & 0 \\ 0 & 0 \end{bmatrix} = \begin{bmatrix} 0(1)+1(0) & 0(0)+1(0) \\ 0(1)+0(0) & 0(0)+0(0) \end{bmatrix} = \begin{bmatrix} 0 & 0 \\ 0 & 0 \end{bmatrix} = 0$

$BA = \begin{bmatrix} 1 & 0 \\ 0 & 0 \end{bmatrix}\begin{bmatrix} 0 & 1 \\ 0 & 0 \end{bmatrix} = \begin{bmatrix} 1(0)+0(0) & 1(1)+0(0) \\ 0(0)+0(0) & 0(1)+0(0) \end{bmatrix} = \begin{bmatrix} 0 & 1 \\ 0 & 0 \end{bmatrix} \neq 0$

69. $2A - B = \begin{bmatrix} 6 & -6 & 0 \\ -4 & 2 & 1 \end{bmatrix} \qquad A + B = \begin{bmatrix} 3 & 0 & 3 \\ -2 & 1 & -4 \end{bmatrix}$

$(2A-B)+(A+B) = 3A = \begin{bmatrix} 6 & -6 & 0 \\ -4 & 2 & 1 \end{bmatrix} + \begin{bmatrix} 3 & 0 & 3 \\ -2 & 1 & -4 \end{bmatrix} = \begin{bmatrix} 9 & -6 & 3 \\ -6 & 3 & -3 \end{bmatrix} \Rightarrow A = \begin{bmatrix} 3 & -2 & 1 \\ -2 & 1 & -1 \end{bmatrix}$

$A + B = \begin{bmatrix} 3 & -2 & 1 \\ -2 & 1 & -1 \end{bmatrix} + B = \begin{bmatrix} 3 & 0 & 3 \\ -2 & 1 & -4 \end{bmatrix} \Rightarrow B - \begin{bmatrix} 3 & 0 & 3 \\ -2 & 1 & -4 \end{bmatrix} - \begin{bmatrix} 3 & -2 & 1 \\ -2 & 1 & -1 \end{bmatrix} = \begin{bmatrix} 0 & 2 & 2 \\ 0 & 0 & -3 \end{bmatrix}$

71. $AB = \begin{bmatrix} 2 & -3 & -5 \\ -1 & 4 & 5 \\ 1 & -3 & -4 \end{bmatrix}\begin{bmatrix} 2 & -2 & -4 \\ -1 & 3 & 4 \\ 1 & -2 & -3 \end{bmatrix}$

$= \begin{bmatrix} 2(2)+(-3)(-1)+(-5)(1) & 2(-2)+(-3)(3)+(-5)(-2) & 2(-4)+(-3)(4)+(-5)(-3) \\ (-1)(2)+4(-1)+5(1) & (-1)(-2)+4(3)+5(-2) & (-1)(-4)+4(4)+5(-3) \\ 1(2)+(-3)(-1)+(-4)(1) & 1(-2)+(-3)(3)+(-4)(-2) & 1(-4)+(-3)(4)+(-4)(-3) \end{bmatrix}$

$= \begin{bmatrix} 2 & -3 & -5 \\ -1 & 4 & 5 \\ 1 & -3 & -4 \end{bmatrix} = A$

(*continued on next page*)

(continued)

$$BA = \begin{bmatrix} 2 & -2 & -4 \\ -1 & 3 & 4 \\ 1 & -2 & -3 \end{bmatrix} \begin{bmatrix} 2 & -3 & -5 \\ -1 & 4 & 5 \\ 1 & -3 & -4 \end{bmatrix}$$

$$= \begin{bmatrix} 2(2)+(-2)(-1)+(-4)(1) & 2(-3)+(-2)(4)+(-4)(-3) & 2(-5)+(-2)(5)+(-4)(-4) \\ (-1)(2)+3(-1)+4(1) & (-1)(-3)+3(4)+4(-3) & (-1)(-5)+3(5)+4(-4) \\ 1(2)+(-2)(-1)+(-3)(1) & 1(-3)+(-2)(4)+(-3)(-3) & 1(-5)+(-2)(5)+(-3)(-4) \end{bmatrix}$$

$$= \begin{bmatrix} 2 & -2 & -4 \\ -1 & 3 & 4 \\ 1 & -2 & -3 \end{bmatrix} = B$$

$$A^2 = \begin{bmatrix} 2 & -3 & -5 \\ -1 & 4 & 5 \\ 1 & -3 & -4 \end{bmatrix} \begin{bmatrix} 2 & -3 & -5 \\ -1 & 4 & 5 \\ 1 & -3 & -4 \end{bmatrix}$$

$$= \begin{bmatrix} 2(2)+(-3)(-1)+(-5)(1) & 2(-3)+(-3)(4)+(-5)(-3) & 2(-5)+(-3)(5)+(-5)(-4) \\ (-1)(2)+4(-1)+5(1) & (-1)(-3)+4(4)+5(-3) & (-1)(-5)+4(5)+5(-4) \\ 1(2)+(-3)(-1)+(-4)(1) & 1(-3)+(-3)(4)+(-4)(-3) & 1(-5)+(-3)(5)+(-4)(-4) \end{bmatrix}$$

$$= \begin{bmatrix} 2 & -3 & -5 \\ -1 & 4 & 5 \\ 1 & -3 & -4 \end{bmatrix} = A$$

73. $A^2 = \begin{bmatrix} 4 & -1 & -4 \\ 3 & 0 & -4 \\ 3 & -1 & -3 \end{bmatrix} \begin{bmatrix} 4 & -1 & -4 \\ 3 & 0 & -4 \\ 3 & -1 & -3 \end{bmatrix}$

$$= \begin{bmatrix} 4(4)+(-1)(3)+(-4)(3) & 4(-1)+(-1)(0)+(-4)(-1) & 4(-4)+(-1)(-4)+(-4)(-3) \\ 3(4)+0(3)+(-4)(3) & 3(-1)+0(0)+(-4)(-1) & 3(-4)+0(-4)+(-4)(-3) \\ 3(4)+(-1)(3)+(-3)(3) & 3(-1)+(-1)(0)+(-3)(-1) & 3(-4)+(-1)(-4)+(-3)(-3) \end{bmatrix}$$

$$= \begin{bmatrix} 1 & 0 & 0 \\ 0 & 1 & 0 \\ 0 & 0 & 1 \end{bmatrix} = I$$

6.2 Critical Thinking/Discussion/Writing

74. a. AB is defined when $n = 5$. The order of AB when this product is defined is $3 \times m$.

 b. BA is defined when $m = 3$. The order of BA when this product is defined is $5 \times n$.

75. $(CA)B$

76. a. $P^2 = \begin{bmatrix} 0.4 & 0.3 & 0.3 \\ 0.6 & 0.3 & 0.1 \\ 0.6 & 0.1 & 0.3 \end{bmatrix} \begin{bmatrix} 0.4 & 0.3 & 0.3 \\ 0.6 & 0.3 & 0.1 \\ 0.6 & 0.1 & 0.3 \end{bmatrix} = \begin{bmatrix} 0.52 & 0.24 & 0.24 \\ 0.48 & 0.28 & 0.24 \\ 0.48 & 0.24 & 0.28 \end{bmatrix} = \begin{bmatrix} \frac{13}{25} & \frac{6}{25} & \frac{6}{25} \\ \frac{12}{25} & \frac{7}{25} & \frac{6}{25} \\ \frac{12}{25} & \frac{6}{25} & \frac{7}{25} \end{bmatrix}$

$$XP^2 = \begin{bmatrix} \frac{1}{3} & \frac{1}{3} & \frac{1}{3} \end{bmatrix} \begin{bmatrix} \frac{13}{25} & \frac{6}{25} & \frac{6}{25} \\ \frac{12}{25} & \frac{7}{25} & \frac{6}{25} \\ \frac{12}{25} & \frac{6}{25} & \frac{7}{25} \end{bmatrix} = \begin{bmatrix} \frac{37}{75} & \frac{19}{75} & \frac{19}{75} \end{bmatrix}$$

b. $P^3 = P^2P = \begin{bmatrix} 0.4 & 0.3 & 0.3 \\ 0.6 & 0.3 & 0.1 \\ 0.6 & 0.1 & 0.3 \end{bmatrix}\begin{bmatrix} 0.52 & 0.24 & 0.24 \\ 0.48 & 0.28 & 0.24 \\ 0.48 & 0.24 & 0.28 \end{bmatrix} = \begin{bmatrix} 0.496 & 0.252 & 0.252 \\ 0.504 & 0.252 & 0.244 \\ 0.504 & 0.244 & 0.252 \end{bmatrix} = \begin{bmatrix} \dfrac{62}{125} & \dfrac{63}{250} & \dfrac{63}{250} \\ \dfrac{63}{125} & \dfrac{63}{250} & \dfrac{61}{250} \\ \dfrac{63}{125} & \dfrac{61}{250} & \dfrac{63}{250} \end{bmatrix}$

$XP^3 = \begin{bmatrix} \dfrac{1}{3} & \dfrac{1}{3} & \dfrac{1}{3} \end{bmatrix}\begin{bmatrix} \dfrac{62}{125} & \dfrac{63}{250} & \dfrac{63}{250} \\ \dfrac{63}{125} & \dfrac{63}{250} & \dfrac{61}{250} \\ \dfrac{63}{125} & \dfrac{61}{250} & \dfrac{63}{250} \end{bmatrix} = \begin{bmatrix} \dfrac{188}{375} & \dfrac{187}{750} & \dfrac{187}{750} \end{bmatrix}$

$P^4 = P^3P = \begin{bmatrix} 0.496 & 0.252 & 0.252 \\ 0.504 & 0.252 & 0.244 \\ 0.504 & 0.244 & 0.252 \end{bmatrix}\begin{bmatrix} 0.4 & 0.3 & 0.3 \\ 0.6 & 0.3 & 0.1 \\ 0.6 & 0.1 & 0.3 \end{bmatrix} = \begin{bmatrix} 0.5008 & 0.2496 & 0.2496 \\ 0.4992 & 0.2512 & 0.2496 \\ 0.4992 & 0.2496 & 0.2512 \end{bmatrix} = \begin{bmatrix} \dfrac{313}{625} & \dfrac{156}{625} & \dfrac{156}{625} \\ \dfrac{312}{625} & \dfrac{157}{625} & \dfrac{156}{625} \\ \dfrac{312}{625} & \dfrac{156}{625} & \dfrac{157}{625} \end{bmatrix}$

$XP^4 = \begin{bmatrix} \dfrac{1}{3} & \dfrac{1}{3} & \dfrac{1}{3} \end{bmatrix}\begin{bmatrix} \dfrac{313}{625} & \dfrac{156}{625} & \dfrac{156}{625} \\ \dfrac{312}{625} & \dfrac{157}{625} & \dfrac{156}{625} \\ \dfrac{312}{625} & \dfrac{156}{625} & \dfrac{157}{625} \end{bmatrix} = \begin{bmatrix} \dfrac{937}{1875} & \dfrac{469}{1875} & \dfrac{469}{1875} \end{bmatrix}$

$P^5 = P^4P = \begin{bmatrix} 0.5008 & 0.2496 & 0.2496 \\ 0.4992 & 0.2512 & 0.2496 \\ 0.4992 & 0.2496 & 0.2512 \end{bmatrix}\begin{bmatrix} 0.4 & 0.3 & 0.3 \\ 0.6 & 0.3 & 0.1 \\ 0.6 & 0.1 & 0.3 \end{bmatrix} = \begin{bmatrix} 0.49984 & 0.25008 & 0.25008 \\ 0.50016 & 0.25008 & 0.24976 \\ 0.50016 & 0.24976 & 0.25008 \end{bmatrix}$

$= \begin{bmatrix} \dfrac{1562}{3125} & \dfrac{1563}{6250} & \dfrac{1563}{6250} \\ \dfrac{1563}{3125} & \dfrac{1563}{6250} & \dfrac{1561}{6250} \\ \dfrac{1563}{3125} & \dfrac{1561}{6250} & \dfrac{1563}{6250} \end{bmatrix}$

$XP^5 = \begin{bmatrix} \dfrac{1}{3} & \dfrac{1}{3} & \dfrac{1}{3} \end{bmatrix}\begin{bmatrix} \dfrac{1562}{3125} & \dfrac{1563}{6250} & \dfrac{1563}{6250} \\ \dfrac{1563}{3125} & \dfrac{1563}{6250} & \dfrac{1561}{6250} \\ \dfrac{1563}{3125} & \dfrac{1561}{6250} & \dfrac{1563}{6250} \end{bmatrix} = \begin{bmatrix} \dfrac{4688}{9375} & \dfrac{4687}{18,750} & \dfrac{4687}{18,750} \end{bmatrix} \Rightarrow$

$XP^n = \begin{bmatrix} \dfrac{1}{2} & \dfrac{1}{4} & \dfrac{1}{4} \end{bmatrix}$

In the long term, the market shares for each company are: *A*, 50%; *B*, 25%; *C*, 25%.

6.2 Maintaining Skills

77. $\left(\dfrac{1}{2}\right)^{-1} = 2$

79. $x^{-1} = \dfrac{1}{8} \Rightarrow x = 8$

81. $\dfrac{1}{4}x - \dfrac{7}{12} = \dfrac{11}{12} - \dfrac{5}{4}x \Rightarrow \dfrac{6}{4}x = \dfrac{18}{12} \Rightarrow$

$x = \dfrac{18}{12} \cdot \dfrac{4}{6} = 1$

83. $\dfrac{2}{4x-1} = \dfrac{3}{4x+1} \Rightarrow 2(4x+1) = 3(4x-1) \Rightarrow 8x+2 = 12x-3 \Rightarrow -4x = -5 \Rightarrow x = \dfrac{5}{4}$

85. $\begin{cases} 3x + 5y = 2 & (1) \\ 6x + 10y = 4 & (2) \end{cases}$

Multiply equation (1) by −2, then add the two equations.
$$-6x - 10y = -4$$
$$\underline{6x + 10y = 4}$$
$$0 = 0$$

The equations are dependent and the system has infinitely many solutions. Solve equation (1) for y in terms of x to find the general solution.
$3x + 5y = 2 \Rightarrow 5y = -3x + 2 \Rightarrow$

$y = -\dfrac{3}{5}x + \dfrac{2}{5} = \dfrac{2-3x}{5}$

Thus, the solution set can be written as $\left\{ \left(x, \dfrac{2-3x}{5} \right) \right\}$.

87. $\begin{cases} 3x + 2y = 6 & (1) \\ 6x + 4y = -13 & (2) \end{cases}$

Multiply equation (1) by −2, then add the equations.
$$-6x - 4y = -12$$
$$\underline{6x + 4y = -13}$$
$$0 = -25 \quad \text{False}$$

The system is inconsistent and the solution set is $\varnothing$.

6.3 The Matrix Inverse

6.3 Practice Problems

1. We need to verify that $AB = BA = I$.

$AB = \begin{bmatrix} 3 & 2 \\ 2 & 1 \end{bmatrix} \begin{bmatrix} -1 & 2 \\ 2 & -3 \end{bmatrix} = \begin{bmatrix} 3(-1)+2(2) & 3(2)+2(-3) \\ 2(-1)+1(2) & 2(2)+1(-3) \end{bmatrix} = \begin{bmatrix} 1 & 0 \\ 0 & 1 \end{bmatrix}$

$BA = \begin{bmatrix} -1 & 2 \\ 2 & -3 \end{bmatrix} \begin{bmatrix} 3 & 2 \\ 2 & 1 \end{bmatrix} = \begin{bmatrix} -1(3)+2(2) & -1(2)+2(1) \\ 2(3)-3(2) & 2(2)-3(1) \end{bmatrix} = \begin{bmatrix} 1 & 0 \\ 0 & 1 \end{bmatrix}$

Thus B is the inverse of A.

2. Suppose A has an inverse B, where $B = \begin{bmatrix} x & y \\ z & w \end{bmatrix}$. Then

$$\begin{bmatrix} 3 & 1 \\ 3 & 1 \end{bmatrix}\begin{bmatrix} x & y \\ z & w \end{bmatrix} = \begin{bmatrix} 1 & 0 \\ 0 & 1 \end{bmatrix} \Rightarrow \begin{bmatrix} 3x+z & 3y+w \\ 3x+z & 3y+w \end{bmatrix} = \begin{bmatrix} 1 & 0 \\ 0 & 1 \end{bmatrix}.$$

Since the matrices are equal, the entries must be equal. So $3x + z = 1$ (in the 1, 1 position) and $3x + z = 0$ (in the 2, 1) position. This is a contradiction, so A does not have an inverse.

3. $[A \mid I] = \begin{bmatrix} 1 & 4 & -2 & 1 & 0 & 0 \\ -1 & 1 & 2 & 0 & 1 & 0 \\ 3 & 7 & -6 & 0 & 0 & 1 \end{bmatrix} \xrightarrow[-3R_1+R_3 \to R_3]{R_1+R_2 \to R_2} \begin{bmatrix} 1 & 4 & -2 & 1 & 0 & 0 \\ 0 & 5 & 0 & 1 & 1 & 0 \\ 0 & -5 & 0 & -3 & 0 & 1 \end{bmatrix}$

$\xrightarrow{R_2+R_3 \to R_3} \begin{bmatrix} 1 & 4 & -2 & 1 & 0 & 0 \\ 0 & 5 & 0 & 1 & 1 & 0 \\ 0 & 0 & 0 & -2 & 1 & 1 \end{bmatrix}$

The inverse of A does not exist.

4. $[A \mid I] = \begin{bmatrix} 1 & 2 & 3 & 1 & 0 & 0 \\ -2 & 3 & 1 & 0 & 1 & 0 \\ 4 & 5 & -2 & 0 & 0 & 1 \end{bmatrix} \xrightarrow[-4R_1+R_3 \to R_3]{2R_1+R_2 \to R_2} \begin{bmatrix} 1 & 2 & 3 & 1 & 0 & 0 \\ 0 & 7 & 7 & 2 & 1 & 0 \\ 0 & -3 & -14 & -4 & 0 & 1 \end{bmatrix}$

$\xrightarrow{\frac{1}{11}(2R_2+R_3) \to R_2} \begin{bmatrix} 1 & 2 & 3 & 1 & 0 & 0 \\ 0 & 1 & 0 & 0 & \frac{2}{11} & \frac{1}{11} \\ 0 & -3 & -14 & -4 & 0 & 1 \end{bmatrix} \xrightarrow[\left(-\frac{1}{14}\right)(3R_2+R_3) \to R_3]{R_1-2R_2 \to R_1} \begin{bmatrix} 1 & 0 & 3 & 1 & -\frac{4}{11} & -\frac{2}{11} \\ 0 & 1 & 0 & 0 & \frac{2}{11} & \frac{1}{11} \\ 0 & 0 & 1 & \frac{2}{7} & -\frac{3}{77} & -\frac{1}{11} \end{bmatrix}$

$\xrightarrow{R_1-3R_3 \to R_1} \begin{bmatrix} 1 & 0 & 3 & \frac{1}{7} & -\frac{19}{77} & \frac{1}{11} \\ 0 & 1 & 0 & 0 & \frac{2}{11} & \frac{1}{11} \\ 0 & 0 & 1 & \frac{2}{7} & -\frac{3}{77} & -\frac{1}{11} \end{bmatrix} \Rightarrow A^{-1} = \begin{bmatrix} \frac{1}{7} & -\frac{19}{77} & \frac{1}{11} \\ 0 & \frac{2}{11} & \frac{1}{11} \\ \frac{2}{7} & -\frac{3}{77} & -\frac{1}{11} \end{bmatrix}$

5. a. $ad - bc = (8)(1) - (2)(4) = 0$

The inverse of matrix A does not exist.

b. $ad - bc = (8)(1) - (-2)(3) = 14$

Thus, matrix B is invertible.

$$B^{-1} = \frac{1}{14}\begin{bmatrix} 1 & 2 \\ -3 & 8 \end{bmatrix} = \begin{bmatrix} \frac{1}{14} & \frac{1}{7} \\ -\frac{3}{14} & \frac{4}{7} \end{bmatrix}$$

6. $\begin{cases} 3x+2y+3z = 9 \\ 3x+ y \quad = 12 \\ x+ \quad z = 6 \end{cases} \Rightarrow AX = B \Rightarrow \begin{bmatrix} 3 & 2 & 3 \\ 3 & 1 & 0 \\ 1 & 0 & 1 \end{bmatrix}\begin{bmatrix} x \\ y \\ z \end{bmatrix} = \begin{bmatrix} 9 \\ 12 \\ 6 \end{bmatrix} \Rightarrow X = A^{-1}B$

Using a graphing calculator, we find that $A^{-1} = \begin{bmatrix} -\dfrac{1}{6} & \dfrac{1}{3} & \dfrac{1}{2} \\ \dfrac{1}{2} & 0 & -\dfrac{3}{2} \\ \dfrac{1}{6} & -\dfrac{1}{3} & \dfrac{1}{2} \end{bmatrix}.$

$X = A^{-1}B \Rightarrow \begin{bmatrix} x \\ y \\ z \end{bmatrix} = \begin{bmatrix} -\dfrac{1}{5} & \dfrac{2}{5} & \dfrac{2}{5} \\ \dfrac{3}{5} & -\dfrac{1}{5} & -\dfrac{6}{5} \\ \dfrac{1}{5} & -\dfrac{2}{5} & \dfrac{3}{5} \end{bmatrix}\begin{bmatrix} 9 \\ 12 \\ 6 \end{bmatrix} \Rightarrow X = \begin{bmatrix} \dfrac{11}{2} \\ -\dfrac{9}{2} \\ \dfrac{1}{2} \end{bmatrix}$

Solution set: $\left\{ \left(\dfrac{11}{2}, -\dfrac{9}{2}, \dfrac{1}{2} \right) \right\}$

7. Following the discussion given in Example 7, we have $X = (I - A)^{-1}D = \dfrac{1}{19}\begin{bmatrix} 36 & 24 \\ 16 & 36 \end{bmatrix}\begin{bmatrix} 800 \\ 3400 \end{bmatrix} = \begin{bmatrix} \dfrac{110,400}{19} \\ \dfrac{135,200}{19} \end{bmatrix}$

To meet the consumer demand for 800 units of energy and 3400 units of food, the energy produced must be $\dfrac{110,400}{19}$ units and food production must be $\dfrac{135,200}{19}$ units.

8. Associate each letter in the phrase with the number representing its position in the alphabet, and partition the numbers into groups of three (inserting two zeros at the right end), forming a 3×6 matrix:

J   A   C   K   I   S   N   O   W   S   A   F   E
[10   1   3]   [11   0   9]   [19   0   14]   [15   23   0]   [19   1   6]   [5   0   0]

$M = \begin{bmatrix} 10 & 11 & 19 & 15 & 19 & 5 \\ 1 & 0 & 0 & 23 & 1 & 0 \\ 3 & 9 & 14 & 0 & 6 & 0 \end{bmatrix}$

Let $A = \begin{bmatrix} 1 & 2 & 3 \\ 1 & 3 & 3 \\ 1 & 2 & 4 \end{bmatrix}$. Then, $AM = \begin{bmatrix} 1 & 2 & 3 \\ 1 & 3 & 3 \\ 1 & 2 & 4 \end{bmatrix}\begin{bmatrix} 10 & 11 & 19 & 15 & 19 & 5 \\ 1 & 0 & 0 & 23 & 1 & 0 \\ 3 & 9 & 14 & 0 & 6 & 0 \end{bmatrix} = \begin{bmatrix} 21 & 38 & 61 & 61 & 39 & 5 \\ 22 & 38 & 61 & 84 & 40 & 5 \\ 24 & 47 & 75 & 61 & 45 & 5 \end{bmatrix}$

The cryptogram is 21 22 24 38 38 47 61 61 75 61 84 61 39 40 45 5 5 5.

6.3 Basic Concepts and Skills

1. For $n \times n$ matrices A and B, if $AB = I$, then B is called the <u>inverse</u> of A.

3. To find the inverse of an invertible matrix A, we transform $[A \mid I]$ by a sequence of row operations into $[I \mid B]$, where $B = \underline{A^{-1}}$.

5. False.

7. $AB = \begin{bmatrix} 1 & 2 \\ 1 & 3 \end{bmatrix}\begin{bmatrix} 3 & -2 \\ -1 & 1 \end{bmatrix} = \begin{bmatrix} 1 & 0 \\ 0 & 1 \end{bmatrix}$

$BA = \begin{bmatrix} 3 & -2 \\ -1 & 1 \end{bmatrix}\begin{bmatrix} 1 & 2 \\ 1 & 3 \end{bmatrix} = \begin{bmatrix} 1 & 0 \\ 0 & 1 \end{bmatrix} \Rightarrow$

B is the inverse of A.

9. $AB = \begin{bmatrix} 3 & 2 \\ 1 & 4 \end{bmatrix} \begin{bmatrix} \dfrac{2}{5} & -\dfrac{1}{5} \\ -\dfrac{1}{10} & \dfrac{3}{10} \end{bmatrix} = \begin{bmatrix} 1 & 0 \\ 0 & 1 \end{bmatrix}$

$BA = \begin{bmatrix} \dfrac{2}{5} & -\dfrac{1}{5} \\ -\dfrac{1}{10} & \dfrac{3}{10} \end{bmatrix} \begin{bmatrix} 3 & 2 \\ 1 & 4 \end{bmatrix} = \begin{bmatrix} 1 & 0 \\ 0 & 1 \end{bmatrix} \Rightarrow$

B is the inverse of A.

11. $AB = \begin{bmatrix} 1 & 0 & -1 \\ -1 & 1 & 1 \end{bmatrix} \begin{bmatrix} 2 & 1 \\ 1 & 1 \\ 1 & 1 \end{bmatrix} = \begin{bmatrix} 1 & 0 \\ 0 & 1 \end{bmatrix}$

$BA = \begin{bmatrix} 2 & 1 \\ 1 & 1 \\ 1 & 1 \end{bmatrix} \begin{bmatrix} 1 & 0 & -1 \\ -1 & 1 & 1 \end{bmatrix} = \begin{bmatrix} 1 & 1 & -1 \\ 0 & 1 & 0 \\ 0 & 1 & 0 \end{bmatrix} \Rightarrow$

B is not the inverse of A.

13. $AB = \begin{bmatrix} 1 & -2 & 1 \\ -8 & 6 & -2 \\ 5 & -3 & 1 \end{bmatrix} \left(\dfrac{1}{2} \begin{bmatrix} 0 & 1 & 2 \\ 1 & 2 & 3 \\ 3 & 1 & 1 \end{bmatrix} \right)$

$= \begin{bmatrix} \dfrac{1}{2} & -1 & -\dfrac{3}{2} \\ 0 & 1 & 0 \\ 0 & 0 & 1 \end{bmatrix} \Rightarrow B$ is not the inverse of A.

15. $AB = \begin{bmatrix} 1 & 1 & 1 \\ 1 & 2 & 3 \\ -1 & 1 & -1 \end{bmatrix} \left(\dfrac{1}{4} \begin{bmatrix} 5 & -2 & -1 \\ 2 & 0 & 2 \\ -3 & 2 & -1 \end{bmatrix} \right)$

$= \begin{bmatrix} 1 & 0 & 0 \\ 0 & 1 & 0 \\ 0 & 0 & 1 \end{bmatrix}$

$BA = \left(\dfrac{1}{4} \begin{bmatrix} 5 & -2 & -1 \\ 2 & 0 & 2 \\ -3 & 2 & -1 \end{bmatrix} \right) \begin{bmatrix} 1 & 1 & 1 \\ 1 & 2 & 3 \\ -1 & 1 & -1 \end{bmatrix}$

$= \begin{bmatrix} 1 & 0 & 0 \\ 0 & 1 & 0 \\ 0 & 0 & 1 \end{bmatrix} \Rightarrow B$ is the inverse of A.

In exercises 17–26, the steps to find the inverse may vary. Be sure to verify that $AA^{-1} = I$.

17. $[A \mid I] = \begin{bmatrix} 2 & 0 & | & 1 & 0 \\ 1 & 3 & | & 0 & 1 \end{bmatrix} \xrightarrow{\frac{1}{2}R_1} \begin{bmatrix} 1 & 0 & | & \dfrac{1}{2} & 0 \\ 1 & 3 & | & 0 & 1 \end{bmatrix}$

$\xrightarrow{R_1 - R_2 \to R_2} \begin{bmatrix} 1 & 0 & | & \dfrac{1}{2} & 0 \\ 0 & -3 & | & \dfrac{1}{2} & -1 \end{bmatrix}$

$\xrightarrow{-\frac{1}{3}R_2} \begin{bmatrix} 1 & 0 & | & \dfrac{1}{2} & 0 \\ 0 & 1 & | & -\dfrac{1}{6} & \dfrac{1}{3} \end{bmatrix} \Rightarrow$

$A^{-1} = \begin{bmatrix} \dfrac{1}{2} & 0 \\ -\dfrac{1}{6} & \dfrac{1}{3} \end{bmatrix}$

19. $[A \mid I] = \begin{bmatrix} 2 & 4 & | & 1 & 0 \\ 3 & 6 & | & 0 & 1 \end{bmatrix}$

$\xrightarrow{\frac{1}{2}R_1 \to R_1} \begin{bmatrix} 1 & 2 & | & \dfrac{1}{2} & 0 \\ 3 & 6 & | & 0 & 1 \end{bmatrix}$

$\xrightarrow{-3R_1 + R_2 \to R_2} \begin{bmatrix} 1 & 2 & | & \dfrac{1}{2} & 0 \\ 0 & 0 & | & -\dfrac{3}{2} & 1 \end{bmatrix} \Rightarrow$

The inverse of A does not exist.

21. $[A \mid I] = \begin{bmatrix} 1 & 6 & 4 & | & 1 & 0 & 0 \\ 0 & 2 & 3 & | & 0 & 1 & 0 \\ 0 & 1 & 2 & | & 0 & 0 & 1 \end{bmatrix}$

$\xrightarrow{2R_3 - R_2 \to R_3} \begin{bmatrix} 1 & 6 & 4 & | & 1 & 0 & 0 \\ 0 & 2 & 3 & | & 0 & 1 & 0 \\ 0 & 0 & 1 & | & 0 & -1 & 2 \end{bmatrix}$

$\xrightarrow{\frac{1}{2}(-3R_3 + R_2) \to R_2} \begin{bmatrix} 1 & 6 & 4 & | & 1 & 0 & 0 \\ 0 & 1 & 0 & | & 0 & 2 & -3 \\ 0 & 0 & 1 & | & 0 & -1 & 2 \end{bmatrix}$

$\xrightarrow[R_1 - 6R_2 \to R_1]{R_1 - 4R_3 \to R_1} \begin{bmatrix} 1 & 0 & 0 & | & 1 & -8 & 10 \\ 0 & 1 & 0 & | & 0 & 2 & -3 \\ 0 & 0 & 1 & | & 0 & -1 & 2 \end{bmatrix}$

$A^{-1} = \begin{bmatrix} 1 & -8 & 10 \\ 0 & 2 & -3 \\ 0 & -1 & 2 \end{bmatrix}$

23. $[A \mid I] = \begin{bmatrix} 2 & 3 & 1 & | & 1 & 0 & 0 \\ 2 & 4 & 1 & | & 0 & 1 & 0 \\ 3 & 7 & 2 & | & 0 & 0 & 1 \end{bmatrix}$

$\xrightarrow{R_2 - R_1 \to R_3} \begin{bmatrix} 2 & 3 & 1 & | & 1 & 0 & 0 \\ 0 & 1 & 0 & | & -1 & 1 & 0 \\ 3 & 7 & 2 & | & 0 & 0 & 1 \end{bmatrix}$

$\xrightarrow{2R_1 - R_3 \to R_1} \begin{bmatrix} 1 & -1 & 0 & | & 2 & 0 & -1 \\ 0 & 1 & 0 & | & -1 & 1 & 0 \\ 3 & 7 & 2 & | & 0 & 0 & 1 \end{bmatrix}$

$\xrightarrow[R_3 - 7R_2 \to R_3]{R_1 + R_2 \to R_1} \begin{bmatrix} 1 & 0 & 0 & | & 1 & 1 & -1 \\ 0 & 1 & 0 & | & -1 & 1 & 0 \\ 3 & 0 & 2 & | & 7 & -7 & 1 \end{bmatrix}$

$\xrightarrow{\frac{1}{2}(R_3 - 3R_1) \to R_3} \begin{bmatrix} 1 & 0 & 0 & | & 1 & 1 & -1 \\ 0 & 1 & 0 & | & -1 & 1 & 0 \\ 0 & 0 & 1 & | & 2 & -5 & 2 \end{bmatrix}$

$A^{-1} = \begin{bmatrix} 1 & 1 & -1 \\ -1 & 1 & 0 \\ 2 & -5 & 2 \end{bmatrix}$

25. $[A \mid I] = \begin{bmatrix} 3 & -3 & 4 & | & 1 & 0 & 0 \\ 2 & -3 & 4 & | & 0 & 1 & 0 \\ 0 & -1 & 1 & | & 0 & 0 & 1 \end{bmatrix}$

$\xrightarrow{R_1 - R_2 \to R_1} \begin{bmatrix} 1 & 0 & 0 & | & 1 & -1 & 0 \\ 2 & -3 & 4 & | & 0 & 1 & 0 \\ 0 & -1 & 1 & | & 0 & 0 & 1 \end{bmatrix}$

$\xrightarrow{2R_1 - R_2 \to R_2} \begin{bmatrix} 1 & 0 & 0 & | & 1 & -1 & 0 \\ 0 & 3 & -4 & | & 2 & -3 & 0 \\ 0 & -1 & 1 & | & 0 & 0 & 1 \end{bmatrix}$

$\xrightarrow{-(R_2 + 4R_3) \to R_2} \begin{bmatrix} 1 & 0 & 0 & | & 1 & -1 & 0 \\ 0 & 1 & 0 & | & -2 & 3 & -4 \\ 0 & -1 & 1 & | & 0 & 0 & 1 \end{bmatrix}$

$\xrightarrow{R_2 + R_3 \to R_3} \begin{bmatrix} 1 & 0 & 0 & | & 1 & -1 & 0 \\ 0 & 1 & 0 & | & -2 & 3 & -4 \\ 0 & 0 & 1 & | & -2 & 3 & -3 \end{bmatrix}$

$A^{-1} = \begin{bmatrix} 1 & -1 & 0 \\ -2 & 3 & -4 \\ -2 & 3 & -3 \end{bmatrix}$

27. $ad - bc = (1)(2) - (0)(3) = 2$

$A^{-1} = \frac{1}{2} \begin{bmatrix} 2 & 0 \\ -3 & 1 \end{bmatrix} = \begin{bmatrix} 1 & 0 \\ -\frac{3}{2} & \frac{1}{2} \end{bmatrix}$

29. $ad - bc = (2)(5) - (-3)(-3) = 1$

$A^{-1} = \begin{bmatrix} 5 & 3 \\ 3 & 2 \end{bmatrix}$

31. $ad - bc = (a)(-a) - (b)(-b) = -a^2 + b^2$

$A^{-1} = \frac{1}{-a^2 + b^2} \begin{bmatrix} -a & b \\ -b & a \end{bmatrix}$ where $a^2 \neq b^2$.

33. $\begin{cases} 2x + 3y = -9 \\ x - 3y = 13 \end{cases} \Rightarrow \begin{bmatrix} 2 & 3 \\ 1 & -3 \end{bmatrix} \begin{bmatrix} x \\ y \end{bmatrix} = \begin{bmatrix} -9 \\ 13 \end{bmatrix}$

35. $\begin{cases} 3x + 2y + z = 8 \\ 2x + y + 3z = 7 \\ x + 3y + 2z = 9 \end{cases} \Rightarrow \begin{bmatrix} 3 & 2 & 1 \\ 2 & 1 & 3 \\ 1 & 3 & 2 \end{bmatrix} \begin{bmatrix} x \\ y \\ z \end{bmatrix} = \begin{bmatrix} 8 \\ 7 \\ 9 \end{bmatrix}$

37. $\begin{bmatrix} 1 & -2 \\ 2 & 1 \end{bmatrix} \begin{bmatrix} x \\ y \end{bmatrix} = \begin{bmatrix} 0 \\ 5 \end{bmatrix} \Rightarrow \begin{cases} x - 2y = 0 \\ 2x + y = 5 \end{cases}$

39. $\begin{bmatrix} 2 & 3 & 1 \\ 5 & 7 & -1 \\ 4 & 3 & 0 \end{bmatrix} \begin{bmatrix} x_1 \\ x_2 \\ x_3 \end{bmatrix} = \begin{bmatrix} -1 \\ 5 \\ 5 \end{bmatrix} \Rightarrow \begin{cases} 2x_1 + 3x_2 + x_3 = -1 \\ 5x_1 + 7x_2 - x_3 = 5 \\ 4x_1 + 3x_2 = 5 \end{cases}$

41. $\begin{bmatrix} 1 & 2 & 5 \\ 2 & 3 & 8 \\ -1 & 1 & 2 \end{bmatrix} \begin{bmatrix} 2 & -1 & -1 \\ 12 & -7 & -2 \\ -5 & 3 & 1 \end{bmatrix} = \begin{bmatrix} 1 & 0 & 0 \\ 0 & 1 & 0 \\ 0 & 0 & 1 \end{bmatrix}$

43. $\begin{cases} x_1 + 2x_2 + 5x_3 = 1 \\ 2x_1 + 3x_2 + 8x_3 = 3 \\ -x_1 + x_2 + 2x_3 = -3 \end{cases} \Rightarrow AX = B \Rightarrow$

$\begin{bmatrix} 1 & 2 & 5 \\ 2 & 3 & 8 \\ -1 & 1 & 2 \end{bmatrix} \begin{bmatrix} x \\ y \\ z \end{bmatrix} = \begin{bmatrix} 1 \\ 3 \\ -3 \end{bmatrix} \Rightarrow X = A^{-1}B \Rightarrow$

$\begin{bmatrix} x \\ y \\ z \end{bmatrix} = \begin{bmatrix} 2 & -1 & -1 \\ 12 & -7 & -2 \\ -5 & 3 & 1 \end{bmatrix} \begin{bmatrix} 1 \\ 3 \\ -3 \end{bmatrix} \Rightarrow \begin{bmatrix} x \\ y \\ z \end{bmatrix} = \begin{bmatrix} 2 \\ -3 \\ 1 \end{bmatrix}$

The solution is $\{(2, -3, 1)\}$.

45. a. $[A \mid I] = \begin{bmatrix} 1 & 1 & 1 & | & 1 & 0 & 0 \\ 1 & 2 & 3 & | & 0 & 1 & 0 \\ 1 & 4 & 9 & | & 0 & 0 & 1 \end{bmatrix}$

$\xrightarrow[R_3 - R_1 \to R_3]{R_2 - R_1 \to R_2} \begin{bmatrix} 1 & 1 & 1 & | & 1 & 0 & 0 \\ 0 & 1 & 2 & | & -1 & 1 & 0 \\ 0 & 3 & 8 & | & -1 & 0 & 1 \end{bmatrix}$

$\xrightarrow[\frac{1}{2}(R_3 - 3R_2) \to R_3]{R_1 - R_2 \to R_1} \begin{bmatrix} 1 & 0 & -1 & | & 2 & -1 & 0 \\ 0 & 1 & 2 & | & -1 & 1 & 0 \\ 0 & 0 & 1 & | & 1 & -\frac{3}{2} & \frac{1}{2} \end{bmatrix}$

(continued on next page)

(*continued*)

$$\xrightarrow[\substack{R_1+R_3\to R_1\\R_2-2R_3\to R_2}]{}\begin{bmatrix}1 & 0 & 0 & 3 & -\dfrac{5}{2} & \dfrac{1}{2}\\[2mm]0 & 1 & 0 & -3 & 4 & -1\\[2mm]0 & 0 & 1 & 1 & -\dfrac{3}{2} & \dfrac{1}{2}\end{bmatrix}$$

$$A^{-1}=\begin{bmatrix}3 & -\dfrac{5}{2} & \dfrac{1}{2}\\[2mm]-3 & 4 & -1\\[2mm]1 & -\dfrac{3}{2} & \dfrac{1}{2}\end{bmatrix}$$

b. $\begin{cases}x+\ y+\ z=6\\x+2y+3z=14\\x+4y+9z=36\end{cases}\Rightarrow AX=B\Rightarrow$

$$\begin{bmatrix}1 & 1 & 1\\1 & 2 & 3\\1 & 4 & 9\end{bmatrix}\begin{bmatrix}x\\y\\z\end{bmatrix}=\begin{bmatrix}6\\14\\36\end{bmatrix}\Rightarrow X=A^{-1}B\Rightarrow$$

$$\begin{bmatrix}x\\y\\z\end{bmatrix}=\begin{bmatrix}3 & -\dfrac{5}{2} & \dfrac{1}{2}\\[2mm]-3 & 4 & -1\\[2mm]1 & -\dfrac{3}{2} & \dfrac{1}{2}\end{bmatrix}\begin{bmatrix}6\\14\\36\end{bmatrix}\Rightarrow\begin{bmatrix}x\\y\\z\end{bmatrix}=\begin{bmatrix}1\\2\\3\end{bmatrix}$$

The solution is $\{(1,2,3)\}$.

6.3 Applying the Concepts

47. $\begin{cases}3x+7y=17\\-5x+4y=13\end{cases}\Rightarrow\begin{bmatrix}3 & 7\\-5 & 4\end{bmatrix}\begin{bmatrix}x\\y\end{bmatrix}=\begin{bmatrix}11\\13\end{bmatrix}$

$$[A\,|\,I]=\begin{bmatrix}3 & 7 & 1 & 0\\-5 & 4 & 0 & 1\end{bmatrix}$$

$$\xrightarrow[]{5R_1+3R_2\to R_2}\begin{bmatrix}3 & 7 & 1 & 0\\0 & 47 & 5 & 3\end{bmatrix}$$

$$\xrightarrow[]{\frac{1}{47}R_2\to R_2}\begin{bmatrix}3 & 7 & 1 & 0\\[1mm]0 & 1 & \dfrac{5}{47} & \dfrac{3}{47}\end{bmatrix}$$

$$\xrightarrow[]{\frac{1}{3}(R_1-7R_2)\to R_1}\begin{bmatrix}1 & 0 & \dfrac{4}{47} & -\dfrac{7}{47}\\[2mm]0 & 1 & \dfrac{5}{47} & \dfrac{3}{47}\end{bmatrix}$$

$$A^{-1}=\begin{bmatrix}\dfrac{4}{47} & -\dfrac{7}{47}\\[2mm]\dfrac{5}{47} & \dfrac{3}{47}\end{bmatrix}$$

$$X=A^{-1}B\Rightarrow\begin{bmatrix}x\\y\end{bmatrix}=\begin{bmatrix}\dfrac{4}{47} & -\dfrac{7}{47}\\[2mm]\dfrac{5}{47} & \dfrac{3}{47}\end{bmatrix}\begin{bmatrix}11\\13\end{bmatrix}=\begin{bmatrix}-1\\2\end{bmatrix}$$

The solution is $\{(-1,2)\}$.

49. $\begin{cases}x+\ y+2z=7\\x-\ y-3z=-6\\2x+3y+\ z=4\end{cases}\Rightarrow\begin{bmatrix}1 & 1 & 2\\1 & -1 & -3\\2 & 3 & 1\end{bmatrix}\begin{bmatrix}x\\y\\z\end{bmatrix}=\begin{bmatrix}7\\-6\\4\end{bmatrix}$

$$[A\,|\,I]=\begin{bmatrix}1 & 1 & 2 & 1 & 0 & 0\\1 & -1 & -3 & 0 & 1 & 0\\2 & 3 & 1 & 0 & 0 & 1\end{bmatrix}$$

$$\xrightarrow[\substack{R_1-R_2\to R_2\\R_3-2R_1\to R_3}]{}\begin{bmatrix}1 & 1 & 2 & 1 & 0 & 0\\0 & 2 & 5 & 1 & -1 & 0\\0 & 1 & -3 & -2 & 0 & 1\end{bmatrix}$$

$$\xrightarrow[\substack{2R_1-R_2\to R_1\\\frac{1}{11}(R_2-2R_3)\to R_3}]{}\begin{bmatrix}2 & 0 & -1 & 1 & 1 & 0\\0 & 2 & 5 & 1 & -1 & 0\\0 & 0 & 1 & \dfrac{5}{11} & -\dfrac{1}{11} & -\dfrac{2}{11}\end{bmatrix}$$

$$\xrightarrow[\substack{\frac{1}{2}(R_1+R_3)\to R_1\\\frac{1}{2}(R_2-5R_3)\to R_2}]{}\begin{bmatrix}1 & 0 & 0 & \dfrac{8}{11} & \dfrac{5}{11} & -\dfrac{1}{11}\\[2mm]0 & 1 & 0 & -\dfrac{7}{11} & -\dfrac{3}{11} & \dfrac{5}{11}\\[2mm]0 & 0 & 1 & \dfrac{5}{11} & -\dfrac{1}{11} & -\dfrac{2}{11}\end{bmatrix}$$

$$A^{-1}=\begin{bmatrix}\dfrac{8}{11} & \dfrac{5}{11} & -\dfrac{1}{11}\\[2mm]-\dfrac{7}{11} & -\dfrac{3}{11} & \dfrac{5}{11}\\[2mm]\dfrac{5}{11} & -\dfrac{1}{11} & -\dfrac{2}{11}\end{bmatrix}$$

$$X=A^{-1}B\Rightarrow$$

$$\begin{bmatrix}x\\y\\z\end{bmatrix}=\begin{bmatrix}\dfrac{8}{11} & \dfrac{5}{11} & -\dfrac{1}{11}\\[2mm]-\dfrac{7}{11} & -\dfrac{3}{11} & \dfrac{5}{11}\\[2mm]\dfrac{5}{11} & -\dfrac{1}{11} & -\dfrac{2}{11}\end{bmatrix}\begin{bmatrix}7\\-6\\4\end{bmatrix}=\begin{bmatrix}2\\-1\\3\end{bmatrix}$$

The solution is $\{(2,-1,3)\}$.

51. $\begin{cases} 2x+2y+3x=7 \\ 5x+3y+5z=3 \\ 3x+5y+\ z=-5 \end{cases} \Rightarrow \begin{bmatrix} 2 & 2 & 3 \\ 5 & 3 & 5 \\ 3 & 5 & 1 \end{bmatrix}\begin{bmatrix} x \\ y \\ z \end{bmatrix}=\begin{bmatrix} 7 \\ 3 \\ -5 \end{bmatrix}$

$[A\,|\,I]=\begin{bmatrix} 2 & 2 & 3 & | & 1 & 0 & 0 \\ 5 & 3 & 5 & | & 0 & 1 & 0 \\ 3 & 5 & 1 & | & 0 & 0 & 1 \end{bmatrix}\xrightarrow[3R_1-2R_3\to R_3]{5R_1-2R_2\to R_2}\begin{bmatrix} 2 & 2 & 3 & | & 1 & 0 & 0 \\ 0 & 4 & 5 & | & 5 & -2 & 0 \\ 0 & -4 & 7 & | & 3 & 0 & -2 \end{bmatrix}$

$\xrightarrow[\frac{1}{12}(R_2+R_3)\to R_3]{R_2-2R_1\to R_1}\begin{bmatrix} -4 & 0 & -1 & | & 3 & -2 & 0 \\ 0 & 4 & 5 & | & 5 & -2 & 0 \\ 0 & 0 & 1 & | & \frac{2}{3} & -\frac{1}{6} & -\frac{1}{6} \end{bmatrix}\xrightarrow[\frac{1}{4}(R_2-5R_3)\to R_2]{-\frac{1}{4}(R_1+R_3)\to R_1}\begin{bmatrix} 1 & 0 & 0 & | & -\frac{11}{12} & \frac{13}{24} & \frac{1}{24} \\ 0 & 1 & 0 & | & \frac{5}{12} & -\frac{7}{24} & \frac{5}{24} \\ 0 & 0 & 1 & | & \frac{2}{3} & -\frac{1}{6} & -\frac{1}{6} \end{bmatrix}$

$A^{-1}=\begin{bmatrix} -\frac{11}{12} & \frac{13}{24} & \frac{1}{24} \\ \frac{5}{12} & -\frac{7}{24} & \frac{5}{24} \\ \frac{2}{3} & -\frac{1}{6} & -\frac{1}{6} \end{bmatrix}$

$X=A^{-1}B\Rightarrow\begin{bmatrix} x \\ y \\ z \end{bmatrix}=\begin{bmatrix} -\frac{11}{12} & \frac{13}{24} & \frac{1}{24} \\ \frac{5}{12} & -\frac{7}{24} & \frac{5}{24} \\ \frac{2}{3} & -\frac{1}{6} & -\frac{1}{6} \end{bmatrix}\begin{bmatrix} 7 \\ 3 \\ -5 \end{bmatrix}=\begin{bmatrix} -5 \\ 1 \\ 5 \end{bmatrix}$

The solution is $\{(-5,\ 1,\ 5)\}$.

53. Let $x=$ the amount invested in a treasury bill, $y=$ the amount invested in bonds, and $z=$ the amount invested in a mutual fund. Then we have

$\begin{cases} x+y+z=90{,}000 \\ 0.03x+0.07y-0.11z=980 \\ 0.03x+0.07y+0.08z=5920 \end{cases} \Rightarrow \begin{bmatrix} 1 & 1 & 1 \\ 0.03 & 0.07 & -0.11 \\ 0.03 & 0.07 & 0.08 \end{bmatrix}\begin{bmatrix} x \\ y \\ z \end{bmatrix}=\begin{bmatrix} 90{,}000 \\ 980 \\ 5920 \end{bmatrix}$

$[A\,|\,I]=\begin{bmatrix} 1 & 1 & 1 & | & 1 & 0 & 0 \\ \frac{3}{100} & \frac{7}{100} & -\frac{11}{100} & | & 0 & 1 & 0 \\ \frac{3}{100} & \frac{7}{100} & \frac{8}{100} & | & 0 & 0 & 1 \end{bmatrix}\xrightarrow[-\frac{3}{100}R_3+R_3\to]{-\frac{3}{100}R_1+R_2\to R_2}\begin{bmatrix} 1 & 1 & 1 & | & 1 & 0 & 0 \\ 0 & \frac{1}{25} & -\frac{7}{50} & | & -\frac{3}{100} & 1 & 0 \\ 0 & \frac{1}{25} & \frac{1}{20} & | & -\frac{3}{100} & 0 & 1 \end{bmatrix}$

$\xrightarrow{-\frac{100}{19}(R_2-R_3)\to R_3}\begin{bmatrix} 1 & 1 & 1 & | & 1 & 0 & 0 \\ 0 & \frac{1}{25} & -\frac{7}{50} & | & -\frac{3}{100} & 1 & 0 \\ 0 & 0 & 1 & | & 0 & -\frac{100}{19} & \frac{100}{19} \end{bmatrix}\xrightarrow{25R_2\to R_2}\begin{bmatrix} 1 & 1 & 1 & | & 1 & 0 & 0 \\ 0 & 1 & -\frac{7}{2} & | & -\frac{3}{4} & 25 & 0 \\ 0 & 0 & 1 & | & 0 & -\frac{100}{19} & \frac{100}{19} \end{bmatrix}$

(continued on next page)

(*continued*)

$$\xrightarrow[\substack{R_1-R_3\to R_1 \\ R_2+\frac{7}{2}R_3\to R_2}]{} \left[\begin{array}{ccc|ccc} 1 & 1 & 0 & 1 & \dfrac{100}{19} & -\dfrac{100}{19} \\[2mm] 0 & 1 & 0 & -\dfrac{3}{4} & \dfrac{125}{19} & \dfrac{350}{19} \\[2mm] 0 & 0 & 1 & 0 & -\dfrac{100}{19} & \dfrac{100}{19} \end{array}\right] \xrightarrow[\ ]{R_1-R_2\to R_1} \left[\begin{array}{ccc|ccc} 1 & 1 & 0 & \dfrac{7}{4} & -\dfrac{25}{19} & -\dfrac{450}{19} \\[2mm] 0 & 1 & 0 & -\dfrac{3}{4} & \dfrac{125}{19} & \dfrac{350}{19} \\[2mm] 0 & 0 & 1 & 0 & -\dfrac{100}{19} & \dfrac{100}{19} \end{array}\right]$$

$$A^{-1} = \left[\begin{array}{ccc} \dfrac{7}{4} & -\dfrac{25}{19} & -\dfrac{450}{19} \\[2mm] -\dfrac{3}{4} & \dfrac{125}{19} & \dfrac{350}{19} \\[2mm] 0 & -\dfrac{100}{19} & \dfrac{100}{19} \end{array}\right] \Rightarrow X = A^{-1}B \Rightarrow \left[\begin{array}{c} x \\ y \\ z \end{array}\right] = \left[\begin{array}{ccc} \dfrac{7}{4} & -\dfrac{25}{19} & -\dfrac{450}{19} \\[2mm] -\dfrac{3}{4} & \dfrac{125}{19} & \dfrac{350}{19} \\[2mm] 0 & -\dfrac{100}{19} & \dfrac{100}{19} \end{array}\right]\left[\begin{array}{c} 90{,}000 \\ 980 \\ 5920 \end{array}\right] = \left[\begin{array}{c} 16{,}000 \\ 48{,}000 \\ 26{,}000 \end{array}\right]$$

Liz invested \$16,000 in a treasury bill, \$48,000 in bonds, and \$26,000 in a mutual fund.

55. $\begin{cases} V - E + R = 2 \\ 2E = 3V \\ 2R = E - 1 \end{cases} \Rightarrow \begin{cases} V - E + R = 2 \\ -3V + 2E = 0 \\ -E + 2R = -1 \end{cases} \Rightarrow \left[\begin{array}{ccc} 1 & -1 & 1 \\ -3 & 2 & 0 \\ 0 & -1 & 2 \end{array}\right]\left[\begin{array}{c} x \\ y \\ z \end{array}\right] = \left[\begin{array}{c} 2 \\ 0 \\ 1 \end{array}\right]$

$$[A \mid I] = \left[\begin{array}{ccc|ccc} 1 & -1 & 1 & 1 & 0 & 0 \\ -3 & 2 & 0 & 0 & 1 & 0 \\ 0 & -1 & 2 & 0 & 0 & 1 \end{array}\right] \xrightarrow{3R_1+R_2\to R_2} \left[\begin{array}{ccc|ccc} 1 & -1 & 1 & 1 & 0 & 0 \\ 0 & -1 & 3 & 3 & 1 & 0 \\ 0 & -1 & 2 & 0 & 0 & 1 \end{array}\right]$$

$$\xrightarrow{-R_2\to R_2} \left[\begin{array}{ccc|ccc} 1 & -1 & 1 & 1 & 0 & 0 \\ 0 & 1 & -3 & -3 & -1 & 0 \\ 0 & -1 & 2 & 0 & 0 & 1 \end{array}\right] \xrightarrow[\substack{-(R_3+R_2)\to R_3}]{R_1+R_2\to R_1} \left[\begin{array}{ccc|ccc} 1 & 0 & -2 & -2 & -1 & 0 \\ 0 & 1 & -3 & -3 & -1 & 0 \\ 0 & 0 & 1 & 3 & 1 & -1 \end{array}\right]$$

$$\xrightarrow[\substack{R_1+2R_3\to R_1 \\ R_2+3R_3\to R_2}]{} \left[\begin{array}{ccc|ccc} 1 & 0 & 0 & 4 & 1 & -2 \\ 0 & 1 & 0 & 6 & 2 & -3 \\ 0 & 0 & 1 & 3 & 1 & -1 \end{array}\right]$$

$$A^{-1} = \left[\begin{array}{ccc} 4 & 1 & -2 \\ 6 & 2 & -3 \\ 3 & 1 & -1 \end{array}\right] \Rightarrow X = A^{-1}B \Rightarrow \left[\begin{array}{c} x \\ y \\ z \end{array}\right] = \left[\begin{array}{ccc} 4 & 1 & -2 \\ 6 & 2 & -3 \\ 3 & 1 & -1 \end{array}\right]\left[\begin{array}{c} 2 \\ 0 \\ 1 \end{array}\right] = \left[\begin{array}{c} 10 \\ 15 \\ 7 \end{array}\right]$$

There are 10 vertices, 15 edges, and 7 regions.

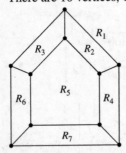

57. $X = \left(\begin{bmatrix} 1 & 0 \\ 0 & 1 \end{bmatrix} - \begin{bmatrix} 0.1 & 0.4 \\ 0.5 & 0.2 \end{bmatrix} \right)^{-1} \begin{bmatrix} 50 \\ 30 \end{bmatrix}$; $(I-A)^{-1} = \begin{bmatrix} \dfrac{9}{10} & -\dfrac{2}{5} \\ -\dfrac{1}{2} & \dfrac{4}{5} \end{bmatrix}^{-1}$

$\begin{bmatrix} \dfrac{9}{10} & -\dfrac{2}{5} & \bigg| & 1 & 0 \\ -\dfrac{1}{2} & \dfrac{4}{5} & \bigg| & 0 & 1 \end{bmatrix} \xrightarrow{\frac{1}{2}R_1 + \frac{9}{10}R_2 \to R_2} \begin{bmatrix} \dfrac{9}{10} & -\dfrac{2}{5} & \bigg| & 1 & 0 \\ 0 & \dfrac{13}{25} & \bigg| & \dfrac{1}{2} & \dfrac{9}{10} \end{bmatrix} \xrightarrow{\frac{13}{25}R_1 + \frac{2}{5}R_2 \to R_1} \begin{bmatrix} \dfrac{117}{250} & 0 & \bigg| & \dfrac{18}{25} & \dfrac{9}{25} \\ 0 & \dfrac{13}{25} & \bigg| & \dfrac{1}{2} & \dfrac{9}{10} \end{bmatrix}$

$\xrightarrow[\frac{25}{13}R_2 \to R_2]{\frac{250}{117}R_1 \to R_1} \begin{bmatrix} 1 & 0 & \bigg| & \dfrac{20}{13} & \dfrac{10}{13} \\ 0 & 1 & \bigg| & \dfrac{25}{26} & \dfrac{45}{26} \end{bmatrix}$

$(I-A)^{-1} = \begin{bmatrix} \dfrac{20}{13} & \dfrac{10}{13} \\ \dfrac{25}{26} & \dfrac{45}{26} \end{bmatrix} \Rightarrow X = (I-A)^{-1}D = \begin{bmatrix} \dfrac{20}{13} & \dfrac{10}{13} \\ \dfrac{25}{26} & \dfrac{45}{26} \end{bmatrix} \begin{bmatrix} 50 \\ 30 \end{bmatrix} = \begin{bmatrix} 100 \\ 100 \end{bmatrix}$

59. $X = \left(\begin{bmatrix} 1 & 0 & 0 \\ 0 & 1 & 0 \\ 0 & 0 & 1 \end{bmatrix} - \begin{bmatrix} 0.2 & 0.2 & 0 \\ 0.1 & 0.1 & 0.3 \\ 0.1 & 0 & 0.2 \end{bmatrix} \right)^{-1} \begin{bmatrix} 400 \\ 600 \\ 800 \end{bmatrix}$

$(I-A)^{-1} = \begin{bmatrix} \dfrac{4}{5} & -\dfrac{1}{5} & 0 \\ -\dfrac{1}{10} & \dfrac{9}{10} & -\dfrac{3}{10} \\ -\dfrac{1}{10} & 0 & \dfrac{4}{5} \end{bmatrix}^{-1} \Rightarrow \begin{bmatrix} \dfrac{4}{5} & -\dfrac{1}{5} & 0 & \bigg| & 1 & 0 & 0 \\ -\dfrac{1}{10} & \dfrac{9}{10} & -\dfrac{3}{10} & \bigg| & 0 & 1 & 0 \\ -\dfrac{1}{10} & 0 & \dfrac{4}{5} & \bigg| & 0 & 0 & 1 \end{bmatrix} \xrightarrow[\frac{1}{10}R_1 + \frac{4}{5}R_3 \to R_3]{\frac{1}{10}R_1 + \frac{4}{5}R_2 \to R_2} \begin{bmatrix} \dfrac{4}{5} & -\dfrac{1}{5} & 0 & \bigg| & 1 & 0 & 0 \\ 0 & \dfrac{7}{10} & -\dfrac{6}{25} & \bigg| & \dfrac{1}{10} & \dfrac{4}{5} & 0 \\ 0 & -\dfrac{1}{50} & \dfrac{16}{25} & \bigg| & \dfrac{1}{10} & 0 & \dfrac{4}{5} \end{bmatrix}$

$\xrightarrow{\frac{625}{277}\left(\frac{1}{50}R_2 + \frac{7}{10}R_3\right) \to R_3} \begin{bmatrix} \dfrac{4}{5} & -\dfrac{1}{5} & 0 & \bigg| & 1 & 0 & 0 \\ 0 & \dfrac{7}{10} & -\dfrac{6}{25} & \bigg| & \dfrac{1}{10} & \dfrac{4}{5} & 0 \\ 0 & 0 & 1 & \bigg| & \dfrac{45}{277} & \dfrac{10}{277} & \dfrac{350}{277} \end{bmatrix}$

$\xrightarrow[\frac{10}{7}\left(\frac{6}{25}R_3 + R_2 \to R_2\right)]{5R_1 \to R_1} \begin{bmatrix} 4 & -1 & 0 & \bigg| & 5 & 0 & 0 \\ 0 & 1 & 0 & \bigg| & \dfrac{55}{277} & \dfrac{320}{277} & \dfrac{120}{277} \\ 0 & 0 & 1 & \bigg| & \dfrac{45}{277} & \dfrac{10}{277} & \dfrac{350}{277} \end{bmatrix} \xrightarrow{\frac{1}{4}(R_1 + R_2) \to R_1} \begin{bmatrix} 1 & 0 & 0 & \bigg| & \dfrac{360}{277} & \dfrac{80}{277} & \dfrac{30}{277} \\ 0 & 1 & 0 & \bigg| & \dfrac{55}{277} & \dfrac{320}{277} & \dfrac{120}{277} \\ 0 & 0 & 1 & \bigg| & \dfrac{45}{277} & \dfrac{10}{277} & \dfrac{350}{277} \end{bmatrix} \Rightarrow$

$(I-A)^{-1} = \begin{bmatrix} \dfrac{360}{277} & \dfrac{80}{277} & \dfrac{30}{277} \\ \dfrac{55}{277} & \dfrac{320}{277} & \dfrac{120}{277} \\ \dfrac{45}{277} & \dfrac{10}{277} & \dfrac{350}{277} \end{bmatrix}$; $X = (I-A)^{-1}D = \begin{bmatrix} \dfrac{360}{277} & \dfrac{80}{277} & \dfrac{30}{277} \\ \dfrac{55}{277} & \dfrac{320}{277} & \dfrac{120}{277} \\ \dfrac{45}{277} & \dfrac{10}{277} & \dfrac{350}{277} \end{bmatrix} \begin{bmatrix} 400 \\ 600 \\ 800 \end{bmatrix} = \begin{bmatrix} \dfrac{216{,}000}{277} \\ \dfrac{310{,}000}{277} \\ \dfrac{304{,}000}{277} \end{bmatrix}$

Answers will vary in exercises 61–64.

61. a. Associate each letter in the phrase with the number representing its position in the alphabet, and partition the numbers into groups of two, forming a 2×9 matrix:

$$M = \begin{bmatrix} 3 & 14 & 15 & 6 & 14 & 3 & 12 & 13 & 15 \\ 1 & 14 & 20 & 9 & 4 & 15 & 15 & 2 & 0 \end{bmatrix} \text{ Let } A = \begin{bmatrix} 1 & 2 \\ 2 & 1 \end{bmatrix} \text{ Then,}$$

$$AM = \begin{bmatrix} 1 & 2 \\ 2 & 1 \end{bmatrix}\begin{bmatrix} 3 & 14 & 15 & 6 & 14 & 3 & 12 & 13 & 15 \\ 1 & 14 & 20 & 9 & 4 & 15 & 15 & 2 & 0 \end{bmatrix} = \begin{bmatrix} 5 & 42 & 55 & 24 & 22 & 33 & 42 & 17 & 15 \\ 7 & 42 & 50 & 21 & 32 & 21 & 39 & 28 & 30 \end{bmatrix}$$

The cryptogram is 5 7 42 42 55 50 24 21 22 32 33 21 42 39 17 28 15 30.

b. $A^{-1} = \begin{bmatrix} 1 & 2 & | & 1 & 0 \\ 2 & 1 & | & 0 & 1 \end{bmatrix} \xrightarrow{\frac{1}{3}(2R_1 - R_2) \to R_2} \begin{bmatrix} 1 & 2 & | & 1 & 0 \\ 0 & 1 & | & \frac{2}{3} & -\frac{1}{3} \end{bmatrix} \xrightarrow{R_1 - 2R_2 \to R_1} \begin{bmatrix} 1 & 0 & | & -\frac{1}{3} & \frac{2}{3} \\ 0 & 1 & | & \frac{2}{3} & -\frac{1}{3} \end{bmatrix} = \begin{bmatrix} -\frac{1}{3} & \frac{2}{3} \\ \frac{2}{3} & -\frac{1}{3} \end{bmatrix}$

$$M = A^{-1}(AM) = \begin{bmatrix} -\frac{1}{3} & \frac{2}{3} \\ \frac{2}{3} & -\frac{1}{3} \end{bmatrix}\begin{bmatrix} 5 & 42 & 55 & 24 & 22 & 33 & 42 & 17 & 15 \\ 7 & 42 & 50 & 21 & 32 & 21 & 39 & 28 & 30 \end{bmatrix}$$

$$= \begin{bmatrix} 3 & 14 & 15 & 6 & 14 & 3 & 12 & 13 & 15 \\ 1 & 14 & 20 & 9 & 4 & 15 & 15 & 2 & 0 \end{bmatrix}$$

63. a. Associate each letter in the phrase with the number representing its position in the alphabet, and partition the numbers into groups of three, forming a 3×6 matrix:

$$M = \begin{bmatrix} 3 & 14 & 6 & 4 & 12 & 2 \\ 1 & 15 & 9 & 3 & 15 & 15 \\ 14 & 20 & 14 & 15 & 13 & 0 \end{bmatrix} \text{ Let } A = \begin{bmatrix} 2 & 1 & 1 \\ 1 & 2 & 1 \\ 1 & 1 & 2 \end{bmatrix} \text{ Then,}$$

$$AM = \begin{bmatrix} 2 & 1 & 1 \\ 1 & 2 & 1 \\ 1 & 1 & 2 \end{bmatrix}\begin{bmatrix} 3 & 14 & 6 & 4 & 12 & 2 \\ 1 & 15 & 9 & 3 & 15 & 15 \\ 14 & 20 & 14 & 15 & 13 & 0 \end{bmatrix} = \begin{bmatrix} 21 & 63 & 35 & 26 & 52 & 19 \\ 19 & 64 & 38 & 25 & 55 & 32 \\ 32 & 69 & 43 & 37 & 53 & 17 \end{bmatrix}$$

The cryptogram is 21 39 32 63 64 69 35 38 43 26 25 37 52 55 53 19 32 17.

b. $A^{-1} = \begin{bmatrix} 2 & 1 & 1 & | & 1 & 0 & 0 \\ 1 & 2 & 1 & | & 0 & 1 & 0 \\ 1 & 1 & 2 & | & 0 & 0 & 1 \end{bmatrix} \xrightarrow[R_1 - 2R_3 \to R_3]{R_1 - 2R_2 \to R_2} \begin{bmatrix} 2 & 1 & 1 & | & 1 & 0 & 0 \\ 0 & -3 & -1 & | & 1 & -2 & 0 \\ 0 & -1 & -3 & | & 1 & 0 & -2 \end{bmatrix}$

$$\xrightarrow{\frac{1}{8}(R_2 - 3R_3) \to R_3} \begin{bmatrix} 2 & 1 & 1 & | & 1 & 0 & 0 \\ 0 & -3 & -1 & | & 1 & -2 & 0 \\ 0 & 0 & 1 & | & -\frac{1}{4} & -\frac{1}{4} & \frac{3}{4} \end{bmatrix} \xrightarrow[-\frac{1}{3}(R_3 + R_2) \to R_2]{R_1 - R_3 \to R_1} \begin{bmatrix} 2 & 1 & 0 & | & \frac{5}{4} & \frac{1}{4} & -\frac{3}{4} \\ 0 & 1 & 0 & | & -\frac{1}{4} & \frac{3}{4} & -\frac{1}{4} \\ 0 & 0 & 1 & | & -\frac{1}{4} & -\frac{1}{4} & \frac{3}{4} \end{bmatrix}$$

$$\xrightarrow{\frac{1}{2}(R_1 - R_2) \to R_1} \begin{bmatrix} 1 & 0 & 0 & | & \frac{3}{4} & -\frac{1}{4} & -\frac{1}{4} \\ 0 & 1 & 0 & | & -\frac{1}{4} & \frac{3}{4} & -\frac{1}{4} \\ 0 & 0 & 1 & | & -\frac{1}{4} & -\frac{1}{4} & \frac{3}{4} \end{bmatrix} = \begin{bmatrix} \frac{3}{4} & -\frac{1}{4} & -\frac{1}{4} \\ -\frac{1}{4} & \frac{3}{4} & -\frac{1}{4} \\ -\frac{1}{4} & -\frac{1}{4} & \frac{3}{4} \end{bmatrix}$$

(continued on next page)

(*continued*)

$$M = A^{-1}(AM) = \begin{bmatrix} \frac{3}{4} & -\frac{1}{4} & -\frac{1}{4} \\ -\frac{1}{4} & \frac{3}{4} & -\frac{1}{4} \\ -\frac{1}{4} & -\frac{1}{4} & \frac{3}{4} \end{bmatrix} \begin{bmatrix} 21 & 63 & 35 & 26 & 52 & 19 \\ 19 & 64 & 38 & 25 & 55 & 32 \\ 32 & 69 & 43 & 37 & 53 & 17 \end{bmatrix} = \begin{bmatrix} 3 & 14 & 6 & 4 & 12 & 2 \\ 1 & 15 & 9 & 3 & 15 & 15 \\ 14 & 20 & 14 & 15 & 13 & 0 \end{bmatrix}$$

6.3 Beyond the Basics

65. a. Yes, by the definition of *inverse* if $AB = I$, then $BA = I$. So $B = A^{-1}$ and $A = B^{-1}$.

b. $\left(A^{-1}\right)^{-1} = A$

67. $I = A^2 B = A(AB)$, so AB is the inverse of A.

69. $ABB^{-1}A^{-1} = AIA^{-1} = AA^{-1} = I$
$B^{-1}A^{-1}AB = B^{-1}IB = B^{-1}B = I \Rightarrow AB$ is invertible and $(AB)^{-1} = B^{-1}A^{-1}$.

71. If A is invertible, then there exists a matrix D such that $AD = I$ and $DA = I$. Then
$AB = AC \Leftrightarrow DAB = DAC \Leftrightarrow IB = IC \Leftrightarrow B = C$.

73. If A is invertible, then $I = A^{-1}A = A^{-1}(AB) = \left(A^{-1}A\right)B = IB = B$.

75. a. $\begin{bmatrix} 3 & 4 \\ 2 & 3 \end{bmatrix}^2 - 6\begin{bmatrix} 3 & 4 \\ 2 & 3 \end{bmatrix} + \begin{bmatrix} 1 & 0 \\ 0 & 1 \end{bmatrix} = \begin{bmatrix} 17 & 24 \\ 12 & 17 \end{bmatrix} - \begin{bmatrix} 18 & 24 \\ 12 & 18 \end{bmatrix} + \begin{bmatrix} 1 & 0 \\ 0 & 1 \end{bmatrix} = \begin{bmatrix} 0 & 0 \\ 0 & 0 \end{bmatrix}$

b. $A^2 - 6A + I = 0 \Rightarrow I = 6A - A^2 \Rightarrow AA^{-1} = 6A - A^2 \Rightarrow A^{-1} = 6I - A$

c. $A^{-1} = 6I - A = \begin{bmatrix} 6 & 0 \\ 0 & 6 \end{bmatrix} - \begin{bmatrix} 3 & 4 \\ 2 & 3 \end{bmatrix} = \begin{bmatrix} 3 & -4 \\ -2 & 3 \end{bmatrix}$

77. $\begin{cases} x - y - z - w = -4 \\ x + y + z - w = 2 \\ 2x + y + z - w = 3 \\ x - y + z - w = -2 \end{cases} \Rightarrow \begin{bmatrix} 1 & -1 & -1 & -1 \\ 1 & 1 & 1 & -1 \\ 2 & 1 & 1 & -1 \\ 1 & -1 & 1 & -1 \end{bmatrix}\begin{bmatrix} x \\ y \\ z \\ w \end{bmatrix} = \begin{bmatrix} -4 \\ 2 \\ 3 \\ -2 \end{bmatrix} \Rightarrow \begin{bmatrix} 1 & -1 & -1 & -1 \\ 1 & 1 & 1 & -1 \\ 2 & 1 & 1 & -1 \\ 1 & -1 & 1 & -1 \end{bmatrix}^{-1}\begin{bmatrix} -4 \\ 2 \\ 3 \\ -2 \end{bmatrix} = \begin{bmatrix} x \\ y \\ z \\ w \end{bmatrix}$

Using a graphing calculator, we have

$$\begin{bmatrix} 1 & -1 & -1 & -1 \\ 1 & 1 & 1 & -1 \\ 2 & 1 & 1 & -1 \\ 1 & -1 & 1 & -1 \end{bmatrix}^{-1} = \begin{bmatrix} 0 & -1 & 1 & 0 \\ 0 & \frac{1}{2} & 0 & -\frac{1}{2} \\ -\frac{1}{2} & 0 & 0 & \frac{1}{2} \\ -\frac{1}{2} & -\frac{3}{2} & 1 & 0 \end{bmatrix}$$

(*continued on next page*)

(*continued*)

$$\begin{bmatrix} 0 & -1 & 1 & 0 \\ 0 & \dfrac{1}{2} & 0 & -\dfrac{1}{2} \\ -\dfrac{1}{2} & 0 & 0 & \dfrac{1}{2} \\ -\dfrac{1}{2} & -\dfrac{3}{2} & 1 & 0 \end{bmatrix} \begin{bmatrix} -4 \\ 2 \\ 3 \\ -2 \end{bmatrix} = \begin{bmatrix} 1 \\ 2 \\ 1 \\ 2 \end{bmatrix}$$

Solution set: $\{(1, 2, 1, 2)\}$

79. a. $ABB^{-1} = AI = A \Rightarrow A = \begin{bmatrix} -1 & 3 \\ 0 & 2 \end{bmatrix}\begin{bmatrix} 1 & 2 \\ 3 & 4 \end{bmatrix} = \begin{bmatrix} 8 & 10 \\ 6 & 8 \end{bmatrix}$

b. $B^{-1}BA = IA = A \Rightarrow A = \begin{bmatrix} 1 & 2 \\ 3 & 4 \end{bmatrix}\begin{bmatrix} -1 & 3 \\ 0 & 2 \end{bmatrix} = \begin{bmatrix} -1 & 7 \\ -3 & 17 \end{bmatrix}$

c. $B^{-1}BAB^{-1} = IAB^{-1} = AB^{-1};$ $\qquad AB^{-1} = \begin{bmatrix} 1 & 2 \\ 3 & 4 \end{bmatrix}\begin{bmatrix} 1 & 1 \\ 1 & 2 \end{bmatrix} = \begin{bmatrix} 3 & 5 \\ 7 & 11 \end{bmatrix}$

$$\begin{bmatrix} a & b \\ c & d \end{bmatrix}\begin{bmatrix} 1 & 2 \\ 3 & 4 \end{bmatrix} = \begin{bmatrix} 3 & 5 \\ 7 & 11 \end{bmatrix} \Rightarrow \begin{bmatrix} a+3b & 2a+4b \\ c+3d & 2c+4d \end{bmatrix} = \begin{bmatrix} 3 & 5 \\ 7 & 11 \end{bmatrix} \Rightarrow \begin{cases} a+3b = 3 \\ 2a+4b = 5 \\ c+3d = 7 \\ 2c+4d = 11 \end{cases} \Rightarrow$$

$$a = \frac{3}{2}, b = \frac{1}{2}, c = \frac{5}{2}, d = \frac{3}{2} \Rightarrow A = \begin{bmatrix} \dfrac{3}{2} & \dfrac{1}{2} \\ \dfrac{5}{2} & \dfrac{3}{2} \end{bmatrix}$$

d. $B^{-1}ABB^{-1} = B^{-1}AI = B^{-1}A \Rightarrow B^{-1}A = \begin{bmatrix} 1 & 1 \\ 1 & 2 \end{bmatrix}\begin{bmatrix} 1 & 2 \\ 3 & 4 \end{bmatrix} = \begin{bmatrix} 4 & 6 \\ 7 & 10 \end{bmatrix}$

$$\begin{bmatrix} 1 & 2 \\ 3 & 4 \end{bmatrix}\begin{bmatrix} a & b \\ c & d \end{bmatrix} = \begin{bmatrix} 4 & 6 \\ 7 & 10 \end{bmatrix} \Rightarrow \begin{bmatrix} a+2c & b+2d \\ 3a+4c & 3b+4d \end{bmatrix} = \begin{bmatrix} 4 & 6 \\ 7 & 10 \end{bmatrix} \Rightarrow \begin{cases} a+2c = 4 \\ b+2d = 6 \\ 3a+4c = 7 \\ 3b+4d = 10 \end{cases} \Rightarrow$$

$$\begin{bmatrix} 1 & 0 & 2 & 0 \\ 0 & 1 & 0 & 2 \\ 3 & 0 & 4 & 0 \\ 0 & 3 & 0 & 4 \end{bmatrix}\begin{bmatrix} a \\ b \\ c \\ d \end{bmatrix} = \begin{bmatrix} 4 \\ 6 \\ 7 \\ 10 \end{bmatrix} \Rightarrow \begin{bmatrix} a \\ b \\ c \\ d \end{bmatrix} = \begin{bmatrix} 1 & 0 & 2 & 0 \\ 0 & 1 & 0 & 2 \\ 3 & 0 & 4 & 0 \\ 0 & 3 & 0 & 4 \end{bmatrix}^{-1}\begin{bmatrix} 4 \\ 6 \\ 7 \\ 10 \end{bmatrix} = \begin{bmatrix} -2 & 0 & 1 & 0 \\ 0 & -2 & 0 & 1 \\ \dfrac{3}{2} & 0 & -\dfrac{1}{2} & 0 \\ 0 & \dfrac{3}{2} & 0 & -\dfrac{1}{2} \end{bmatrix}\begin{bmatrix} 4 \\ 6 \\ 7 \\ 10 \end{bmatrix} = \begin{bmatrix} -1 \\ -2 \\ \dfrac{5}{2} \\ 4 \end{bmatrix} \Rightarrow$$

$$A = \begin{bmatrix} -1 & -2 \\ \dfrac{5}{2} & 4 \end{bmatrix}$$

6.3 Critical Thinking/Discussion/Writing

81. a. True. $A^2B = AAB = ABA = BAA = BA^2$

b. False. For example, if $A = I$ and $B = -I$, then $A + B = \begin{bmatrix} 1 & 0 \\ 0 & 1 \end{bmatrix} + \begin{bmatrix} -1 & 0 \\ 0 & -1 \end{bmatrix} = \begin{bmatrix} 0 & 0 \\ 0 & 0 \end{bmatrix}$, which is not invertible.

82. Let $A = \begin{bmatrix} 0 & -1 \\ 0 & 0 \end{bmatrix}$. Then $A^2 = \begin{bmatrix} 0 & 0 \\ 0 & 0 \end{bmatrix}$.

$$I + A = \begin{bmatrix} 1 & 0 \\ 0 & 1 \end{bmatrix} + \begin{bmatrix} 0 & -1 \\ 0 & 0 \end{bmatrix} = \begin{bmatrix} 1 & -1 \\ 0 & 1 \end{bmatrix}$$

$$(I + A)^{-1} = \begin{bmatrix} 1 & 1 \\ 0 & 1 \end{bmatrix}$$

$$I - A = \begin{bmatrix} 1 & 0 \\ 0 & 1 \end{bmatrix} - \begin{bmatrix} 0 & -1 \\ 0 & 0 \end{bmatrix} = \begin{bmatrix} 1 & 1 \\ 0 & 1 \end{bmatrix}$$

83. True. A matrix has an inverse if and only if it is square.

6.3 Maintaining Skills

85. 1

87. −1

89. $\begin{bmatrix} -1 \\ -3 \end{bmatrix} \begin{bmatrix} 5 & 0 \end{bmatrix} = \begin{bmatrix} -1(5) & -1(0) \\ -3(5) & -3(0) \end{bmatrix} = \begin{bmatrix} -5 & 0 \\ -15 & 0 \end{bmatrix}$

91. $(-1)\begin{bmatrix} -6 & 7 \\ 4 & -1 \end{bmatrix} = \begin{bmatrix} 6 & -7 \\ -4 & 1 \end{bmatrix}$

In exercises 92–95, be sure to check the solution in the original equations.

93. $\begin{cases} 16x - 9y = -5 & (1) \\ 10x + 18y = -11 & (2) \end{cases}$

Multiply equation (1) by 2, then add the resulting equation and equation (2).

$32x - 18y = -10$
$\underline{10x + 18y = -11}$
$\,42x = -21 \Rightarrow x = -\dfrac{1}{2}$

Substitute $x = -\dfrac{1}{2}$ in equation (2) and solve for y.

$10\left(-\dfrac{1}{2}\right) + 18y = -11 \Rightarrow -5 + 18y = -11 \Rightarrow$

$18y = -6 \Rightarrow y = -\dfrac{1}{3}$

Solution set: $\left\{ \left(-\dfrac{1}{2}, -\dfrac{1}{3} \right) \right\}$

95. $\begin{cases} 2x - y + 2z = 3 & (1) \\ 2x + 2y - z = 0 & (2) \\ -x + 2y + 2z = -12 & (3) \end{cases}$

Write the augmented matrix.

$$\begin{bmatrix} 2 & -1 & 2 & | & 3 \\ 2 & 2 & -1 & | & 0 \\ -1 & 2 & 2 & | & -12 \end{bmatrix}$$

Now use Gauss-Jordan elimination to solve the system.

$$\begin{bmatrix} 2 & -1 & 2 & | & 3 \\ 2 & 2 & -1 & | & 0 \\ -1 & 2 & 2 & | & -12 \end{bmatrix}$$

$$\xrightarrow[R_1 + 2R_3 \to R_3]{R_2 - R_1 \to R_2} \begin{bmatrix} 2 & -1 & 2 & | & 3 \\ 0 & 3 & -3 & | & -3 \\ 0 & 3 & 6 & | & -21 \end{bmatrix}$$

$$\xrightarrow{\frac{1}{9}(R_3 - R_2) \to R_3} \begin{bmatrix} 2 & -1 & 2 & | & 3 \\ 0 & 3 & -3 & | & -3 \\ 0 & 0 & 1 & | & -2 \end{bmatrix}$$

$$\xrightarrow[\frac{1}{3}(R_2 + 3R_3) \to R_2]{R_1 - 2R_3 \to R_1} \begin{bmatrix} 2 & -1 & 0 & | & 7 \\ 0 & 1 & 0 & | & -3 \\ 0 & 0 & 1 & | & -2 \end{bmatrix}$$

$$\xrightarrow{\frac{1}{2}(R_1 + R_2) \to R_1} \begin{bmatrix} 1 & 0 & 0 & | & 2 \\ 0 & 1 & 0 & | & -3 \\ 0 & 0 & 1 & | & -2 \end{bmatrix}$$

Solution set: $\{(2, -3, -2)\}$

6.4 Determinants and Cramer's Rule

6.4 Practice Problems

1. a. $\begin{vmatrix} 5 & 1 \\ 3 & -7 \end{vmatrix} = 5(-7) - 1(3) = -38$

b. $\begin{vmatrix} 2 & -9 \\ -4 & 18 \end{vmatrix} = 2(18) - (-9)(-4) = 0$

2. a. $M_{11} = \begin{bmatrix} 3 & -1 & 2 \\ 4 & 5 & 6 \\ 7 & 1 & 2 \end{bmatrix} = \begin{vmatrix} 5 & 6 \\ 1 & 2 \end{vmatrix}$
$= (5)(2) - 6(1) = 4$

$M_{23} = \begin{bmatrix} 3 & -1 & 2 \\ 4 & 5 & 6 \\ 7 & 1 & 2 \end{bmatrix} = \begin{vmatrix} 3 & -1 \\ 7 & 1 \end{vmatrix}$
$= 3(1) - (-1)(7) = 10$

$M_{32} = \begin{bmatrix} 3 & -1 & 2 \\ 4 & 5 & 6 \\ 7 & 1 & 2 \end{bmatrix} = \begin{vmatrix} 3 & 2 \\ 4 & 6 \end{vmatrix}$
$= 3(6) - 2(4) = 10$

b. $A_{11} = (-1)^{1+1} M_{11} = 4$

$A_{23} = (-1)^{2+3} M_{23} = -10$

$A_{32} = (-1)^{3+2} M_{32} = -10$

3. Expand by the third row:

$$\begin{vmatrix} 2 & -3 & 7 \\ -2 & -1 & 9 \\ 0 & 2 & -9 \end{vmatrix}$$

$$= a_{31}A_{31} + a_{32}A_{32} + a_{33}A_{33}$$

$$= 0 + 2(-1)^{3+2}\begin{vmatrix} 2 & 7 \\ -2 & 9 \end{vmatrix} - 9(-1)^{3+3}\begin{vmatrix} 2 & -3 \\ -2 & -1 \end{vmatrix}$$

$$= 0 - 2(32) - 9(-8) = 8$$

4. $D = \begin{vmatrix} 2 & 3 \\ 5 & 9 \end{vmatrix} = 3,\ D_x = \begin{vmatrix} 7 & 3 \\ 4 & 9 \end{vmatrix} = 51$

$D_y = \begin{vmatrix} 2 & 7 \\ 5 & 4 \end{vmatrix} = -27$

$x = \dfrac{D_x}{D} = \dfrac{51}{3} = 17;\ y = \dfrac{D_y}{D} = \dfrac{-27}{3} = -9$

The solution is $\{(17, -9)\}$.

5. $D = \begin{vmatrix} 3 & 2 & 1 \\ 4 & 3 & 1 \\ 5 & 1 & 1 \end{vmatrix} = -3,\ D_x = \begin{vmatrix} 4 & 2 & 1 \\ 5 & 3 & 1 \\ 9 & 1 & 1 \end{vmatrix} = -6$

$D_y = \begin{vmatrix} 3 & 4 & 1 \\ 4 & 5 & 1 \\ 5 & 9 & 1 \end{vmatrix} = 3,\ D_z = \begin{vmatrix} 3 & 2 & 4 \\ 4 & 3 & 5 \\ 5 & 1 & 9 \end{vmatrix} = 0$

$x = \dfrac{D_x}{D} = \dfrac{-6}{-3} = 2;\ y = \dfrac{D_y}{D} = \dfrac{3}{-3} = -1$

$z = \dfrac{D_z}{D} = \dfrac{0}{-3} = 0$

Solution set: $\{(2, -1, 0)\}$

6.4 Basic Concepts and Skills

1. The determinant of $\begin{bmatrix} a & b \\ c & d \end{bmatrix}$ is <u>$ad - bc$</u>.

3. To expand an $n \times n$ determinant, you multiply each element of some row (or column) by its <u>cofactor</u> and add the result.

5. False.

7. $\begin{vmatrix} 2 & 3 \\ 4 & 5 \end{vmatrix} = 2(5) - 4(3) = -2$

9. $\begin{vmatrix} 4 & -2 \\ 3 & -3 \end{vmatrix} = 4(-3) - (-2)(3) = -6$

11. $\begin{vmatrix} -1 & -3 \\ -4 & -5 \end{vmatrix} = (-1)(-5) - (-3)(-4) = -7$

13. $\begin{vmatrix} \dfrac{3}{8} & \dfrac{1}{2} \\ -\dfrac{1}{9} & 5 \end{vmatrix} = \left(\dfrac{3}{8}\right)(5) - \left(\dfrac{1}{2}\right)\left(-\dfrac{1}{9}\right) = \dfrac{139}{72}$

15. $\begin{vmatrix} \sqrt{a} & \sqrt{b} \\ \sqrt{b} & \sqrt{a} \end{vmatrix} = a - b$

17. a. $M_{21} = \begin{bmatrix} 2 & -3 & 4 \\ 1 & -1 & 2 \\ 0 & 1 & 2 \end{bmatrix} = \begin{vmatrix} -3 & 4 \\ 1 & 2 \end{vmatrix}$

$= (-3)(2) - 4(1) = -10$

b. $M_{23} = \begin{bmatrix} 2 & -3 & 4 \\ 1 & -1 & 2 \\ 0 & 1 & 2 \end{bmatrix} = \begin{vmatrix} 2 & -3 \\ 0 & 1 \end{vmatrix} = 2(1) - 0 = 2$

c. $M_{32} = \begin{bmatrix} 2 & -3 & 4 \\ 1 & -1 & 2 \\ 0 & 1 & 2 \end{bmatrix} = \begin{vmatrix} 2 & 4 \\ 1 & 2 \end{vmatrix} = 2(2) - 4(1) = 0$

19. a. $M_{11} = \begin{bmatrix} 2 & -3 & 4 \\ 1 & -1 & 2 \\ 0 & 1 & 2 \end{bmatrix} = \begin{vmatrix} -1 & 2 \\ 1 & 2 \end{vmatrix}$

$= (-1)(2) - 2(1) = -4$

b. $M_{22} = \begin{bmatrix} 2 & -3 & 4 \\ 1 & -1 & 2 \\ 0 & 1 & 2 \end{bmatrix} = \begin{vmatrix} 2 & 4 \\ 0 & 2 \end{vmatrix} = 4$

c. $M_{31} = \begin{bmatrix} 2 & -3 & 4 \\ 1 & -1 & 2 \\ 0 & 1 & 2 \end{bmatrix} = \begin{vmatrix} -3 & 4 \\ -1 & 2 \end{vmatrix}$

$= (-3)(2) - 4(-1) = -2$

21. Expand by the third row:

$$\begin{vmatrix} 1 & 0 & -1 \\ 0 & 2 & 2 \\ -1 & 0 & 0 \end{vmatrix} = a_{31}A_{31} + a_{32}A_{32} + a_{33}A_{33}$$

$$= -1(-1)^{3+1}\begin{vmatrix} 0 & -1 \\ 2 & 2 \end{vmatrix} + 0 + 0 = -2$$

23. Expand by the third row:

$$\begin{vmatrix} 1 & 2 & 3 \\ 0 & 3 & 4 \\ 0 & 0 & 4 \end{vmatrix} = a_{31}A_{31} + a_{32}A_{32} + a_{33}A_{33}$$

$$= 0 + 0 + 4(-1)^{3+3}\begin{vmatrix} 1 & 2 \\ 0 & 3 \end{vmatrix} = 12$$

25. Expand by the first row:

$$\begin{vmatrix} 1 & 0 & 0 \\ 2 & 0 & 0 \\ 3 & 4 & 5 \end{vmatrix} = a_{11}A_{11} + a_{12}A_{12} + a_{13}A_{13}$$

$$= 1(-1)^{1+1}\begin{vmatrix} 0 & 0 \\ 4 & 5 \end{vmatrix} + 0 + 0 = 0$$

27. Expand by the first row:

$$\begin{vmatrix} 1 & 6 & 0 \\ 2 & 5 & 3 \\ 3 & 4 & 0 \end{vmatrix} = a_{11}A_{11} + a_{12}A_{12} + a_{13}A_{13}$$

$$= 1(-1)^{1+1}\begin{vmatrix} 5 & 3 \\ 4 & 0 \end{vmatrix} + 6(-1)^{1+2}\begin{vmatrix} 2 & 3 \\ 3 & 0 \end{vmatrix} + 0$$

$$= -12 - 6(-9) = 42$$

29. Expand by the first row:

$$\begin{vmatrix} 3 & 4 & 1 \\ 1 & 4 & 3 \\ 4 & 3 & 1 \end{vmatrix} = a_{11}A_{11} + a_{12}A_{12} + a_{13}A_{13}$$

$$= 3(-1)^{1+1}\begin{vmatrix} 4 & 3 \\ 3 & 1 \end{vmatrix} + 4(-1)^{1+2}\begin{vmatrix} 1 & 3 \\ 4 & 1 \end{vmatrix}$$

$$+ 1(-1)^{1+3}\begin{vmatrix} 1 & 4 \\ 4 & 3 \end{vmatrix}$$

$$= 3(-5) - 4(-11) + 1(-13) = 16$$

31. Expand by the first row:

$$\begin{vmatrix} 0 & 1 & 6 \\ 1 & 0 & 4 \\ 8 & 3 & 1 \end{vmatrix} = a_{11}A_{11} + a_{12}A_{12} + a_{13}A_{13}$$

$$= 0 + 1(-1)^{1+2}\begin{vmatrix} 1 & 4 \\ 8 & 1 \end{vmatrix} + 6(-1)^{1+3}\begin{vmatrix} 1 & 0 \\ 8 & 3 \end{vmatrix}$$

$$= -1(-31) + 6(3) = 49$$

33. Expand by the first row:

$$\begin{vmatrix} a & b & 0 \\ 0 & a & b \\ b & 0 & a \end{vmatrix} = a_{11}A_{11} + a_{12}A_{12} + a_{13}A_{13}$$

$$= a(-1)^{1+1}\begin{vmatrix} a & b \\ 0 & a \end{vmatrix} + b(-1)^{1+2}\begin{vmatrix} 0 & b \\ b & a \end{vmatrix} + 0$$

$$= a(a^2) - b(-b^2) = a^3 + b^3$$

35. Expand by the first row:

$$\begin{vmatrix} a & b & c \\ c & a & b \\ b & c & a \end{vmatrix} = a_{11}A_{11} + a_{12}A_{12} + a_{13}A_{13}$$

$$= a(-1)^{1+1}\begin{vmatrix} a & b \\ c & a \end{vmatrix} + b(-1)^{1+2}\begin{vmatrix} c & b \\ b & a \end{vmatrix}$$

$$+ c(-1)^{1+3}\begin{vmatrix} c & a \\ b & c \end{vmatrix}$$

$$= a(a^2 - bc) - b(ac - b^2) + c(c^2 - ab)$$

$$= a^3 + b^3 + c^3 - 3abc$$

37. $D = \begin{vmatrix} 1 & 1 \\ 1 & -1 \end{vmatrix} = -2, D_x = \begin{vmatrix} 8 & 1 \\ -2 & -1 \end{vmatrix} = -6$

$$D_y = \begin{vmatrix} 1 & 8 \\ 1 & -2 \end{vmatrix} = -10$$

$$x = \frac{-6}{-2} = 3, y = \frac{-10}{-2} = 5$$

The solution is $\{(3, 5)\}$.

39. $D = \begin{vmatrix} 5 & 3 \\ 2 & 1 \end{vmatrix} = -1, D_x = \begin{vmatrix} 11 & 3 \\ 4 & 1 \end{vmatrix} = -1$

$$D_y = \begin{vmatrix} 5 & 11 \\ 2 & 4 \end{vmatrix} = -2$$

$$x = \frac{-1}{-1} = 1, y = \frac{-2}{-1} = 2$$

The solution is $\{(1, 2)\}$.

41. $D = \begin{vmatrix} 2 & 9 \\ 3 & -2 \end{vmatrix} = -31, D_x = \begin{vmatrix} 4 & 9 \\ 6 & -2 \end{vmatrix} = -62$

$$D_y = \begin{vmatrix} 2 & 4 \\ 3 & 6 \end{vmatrix} = 0$$

$$x = \frac{-62}{-31} = 2, y = \frac{0}{-31} = 0$$

The solution is $\{(2, 0)\}$.

43. $D = \begin{vmatrix} 2 & -3 \\ 4 & -6 \end{vmatrix} = 0 \Rightarrow$ there is not a unique

solution. $2x - 3y = 4 \Rightarrow x = \frac{3}{2}y + 2$.

The solution is $\left\{ \left(\frac{3}{2}y + 2, y \right) \right\}$.

45. Let $u = \dfrac{1}{x}$ and $v = \dfrac{1}{y}$. Then $\begin{cases} \dfrac{2}{x} + \dfrac{3}{y} = 2 \\ \dfrac{5}{x} + \dfrac{8}{y} = \dfrac{31}{6} \end{cases} \Rightarrow$

$\begin{cases} 2u + 3v = 2 \\ 5u + 8v = \dfrac{31}{6} \end{cases} \Rightarrow D = \begin{vmatrix} 2 & 3 \\ 5 & 8 \end{vmatrix} = 1,$

$D_u = \begin{vmatrix} 2 & 3 \\ \frac{31}{6} & 8 \end{vmatrix} = \dfrac{1}{2}, D_v = \begin{vmatrix} 2 & 2 \\ 5 & \frac{31}{6} \end{vmatrix} = \dfrac{1}{3} \Rightarrow$

$u = \dfrac{1}{2}, v = \dfrac{1}{3} \Rightarrow x = 2, y = 3$

The solution is $\{(2, 3)\}$.

47. $D = \begin{vmatrix} 1 & -2 & 1 \\ 3 & 1 & -1 \\ 0 & 1 & 1 \end{vmatrix} = 11, D_x = \begin{vmatrix} -1 & -2 & 1 \\ 4 & 1 & -1 \\ 1 & 1 & 1 \end{vmatrix} = 11$

$D_y = \begin{vmatrix} 1 & -1 & 1 \\ 3 & 4 & -1 \\ 0 & 1 & 1 \end{vmatrix} = 11, D_z = \begin{vmatrix} 1 & -2 & -1 \\ 3 & 1 & 4 \\ 0 & 1 & 1 \end{vmatrix} = 0$

$x = \dfrac{11}{11} = 1, y = \dfrac{11}{11} = 1, z = \dfrac{0}{11} = 0$

The solution is $\{(1, 1, 0)\}$.

49. $D = \begin{vmatrix} 1 & 1 & -1 \\ 2 & 3 & 1 \\ 0 & 2 & 1 \end{vmatrix} = -5, D_x = \begin{vmatrix} -3 & 1 & -1 \\ 2 & 3 & 1 \\ 1 & 2 & 1 \end{vmatrix} = -5$

$D_y = \begin{vmatrix} 1 & -3 & -1 \\ 2 & 2 & 1 \\ 0 & 1 & 1 \end{vmatrix} = 5, D_z = \begin{vmatrix} 1 & 1 & -3 \\ 2 & 3 & 2 \\ 0 & 2 & 1 \end{vmatrix} = -15$

$x = \dfrac{-5}{-5} = 1, y = \dfrac{5}{-5} = -1, z = \dfrac{-15}{-5} = 3$

The solution is $\{(1, -1, 3)\}$.

51. $D = \begin{vmatrix} 1 & 1 & 1 \\ 2 & -3 & 5 \\ 1 & 2 & -4 \end{vmatrix} = 22, D_x = \begin{vmatrix} 3 & 1 & 1 \\ 4 & -3 & 5 \\ -1 & 2 & -4 \end{vmatrix} = 22$

$D_y = \begin{vmatrix} 1 & 3 & 1 \\ 2 & 4 & 5 \\ 1 & -1 & -4 \end{vmatrix} = 22, D_z = \begin{vmatrix} 1 & 1 & 3 \\ 2 & -3 & 4 \\ 1 & 2 & -1 \end{vmatrix} = 22$

$x = \dfrac{22}{22} = 1, y = \dfrac{22}{22} = 1, z = \dfrac{22}{22} = 1$

The solution is $\{(1, 1, 1)\}$.

53. $D = \begin{vmatrix} 2 & -3 & 5 \\ 3 & 5 & -2 \\ 1 & 2 & -3 \end{vmatrix} = -38$

$D_x = \begin{vmatrix} 11 & -3 & 5 \\ 7 & 5 & -2 \\ -4 & 2 & -3 \end{vmatrix} = -38$

$D_y = \begin{vmatrix} 2 & 11 & 5 \\ 3 & 7 & -2 \\ 1 & -4 & -3 \end{vmatrix} = -76$

$D_z = \begin{vmatrix} 2 & -3 & 11 \\ 3 & 5 & 7 \\ 1 & 2 & -4 \end{vmatrix} = -114$

$x = \dfrac{-38}{-38} = 1, y = \dfrac{-76}{-38} = 2, z = \dfrac{-114}{-38} = 3$

The solution is $\{(1, 2, 3)\}$.

55. $D = \begin{vmatrix} 5 & 2 & 1 \\ 2 & 1 & 3 \\ 3 & 2 & 4 \end{vmatrix} = -7, D_x = \begin{vmatrix} 12 & 2 & 1 \\ 13 & 1 & 3 \\ 19 & 2 & 4 \end{vmatrix} = -7$

$D_y = \begin{vmatrix} 5 & 12 & 1 \\ 2 & 13 & 3 \\ 3 & 19 & 4 \end{vmatrix} = -14, D_z = \begin{vmatrix} 5 & 2 & 12 \\ 2 & 1 & 13 \\ 3 & 2 & 19 \end{vmatrix} = -21$

$x = \dfrac{-7}{-7} = 1, y = \dfrac{-14}{-7} = 2, z = \dfrac{-21}{-7} = 3$

The solution is $\{(1, 2, 3)\}$.

6.4 Applying the Concepts

57. $A = |D| = \dfrac{1}{2} \begin{vmatrix} 1 & 2 & 1 \\ -3 & 4 & 1 \\ 4 & 6 & 1 \end{vmatrix} = 11$

59. $A = |D| = \dfrac{1}{2} \begin{vmatrix} -2 & 1 & 1 \\ -3 & -5 & 1 \\ 2 & 4 & 1 \end{vmatrix} = \dfrac{21}{2} = 10.5$

61. $\begin{vmatrix} 0 & 3 & 1 \\ -1 & 1 & 1 \\ 2 & 7 & 1 \end{vmatrix} = 0 \Rightarrow$ the points are collinear.

63. $\begin{vmatrix} 0 & -4 & 1 \\ 3 & -2 & 1 \\ 1 & -4 & 1 \end{vmatrix} = -2 \Rightarrow$ the points are not collinear.

65. $\begin{vmatrix} x & y & 1 \\ -1 & -1 & 1 \\ 1 & 3 & 1 \end{vmatrix} = 0 \Rightarrow x(-1)^{1+1}\begin{vmatrix} -1 & 1 \\ 3 & 1 \end{vmatrix} + y(-1)^{1+2}\begin{vmatrix} -1 & 1 \\ 1 & 1 \end{vmatrix} + 1(-1)^{1+3}\begin{vmatrix} -1 & -1 \\ 1 & 3 \end{vmatrix} = 0 \Rightarrow -4x + 2y - 2 = 0 \Rightarrow y = 2x + 1$

67. $\begin{vmatrix} x & y & 1 \\ 0 & \frac{1}{3} & 1 \\ 1 & 1 & 1 \end{vmatrix} = 0 \Rightarrow x(-1)^{1+1}\begin{vmatrix} \frac{1}{3} & 1 \\ 1 & 1 \end{vmatrix} + y(-1)^{1+2}\begin{vmatrix} 0 & 1 \\ 1 & 1 \end{vmatrix} + 1(-1)^{1+3}\begin{vmatrix} 0 & \frac{1}{3} \\ 1 & 1 \end{vmatrix} = 0 \Rightarrow -\frac{2}{3}x + y - \frac{1}{3} = 0 \Rightarrow y = \frac{2}{3}x + \frac{1}{3}$

6.4 Beyond the Basics

69. a. $\begin{vmatrix} 0 & 0 \\ 2 & 5 \end{vmatrix} = 0(5) - 2(0) = 0$

b. $\begin{vmatrix} 1 & 0 \\ 3 & 0 \end{vmatrix} = 1(0) - 3(0) = 0$

c. Expand by the second row: $\begin{vmatrix} 1 & -2 & -3 \\ 0 & 0 & 0 \\ 4 & 4 & -7 \end{vmatrix} = 0(-1)^{2+1}\begin{vmatrix} -2 & -3 \\ 4 & -7 \end{vmatrix} + 0(-1)^{2+2}\begin{vmatrix} 1 & -3 \\ 4 & -7 \end{vmatrix} + 0(-1)^{2+3}\begin{vmatrix} 1 & -2 \\ 4 & 4 \end{vmatrix} = 0$

d. Expand by the third column: $\begin{vmatrix} 4 & 5 & 0 \\ 6 & -7 & 0 \\ 8 & 15 & 0 \end{vmatrix} = 0(-1)^{1+3}\begin{vmatrix} 6 & -7 \\ 8 & 15 \end{vmatrix} + 0(-1)^{2+3}\begin{vmatrix} 4 & 5 \\ 8 & 15 \end{vmatrix} + 0(-1)^{3+3}\begin{vmatrix} 4 & 5 \\ 6 & -7 \end{vmatrix} = 0$

71. a. Expand by the second row: $\begin{vmatrix} 2 & 3 & 5 \\ -1 & 4 & 8 \\ 2 & 3 & 5 \end{vmatrix} = -1(-1)^{2+1}\begin{vmatrix} 3 & 5 \\ 3 & 5 \end{vmatrix} + 4(-1)^{2+2}\begin{vmatrix} 2 & 5 \\ 2 & 5 \end{vmatrix} + 8(-1)^{2+3}\begin{vmatrix} 2 & 5 \\ 2 & 5 \end{vmatrix} = 0$

b. Expand by the second column: $\begin{vmatrix} 3 & 4 & 3 \\ 1 & 2 & 1 \\ -1 & 6 & -1 \end{vmatrix} = 4(-1)^{1+2}\begin{vmatrix} 1 & 1 \\ -1 & -1 \end{vmatrix} + 2(-1)^{2+2}\begin{vmatrix} 3 & 3 \\ -1 & -1 \end{vmatrix} + 6(-1)^{3+2}\begin{vmatrix} 3 & 3 \\ 1 & 1 \end{vmatrix} = 0$

73. $\begin{vmatrix} 2 & -3 & 4 \\ -4 & 7 & -8 \\ 5 & -1 & 3 \end{vmatrix} = -14, \quad \begin{vmatrix} 2 & -3 & 4 \\ 0 & 1 & 0 \\ 5 & -1 & 3 \end{vmatrix} = -14$

75. $\begin{vmatrix} 2 & 0 & -3 & 4 \\ 0 & 1 & 0 & 5 \\ 5 & 0 & -9 & 8 \\ 1 & 2 & 0 & 7 \end{vmatrix} \xrightarrow[R_3 - 5R_4 \to R_3]{R_1 - 2R_4 \to R_1} \begin{vmatrix} 0 & -4 & -3 & -10 \\ 0 & 1 & 0 & 5 \\ 0 & -10 & -9 & -27 \\ 1 & 2 & 0 & 7 \end{vmatrix}$

Expand by column 1:

$\begin{vmatrix} 0 & -4 & -3 & -10 \\ 0 & 1 & 0 & 5 \\ 0 & -10 & -9 & -27 \\ 1 & 2 & 0 & 7 \end{vmatrix} = 0 + 0 + 0 + 1(-1)^{4+1}\begin{vmatrix} -4 & -3 & -10 \\ 1 & 0 & 5 \\ -10 & -9 & -27 \end{vmatrix} = 21$

77. $\begin{vmatrix} 5 & 7 & 1 & 2 \\ 6 & 8 & 9 & 3 \\ 24 & 22 & 6 & 10 \\ 21 & 17 & 7 & 10 \end{vmatrix} \xrightarrow[\substack{-6R_1+R_3\to R_3 \\ -7R_1+R_4\to R_4}]{-9R_1+R_2\to R_2} \begin{vmatrix} 5 & 7 & 1 & 2 \\ -39 & -55 & 0 & -15 \\ -6 & -20 & 0 & -2 \\ -14 & -32 & 0 & -4 \end{vmatrix}$

Expand by column 3: $\begin{vmatrix} 5 & 7 & 1 & 2 \\ -39 & -55 & 0 & -15 \\ -6 & -20 & 0 & -2 \\ -14 & -32 & 0 & -4 \end{vmatrix} = 1(-1)^{1+3} \begin{vmatrix} -39 & -55 & -15 \\ -6 & -20 & -2 \\ -14 & -32 & -4 \end{vmatrix} = 476$

79. $\begin{vmatrix} 3 & 2 \\ 6 & x \end{vmatrix} = 0 \Rightarrow 3x - 12 = 0 \Rightarrow x = 4$

81. $\begin{vmatrix} x & -2 \\ 1 & x-1 \end{vmatrix} = 0 \Rightarrow x^2 - x + 2 = 0 \Rightarrow x = \dfrac{1 \pm \sqrt{1-8}}{2} = \dfrac{1 \pm i\sqrt{7}}{2}$

83. Expand by the second row:

$\begin{vmatrix} 1 & -3 & 1 \\ 4 & 7 & x \\ 0 & 2 & 2 \end{vmatrix} = 0 \Rightarrow 4(-1)^{2+1} \begin{vmatrix} -3 & 1 \\ 2 & 2 \end{vmatrix} + 7(-1)^{2+2} \begin{vmatrix} 1 & 1 \\ 0 & 2 \end{vmatrix} + x(-1)^{2+3} \begin{vmatrix} 1 & -3 \\ 0 & 2 \end{vmatrix} = 0 \Rightarrow$

$-4(-8) + 7(2) - 2x = 46 - 2x = 0 \Rightarrow x = 23$

85. Expand by the second row:

$\begin{vmatrix} x & 0 & 1 \\ 0 & x & 0 \\ 1 & 0 & x \end{vmatrix} = 0 \Rightarrow 0 + x(-1)^{2+2} \begin{vmatrix} x & 1 \\ 1 & x \end{vmatrix} + 0 \Rightarrow x(x^2 - 1) = 0 \Rightarrow x(x-1)(x+1) = 0 \Rightarrow x = 0 \text{ or } x = 1 \text{ or } x = -1$

87. $\dfrac{1}{2} \begin{vmatrix} -4 & 2 & 1 \\ 0 & k & 1 \\ -2 & k & 1 \end{vmatrix} = 28 \Rightarrow 0 + k(-1)^{2+2} \begin{vmatrix} -4 & 1 \\ -2 & 1 \end{vmatrix} + 1(-1)^{2+3} \begin{vmatrix} -4 & 2 \\ -2 & k \end{vmatrix} = 56 \Rightarrow -2k - (-4k + 4) = 2k - 4 = 56 \Rightarrow k = 30$

6.4 Critical Thinking/Discussion/Writing

89. Let $u = \dfrac{1}{x-2}$ and $v = \dfrac{1}{y+1}$. Then $\begin{cases} \dfrac{1}{x-2} + \dfrac{3}{y+1} = 13 \\ \dfrac{4}{x-2} - \dfrac{5}{y+1} = 1 \end{cases} \Rightarrow \begin{cases} u + 3v = 13 \\ 4u - 5v = 1 \end{cases}$.

$D = \begin{vmatrix} 1 & 3 \\ 4 & -5 \end{vmatrix} = -17, \quad D_u = \begin{vmatrix} 13 & 3 \\ 1 & -5 \end{vmatrix} = -68, \quad D_v = \begin{vmatrix} 1 & 13 \\ 4 & 1 \end{vmatrix} = -51$

$u = \dfrac{-68}{-14}, \quad v = \dfrac{-51}{-17} = 3$

$4 = \dfrac{1}{x-2} \Rightarrow 4x - 8 = 1 \Rightarrow x = \dfrac{9}{4}; \quad 3 = \dfrac{1}{y+1} \Rightarrow 3y + 3 = 1 \Rightarrow y = -\dfrac{2}{3}$

The solution is $\left\{ \left(\dfrac{9}{4}, -\dfrac{2}{3} \right) \right\}$.

90. Let $u = \dfrac{1}{x+1}$ and $v = \dfrac{1}{y-1}$. Then

$$\begin{cases} \dfrac{6}{x+1} + \dfrac{4}{y-1} = 7 \\ \dfrac{8}{x+1} + \dfrac{5}{y-1} = 9 \end{cases} \Rightarrow \begin{cases} 6u + 4v = 7 \\ 8u + 5v = 9 \end{cases}$$

$$D = \begin{vmatrix} 6 & 4 \\ 8 & 5 \end{vmatrix} = -2, D_u = \begin{vmatrix} 7 & 4 \\ 9 & 5 \end{vmatrix} = -1$$

$$D_y = \begin{vmatrix} 6 & 7 \\ 8 & 9 \end{vmatrix} = -2$$

$$u = \frac{-1}{-2} = \frac{1}{2}, v = \frac{-2}{-2} = 1$$

$$\frac{1}{2} = \frac{1}{x+1} \Rightarrow x+1 = 2 \Rightarrow x = 1$$

$$1 = \frac{1}{y-1} \Rightarrow y-1 = 1 \Rightarrow y = 2$$

The solution is $\{(1, 2)\}$.

91. Let $u = \dfrac{1}{3^x}$ and $v = \dfrac{1}{4^y}$. Then

$$\begin{cases} \dfrac{1}{3^x} + \dfrac{4}{4^y} = 25 \\ \dfrac{2}{3^x} - \dfrac{1}{4^y} = 14 \end{cases} \Rightarrow \begin{cases} u + 4v = 25 \\ 2u - v = 14 \end{cases}$$

$$D = \begin{vmatrix} 1 & 4 \\ 2 & -1 \end{vmatrix} = -9, D_u = \begin{vmatrix} 25 & 4 \\ 14 & -1 \end{vmatrix} = -81$$

$$D_y = \begin{vmatrix} 1 & 25 \\ 2 & 14 \end{vmatrix} = -36$$

$$u = \frac{-81}{-9} = 9, v = \frac{-36}{-9} = 4$$

$$9 = \frac{1}{3^x} \Rightarrow 3^x = \frac{1}{9} \Rightarrow x = -2$$

$$4 = \frac{1}{4^y} \Rightarrow 4^y = \frac{1}{4} \Rightarrow y = -1$$

The solution is $\{(-2, -1)\}$.

92. We use the change of base formula along with the power rule of logarithms. See section 4.3.

a. Recall that $\log 8 = \log\left(2^3\right) = 3\log 2$.

$$\begin{vmatrix} \log_2 3 & \log_8 3 \\ \log_3 4 & \log_3 4 \end{vmatrix}$$

$$= \log_2 3 \cdot \log_3 4 - \log_8 3 \cdot \log_3 4$$

$$= \frac{\log 3 \cdot \log 4}{\log 2 \cdot \log 3} - \frac{\log 3 \cdot \log 4}{\log 8 \cdot \log 3}$$

$$= \frac{\log 3 \cdot \log 4}{\log 2 \cdot \log 3} - \frac{\log 3 \cdot \log 4}{3\log 2 \cdot \log 3}$$

$$= \frac{\log 4}{\log 2} - \frac{\log 4}{3\log 2}$$

$$= \frac{3\log 4 - \log 4}{3\log 2} = \frac{2\log 4}{3\log 2}$$

$$= \frac{2\log\left(2^2\right)}{3\log 2} = \frac{4\log 2}{3\log 2} = \frac{4}{3}$$

b.
$$\begin{vmatrix} \log_3 512 & \log_4 3 \\ \log_3 8 & \log_4 9 \end{vmatrix}$$

$$= \log_3 512 \cdot \log_4 9 - \log_4 3 \cdot \log_3 8$$

$$= \frac{\log 512 \cdot \log 9}{\log 3 \cdot \log 4} - \frac{\log 3 \cdot \log 8}{\log 4 \cdot \log 3}$$

$$= \frac{\log\left(2 \cdot 16^2\right) \cdot 2\log 3}{\log 3 \cdot \log 4} - \frac{3\log 2}{\log 4}$$

$$= \frac{(\log 2 + 2\log 16) \cdot 2}{2\log 2} - \frac{3\log 2}{2\log 2}$$

$$= \frac{\log 2 + 2\log\left(2^4\right)}{\log 2} - \frac{3}{2}$$

$$= \frac{\log 2 + 8\log 2}{\log 2} - \frac{3}{2}$$

$$= \frac{9\log 2}{\log 2} - \frac{3}{2} = 9 - \frac{3}{2} = \frac{15}{2}$$

93. The system has no solution when $D = 0$.

$$D = \begin{vmatrix} k & 3 & -1 \\ 1 & 2 & 1 \\ -k & 1 & 2 \end{vmatrix}$$

$$= -1\begin{vmatrix} 3 & -1 \\ 1 & 2 \end{vmatrix} + 2\begin{vmatrix} k & -1 \\ -k & 2 \end{vmatrix} - 1\begin{vmatrix} k & 3 \\ -k & 1 \end{vmatrix}$$

$$= -7 + 2(2k - k) - (k + 3k)$$

$$= -7 + 2k - 4k = -7 - 2k$$

$$-7 - 2k = 0 \Rightarrow k = -\frac{7}{2}$$

The system has no solution for $k = -\dfrac{7}{2}$.

94. $D = \begin{vmatrix} 2 & a & 6 \\ 1 & 2 & b \\ 1 & 1 & 3 \end{vmatrix} = \begin{vmatrix} a & 6 \\ 2 & b \end{vmatrix} - \begin{vmatrix} 2 & 6 \\ 1 & b \end{vmatrix} + 3\begin{vmatrix} 2 & a \\ 1 & 2 \end{vmatrix}$

$$= (ab - 12) - (2b - 6) + 3(4 - a)$$

$$= ab - 3a - 2b + 6$$

The system of equations has a unique solution when $D \neq 0$.

$$ab - 3a - 2b + 6 = 0 \Rightarrow ab - 3a = 2b - 6 \Rightarrow$$
$$a(b - 3) = 2(b - 3) \Rightarrow a = 2, b \neq 3$$

So, the system has a unique solution if $a \neq 2$ and $b \neq 3$.

(continued on next page)

(continued)

If $a = 2$, then row 1 = 2 row 3, so there are infinitely many solutions.

If $a = b = 3$, then $D = 0$, but all rows are different, so there is no solution.

6.4 Maintaining Skills

You may want to refer to sections 1.2, 8.3, and 8.4.

95. $x^2 = 8y \Rightarrow \dfrac{1}{8}x^2 = y$

97. $(x-2)^2 = 6(y-3) \Rightarrow x^2 - 4x + 4 = 6y - 18 \Rightarrow$

$x^2 - 4x + 22 = 6y \Rightarrow \dfrac{1}{6}x^2 - \dfrac{2}{3}x + \dfrac{11}{3} = y$

99. $-2 = a(-2)^2 \Rightarrow -2 = 4a \Rightarrow -\dfrac{1}{2} = a$

The function is $y = -\dfrac{1}{2}x^2$.

101. The formula for the x-value of the vertex gives

$x = -\dfrac{b}{2a} = -3 \Rightarrow b = 6a$

Using (0, 10) gives

$10 = a(0^2) + b(0) + c \Rightarrow c = 10$.

Substituting $(-3, -8)$, $b = 6a$, and $c = 10$ gives

$-8 = a(-3)^2 + 6a(-3) + 10 \Rightarrow$

$-8 = 9a - 18a + 10 \Rightarrow -18 = -9a \Rightarrow 2 = a \Rightarrow$

$b = 12$

The function is $y = 2x^2 + 12x + 10$.

103. $f(x) = 2x^2 + 4x - 6$

a. Opens up since $a > 0$.

b. $x = -\dfrac{b}{2a} = -\dfrac{4}{2(2)} = -1$

$f(-1) = 2(-1)^2 + 4(-1) - 6 = -8$

The vertex is $(-1, -8)$.

c. The axis of symmetry is $x = -1$.

d. To find the x-intercepts, let $y = 0$ and solve for x.

$2x^2 + 4x - 6 = 0 \Rightarrow 2(x^2 + 2x - 3) = 0 \Rightarrow$

$(x+3)(x-1) = 0 \Rightarrow x = -3, \ x = 1$

The x-intercepts are -3 and 1.

e. To find the y-intercept, let $x = 0$ and solve for y.

$y = 2(0)^2 + 4(0) - 6 = -6$

The y-intercept is -6.

Chapter 6 Review Exercises

Basic Skills and Concepts

1. 1×4

3. 3×2

5. $a_{12} = -1, a_{14} = -4, a_{23} = 3, a_{21} = 5$

7. $\begin{bmatrix} 2 & -3 & | & 7 \\ 3 & 1 & | & 6 \end{bmatrix}$

9. Answers may vary.

$\begin{bmatrix} 0 & 1 & 2 & 1 \\ 2 & 0 & 3 & 4 \\ 1 & -2 & 1 & 7 \end{bmatrix} \xrightarrow{R_1 \leftrightarrow R_3} \begin{bmatrix} 1 & -2 & 1 & 7 \\ 2 & 0 & 3 & 4 \\ 0 & 1 & 2 & 1 \end{bmatrix}$

$\xrightarrow{2R_1 - R_2 \to R_2} \begin{bmatrix} 1 & -2 & 1 & 7 \\ 0 & -4 & -1 & 10 \\ 0 & 1 & 2 & 1 \end{bmatrix}$

$\xrightarrow{R_2 \leftrightarrow R_3} \begin{bmatrix} 1 & -2 & 1 & 7 \\ 0 & 1 & 2 & 1 \\ 0 & -4 & -1 & 10 \end{bmatrix}$

$\xrightarrow{\frac{1}{7}(4R_2 + R_3) \to R_3} \begin{bmatrix} 1 & -2 & 1 & 7 \\ 0 & 1 & 2 & 1 \\ 0 & 0 & 1 & 2 \end{bmatrix}$

11. $\begin{bmatrix} 3 & 1 & 3 & 1 \\ 2 & 1 & 1 & 1 \\ 1 & -1 & -1 & 0 \end{bmatrix} \xrightarrow{R_1 \leftrightarrow R_3} \begin{bmatrix} 1 & -1 & -1 & 0 \\ 2 & 1 & 1 & 1 \\ 3 & 1 & 3 & 1 \end{bmatrix}$

$\xrightarrow[-3R_1 + R_3 \to R_3]{-2R_1 + R_2 \to R_2} \begin{bmatrix} 1 & -1 & -1 & 0 \\ 0 & 3 & 3 & 1 \\ 0 & 4 & 6 & 1 \end{bmatrix}$

$\xrightarrow{-\frac{1}{2}(-2R_2 + R_3) \to R_2} \begin{bmatrix} 1 & -1 & -1 & 0 \\ 0 & 1 & 0 & \frac{1}{2} \\ 0 & 4 & 6 & 1 \end{bmatrix}$

$\xrightarrow[\frac{1}{6}(R_3 - 4R_2) \to R_3]{R_1 + R_2 \to R_1} \begin{bmatrix} 1 & 0 & -1 & \frac{1}{2} \\ 0 & 1 & 0 & \frac{1}{2} \\ 0 & 0 & 1 & -\frac{1}{6} \end{bmatrix}$

(continued on next page)

(continued)

$$\xrightarrow{R_1+R_3\to R_1}\begin{bmatrix}1&0&0&\frac{1}{3}\\0&1&0&\frac{1}{2}\\0&0&1&-\frac{1}{6}\end{bmatrix}$$

13. $\begin{cases}x+y-z=0\\2x+y-2z=3\\3x-2y+3z=9\end{cases}\Rightarrow\begin{bmatrix}1&1&-1&0\\2&1&-2&3\\3&-2&3&9\end{bmatrix}$

$$\xrightarrow[3R_1-R_3\to R_3]{2R_1-R_2\to R_2}\begin{bmatrix}1&1&-1&0\\0&1&0&-3\\0&5&-6&-9\end{bmatrix}$$

$$\xrightarrow[-\frac{1}{6}(R_3-5R_2)\to R_3]{R_1-R_2\to R_1}\begin{bmatrix}1&0&-1&3\\0&1&0&-3\\0&0&1&-1\end{bmatrix}$$

$$\xrightarrow{R_1+R_3\to R_1}\begin{bmatrix}1&0&0&2\\0&1&0&-3\\0&0&1&-1\end{bmatrix}$$

The solution is $\{(2,-3,-1)\}$.

15. $\begin{cases}x-2y+3z=-2\\2x-3y+z=9\\3x-y+2z=5\end{cases}\Rightarrow\begin{bmatrix}1&-2&3&-2\\2&-3&1&9\\3&-1&2&5\end{bmatrix}$

$$\xrightarrow[R_3-3R_1\to R_3]{R_2-2R_1\to R_2}\begin{bmatrix}1&-2&3&-2\\0&1&-5&13\\0&5&-7&11\end{bmatrix}$$

$$\xrightarrow{\frac{1}{18}(R_3-5R_2)\to R_3}\begin{bmatrix}1&-2&3&-2\\0&1&-5&13\\0&0&1&-3\end{bmatrix}$$

$z=-3;\ y-5(-3)=13\Rightarrow y=-2$
$x-2(-2)+3(-3)=-2\Rightarrow x=3$
The solution is $\{(3,-2,-3)\}$.

17. $\begin{cases}x-2y-2z=11\\3x+4y-z=-2\\4x+5y+7z=7\end{cases}\Rightarrow\begin{bmatrix}1&-2&-2&1\\3&4&-1&-2\\4&5&7&7\end{bmatrix}$

$$\xrightarrow[R_3-4R_1\to R_3]{\frac{1}{5}(R_2-3R_1)\to R_2}\begin{bmatrix}1&-2&-2&1\\0&2&1&-7\\0&13&15&-37\end{bmatrix}$$

$$\xrightarrow[\frac{2}{17}(R_3-\frac{13}{2}R_2)\to R_3]{R_1+R_2\to R_1}\begin{bmatrix}1&0&-1&4\\0&2&1&-7\\0&0&1&1\end{bmatrix}$$

$$\xrightarrow[\frac{1}{2}(R_2-R_3)\to R_2]{R_1+R_3\to R_1}\begin{bmatrix}1&0&0&5\\0&1&0&-4\\0&0&1&1\end{bmatrix}$$

The solution is $\{(5,-4,1)\}$.

19. $\begin{cases}2x-y+3z=4\\x+3y+3z=-2\\3x+2y-6z=6\end{cases}\Rightarrow\begin{bmatrix}2&-1&3&4\\1&3&3&-2\\3&2&-6&6\end{bmatrix}$

$$\xrightarrow{R_1\leftrightarrow R_2}\begin{bmatrix}1&3&3&-2\\2&-1&3&4\\3&2&-6&6\end{bmatrix}$$

$$\xrightarrow[R_3-3R_1\to R_3]{2R_1-R_2\to R_2}\begin{bmatrix}1&3&3&-2\\0&7&3&-8\\0&-7&-15&12\end{bmatrix}$$

$$\xrightarrow{-\frac{1}{12}(R_2+R_3)\to R_3}\begin{bmatrix}1&3&3&-2\\0&7&3&-8\\0&0&1&-\frac{1}{3}\end{bmatrix}$$

$$\xrightarrow[\frac{1}{7}(R_2-3R_3)\to R_2]{R_1-3R_3\to R_1}\begin{bmatrix}1&3&0&-1\\0&1&0&-1\\0&0&1&-\frac{1}{3}\end{bmatrix}$$

$$\xrightarrow{R_1-3R_2\to R_1}\begin{bmatrix}1&0&0&2\\0&1&0&-1\\0&0&1&-\frac{1}{3}\end{bmatrix}$$

The solution is $\left\{\left(2,-1,-\frac{1}{3}\right)\right\}$.

21. $\begin{bmatrix}x-y&0\\1&x+y\end{bmatrix}=\begin{bmatrix}1&0\\1&3\end{bmatrix}\Rightarrow\begin{cases}x-y=1\\x+y=3\end{cases}\Rightarrow$
$2x=4\Rightarrow x=2,\ y=1$

23. a. $\begin{bmatrix}1&2\\-3&4\end{bmatrix}+\begin{bmatrix}2&-3\\-5&6\end{bmatrix}=\begin{bmatrix}3&-1\\-8&10\end{bmatrix}$

b. $\begin{bmatrix}1&2\\-3&4\end{bmatrix}-\begin{bmatrix}2&-3\\-5&6\end{bmatrix}=\begin{bmatrix}-1&5\\2&-2\end{bmatrix}$

c. $2\begin{bmatrix}1&2\\-3&4\end{bmatrix}=\begin{bmatrix}2&4\\-6&8\end{bmatrix}$

d. $-3\begin{bmatrix}2&-3\\-5&6\end{bmatrix}=\begin{bmatrix}-6&9\\15&-18\end{bmatrix}$

e. $2\begin{bmatrix} 1 & 2 \\ -3 & 4 \end{bmatrix} - 3\begin{bmatrix} 2 & -3 \\ -5 & 6 \end{bmatrix}$

$\quad = \begin{bmatrix} 2 & 4 \\ -6 & 8 \end{bmatrix} - \begin{bmatrix} 6 & -9 \\ -15 & 18 \end{bmatrix} = \begin{bmatrix} -4 & 13 \\ 9 & -10 \end{bmatrix}$

25. $3A + 2B - 3X = 0 \Rightarrow 3\begin{bmatrix} 1 & 2 \\ -3 & 4 \end{bmatrix} + 2\begin{bmatrix} 2 & -3 \\ -5 & 6 \end{bmatrix} = 3X \Rightarrow \begin{bmatrix} 3 & 6 \\ -9 & 12 \end{bmatrix} + \begin{bmatrix} 4 & -6 \\ -10 & 12 \end{bmatrix} = 3X \Rightarrow \begin{bmatrix} 7 & 0 \\ -19 & 24 \end{bmatrix} = 3X \Rightarrow$

$\quad X = \begin{bmatrix} \dfrac{7}{3} & 0 \\ -\dfrac{19}{3} & 8 \end{bmatrix}$

27. a. $AB = \begin{bmatrix} 0 & 1 \\ 2 & 3 \end{bmatrix}\begin{bmatrix} -1 & -1 \\ -3 & 4 \end{bmatrix} = \begin{bmatrix} 0(-1)+1(-3) & 0(-1)+1(4) \\ 2(-1)+3(-3) & 2(-1)+3(4) \end{bmatrix} = \begin{bmatrix} -3 & 4 \\ -11 & 10 \end{bmatrix}$

b. $BA = \begin{bmatrix} -1 & -1 \\ -3 & 4 \end{bmatrix}\begin{bmatrix} 0 & 1 \\ 2 & 3 \end{bmatrix} = \begin{bmatrix} -1(0)-1(2) & -1(1)-1(3) \\ -3(0)+4(2) & -3(1)+4(3) \end{bmatrix} = \begin{bmatrix} -2 & -4 \\ 8 & 9 \end{bmatrix}$

29. a. $AB = \begin{bmatrix} 1 & 2 & -1 \end{bmatrix}\begin{bmatrix} 2 \\ 3 \\ 1 \end{bmatrix} = \begin{bmatrix} 1(2)+2(3)-1(1) \end{bmatrix} = \begin{bmatrix} 7 \end{bmatrix}$

b. $BA = \begin{bmatrix} 2 \\ 3 \\ 1 \end{bmatrix}\begin{bmatrix} 1 & 2 & -1 \end{bmatrix} = \begin{bmatrix} 2(1) & 2(2) & 2(-1) \\ 3(1) & 3(2) & 3(-1) \\ 1(1) & 1(2) & 1(-1) \end{bmatrix} = \begin{bmatrix} 2 & 4 & -2 \\ 3 & 6 & -3 \\ 1 & 2 & -1 \end{bmatrix}$

31. $\begin{bmatrix} 1 & 2 \\ 3 & -1 \end{bmatrix}A = \begin{bmatrix} 5 & 6 \\ 1 & -3 \end{bmatrix} \Rightarrow A = \begin{bmatrix} 1 & 2 \\ 3 & -1 \end{bmatrix}^{-1}\begin{bmatrix} 5 & 6 \\ 1 & -3 \end{bmatrix}$

$\begin{bmatrix} 1 & 2 \\ 3 & -1 \end{bmatrix}^{-1} = \dfrac{1}{-1-6}\begin{bmatrix} -1 & -2 \\ -3 & 1 \end{bmatrix} = \begin{bmatrix} \dfrac{1}{7} & \dfrac{2}{7} \\ \dfrac{3}{7} & -\dfrac{1}{7} \end{bmatrix}$

$A = \begin{bmatrix} \dfrac{1}{7} & \dfrac{2}{7} \\ \dfrac{3}{7} & -\dfrac{1}{7} \end{bmatrix}\begin{bmatrix} 5 & 6 \\ 1 & -3 \end{bmatrix} = \begin{bmatrix} 1 & 0 \\ 2 & 3 \end{bmatrix}$

33. $\begin{bmatrix} 5 & 3 \\ 4 & 2 \end{bmatrix}A = \begin{bmatrix} 1 & 0 \\ 0 & 1 \end{bmatrix} \Rightarrow A = \begin{bmatrix} 5 & 3 \\ 4 & 2 \end{bmatrix}^{-1}\begin{bmatrix} 1 & 0 \\ 0 & 1 \end{bmatrix}$

$\begin{bmatrix} 5 & 3 \\ 4 & 2 \end{bmatrix}^{-1} = \dfrac{1}{10-12}\begin{bmatrix} 2 & -3 \\ -4 & 5 \end{bmatrix} = \begin{bmatrix} -1 & \dfrac{3}{2} \\ 2 & -\dfrac{5}{2} \end{bmatrix}$

$A = \begin{bmatrix} -1 & \dfrac{3}{2} \\ 2 & -\dfrac{5}{2} \end{bmatrix}\begin{bmatrix} 1 & 0 \\ 0 & 1 \end{bmatrix} = \begin{bmatrix} -1 & \dfrac{3}{2} \\ 2 & -\dfrac{5}{2} \end{bmatrix}$

35. a. $A^{-1} = \dfrac{1}{0-(-1)}\begin{bmatrix} 3 & -1 \\ 1 & 0 \end{bmatrix} = \begin{bmatrix} 3 & -1 \\ 1 & 0 \end{bmatrix}$

b. $\left(A^2\right)^{-1} = \left(\begin{bmatrix} 0 & 1 \\ -1 & 3 \end{bmatrix}^2\right)^{-1} = \begin{bmatrix} -1 & 3 \\ -3 & 8 \end{bmatrix}^{-1}$

$\quad = \dfrac{1}{-8+9}\begin{bmatrix} 8 & -3 \\ 3 & -1 \end{bmatrix} = \begin{bmatrix} 8 & -3 \\ 3 & -1 \end{bmatrix}$

c. $\left(A^{-1}\right)^2 = \begin{bmatrix} 3 & -1 \\ 1 & 0 \end{bmatrix}\begin{bmatrix} 3 & -1 \\ 1 & 0 \end{bmatrix} = \begin{bmatrix} 8 & -3 \\ 3 & -1 \end{bmatrix}$

d. $\left(A^2\right)\left(A^{-1}\right)^2 = \begin{bmatrix} -1 & 3 \\ -3 & 8 \end{bmatrix}\begin{bmatrix} 8 & -3 \\ 3 & -1 \end{bmatrix} = \begin{bmatrix} 1 & 0 \\ 0 & 1 \end{bmatrix}$

So $\left(A^{-1}\right)^2$ is the inverse of A^2.

c. $\left(A^{-1}\right)^2 = \begin{bmatrix} 3 & -4 \\ -2 & 3 \end{bmatrix}\begin{bmatrix} 3 & -4 \\ -2 & 3 \end{bmatrix} = \begin{bmatrix} 17 & -24 \\ -12 & 17 \end{bmatrix}$

d. $\left(A^2\right)\left(A^{-1}\right)^2 = \begin{bmatrix} 17 & 24 \\ 12 & 17 \end{bmatrix}\begin{bmatrix} 17 & -24 \\ -12 & 17 \end{bmatrix} = \begin{bmatrix} 1 & 0 \\ 0 & 1 \end{bmatrix}$

So $\left(A^{-1}\right)^2$ is the inverse of A^2.

For exercises 37–40, we show that $AB = I$. You should also show that $BA = I$ in order to show that B is the inverse of A.

37. $AB = \begin{bmatrix} 7 & 6 \\ 6 & 5 \end{bmatrix}\begin{bmatrix} -5 & 6 \\ 6 & -7 \end{bmatrix} = \begin{bmatrix} 1 & 0 \\ 0 & 1 \end{bmatrix}$

39. $AB = \begin{bmatrix} 2 & -1 & 0 \\ 1 & 0 & 4 \\ 1 & -1 & 1 \end{bmatrix}\left(\dfrac{1}{5}\right)\begin{bmatrix} 4 & 1 & -4 \\ 3 & 2 & -8 \\ -1 & 1 & 1 \end{bmatrix} = \left(\dfrac{1}{5}\right)\begin{bmatrix} 2 & -1 & 0 \\ 1 & 0 & 4 \\ 1 & -1 & 1 \end{bmatrix}\begin{bmatrix} 4 & 1 & -4 \\ 3 & 2 & -8 \\ -1 & 1 & 1 \end{bmatrix} = \left(\dfrac{1}{5}\right)\begin{bmatrix} 5 & 0 & 0 \\ 0 & 5 & 0 \\ 0 & 0 & 5 \end{bmatrix} = \begin{bmatrix} 1 & 0 & 0 \\ 0 & 1 & 0 \\ 0 & 0 & 1 \end{bmatrix}$

41. $\begin{bmatrix} 3 & 1 \\ 2 & 4 \end{bmatrix}^{-1} = \dfrac{1}{12-2}\begin{bmatrix} 4 & -1 \\ -2 & 3 \end{bmatrix} = \begin{bmatrix} \dfrac{2}{5} & -\dfrac{1}{10} \\ -\dfrac{1}{5} & \dfrac{3}{10} \end{bmatrix}$

43. $\begin{bmatrix} 1 & 2 & -2 & | & 1 & 0 & 0 \\ -1 & 3 & 0 & | & 0 & 1 & 0 \\ 0 & -2 & 1 & | & 0 & 0 & 1 \end{bmatrix} \xrightarrow{R_1 + R_2 \to R_2} \begin{bmatrix} 1 & 2 & -2 & | & 1 & 0 & 0 \\ 0 & 5 & -2 & | & 1 & 1 & 0 \\ 0 & -2 & 1 & | & 0 & 0 & 1 \end{bmatrix} \xrightarrow[R_2 + 2R_3 \to R_2]{R_1 + R_3 \to R_1} \begin{bmatrix} 1 & 0 & -1 & | & 1 & 0 & 1 \\ 0 & 1 & 0 & | & 1 & 1 & 2 \\ 0 & -2 & 1 & | & 0 & 0 & 1 \end{bmatrix}$

$\xrightarrow{2R_2 + R_3 \to R_3} \begin{bmatrix} 1 & 0 & -1 & | & 1 & 0 & 1 \\ 0 & 1 & 0 & | & 1 & 1 & 2 \\ 0 & 0 & 1 & | & 2 & 2 & 5 \end{bmatrix} \xrightarrow{R_1 + R_3 \to R_1} \begin{bmatrix} 1 & 0 & 0 & | & 3 & 2 & 6 \\ 0 & 1 & 0 & | & 1 & 1 & 2 \\ 0 & 0 & 1 & | & 2 & 2 & 5 \end{bmatrix}$

Thus, $\begin{bmatrix} 1 & 2 & -2 \\ -1 & 3 & 0 \\ 0 & -2 & 1 \end{bmatrix}^{-1} = \begin{bmatrix} 3 & 2 & 6 \\ 1 & 1 & 2 \\ 2 & 2 & 5 \end{bmatrix}$

45. $\begin{cases} x + 3y = 7 \\ 2x + 5y = 4 \end{cases} \Rightarrow \begin{bmatrix} 1 & 3 \\ 2 & 5 \end{bmatrix}\begin{bmatrix} x \\ y \end{bmatrix} = \begin{bmatrix} 7 \\ 4 \end{bmatrix} \Rightarrow \begin{bmatrix} x \\ y \end{bmatrix} = \begin{bmatrix} 1 & 3 \\ 2 & 5 \end{bmatrix}^{-1}\begin{bmatrix} 7 \\ 4 \end{bmatrix} = \begin{bmatrix} -5 & 3 \\ 2 & -1 \end{bmatrix}\begin{bmatrix} 7 \\ 4 \end{bmatrix} = \begin{bmatrix} -23 \\ 10 \end{bmatrix}$

The solution is $\{(-23, 10)\}$.

47. $\begin{cases} x + 3y + 3z = 3 \\ x + 4y + 3z = 5 \\ x + 3y + 4z = 6 \end{cases} \Rightarrow \begin{bmatrix} 1 & 3 & 3 \\ 1 & 4 & 3 \\ 1 & 3 & 4 \end{bmatrix}\begin{bmatrix} x \\ y \\ z \end{bmatrix} = \begin{bmatrix} 3 \\ 5 \\ 6 \end{bmatrix} \Rightarrow \begin{bmatrix} x \\ y \\ z \end{bmatrix} = \begin{bmatrix} 1 & 3 & 3 \\ 1 & 4 & 3 \\ 1 & 3 & 4 \end{bmatrix}^{-1}\begin{bmatrix} 3 \\ 5 \\ 6 \end{bmatrix} = \begin{bmatrix} 7 & -3 & -3 \\ -1 & 1 & 0 \\ -1 & 0 & 1 \end{bmatrix}\begin{bmatrix} 3 \\ 5 \\ 6 \end{bmatrix} = \begin{bmatrix} -12 \\ 2 \\ 3 \end{bmatrix}$

The solution is $\{(-12, 2, 3)\}$.

49. $\begin{cases} x - y + z = 3 \\ 4x + 2y = 5 \\ 7x - y - z = 6 \end{cases} \Rightarrow \begin{bmatrix} 1 & -1 & 1 \\ 4 & 2 & 0 \\ 7 & -1 & -1 \end{bmatrix}\begin{bmatrix} x \\ y \\ z \end{bmatrix} = \begin{bmatrix} 3 \\ 5 \\ 6 \end{bmatrix} \Rightarrow \begin{bmatrix} x \\ y \\ z \end{bmatrix} = \begin{bmatrix} 1 & -1 & 1 \\ 4 & 2 & 0 \\ 7 & -1 & -1 \end{bmatrix}^{-1}\begin{bmatrix} 3 \\ 5 \\ 6 \end{bmatrix} = \begin{bmatrix} \dfrac{1}{12} & \dfrac{1}{12} & \dfrac{1}{12} \\ -\dfrac{1}{6} & \dfrac{1}{3} & -\dfrac{1}{6} \\ \dfrac{3}{4} & \dfrac{1}{4} & -\dfrac{1}{4} \end{bmatrix}\begin{bmatrix} 3 \\ 5 \\ 6 \end{bmatrix} = \begin{bmatrix} \dfrac{7}{6} \\ \dfrac{1}{6} \\ 2 \end{bmatrix}$

The solution is $\left\{\left(\dfrac{7}{6}, \dfrac{1}{6}, 2\right)\right\}$.

51. $\begin{vmatrix} 2 & -5 \\ 3 & 4 \end{vmatrix} = 2(4) - (-5)(3) = 23$

53. $\begin{vmatrix} 12 & 21 \\ 4 & 7 \end{vmatrix} = 12(7) - 21(4) = 0$

55. a. $A = \begin{bmatrix} 4 & -1 & 2 \\ -2 & -3 & 5 \\ 0 & 2 & -4 \end{bmatrix}$

$M_{12} = \begin{vmatrix} -2 & 5 \\ 0 & -4 \end{vmatrix} = (-2)(-4) - 5(0) = 8 \qquad M_{23} = \begin{vmatrix} 4 & -1 \\ 0 & 2 \end{vmatrix} = 4(2) - (-1)(0) = 8$

$M_{22} = \begin{vmatrix} 4 & 2 \\ 0 & -4 \end{vmatrix} = 4(-4) - 2(0) = -16$

b. $A_{12} = (-1)^{1+2}M_{12} = -8 \qquad A_{23} = (-1)^{2+3}M_{23} = -8 \qquad A_{22} = (-1)^{2+2}M_{22} = -16$

57. a. $\begin{vmatrix} 1 & -2 & 3 \\ 4 & -1 & -2 \\ -2 & 1 & 5 \end{vmatrix} = a_{21}A_{21} + a_{22}A_{22} + a_{23}A_{23} = 4(-1)^{2+1}\begin{vmatrix} -2 & 3 \\ 1 & 5 \end{vmatrix} - 1(-1)^{2+2}\begin{vmatrix} 1 & 3 \\ -2 & 5 \end{vmatrix} - 2(-1)^{2+3}\begin{vmatrix} 1 & -2 \\ -2 & 1 \end{vmatrix}$

$= -4(-13) - 1(11) + 2(-3) = 35$

b. $\begin{vmatrix} 1 & -2 & 3 \\ 4 & -1 & -2 \\ -2 & 1 & 5 \end{vmatrix} = a_{13}A_{13} + a_{23}A_{23} + a_{33}A_{33} = 3(-1)^{1+3}\begin{vmatrix} 4 & -1 \\ -2 & 1 \end{vmatrix} - 2(-1)^{2+3}\begin{vmatrix} 1 & -2 \\ -2 & 1 \end{vmatrix} + 5(-1)^{3+3}\begin{vmatrix} 1 & -2 \\ 4 & -1 \end{vmatrix}$

$= 3(2) + 2(-3) + 5(7) = 35$

59. Expand by the first column: $\begin{vmatrix} 1 & 2 & 0 \\ 0 & 1 & 2 \\ 1 & 0 & 2 \end{vmatrix} = 1(-1)^{1+1}\begin{vmatrix} 1 & 2 \\ 0 & 2 \end{vmatrix} + 0 + 1(-1)^{3+1}\begin{vmatrix} 2 & 0 \\ 1 & 2 \end{vmatrix} = 2 + 4 = 6$

61. $D = \begin{vmatrix} 5 & 3 \\ 2 & 1 \end{vmatrix} = -1, \; D_x = \begin{vmatrix} 11 & 3 \\ 4 & 1 \end{vmatrix} = -1, \; D_y = \begin{vmatrix} 5 & 11 \\ 2 & 4 \end{vmatrix} = -2$

$x = \dfrac{-1}{-1} = 1, y = \dfrac{-2}{-1} = 2$

The solution is $\{(1, 2)\}$.

63. $D = \begin{vmatrix} 3 & 1 & -1 \\ 1 & 3 & -1 \\ 1 & 1 & -3 \end{vmatrix} = -20, \; D_x = \begin{vmatrix} 14 & 1 & -1 \\ 16 & 3 & -1 \\ -10 & 1 & -3 \end{vmatrix} = -100, \; D_y = \begin{vmatrix} 3 & 14 & -1 \\ 1 & 16 & -1 \\ 1 & -10 & -3 \end{vmatrix} = -120, \; D_z = \begin{vmatrix} 3 & 1 & 14 \\ 1 & 3 & 16 \\ 1 & 1 & -10 \end{vmatrix} = -140$

$x = \dfrac{-100}{-20} = 5, y = \dfrac{-120}{-20} = 6, z = \dfrac{-140}{-20} = 7$

The solution is $\{(5, 6, 7)\}$.

65. Expand by the third row:

$\begin{vmatrix} 1 & 2 & 4 \\ -3 & 5 & 7 \\ 1 & x & 4 \end{vmatrix} = 0 \Rightarrow 1(-1)^{3+1}\begin{vmatrix} 2 & 4 \\ 5 & 7 \end{vmatrix} + x(-1)^{3+2}\begin{vmatrix} 1 & 4 \\ -3 & 7 \end{vmatrix} + 4(-1)^{3+3}\begin{vmatrix} 1 & 2 \\ -3 & 5 \end{vmatrix} = 0 \Rightarrow -6 - 19x + 4(11) = 0 \Rightarrow x = 2$

67. $\begin{vmatrix} x & y & 1 \\ 2 & 3 & 1 \\ 1 & -4 & 1 \end{vmatrix} = 0 \Rightarrow x(-1)^{1+1}\begin{vmatrix} 3 & 1 \\ -4 & 1 \end{vmatrix} + y(-1)^{1+2}\begin{vmatrix} 2 & 1 \\ 1 & 1 \end{vmatrix} + 1(-1)^{1+3}\begin{vmatrix} 2 & 3 \\ 1 & -4 \end{vmatrix} = 0 \Rightarrow 7x - y - 11 = 0 \Rightarrow y = 7x - 11$

Using the given points to find the slope and y-intercept, we have $m = \dfrac{3 - (-4)}{2 - 1} = 7$ and

$3 = 7(2) + b \Rightarrow b = -11$.

Applying the Concepts

69.

$$\begin{matrix} & A & B & C \\ \text{Method 1} \\ \text{Method 2} \\ \text{Method 3} \end{matrix} \begin{bmatrix} 4 & 8 & 2 \\ 5 & 7 & 1 \\ 5 & 4 & 8 \end{bmatrix} \begin{bmatrix} 10 \\ 4 \\ 6 \end{bmatrix} = \begin{bmatrix} 84 \\ 84 \\ 114 \end{bmatrix} \Rightarrow \text{Method 3 is the most profitable.}$$

71. Let x = the speed of the plane. Let y = the velocity of the wind. Using Gauss-Jordan elimination, we have

$$\begin{cases} 3(x+y) = 1680 \\ 3.5(x-y) = 1680 \end{cases} \Rightarrow \begin{cases} 3x + 3y = 1680 \\ 3.5x - 3.5y = 1680 \end{cases} \Rightarrow$$

$$\begin{bmatrix} 3 & 3 & | & 1680 \\ 3.5 & -3.5 & | & 1680 \end{bmatrix}$$

$$\xrightarrow[\frac{1}{3.5}R_2 \to R_2]{\frac{1}{3}R_1 \to R_1} \begin{bmatrix} 1 & 1 & | & 560 \\ 1 & -1 & | & 480 \end{bmatrix}$$

$$\xrightarrow[\frac{1}{2}(R_1 - R_2) \to R_2]{\frac{1}{2}(R_1 + R_2) \to R_1} \begin{bmatrix} 1 & 0 & | & 520 \\ 0 & 1 & | & 40 \end{bmatrix} \Rightarrow$$

$x = 520, y = 40$

The plane is traveling at 520 mph and the wind velocity is 40 mph.

73. Let x = Andrew's amount. Let y = Bonnie's amount. Let z = Chauncie's amount. Using Gaussian elimination, we have

$$\begin{cases} x+y+z = 320 \\ x = 2z \\ y+z = x-20 \end{cases} \Rightarrow \begin{cases} x+y+z = 320 \\ x \quad\;\; -2z = 0 \\ -x+y+z = -20 \end{cases} \Rightarrow$$

$$\begin{vmatrix} 1 & 1 & 1 & | & 320 \\ 1 & 0 & -2 & | & 0 \\ -1 & 1 & 1 & | & -20 \end{vmatrix}$$

$$\xrightarrow[\frac{1}{2}(R_1 + R_3) \to R_3]{R_1 - R_2 \to R_2} \begin{vmatrix} 1 & 1 & 1 & | & 320 \\ 0 & 1 & 3 & | & 320 \\ 0 & 1 & 1 & | & 150 \end{vmatrix}$$

$$\xrightarrow[]{\frac{1}{2}(R_2 - R_3) \to R_3} \begin{vmatrix} 1 & 1 & 1 & | & 320 \\ 0 & 1 & 3 & | & 320 \\ 0 & 0 & 1 & | & 85 \end{vmatrix} \Rightarrow$$

$z = 85, y + 3(85) = 320 \Rightarrow y = 65$
$x + 65 + 85 = 320 \Rightarrow x = 170$
Andrew has \$170, Bonnie has \$65, and Chauncie has \$85.

75. Let x = the number of registered nurses. Let y = the number of licensed practical nurses. Let z = the number of nurse's aids. Using Gaussian elimination, we have

$$\begin{cases} 75x + 20y + 30z = 4850 \\ 3x + 5y + 4z = 530 \\ x + y + z = 130 \end{cases} \Rightarrow$$

$$\begin{bmatrix} 75 & 20 & 30 & | & 4850 \\ 3 & 5 & 4 & | & 530 \\ 1 & 1 & 1 & | & 130 \end{bmatrix}$$

$$\xrightarrow[]{R_1 \leftrightarrow R_3} \begin{bmatrix} 1 & 1 & 1 & | & 130 \\ 3 & 5 & 4 & | & 530 \\ 75 & 20 & 30 & | & 4850 \end{bmatrix}$$

$$\xrightarrow[75R_1 - R_3 \to R_3]{\frac{1}{2}(R_2 - 3R_1) \to R_2} \begin{bmatrix} 1 & 1 & 1 & | & 130 \\ 0 & 1 & \frac{1}{2} & | & 70 \\ 0 & 55 & 45 & | & 4900 \end{bmatrix}$$

$$\xrightarrow[]{\frac{2}{35}(R_3 - 55R_2) \to R_3} \begin{bmatrix} 1 & 1 & 1 & | & 130 \\ 0 & 1 & \frac{1}{2} & | & 70 \\ 0 & 0 & 1 & | & 60 \end{bmatrix} \Rightarrow$$

$z = 60, y + \frac{1}{2}(60) = 70 \Rightarrow y = 40,$
$x + 60 + 40 = 130 \Rightarrow 30$
There are 30 registered nurses, 40 licensed practical nurses, and 60 nurses aides.

Chapter 6 Practice Test A

1. $5 \times 4, a_{43} = 9$

2. $\begin{bmatrix} 7 & -3 & 9 & | & 5 \\ -2 & 4 & 3 & | & -12 \\ 8 & -5 & 1 & | & -9 \end{bmatrix}$

3. $\begin{cases} 4x \quad\quad - z = -3 \\ x + 3y \quad\;\; = 9 \\ 2x + 7y + 5z = 8 \end{cases}$

4. $x - 5 = 5 \Rightarrow x = 10; y + 3 = 13 \Rightarrow y = 10; z = 0$
The solution is $\{(10, 10, 0)\}$.

5. $\begin{cases} x + 2y + z = 6 \\ x + y - z = 7 \\ 2x - y + 2z = -3 \end{cases} \Rightarrow \begin{bmatrix} 1 & 2 & 1 & | & 6 \\ 1 & 1 & -1 & | & 7 \\ 2 & -1 & 2 & | & -3 \end{bmatrix}$

$\xrightarrow[\frac{1}{5}(2R_1 - R_3) \to R_3]{R_1 - R_2 \to R_2} \begin{bmatrix} 1 & 2 & 1 & | & 6 \\ 0 & 1 & 2 & | & -1 \\ 0 & 1 & 0 & | & 3 \end{bmatrix}$

$\xrightarrow{\frac{1}{2}(R_2 - R_3) \to R_3} \begin{bmatrix} 1 & 2 & 1 & | & 6 \\ 0 & 1 & 2 & | & -1 \\ 0 & 0 & 1 & | & -2 \end{bmatrix}$

$\xrightarrow[R_2 - 2R_3 \to R_2]{R_1 - R_3 \to R_1} \begin{bmatrix} 1 & 2 & 0 & | & 8 \\ 0 & 1 & 0 & | & 3 \\ 0 & 0 & 1 & | & -2 \end{bmatrix}$

$\xrightarrow{R_1 - 2R_2 \to R_1} \begin{bmatrix} 1 & 0 & 0 & | & 2 \\ 0 & 1 & 0 & | & 3 \\ 0 & 0 & 1 & | & -2 \end{bmatrix}$

The solution is $\{(2, 3, -2)\}$.

6. $\begin{cases} 2x + y - 4z = 6 \\ -x + 3y - z = -2 \\ 2x - 6y + 2z = 4 \end{cases} \Rightarrow \begin{bmatrix} 2 & 1 & -4 & | & 6 \\ -1 & 3 & -1 & | & -2 \\ 2 & -6 & 2 & | & 4 \end{bmatrix}$

$\xrightarrow[R_1 - R_3 \to R_3]{R_1 + 2R_2 \to R_2} \begin{bmatrix} 2 & 1 & -4 & | & 6 \\ 0 & 7 & -6 & | & 2 \\ 0 & 7 & -6 & | & 2 \end{bmatrix}$

$\xrightarrow[R_2 - R_3 \to R_3]{\frac{1}{14}(7R_1 - R_2) \to R_1} \begin{bmatrix} 1 & 0 & -\frac{11}{7} & | & \frac{20}{7} \\ 0 & 7 & -6 & | & 2 \\ 0 & 0 & 0 & | & 0 \end{bmatrix}$

$\xrightarrow{\frac{1}{7}R_2 \to R_2} \begin{bmatrix} 1 & 0 & -\frac{11}{7} & | & \frac{20}{7} \\ 0 & 1 & -\frac{6}{7} & | & \frac{2}{7} \\ 0 & 0 & 0 & | & 0 \end{bmatrix} \Rightarrow$

$x - \frac{11}{7}z = \frac{20}{7} \Rightarrow x = \frac{11}{7}z + \frac{20}{7}$

$y - \frac{6}{7}z = \frac{2}{7} \Rightarrow y = \frac{6}{7}z + \frac{2}{7}$

The solution is $\left\{ \left(\frac{11}{7}z + \frac{20}{7}, \frac{6}{7}z + \frac{2}{7}, z \right) \right\}$.

7. $A - B = \begin{bmatrix} 5 & -2 \\ 4 & 0 \\ 7 & 6 \end{bmatrix} - \begin{bmatrix} -3 & -1 \\ 0 & -8 \\ -4 & 6 \end{bmatrix} = \begin{bmatrix} 8 & -1 \\ 4 & 8 \\ 11 & 0 \end{bmatrix}$

8. $AB = \begin{bmatrix} 3 & -7 & 2 \end{bmatrix} \begin{bmatrix} 0 \\ 1 \\ 4 \end{bmatrix} = [3(0) - 7(1) + 2(4)] = [1]$

9. The product AB is not defined.

10. $2A = 2\begin{bmatrix} -3 & 1 & 0 \\ 5 & 7 & 2 \end{bmatrix} = \begin{bmatrix} -6 & 2 & 0 \\ 10 & 14 & 4 \end{bmatrix}$

11. $A + BA = \begin{bmatrix} -3 & 1 & 0 \\ 5 & 7 & 2 \end{bmatrix} + \begin{bmatrix} -1 & 4 \\ 8 & 2 \end{bmatrix} \begin{bmatrix} -3 & 1 & 0 \\ 5 & 7 & 2 \end{bmatrix}$

$= \begin{bmatrix} -3 & 1 & 0 \\ 5 & 7 & 2 \end{bmatrix} + \begin{bmatrix} 23 & 27 & 8 \\ -14 & 22 & 4 \end{bmatrix}$

$= \begin{bmatrix} 20 & 28 & 8 \\ -9 & 29 & 6 \end{bmatrix}$

12. $C^2 = \begin{bmatrix} 1 & 5 \\ 0 & 4 \end{bmatrix} \begin{bmatrix} 1 & 5 \\ 0 & 4 \end{bmatrix} = \begin{bmatrix} 1 & 25 \\ 0 & 16 \end{bmatrix}$

13. $C^{-1} = \begin{bmatrix} 1 & 5 \\ 0 & 4 \end{bmatrix} = \frac{1}{4} \begin{bmatrix} 4 & -5 \\ 0 & 1 \end{bmatrix} = \begin{bmatrix} 1 & -\frac{5}{4} \\ 0 & \frac{1}{4} \end{bmatrix}$

14. $\begin{bmatrix} 2 & 1 & 3 \\ 1 & 2 & -1 \\ 3 & 1 & 5 \end{bmatrix}^{-1} = \begin{bmatrix} 2 & 1 & 3 & | & 1 & 0 & 0 \\ 1 & 2 & -1 & | & 0 & 1 & 0 \\ 3 & 1 & 5 & | & 0 & 0 & 1 \end{bmatrix}$

$\xrightarrow{R_1 \leftrightarrow R_2} \begin{bmatrix} 1 & 2 & -1 & | & 0 & 1 & 0 \\ 2 & 1 & 3 & | & 1 & 0 & 0 \\ 3 & 1 & 5 & | & 0 & 0 & 1 \end{bmatrix}$

$\xrightarrow[3R_1 - R_3 \to R_3]{2R_1 - R_2 \to R_2} \begin{bmatrix} 1 & 2 & -1 & | & 0 & 1 & 0 \\ 0 & 3 & -5 & | & -1 & 2 & 0 \\ 0 & 5 & -8 & | & 0 & 3 & -1 \end{bmatrix}$

$\xrightarrow[3R_3 - 5R_2 \to R_3]{5R_1 - R_2 \to R_1} \begin{bmatrix} 5 & 7 & 0 & | & 1 & 3 & 0 \\ 0 & 3 & -5 & | & -1 & 2 & 0 \\ 0 & 0 & 1 & | & 5 & -1 & -3 \end{bmatrix}$

$\xrightarrow{\frac{1}{3}(R_2 + 5R_3) \to R_2} \begin{bmatrix} 5 & 7 & 0 & | & 1 & 3 & 0 \\ 0 & 1 & 0 & | & 8 & -1 & -5 \\ 0 & 0 & 1 & | & 5 & -1 & -3 \end{bmatrix}$

$\xrightarrow{\frac{1}{5}(R_1 - 7R_2) \to R_1} \begin{bmatrix} 1 & 0 & 0 & | & -11 & 2 & 7 \\ 0 & 1 & 0 & | & 8 & -1 & -5 \\ 0 & 0 & 1 & | & 5 & -1 & -3 \end{bmatrix}$

$\begin{bmatrix} 2 & 1 & 3 \\ 1 & 2 & -1 \\ 3 & 1 & 5 \end{bmatrix}^{-1} = \begin{bmatrix} -11 & 2 & 7 \\ 8 & -1 & -5 \\ 5 & -1 & -3 \end{bmatrix}$

15. $\begin{bmatrix} 5 & 2 \\ 3 & 1 \end{bmatrix}\begin{bmatrix} x \\ y \end{bmatrix} = \begin{bmatrix} 32 \\ 18 \end{bmatrix}$

16. $\begin{cases} 12x - 3y = 5 \\ -2x + 7y = -9 \end{cases}$

17. $\begin{bmatrix} 1 & 5 & -2 \\ 4 & -2 & 7 \end{bmatrix} - 5X = 2\begin{bmatrix} 2 & 5 & -11 \\ 18 & 8 & 11 \end{bmatrix} \Rightarrow$

$\begin{bmatrix} 1 & 5 & -2 \\ 4 & -2 & 7 \end{bmatrix} - 5X = \begin{bmatrix} 4 & 10 & -22 \\ 36 & 16 & 22 \end{bmatrix} \Rightarrow$

$\begin{bmatrix} 1 & 5 & -2 \\ 4 & -2 & 7 \end{bmatrix} - \begin{bmatrix} 4 & 10 & -22 \\ 36 & 16 & 22 \end{bmatrix} = 5X \Rightarrow$

$\begin{bmatrix} -3 & -5 & 20 \\ -32 & -18 & -15 \end{bmatrix} = 5X \Rightarrow$

$\begin{bmatrix} -\dfrac{3}{5} & -1 & 4 \\ -\dfrac{32}{5} & -\dfrac{18}{5} & -3 \end{bmatrix} = X$

18. $\begin{vmatrix} \dfrac{1}{2} & -\dfrac{1}{4} \\ \dfrac{1}{2} & \dfrac{3}{4} \end{vmatrix} = \left(\dfrac{1}{2}\right)\left(\dfrac{3}{4}\right) - \left(-\dfrac{1}{4}\right)\left(\dfrac{1}{2}\right) = \dfrac{1}{2}$

19. Expand by the second row:

$\begin{vmatrix} 1 & 3 & 5 \\ 2 & 0 & 10 \\ -3 & 1 & -15 \end{vmatrix}$

$= 2(-1)^{2+1}\begin{vmatrix} 3 & 5 \\ 1 & -15 \end{vmatrix} + 0 + 10(-1)^{2+3}\begin{vmatrix} 1 & 3 \\ -3 & 1 \end{vmatrix}$

$= -2(-50) - 10(10) = 0$

20. $\begin{cases} 2x - y + z = 3 \\ x + y + z = 6 \\ 4x + 3y - 2z = 4 \end{cases} \Rightarrow D = \begin{vmatrix} 2 & -1 & 1 \\ 1 & 1 & 1 \\ 4 & 3 & -2 \end{vmatrix}$

$D_x = \begin{vmatrix} 3 & -1 & 1 \\ 6 & 1 & 1 \\ 4 & 3 & -2 \end{vmatrix}, D_y = \begin{vmatrix} 2 & 3 & 1 \\ 1 & 6 & 1 \\ 4 & 4 & -2 \end{vmatrix},$

$D_z = \begin{vmatrix} 2 & -1 & 3 \\ 1 & 1 & 6 \\ 4 & 3 & 4 \end{vmatrix}$

$x = \dfrac{\begin{vmatrix} 3 & -1 & 1 \\ 6 & 1 & 1 \\ 4 & 3 & -2 \end{vmatrix}}{\begin{vmatrix} 2 & -1 & 1 \\ 1 & 1 & 1 \\ 4 & 3 & -2 \end{vmatrix}}, y = \dfrac{\begin{vmatrix} 2 & 3 & 1 \\ 1 & 6 & 1 \\ 4 & 4 & -2 \end{vmatrix}}{\begin{vmatrix} 2 & -1 & 1 \\ 1 & 1 & 1 \\ 4 & 3 & -2 \end{vmatrix}},$

$z = \dfrac{\begin{vmatrix} 2 & -1 & 3 \\ 1 & 1 & 6 \\ 4 & 3 & 4 \end{vmatrix}}{\begin{vmatrix} 2 & -1 & 1 \\ 1 & 1 & 1 \\ 4 & 3 & -2 \end{vmatrix}}$

Chapter 6 Practice Test B

1. D   **2.** D   **3.** D

4. $x + 3 = 9 \Rightarrow x = 6; y + 4 = 2 \Rightarrow y = -2; z = 3$

The answer is C.

5. $\begin{cases} 2x + y = 15 \\ 2y + z = 25 \\ 2z + x = 26 \end{cases} \Rightarrow \begin{bmatrix} 2 & 1 & 0 & | & 15 \\ 0 & 2 & 1 & | & 25 \\ 1 & 0 & 2 & | & 26 \end{bmatrix}$

$\xrightarrow{R_1 - 2R_3 \to R_3} \begin{bmatrix} 2 & 1 & 0 & | & 15 \\ 0 & 2 & 1 & | & 25 \\ 0 & 1 & -4 & | & -37 \end{bmatrix}$

$\xrightarrow{\frac{1}{9}(R_2 - 2R_3) \to R_3} \begin{bmatrix} 2 & 1 & 0 & | & 15 \\ 0 & 2 & 1 & | & 25 \\ 0 & 0 & 1 & | & 11 \end{bmatrix} \Rightarrow z = 11$

The answer is C.

6. $\begin{cases} 2x + y = 17 \\ y + 2z = 15 \\ x + z = 9 \end{cases} \Rightarrow \begin{bmatrix} 2 & 1 & 0 & | & 17 \\ 0 & 1 & 2 & | & 15 \\ 1 & 0 & 1 & | & 9 \end{bmatrix}$

$\xrightarrow{2R_3 - R_1 \to R_3} \begin{bmatrix} 2 & 1 & 0 & | & 17 \\ 0 & 1 & 2 & | & 15 \\ 0 & -1 & 2 & | & 1 \end{bmatrix}$

$\xrightarrow[\frac{1}{4}(R_2 + R_3) \to R_3]{\frac{1}{2}(R_1 + R_3) \to R_1} \begin{bmatrix} 1 & 0 & 1 & | & 9 \\ 0 & 1 & 2 & | & 15 \\ 0 & 0 & 1 & | & 4 \end{bmatrix}$

$\xrightarrow[R_2 - 2R_3 \to R_2]{R_1 - R_3 \to R_1} \begin{bmatrix} 1 & 0 & 0 & | & 5 \\ 0 & 1 & 0 & | & 7 \\ 0 & 0 & 1 & | & 4 \end{bmatrix} \Rightarrow$

$x = 5, y = 7, z = 4 \Rightarrow 4(5) + 3(7) + 4 = 45$

The answer is D.

7. $\begin{bmatrix} -1 & 4 \\ 0 & 4 \\ 8 & -4 \end{bmatrix} - \begin{bmatrix} 7 & 2 \\ 17 & 4 \\ 2 & 2 \end{bmatrix} = \begin{bmatrix} -8 & 2 \\ -17 & 0 \\ 6 & -6 \end{bmatrix}$

The answer is C.

8. $AB = \begin{bmatrix} -8 & 2 & 9 \end{bmatrix} \begin{bmatrix} 3 \\ 0 \\ -3 \end{bmatrix} = \begin{bmatrix} -8(3) + 2(0) + 9(-3) \end{bmatrix}$

$= \begin{bmatrix} -51 \end{bmatrix}$. The answer is B.

9. AB is not defined. The answer is A.

10. $2A = 2\begin{bmatrix} 2 & 1 & -3 \\ -5 & 2 & 1 \end{bmatrix} = \begin{bmatrix} 4 & 2 & -6 \\ -10 & 4 & 2 \end{bmatrix}$

The answer is B.

11. $A + BA = \begin{bmatrix} 2 & 1 & -3 \\ -5 & 2 & 1 \end{bmatrix} + \begin{bmatrix} -3 & 7 \\ 2 & 4 \end{bmatrix}\begin{bmatrix} 2 & 1 & -3 \\ -5 & 2 & 1 \end{bmatrix}$

$= \begin{bmatrix} 2 & 1 & -3 \\ -5 & 2 & 1 \end{bmatrix} + \begin{bmatrix} -41 & 11 & 16 \\ -16 & 10 & -2 \end{bmatrix}$

$= \begin{bmatrix} -39 & 12 & 13 \\ -21 & 12 & -1 \end{bmatrix}$

The answer is C.

12. $C^2 = \begin{bmatrix} 5 & 4 \\ 1 & 0 \end{bmatrix}\begin{bmatrix} 5 & 4 \\ 1 & 0 \end{bmatrix} = \begin{bmatrix} 29 & 20 \\ 5 & 4 \end{bmatrix}$

The answer is C.

13. $C^{-1} = \begin{bmatrix} 5 & 4 \\ 1 & 0 \end{bmatrix}^{-1} = -\frac{1}{4}\begin{bmatrix} 0 & -4 \\ -1 & 5 \end{bmatrix} = \begin{bmatrix} 0 & 1 \\ \frac{1}{4} & -\frac{5}{4} \end{bmatrix}$

The answer is B.

14. $\begin{bmatrix} 1 & 0 & 0 \\ 2 & 1 & 0 \\ 3 & -4 & 1 \end{bmatrix}^{-1} = \begin{bmatrix} 1 & 0 & 0 & 1 & 0 & 0 \\ 2 & 1 & 0 & 0 & 1 & 0 \\ 3 & -4 & 1 & 0 & 0 & 1 \end{bmatrix}$

$\xrightarrow[3R_1 - R_3 \to R_3]{R_2 - 2R_1 \to R_2} \begin{bmatrix} 1 & 0 & 0 & 1 & 0 & 0 \\ 0 & 1 & 0 & -2 & 1 & 0 \\ 0 & 4 & -1 & 3 & 0 & -1 \end{bmatrix}$

$\xrightarrow{4R_2 - R_3 \to R_3} \begin{bmatrix} 1 & 0 & 0 & 1 & 0 & 0 \\ 0 & 1 & 0 & -2 & 1 & 0 \\ 0 & 0 & 1 & -11 & 4 & 1 \end{bmatrix} \Rightarrow$

$\begin{bmatrix} 1 & 0 & 0 \\ 2 & 1 & 0 \\ 3 & -4 & 1 \end{bmatrix}^{-1} = \begin{bmatrix} 1 & 0 & 0 \\ -2 & 1 & 0 \\ -11 & 4 & 1 \end{bmatrix}$

The answer is D.

15. A

16. A

17. $\begin{bmatrix} -2 & -3 & 1 \\ -5 & 3 & -2 \end{bmatrix} - 3X = -5\begin{bmatrix} -1 & -2 & -1 \\ 1 & 0 & 1 \end{bmatrix} \Rightarrow$

$\begin{bmatrix} -2 & -3 & 1 \\ -5 & 3 & -2 \end{bmatrix} - 3X = \begin{bmatrix} 5 & 10 & 5 \\ -5 & 0 & -5 \end{bmatrix} \Rightarrow$

$\begin{bmatrix} -2 & -3 & 1 \\ -5 & 3 & -2 \end{bmatrix} - \begin{bmatrix} 5 & 10 & 5 \\ -5 & 0 & -5 \end{bmatrix} = 3X \Rightarrow$

$3X = \begin{bmatrix} -7 & -13 & -4 \\ 0 & 3 & 3 \end{bmatrix} \Rightarrow$

$X = \begin{bmatrix} -\frac{7}{3} & -\frac{13}{3} & -\frac{4}{3} \\ 0 & 1 & 1 \end{bmatrix}$

The answer is B.

18. $\begin{vmatrix} -8 & 5 \\ -4 & -1 \end{vmatrix} = (-8)(-1) - 5(-4) = 28$

The answer is D.

19. $\begin{vmatrix} 2 & 3 & -2 \\ 3 & 0 & -3 \\ -3 & 0 & -5 \end{vmatrix} = 3(-1)^{1+2}\begin{vmatrix} 3 & -3 \\ -3 & -5 \end{vmatrix} = -3(-24) = 72$

The answer is C.

20. $\begin{cases} x + y + z = -6 \\ x - y + 3z = -22 \\ 2x + y + z = -10 \end{cases} \Rightarrow D = \begin{vmatrix} 1 & 1 & 1 \\ 1 & -1 & 3 \\ 2 & 1 & 1 \end{vmatrix} = 4$

$D_x = \begin{vmatrix} -6 & 1 & 1 \\ -22 & -1 & 3 \\ -10 & 1 & 1 \end{vmatrix} = -16,\ D_y = \begin{vmatrix} 1 & -6 & 1 \\ 1 & -22 & 3 \\ 2 & -10 & 1 \end{vmatrix} = 12$

$D_z = \begin{vmatrix} 1 & 1 & -6 \\ 1 & -1 & -22 \\ 2 & 1 & -10 \end{vmatrix} = -20$

$x = \frac{-16}{4} = -4,\ y = \frac{12}{4} = 3,\ z = \frac{-20}{4} = -5$

The answer is B.

Cumulative Review Exercises (Chapters P–6)

1. $\sqrt{(-1-2)^2 + (y - (-3))^2} = 5 \Rightarrow$
$9 + (y+3)^2 = 25 \Rightarrow y^2 + 6y - 7 = 0 \Rightarrow$
$(y+7)(y-1) = 0 \Rightarrow y = -7$ or $y = 1$

2. $\frac{4}{x-1} - \frac{3}{x+2} = \frac{18}{(x+2)(x-1)} \Rightarrow$
$4(x+2) - 3(x-1) = 18 \Rightarrow x + 11 = 18 \Rightarrow x = 7$

3. $|2x - 5| = 3 \Rightarrow 2x - 5 = 3$ or $2x - 5 = -3 \Rightarrow$
$x = 4$ or $x = 1$

4. $4x^2 = 8x - 13 \Rightarrow 4x^2 - 8x + 13 = 0 \Rightarrow$

$x = \dfrac{8 \pm \sqrt{64 - 4(4)(13)}}{2(4)} = \dfrac{8 \pm \sqrt{-144}}{8} = 1 \pm \dfrac{3}{2}i$

5. $\left(\dfrac{3x-1}{x+5}\right)^2 - 3\left(\dfrac{3x-1}{x+5}\right) - 28 = 0$

Let $u = \dfrac{3x-1}{x+5}$. Then we have $u^2 - 3u - 28 = 0 \Rightarrow$

$(u+4)(u-7) = 0 \Rightarrow u = -4$ or $u = 7$

$\dfrac{3x-1}{x+5} = -4 \Rightarrow 3x - 1 = -4x - 20 \Rightarrow x = -\dfrac{19}{7}$

$\dfrac{3x-1}{x+5} = 7 \Rightarrow 3x - 1 = 7x + 35 \Rightarrow x = -9$

The solution is $\left\{ -\dfrac{19}{7}, -9 \right\}$.

6. $\log_2 |x| + \log_2 |x+6| = 4 \Rightarrow$

$\log_2 (|x| \cdot |x+6|) = 4 \Rightarrow 2^4 = |x^2 + 6x| \Rightarrow$

$x^2 + 6x = 16$ or $x^2 + 6x = -16$

$x^2 + 6x - 16 = 0 \Rightarrow (x+8)(x-2) = 0 \Rightarrow$

$x = -8$ or $x = 2$

$x^2 + 6x + 16 = 0 \Rightarrow$

$x = \dfrac{-6 \pm \sqrt{36 - 64}}{2} = -3 \pm i\sqrt{7}$ (reject this)

The solution is $\{-8, 2\}$.

7. Solve $x + 2 = 0 \Rightarrow x = -2$ and

$2x - 1 = 0 \Rightarrow x = \dfrac{1}{2}$. The intervals to be tested

are $(-\infty, -2), \left(-2, \dfrac{1}{2}\right)$, and $\left(\dfrac{1}{2}, \infty\right)$.

Interval	Test point	Value of $\dfrac{x+2}{2x-1}$	Result
$(-\infty, -2)$	-3	$1/7$	$+$
$(-2, 1/2)$	0	-2	$-$
$(1/2, \infty)$	1	3	$+$

The solution set is $(-\infty, -2) \cup \left(\dfrac{1}{2}, \infty\right)$.

8. Solve the associated equation:

$x^2 - 7x + 6 = 0 \Rightarrow (x-6)(x-1) = 0 \Rightarrow x = 6$

or $x = 1$. The intervals are $(-\infty, 1], [1, 6]$, and $[6, \infty)$.

Interval	Test point	Value of $x^2 - 7x + 6$	Result
$(-\infty, 1]$	0	6	$+$
$[1, 6]$	2	-4	$-$
$(6, \infty)$	7	6	$+$

The solution set is $[1, 6]$.

9. The factors of the constant term are $\{\pm 1, \pm 3\}$. The factors of the leading coefficient are $\{\pm 1, \pm 2, \pm 4\}$. The possible rational zeros are $\left\{ \pm 1, \pm \dfrac{1}{2}, \pm \dfrac{1}{4}, \pm 3, \pm \dfrac{3}{2}, \pm \dfrac{3}{4} \right\}$.

10. Using synthetic division, we have

$$\begin{array}{r|rrrr} \frac{1}{2} & 4 & 8 & -11 & 3 \\ & & 2 & 5 & -3 \\ \hline & 4 & 10 & -6 & 0 \end{array}$$

$4x^3 + 8x^2 - 11x + 3 = \left(x - \dfrac{1}{2}\right)\left(4x^2 + 10x - 6\right)$

The zeros of the depressed function $4x^2 + 10x - 6$ are also zeros of the original function.

$4x^2 + 10x - 6 = 0 \Rightarrow 2(x+3)(2x-1) = 0 \Rightarrow$

$x = -3$ or $x = \dfrac{1}{2}$.

So $\dfrac{1}{2}$ is a zero of multiplicity 2.

11. $h = kr^3 \Rightarrow 10{,}125 = k(15^3) \Rightarrow k = 3$

$h = 3(20^3) = 24{,}000$ horsepower

12. Let $x =$ the number of acres to be annexed. Then

$0.12(400 + x) = 0.02(400) + 0.2x \Rightarrow$

$48 + 0.12x = 8 + 0.2x \Rightarrow x = 500$ square miles

13. Using substitution, we have

$\begin{cases} 3x + y = 2 \\ 4x + 5y = -1 \end{cases} \Rightarrow y = -3x + 2$

$4x + 5(-3x + 2) = -1 \Rightarrow -11x + 10 = -1 \Rightarrow$

$x = 1; y = -3(1) + 2 = -1$

The solution is $\{(1, -1)\}$.

14. Using elimination, we have

$$\begin{cases} 2x+y=5 \\ y^2-2y=-3x+5 \end{cases} \Rightarrow \begin{cases} 2x+\qquad y=5 \\ 3x+y^2-2y=5 \end{cases} \Rightarrow$$

$$\begin{cases} -6x-\qquad 3y=-15 \\ 6x+2y^2-4y=10 \end{cases} \Rightarrow 2y^2-7y=-5 \Rightarrow$$

$$2y^2-7y+5=0 \Rightarrow (y-1)(2y-5)=0 \Rightarrow$$

$$y=1 \text{ or } y=\frac{5}{2}$$

$$2x+1=5 \Rightarrow x=2$$

$$2x+\frac{5}{2}=5 \Rightarrow x=\frac{5}{4}$$

The solution is $\left\{(2,1),\left(\frac{5}{4},\frac{5}{2}\right)\right\}$.

15. Shift the graph of $f(x)=|x|$ one unit left, stretch by a factor of 3, then shift the resulting graph two units up.

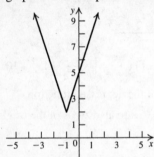

16.

$$\begin{bmatrix} 1 & 2 & -2 \\ -1 & 3 & 0 \\ 0 & -2 & 1 \end{bmatrix}^{-1} = \begin{bmatrix} 1 & 2 & -2 & | & 1 & 0 & 0 \\ -1 & 3 & 0 & | & 0 & 1 & 0 \\ 0 & -2 & 1 & | & 0 & 0 & 1 \end{bmatrix}$$

$$\xrightarrow{R_1+R_2 \to R_2} \begin{bmatrix} 1 & 2 & -2 & | & 1 & 0 & 0 \\ 0 & 5 & -2 & | & 1 & 1 & 0 \\ 0 & -2 & 1 & | & 0 & 0 & 1 \end{bmatrix}$$

$$\xrightarrow[2R_2+5R_3 \to R_3]{R_1+R_3 \to R_1} \begin{bmatrix} 1 & 0 & -1 & | & 1 & 0 & 1 \\ 0 & 5 & -2 & | & 1 & 1 & 0 \\ 0 & 0 & 1 & | & 2 & 2 & 5 \end{bmatrix}$$

$$\xrightarrow[\frac{1}{5}(R_2+2R_3) \to R_2]{R_1+R_3 \to R_1} \begin{bmatrix} 1 & 0 & 0 & | & 3 & 2 & 6 \\ 0 & 1 & 0 & | & 1 & 1 & 2 \\ 0 & 0 & 1 & | & 2 & 2 & 5 \end{bmatrix} \Rightarrow$$

$$\begin{bmatrix} 1 & 2 & -2 \\ -1 & 3 & 0 \\ 0 & -2 & 1 \end{bmatrix}^{-1} = \begin{bmatrix} 3 & 2 & 6 \\ 1 & 1 & 2 \\ 2 & 2 & 5 \end{bmatrix}$$

17. $\begin{bmatrix} 1 & 2 & -2 \\ -1 & 3 & 0 \\ 0 & -2 & 1 \end{bmatrix}\begin{bmatrix} x \\ y \\ z \end{bmatrix} = \begin{bmatrix} 5 \\ 2 \\ -3 \end{bmatrix} \Rightarrow$

$$\begin{bmatrix} x \\ y \\ z \end{bmatrix} = \begin{bmatrix} 1 & 2 & -2 \\ -1 & 3 & 0 \\ 0 & -2 & 1 \end{bmatrix}^{-1}\begin{bmatrix} 5 \\ 2 \\ -3 \end{bmatrix}$$

$$= \begin{bmatrix} 3 & 2 & 6 \\ 1 & 1 & 2 \\ 2 & 2 & 5 \end{bmatrix}\begin{bmatrix} 5 \\ 2 \\ -3 \end{bmatrix} = \begin{bmatrix} 1 \\ 1 \\ -1 \end{bmatrix}$$

The solution is $\{(1, 1, -1)\}$.

18. a. $F(x)=(x+2)^2+3(x+2)-1=x^2+7x+9$

b. $F(4)=4^2+7(4)+9=53$

19. $y=\dfrac{x}{x+4}$. Switch the variables, and then

solve for y to find $f^{-1}(x)$: $x=\dfrac{y}{y+4} \Rightarrow$

$$xy+4x=y \Rightarrow xy-y=-4x \Rightarrow$$

$$y(x-1)=-4x \Rightarrow y=-\frac{4x}{x-1} \Rightarrow$$

$$f^{-1}(x)=-\frac{4x}{x-1}$$

20. Domain: $(-\infty,-4)\cup(-4,\infty)$

Range: $(-\infty,1)\cup(1,\infty)$

Chapter 7 Conic Sections

7.2 The Parabola

7.2 Practice Problems

1. a.

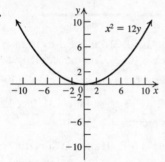

Vertex: (0, 0)

$x^2 = 12y = 4ay \Rightarrow a = 3$, so the focus is (0, 3).

The directrix is $y = -3$.

The axis is the y-axis.

b.

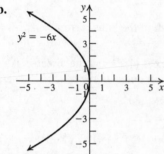

Vertex: (0, 0)

$y^2 = -6x = 4ax \Rightarrow a = -\dfrac{3}{2}$, so the focus is $\left(-\dfrac{3}{2}, 0\right)$.

The directrix is $x = \dfrac{3}{2}$.

The axis is the x-axis.

2. a. The vertex is (0, 0) and the focus is (0, 2), so the graph opens up. The general form is $x^2 = 4ay$.

$a = 2 \Rightarrow x^2 = 4(2)y \Rightarrow x^2 = 8y$.

b. Since the vertex is (0, 0), the axis of the parabola is the x-axis, and the parabola passes through (1, 2) which is to the right of the vertex, the parabola opens to the right and the general form is $y^2 = 4ax$.

We need to solve for a:

$2^2 = 4a(1) \Rightarrow 4 = 4a \Rightarrow 1 = a$. The equation is $y^2 = 4x$.

3. Rewrite the equation as $2\left(x^2 - 4x\right) = y - 7$, then complete the square to put the equation into standard form:

$$2x^2 - 8x - y + 7 = 0$$
$$2x^2 - 8x = y - 7$$
$$2\left(x^2 - 4x\right) = y - 7$$
$$2\left(x^2 - 4x + 4\right) = y - 7 + 2(4)$$
$$2(x - 2)^2 = y + 1$$
$$(x - 2)^2 = \frac{1}{2}(y + 1)$$

$4a = \dfrac{1}{2} \Rightarrow a = \dfrac{1}{8}$; vertex $= (h, k) = (2, -1)$

The parabola opens up, so the focus is at

$(h, k + a) = \left(2, -1 + \dfrac{1}{8}\right) = \left(2, -\dfrac{7}{8}\right)$. The

directrix is located at

$y = k - a = -1 - \dfrac{1}{8} = -\dfrac{9}{8}$.

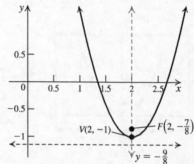

4. The equation is $x^2 = 4ay \Rightarrow x^2 = 4(7.3)y \Rightarrow$ $x^2 = 29.2y$. To find the thickness y of the mirror at the edge, substitute $x = 1.5$ (half the diameter) in the equation and solve for y.

$1.5^2 = 29.2y \Rightarrow y \approx 0.077055$

The mirror is about 0.077055 in. thick at the edge.

7.2 Basic Concepts Skills

1. A parabola is the set of all points P in the plane that are equidistant from a fixed line called the <u>directrix</u> and a fixed point not on the line called the <u>focus</u>.

3. The point at which the axis intersects the parabola is called the <u>vertex</u> of the parabola.

5. True

7. $x^2 = 2y = 4ay \Rightarrow a = \dfrac{1}{2}$

   Focus: $\left(0, \dfrac{1}{2}\right)$, directrix: $y = -\dfrac{1}{2}$, graph (e)

9. $16x^2 = -9y \Rightarrow x^2 = -\dfrac{9}{16}y = 4ay \Rightarrow a = -\dfrac{9}{64}$

   Focus: $\left(0, -\dfrac{9}{64}\right)$, directrix: $y = \dfrac{9}{64}$, graph (d)

11. $y^2 = 2x = 4ax \Rightarrow a = \dfrac{1}{2}$

    Focus: $\left(\dfrac{1}{2}, 0\right)$, directrix: $x = -\dfrac{1}{2}$, graph (g)

13. $9y^2 = -16x \Rightarrow y^2 = -\dfrac{16}{9}x = 4ax \Rightarrow a = -\dfrac{4}{9}$

    Focus: $\left(-\dfrac{4}{9}, 0\right)$, directrix: $x = \dfrac{4}{9}$, graph (f)

15.

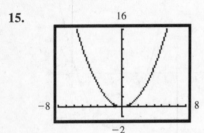

17.

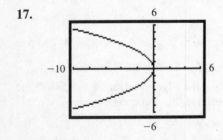

19.

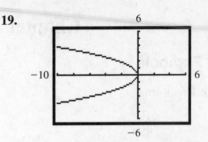

21. The focus is (0, 2) and the directrix is $y = 4$, so the vertex is (0, 3), and the graph opens down. The general form is $(x-h)^2 = -4a(y-k)$.

    $a = 3 - 2 = 1 \Rightarrow x^2 = -4(y-3)$. The length of the latus rectum $= 4(1) = 4$.

23. The focus is (−2, 0) and the directrix is $x = 3$, so the vertex is $(1/2, 0)$, and the graph opens to the left. The general form is

    $(y-k)^2 = -4a(x-h)$. $a = \dfrac{1}{2} - (-2) = \dfrac{5}{2} \Rightarrow$

    $y^2 = -4\left(\dfrac{5}{2}\right)\left(x - \dfrac{1}{2}\right) \Rightarrow y^2 = -10\left(x - \dfrac{1}{2}\right)$.

    The length of the latus rectum $= 4\left(\dfrac{5}{2}\right) = 10$.

25. The vertex is (1, 1) and the directrix is $x = 3$, so the graph opens to the left. The general form is $(y-k)^2 = -4a(x-h)$.

    $a = 3 - 1 = 2 \Rightarrow$

    $(y-1)^2 = -4(2)(x-1) \Rightarrow (y-1)^2 = -8(x-1)$.

    The length of the latus rectum $= 4(2) = 8$.

27. The vertex is (1, 1) and the directrix is $y = -3$, so the graph opens up. The general form is

    $(x-h)^2 = 4a(y-k)$. $a = 1 - (-3) = 4 \Rightarrow$

    $(x-1)^2 = 4(4)(y-1) \Rightarrow (x-1)^2 = 16(y-1)$.

    The length of the latus rectum $= 4(4) = 16$.

29. The vertex is (1, 0) and the focus is (3, 0), so the graph opens to the right. The general form is $(y-k)^2 = 4a(x-h)$. $a = 3 - 1 = 2 \Rightarrow$

    $(y-0)^2 = 4(2)(x-1) \Rightarrow y^2 = 8(x-1)$. The length of the latus rectum $= 4(2) = 8$.

31. The vertex is (0, 1) and the focus is (0, −2), so the graph opens down. The general form is

    $(x-h)^2 = -4a(y-k)$. $a = 1 - (-2) = 3 \Rightarrow$

    $(x-0)^2 = -4(3)(y-1) \Rightarrow x^2 = -12(y-1)$.

    The length of the latus rectum $= 4(3) = 12$.

33. The vertex is (2, 3) and the directrix is $x = 4$, so the graph opens to the left. The general form is $(y - k)^2 = -4a(x - h)$. $a = 4 - 2 = 2 \Rightarrow$

$(y - 3)^2 = -4(2)(x - 2) \Rightarrow (y - 3)^2 = -8(x - 2)$. The length of the latus rectum = $4(2) = 8$.

35. The vertex is (2, 3) and the directrix is $y = 1$ so the graph opens up. The general form is $(x - h)^2 = 4a(y - k)$. $a = 3 - 1 = 2 \Rightarrow$

$(x - 2)^2 = 4(2)(y - 3) \Rightarrow (x - 2)^2 = 8(y - 3)$. The length of the latus rectum = $4(2) = 8$.

37.

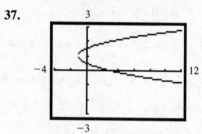

39.

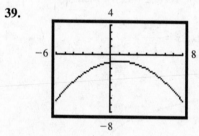

41. $4a = 2 \Rightarrow a = 1/2$. The graph opens to the right, so the focus is $(h + a, k)$, and the directrix is $x = h - a$. Vertex: (−1, 1), focus $(-1/2, 1)$, directrix: $x = -3/2$.

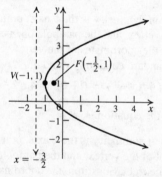

43. $4a = 3 \Rightarrow a = 3/4$. The graph opens up, so the focus is $(h, k + a)$, and the directrix is $y = k - a$. Vertex: (−2, 2), focus $(-2, 11/4)$, directrix: $y = 5/4$.

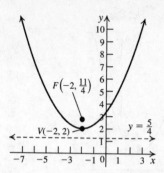

45. $4a = 6 \Rightarrow a = 3/2$. The graph opens to the left, so the focus is $(h - a, k)$, and the directrix is $x = h + a$. Vertex: (2, −1), focus $(1/2, -1)$, directrix: $x = 7/2$.

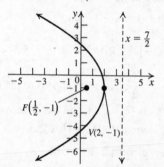

47. $4a = 10 \Rightarrow a = 5/2$. The graph opens down, so the focus is $(h, k - a)$, and the directrix is $y = k + a$. Vertex: (1, 3), focus $(1, 1/2)$, directrix: $y = 11/2$.

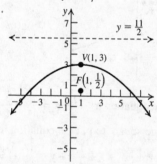

49. Rewrite the equation as $y - 2 = x^2 + 2x$, then complete the square to put the equation into standard form: $y - 2 + 1 = x^2 + 2x + 1 \Rightarrow$

$y - 1 = (x + 1)^2 \Rightarrow 4a = 1 \Rightarrow a = 1/4$. The graph opens up, so the focus is $(h, k + a)$, and the directrix is $y = k - a$. Vertex: (−1, 1), focus $(-1, 5/4)$, directrix: $y = 3/4$.

(continued on next page)

(continued)

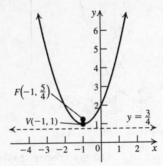

51. Rewrite the equation as $2(y^2 + 2y) = 2x - 1$, then complete the square to put the equation into standard form:

$$2(y^2 + 2y + 1) = 2x - 1 + 2 \Rightarrow$$

$$2(y+1)^2 = 2\left(x + \frac{1}{2}\right) \Rightarrow (y+1)^2 = x + \frac{1}{2} \Rightarrow$$

$4a = 1 \Rightarrow a = \frac{1}{4}$. The graph opens to the right, so the focus is $(h + a, k)$, and the directrix is $x = h - a$. Vertex: $(-1/2, -1)$, focus $(-1/4, -1)$, directrix: $x = -3/4$.

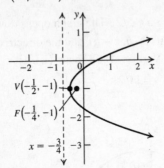

53. Rewrite the equation as $x^2 + x = -y - \frac{5}{4}$, then complete the square to put the equation into standard form: $x^2 + x + \frac{1}{4} = -y - \frac{5}{4} + \frac{1}{4} \Rightarrow$

$\left(x + \frac{1}{2}\right)^2 = -(y+1) \Rightarrow 4a = 1 \Rightarrow a = \frac{1}{4}$. The graph opens down, so the focus is $(h, k - a)$, and the directrix is $y = k + a$. Vertex: $(-1/2, -1)$, focus $(-1/2, -5/4)$, directrix: $y = -3/4$.

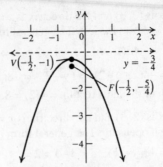

55. Complete the square to put the equation into standard form:

$$x + 2 - 48 = -3\left(y^2 - 8y + 16\right) \Rightarrow$$

$$x - 46 = -3(y-4)^2 \Rightarrow -\frac{1}{3}(x - 46) = (y - 4)^2 \Rightarrow$$

$4a = \frac{1}{3} \Rightarrow a = \frac{1}{12}$. The graph opens to the left, so the focus is $(h - a, k)$, and the directrix is $x = h + a$. Vertex: $(46, 4)$, focus $\left(\frac{551}{12}, 4\right)$,

directrix: $x = \frac{553}{12}$.

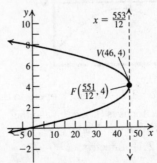

57. If the axis of symmetry is the y-axis, point P is above the vertex, and the parabola opens up. Substitute the coordinates of the vertex and P into the standard equation to find a:

$$(x - h)^2 = 4a(y - k) \Rightarrow (1 - 0)^2 = 4a(2 - 0) \Rightarrow$$

$a = \frac{1}{8} \Rightarrow$ the equation is $x^2 = \frac{1}{2}y$.

If the axis of symmetry is the x-axis, point P is to the right of the vertex, and the parabola opens to the right. Substitute the coordinates of the vertex and P into the standard equation to find a: $(y - k)^2 = 4a(x - h) \Rightarrow$

$$(2 - 0)^2 = 4a(1 - 0) \Rightarrow a = 1 \Rightarrow y^2 = 4x.$$

59. If the axis of symmetry is parallel to the *y*-axis, point *P* is above the vertex, and the parabola opens up. Substitute the coordinates of the vertex and *P* into the standard equation to find *a*: $(x-h)^2 = 4a(y-k) \Rightarrow$

$(2-0)^2 = 4a(3-1) \Rightarrow a = 1/2 \Rightarrow$ the

equation is $x^2 = 2(y-1)$.

If the axis of symmetry is parallel to the *x*-axis, point *P* is to the right of the vertex, and the parabola opens to the right. Substitute the coordinates of the vertex and *P* into the standard equation to find *a*:

$(y-k)^2 = 4a(x-h) \Rightarrow$

$(3-1)^2 = 4a(2-0) \Rightarrow a = 1/2 \Rightarrow$ the

equation is $(y-1)^2 = 2x$.

61. If the axis of symmetry is parallel to the *y*-axis, point *P* is below the vertex, and the parabola opens down. Substitute the coordinates of the vertex and *P* into the standard equation to find *a*:

$(x-h)^2 = -4a(y-k) \Rightarrow$

$(-3-(-2))^2 = -4a(0-1) \Rightarrow a = \dfrac{1}{4} \Rightarrow$ the

equation is $(x+2)^2 = -(y-1)$.

If the axis of symmetry is parallel to the *x*-axis, point *P* is to the left of the vertex, and the parabola opens to the left. Substitute the coordinates of the vertex and *P* into the standard equation to find *a*:

$(y-k)^2 = -4a(x-h) \Rightarrow$

$(0-1)^2 = -4a(-3-(-2)) \Rightarrow a = \dfrac{1}{4} \Rightarrow$ the

equation is $(y-1)^2 = -(x+2)$.

63. If the axis of symmetry is parallel to the *y*-axis, point *P* is above the vertex, and the parabola opens up. Substitute the coordinates of the vertex and *P* into the standard equation to find *a*: $(x-h)^2 = 4a(y-k) \Rightarrow$

$(0-(-1))^2 = 4a(2-1) \Rightarrow a = 1/4 \Rightarrow$ the

equation is $(x+1)^2 = y-1$.

If the axis of symmetry is parallel to the *x*-axis, point *P* is to the right of the vertex, and the parabola opens to the right. Substitute the coordinates of the vertex and *P* into the standard equation to find *a*:

$(y-k)^2 = 4a(x-h) \Rightarrow$

$(2-1)^2 = 4a(0-(-1)) \Rightarrow a = \dfrac{1}{4} \Rightarrow$ the

equation is $(y-1)^2 = x+1$.

7.2 Applying the Concepts

65. If we sketch the parabola so that the focus is on the *y*-axis, and the vertex is at (0, 0), then the points (20, 20) and (−20, 20) must lie on the parabola.

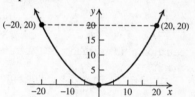

$x^2 = 4ay \Rightarrow 20^2 = 4a(20) \Rightarrow a = 5 \Rightarrow$ the

receptor should be placed at the focus 5 inches from the vertex.

67. If we sketch the parabola so that the focus is on the *y*-axis, and the vertex is at (0, 0), then the points (9, 6) and (−9, 6) must lie on the parabola.

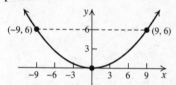

$x^2 = 4ay \Rightarrow 9^2 = 4a(6) \Rightarrow a = \dfrac{81}{24} = \dfrac{27}{8} \Rightarrow$

the heating element should be placed at the

focus $3\dfrac{3}{8}$ feet from the vertex.

69. $x = 4y^2 \Rightarrow \dfrac{x}{4} = y^2 \Rightarrow 4a = \dfrac{1}{4} \Rightarrow a = \dfrac{1}{16} \Rightarrow$

the bulb should be placed at the focus,

$\left(\dfrac{1}{16}, 0\right)$.

71. $y = 4x^2 \Rightarrow \dfrac{1}{4}y = x^2 \Rightarrow 4a = \dfrac{1}{4} \Rightarrow a = \dfrac{1}{16} \Rightarrow$

the microphone should be placed at the focus,

$\left(0, \dfrac{1}{16}\right)$.

73. If we sketch the parabola so that the focus is on the y-axis, the roadbed is the x-axis, and the vertex is at $(0, 0)$, then the points $(400, 120)$ and $(-400, 120)$ must lie on the parabola.

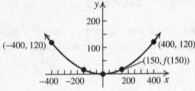

$x^2 = 4ay \Rightarrow 400^2 = 4a(120) \Rightarrow a = \dfrac{1000}{3} \Rightarrow$

the equation of the parabola is

$x^2 = \dfrac{4000}{3}y \Rightarrow y = \dfrac{3}{4000}x^2$. The point on

the cable 250 feet from the tower has coordinates $\left(150, f(150)\right)$. The length of the

cable is $f(150) = \dfrac{135}{8} = 16.875$ feet.

75. The vertex of the parabola occurs at $(35, 30)$. The parabola also passes through the origin.

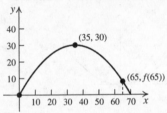

The standard form of the equation is

$(x - 35)^2 = -4a(y - 30)$. Substitute $(0, 0)$ for (x, y) and solve for a:

$(0 - 35)^2 = -4a(0 - 30) \Rightarrow a = \dfrac{245}{24}$. The

equation of the parabola is

$(x - 35)^2 = -\dfrac{245}{6}(y - 30) \Rightarrow$

$y = -\dfrac{6}{245}(x - 35)^2 + 30$.

Now find $f(65) = \dfrac{390}{49} \approx 7.96$ yards.

77. Rewrite the equation as

$10y - 500 = x^2 - 30x$, then complete the square to put the equation in standard form:

$10y - 500 + 225 = x^2 - 30x + 225 \Rightarrow$

$10(y - 27.5) = (x - 15)^2$ So, the vertex is $(15, 27.5)$. The output is 15 tons at a cost of $27.50.

7.2 Beyond the Basics

79. Solve the system $\begin{cases} 2x - 3y = -16 \\ y^2 = 16x \end{cases}$ using

substitution: $2x - 3y = -16 \Rightarrow x = \dfrac{3}{2}y - 8 \Rightarrow$

$y^2 = 16\left(\dfrac{3}{2}y - 8\right) \Rightarrow y^2 = 24y - 128 \Rightarrow$

$y^2 - 24y + 128 = 0 \Rightarrow (y - 8)(y - 16) = 0 \Rightarrow$
$y = 8$ or $y = 16$.
$2x - 3(8) + 16 = 0 \Rightarrow x = 4$
$2x - 3(16) + 16 = 0 \Rightarrow x = 16$. The line and the parabola intersect at $(16, 16)$ and $(4, 8)$. Verify using a graphing calculator.

81. The vertex is the midpoint of the perpendicular segment connecting the directrix with the focus. The slope of the line connecting the vertex and the focus is

$\dfrac{6 - 2}{-6 - (-2)} = \dfrac{4}{-4} = -1$. The equation of that

line is $y - 2 = -1(x + 2) \Rightarrow y = -x$.
The distance between the focus and the vertex

is $\sqrt{(-6 - (-2))^2 + (6 - 2)^2} = \sqrt{32} = 4\sqrt{2}$, so

the distance from the vertex to the directrix is
also $4\sqrt{2}$. Let (x, y) be the point on the
directrix on the line $y = -x - 4$. Then

$\sqrt{(x - (-2))^2 + (y - 2)^2} = 4\sqrt{2} \Rightarrow$

$(x + 2)^2 + (y - 2)^2 = 32$. Substituting

$y = -x$, we have

$(x + 2)^2 + (-x - 2)^2 = 2x^2 + 8x + 8 = 32 \Rightarrow$

$2(x^2 + 4x - 12) = 0 \Rightarrow x = -6$ or $x = 2$. The

point we are looking for is $(2, -2)$. The slope of the directrix is 1. The equation of the directrix is $y + 2 = x - 2 \Rightarrow y = x - 4$.

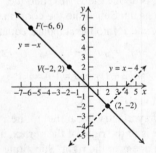

83. The latus rectum passes through the focus and is perpendicular to the axis of the parabola. The focus is the midpoint of the latus rectum, so the focus is (3, 1). The parabola opens to the right or to the left. Then $(h, 1) =$ the coordinates of the vertex, and $a = 3 - h$. Substituting the coordinates of the point (3, 5) into the general form of the equation of a parabola, we have

$(5-1)^2 = 4(3-h)(3-h) \Rightarrow h = 1$ or $h = 5$. If $h = 1$, the vertex of the parabola is (1, 1), $a = 2$, and the parabola opens to the right. Its equation is $(y-1)^2 = 8(x-1)$.

If $h = 5$, the vertex of the parabola is (5, 1), $a = -2$, and the parabola opens to the left. Its equation is $(y-1)^2 = -8(x-5)$.

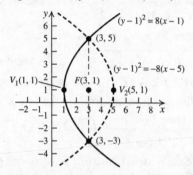

85. The parabola opens up or down, so the equation is of the form $(x-h)^2 = \pm 4a(y-k)$. Substitute the coordinates of the given points into the equation $ax^2 + bx + c = y$, and solve the system

$$\begin{cases} a(0^2) + b(0) + c = 5 \\ a(1^2) + b(1) + c = 4 \\ a(2^2) + b(2) + c = 7 \end{cases} \Rightarrow a = 2, b = -3, c = 5.$$

Now rewrite the equation $2x^2 - 3x + 5 = y$ as

$2\left(x^2 - \dfrac{3}{2}x\right) = y - 5$ and complete the square:

$2\left(x^2 - \dfrac{3}{2}x + \dfrac{9}{16}\right) = y - 5 + \dfrac{9}{8} \Rightarrow$

$\left(x - \dfrac{3}{4}\right)^2 = \dfrac{1}{2}\left(y - \dfrac{31}{8}\right).$

87. Rewrite the equation as $x^2 - 8x = -2y - 4$, and then complete the square to put the equation in standard form:

$x^2 - 8x + 16 = -2y - 4 + 16 \Rightarrow$

$(x-4)^2 = -2(y-6)$. The vertex is (4, 6).

$4a = -2 \Rightarrow a = -\dfrac{1}{2}$, so the focus is $\left(4, \dfrac{11}{2}\right)$.

The directrix is $y = \dfrac{13}{2}$. The axis of the parabola is $x = 4$.

89. Rewrite the equation as $y^2 - 6y = -3x - 15$, and then complete the square to put the equation in standard form:

$y^2 - 6y + 9 = -3x - 15 + 9 \Rightarrow$

$(y-3)^2 = -3(x+2)$. The vertex is (−2, 3).

$4a = -3 \Rightarrow a = -\dfrac{3}{4}$, so the focus is $\left(-\dfrac{11}{4}, 3\right)$,

and the directrix is $x = -\dfrac{5}{4}$. The axis is $y = 3$.

91. Step 1: Let $m =$ the slope of the tangent line. Then the equation of the tangent line is

$y - 3 = m(x - 1) \Rightarrow y = mx - m + 3$.

Step 2: $y = 3x^2 \Rightarrow mx - m + 3 = 3x^2$.

Step 3:

$mx - m + 3 = 3x^2 \Rightarrow 3x^2 - mx + m - 3 = 0 \Rightarrow$
$a = 3, b = -m, c = m - 3$
$b^2 - 4ac = (-m)^2 - 4(3)(m-3)$, so
$b^2 - 4ac = 0 \Rightarrow (-m)^2 - 4(3)(m-3) = 0 \Rightarrow$
$m^2 - 12m + 36 = 0 \Rightarrow (m-6)^2 = 0 \Rightarrow m = 6$

Step 4: The equation of the tangent line is

$y - 3 = 6(x-1) \Rightarrow y = 6x - 3$.

93. Step 1: Let $m =$ the slope of the tangent line. Then the equation of the tangent line is

$y - y_1 = m(x - x_1) \Rightarrow \dfrac{y - y_1}{m} + x_1 = x$.

Step 2: $x = 4ay^2 \Rightarrow \dfrac{y - y_1}{m} + x_1 = 4ay^2$.

$x_1 = 4ay_1^2 \Rightarrow \dfrac{y - y_1}{m} + x_1 = 4ay^2 \Rightarrow$

$\dfrac{y - y_1}{m} + 4ay_1^2 = 4ay^2$

Step 3: $\dfrac{y - y_1}{m} + 4ay_1^2 = 4ay^2 \Rightarrow$

$4ay^2 - \dfrac{y}{m} + \left(\dfrac{y_1}{m} - 4ay_1^2\right) = 0 \Rightarrow$

$4ay^2 - \dfrac{y}{m} + \left(\dfrac{y_1 - 4may_1^2}{m}\right) = 0$

$a = 4a,\ b = -\dfrac{1}{m},\ c = \dfrac{y_1}{m} - 4ay_1^2$

(continued on next page)

(*continued*)

$$b^2 - 4ac = \left(-\frac{1}{m}\right)^2 - 4(4a)\left(\frac{y_1}{m} - 4ay_1^2\right), \text{ so}$$

$$b^2 - 4ac = 0 \Rightarrow \frac{1}{m^2} - \frac{16ay_1}{m} + 64a^2y_1^2 = 0 \Rightarrow$$

$$\left(\frac{1}{m} - 8ay_1\right)^2 = 0 \Rightarrow \frac{1}{m} = 8ay_1 \Rightarrow m = \frac{1}{8ay_1}$$

Step 4: The equation of the tangent line is

$$y - y_1 = \frac{1}{8ay_1}(x - x_1) \Rightarrow$$

$$y - y_1 = \frac{1}{8ay_1}x - \frac{1}{8ay_1}x_1 \Rightarrow$$

$$y - y_1 = \frac{1}{8ay_1}x - \frac{1}{8ay_1}\left(4ay_1^2\right) \Rightarrow$$

$$y - y_1 = \frac{1}{8ay_1}x - \frac{1}{2}y_1 \Rightarrow y = \frac{1}{8ay_1}x + \frac{1}{2}y_1$$

7.2 Critical Thinking/Discussion/Writing

95. a. No. A parabola that opens to the right or left is not the graph of a function because it does not pass the vertical line test.

b. A parabola is always a function if the directrix is parallel to the x-axis, so the slope of the directrix is 0.

96.

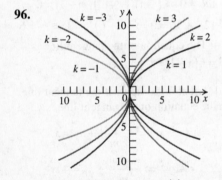

The parabola widens as $|k|$ increases.

97. False. Its axis is parallel to the x-axis.

98. For a point $P(x_1, y_1)$ on a parabola, the distance from P to the focus is the same as the distance from P to the directrix. Using the given formula, the distance from P to the directrix is $\dfrac{|3x_1 + 4y_1 - 7|}{\sqrt{3^2 + 4^2}} = \dfrac{|3x_1 + 4y_1 - 7|}{5}$.

The distance from P to the focus is

$$\sqrt{(x_1 - 4)^2 + (y_1 - 5)^2}.$$

Find the equation of the parabola by setting $\dfrac{|3x_1 + 4y_1 - 7|}{5} = \sqrt{(x_1 - 4)^2 + (y_1 - 5)^2}$ and simplifying:

$$16x^2 + 9y^2 - 24xy - 158x - 194y + 976 = 0$$

The axis is perpendicular to the directrix and passes through the focus. The slope of the directrix is $-\dfrac{3}{4}$, so the slope of the axis is $\dfrac{4}{3}$. The equation of the line through (4, 5) with slope $\dfrac{4}{3}$ is $y - 5 = \dfrac{4}{3}(x - 4) \Rightarrow y = \dfrac{4}{3}x - \dfrac{1}{3}$.

99. Using the formula given in group project 1, the distance from the vertex to the directrix is

$$\frac{|3(6) - 5(-3) + 1|}{\sqrt{3^2 + (-5)^2}} = \frac{34}{\sqrt{34}} = \sqrt{34}. \text{ The distance}$$

from the focus (x, y) to the vertex is

(1) $\sqrt{(x - 6)^2 + (y + 3)^2} = \sqrt{34}$. The distance from the focus (x, y) to the directrix is twice the distance from the focus to the vertex:

(2) $\dfrac{|3x - 5y + 1|}{\sqrt{3^2 + (-5)^2}} = 2\sqrt{34} \Rightarrow |3x - 5y + 1| = 68.$

Solve the system consisting of equations (1) and (2) to find the coordinates of the focus:

$$\begin{cases} (x - 6)^2 + (y + 3)^2 = 34 \\ 3x - 5y = 67 \end{cases} \Rightarrow$$

$$\begin{cases} x^2 + y^2 - 12x + 6y + 11 = 0 \\ x = \dfrac{5}{3}y + \dfrac{67}{3} \end{cases} \Rightarrow$$

$$y^2 + 16y + 64 = 0 \Rightarrow y = -8$$

$$3x - 5(-8) = 67 \Rightarrow x = 9$$

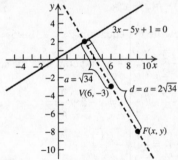

The focus is at (9, −8).

7.2 Maintaining Skills

101. $d = \sqrt{\left(-1-(-1)\right)^2 + \left(-5-2\right)^2} = \sqrt{0 + (-7)^2}$
$= \sqrt{49} = 7$

103. $(2x+3)^2 = 4x^2 + 12x + 9$

105. To find the *x*-intercepts, let $y = 0$ and solve for *x*.
$x^2 + 8x + 0^2 - 6(0) + 16 = 0 \Rightarrow$
$x^2 + 8x + 16 = 0 \Rightarrow (x+4)^2 = 0 \Rightarrow x = 4$

To find the *y*-intercept, let $x = 0$ and solve for *y*.
$0^2 + 8(0) + y^2 - 6y + 16 = 0 \Rightarrow$
$y^2 - 6y + 16 = 0$
Use the quadratic formula to solve for *y*.
$y = \dfrac{-b \pm \sqrt{b^2 - 4ac}}{2a}$
$= \dfrac{-(-6) \pm \sqrt{(-6)^2 - 4(1)(16)}}{2(1)}$
$= \dfrac{6 \pm \sqrt{36 - 64}}{2} = \dfrac{6 \pm \sqrt{-28}}{2}$, which is not
a real number.
The *x*-intercept is 4. There are no *y*-intercepts.

107. $2 + \sqrt{3x-5} = x - 1 \Rightarrow \sqrt{3x-5} = x - 3 \Rightarrow$
$3x - 5 = (x-3)^2 \Rightarrow 3x - 5 = x^2 - 6x + 9 \Rightarrow$
$0 = x^2 - 9x + 14 \Rightarrow 0 = (x-2)(x-7) \Rightarrow$
$x = 2, \ x = 7$
When we check $x = 2$ in the original equation, we have
$2 + \sqrt{3(2)-5} = 2 - 1 \Rightarrow 2 + \sqrt{1} = 2 - 1 \Rightarrow$
$3 = 1$, which is false. Thus the solution set is $\{7\}$.

109. $4x^2 - 8x + 14 = 4\left(x^2 - 2x\right) + 14$
$= 4\left(x^2 - 2x + 1\right) + (14 - 4)$
$= 4(x-1)^2 + 10$

111. The vertex of $y = (x+2)^2 + 3$ is $(-2, 3)$. The vertex of the new parabola is $(1, -2)$, so the new parabola is shifted one unit to the left and one unit down. The equation of the new parabola is $y = (x+1)^2 + 2$.

7.3   The Ellipse

7.3   Practice Problems

1. Since the foci are $(0, -8)$ and $(0, 8)$, the major axis is on the *y*-axis, and $c = 8$. One vertex is $(0, 10)$, so the other vertex is $(0, -10)$, and $a = 10$.
$b^2 = a^2 - c^2 \Rightarrow b^2 = 10^2 - 8^2 \Rightarrow b = 6$.
Thus, the equation is $\dfrac{x^2}{36} + \dfrac{y^2}{100} = 1$.

2. $4x^2 + y^2 = 16 \Rightarrow \dfrac{x^2}{4} + \dfrac{y^2}{16} = 1 \Rightarrow$
$a^2 = 16 \Rightarrow a = 4; \ b^2 = 4 \Rightarrow b = 2$

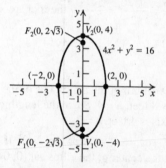

The length of the major axis $= 2a = 8$.
The length of the minor axis $= 2b = 4$.

3. Since the foci $(2, -3)$ and $(2, 5)$ lie on the vertical line $x = 2$, the ellipse is a vertical ellipse. The center of the ellipse is at
$\left(2, \dfrac{-3+5}{2}\right) = (2, 1) = (h, k)$. Since the major axis has length 10, the vertices are 5 units from the center, so $a = 5$. The foci are 4 units from the center, so $c = 4$.
$b^2 = a^2 - c^2 \Rightarrow b^2 = 5^2 - 4^2 \Rightarrow b^2 = 9$.
Thus, the equation is $\dfrac{(x-2)^2}{9} + \dfrac{(y-1)^2}{25} = 1$.

4. Rewrite the equation as
$\left(x^2 - 6x\right) + 4\left(y^2 + 2y\right) = 29$, then complete both squares and write the equation in standard form:
$\left(x^2 - 6x\right) + 4\left(y^2 + 2y\right) = 29$
$\left(x^2 - 6x + 9\right) + 4\left(y^2 + 2y + 1\right) = 29 + 9 + 4$
$(x-3)^2 + 4(y+1)^2 = 42$

(*continued on next page*)

(continued)

$$\frac{(x-3)^2}{42} + \frac{2(y+1)^2}{21} = 1$$

$$\frac{(x-3)^2}{42} + \frac{(y+1)^2}{21/2} = 1$$

The center is $(3, -1)$.

$a^2 = 42 \Rightarrow a = \sqrt{42}$, so the vertices are

$\left(3 - \sqrt{42}, -1\right)$ and

$\left(3 + \sqrt{42}, -1\right)$.

$$b^2 = a^2 - c^2 \Rightarrow \frac{21}{2} = 42 - c^2 \Rightarrow c^2 = \frac{63}{2} \Rightarrow$$

$$c = \sqrt{\frac{63}{2}} = \frac{\sqrt{126}}{2} = \frac{3\sqrt{14}}{2}, \text{ the foci are}$$

$$\left(3 - \frac{3\sqrt{14}}{2}, -1\right) \text{ and } \left(3 + \frac{3\sqrt{14}}{2}, -1\right).$$

5. Since the length of the major axis of the ellipse is 8 feet, $a = 4$. Since the length of the minor axis is 4 feet, $b = 2$.

   $2^2 = 4^2 - c^2 \Rightarrow c^2 = 12 \Rightarrow c = 2\sqrt{3}$. If we position the center of the ellipse at $(0, 0)$ and the major axis along the x-axis, the foci of the ellipse are $\left(-2\sqrt{3}, 0\right)$ and $\left(2\sqrt{3}, 0\right)$. The distance between the two foci is

   $4\sqrt{3} \approx 6.9282$ feet. Thus the stone should be positioned 6.9282 feet from the source.

7.3 Basic Concepts Skills

1. An ellipse is the set of all points in the plane, the <u>sum</u> of whose distances from two fixed points is a constant.

3. The standard equation of an ellipse with center $(0,0)$, vertices $(\pm a, 0)$, foci $(\pm c, 0)$ is

   $\frac{x^2}{a^2} + \frac{y^2}{b^2} = 1$, where $b^2 = \underline{a^2 - c^2}$.

5. True

7. $a^2 = 16 \Rightarrow$ the vertices are $(4, 0)$ and $(-4, 0)$.
   $b^2 = a^2 - c^2 \Rightarrow 4 = 16 - c^2 \Rightarrow c = 2\sqrt{3} \Rightarrow$
   the foci are $\left(2\sqrt{3}, 0\right)$ and $\left(-2\sqrt{3}, 0\right)$.

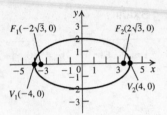

9. $a^2 = 9 \Rightarrow$ the vertices are $(3, 0)$ and $(-3, 0)$.
   $b^2 = a^2 - c^2 \Rightarrow 1 = 9 - c^2 \Rightarrow c = 2\sqrt{2} \Rightarrow$
   the foci are $\left(2\sqrt{2}, 0\right)$ and $\left(-2\sqrt{2}, 0\right)$.

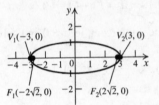

11. $a^2 = 25 \Rightarrow$ the vertices are $(5, 0)$ and $(-5, 0)$.
    $b^2 = a^2 - c^2 \Rightarrow 16 = 25 - c^2 \Rightarrow c = 3 \Rightarrow$
    the foci are $(3, 0)$ and $(-3, 0)$.

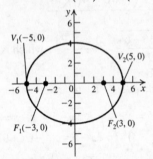

13. $a^2 = 36 \Rightarrow$ the vertices are $(0, 6)$ and $(0, -6)$.
    $b^2 = a^2 - c^2 \Rightarrow 16 = 36 - c^2 \Rightarrow c = 2\sqrt{5} \Rightarrow$
    the foci are $\left(0, 2\sqrt{5}\right)$ and $\left(0, -2\sqrt{5}\right)$.

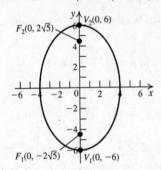

15. A circle with radius 2, centered at the origin.

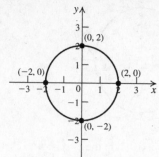

17. $x^2 + 4y^2 = 4 \Rightarrow \dfrac{x^2}{4} + y^2 = 1.$ $a^2 = 4 \Rightarrow$ the vertices are $(2, 0)$ and $(-2, 0)$.
$b^2 = a^2 - c^2 \Rightarrow 1 = 4 - c^2 \Rightarrow c = \sqrt{3} \Rightarrow$ the foci are $\left(\sqrt{3}, 0\right)$ and $\left(-\sqrt{3}, 0\right)$.

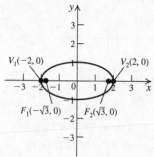

19. $9x^2 + 4y^2 = 36 \Rightarrow \dfrac{x^2}{4} + \dfrac{y^2}{9} = 1.$ $a^2 = 9 \Rightarrow$ the vertices are $(0, 3)$ and $(0, -3)$.
$b^2 = a^2 - c^2 \Rightarrow 4 = 9 - c^2 \Rightarrow c = \pm\sqrt{5} \Rightarrow$ the foci are $\left(0, \sqrt{5}\right)$ and $\left(0, -\sqrt{5}\right)$.

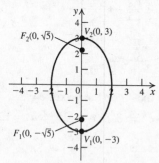

21. $3x^2 + 4y^2 = 12 \Rightarrow \dfrac{x^2}{4} + \dfrac{y^2}{3} = 1.$ $a^2 = 4 \Rightarrow$ the vertices are $(2, 0)$ and $(-2, 0)$.
$b^2 = a^2 - c^2 \Rightarrow 3 = 4 - c^2 \Rightarrow c = 1 \Rightarrow$ the foci are $(1, 0)$ and $(-1, 0)$.

23. $5x^2 - 10 = -2y^2 \Rightarrow 5x^2 + 2y^2 = 10 \Rightarrow$
$\dfrac{x^2}{2} + \dfrac{y^2}{5} = 1.$ $a^2 = 5 \Rightarrow$ the vertices are $\left(0, \sqrt{5}\right)$ and $\left(0, -\sqrt{5}\right)$.
$b^2 = a^2 - c^2 \Rightarrow 2 = 5 - c^2 \Rightarrow c = \sqrt{3} \Rightarrow$ the foci are $\left(0, \sqrt{3}\right)$ and $\left(0, -\sqrt{3}\right)$.

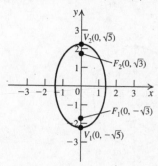

25. $2x^2 + 3y^2 = 7 \Rightarrow \dfrac{x^2}{7/2} + \dfrac{y^2}{7/3} = 1 \Rightarrow a^2 = \dfrac{7}{2} \Rightarrow$
the vertices are $\left(\dfrac{\sqrt{14}}{2}, 0\right)$ and $\left(-\dfrac{\sqrt{14}}{2}, 0\right)$.
$b^2 = a^2 - c^2 \Rightarrow \dfrac{7}{3} = \dfrac{7}{2} - c^2 \Rightarrow c = \dfrac{\sqrt{42}}{6} \Rightarrow$
the foci are $\left(\dfrac{\sqrt{42}}{6}, 0\right)$ and $\left(-\dfrac{\sqrt{42}}{6}, 0\right)$.

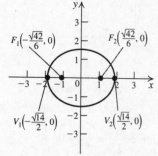

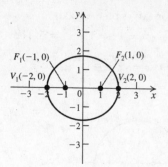

27. The major axis is on the x-axis.
$b^2 = a^2 - c^2 \Rightarrow b^2 = 3^2 - 1^2 = 8$. The
equation is $\dfrac{x^2}{9} + \dfrac{y^2}{8} = 1$.

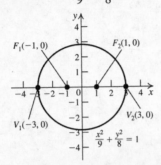

29. The major axis is on the y-axis.
$b^2 = a^2 - c^2 \Rightarrow b^2 = 4^2 - 2^2 = 12$. The
equation is $\dfrac{x^2}{12} + \dfrac{y^2}{16} = 1$.

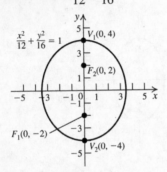

31. The major axis is on the x-axis.
$b^2 = a^2 - c^2 \Rightarrow 3^2 = a^2 - 4^2 \Rightarrow a^2 = 25$.
The equation is $\dfrac{x^2}{25} + \dfrac{y^2}{9} = 1$.

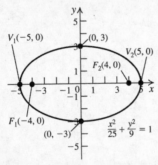

33. The major axis is on the y-axis.
$b^2 = a^2 - c^2 \Rightarrow 4^2 = a^2 - 2^2 \Rightarrow a^2 = 20$.
The equation is $\dfrac{x^2}{16} + \dfrac{y^2}{20} = 1$.

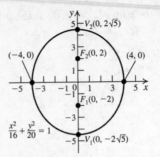

35. The major axis is on the y-axis.
Major axis length $= 10 \Rightarrow a = 5$ and minor
axis length $= 6 \Rightarrow b = 3$. The equation is
$\dfrac{x^2}{9} + \dfrac{y^2}{25} = 1$.

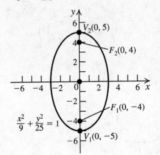

37. The major axis is on the x-axis. $a = 6$.
$b^2 = a^2 - c^2 \Rightarrow b^2 = 6^2 - 3^2 = 27$.
The equation is $\dfrac{x^2}{36} + \dfrac{y^2}{27} = 1$.

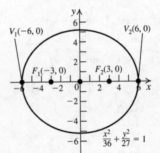

39. The major axis is on the y-axis. $c = 2$.
$b^2 = a^2 - c^2 \Rightarrow 3^2 = a^2 - 2^2 \Rightarrow a^2 = 13$.
The equation is $\dfrac{x^2}{9} + \dfrac{y^2}{13} = 1$.

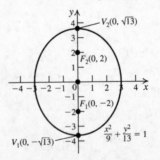

41. Center: (1, 1); $a = 3 \Rightarrow$ vertices: (1, 4) and (1, −2). $b^2 = a^2 - c^2 \Rightarrow 4 = 9 - c^2 \Rightarrow$ $c = \sqrt{5}$. The foci are $\left(1, 1 + \sqrt{5}\right)$ and $\left(1, 1 - \sqrt{5}\right)$.

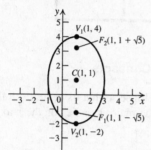

$a = \sqrt{5} \Rightarrow$ vertices: $\left(-3 + \sqrt{5}, 1\right)$ and $\left(-3 - \sqrt{5}, 1\right)$. $b^2 = a^2 - c^2 \Rightarrow 4 = 5 - c^2 \Rightarrow$ $c = \sqrt{5}$. The foci are $(-2, 1)$ and $(-4, 1)$.

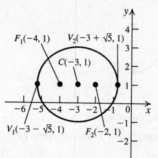

43. Center: (0, −3); $a = 4 \Rightarrow$ vertices: (4, −3) and (−4, −3). $b^2 = a^2 - c^2 \Rightarrow 4 = 16 - c^2 \Rightarrow$ $c = 2\sqrt{3}$. The foci are $\left(2\sqrt{3}, -3\right)$ and $\left(-2\sqrt{3}, -3\right)$.

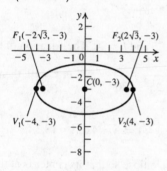

45. $3(x-1)^2 + 4(y+2)^2 = 12 \Rightarrow$ $\dfrac{(x-1)^2}{4} + \dfrac{(y+2)^2}{3} = 1 \Rightarrow$ center: (1, −2). $a = 2 \Rightarrow$ vertices: (3, −2) and (−1, −2). $b^2 = a^2 - c^2 \Rightarrow 3 = 4 - c^2 \Rightarrow c = 1$. The foci are (2, −2), (0, −2).

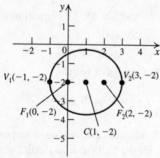

47. $4(x+3)^2 + 5(y-1)^2 = 20 \Rightarrow$ $\dfrac{(x+3)^2}{5} + \dfrac{(y-1)^2}{4} = 1 \Rightarrow$ center: (−3, 1).

49. Rewrite the equation as $5(x^2 + 2x) + 9(y^2 - 4y) = 4$, then complete both squares:
$5(x^2 + 2x + 1) + 9(y^2 - 4y + 4) = 4 + 5 + 36 \Rightarrow$
$5(x+1)^2 + 9(y-2)^2 = 45 \Rightarrow$
$\dfrac{(x+1)^2}{9} + \dfrac{(y-2)^2}{5} = 1 \Rightarrow$ center: $(-1, 2)$.
$a = 3 \Rightarrow$ vertices: $(2, 2)$ and $(-4, 2)$.
$b^2 = a^2 - c^2 \Rightarrow 5 = 9 - c^2 \Rightarrow c = 2$.
The foci are (1, 2) and (−3, 2).

51. Rewrite the equation as $9(x^2 + 4x) + 5(y^2 - 8y) = -71$, then complete both squares: $9(x^2 + 4x + 4) + 5(y^2 - 8y + 16)$
$= -71 + 36 + 80 \Rightarrow$
$9(x+2)^2 + 5(y-4)^2 = 45 \Rightarrow$
$\dfrac{(x+2)^2}{5} + \dfrac{(y-4)^2}{9} = 1 \Rightarrow$ center: $(-2, 4)$.
$a = 3 \Rightarrow$ vertices: $(-2, 7)$ and $(-2, 1)$.
$b^2 = a^2 - c^2 \Rightarrow 5 = 9 - c^2 \Rightarrow c = 2$. The foci are (−2, 2) and (−2, 6).

(*continued on next page*)

(continued)

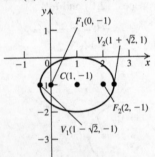

53. Rewrite the equation as

$(x^2 - 2x) + 2(y^2 + 2y) = -1$, then complete

both squares:

$(x^2 - 2x + 1) + 2(y^2 + 2y + 1) = -1 + 1 + 2 \Rightarrow$

$(x-1)^2 + 2(y+1)^2 = 2 \Rightarrow$

$\dfrac{(x-1)^2}{2} + (y+1)^2 = 1 \Rightarrow$ center: $(1, -1)$.

$a = \sqrt{2} \Rightarrow$ vertices: $\left(1 + \sqrt{2}, -1\right)$ and $\left(1 - \sqrt{2}, -1\right)$.

$b^2 = a^2 - c^2 \Rightarrow 1 = 2 - c^2 \Rightarrow c = 1$. The foci

are $(2, -1)$ and $(0, -1)$.

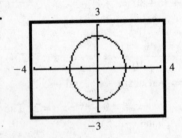

55. Rewrite the equation as

$2(x^2 - 2x) + 9(y^2 + 2y) = -12$, then complete

both squares:

$2(x^2 - 2x + 1) + 9(y^2 + 2y + 1) = -12 + 2 + 9 \Rightarrow$

$2(x-1)^2 + 9(y+1)^2 = -1 \Rightarrow$ there is no

graph.

57.

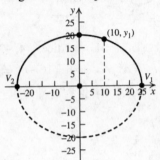

59.

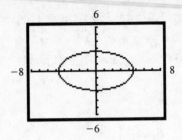

7.3 Applying the Concepts

61. Sketch the ellipse so that its center is at the
origin and the major axis lies on the *x*-axis.

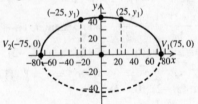

Then $a = 25$ and $b = 20$. The equation of the

ellipse is $\dfrac{x^2}{25^2} + \dfrac{y^2}{20^2} = 1$. Let $x = 10$, then

solve for *y*:

$\dfrac{10^2}{25^2} + \dfrac{y^2}{20^2} = 1 \Rightarrow y = 4\sqrt{21} \approx 18.3$ ft.

63. Sketch the ellipse so that its center is at the
origin and the major axis lies on the *x*-axis.

Then $a = 75$ and $b = 45$. The equation of the

ellipse is $\dfrac{x^2}{75^2} + \dfrac{y^2}{45^2} = 1$.

Let $x = 25$, then solve for *y*:

$\dfrac{25^2}{75^2} + \dfrac{y^2}{45^2} = 1 \Rightarrow y = 30\sqrt{2} \approx 42.4$ m.

65. The bet should not be accepted. The pool
shark can hit the ball from any point straight
into the pocket, or he can shoot it through the
other focus and it will fall into the pocket
because of the reflecting property.

67. $a = 125$ and $b = 80$, so $b^2 = a^2 - c^2 \Rightarrow$
$80^2 = 125^2 - c^2 \Rightarrow c = 15\sqrt{41} \approx 96.05$ feet
from the center or $125 - 96.05 \approx 29$ feet from
the wall on the opposite side.

69. Using the hints given in exercise 58, we have
$a = \dfrac{91.38 + 94.54}{2} = 92.96 \Rightarrow a^2 = 8641.5616$
and $c = 92.96 - 91.38 = 1.58$.
$b^2 = 92.96^2 - 1.58^2 = 8639.0652$. The
equation is $\dfrac{x^2}{8641.5616} + \dfrac{y^2}{8639.0652} = 1$.

71. Using the hints given in exercise 58, we have
$a = \dfrac{837.05 + 936.37}{2} = 886.71 \Rightarrow$
$a^2 = 786,254.6241$ and
$c = 886.71 - 837.05 = 49.66$.
$b^2 = 886.71^2 - 49.66^2 = 783,788.5085$.
The equation is
$\dfrac{x^2}{786,254.6241} + \dfrac{y^2}{783,788.5085} = 1$.

73. $a = \dfrac{5.39 \times 10^9}{2} = 2.695 \times 10^9$ and
$b = \dfrac{1.36 \times 10^9}{2} = 6.8 \times 10^8$. So
$b^2 = a^2 - c^2 \Rightarrow$
$(6.8 \times 10^8)^2 = (2.695 \times 10^9)^2 - c^2 \Rightarrow$
$c \approx 2.6078 \times 10^9$. The perihelion $= a - c$
$\approx 8.72 \times 10^7$ km. The aphelion $=$
$a + c \approx 5.3028 \times 10^9$ km.

7.3 Beyond the Basics

75. The center is $(0, 0)$ and $a = 5$, so the equation
is either $\dfrac{x^2}{5^2} + \dfrac{y^2}{b^2} = 1$ or $\dfrac{x^2}{b^2} + \dfrac{y^2}{5^2} = 1$.
Substitute the coordinates of the point for x
and y, then solve for b:
$\dfrac{(-3)^2}{5^2} + \dfrac{(16/5)^2}{b^2} = 1 \Rightarrow 9b^2 + 256 = 25b^2 \Rightarrow$
$b^2 = 16 \Rightarrow b = 4$.
$\dfrac{(-3)^2}{b^2} + \dfrac{(16/5)^2}{5^2} = 1 \Rightarrow$
$5625 + 256b^2 = 625b^2 \Rightarrow b^2 = \dfrac{625}{41}$.

The equations are either $\dfrac{x^2}{25} + \dfrac{y^2}{16} = 1$ or
$\dfrac{x^2}{625/41} + \dfrac{y^2}{25} = 1$.

77. Substitute the coordinates of the points into
the equation of the ellipse and then solve the
system: $\begin{cases} \dfrac{2^2}{b^2} + \dfrac{1^2}{a^2} = 1 \\ \dfrac{1^2}{b^2} + \dfrac{(-3)^2}{a^2} = 1 \end{cases}$.

Let $u = \dfrac{1}{a^2}$ and $v = \dfrac{1}{b^2}$. Then
$\begin{cases} \dfrac{2^2}{b^2} + \dfrac{1^2}{a^2} = 1 \\ \dfrac{1^2}{b^2} + \dfrac{(-3)^2}{a^2} = 1 \end{cases} \Rightarrow \begin{cases} 4v + u = 1 \\ v + 9u = 1 \end{cases} \Rightarrow$

$u = \dfrac{3}{35}, v = \dfrac{8}{35} \Rightarrow \dfrac{1}{a^2} = \dfrac{3}{35} \Rightarrow a = \dfrac{\sqrt{105}}{3}$ and
$\dfrac{1}{b^2} = \dfrac{8}{35} \Rightarrow b = \dfrac{\sqrt{70}}{4}$.

The length of the major axis is $\dfrac{2\sqrt{105}}{3}$. The
length of the minor axis is $\dfrac{\sqrt{70}}{2}$.

79. When $e = 0$, the ellipse becomes a circle.

81. $20x^2 + 36y^2 = 720 \Rightarrow \dfrac{x^2}{36} + \dfrac{y^2}{20} = 1$
$a = 6, b^2 = a^2 - c^2 \Rightarrow 20 = 36 - c^2 \Rightarrow c = 4$
$e = \dfrac{4}{6} = \dfrac{2}{3}$

83. $a = 5, b^2 = a^2 - c^2 \Rightarrow 9 = 25 - c^2 \Rightarrow c = 4$
$e = \dfrac{4}{5}$

85. $2c = 4 \Rightarrow c = 2; e = \dfrac{c}{a} \Rightarrow \dfrac{1}{2} = \dfrac{2}{a} \Rightarrow a = 4$
$b^2 = a^2 - c^2 \Rightarrow b^2 = 16 - 4 = 12$
The equation is $\dfrac{x^2}{16} + \dfrac{y^2}{12} = 1$.

87. The distance from P to the point $(4, 0)$ is $\sqrt{(x-4)^2 + y^2}$, and the distance from P to the line is $|x-16|$. So, the equation of the path of P is $\sqrt{(x-4)^2 + y^2} = \frac{1}{2}|x-16| \Rightarrow$

$(x-4)^2 + y^2 = \frac{1}{4}(x-16)^2 \Rightarrow$

$\frac{3x^2}{4} + y^2 = 48 \Rightarrow \frac{x^2}{64} + \frac{y^2}{48} = 1 \Rightarrow$

$a = 8, 48 = 64 - c^2 \Rightarrow c = 4 \Rightarrow e = \frac{1}{2}.$

89. Assume that the center of the ellipse is at the origin, the major axis is on the x-axis, and the minor axis is on the y-axis. So the equation is $\frac{x^2}{a^2} + \frac{y^2}{b^2} = 1$. The length of the latus rectum is the difference in the y-coordinates of the two points so the ellipse with x-coordinate c. Letting $x = c$,

$\frac{c^2}{a^2} + \frac{y^2}{b^2} = 1 \Rightarrow \frac{a^2 - b^2}{a^2} + \frac{y^2}{b^2} = 1 \Rightarrow$

$y = \pm \frac{b^2}{a}$. So the length of the latus rectum is $\frac{2b^2}{a}$.

91. Using substitution, we have

$\begin{cases} x + 3y = -2 \\ 4x^2 + 3y^2 = 7 \end{cases} \Rightarrow \begin{cases} x = -3y - 2 \\ 4x^2 + 3y^2 = 7 \end{cases} \Rightarrow$

$4(-3y-2)^2 + 3y^2 = 7 \Rightarrow$

$39y^2 + 48y + 9 = 0 \Rightarrow y = -1 \text{ or } y = -\frac{3}{13}$

$x + 3(-1) = -2 \Rightarrow x = 1$

$x + 3\left(-\frac{3}{13}\right) = -2 \Rightarrow x = -\frac{17}{13}$

The points of intersection are $\left(-\frac{17}{13}, -\frac{3}{13}\right)$ and $(1, -1)$.

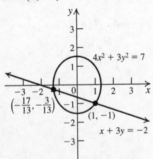

93. Using elimination, we have

$\begin{cases} x^2 + y^2 = 20 \\ 9x^2 + y^2 = 36 \end{cases} \Rightarrow 8x^2 = 16 \Rightarrow x = \pm\sqrt{2}$

$2 + y^2 = 20 \Rightarrow y = \pm 3\sqrt{2}$. The points of intersection are $\left(-\sqrt{2}, -3\sqrt{2}\right), \left(-\sqrt{2}, 3\sqrt{2}\right),$ $\left(\sqrt{2}, -3\sqrt{2}\right),$ and $\left(\sqrt{2}, 3\sqrt{2}\right)$.

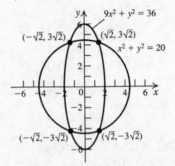

95. $\frac{x^2}{a^2} + \frac{y^2}{b^2} = 1 \Rightarrow \frac{x^2}{(5/2)^2} + \frac{y^2}{5^2} = 1 \Rightarrow$

$\frac{4x^2}{25} + \frac{y^2}{25} = 1 \Rightarrow 4x^2 + y^2 = 25 \Rightarrow$

$y^2 = 25 - 4x^2$

The tangent line equation is $y = m(x-2) + 3$.

Solve $25 - 4x^2 = [m(x-2)+3]^2$ to find m.

$25 - 4x^2 = m^2(x-2)^2 + 6m(x-2) + 9 \Rightarrow$

$(4 + m^2)x^2 + (6m - 4m^2)x$
$\qquad + (4m^2 - 12m - 16) = 0$

For a unique solution, the discriminant must be zero.

$(6m - 4m^2)^2 - 4(4 + m^2)(4m^2 - 12m - 16) = 0$

$\qquad\qquad 36m^2 + 192m + 256 = 0$

$\qquad\qquad 9m^2 + 48m + 64 = 0$

$\qquad\qquad (3m + 8)^2 = 0$

$\qquad\qquad m = -\frac{8}{3}$

The tangent line as equation

$y = -\frac{8}{3}(x-2) + 3 = -\frac{8}{3}x + \frac{25}{3}.$

7.3 Critical Thinking/Discussion/Writing

97. There are four circles, one in each quadrant:
$(x-3)^2 + (y-3)^2 = 9, (x+3)^2 + (y-3)^2 = 9,$
$(x-3)^2 + (y+3)^2 = 9,$ and
$(x+3)^2 + (y+3)^2 = 9.$

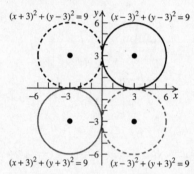

98. There are eight ellipses, two in each quadrant:
$$\frac{(x-3)^2}{9} + \frac{(y-2)^2}{4} = 1, \frac{(x-2)^2}{4} + \frac{(y-3)^2}{9} = 1,$$
$$\frac{(x+3)^2}{9} + \frac{(y-2)^2}{4} = 1, \frac{(x+2)^2}{4} + \frac{(y-3)^2}{9} = 1,$$
$$\frac{(x+3)^2}{9} + \frac{(y+2)^2}{4} = 1, \frac{(x+2)^2}{4} + \frac{(y+3)^2}{9} = 1,$$
$$\frac{(x-3)^2}{9} + \frac{(y+2)^2}{4} = 1, \frac{(x-2)^2}{4} + \frac{(y+3)^2}{9} = 1$$

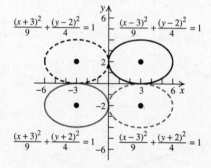

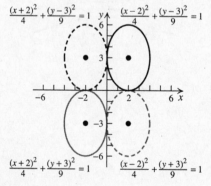

7.3 Maintaining Skills

99.
$$x^2 + 6x + y^2 - 2y - 5 = 0$$
$$\left(x^2 + 6x\right) + \left(y^2 - 2y\right) = 5$$
$$\left(x^2 + 6x + 9\right) + \left(y^2 - 2y + 1\right) = 5 + 9 + 1$$
$$(x+3)^2 + (y-1)^2 = 15$$
Center: $(-3, 1)$
Radius: $\sqrt{15}$

101. To find the *x*-intercepts, let $y = 0$ and solve for *x*.
$$x^2 + 0^2 = 25 \Rightarrow x^2 = 25 \Rightarrow x = \pm 5$$
To find the *y*-intercept, let $x = 0$ and solve for *y*.
$$0^2 + y^2 = 25 \Rightarrow y^2 = 25 \Rightarrow y = \pm 25$$
The *x*-intercepts are −5 and 5. The *y*-intercepts are −5 and 5.

103. $y = -4x + 11$

105. $m = \dfrac{-5 - 7}{-1 - (-3)} = \dfrac{-12}{2} = -6$
$$y - 7 = -6\left(x - (-3)\right) \Rightarrow y - 7 = -6\left(x + 3\right) \Rightarrow$$
$$y = -6x - 18 + 7 = -6x - 11$$

Review Section 3.6 for the procedures to find the asymptotes of a rational function.

107. $\dfrac{2x^2 - 3x + 1}{x - 2}$
$x - 2 = 0 \Rightarrow x = 2 \Rightarrow$ the vertical asymptote is $x = 2$. Since the largest exponent in the numerator is greater than the largest exponent in the denominator, there is no horizontal asymptote.

$$\begin{array}{r} 2x+1 \\ x-2\overline{\smash{\big)}\,2x^2-3x+1} \\ \underline{2x^2-4x} \\ x+1 \\ \underline{x-2} \\ -3 \end{array}$$
The oblique asymptote is $y = 2x + 1$.

7.4 The Hyperbola

7.4 Practice Problems

1. The transverse axis is on the y-axis.

2. Rewrite the equation in standard form to determine a and b. $x^2 - 4y^2 = 8 \Rightarrow$

$$\frac{x^2}{8} - \frac{y^2}{2} = 1 \Rightarrow a^2 = 8 \Rightarrow a = 2\sqrt{2} \text{ and}$$

$b^2 = 2 \Rightarrow b = \sqrt{2}.$

The transverse axis of the hyperbola is along the x-axis, so the vertices are $\left(-2\sqrt{2}, 0\right)$ and $\left(2\sqrt{2}, 0\right)$.

To find the foci, we need c:

$c^2 = a^2 + b^2 \Rightarrow c^2 = 8 + 2 \Rightarrow c = \sqrt{10}$. The foci are $\left(-\sqrt{10}, 0\right)$ and $\left(\sqrt{10}, 0\right)$.

3. Since the foci of the hyperbola, $(0, -6)$ and $(0, 6)$ lie on the y-axis, the transverse axis also lies on the y-axis, and $c = 6$. The center of the hyperbola is $(0, 0)$, so the standard form of this hyperbola is $\frac{y^2}{a^2} - \frac{x^2}{b^2} = 1$.

The vertices are $(0, -4)$ and $(0, 4)$, so $a = 4$.
$c^2 = a^2 + b^2 \Rightarrow 6^2 = 4^2 + b^2 \Rightarrow b^2 = 20$.

The equation is $\frac{y^2}{16} - \frac{x^2}{20} = 1$.

4. The hyperbola $\frac{y^2}{4} - \frac{x^2}{9} = 1$ is of the form

$\frac{y^2}{a^2} - \frac{x^2}{b^2} = 1$, so $b = 3$ and $a = 2$. The

asymptotes are of the form $y = \frac{a}{b}x$ and

$y = -\frac{a}{b}x$. They are $y = \frac{2}{3}x$ and $y = -\frac{2}{3}x$.

5. a. $25x^2 - 4y^2 = 100 \Rightarrow \frac{x^2}{4} - \frac{y^2}{25} = 1 \Rightarrow a = 2,$

$b = 5.$
$c^2 = a^2 + b^2 \Rightarrow c^2 = 4 + 25 \Rightarrow c = \sqrt{29}.$
The vertices are $(2, 0)$ and $(-2, 0)$. The endpoints of the conjugate axis are $(0, 5)$ and $(0, -5)$, and the foci are $\left(-\sqrt{29}, 0\right)$ and $\left(\sqrt{29}, 0\right)$.

The asymptotes are $y = \frac{b}{a}x = \frac{5}{2}x$ and

$y = -\frac{b}{a}x = -\frac{5}{2}x.$

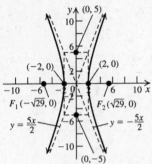

b. $9y^2 - x^2 = 1 \Rightarrow \frac{y^2}{1/9} - \frac{x^2}{1} = 1 \Rightarrow a^2 = \frac{1}{9} \Rightarrow$

$a = \frac{1}{3}$ and $b = 1$.

$c^2 = a^2 + b^2 \Rightarrow c^2 = \frac{1}{9} + 1 \Rightarrow c = \frac{\sqrt{10}}{3}.$

The vertices are $\left(0, \frac{1}{3}\right)$ and $\left(0, -\frac{1}{3}\right)$.

The endpoints of the conjugate axis are $(1, 0)$ and $(-1, 0)$, and the foci are

$\left(0, \frac{\sqrt{10}}{3}\right)$ and $\left(0, -\frac{\sqrt{10}}{3}\right)$.

The asymptotes are $y = \frac{a}{b}x = \frac{1/3}{1}x = \frac{1}{3}x$

and $y = -\frac{a}{b}x = -\frac{1/3}{1}x = -\frac{1}{3}x.$

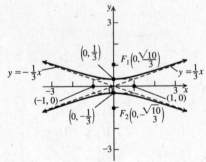

6. a. $\dfrac{(x+1)^2}{4} - \dfrac{(y-1)^2}{16} = 1 \Rightarrow a = 2,\ b = 4,$ and

center $(-1, 1)$. The vertices are

$(-1-2, 1) = (-3, 1)$ and $(-1+2, 1) = (1, 1).$

The endpoints of the conjugate axes are

$(-1, 1-4) = (-1, -3)$ and

$(-1, 1+4) = (-1, 5).$ The asymptotes are

$$y - k = \pm\frac{b}{a}(x-h) \Rightarrow y - 1 = \pm 2(x+1).$$

$$c^2 = a^2 + b^2 \Rightarrow c^2 = 4 + 16 \Rightarrow c = \pm\sqrt{20}$$

The foci are $\left(-\sqrt{20}-1,\ 1\right)$ and

$\left(\sqrt{20}-1,\ 1\right).$

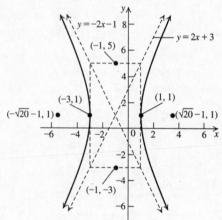

b. $\dfrac{(y-1)^2}{4} - \dfrac{(x+1)^2}{9} = 1 \Rightarrow a = 2,\ b = 3,$ and

center $(-1, 1).$ The vertices are

$(-1, 1-2) = (-1, -1)$ and

$(-1, 1+2) = (-1, 3).$ The endpoints of the

conjugate axes are $(-1-3, 1) = (-4, 1)$ and

$(-1+3, 1) = (2, 1)$ The asymptotes are

$$y - k = \pm\frac{a}{b}(x-h) \Rightarrow y - 1 = \pm\frac{2}{3}(x+1).$$

$$c^2 = a^2 + b^2 \Rightarrow c^2 = 4 + 9 \Rightarrow c = \pm\sqrt{13}$$

The foci are $\left(-1,\ 1+\sqrt{13}\right)$ and

$\left(-1,\ 1-\sqrt{13}\right).$

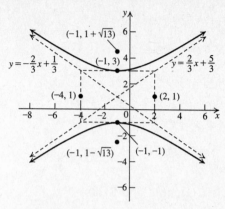

7. Complete the square to put the equation in standard form:

$$\left(x^2 - 2x\right) - 4\left(y^2 - 4y\right) = 20$$

$$\left(x^2 - 2x + 1\right) - 4\left(y^2 - 4y + 4\right) = 20 + 1 - 16$$

$$(x-1)^2 - 4(y-2)^2 = 5$$

$$\frac{(x-1)^2}{5} - \frac{4(y-2)^2}{5} = 1$$

$$\frac{(x-1)^2}{5} - \frac{(y-2)^2}{5/4} = 1$$

The center is $(1, 2),\ a = \sqrt{5}$ and $b = \dfrac{\sqrt{5}}{2}.$

The vertices are $\left(1+\sqrt{5},\ 2\right)$ and $\left(1-\sqrt{5},\ 2\right).$

The endpoints of the conjugate axis are

$\left(1,\ 2+\dfrac{\sqrt{5}}{2}\right)$ and $\left(1,\ 2-\dfrac{\sqrt{5}}{2}\right).$

The asymptotes are $y - k = \pm\dfrac{b}{a}(x-h) \Rightarrow$

$$y - 2 = \pm\frac{\sqrt{5}/2}{\sqrt{5}}(x-1) \Rightarrow y - 2 = \pm\frac{1}{2}(x-1).$$

$$c^2 = a^2 + b^2 \Rightarrow c^2 = 5 + \frac{5}{4} = \frac{25}{4} \Rightarrow c = \frac{5}{2}.$$

The foci are $\left(1+\dfrac{5}{2},\ 2\right) = \left(\dfrac{7}{2},\ 2\right)$ and

$\left(1-\dfrac{5}{2},\ 2\right) = \left(-\dfrac{3}{2},\ 2\right).$

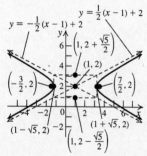

8. The hyperbola has foci $(-150, 0)$ and $(150, 0)$, so $c = 150$. The difference of the distances from the boat to A and B is 260 miles, so $a = 130$.

$$c^2 = a^2 + b^2 \Rightarrow$$
$$150^2 = 130^2 + b^2 \Rightarrow b^2 = 5600.$$
The equation of the hyperbola is
$$\frac{x^2}{130^2} - \frac{y^2}{5600} = 1 \Rightarrow \frac{x^2}{16,900} - \frac{y^2}{5600} = 1.$$

7.4 Basic Concepts and Skills

1. A hyperbola is a set of all points in the plane, the absolute value of the <u>difference</u> of whose distances form two fixed points is constant.

3. The standard equation of the hyperbola with center $(0, 0)$, vertices $(\pm a, 0)$, and foci $(\pm c, 0)$ is $\dfrac{x^2}{a^2} - \dfrac{y^2}{b^2} = 1$, where $b^2 = \underline{c^2 - a^2}$.

5. True

7. g **9.** h

11. d **13.** b

15. $a^2 = 1 \Rightarrow$ the vertices are $(1, 0)$ and $(-1, 0)$.
$$c^2 = a^2 + b^2 \Rightarrow c^2 = 1 + 4 \Rightarrow c = \sqrt{5} \Rightarrow$$
the foci are $\left(\sqrt{5}, 0\right)$ and $\left(-\sqrt{5}, 0\right)$.

Transverse axis: x-axis. Hyperbola opens left and right. Vertices of the fundamental rectangle: $(1, 2), (-1, 2), (-1, -2), (1, -2)$.
Asymptotes: $y = \pm 2x$.

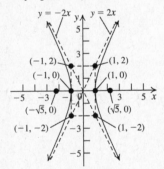

17. $x^2 - y^2 = -1 \Rightarrow y^2 - x^2 = 1 \Rightarrow a^2 = 1 \Rightarrow$ the vertices are $(0, 1)$ and $(0, -1)$.
$$c^2 = a^2 + b^2 \Rightarrow c^2 = 1 + 1 \Rightarrow c = \sqrt{2} \Rightarrow$$
the foci are $\left(0, \sqrt{2}\right)$ and $\left(0, -\sqrt{2}\right)$.

Transverse axis: y-axis. Hyperbola opens up and down. Vertices of the fundamental rectangle: $(1, 1), (-1, 1), (-1, -1), (1, -1)$.
Asymptotes: $y = \pm x$.

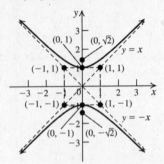

19. $9y^2 - x^2 = 36 \Rightarrow \dfrac{y^2}{4} - \dfrac{x^2}{36} = 1 \Rightarrow a^2 = 4 \Rightarrow$
the vertices are $(0, 2)$ and $(0, -2)$.
$$c^2 = a^2 + b^2 \Rightarrow c^2 = 4 + 36 \Rightarrow c = 2\sqrt{10} \Rightarrow$$
the foci are $\left(0, 2\sqrt{10}\right)$ and $\left(0, -2\sqrt{10}\right)$.

Transverse axis: y-axis. Hyperbola opens up and down. Vertices of the fundamental rectangle: $(6, 2), (-6, 2), (-6, -2), (6, -2)$.
Asymptotes: $y = \pm \dfrac{a}{b}x = \pm \dfrac{1}{3}x$

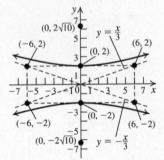

21. $4x^2 - 9y^2 - 36 = 0 \Rightarrow \dfrac{x^2}{9} - \dfrac{y^2}{4} = 1 \Rightarrow$
$a^2 = 9 \Rightarrow$ the vertices are $(3, 0)$ and $(-3, 0)$.
$$c^2 = a^2 + b^2 \Rightarrow c^2 = 9 + 4 \Rightarrow c = \sqrt{13} \Rightarrow$$
the foci are $\left(\sqrt{13}, 0\right)$ and $\left(-\sqrt{13}, 0\right)$.

Transverse axis: x-axis. Hyperbola opens left and right. Vertices of the fundamental rectangle: $(3, 2), (-3, 2), (-3, -2), (3, -2)$.
Asymptotes: $y = \pm \dfrac{b}{a}x = \pm \dfrac{2}{3}x$

(continued on next page)

(*continued*)

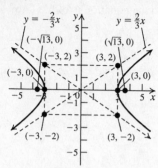

23. $y = \pm\sqrt{4x^2 + 1} \Rightarrow y^2 - \dfrac{x^2}{1/4} = 1 \Rightarrow a^2 = 1 \Rightarrow$

the vertices are (0, 1) and (0, –1).

$c^2 = a^2 + b^2 \Rightarrow c^2 = 1 + \dfrac{1}{4} \Rightarrow c = \dfrac{\sqrt{5}}{2} \Rightarrow$ the

foci are $\left(0, \dfrac{\sqrt{5}}{2}\right)$ and $\left(0, -\dfrac{\sqrt{5}}{2}\right)$. Transverse

axis: *y*-axis. Hyperbola opens up and down.
Vertices of the fundamental rectangle:

$\left(\dfrac{1}{2}, 1\right), \left(-\dfrac{1}{2}, 1\right), \left(-\dfrac{1}{2}, -1\right)$, and $\left(\dfrac{1}{2}, -1\right)$.

Asymptotes: $y = \pm\dfrac{a}{b}x = \pm 2x$

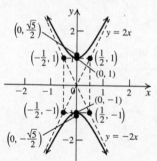

25. $y = \pm\sqrt{9x^2 - 1} \Rightarrow 9x^2 - y^2 = 1 \Rightarrow$

$\dfrac{x^2}{1/9} - y^2 = 1 \Rightarrow a^2 = \dfrac{1}{9} \Rightarrow$ the vertices are

$\left(\dfrac{1}{3}, 0\right)$ and $\left(-\dfrac{1}{3}, 0\right)$. $c^2 = a^2 + b^2 \Rightarrow$

$c^2 = \dfrac{1}{9} + 1 \Rightarrow c = \dfrac{\sqrt{10}}{3} \Rightarrow$ the foci are

$\left(\dfrac{\sqrt{10}}{3}, 0\right)$ and $\left(-\dfrac{\sqrt{10}}{3}, 0\right)$.

Transverse axis: *x*-axis. Hyperbola opens left
and right. Vertices of the fundamental

rectangle: $\left(\dfrac{1}{3}, 1\right), \left(-\dfrac{1}{3}, 1\right), \left(-\dfrac{1}{3}, -1\right)$, and

$\left(\dfrac{1}{3}, -1\right)$. Asymptotes: $y = \pm\dfrac{b}{a}x \Rightarrow y = \pm 3x$

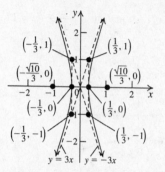

27. The transverse axis is the *x*-axis and the center
is (0, 0), so the equation is of the form

$\dfrac{x^2}{a^2} - \dfrac{y^2}{b^2} = 1$.

$a = 2, c = 3 \Rightarrow 9 = 4 + b^2 \Rightarrow b^2 = 5$. The

equation is $\dfrac{x^2}{4} - \dfrac{y^2}{5} = 1$.

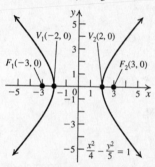

29. The transverse axis is the *y*-axis and the center
is (0, 0), so the equation is of the form

$\dfrac{y^2}{a^2} - \dfrac{x^2}{b^2} = 1$.

$a = 4, c = 6 \Rightarrow 36 = 16 + b^2 \Rightarrow b^2 = 20$. The

equation is $\dfrac{y^2}{16} - \dfrac{x^2}{20} = 1$.

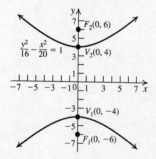

31. The transverse axis is the y-axis and the center is $(0, 0)$, so the equation is of the form

$$\frac{y^2}{a^2} - \frac{x^2}{b^2} = 1. \ a = 2, c = 5 \Rightarrow 25 = 4 + b^2 \Rightarrow$$

$b^2 = 21$. The equation is $\dfrac{y^2}{4} - \dfrac{x^2}{21} = 1$.

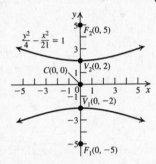

33. The transverse axis is the x-axis and the center is $(0, 0)$, so the equation is of the form

$$\frac{x^2}{a^2} - \frac{y^2}{b^2} = 1. \ a = 1, c = 5 \Rightarrow 25 = 1 + b^2 \Rightarrow$$

$b^2 = 24$. The equation is $x^2 - \dfrac{y^2}{24} = 1$.

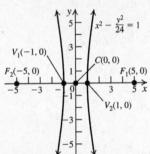

35. The transverse axis is the y-axis and the center is $(0, 0)$, so the equation is of the form

$$\frac{y^2}{a^2} - \frac{x^2}{b^2} = 1.$$

Length of the transverse axis $= 6 \Rightarrow a = 3$.

$c = 5 \Rightarrow 25 = 9 + b^2 \Rightarrow b^2 = 16$.

The equation is $\dfrac{y^2}{9} - \dfrac{x^2}{16} = 1$.

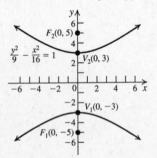

37. The transverse axis is the x-axis and the center is $(0, 0)$, so the equation is of the form

$$\frac{x^2}{a^2} - \frac{y^2}{b^2} = 1. \ y = 2x = \frac{b}{a}x \Rightarrow b = 2a.$$

$c^2 = a^2 + b^2 \Rightarrow 5 = a^2 + (2a)^2 \Rightarrow a = 1, b = 2.$

The equation is $x^2 - \dfrac{y^2}{4} = 1$.

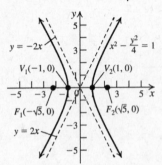

39. The transverse axis is the y-axis and the center is $(0, 0)$, so the equation is of the form

$$\frac{y^2}{a^2} - \frac{x^2}{b^2} = 1. \ y = x = \frac{a}{b}x \Rightarrow a = b.$$

$a = 4 \Rightarrow b = 4$. The equation is $\dfrac{y^2}{16} - \dfrac{x^2}{16} = 1$.

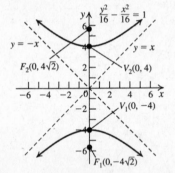

41.

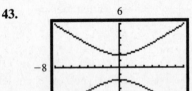

43.

45.

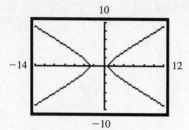

47.

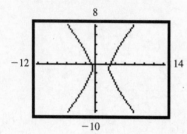

49. Center: $(1, -1)$; vertices: $(4, -1)$, $(-2, -1)$; transverse axis: $y = -1$; asymptotes:

$$y - k = \pm\frac{b}{a}(x-h) \Rightarrow y + 1 = \pm\frac{4}{3}(x-1)$$

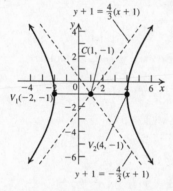

51. Center: $(-2, 0)$; vertices: $(3, 0)$, $(-7, 0)$; transverse axis: x-axis $(y = 0)$; asymptotes:

$$y - k = \pm\frac{b}{a}(x-h) \Rightarrow y = \pm\frac{7}{5}(x+2)$$

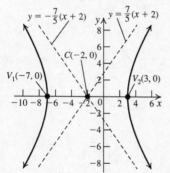

53. Center: $(-4, -3)$; vertices: $(1, -3)$, $(-9, -3)$; transverse axis: $y = -3$; asymptotes:

$$y - k = \pm\frac{b}{a}(x-h) \Rightarrow y + 3 = \pm\frac{7}{5}(x+4).$$

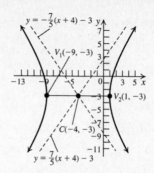

55. $4x^2 - (y+1)^2 = 25 \Rightarrow \dfrac{x^2}{25/4} - \dfrac{(y+1)^2}{25} = 1 \Rightarrow$

center: $(0, -1)$; vertices: $\left(\dfrac{5}{2}, -1\right), \left(-\dfrac{5}{2}, -1\right)$;

transverse axis: $y = -1$; asymptotes:

$$y - k = \pm\frac{b}{a}(x-h) \Rightarrow y + 1 = \pm 2x.$$

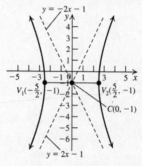

57. $(y+1)^2 - 9(x-2)^2 = 25 \Rightarrow$

$\dfrac{(y+1)^2}{25} - \dfrac{(x-2)^2}{25/9} = 1 \Rightarrow$ center: $(2, -1)$;

vertices: $(2, 4)$, $(2, -6)$;

transverse axis: $x = 2$; asymptotes:

$$y - k = \pm\frac{a}{b}(x-h) \Rightarrow y + 1 = \pm 3(x-2).$$

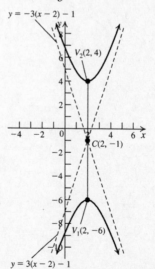

59. Complete the square to put the equation in standard form: $x^2 - y^2 + 6x = 36 \Rightarrow$

$x^2 + 6x + 9 - y^2 = 36 + 9 \Rightarrow$

$\dfrac{(x+3)^2}{45} - \dfrac{y^2}{45} = 1 \Rightarrow$ center: $(-3, 0)$; vertices:

$\left(-3 + 3\sqrt{5}, 0\right), \left(-3 - 3\sqrt{5}, 0\right)$; transverse axis:

x-axis $(y = 0)$; asymptotes:

$y - k = \pm\dfrac{b}{a}(x - h) \Rightarrow y = \pm(x + 3)$.

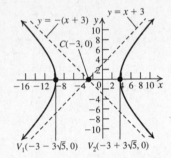

61. Complete the square to put the equation in standard form: $x^2 - 4y^2 - 4x = 0 \Rightarrow$

$x^2 - 4x + 4 - 4y^2 = 0 + 4 \Rightarrow$

$\dfrac{(x-2)^2}{4} - y^2 = 1 \Rightarrow$ center: $(2, 0)$; vertices:

$(4, 0), (0, 0)$; transverse axis: x-axis $(y = 0)$;

asymptotes: $y - k = \pm\dfrac{b}{a}(x - h) \Rightarrow$

$y = \pm\dfrac{1}{2}(x - 2)$

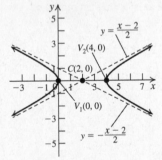

63. Complete the square to put the equation in standard form: $2x^2 - y^2 + 12x - 8y + 3 = 0 \Rightarrow$

$2(x^2 + 6x + 9) - (y^2 + 8y + 16) = -3 + 18 - 16 \Rightarrow$

$(y+4)^2 - \dfrac{(x+3)^2}{1/2} = 1 \Rightarrow$ center: $(-3, -4)$;

vertices: $(-3, -3), (-3, -5)$; transverse axis:

$x = -3$; asymptotes: $y - k = \pm\dfrac{a}{b}(x - h) \Rightarrow$

$y + 4 = \pm\sqrt{2}(x + 3)$

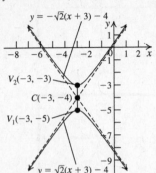

65. Complete the square to put the equation in standard form:

$3x^2 - 18x - 2y^2 - 8y + 1 = 0 \Rightarrow$

$3(x^2 - 6x + 9) - 2(y^2 + 4y + 4) = -1 + 27 - 8 \Rightarrow$

$\dfrac{(x-3)^2}{6} - \dfrac{(y+2)^2}{9} = 1 \Rightarrow$ center: $(3, -2)$;

vertices: $\left(3 + \sqrt{6}, -2\right), \left(3 - \sqrt{6}, -2\right)$;

transverse axis: $y = -2$; asymptotes:

$y - k = \pm\dfrac{b}{a}(x - h) \Rightarrow y + 2 = \pm\dfrac{\sqrt{6}}{2}(x - 3)$

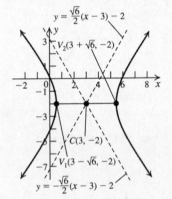

67. Complete the square to put the equation in standard form: $y^2 + 2\sqrt{2} - x^2 + 2\sqrt{2}x = 1 \Rightarrow$

$y^2 - \left(x^2 - 2\sqrt{2}x + 2\right) = 1 - 2\sqrt{2} - 2 \Rightarrow$

$\dfrac{(x-\sqrt{2})^2}{1 + 2\sqrt{2}} - \dfrac{y^2}{1 + 2\sqrt{2}} = 1 \Rightarrow$ center: $\left(\sqrt{2}, 0\right)$;

vertices: $\left(\sqrt{2} + \sqrt{1 + 2\sqrt{2}}, 0\right)$,

$\left(\sqrt{2} - \sqrt{1 + 2\sqrt{2}}, 0\right)$; transverse axis: x-axis

$(y = 0)$

(continued on next page)

(*continued*)

asymptotes: $y - k = \pm \dfrac{b}{a}(x - h) \Rightarrow$

$y = \pm\left(x - \sqrt{2}\right)$

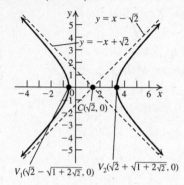

69. $x^2 - 6x + 12y + 33 = 0 \Rightarrow$

$x^2 - 6x + 9 = -12y - 33 + 9 \Rightarrow$

$(x - 3)^2 = -12(y + 2) \Rightarrow$ the conic is a

parabola.

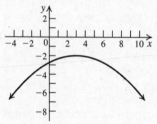

71. $y^2 - 9x^2 = -1 \Rightarrow \dfrac{x^2}{1/9} - y^2 = 1 \Rightarrow$ the conic is

a hyperbola.

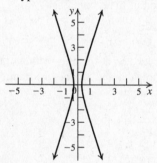

73. $x^2 + y^2 - 4x + 8y = 16 \Rightarrow$

$x^2 - 4x + 4 + y^2 + 8y + 16 = 16 + 4 + 16 \Rightarrow$

$(x - 2)^2 + (y + 4)^2 = 36 \Rightarrow$ the conic is a

circle.

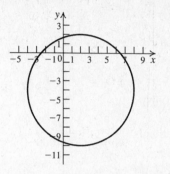

75. $2x^2 - 4x + 3y + 8 = 0 \Rightarrow$

$2(x^2 - 2x + 1) = -3y - 8 + 2 \Rightarrow$

$2(x - 1)^2 = -3(y + 2) \Rightarrow$ the conic is a

parabola.

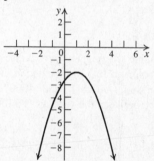

77. $4x^2 + 9y^2 + 8x - 54y + 49 = 0 \Rightarrow$

$4(x^2 + 2x + 1) + 9(y^2 - 6y + 9) = -49 + 4 + 81 \Rightarrow$

$\dfrac{(x + 1)^2}{9} + \dfrac{(y - 3)^2}{4} = 1 \Rightarrow$ the conic is an

ellipse.

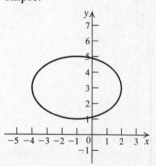

7.4 Applying the Concepts

79. A hyperbola is the set of all points in the plane
   whose distances from two fixed points have a
   constant difference. The difference in the
   distances of the location of the explosion from
   point *A* to point *B* is 600 meters, so *A* and *B*
   are the foci of the hyperbola. Let the
   coordinates of *A* be (−500, 0) and those of *B*
   be (500, 0), so $c = 500$.

   (*continued on next page*)

(continued)

The distance between V_1 and V_2 is 600, so $a = 300$.

$c^2 = a^2 + b^2 \Rightarrow 500^2 = 300^2 + b^2 \Rightarrow$
$b^2 = 160,000$.

The equation is $\dfrac{x^2}{90,000} - \dfrac{y^2}{160,000} = 1$.

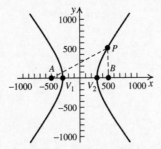

81. Using the same reasoning as in exercises 79 and 80, Nicole and Juan are located at the foci of the hyperbola, $(-4000, 0)$ and $(4000, 0)$. The distance between the vertices is 3300, so $a = 1650$. $c^2 = a^2 + b^2 \Rightarrow$
$4000^2 = 1650^2 + b^2 \Rightarrow b^2 = 13,277,500$.

The equation is $\dfrac{x^2}{2,722,500} - \dfrac{y^2}{13,277,500} = 1$.

Because Nicole hears the thunder first, the graph consists only of the part of the hyperbola closest to Nicole:

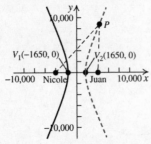

83. Let the coordinates of A and B be $(-150, 0)$ and $(150, 0)$, respectively. At 300,000 km per sec, the difference in the distances the signals travel is 150 km, so $a = 75$. $c^2 = a^2 + b^2 \Rightarrow$
$150^2 = 75^2 + b^2 \Rightarrow b^2 = 16,875$. The

equation is $\dfrac{x^2}{5625} - \dfrac{y^2}{16,875} = 1$.

85. The bullet reaches the target in
$t = \dfrac{d}{r} = \dfrac{1600 \text{ ft}}{2000 \text{ ft/sec}} = \dfrac{4}{5}$ second. The person
hears the crack of the gun and the thud of the bullet at the same time. So,

$\dfrac{d_1}{1100} = \dfrac{4}{5} + \dfrac{d_2}{1100} \Rightarrow \dfrac{d_1}{1100} - \dfrac{d_2}{1100} = \dfrac{4}{5} \Rightarrow$

$d_1 - d_2 = \dfrac{4}{5}(1100) = 880$, a constant. So P

lies on a hyperbola with $a = 440$. The foci are at $(-800, 0)$ and $(800, 0)$, so
$b^2 = c^2 - a^2 \Rightarrow$
$b^2 = 800^2 - 440^2 = 446,400$
The equation of the hyperbola is

$\dfrac{x^2}{193,600} - \dfrac{y^2}{446,400} = 1$.

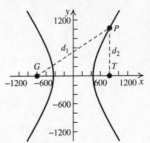

87. A and B are the foci of hyperbola 1, and the difference of the distances is 120, so $c = 100$ and $a = 60$. $c^2 = a^2 + b^2 \Rightarrow$
$100^2 = 60^2 + b^2 \Rightarrow b^2 = 80^2$. The equation of

hyperbola 1 is $\dfrac{x^2}{3600} - \dfrac{y^2}{6400} = 1$. C and D are

the foci of hyperbola 2, and the difference of the distances is 80, so $c = 150$ and $a = 40$.
$c^2 = a^2 + b^2 \Rightarrow 150^2 = 40^2 + b^2 \Rightarrow$
$b^2 = 20,900$.
The equation of hyperbola 2 is

$\dfrac{y^2}{1600} - \dfrac{x^2}{20,900} = 1$.

The ship is located at the intersection of the two hyperbolas, in the second quadrant. Solve the system

$$\begin{cases} \dfrac{x^2}{3600} - \dfrac{y^2}{6400} = 1 \\ \dfrac{y^2}{1600} - \dfrac{x^2}{20,900} = 1 \end{cases}$$

or use a graphing calculator to find the intersection. Since the ship is in the second quadrant, graph only those portions of the curve in that quadrant.

(continued on next page)

(continued)

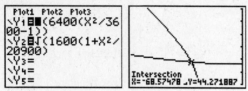

The ship is at approximately
(−68.5748, 44.2719).

7.4 Beyond the Basics

89. For all three parts, $c = 5$ and the transverse axis is the x-axis.

  a. $a = 1$, so $b^2 = 25 - 1 = 24$. The equation is $x^2 - \dfrac{y^2}{24} = 1$.

  b. $a = 2$, so $b^2 = 25 - 4 = 21$. The equation is $\dfrac{x^2}{4} - \dfrac{y^2}{21} = 1$.

  c. $a = 4$, so $b^2 = 25 - 16 = 9$. The equation is $\dfrac{x^2}{16} - \dfrac{y^2}{9} = 1$.

91. For all three parts, $a = 1$ and the transverse axis is the y-axis.

  a. $c = 6$, so $b^2 = 36 - 1 = 35$. The equation is $y^2 - \dfrac{x^2}{35} = 1$.

  b. $c = 4$, so $b^2 = 16 - 1 = 15$. The equation is $y^2 - \dfrac{x^2}{15} = 1$.

  c. $c = 3$, so $b^2 = 9 - 1 = 8$. The equation is $y^2 - \dfrac{x^2}{8} = 1$.

93. $\sqrt{(x+c)^2 + y^2} - \sqrt{(x-c)^2 + y^2} = \pm 2a \Rightarrow \sqrt{(x+c)^2 + y^2} = \pm 2a + \sqrt{(x-c)^2 + y^2} \Rightarrow$

$(x+c)^2 + y^2 = 4a^2 \pm 4a\sqrt{(x-c)^2 + y^2} + (x-c)^2 + y^2 \Rightarrow$

$x^2 + 2xc + c^2 + y^2 = 4a^2 \pm 4a\sqrt{(x-c)^2 + y^2} + x^2 - 2xc + c^2 + y^2 \Rightarrow$

$xc = a^2 \pm a\sqrt{(x-c)^2 + y^2} \Rightarrow xc - a^2 = \pm a\sqrt{(x-c)^2 + y^2} \Rightarrow a^4 - 2a^2 xc + x^2 c^2 = a^2(x^2 - 2xc + c^2 + y^2) \Rightarrow$

$a^4 + x^2 c^2 = a^2 x^2 + a^2 c^2 + a^2 y^2 \Rightarrow x^2 c^2 - a^2 x^2 - a^2 y^2 = -a^4 + a^2 c^2 \Rightarrow (c^2 - a^2)x^2 - a^2 y^2 = a^2(c^2 - a^2)$

95. Assume that the transverse axis is the x-axis. Then the equation of the hyperbola is $\dfrac{x^2}{a^2} - \dfrac{y^2}{b^2} = 1$. The hyperbola is equilateral, so $a = b$, and the equations of the asymptotes are $y = \pm x$. These lines are perpendicular to each other. Similarly, it can be shown that the asymptotes are perpendicular to each other if the transverse axis is the y-axis.

97. Since $b > 0$ and $c^2 = a^2 + b^2$, it follows that $c^2 > a^2$ and $c > a$. So, $e > 1$. When $e = 1$, the hyperbola becomes the union of two rays.

99. $36x^2 - 25y^2 = 900 \Rightarrow \dfrac{x^2}{25} - \dfrac{y^2}{36} = 1$

$a = 5, b = 6 \Rightarrow c^2 = 25 + 36 \Rightarrow c = \sqrt{61}$

$e = \dfrac{c}{a} = \dfrac{\sqrt{61}}{5}$.

Length of latus rectum $= \dfrac{2b^2}{a} = \dfrac{72}{5}$.

101. $8(x-1)^2 - (y+2)^2 = 2 \Rightarrow$

$$\frac{(x-1)^2}{1/4} - \frac{(y+2)^2}{2} = 1$$

$$a = \frac{1}{2}, b = \sqrt{2} \Rightarrow c^2 = \frac{1}{4} + 2 \Rightarrow c = \frac{3}{2} \Rightarrow$$

$$e = \frac{c}{a} = 3$$

Length of latus rectum = $\dfrac{2b^2}{a} = 8$.

103. $e = \dfrac{c}{a} \Rightarrow e^2 = \dfrac{c^2}{a^2} \Rightarrow e^2 = \dfrac{a^2 + b^2}{a^2} = 1 + \dfrac{b^2}{a^2} \Rightarrow$

$$e^2 - 1 = \frac{b^2}{a^2} \Rightarrow a^2(e^2 - 1) = b^2$$

105. The equation of the path is

$\sqrt{(x-3)^2 + y^2} = 2|x+1|$. Square both sides, simplify, complete the square, and write the equation in standard form:

$(x-3)^2 + y^2 = 4(x+1)^2 \Rightarrow$

$x^2 - 6x + 9 + y^2 = 4x^2 + 8x + 4 \Rightarrow$

$3x^2 + 14x - y^2 = 5 \Rightarrow$

$$3\left(x^2 + \frac{14}{3}x + \frac{49}{9}\right) - y^2 = 5 + \frac{49}{3} \Rightarrow$$

$$3\left(x + \frac{7}{3}\right)^2 - y^2 = \frac{64}{3} \Rightarrow \frac{\left(x + \frac{7}{3}\right)^2}{64/9} - \frac{y^2}{64/3} = 1$$

$$c = \sqrt{\frac{64}{9} + \frac{64}{3}} = \frac{16}{3}; a = \frac{8}{3}; e = \frac{c}{a} = 2$$

107. Solve using substitution:

$$\begin{cases} y - 2x - 20 = 0 \\ y^2 - 4x^2 = 36 \end{cases} \Rightarrow \begin{cases} y = 2x + 20 \\ y^2 - 4x^2 = 36 \end{cases} \Rightarrow$$

$(2x + 20)^2 - 4x^2 = 36 \Rightarrow 80x + 400 = 36 \Rightarrow$

$$x = -\frac{91}{20}; y = 2\left(-\frac{91}{20}\right) + 20 = \frac{109}{10}$$

The only point of intersection is $\left(-\dfrac{91}{20}, \dfrac{109}{10}\right)$.

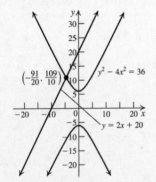

109. Solve using elimination:

$$\begin{cases} x^2 + y^2 = 15 \\ x^2 - y^2 = 1 \end{cases} \Rightarrow 2x^2 = 16 \Rightarrow x = \pm 2\sqrt{2}$$

$$\left(\pm 2\sqrt{2}\right)^2 + y^2 = 15 \Rightarrow y^2 = 7 \Rightarrow y = \pm\sqrt{7}$$

The points of intersection are $\left(2\sqrt{2}, \sqrt{7}\right)$,

$\left(2\sqrt{2}, -\sqrt{7}\right), \left(-2\sqrt{2}, -\sqrt{7}\right)$, and

$\left(-2\sqrt{2}, \sqrt{7}\right)$.

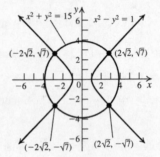

111. Solve using substitution:

$$\begin{cases} 9x^2 + 4y^2 = 72 \\ x^2 - 9y^2 = -77 \end{cases} \Rightarrow \begin{cases} 9x^2 + 4y^2 = 72 \\ x^2 = 9y^2 - 77 \end{cases} \Rightarrow$$

$9(9y^2 - 77) + 4y^2 = 72 \Rightarrow$

$85y^2 - 693 = 72 \Rightarrow y = \pm 3$

$x^2 = 9(9) - 77 \Rightarrow x = \pm 2$

The points of intersection are $(-2, -3)$, $(-2, 3)$, $(2, -3)$, and $(2, 3)$.

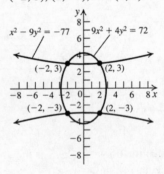

113. The equation of the hyperbola is

$$\frac{x^2}{5/3} - \frac{y^2}{5/7} = 1 \Rightarrow \frac{3x^2}{5} - \frac{7y^2}{5} = 1 \Rightarrow 3x^2 - 7y^2 = 5 \Rightarrow 7y^2 = 3x^2 - 5$$

Step 1: Let m = the slope of the tangent line. Then the equation of the tangent line is

$$y - 1 = m(x - 2) \Rightarrow y = m(x - 2) + 1$$

Step 2: $7y^2 = 7\left[m(x-2)+1\right]^2 = 7\left[m^2(x-2)^2 + 2m(x-2) + 1\right] = 7m^2(x-2)^2 + 14m(x-2) + 7$

Step 3: $3x^2 - 5 = 7m^2(x-2)^2 + 14m(x-2) + 7 \Rightarrow 3x^2 - 5 = 7m^2\left(x^2 - 4x + 4\right) + 14mx - 28m + 7 \Rightarrow$

$$3x^2 - 5 = 7m^2x^2 - 28m^2x + 28m^2 + 14mx - 28m + 7 \Rightarrow$$

$$3x^2 - 7m^2x^2 + 28m^2x - 14mx - 28m^2 + 28m - 12 = 0 \Rightarrow$$

$$\left(3 - 7m^2\right)x^2 + \left(28m^2 - 14m\right)x - \left(28m^2 - 28m + 12\right) = 0$$

Now set the discriminant equal to 0 and solve for x:

$$\left(28m^2 - 14m\right)^2 - 4\left(3 - 7m^2\right)\left[-\left(28m^2 - 28m + 12\right)\right] = 0 \Rightarrow$$

$$784m^4 - 784m^3 + 196m^2 - 784m^4 + 784m^3 - 336m + 144 = 0 \Rightarrow$$

$$196m^2 - 336m + 144 = 0 \Rightarrow 49m^2 - 84m + 36 = 0 \Rightarrow (7m - 6)^2 = 0 \Rightarrow m = \frac{6}{7}$$

Step 4: The equation of the tangent line is $y - 1 = \frac{6}{7}(x - 2) \Rightarrow y = \frac{6}{7}x - \frac{5}{7}$.

7.4 Critical Thinking/Writing/Discussion

115. a. $4x^2 - 9y^2 = 0 \Rightarrow (2x + 3y)(2x - 3y) = 0$.

The graph consists of two lines, $y = \pm\frac{2}{3}x$.

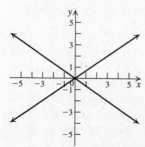

b. All the variable terms on the left side of the equation are always nonnegative, so there are no x or y values that would satisfy the equation. There is no graph.

c. Since both x^2 and y^2 are nonnegative, the only solution of the equation is (0, 0).

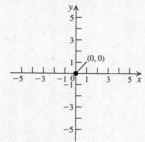

d. Complete the square:

$$x^2 + y^2 - 4x + 8y = -20 \Rightarrow$$

$$(x^2 - 4x + 4) + (y^2 + 8y + 16) = -20 + 4 + 16 \Rightarrow$$

$$(x - 2)^2 + (y + 4)^2 = 0 \Rightarrow x = 2 \text{ and } y = -4.$$

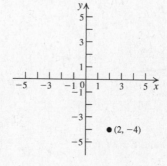

e. $y^2 - 2x^2 = 0 \Rightarrow y^2 = 2x^2 \Rightarrow y = \pm x\sqrt{2}$.

The graph consists of two lines.

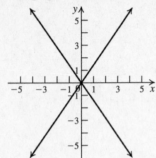

116. $Ax^2 + Cy^2 + Dx + Ey + F = 0$, $AC \neq 0$

Rewriting the equation, we have

$$A\left(x^2 + \frac{D}{A}x + \frac{D^2}{4A^2}\right) + C\left(y^2 + \frac{E}{C}y + \frac{E^2}{4C^2}\right)$$

$$= \frac{D^2}{4A} + \frac{E^2}{4C} - F \Rightarrow$$

$$A\left(x + \frac{D}{2A}\right)^2 + C\left(y + \frac{E}{2C}\right)^2 = \frac{D^2}{4A} + \frac{E^2}{4C} - F.$$

If the term on the right side is not zero, then the equation can be rewritten as

$$\frac{A\left(x + \dfrac{D}{2A}\right)^2}{\dfrac{D^2}{4A} + \dfrac{E^2}{4C} - F} + \frac{C\left(y + \dfrac{E}{2C}\right)^2}{\dfrac{D^2}{4A} + \dfrac{E^2}{4C} - F} = 1 \Rightarrow$$

$$\frac{\left(x + \dfrac{D}{2A}\right)^2}{\dfrac{D^2}{4A^2} + \dfrac{E^2}{4AC} - \dfrac{F}{A}} + \frac{\left(y + \dfrac{E}{2C}\right)^2}{\dfrac{D^2}{4AC} + \dfrac{E^2}{4C^2} - \dfrac{F}{C}} = 1$$

(i) If $A > 0$ and $C > 0$, and if $\dfrac{D^2}{4A} + \dfrac{E^2}{4C} - F = 0$,

the graph consists of one point,

$\left(-\dfrac{D}{2A}, -\dfrac{E}{2C}\right)$. If $\dfrac{D^2}{4A} + \dfrac{E^2}{4C} - F > 0$, the

graph is an ellipse. If $\dfrac{D^2}{4A} + \dfrac{E^2}{4C} - F < 0$, then

there is no graph.

(ii) If $A > 0$ and $C < 0$, and if $\dfrac{D^2}{4A} + \dfrac{E^2}{4C} - F = 0$,

the graph consists of two lines,

$\sqrt{A}\left(x + \dfrac{D}{2A}\right) = \pm\sqrt{-C}\left(y + \dfrac{E}{2C}\right)$. If

$\dfrac{D^2}{4A} + \dfrac{E^2}{4C} - F > 0$, the graph is a hyperbola

with a horizontal transverse axis. If

$\dfrac{D^2}{4A} + \dfrac{E^2}{4C} - F < 0$, then the graph is a

hyperbola with a vertical transverse axis.

(iii) If $A < 0$ and $C > 0$, and if $\dfrac{D^2}{4A} + \dfrac{E^2}{4C} - F = 0$,

the graph consists of two lines,

$\sqrt{-A}\left(x + \dfrac{D}{2A}\right) = \pm\sqrt{C}\left(y + \dfrac{E}{2C}\right)$. If

$\dfrac{D^2}{4A} + \dfrac{E^2}{4C} - F > 0$, the graph is a hyperbola

with a vertical transverse axis. If

$\dfrac{D^2}{4A} + \dfrac{E^2}{4C} - F < 0$, then the graph is a

hyperbola with a horizontal transverse axis.

(iv) If $A < 0$ and $C < 0$, and if $\dfrac{D^2}{4A} + \dfrac{E^2}{4C} - F = 0$,

the graph consists of one point,

$\left(-\dfrac{D}{2A}, -\dfrac{E}{2C}\right)$. If $\dfrac{D^2}{4A} + \dfrac{E^2}{4C} - F > 0$, there is

no graph. If $\dfrac{D^2}{4A} + \dfrac{E^2}{4C} - F < 0$, then the

graph is an ellipse.

117. The hyperbolas $x^2 - y^2 = -1$ and

$x^2 - y^2 = 1$ have the same asymptotes
because $a = b = 1$ for both graphs. The
asymptotes are $y = x$ and $y = -x$. Note that the
graphs are the reflections of each other across
$y = x$ and across $y = -x$. Also note that the
graph of one can be obtained by rotating the
graph of the other about the origin 90°.

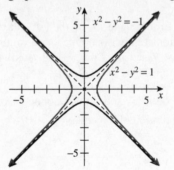

7.4 Maintaining Skills

119. $g(10) = 2(10) + 1 = 21$

121. $f(6) = \dfrac{16(6)}{6^2 + 12} = \dfrac{96}{48} = 2$

123. $(-1)^8 \left(\dfrac{1}{8+1} \right) = \dfrac{1}{9}$

125. $(-1)^{7-2} (2(7)+1) + (-1)^7 = -1(15) + (-1)$
$= -16$

127. $\dfrac{a \cdot 2 \cdot 3 \cdot 4 \cdot 5 \cdot 6 \cdot 7 \cdot 8 \cdot 9 \cdot 10}{3 \cdot 5 \cdot 7 \cdot 8 \cdot 9 \cdot 10 \cdot a} = 2 \cdot 4 \cdot 6 = 48$

129. $3a - 2b + 6c - 5d = 3a + 6c - 2b - 5d$
$= (3a + 6c) - (2b + 5d)$
True

Chapter 7 Review Exercises

1. $y^2 = -6x \Rightarrow -6 = 4a \Rightarrow a = -\dfrac{3}{2}$

Vertex: (0, 0), focus: $\left(-\dfrac{3}{2}, 0 \right)$, axis: x-axis,

directrix: $x = \dfrac{3}{2}$.

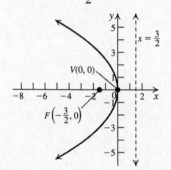

3. $x^2 = 7y \Rightarrow 7 = 4a \Rightarrow \dfrac{7}{4} = a$.

Vertex: (0, 0), focus: $\left(0, \dfrac{7}{4} \right)$, axis: y-axis,

directrix: $y = -\dfrac{7}{4}$.

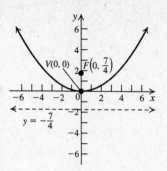

5. $(x-2)^2 = -(y+3) \Rightarrow -1 = 4a \Rightarrow -\dfrac{1}{4} = a$.

Vertex: (2, −3), focus: $\left(2, -\dfrac{13}{4} \right)$, axis: $x = 2$,

directrix: $y = -\dfrac{11}{4}$.

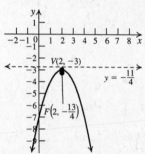

7. Rearrange the equation and complete the square to put the equation in standard form:

$y^2 = -4y + 2x + 1 \Rightarrow y^2 + 4y + 4 = 2x + 1 + 4 \Rightarrow$

$(y+2)^2 = 2 \left(x + \dfrac{5}{2} \right) \Rightarrow 2 = 4a \Rightarrow a = \dfrac{1}{2}$.

Vertex: $\left(-\dfrac{5}{2}, -2 \right)$, focus: (−2, −2),

axis: $y = -2$, directrix: $x = -3$.

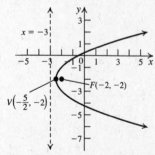

9. The vertex is (0, 0) and the focus is (−3, 0), so the graph opens to the left. The general form is $(y-k)^2 = -4a(x-h)$.

$a = -3 \Rightarrow y^2 = -12x$.

11. The focus is (0, 4) and the directrix is $y = -4$, so the vertex is (0, 0), and the graph opens up. The general form is $(x - h)^2 = 4a(y - k)$.

$a = 4 \Rightarrow x^2 = 16y$.

13. $a^2 = 25 \Rightarrow$ the vertices are (5, 0) and (−5, 0).

$b^2 = a^2 - c^2 \Rightarrow 4 = 25 - c^2 \Rightarrow c = \pm\sqrt{21} \Rightarrow$ the foci are $\left(\sqrt{21}, 0\right)$ and $\left(-\sqrt{21}, 0\right)$.

Endpoints of the minor axis: (0, −2) and (0, 2).

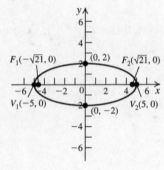

15. $4x^2 + y^2 = 4 \Rightarrow x^2 + \dfrac{y^2}{4} = 1$.

$a^2 = 4 \Rightarrow$ the vertices are (0, 2) and (0, −2).

$b^2 = a^2 - c^2 \Rightarrow 1 = 4 - c^2 \Rightarrow c = \pm\sqrt{3} \Rightarrow$ the foci are $\left(0, \sqrt{3}\right)$ and $\left(0, -\sqrt{3}\right)$.

Endpoints of the minor axis: (1, 0) and (−1, 0).

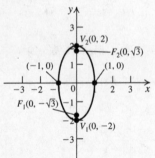

17. $16(x + 1)^2 + 9(y + 4)^2 = 144 \Rightarrow$

$\dfrac{(x + 1)^2}{9} + \dfrac{(y + 4)^2}{16} = 1$.

The center is (−1, −4) and $a^2 = 16 \Rightarrow$ the vertices are (−1, −8) and (−1, 0).

$b^2 = a^2 - c^2 \Rightarrow 9 = 16 - c^2 \Rightarrow c = \pm\sqrt{7}$.

The foci are $\left(-1, -4 - \sqrt{7}\right)$ and $\left(-1, -4 + \sqrt{7}\right)$.

Endpoints of the minor axis: (−4, −4) and (2, −4).

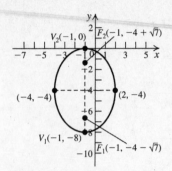

19. Rearrange the equation and complete the square to put the equation in standard form:

$x^2 + 9y^2 + 2x - 18y + 1 = 0 \Rightarrow$
$x^2 + 2x + 1 + 9(y^2 - 2y + 1) = -1 + 1 + 9 \Rightarrow$
$(x + 1)^2 + 9(y - 1)^2 = 9 \Rightarrow \dfrac{(x + 1)^2}{9} + (y - 1)^2 = 1$

The center is (−1, 1) and $a^2 = 9 \Rightarrow$ the vertices are (−4, 1) and (2, 1).

$b^2 = a^2 - c^2 \Rightarrow 1 = 9 - c^2 \Rightarrow c = \pm 2\sqrt{2}$

The foci are $\left(-1 - 2\sqrt{2}, 1\right)$ and $\left(-1 + 2\sqrt{2}, 1\right)$.

Endpoints of the minor axis: (−1, 2) and (−1, 0).

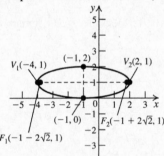

21. $a = 4$, $b = 2$, major axis: x-axis. $\dfrac{x^2}{16} + \dfrac{y^2}{4} = 1$

23. $a = 10$, center (0, 0), major axis: x-axis, $c = 5$, so $b^2 = 100 - 25 = 75$. $\dfrac{x^2}{100} + \dfrac{y^2}{75} = 1$

25. $a^2 = 16 \Rightarrow$ the vertices are (0, 4) and (0, −4).

$c^2 = a^2 + b^2 \Rightarrow c^2 = 16 + 4 \Rightarrow c = \pm 2\sqrt{5} \Rightarrow$ the foci are $\left(0, 2\sqrt{5}\right)$ and $\left(0, -2\sqrt{5}\right)$.

Asymptotes: $y = \pm\dfrac{a}{b}x = \pm 2x$.

(continued on next page)

(*continued*)

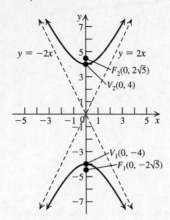

27. $8x^2 - y^2 = 8 \Rightarrow x^2 - \dfrac{y^2}{8} = 1 \Rightarrow a^2 = 1 \Rightarrow$ the

vertices are $(-1, 0)$ and $(1, 0)$.

$c^2 = a^2 + b^2 \Rightarrow c^2 = 1 + 8 \Rightarrow c = \pm 3 \Rightarrow$ the
foci are $(3, 0)$ and $(-3, 0)$. Asymptotes:

$y = \pm \dfrac{b}{a} x = \pm 2\sqrt{2} x$.

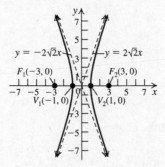

29. The center is $(-2, 3)$. $a^2 = 9 \Rightarrow$ the vertices
are $(1, 3)$ and $(-5, 3)$.

$c^2 = a^2 + b^2 \Rightarrow c^2 = 9 + 4 \Rightarrow c = \pm\sqrt{13} \Rightarrow$
the foci are $\left(-2 + \sqrt{13}, 3\right)$ and $\left(-2 - \sqrt{13}, 3\right)$.

Asymptotes:

$y - k = \pm \dfrac{b}{a}(x - h) \Rightarrow y - 3 = \pm \dfrac{2}{3}(x + 2)$.

31. Rearrange the equation and complete the
square to put the equation in standard form:

$4y^2 - x^2 + 40y - 4x + 60 = 0 \Rightarrow$

$4(y^2 + 10y + 25) - (x^2 + 4x + 4)$
$\qquad\qquad = -60 + 100 - 4$

$4(y + 5)^2 - (x + 2)^2 = 36 \Rightarrow$

$\dfrac{(y + 5)^2}{9} - \dfrac{(x + 2)^2}{36} = 1.$

The center is $(-2, -5)$. $a^2 = 9 \Rightarrow$ the vertices
are $(-2, -5 - 3) = (-2, -8)$ and
$(-2, -5 + 3) = (-2, -2)$.

$c^2 = a^2 + b^2 \Rightarrow c^2 = 9 + 36 \Rightarrow c = \pm 3\sqrt{5} \Rightarrow$
the foci are $\left(-2, -5 + 3\sqrt{5}\right)$ and

$\left(-2, -5 - 3\sqrt{5}\right)$. Asymptotes:

$y - k = \pm \dfrac{a}{b}(x - h) \Rightarrow y + 5 = \pm \dfrac{1}{2}(x + 2)$.

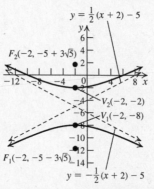

33. The vertices are $(\pm 1, 0)$ and the foci are
$(\pm 2, 0)$, so the center is $(0, 0)$, $a = 1$, $c = 2$,
and the transverse axis is the x-axis.

$c^2 = a^2 + b^2 \Rightarrow 4 = 1 + b^2 \Rightarrow b^2 = 3$. The

equation is $x^2 - \dfrac{y^2}{3} = 1$.

35. The vertices are $(\pm 2, 0)$ so the center is $(0, 0)$,
$a = 2$, and the transverse axis is the x-axis.

The asymptotes are $y = \pm 3x \Rightarrow \pm \dfrac{b}{a} = \pm 3 \Rightarrow$

$\dfrac{b}{2} = 3 \Rightarrow b = 6$. The equation is $\dfrac{x^2}{4} - \dfrac{y^2}{36} = 1$.

37. Hyperbola. In standard form, the equation is
$$\frac{x^2}{4} - \frac{y^2}{5} = 1.$$

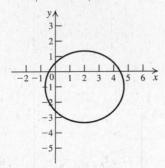

39. Ellipse. In standard form, the equation is
$$\frac{(x-2)^2}{22/3} + \frac{(y+1)^2}{11/2} = 1.$$

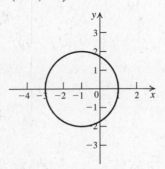

41. Circle. In standard form, the equation is
$$(x+1)^2 + y^2 = 4.$$

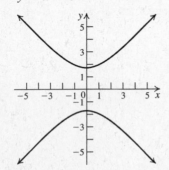

43. Hyperbola. In standard form, the equation is
$$y^2 - x^2 = 3.$$

45. Circle. In standard form, the equation is
$$(x-1)^2 + (y+2)^2 = \frac{10}{3}$$

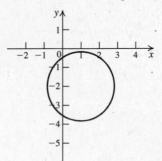

47. Ellipse. In standard form, the equation is
$$\frac{x^2}{4} + \frac{y^2}{9/2} = 1.$$

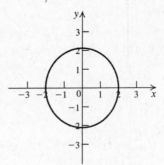

49. $4x^2 + 9y^2 = 36 \Rightarrow \dfrac{x^2}{9} + \dfrac{y^2}{4} = 1 \Rightarrow$ the vertices of the ellipse are $(3, 0)$ and $(-3, 0)$.
$b^2 = a^2 - c^2 \Rightarrow 4 = 9 - c^2 \Rightarrow c = \pm\sqrt{5} \Rightarrow$ the foci are $\left(\sqrt{5}, 0\right)$ and $\left(-\sqrt{5}, 0\right)$. For the hyperbola, the transverse axis is the x-axis, $a = \sqrt{5}$, and $c = 3$, so $c^2 = a^2 + b^2 \Rightarrow$ $9 = 5 + b^2 \Rightarrow b^2 = 4$. The equation of the hyperbola is $\dfrac{x^2}{5} - \dfrac{y^2}{4} = 1$.

51. Solve using substitution:
$$\begin{cases} x^2 - 4y^2 = 36 \\ x - 2y - 20 = 0 \end{cases} \Rightarrow \begin{cases} x^2 - 4y^2 = 36 \\ x = 2y + 20 \end{cases} \Rightarrow$$
$(2y + 20)^2 - 4y^2 = 36 \Rightarrow 80y + 400 = 36 \Rightarrow$
$y = -\dfrac{91}{20}; \; x = 2\left(-\dfrac{91}{20}\right) + 20 = \dfrac{109}{10}$

The only point of intersection is $\left(\dfrac{109}{10}, -\dfrac{91}{20}\right)$.

(continued on next page)

(*continued*)

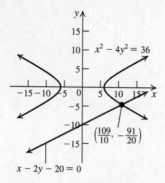

$x^2 - 4y^2 = 36$

$\left(\frac{109}{10}, -\frac{91}{20}\right)$

$x - 2y - 20 = 0$

53. Solve using elimination:

$$\begin{cases} 3x^2 - 7y^2 = 5 \\ 9y^2 - 2x^2 = 1 \end{cases} \Rightarrow \begin{cases} 6x^2 - 14y^2 = 10 \\ -6x^2 + 27y^2 = 3 \end{cases} \Rightarrow$$

$13y^2 = 13 \Rightarrow y = \pm 1$

$3x^2 - 7 = 5 \Rightarrow x^2 = 4 \Rightarrow x = \pm 2$

The points of intersection are $(-2, -1)$, $(-2, 1)$, $(2, -1)$, and $(2, 1)$.

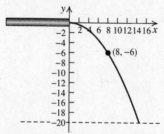

$9y^2 - 2x^2 = 1$ $\quad 3x^2 - 7y^2 = 5$

$(-2, 1)$ $(2, 1)$

$(-2, -1)$ $(2, -1)$

55. Let the end of the pipe, and therefore the vertex of the parabola, be at $(0, 0)$. Then the ground is at $y = -20$. The water goes through the point $(8, -6)$.

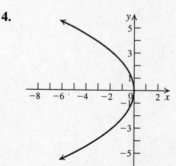

$(8, -6)$

The equation is of the form $y = -4ax^2$.
Substitute $(8, -6)$ into the equation and solve

for a: $-6 = -4a(64) \Rightarrow a = \dfrac{3}{128}$ and the

equation of the parabola is $y = -\dfrac{3}{32}x^2$.

Let $y = -20$ and solve for x:

$-20 = -\dfrac{3}{32}x^2 \Rightarrow x^2 = \dfrac{640}{3} \Rightarrow x = \pm 14.61$.

We are looking for the positive solution. The water hits the ground 14.6 feet from the end of the pipe.

57. The equation of the ellipse is $\dfrac{x^2}{36} + \dfrac{y^2}{49/4} = 1$.

At $x = 4$, $y = \dfrac{7\sqrt{5}}{6} \approx 2.61$. (Reject the

negative solution.)
This is the radius of the cross section. The circumference $= 2\pi(2.61) \approx 16.4$ inches.

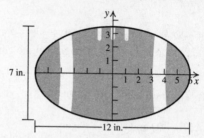

7 in.

12 in.

Chapter 7 Practice Test A

1. The focus is at $(0, 12)$ and the directrix is $x = -12$, so the vertex is at $(-6, 12)$ and the parabola opens to the right. The general form is $(y - k)^2 = 4a(x - h)$. $a = 0 - (-6) = 6 \Rightarrow$

$(y - 12)^2 = 4(6)(x + 6) = 24(x + 6)$.

2. $y^2 - 2y + 8x + 25 = 0 \Rightarrow$

$y^2 - 2y + 1 = -8x - 25 + 1 \Rightarrow$

$(y - 1)^2 = -8(x + 3)$

3. $x^2 = -9y = 4ay \Rightarrow a = -\dfrac{9}{4}$

Focus: $\left(0, -\dfrac{9}{4}\right)$, directrix: $y = \dfrac{9}{4}$.

4.

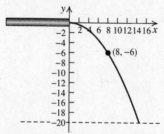

5. $(x + 2)^2 = -8(y - 1) = 4a(y - 1) \Rightarrow a = -2$.
Vertex: $(-2, 1)$, focus: $(-2, -1)$, directrix: $y = 3$.

6. Let the vertex be at (0, 0). Then the parabola goes through the point (3, 2). The general form of the equation is $(x-h)^2 = 4a(y-k)$. Substitute the values for x, y, h, and k, then solve for a: $3^2 = 4a(2) \Rightarrow a = \dfrac{9}{8}$. The equation of the parabola is $x^2 = \dfrac{9}{2}y$. Now find the value of y for $x = 1$:

   $1 = \dfrac{9}{2}y \Rightarrow y = \dfrac{2}{9} \approx 0.22$ feet.

7. Vertex (3, −1) and directrix $x = -3 \Rightarrow a = 6$. The parabola opens to the right, so the general form of the equation is $(y-k)^2 = 4a(x-h)$. The equation is $(y+1)^2 = 24(x-3)$.

8. Foci (0, −2) and (0, 2) $\Rightarrow c = 2$. Vertices (0, −4) and (0, 4) $\Rightarrow a = 4$. $b^2 = a^2 - c^2 \Rightarrow b^2 = 4^2 - 2^2 = 12$. The equation is $\dfrac{x^2}{12} + \dfrac{y^2}{16} = 1$.

9. Foci (−2, 0) and (2, 0) $\Rightarrow c = 2$. y-intercepts −5 and 5 $\Rightarrow b = 5$. $b^2 = a^2 - c^2 \Rightarrow 25 = a^2 - 2^2 \Rightarrow a^2 = 29$. The equation is $\dfrac{x^2}{29} + \dfrac{y^2}{25} = 1$.

10. $a = 9$, $b = 2$. The equation is $\dfrac{x^2}{81} + \dfrac{y^2}{4} = 1$.

11.

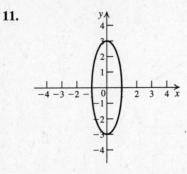

12. $49y^2 - x^2 = 49 \Rightarrow y^2 - \dfrac{x^2}{49} = 1 \Rightarrow$

   $a^2 = 1, b^2 = 49$. The vertices are at (0, 1) and (0, −1).

   $c^2 = a^2 + b^2 \Rightarrow c^2 = 1 + 49 \Rightarrow c = \pm 5\sqrt{2}$. The foci are at $\left(0, 5\sqrt{2}\right)$ and $\left(0, -5\sqrt{2}\right)$.

13.

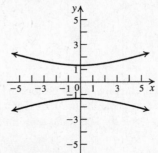

14. The center is (0, 0) and the transverse axis is the y-axis, so the equation is of the form $\dfrac{y^2}{a^2} - \dfrac{x^2}{b^2} = 1$.

   $a = 6, c = \sqrt{45} \Rightarrow 45 = 36 + b^2 \Rightarrow b^2 = 9$.

   The equation is $\dfrac{y^2}{36} - \dfrac{x^2}{9} = 1$.

15. $y^2 - x^2 + 2x = 2 \Rightarrow$
   $y^2 - (x^2 - 2x + 1) = 2 - 1 \Rightarrow y^2 - (x-1)^2 = 1$

16. $x^2 - 2x - 4y^2 - 16y = 19 \Rightarrow$
   $(x^2 - 2x + 1) - 4(y^2 + 4y + 4) = 19 + 1 - 16 \Rightarrow$
   $(x-1)^2 - 4(y+2)^2 = 4 \Rightarrow$
   $\dfrac{(x-1)^2}{4} - (y+2)^2 = 1$

17. Hyperbola

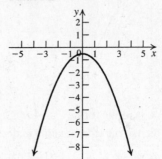

18. Parabola

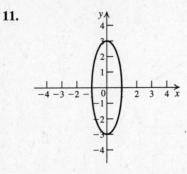

19. Circle

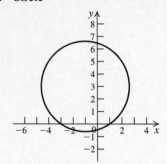

20. Ellipse

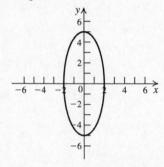

Chapter 7 Practice Test B

1. The focus is at $(-10, 0)$ and the directrix is $x = 10$, so the vertex is at $(0, 0)$ and the parabola opens to the left. The general form is
$$(y-k)^2 = -4a(x-h).$$
$a = 0 - (-10) = 10 \Rightarrow y^2 = -4(10)x = -40x$.
The answer is B.

2. $y^2 - 4y - 5x + 24 = 0 \Rightarrow$
$y^2 - 4y + 4 = 5x - 24 + 4 \Rightarrow (y-2)^2 = 5(x-4)$
. The answer is B.

3. $x = 7y^2 \Rightarrow \dfrac{x}{7} = y^2 \Rightarrow \dfrac{1}{7} = 4a \Rightarrow \dfrac{1}{28} = a \Rightarrow$ the

focus at $\left(\dfrac{1}{28}, 0\right)$ and the directrix is $x = -\dfrac{1}{28}$.

The answer is A.

4. The answer is A.

5. $(x-3)^2 = 12(y-1) \Rightarrow$ the vertex is $(3, 1)$.
$12 = 4a \Rightarrow a = 3 \Rightarrow$ the focus is at $(3, 4)$, and
the directrix is $y = -2$. The answer is A.

6. Let the vertex be located at $(0, 25)$ and one end of the base is at $(90, 0)$. The general form of the equation is $(x-h)^2 = -4a(y-k)$. Substitute the values for x, y, h, and k, then solve for a:

7. Vertex $(-2, 1)$ and directrix $x = 2 \Rightarrow a = 4$.
The parabola opens to the left, so the general form of the equation is
$$(y-k)^2 = -4a(x-h).$$
The equation is $(y-1)^2 = -16(x+2)$.
The answer is A.

$(90-0)^2 = -4a(0-25) \Rightarrow a = 81$
The equation of the parabola is
$x^2 = -4(81)(y-25) \Rightarrow x^2 = -324(y-25)$.
Find the value of y for
$x = 45$: $45^2 = -324(y-25) \Rightarrow y = 18.75$.
The answer is D.

8. Foci $(-3, 0)$ and $(3, 0) \Rightarrow c = 3$.
Vertices $(-5, 0)$ and $(5, 0) \Rightarrow a = 5$.
$b^2 = a^2 - c^2 \Rightarrow b^2 = 5^2 - 3^2 = 16$
The equation is $\dfrac{x^2}{25} + \dfrac{y^2}{16} = 1$.
The answer is B.

9. Foci $(0, -3)$ and $(0, 3) \Rightarrow c = 3$. y-intercepts
-7 and $7 \Rightarrow a = 7$. $b^2 = a^2 - c^2 \Rightarrow$
$49 = 3^2 - b^2 \Rightarrow b^2 = 40$. The equation is
$\dfrac{x^2}{40} + \dfrac{y^2}{49} = 1$. The answer is D.

10. $a = 8$, $b = 4$. The equation is $\dfrac{x^2}{16} + \dfrac{y^2}{64} = 1$.
The answer is A.

11. $9(x-1)^2 + 4(y-2)^2 = 36 \Rightarrow$
$\dfrac{(x-1)^2}{4} + \dfrac{(y-2)^2}{9} = 1$. The answer is A.

12. $a = 11$, $b = 2$. The vertices are at $(11, 0)$ and
$(-11, 0)$. $c^2 = a^2 + b^2 \Rightarrow c^2 = 121 + 4 \Rightarrow$
$c = 5\sqrt{5}$. The foci are at $\left(-5\sqrt{5}, 0\right)$ and
$\left(5\sqrt{5}, 0\right)$. The answer is C.

13. The answer is B.

14. The center is (0, 0) and the transverse axis is the y-axis, so the equation is of the form

$$\frac{y^2}{a^2} - \frac{x^2}{b^2} = 1.$$

$$a = 5, c = 10 \Rightarrow 100 = 25 + b^2 \Rightarrow b^2 = 75.$$

The equation is $\dfrac{y^2}{25} - \dfrac{x^2}{75} = 1$.

The answer is B.

15. $y^2 - 4x^2 - 2y - 16x - 19 = 0 \Rightarrow$

$$y^2 - 2y + 1 - 4(x^2 + 4x + 4) = 19 + 1 - 16 \Rightarrow$$

$$(y-1)^2 - 4(x+2)^2 = 4 \Rightarrow$$

$$\frac{(y-1)^2}{4} - (x+2)^2 = 1. \text{ The answer is C.}$$

16. $4y^2 - 9x^2 - 16y - 36x - 56 = 0 \Rightarrow$

$$4(y^2 - 4y + 4) - 9(x^2 + 4x + 4)$$
$$= 56 + 16 - 36 \Rightarrow$$

$$4(y-2)^2 - 9(x+2)^2 = 36 \Rightarrow$$

$$\frac{(y-2)^2}{9} - \frac{(x+2)^2}{4} = 1. \text{ The answer is C.}$$

17. C **18.** A **19.** D **20.** D

Cumulative Review Exercises (Chapters P–7)

1. $\dfrac{\left((x+h)^2 - 3(x+h) + 2\right) - \left(x^2 - 3x + 2\right)}{h}$

$$= \frac{x^2 + 2xh + h^2 - 3x - 3h + 2 - x^2 + 3x - 2}{h}$$

$$= \frac{h^2 + 2xh - 3h}{h} = h + 2x - 3$$

2.

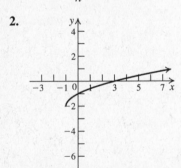

3. Switch the variables, then solve for y:

$$y = 2x - 3 \Rightarrow x = 2y - 3 \Rightarrow \frac{x+3}{2} = y = f^{-1}(x)$$

$$f(f^{-1}(x)) = 2\left(\frac{x+3}{2}\right) - 3 = x$$

4.

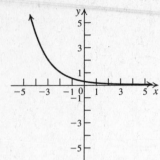

5. $\log_5(x-1) + \log_5(x-2) = 3\log_5 \sqrt[3]{6} \Rightarrow$

$$\log_5\left((x-1)(x-2)\right) = \log_5\left(\sqrt[3]{6}\right)^3 \Rightarrow$$

$$x^2 - 3x + 2 = 6 \Rightarrow x^2 - 3x - 4 = 0 \Rightarrow$$

$(x-4)(x+1) = 0 \Rightarrow x = 4 \text{ or } x = -1.$ (Reject the negative solution.) The solution is $\{4\}$.

6. $\log_a \sqrt[3]{x\sqrt{yz}} = \log_a\left(x\sqrt{yz}\right)^{1/3}$

$$= \frac{1}{3}\log_a x + \frac{1}{3}\log_a \sqrt{yz}$$

$$= \frac{1}{3}\log_a x + \frac{1}{3}\log_a (yz)^{1/2}$$

$$= \frac{1}{3}\log_a x + \frac{1}{3} \cdot \frac{1}{2}\log_a (yz)$$

$$= \frac{1}{3}\log_a x + \frac{1}{6}\log_a y + \frac{1}{6}\log_a z$$

7. $\dfrac{x}{x-2} \geq 1 \Rightarrow \dfrac{x}{x-2} - 1 \geq 0 \Rightarrow \dfrac{2}{x-2} \geq 0.$

$x - 2 > 0 \Rightarrow x > 2.$ The solution is $(2, \infty)$.

8. $I = \dfrac{V}{R}(1 - e^{-0.3t}) \Rightarrow 1 - \dfrac{IR}{V} = e^{-0.3t} \Rightarrow$

$$\ln\left(1 - \frac{IR}{V}\right) = -0.3t \Rightarrow t = -\frac{\ln\left(1 - \dfrac{IR}{V}\right)}{0.3}$$

9. Solve using elimination:

$$\begin{cases} 1.4x - 0.5y = 1.3 \\ 0.4x + 1.1y = 4.1 \end{cases} \Rightarrow \begin{cases} 0.56x - 0.2y = 0.52 \\ -0.56x - 1.54y = -5.74 \end{cases} \Rightarrow$$

$-1.74y = -5.22 \Rightarrow y = 3$

$0.4x + 1.1(3) = 4.1 \Rightarrow x = 2$

The solution is $\{(2, 3)\}$.

10. Switch the first and second equations:

$$\begin{cases} 2x + y - 4z = 3 \\ x - 2y + 3z = 4 \\ -3x + 4y - z = -2 \end{cases} \Rightarrow \begin{cases} x - 2y + 3z = 4 \\ 2x + y - 4z = 3 \\ -3x + 4y - z = -2 \end{cases}.$$

Multiply the first equation by -2, add the result to the second equation, and replace the second equation with the new equation:

(continued on next page)

(*continued*)

$$\begin{cases} x-2y+3z=4 \\ 2x+y-4z=3 \\ -3x+4y-z=-2 \end{cases} \Rightarrow \begin{cases} x-2y+3z=4 \\ 5y-10z=-5 \\ -3x+4y-z=-2 \end{cases}$$

Divide the second equation by 5 and replace the equation with the result. Then multiply the first equation by 3, add the result to the third equation, and replace the third equation with the new equation.

$$\begin{cases} x-2y+3z=4 \\ 5y-10z=-5 \\ -3x+4y-z=-2 \end{cases} \Rightarrow \begin{cases} x-2y+3z=4 \\ y-2z=-1 \\ -2y+8z=10 \end{cases}.$$

Divide the third equation by -2 and replace the equation with the result:

$$\begin{cases} x-2y+3z=4 \\ y-2z=-1 \\ -2y+8z=10 \end{cases} \Rightarrow \begin{cases} x-2y+3z=4 \\ y-2z=-1 \\ y-4z=-5 \end{cases}.$$

Subtract the third equation from the second equation and solve for z:

$$\begin{cases} x-2y+3z=4 \\ y-2z=-1 \\ y-4z=-5 \end{cases} \Rightarrow \begin{cases} x-2y+3z=4 \\ y-2z=-1 \\ z=2 \end{cases}$$

Substitute $z=2$ into the second equation and solve for y: $y-2(2)=-1 \Rightarrow y=3$. Substitute the values for y and z into the first equation and solve for x: $x-2(3)+3(2)=4 \Rightarrow x=4$. The solution is $\{(4, 3, 2)\}$.

11. Solve using substitution:

$$\begin{cases} y=2-\log x \\ y-\log(x+3)=1 \end{cases} \Rightarrow$$
$$2-\log x - \log(x+3)=1 \Rightarrow$$
$$\log x + \log(x+3)=1 \Rightarrow \log\big(x(x+3)\big)=1 \Rightarrow$$
$$x^2+3x=10 \Rightarrow x^2+3x-10=0 \Rightarrow$$
$$(x+5)(x-2)=0 \Rightarrow x=-5 \text{ or } x=2$$

(Reject the negative solution.)
The solution is $\{(2, 2-\log 2)\}$.

12. Solve using substitution:

$$\begin{cases} y=x^2-1 \\ 3x^2+8y^2=8 \end{cases} \Rightarrow \begin{cases} y+1=x^2 \\ 3x^2+8y^2=8 \end{cases} \Rightarrow$$
$$3(y+1)+8y^2=8 \Rightarrow 8y^2+3y-5=0 \Rightarrow$$
$$(y+1)(8y-5)=0 \Rightarrow y=-1 \text{ or } y=\frac{5}{8}$$

$$-1=x^2-1 \Rightarrow x^2=0 \Rightarrow x=0$$
$$\frac{5}{8}=x^2-1 \Rightarrow \frac{13}{8}=x^2 \Rightarrow x=\pm\frac{\sqrt{26}}{4}$$

The solutions are $\left\{(0,-1), \left(\dfrac{\sqrt{26}}{4}, \dfrac{5}{8}\right), \left(-\dfrac{\sqrt{26}}{4}, \dfrac{5}{8}\right)\right\}.$

13. Expand by the first column:

$$\begin{vmatrix} 1 & 4 & 7 \\ 2 & 5 & 8 \\ 3 & 6 & 9 \end{vmatrix} = a_{11}A_{11}+a_{21}A_{21}+a_{31}A_{31}$$

$$=1(-1)^{1+1}\begin{vmatrix} 5 & 8 \\ 6 & 9 \end{vmatrix}+2(-1)^{2+1}\begin{vmatrix} 4 & 7 \\ 6 & 9 \end{vmatrix}$$
$$+3(-1)^{3+1}\begin{vmatrix} 4 & 7 \\ 5 & 8 \end{vmatrix}$$
$$=-3-2(-6)+3(-3)=0$$

14. $\begin{cases} 2x-3y=-4 \\ 5x+7y=1 \end{cases} \Rightarrow D=\begin{vmatrix} 2 & -3 \\ 5 & 7 \end{vmatrix}=29$

$$D_x=\begin{vmatrix} -4 & -3 \\ 1 & 7 \end{vmatrix}=-25, D_y=\begin{vmatrix} 2 & -4 \\ 5 & 1 \end{vmatrix}=22$$
$$x=-\frac{25}{29}, y=\frac{22}{29}$$

15. $A=\begin{vmatrix} 3 & -2 \\ -5 & 4 \end{vmatrix} \Rightarrow A^{-1}=\dfrac{1}{12-10}\begin{vmatrix} 4 & 2 \\ 5 & 3 \end{vmatrix}=\begin{vmatrix} 2 & 1 \\ \dfrac{5}{2} & \dfrac{3}{2} \end{vmatrix}$

16. To find the point of intersection, solve the system

$$\begin{cases} x+2y-3=0 \\ 3x+4y-5=0 \end{cases} \Rightarrow \begin{cases} -2x-4y+6=0 \\ 3x+4y-5=0 \end{cases} \Rightarrow$$
$$x=-1; -1+2y-3=0 \Rightarrow y=2$$

The point of intersection is $(-1, 2)$. The line

$$x-3y+5=0 \Rightarrow y=\frac{1}{3}x+\frac{5}{3} \Rightarrow \text{ its slope is}$$

$1/3$. The slope of the perpendicular is -3. The equation of the line through $(-1, 2)$ with slope -3 is $y-2=-3(x+1) \Rightarrow y=-3x-1$.

17. Let $u=x^2$. Then the equation becomes

$$2u^2-5u+3=0 \Rightarrow (2u-3)(u-1)=0 \Rightarrow$$
$$u=\frac{3}{2} \text{ or } u=1. \text{ Now solve for } x:$$
$$x^2=\frac{3}{2} \Rightarrow x=\pm\frac{\sqrt{6}}{2}; x^2=1 \Rightarrow x=\pm 1.$$

The solutions are $\left\{\pm\dfrac{\sqrt{6}}{2}, \pm 1\right\}.$

18. Hyperbola

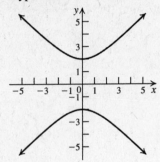

19. Circle

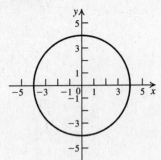

20.

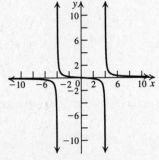

Chapter 8 Further Topics in Algebra

8.1 Sequences and Series

8.1 Practice Problems

1. $a_n = -2^n \Rightarrow a_1 = -2^1 = -2; \; a_2 = -2^2 = -4;$
$a_3 = -2^3 = -8; \; a_4 = -2^4 = -16$

2.

$n:$	1	2	3	4	5	...	n
term:	0	$-\dfrac{1}{2}$	$\dfrac{2}{3}$	$-\dfrac{3}{4}$	$\dfrac{4}{5}$	...	a_n

Since the terms alternate, and the negative terms are the even-numbered terms, use the factor $(-1)^{n+1}$ to alternate the signs. The absolute value of each term is $\dfrac{n-1}{n} = 1 - \dfrac{1}{n}$, so the general term is $(-1)^{n+1}\left(1 - \dfrac{1}{n}\right)$.

3. $a_1 = -3, \; a_{n+1} = 2a_n + 5$
$a_2 = 2a_1 + 5 = 2(-3) + 5 = -1$
$a_3 = 2a_2 + 5 = 2(-1) + 5 = 3$
$a_4 = 2a_3 + 5 = 2(3) + 5 = 11$
$a_5 = 2a_4 + 5 = 2(11) + 5 = 27$

4. The female bee is a_0. Since female bees have two parents, $a_1 = 2$. The female parent has two parents and the male parent has one, so $a_2 = 3$. There are two females and one male in this generation, so the next generation has three females and two males. We can use a tree to represent this:

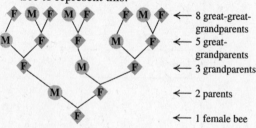

← 8 great-great-grandparents
← 5 great-grandparents
← 3 grandparents
← 2 parents
← 1 female bee

$a_k = a_{k-2} + a_{k-1}$ for $k \ge 3$.

5. a. $\dfrac{13!}{12!} = \dfrac{13 \cdot 12!}{12!} = 13$

b. $\dfrac{n!}{(n-3)!} = \dfrac{n(n-1)(n-2)(n-3)!}{(n-3)!}$
$= n(n-1)(n-2)$

6. $a_n = \dfrac{(-1)^n 2^n}{n!} \Rightarrow a_1 = \dfrac{(-1)^1 2^1}{1!} = -2$

$a_2 = \dfrac{(-1)^2 2^2}{2!} = \dfrac{4}{2 \cdot 1} = 2$

$a_3 = \dfrac{(-1)^3 2^3}{3!} = \dfrac{-8}{3 \cdot 2 \cdot 1} = -\dfrac{8}{6} = -\dfrac{4}{3}$

$a_4 = \dfrac{(-1)^4 2^4}{4!} = \dfrac{16}{4 \cdot 3 \cdot 2 \cdot 1} = \dfrac{16}{24} = \dfrac{2}{3}$

$a_5 = \dfrac{(-1)^5 2^5}{5!} = \dfrac{-32}{5 \cdot 4 \cdot 3 \cdot 2 \cdot 1} = -\dfrac{32}{120} = -\dfrac{4}{15}$

7. $\displaystyle\sum_{k=0}^{3} (-1)^k k!$
$= (-1)^0 (0!) + (-1)^1 (1!) + (-1)^2 (2!) + (-1)^3 (3!)$
$= 1 + (-1) + (2 \cdot 1) + (-1)(3 \cdot 2 \cdot 1) = -4$

8. $2 - 4 + 6 - 8 + 10 - 12 + 14$ alternates addition and subtraction of consecutive even integers from 2 to 14. The even integers can be represented as $2k$ with k ranging from 1 to 7. We alternate the signs using the factor $(-1)^{k+1}$. Thus, the expression is

$\displaystyle\sum_{k=1}^{7} (-1)^{k+1}(2k)$.

8.1 Basic Concepts and Skills

1. An infinite sequence is a function whose domain is the set of <u>positive integers</u>.

3. By definition, $0! = \underline{1}$.

5. False. $a_1 = 3, \; a_2 = a_1^2 = 3^2 = 9,$
$a_3 = a_2^2 = 9^2 = 81$

7. $a_1 = 3(1) - 2 = 1, \; a_2 = 3(2) - 2 = 4,$
$a_3 = 3(3) - 2 = 7, \; a_4 = 3(4) - 2 = 10$

9. $a_1 = 1 - \dfrac{1}{1} = 0, \; a_2 = 1 - \dfrac{1}{2} = \dfrac{1}{2},$
$a_3 = 1 - \dfrac{1}{3} = \dfrac{2}{3}, \; a_4 = 1 - \dfrac{1}{4} = \dfrac{3}{4}$

11. $a_1 = -1^2 = -1, \; a_2 = -2^2 = -4,$
$a_3 = -3^2 = -9, \; a_4 = -4^2 = -16$

13. $a_1 = \dfrac{2(1)}{1+1} = 1, a_2 = \dfrac{2(2)}{2+1} = \dfrac{4}{3}, a_3 = \dfrac{2(3)}{3+1} = \dfrac{3}{2},$

$a_4 = \dfrac{2(4)}{4+1} = \dfrac{8}{5}$

15. $a_1 = (-1)^{1+1} = 1, a_2 = (-1)^{2+1} = -1,$

$a_3 = (-1)^{3+1} = 1, a_4 = (-1)^{4+1} = -1$

17. $a_1 = 3 - \dfrac{1}{2^1} = \dfrac{5}{2}, a_2 = 3 - \dfrac{1}{2^2} = \dfrac{11}{4},$

$a_3 = 3 - \dfrac{1}{2^3} = \dfrac{23}{8}, a_4 = 3 - \dfrac{1}{2^4} = \dfrac{47}{16}$

19. $a_1 = a_2 = a_3 = a_4 = 0.6$

21. $a_1 = \dfrac{(-1)^1}{1!} = -1, a_2 = \dfrac{(-1)^2}{2!} = \dfrac{1}{2},$

$a_3 = \dfrac{(-1)^3}{3!} = -\dfrac{1}{6}, a_4 = \dfrac{(-1)^4}{4!} = \dfrac{1}{24}$

23. $a_1 = (-1)^1 3^{-1} = -\dfrac{1}{3}, a_2 = (-1)^2 3^{-2} = \dfrac{1}{9},$

$a_3 = (-1)^3 3^{-3} = -\dfrac{1}{27}, a_4 = (-1)^4 3^{-4} = \dfrac{1}{81}$

25. $a_1 = \dfrac{e^1}{2(1)} = \dfrac{e}{2}, a_2 = \dfrac{e^2}{2(2)} = \dfrac{e^2}{4}, a_3 = \dfrac{e^3}{2(3)} = \dfrac{e^3}{6},$

$a_4 = \dfrac{e^4}{2(4)} = \dfrac{e^4}{8}$

27. $a_n = 3n - 2$

29. $a_n = \dfrac{1}{n+1}$

31. $a_n = (-1)^{n+1}(2)$

33. $a_n = \dfrac{3^{n-1}}{2^n}$

35. $a_n = n(n+1)$

37. $a_n = 2 - \dfrac{(-1)^n}{n+1}$

39. $a_n = \dfrac{3^{n+1}}{n+1}$

41. $a_1 = 2, a_2 = 2 + 3 = 5, a_3 = 5 + 3 = 8,$
$a_4 = 8 + 3 = 11, a_5 = 11 + 3 = 14$

43. $a_1 = 3, a_2 = 2(3) = 6, a_3 = 2(6) = 12,$
$a_4 = 2(12) = 24, a_5 = 2(24) = 48$

45. $a_1 = 7, a_2 = -2(7) + 3 = -11,$
$a_3 = -2(-11) + 3 = 25, a_4 = -2(25) + 3 = -47,$
$a_5 = -2(-47) + 3 = 97$

47. $a_1 = 2, a_2 = \dfrac{1}{2}, a_3 = \dfrac{1}{1/2} = 2, a_4 = \dfrac{1}{2},$

$a_5 = \dfrac{1}{1/2} = 2$

49. $a_1 = 25, a_2 = \dfrac{(-1)^1}{5(25)} = -\dfrac{1}{125},$

$a_3 = \dfrac{(-1)^2}{5(-1/125)} = -25, a_4 = \dfrac{(-1)^3}{5(-25)} = \dfrac{1}{125},$

$a_5 = \dfrac{(-1)^4}{5(1/125)} = 25$

In exercises 51–56, we show only the first screen to determine the terms of the sequence before we list the ten terms.

51. a.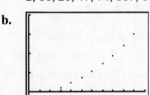

2, 11, 26, 47, 74, 107, 146, 191, 242, 299

b.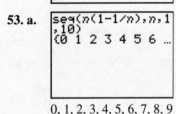

53. a. seq($n(1-1/n),n,1,10$)
{0 1 2 3 4 5 6 …

0, 1, 2, 3, 4, 5, 6, 7, 8, 9

b.

55. a.

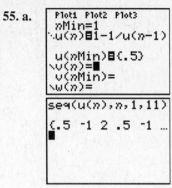

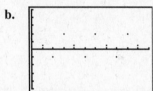

$$0.5, -1, 2, 0.5, -1, 2, 0.5, -1, 2, 0.5$$

b.

57. $\dfrac{3!}{5!} = \dfrac{3!}{5 \cdot 4 \cdot 3!} = \dfrac{1}{20}$

59. $\dfrac{12!}{11!} = \dfrac{12 \cdot 11!}{11!} = 12$

61. $\dfrac{n!}{(n+1)!} = \dfrac{n!}{(n+1) \cdot n!} = \dfrac{1}{n+1}$

63. $\dfrac{(2n+1)!}{(2n)!} = \dfrac{(2n+1) \cdot (2n)!}{(2n)!} = 2n+1$

65. $\displaystyle\sum_{k=1}^{7} 5 = 5+5+5+5+5+5+5 = 35$

67. $\displaystyle\sum_{j=0}^{5} j^2 = 0^2 + 1^2 + 2^2 + 3^2 + 4^2 + 5^2 = 55$

69. $\displaystyle\sum_{i=1}^{5} (2i-1) = (2(1)-1) + (2(2)-1) + (2(3)-1)$
$$+ (2(4)-1) + (2(5)-1) = 25$$

71. $\displaystyle\sum_{j=3}^{7} \dfrac{j+1}{j} = \dfrac{4}{3} + \dfrac{5}{4} + \dfrac{6}{5} + \dfrac{7}{6} + \dfrac{8}{7} = \dfrac{853}{140}$

73. $\displaystyle\sum_{i=2}^{6} (-1)^i 3^{i-1}$
$$= (-1)^2 (3^1) + (-1)^3 (3^2) + (-1)^4 (3^3)$$
$$+ (-1)^5 (3^4) + (-1)^6 (3^5)$$
$$= 183$$

75. $\displaystyle\sum_{k=4}^{7} (2 - k^2)$
$$= \left(2 - 4^2\right) + \left(2 - 5^2\right) + \left(2 - 6^2\right) + \left(2 - 7^2\right)$$
$$= -118$$

77. $\displaystyle\sum_{k=1}^{51} 2k - 1$ **79.** $\displaystyle\sum_{k=1}^{11} \dfrac{1}{5k}$

81. $\displaystyle\sum_{k=1}^{50} \dfrac{(-1)^{k+1}}{k}$ **83.** $\displaystyle\sum_{k=1}^{10} \dfrac{k}{k+1}$

85.
```
sum(seq(12n²,n,1
,10))
              4620
```

87.
```
sum(seq(7/(1-n²)
,n,5,30))
        -1.345430108
```

89.
```
sum(seq((-1)^n(n
),n,1,100))
                50
```

8.1 Applying the Concepts

91. a. Use a table to determine the pattern. It appears that each second, the body falls 32 feet per second faster than it did in the second before.

Time	Distance each second
1	16
2	48
3	80
4	112
5	144
6	176
7	208

b. $16 + 32(n-1) = 32n - 16$ feet

93. The fine is $50 plus $25 for each additional day the work is not done. So on the ninth day, she will be fined $50 + 8($25) = $250.

95. In three years, there are 6 six-month periods. For the first six-month period, cell phone usage was 600 minutes per month. For the next 5 six-month periods, the usage doubled during each period. At the end of three years, cell phone use was $600 \cdot 2^5 = 19,200$ minutes per month.

97. a.
$$A_1 = 10,000\left(1+\frac{0.06}{2}\right)^1 = \$10,300$$

$$A_2 = 10,000\left(1+\frac{0.06}{2}\right)^2 = \$10,609$$

$$A_3 = 10,000\left(1+\frac{0.06}{2}\right)^3 = \$10,927.27$$

$$A_4 = 10,000\left(1+\frac{0.06}{2}\right)^4 = \$11,255.09$$

$$A_5 = 10,000\left(1+\frac{0.06}{2}\right)^5 = \$11,592.74$$

$$A_6 = 10,000\left(1+\frac{0.06}{2}\right)^6 = \$11,940.52$$

b. The interest is compounded semiannually, so in eight years, there are 16 compounding periods.

$$A_{16} = 100\left(1+\frac{0.08}{4}\right)^{16} = \$16,047.10$$

99.
$A_1 = (100,000)(1.05) = \$105,000$
$A_2 = (105,000)(1.05) = \$110,250$
$A_3 = (110,250)(1.05) = \$115,762.50$
$A_4 = (115,762.50)(1.05) = \$121,550.63$
$A_5 = (121,550.63)(1.05) = \$127,628.16$
$A_6 = (127,628.16)(1.05) = \$134,009.56$
$A_7 = (134,009.56)(1.05) = \$140,710.04$

The formula is $100,000\left(1.05^n\right)$.

8.1 Beyond the Basics

101. $a_1 = 2^{1/2}, a_2 = \left(2 \cdot 2^{1/2}\right)^{1/2} = 2^{3/4},$

$a_3 = \left(2 \cdot 2^{3/4}\right)^{1/2} = 2^{7/8},$

$a_4 = \left(2 \cdot 2^{7/8}\right)^{1/2} = 2^{15/16},$

$a_5 = \left(2 \cdot 2^{15/16}\right)^{1/2} = 2^{31/32}$

$a_n = 2^{(2^n-1)/2^n} = 2^{1-1/2^n}$

103. a. $a_1 = 1, a_2 = \dfrac{1}{2}, a_3 = \dfrac{1}{4}, a_4 = \dfrac{1}{8}, a_5 = \dfrac{1}{16}$

b. $a_n = \dfrac{1}{2^{n-1}}$

105. $a_1 = a_2 = 1,$
$a_3 = a_{a_2} + a_{3-a_2} = a_1 + a_2 = 1+1 = 2,$
$a_4 = a_{a_3} + a_{4-a_3} = a_2 + a_2 = 1+1 = 2,$
$a_5 = a_{a_4} + a_{5-a_4} = a_2 + a_3 = 1+2 = 3,$
$a_6 = a_{a_5} + a_{6-a_5} = a_3 + a_3 = 2+2 = 4,$
$a_7 = a_{a_6} + a_{7-a_6} = a_4 + a_3 = 2+2 = 4,$
$a_8 = a_{a_7} + a_{8-a_7} = a_4 + a_4 = 2+2 = 4,$
$a_9 = a_{a_8} + a_{9-a_8} = a_4 + a_5 = 2+3 = 5,$
$a_{10} = a_{a_9} + a_{10-a_9} = a_5 + a_5 = 3+3 = 6$

107. $\displaystyle\sum_{n=1}^{20} n^2 = \sum_{m=0}^{19} a_m$

Write the first five terms to determine the pattern: $\displaystyle\sum_{n=1}^{1} n^2 = 1 = \sum_{m=0}^{0} a_m = a_0 \Rightarrow a_0 = 1,$

$\displaystyle\sum_{n=1}^{2} n^2 = 1 + 2^2 = 5 = \sum_{m=0}^{1} a_m = a_0 + a_1 \Rightarrow$
$5 = 1 + a_1 \Rightarrow a_1 = 4 = 2^2.$

$\displaystyle\sum_{n=1}^{3} n^2 = 1 + 2^2 + 3^2 = 14$

$\displaystyle = \sum_{m=0}^{2} a_m = a_0 + a_1 + a_2 \Rightarrow$
$14 = 1 + 4 + a_2 \Rightarrow a_2 = 9 = 3^2$

$\displaystyle\sum_{n=1}^{4} n^2 = 1 + 2^2 + 3^2 + 4^2 = 30$

$\displaystyle = \sum_{m=0}^{3} a_m = a_0 + a_1 + a_2 + a_3 \Rightarrow$
$30 = 1 + 4 + 9 + a_3 \Rightarrow a_3 = 16 = 4^2$

$\displaystyle\sum_{n=1}^{5} n^2 = 1 + 2^2 + 3^2 + 4^2 + 5^2 = 55$

$\displaystyle = \sum_{m=0}^{4} a_m = a_0 + a_1 + a_2 + a_3 + a_4 \Rightarrow$
$55 = 1 + 4 + 9 + 16 + a_4 \Rightarrow a_4 = 25 = 5^2$

Following the pattern, we see that
$a_m = (m+1)^2.$

109. $\displaystyle\sum_{n=0}^{10} n^3 = \sum_{m=p}^{q} (m-2)^3 \Rightarrow n^3 = (m-2)^3 \Rightarrow$
$n = m-2 \Rightarrow m = n+2$. So, $p = 2$ and $q = 12$.

111. $7! - 2(5!) = (7 \cdot 6)(5!) - 2(5!) = (42 - 2)(5!) = 40(5!)$

113. $\dfrac{1}{5!} + \dfrac{1}{6!} = \dfrac{x}{7!} \Rightarrow \dfrac{1}{5!} + \dfrac{1}{6 \cdot 5!} = \dfrac{x}{7 \cdot 6 \cdot 5!}$

Multiply both sides of the equation by $7! = 7 \cdot 6 \cdot 5!$ and simplify. Then solve for x.

$\dfrac{7 \cdot 6 \cdot 5!}{5!} + \dfrac{7 \cdot 6 \cdot 5!}{6 \cdot 5!} = \dfrac{7 \cdot 6 \cdot 5! x}{7 \cdot 6 \cdot 5!} \Rightarrow 42 + 7 = x \Rightarrow x = 49$

8.1 Critical Thinking/Discussion/Writing

115. $n^2 + n = (n-1)(n-2)(n-3)(n-4)(n-5) + n^2 + n \Rightarrow 0 = (n-1)(n-2)(n-3)(n-4)(n-5) \Rightarrow n = 1, 2, 3, 4, 5$, so $k = 5$ is the upper limit.

116. If $a_0 = 13$, then $a_1 = 3(13) + 1 = 40$, $a_2 = a_1/2 = 40/2 = 20, a_3 = a_2/2 = 20/2 = 10$, $a_4 = a_3/2 = 10/2 = 5$, $a_5 = 3a_4 + 1 = 3(5) + 1 = 16$, $a_6 = a_5/2 = 16/2 = 8, a_7 = a_6/2 = 8/2 = 4$, $a_8 = a_7/2 = 4/2 = 2$, . $a_9 = a_8/2 = 2/2 = 1$

So the conjecture is true for $a_0 = 13$.

117. $\dfrac{(2n)!}{n!} = \dfrac{(2n)(2n-1)(2n-2)(2n-3)(2n-4)(2n-5)\cdots}{n(n-1)(n-2)\cdots} = \dfrac{2n(2n-1)2(n-1)(2n-3)2(n-2)(2n-5)\cdots}{n(n-1)(n-2)\cdots}$

$= 2^n (2n-1)(2n-3)(2n-5)\cdots$

Note that the odd numbers are represented by $2n - 1, 2n - 3, 2n - 5$, etc. Therefore,

$\dfrac{(2n)!}{n!} = 2^n (2n-1)(2n-3)(2n-5)\cdots = 2^n \left[1 \cdot 3 \cdot 5 \cdot (2n-1)\right]$.

118. $n!$ is divisible by all integers between 2 and n (from the definition of $n!$). So, if we divide $(n! + 1)$ by any integer between 2 and n, we will end up with remainder 1. Hence, $(n! + 1)$ is not divisible by any integer between 2 and n.

8.1 Maintaining Skills

119. $a_2 - a_1 = 5 - 1 = 4$
$a_3 - a_2 = 9 - 5 = 4$
$a_4 - a_3 = 13 - 9 = 4$
$a_5 - a_4 = 17 - 13 = 4$

121. $a_2 - a_1 = 1 - 6 = -5$
$a_3 - a_2 = -4 - 1 = -5$
$a_4 - a_3 = -9 - (-4) = -5$
$a_5 - a_4 = -14 - (-9) = -5$

123. $a_2 = a_1 + d = 3 + 4 = 7$
$a_3 = a_2 + d = 7 + 4 = 11$
$a_4 = a_3 + d = 11 + 4 = 15$
$a_5 = a_4 + d = 15 + 4 = 19$

125. $a_2 = a_1 + d = 7 + (-3) = 4$
$a_3 = a_2 + d = 4 + (-3) = 1$
$a_4 = a_3 + d = 1 + (-3) = -2$
$a_5 = a_4 + d = -2 + (-3) = -5$

127. $a_n = 3n + 5$
$a_1 = 3(1) + 5 = 8$
$a_2 = 3(2) + 5 = 11$
$a_3 = 3(3) + 5 = 14$
$a_4 = 3(4) + 5 = 17$
$a_5 = 3(5) + 5 = 20$
$a_2 - a_1 = 11 - 8 = 3$
$a_3 - a_2 = 14 - 11 = 3$
$a_4 - a_3 = 17 - 14 = 3$
$a_5 - a_4 = 20 - 17 = 3$
Note that
$a_n - a_{n-1} = (3n + 5) - (3(n-1) + 1)$
$= (3n + 5) - (3n - 3 + 1)$
$= (3n + 5) - (3n - 2) = 3$

129. $a_n = -2n - 6$

$a_1 = -2(1) - 6 = -8$

$a_2 = -2(2) - 6 = -10$

$a_3 = -2(3) - 6 = -12$

$a_4 = -2(4) - 6 = -14$

$a_5 = -2(5) - 6 = -16$

$a_2 - a_1 = -10 - (-8) = -2$

$a_3 - a_2 = -12 - (-10) = -2$

$a_4 - a_3 = -14 - (-12) = -2$

$a_5 - a_4 = -16 - (-14) = -2$

Note that

$a_n - a_{n-1} = (-2n - 6) - (-2(n-1) - 6)$

$= (-2n - 6) - (-2n + 2 - 6)$

$= (-2n - 6) - (-2n - 4) = -2$

131. $a_n = 2n - 5$

$a_{n+1} = 2(n+1) - 5 = 2n + 2 - 5 = 2n - 3$

$a_{n-1} = 2(n-1) - 5 = 2n - 2 - 5 = 2n - 7$

8.2 Arithmetic Sequences; Partial Sums

8.2 Practice Problems

1. $(-2) - 3 = -5$, $-7 - (-2) = -5$, etc. The common difference is -5.

2. $a_1 = -3$ and $d = 1 - (-3) = 4$, so the expression is

$a_n = -3 + 4(n-1) = -3 + 4n - 4 = 4n - 7$.

3. $a_4 = a_1 + d(4-1) \Rightarrow 41 = a_1 + 3d$

$a_{15} = a_1 + d(15-1) \Rightarrow 8 = a_1 + 14d$

$\begin{cases} a_1 + 3d = 41 \\ a_1 + 14d = 8 \end{cases} \Rightarrow -11d = 33 \Rightarrow d = -3, a_1 = 50$

$a_n = 50 - 3(n-1) = -3n + 53$

4. $d = \dfrac{5}{6} - \dfrac{2}{3} = \dfrac{1}{6}$; $n = 10$; $a_1 = \dfrac{2}{3}$; $a_{10} = \dfrac{13}{6}$

$S_{10} = 10\left(\dfrac{\frac{2}{3} + \frac{13}{6}}{2}\right) = \dfrac{85}{6}$

5. $d = 32$; $n = 5$; $a_{11} = 32(11) - 16 = 336$;

$a_{15} = 32(15) - 16 = 464$

$S = 5\left(\dfrac{336 + 464}{2}\right) = 2000$

The object fell 2000 feet during the 11th through 15th seconds.

8.2 Basic Concepts and Skills

1. If 5 is the common difference of an arithmetic sequence with general term a_n, then

$a_{17} - a_{16} = \underline{5}$.

3. If 14 is the term immediately following the sequence term 17 is an arithmetic sequence, then the common difference is $\underline{-3}$.

5. False. The common difference of an arithmetic sequence may be positive or negative.

7. The sequence is arithmetic. $a_1 = 1, d = 1$

9. The sequence is arithmetic. $a_1 = 2, d = 3$

11. The sequence is not arithmetic.

13. The sequence is not arithmetic.

15. The sequence is arithmetic. $a_1 = 0.6, d = -0.4$

17. Write the first four terms of the sequence:

$a_1 = 2(1) + 6 = 8, a_2 = 2(2) + 6 = 10,$

$a_3 = 2(3) + 6 = 12, a_4 = 2(4) + 6 = 14$.

It appears that the sequence is arithmetic, with a common difference of 2. Verify as follows:

$d = a_{n+1} - a_n = (2(n+1) + 6) - (2n + 6) = 2$.

19. Write the first four terms of the sequence:

$a_1 = 1 - 1^2 = 0, a_2 = 1 - 2^2 = -3,$

$a_3 = 1 - 3^2 = -8, a_4 = 1 - 4^2 = -15$.

There is no common difference, so the sequence is not arithmetic.

21. $d = 3, a_1 = 5 \Rightarrow a_n = 5 + 3(n-1) = 3n + 2$

23. $d = -5, a_1 = 11 \Rightarrow a_n = 11 - 5(n-1) = 16 - 5n$

25. $d = -\dfrac{1}{4}, a_1 = \dfrac{1}{2} \Rightarrow a_n = \dfrac{1}{2} - \dfrac{1}{4}(n-1) = \dfrac{3-n}{4}$

27. $d = -\dfrac{2}{5}, a_1 = -\dfrac{3}{5} \Rightarrow$

$a_n = -\dfrac{3}{5} - \dfrac{2}{5}(n-1) = \dfrac{-1-2n}{5} = -\dfrac{2n+1}{5}$

29. $d = 3, a_1 = e \Rightarrow a_n = e + 3(n-1) = e + 3n - 3$

31. $a_4 = a_1 + d(4-1) \Rightarrow 21 = a_1 + 3d$

$a_{10} = a_1 + d(10-1) \Rightarrow 60 = a_1 + 9d$

$\begin{cases} a_1 + 3d = 21 \\ a_1 + 9d = 60 \end{cases} \Rightarrow -6d = -39 \Rightarrow d = \dfrac{13}{2}, a_1 = \dfrac{3}{2}$

$a_n = \dfrac{3}{2} + \dfrac{13}{2}(n-1) = \dfrac{13}{2}n - 5$

33. $a_7 = a_1 + d(7-1) \Rightarrow 8 = a_1 + 6d$
$a_{15} = a_1 + d(15-1) \Rightarrow -8 = a_1 + 14d$

$\begin{cases} a_1 + 6d = 8 \\ a_1 + 14d = -8 \end{cases} \Rightarrow 8d = -16 \Rightarrow d = -2, a_1 = 20$

$a_n = 20 - 2(n-1) = 22 - 2n$

35. $a_3 = a_1 + d(3-1) \Rightarrow 7 = a_1 + 2d$
$a_{23} = a_1 + d(23-1) \Rightarrow 17 = a_1 + 22d$

$\begin{cases} a_1 + 2d = 7 \\ a_1 + 22d = 17 \end{cases} \Rightarrow 20d = 10 \Rightarrow d = \frac{1}{2}, a_1 = 6$

$a_n = 6 + \frac{1}{2}(n-1) = \frac{n+11}{2}$

37. $n = 50; S = 50\left(\frac{1+50}{2}\right) = 1275$

39. $d = 2, a_1 = 1, a_n = 99 \Rightarrow 99 = 1 + 2(n-1) \Rightarrow$

$n = 50; S = 50\left(\frac{1+99}{2}\right) = 2500$

41. $d = 3, a_1 = 3, a_n = 300 \Rightarrow 300 = 3 + 3(n-1) \Rightarrow$

$n = 100; S = 100\left(\frac{3+300}{2}\right) = 15,150$

43. $d = -3, a_1 = 2, a_n = -34 \Rightarrow$
$-34 = 2 - 3(n-1) \Rightarrow n = 13$

$S = 13\left(\frac{2-34}{2}\right) = -208$

45. $d = \frac{2}{3}, a_1 = \frac{1}{3}, a_n = 7 \Rightarrow$

$7 = \frac{1}{3} + \frac{2}{3}(n-1) \Rightarrow n = 11$

$S = 11\left(\frac{1/3+7}{2}\right) = \frac{121}{3}$

47. $d = 5, n = 50, a_1 = 2 \Rightarrow a_n = 2 + 5(50-1) = 247$

$S = 50\left(\frac{2+247}{2}\right) = 6225$

49. $d = 4, n = 20, a_1 = -15 \Rightarrow$
$a_n = -15 + 4(20-1) = 61$

$S = 20\left(\frac{-15+61}{2}\right) = 460$

51. $d = 0.2, n = 100, a_1 = 3.5 \Rightarrow$
$a_n = 3.5 + 0.2(100-1) = 23.3$

$S = 100\left(\frac{3.5+23.3}{2}\right) = 1340$

In exercises 53–58, use the formula for finding the terms in an arithmetic sequence, $a_n = a_1 + d(n-1)$.

53. $a_n = 75, a_1 = 1, d = 2$
$75 = 1 + 2(n-1) \Rightarrow 37 = n - 1 \Rightarrow n = 38$

55. $a_n = 95, a_1 = -1, d = 4$
$95 = -1 + 4(n-1) \Rightarrow 24 = n - 1 \Rightarrow n = 25$

57. $a_n = 50\sqrt{3}, a_1 = 2\sqrt{3}, d = 2\sqrt{3}$
$50\sqrt{3} = 2\sqrt{3} + 2\sqrt{3}(n-1) \Rightarrow$
$48\sqrt{3} = 2\sqrt{3}(n-1) \Rightarrow 24 = n - 1 \Rightarrow n = 25$

8.2 Applying the Concepts

59. $d = 2, n = 30, a_1 = 10 \Rightarrow a_n = 10 + 2(30-1) = 68$

$S = 30\left(\frac{10+68}{2}\right) = 1170$

61. $d = 25, n = 30, a_1 = 50 \Rightarrow$
$a_{30} = 50 + 25(30-1) = 775$

$S = 30\left(\frac{50+775}{2}\right) = \$12,375$

63. Antonio has 17 different hourly wages over the four years (his original wage plus 16 raises), so $n = 17$. $a_1 = \$12.75, d = 0.25 \Rightarrow$

$a_{13} = 12.75 + 0.25(17-1) = \$16.75$.

65. $n = 25, a_1 = 20, d = 2 \Rightarrow$
$a_{25} = 20 + 2(25-1) = 68$

$S = 25\left(\frac{20+68}{2}\right) = 1100$.

There are 1100 seats in the theater.

67. $n = 17, a_1 = 40, a_{28} = 8 \Rightarrow$

$S = 17\left(\frac{40+8}{2}\right) = 408$

There are 408 boxes in the stack.

69. Using the formula for the sum of an arithmetic sequence with $a_1 = 3$, we have

$192 = n\left(\frac{3+a_n}{2}\right) \Rightarrow 384 = 3n + na_n$. The last

term in the sequence, given $d = 6$, is
$a_n = 3 + 6(n-1) = 6n - 3$. Solve the system:

$\begin{cases} 3n + na_n = 384 \\ a_n = 6n - 3 \end{cases} \Rightarrow 3n + n(6n-3) = 384 \Rightarrow$

$6n^2 = 384 \Rightarrow n^2 = 64 \Rightarrow n = \pm 8$ (Reject the negative solution.) The flea market was open for 8 days.

8.2 Beyond the Basics

71. $a_n = \log\left(\dfrac{a^n}{b^{n-1}}\right),\ a_{n+1} = \log\left(\dfrac{a^{n+1}}{b^n}\right)$

$a_{n+1} - a_n = \log\left(\dfrac{a^{n+1}}{b^n}\right) - \log\left(\dfrac{a^n}{b^{n-1}}\right)$

$= \log\left(\dfrac{\dfrac{a^{n+1}}{b^n}}{\dfrac{a^n}{b^{n-1}}}\right) = \log\left(\dfrac{a^{n+1}}{b^n} \cdot \dfrac{b^{n-1}}{a^n}\right)$

$= \log\dfrac{a}{b}$

73. $S_n = 3n^2 + 4n$

$S_{n-1} = 3(n-1)^2 + 4(n-1)$
$= 3(n^2 - 2n + 1) + 4n - 4$
$= 3n^2 - 6n + 3 + 4n - 4$
$= 3n^2 - 2n - 1$

$a_n = S_n - S_{n-1} = (3n^2 + 4n) - (3n^2 - 2n - 1)$
$= 6n + 1$

$d = a_n - a_{n-1} = (6n + 1) - (6(n-1) + 1)$
$= (6n + 1) - (6n - 5) = 6$

Since $d = 6$ is a constant, the sequence a_n is an arithmetic sequence.

75. 45 is divisible by 3 and the largest number less than 100 that is divisible by 3 is 99, so $a_1 = 45$, $d = 3, a_n = 99 \Rightarrow 99 = 45 + 3(n-1) \Rightarrow n = 19.$

$S = 19\left(\dfrac{45 + 99}{2}\right) = 1368.$

77. The sequence of the reciprocals is
$2, \dfrac{5}{3}, \dfrac{4}{3}, 1, \ldots$ The common difference is $-\dfrac{1}{3}$,
so the sequence is arithmetic, and the original sequence is harmonic.

79. $m = a + d$ and $b = a + 2d$. So
$\dfrac{a + b}{2} = \dfrac{a + a + 2d}{2} = \dfrac{2a + 2d}{2} = a + d = m$

8.2 Critical Thinking/Discussion/Writing

80. $a = a_1 = 1, b = a_n = 30 \Rightarrow$

$S = 465 = n\left(\dfrac{1 + 30}{2}\right) \Rightarrow n = 30$. So, there are

30 terms in the sequence and $k = 28$.
$30 = 1 + d(30 - 1) \Rightarrow d = 1.$

81. $a_1 = 10, a_{21} = a_1 + d(21 - 1) \Rightarrow 0 = 10 + 20d \Rightarrow$

$d = -\dfrac{1}{2} \Rightarrow a_n = 10 - \dfrac{1}{2}(n - 1) = \dfrac{21 - n}{2}$

82. First term $= -a_1$; difference $= -d$.

83. There are $100 - 22 + 1 = 79$ terms in the series.

84. In the set of counting numbers, $a_n = n$.

$12,403 = n\left(\dfrac{1 + n}{2}\right) \Rightarrow 24,806 = n^2 + n \Rightarrow$

$n^2 + n - 24,806 = 0 \Rightarrow (n - 157)(n + 158) = 0 \Rightarrow$

$n = -158$ (reject this) or $n = 157.$
The sum of the first 157 counting number is 12,403.

8.2 Maintaining Skills

85. $\dfrac{a_2}{a_1} = \dfrac{6}{3} = 2$

$\dfrac{a_3}{a_2} = \dfrac{12}{6} = 2$

$\dfrac{a_4}{a_3} = \dfrac{24}{12} = 2$

$\dfrac{a_5}{a_4} = \dfrac{48}{24} = 2$

87. $\dfrac{a_2}{a_1} = \dfrac{-12}{4} = -3$

$\dfrac{a_3}{a_2} = \dfrac{36}{-12} = -3$

$\dfrac{a_4}{a_3} = \dfrac{-108}{36} = -3$

$\dfrac{a_5}{a_4} = \dfrac{324}{-108} = -3$

89. $a_2 = a_1 r = 2 \cdot 3 = 6$
$a_3 = a_2 r = 6 \cdot 3 = 18$
$a_4 = a_3 r = 18 \cdot 3 = 54$
$a_5 = a_4 r = 54 \cdot 3 = 162$

91. $a_2 = a_1 r = 1(-2) = -2$
$a_3 = a_2 r = -2(-2) = 4$
$a_4 = a_3 r = 4(-2) = -8$
$a_5 = a_4 r = -8(-2) = 16$

93. $a_n = 5 \cdot 2^n$
$a_1 = 5 \cdot 2^1 = 10$
$a_2 = 5 \cdot 2^2 = 20$
$a_3 = 5 \cdot 2^3 = 40$
$a_4 = 5 \cdot 2^4 = 80$
$a_5 = 5 \cdot 2^5 = 160$

$\dfrac{a_2}{a_1} = \dfrac{20}{10} = 2$

$\dfrac{a_3}{a_2} = \dfrac{40}{20} = 2$

$\dfrac{a_4}{a_3} = \dfrac{80}{40} = 2$

$\dfrac{a_5}{a_4} = \dfrac{160}{80} = 2$

Note that

$\dfrac{a_{n+1}}{a_n} = \dfrac{5 \cdot 2^{n+1}}{5 \cdot 2^n} = \dfrac{5 \cdot 2^n \cdot 2}{5 \cdot 2^n} = 2$

95. $a_n = 2(-3)^n$

$a_1 = 2(-3)^1 = -6$

$a_2 = 2(-3)^2 = 18$

$a_3 = 2(-3)^3 = -54$

$a_4 = 2(-3)^4 = 162$

$a_5 = 2(-3)^5 = -486$

$\dfrac{a_2}{a_1} = \dfrac{18}{-6} = -3$

$\dfrac{a_3}{a_2} = \dfrac{-54}{18} = -3$

$\dfrac{a_4}{a_3} = \dfrac{162}{-54} = -3$

$\dfrac{a_5}{a_4} = \dfrac{-486}{162} = -3$

Note that

$$\dfrac{a_{n+1}}{a_n} = \dfrac{2(-3)^{n+1}}{2(-3)^n} = \dfrac{2(-3)^n(-3)}{2(-3)^n} = -3$$

97. $a_n = -3 \cdot 2^n$

$a_{n+1} = -3 \cdot 2^{n+1}$

$a_{n-1} = -3 \cdot 2^{n-1}$

8.3 Geometric Sequences and Series

8.3 Practice Problems

1. $r = \dfrac{18}{6} = 3$

2. The first four terms of the sequence are

$\dfrac{3}{2}, \left(\dfrac{3}{2}\right)^2 = \dfrac{9}{4}, \left(\dfrac{3}{2}\right)^3 = \dfrac{27}{8}, \left(\dfrac{3}{2}\right)^4 = \dfrac{81}{16}.$

Since the common ratio is $\dfrac{3}{2}$, the sequence is geometric.

3. a. $a_1 = 2$

b. $r = \dfrac{6/5}{2} = \dfrac{3}{5}$

c. $a_n = 2 \cdot \left(\dfrac{3}{5}\right)^{n-1}$

4. $a_{18} = a_1 r^{18-1} = 7(1.5^{17}) \approx 6896.8288$

5. $a_1 = \dfrac{1}{9}; \ r = \dfrac{1/3}{1/9} = 3$

$a_n = a_1 r^{n-1} \Rightarrow \dfrac{1}{9} \cdot 3^{n-1} = 243 \Rightarrow$

$3^{n-1} = 2187 \Rightarrow 3^{n-1} = 3^7 \Rightarrow$

$n - 1 = 7 \Rightarrow n = 8$

There are 8 terms in the sequence.

6. $a_1 = 3(0.4)^1 = 1.2; \ r = 0.4$

$S_{17} = \displaystyle\sum_{i=1}^{17} 3(0.4)^i = 1.2\left(\dfrac{1 - 0.4^{17}}{1 - 0.4}\right) \approx 2$

7. $A = 1500\left[\dfrac{\left(1 + \dfrac{0.045}{1}\right)^{(1)(30)} - 1}{\dfrac{0.045}{1}}\right]$

$= \$91,510.60$

8. $a_1 = 3, \ r = \dfrac{6/3}{3} = \dfrac{2}{3}$

Since $|r| < 1,$ we can use the formula for the sum of an infinite geometric series.

$S = \dfrac{a_1}{1 - r} = \dfrac{3}{1 - \dfrac{2}{3}} = 9$

9. $\displaystyle\sum_{i=1}^{\infty} (10,000,000)(0.85)^{i-1} = \dfrac{a_1}{1 - r}$

$= \dfrac{10,000,000}{1 - .85}$

$\approx \$66,666,666.67$

8.3 Basic Concepts and Skills

1. If 5 is the common ratio of a geometric sequence with general term $a_n,$ then

$\dfrac{a_{63}}{a_{62}} = \underline{5}.$

3. If 24 is the term immediately following the sequence term 8 is a geometric sequence, then the common ratio is $\underline{3}$.

5. True

7. The sequence is geometric. $a_1 = 3, r = 2$

9. The sequence is not geometric.

11. The sequence is geometric. $a_1 = 1, r = -3$

13. The sequence is geometric. $a_1 = 7, r = -1$

15. The sequence is geometric. $a_1 = 9, r = \dfrac{1}{3}$

17. The sequence is geometric. $a_1 = -\dfrac{1}{2}, r = -\dfrac{1}{2}$

19. The sequence is geometric. $a_1 = 1, r = 2$

21. The sequence is not geometric.

23. The sequence is geometric. $a_1 = \dfrac{1}{3}, r = \dfrac{1}{3}$

25. The sequence is geometric. $a_1 = \sqrt{5}, r = \sqrt{5}$

27. $a_1 = 2, r = 5, a_n = 2 \cdot 5^{n-1}$

29. $a_1 = 5, r = \dfrac{2}{3}, a_n = 5\left(\dfrac{2}{3}\right)^{n-1}$

31. $a_1 = 0.2, r = -3, a_n = 0.2\,(-3)^{n-1}$

33. $a_1 = \pi^4, r = \pi^2, a_n = \pi^4 (\pi^2)^{n-1} = \pi^{2n+2}$

35. $a_7 = a_1 r^{7-1} = 5(2^6) = 320$

37. $a_{10} = a_1 r^{10-1} = 3(-2)^9 = -1536$

39. $a_6 = a_1 r^{6-1} = \dfrac{1}{16}(3)^5 = \dfrac{243}{16}$

41. $a_9 = a_1 r^{9-1} = -1\left(\dfrac{5}{2}\right)^8 = -\dfrac{390,625}{256}$

43. $a_{20} = a_1 r^{20-1} = 500\left(-\dfrac{1}{2}\right)^{19} = -\dfrac{125}{131,072}$

45. $a_1 = 5;\ r = \dfrac{10}{5} = 2$

$a_n = a_1 r^{n-1} \Rightarrow 5 \cdot 2^{n-1} = 5120 \Rightarrow$
$2^{n-1} = 1024 \Rightarrow 2^{n-1} = 2^{10} \Rightarrow$
$n - 1 = 10 \Rightarrow n = 11$
There are 11 terms in the sequence.

47. $a_1 = 16;\ r = \dfrac{8}{16} = \dfrac{1}{2}$

$a_n = a_1 r^{n-1} \Rightarrow 16 \cdot \left(\dfrac{1}{2}\right)^{n-1} = \dfrac{1}{32} \Rightarrow$
$\left(\dfrac{1}{2}\right)^{n-1} = \dfrac{1}{16 \cdot 32} \Rightarrow \left(\dfrac{1}{2}\right)^{n-1} = \dfrac{1}{2^4 \cdot 2^5} = \dfrac{1}{2^9} \Rightarrow$
$n - 1 = 9 \Rightarrow n = 10$
There are 10 terms in the sequence.

49. $a_1 = 18;\ r = \dfrac{-12}{18} = -\dfrac{2}{3}$

$a_n = a_1 r^{n-1} \Rightarrow 18 \cdot \left(-\dfrac{2}{3}\right)^{n-1} = \dfrac{512}{729} \Rightarrow$
$\left(-\dfrac{2}{3}\right)^{n-1} = \dfrac{512}{729 \cdot 18} \Rightarrow$
$\left(-\dfrac{2}{3}\right)^{n-1} = \dfrac{2^9}{3^6 \cdot 3^2 \cdot 2} = \dfrac{2^8}{3^8} = \left(\dfrac{2}{3}\right)^8 = \left(-\dfrac{2}{3}\right)^8 \Rightarrow$
$n - 1 = 8 \Rightarrow n = 9$
There are 9 terms in the sequence.

51. $r = 5, S_{10} = \dfrac{(1/10)\,(1 - 5^{10})}{1 - 5} = \dfrac{1,220,703}{5}$

53. $r = -5, S_{12} = \dfrac{(1/25)(1 - 5^{12})}{1 - (-5)} = -\dfrac{40,690,104}{25}$

55. $r = \dfrac{1}{4}, S_8 = \dfrac{5(1 - (1/4)^8)}{1 - 1/4} = \dfrac{109,225}{16,384}$

57. $\displaystyle\sum_{i=1}^{5}\left(\dfrac{1}{2}\right)^{i-1} = \dfrac{(1)(1 - (1/2)^5)}{1 - 1/2} = \dfrac{31}{16}$

59. $a_1 = 3, r = \dfrac{2}{3}$

$\displaystyle\sum_{i=1}^{8} 3\left(\dfrac{2}{3}\right)^{i-1} = \dfrac{3(1 - (2/3)^8)}{1 - (2/3)} = \dfrac{6305}{729}$

61. $a_1 = \dfrac{1}{4}, r = 2$

$\displaystyle\sum_{i=3}^{10}\dfrac{2^{i-1}}{4} = \sum_{i=1}^{10}\dfrac{2^{i-1}}{4} - \sum_{i=1}^{2}\dfrac{2^{i-1}}{4}$

$= \dfrac{(1/4)(1 - 2^{10})}{1 - 2} - \dfrac{(1/4)(1 - 2^2)}{1 - 2}$

$= \dfrac{1023}{4} - \dfrac{3}{4} = 255$

63. $a_1 = -\dfrac{3}{5}, r = -\dfrac{5}{2}$

$\displaystyle\sum_{i=1}^{20}\left(-\dfrac{3}{5}\right)\left(-\dfrac{5}{2}\right)^{i-1} = \dfrac{(-3/5)(1 - (-5/2)^{20})}{1 - (-5/2)}$

$= \dfrac{3(5^{20} - 2^{20})}{35(2^{19})}$

65. $a_1 = \dfrac{1}{5};\ r = \dfrac{1}{2}$

$\dfrac{1}{320} = \dfrac{1}{5}\left(\dfrac{1}{2}\right)^{n-1} \Rightarrow \dfrac{1}{64} = \left(\dfrac{1}{2}\right)^{n-1} \Rightarrow$
$\left(\dfrac{1}{2}\right)^6 = \left(\dfrac{1}{2}\right)^{n-1} \Rightarrow 6 = n - 1 \Rightarrow n = 7$

Thus, the sum is $\displaystyle\sum_{n=1}^{7}\dfrac{1}{5}\left(\dfrac{1}{2}\right)^{n-1}$.

```
sum(seq((1/5)(1/
2)^(n-1),n,1,7)
           .396875
Ans▶Frac
           127/320
■
```

67.

```
sum(seq(3^(n-2),
n,1,10)
       9841.333333
```

69.

```
sum(seq((-1)^(n-
1)*3^(2-n),n,1,1
0)
       2.249961896
```

71. $a_1 = \dfrac{1}{3}, r = \dfrac{1}{3}$

$$\dfrac{1}{3} + \dfrac{1}{9} + \dfrac{1}{27} + \dfrac{1}{81} + \cdots = \dfrac{1/3}{1-(1/3)} = \dfrac{1}{2}$$

73. $a_1 = -\dfrac{1}{2}, r = -\dfrac{1}{2}$

$$-\dfrac{1}{2} + \dfrac{1}{4} - \dfrac{1}{8} + \dfrac{1}{16} + \cdots = \dfrac{-1/2}{1-(-1/2)} = -\dfrac{1}{3}$$

75. $a_1 = 8, r = -\dfrac{1}{4}$

$$8 - 2 + \dfrac{1}{2} - \dfrac{1}{8} + \cdots = \dfrac{8}{1-(-1/4)} = \dfrac{32}{5}$$

77. $a_1 = 5, r = \dfrac{1}{3}; \displaystyle\sum_{n=0}^{\infty} 5\left(\dfrac{1}{3}\right)^n = \dfrac{5}{1-(1/3)} = \dfrac{15}{2}$

79. $a_1 = 1, r = -\dfrac{1}{4}; \displaystyle\sum_{n=0}^{\infty} \left(-\dfrac{1}{4}\right)^n = \dfrac{1}{1-(-1/4)} = \dfrac{4}{5}$

8.3 Applying the Concepts

81. $a_5 = 20,000(1.03^5) = 23,185$

83. $a_{36} = 100\left(\dfrac{\left(1+\dfrac{0.06}{12}\right)^{36} - 1}{\dfrac{0.06}{12}}\right) = \$3933.61$

85. $a_{10} = 2^{10} = 1024$.

A person has 1024 ancestors in the tenth generation back.

87. $a_1 = \$36,000, r = 1.05$

$$S_{20} = \dfrac{36,000(1-1.05^{20})}{1-1.05} = \$1,190,374.35$$

89. $a_1 = 56.25, r = 0.8$

$$S_5 = \dfrac{56.25(1-0.8^5)}{1-0.8} = 189.09 \text{ cm}$$

91. $D = 5 + 2(5)\left(\dfrac{3}{5}\right) + 2(5)\left(\dfrac{3}{5}\right)^2 + 2(5)\left(\dfrac{3}{5}\right)^3 + \cdots$

$$= 5 + 10\left(\dfrac{3}{5}\right) + 10\left(\dfrac{3}{5}\right)^2 + 10\left(\dfrac{3}{5}\right)^3 + \cdots$$

$$= 5 + \dfrac{6}{1-(3/5)} = 20 \text{ m}$$

8.3 Beyond the Basics

93. The area of each square is 1/2 the area of the next larger square. However, we are including the shaded regions only, so $r = -1/2$. (This will eliminate the white squares.)

$$\sum_{i=1}^{\infty} \left(-\dfrac{1}{2}\right)^{i-1} = \dfrac{1}{1-(-1/2)} = \dfrac{2}{3}.$$

95. Since $a_1, a_2, a_3, \ldots$ is a geometric sequence, $a_n = a_{n-1}r$ for every $n \geq 1$, where r is the common ratio. Taking the logarithm of both sides, we have $\ln a_n = \ln(a_{n-1}r)$ $= \ln a_{n-1} + \ln r$. So the sequence $\ln a_1, \ln a_2, \ln a_3, \ldots$ is arithmetic with common difference $\ln r$.

97. Since $a_1, a_2, a_3, \ldots$ is a geometric sequence, $a_n = a_{n-1}r$ for every $n \geq 1$, where r is the common ratio. Taking the reciprocal of both sides, we have $\dfrac{1}{a_n} = \dfrac{1}{a_{n-1}r} = \dfrac{1}{a_{n-1}} \cdot \dfrac{1}{r}$. So the sequence $\dfrac{1}{a_1}, \dfrac{1}{a_2}, \dfrac{1}{a_3}, \cdots$ is geometric with common ratio $1/r$.

99. Since $a_n = \dfrac{2a_{n-1}}{x}$ for every $n \geq 1$, the sequence is geometric with common ratio $2/x$.

101. Since $a_n = -\dfrac{a_{n-1}}{y}$ for every $n \geq 1$, the sequence is geometric with common ratio $-\dfrac{1}{y}$.

8.3 Critical Thinking/Discussion/Writing

103. $a_{1001} = 10 + 1000(2.7) = 2710$.

$b_{1001} = 10(2^{1000})$, so b_{1001} is larger.

104. $\dfrac{a_1\left(1-\left(\dfrac{1}{2}\right)^{10}\right)}{1-\dfrac{1}{2}} = 1023 \Rightarrow a_1 = 512$

105. $5 + 5(2) + 5(2^2) + \cdots + 5(2^{15}) = \displaystyle\sum_{i=0}^{15} 5(2^i)$ or

$\displaystyle\sum_{i=1}^{16} 5(2^{i-1})$.

106. Let $\dfrac{a}{r}, a,$ and ar be the three terms of the

geometric sequence. Then we have

$\begin{cases} \dfrac{a}{r} + a + ar = 35 \\ \left(\dfrac{a}{r}\right)a(ar) = 1000 \end{cases} \Rightarrow \begin{cases} a + ar + ar^2 = 35r \\ a^3 = 1000 \end{cases} \Rightarrow$

$a = 10$

$10 + 10r + 10r^2 = 35r \Rightarrow 2r^2 - 5r + 2 = 0 \Rightarrow$

$(2r-1)(r-2) \Rightarrow r = \dfrac{1}{2}$ or $r = 2$.

If $r = \dfrac{1}{2}$, then the three numbers are 20, 10,

and 5. If $r = 2$, then the numbers are also 5, 10,
and 20.

107. Let $a, a+d,$ and $a+2d$ be the three terms of
the arithmetic sequence. Then
$a + (a+d) + (a+2d) = 3a + 3d = 15 \Rightarrow$
$a + d = 5$.
The geometric sequence is $a+1, a+d+4,$ and

$a + 2d + 19$, and $\dfrac{a+d+4}{a+1} = \dfrac{a+2d+19}{a+d+4}$.

Use substitution to solve the system:

$\begin{cases} a + d = 5 \\ \dfrac{a+d+4}{a+1} = \dfrac{a+2d+19}{a+d+4} \end{cases} \Rightarrow$

$\begin{cases} d = 5 - a \\ \dfrac{a+d+4}{a+1} = \dfrac{a+2d+19}{a+d+4} \end{cases} \Rightarrow$

$\dfrac{a+(5-a)+4}{a+1} = \dfrac{a+2(5-a)+19}{a+(5-a)+4} \Rightarrow$

$\dfrac{9}{a+1} = \dfrac{29-a}{9} \Rightarrow 29 + 28a - a^2 = 81 \Rightarrow$

$-(a^2 - 28a + 52) = 0 \Rightarrow (a-2)(a-26) = 0 \Rightarrow$
$a = 2$ or $a = 26$.
If $a = 2$, then $d = 3$, and the arithmetic sequence
is 2, 5, 8. If $a = 26$, then $d = -21$, and the
sequence is 26, 5, -16.

108. Let $a, ar,$ and ar^2 be the three terms of the
geometric sequence. So,
$a \cdot ar \cdot ar^2 = a^3 r^3 = 1000 \Rightarrow ar = 10$.
The arithmetic sequence is $a, ar + 6,$ and

$ar^2 + 7$. Then,
$(ar + 6) - a = (ar^2 + 7) - (ar + 6)$.

Substituting $ar = 10$ gives
$(10 + 6) - a = (10r + 7) - (10 + 6) \Rightarrow$
$16 - a = 10r - 9 \Rightarrow 25 = 10r + a$

$ar = 10 \Rightarrow r = \dfrac{10}{a}$

Substituting, we have

$25 = 10\left(\dfrac{10}{a}\right) + a \Rightarrow 25a = 100 + a^2 \Rightarrow$

$a^2 - 25a + 100 = 0 \Rightarrow (a-5)(a-20) = 0 \Rightarrow$
$a = 5, a = 20$

If $a = 5$, then we have $5r = 10 \Rightarrow r = 2$, and
the geometric sequence is 5, 10, 20.

If $a = 20$, then we have $20r = 10 \Rightarrow r = \dfrac{1}{2}$, and

the geometric sequence is 20, 10, 5.

8.3 Maintaining Skills

109. $P_n : n(n+1)$ is even.
$P_3 : 3(3+1)$ is even.　　　True
$P_4 : 4(4+1)$ is even.　　　True

111. $P_n : 1 + 2 + 3 + \cdots + n = \dfrac{n(n+1)}{2}$

$P_3 : 1 + 2 + 3 = \dfrac{3(3+1)}{2}$　　　True

$P_4 : 1 + 2 + 3 + 4 = \dfrac{4(4+1)}{2}$　　　True

113. $P_n : n^2 + n$ is divisible by 2.
$P_{n+1} : (n+1)^3 + (n+1)$ is divisible by 2.

115. $P_n : 3^n > 5n$
$P_{n+1} : 3^{n+1} > 5(n+1)$

117. $\dfrac{k(k+1)}{2} + (k+1) = \dfrac{k(k+1)}{2} + \dfrac{2(k+1)}{2}$

$= \dfrac{k^2 + k + 2k + 2}{2}$

$= \dfrac{k^2 + 3k + 2}{2}$

$= \dfrac{(k+1)(k+2)}{2}$

8.4 Mathematical Induction

8.4 Practice Problems

1. $P_{k+1} : \left[(k+1)+3\right]^2 > (k+1)^2 + 9 \Rightarrow$
   $(k+4)^2 > (k+1)^2 + 9$

2. For $n = 1 : 1 = \dfrac{1(1+1)}{2}$ is true. Assume that it is

   true for $n = k : 1 + 2 + 3 + \ldots + k = \dfrac{k(k+1)}{2}$.

   Then for $n = k+1 : 1 + 2 + 3 + \ldots + k + (k+1)$

   $= \dfrac{k(k+1)}{2} + (k+1) = \dfrac{k^2 + k + 2k + 2}{2}$

   $= \dfrac{(k+1)(k+2)}{2}$, which is exactly the

   statement for $n = k+1$. Therefore the formula
   is true for all natural numbers.

3. For $n = 1 : 3^1 > 1$ is true. Assume that the
   inequality is true for $n = k : 3^k > k$. Then, we
   must use this fact to prove that for
   $n = k+1 : 3^{k+1} > k+1$.

   $3^k > k \Rightarrow 3^k \cdot 3 > 3 \cdot k \Rightarrow 3^{k+1} > 3k > k+1$,
   which is exactly the statement for $n = k+1$.
   Therefore the formula is true for all natural
   numbers.

8.4 Basic Concepts and Skills

1. Mathematical induction can only be used to
   prove statements about the <u>natural numbers</u>.

3. The second step in a mathematical induction
   proof is to assume that P_k is true for a natural
   number k, and then show that <u>P_{k+1} is true</u>.

5. False. The number e is a constant. The
   statement $e^n \geq n$ can be proven by
   mathematical induction for all natural
   numbers n.

7. $P_{k+1} : ((k+1)+1)^2 - 2(k+1)$
   $= k^2 + 4k + 4 - 2k - 2$
   $= k^2 + 2k + 2 = (k+1)^2 + 1$

9. $P_{k+1} : 2^{k+1} > 5(k+1)$

11. For $n = 1 : 2 = 2$ is true. Assume that it is true
    for $n = k$:
    $2 + 4 + 6 + \ldots + 2k = k^2 + k = k(k+1)$

Then for $n = k+1$:
$2 + 4 + 6 + \ldots + 2k + 2(k+1)$
$\quad = k(k+1) + 2(k+1) = k^2 + 3k + 2$
$\quad = (k+1)(k+2)$
which is exactly the statement for $n = k+1$.
Therefore the formula is true for all natural
numbers.

13. For $n = 1 : 4 = 2(1)(1+1)$ is true. Assume that
    it is true for $n = k$:
    $4 + 8 + 12 + \ldots + 4k = 2k(k+1)$

    Then for $n = k+1$:
    $4 + 8 + 12 + \ldots + 4k + (4k+4)$
    $\quad = 2k(k+1) + (4k+4) = 2k^2 + 2k + 4k + 4$
    $\quad = 2(k^2 + 3k + 2) = 2(k+1)(k+2)$
    which is exactly the statement for $n = k+1$.
    Therefore the formula is true for all natural
    numbers.

15. For $n = 1 : 1 = 1(2-1)$ is true. Assume that it is
    true for $n = k$:
    $1 + 5 + 9 + \ldots + (4k-3) = k(2k-1)$

    Then for $n = k+1$:
    $1 + 5 + 9 + \ldots + (4k-3) + (4k+1)$
    $\quad = k(2k-1) + (4k+1) = 2k^2 - k + 4k + 1$
    $\quad = 2k^2 + 3k + 1 = (k+1)(2k+1)$
    $\quad = (k+1)(2(k+1)-1)$
    which is exactly the statement for $n = k+1$.
    Therefore the formula is true for all natural
    numbers.

17. For $n = 1 : 3 = \dfrac{3(3-1)}{2}$ is true. Assume that it is

    true for $n = k$:

    $3 + 9 + 27 + \ldots + 3^k = \dfrac{3(3^k - 1)}{2}$

    Then for $n = k+1$:

    $3 + 9 + 27 + \ldots + 3^k + 3^{k+1} = \dfrac{3(3^k - 1)}{2} + 3^{k+1}$

    $\qquad = \dfrac{3^{k+1} - 3 + 2(3^{k+1})}{2}$

    $\qquad = \dfrac{3(3^{k+1}) - 3}{2}$

    $\qquad = \dfrac{3(3^{k+1} - 1)}{2}$

    which is exactly the statement for $n = k+1$.
    Therefore the formula is true for all natural
    numbers.

19. For $n = 1 : \dfrac{1}{(1)(2)} = \dfrac{1}{1+1}$ is true. Assume that it

is true for $n = k$:

$$\frac{1}{(1)(2)} + \frac{1}{(2)(3)} + \frac{1}{(3)(4)} + \ldots + \frac{1}{k(k+1)} = \frac{k}{k+1}.$$

Then for $n = k + 1$:

$$\frac{1}{(1)(2)} + \frac{1}{(2)(3)} + \frac{1}{(3)(4)} + \cdots$$

$$+ \frac{1}{k(k+1)} + \frac{1}{(k+1)(k+2)}$$

$$= \frac{k}{k+1} + \frac{1}{(k+1)(k+2)} = \frac{k^2 + 2k + 1}{(k+1)(k+2)}$$

$$= \frac{(k+1)^2}{(k+1)(k+2)} = \frac{k+1}{k+2}$$

which is exactly the statement for $n = k + 1$. Therefore the formula is true for all natural numbers.

21. For $n = 1 : 2 \le 2$ is true. Assume that it is true

for $n = k : 2 \le 2^k$. Then for $n = k + 1$:

$$2 \le 2^k \Rightarrow 2 \cdot 2 \le 2(2^k) \Rightarrow 4 \le 2^{k+1} \Rightarrow 2 \le 2^{k+1}$$

Therefore the formula is true for all natural numbers.

23. For $n = 1 : 1(1+2) < (1+1)^2$ is true. Assume

that it is true for $n = k : k(k+2) < (k+1)^2$.
Then for $n = k + 1$:

$$(k+1)(k+1+2) = (k+1)(k+3) < (k+1+1)^2 \Rightarrow$$

$$k^2 + 4k + 3 < k^2 + 4k + 4 \Rightarrow$$

$$k^2 + 2k + 2k + 3 < k^2 + 2k + 1 + 2k + 3 \Rightarrow$$

$$k(k+2) + (2k+3) < (k+1)^2 + (2k+3) \Rightarrow$$

$$k(k+2) < (k+1)^2, \text{ which is exactly the}$$

statement for $n = k + 1$. Therefore the formula is true for all natural numbers.

25. For $n = 1 : \dfrac{1!}{1} = (1-1)!$ is true. Assume that it

is true for $n = k : \dfrac{k!}{k} = (k-1)!$. Then for

$n = k + 1 : \dfrac{(k+1)!}{k+1} = \dfrac{(k+1)k!}{k+1} = k!$, which is

exactly the statement for $n = k + 1$. Therefore the formula is true for all natural numbers.

27. For $n = 1 : 1^2 = \dfrac{1(1+1)(2+1)}{6}$ is true. Assume

that it is true for $n = k$:

$$1^2 + 2^2 + 3^2 + \cdots + k^2 = \frac{k(k+1)(2k+1)}{6}.$$

Then for $n = k + 1$:

$$1^2 + 2^2 + 3^2 + \ldots + k^2 + (k+1)^2$$

$$= \frac{k(k+1)(2k+1)}{6} + (k+1)^2$$

$$= \frac{k(k+1)(2k+1) + 6(k+1)^2}{6}$$

$$= \frac{(k+1)\big(k(2k+1) + 6(k+1)\big)}{6}$$

$$= \frac{(k+1)(2k^2 + 7k + 6)}{6}$$

$$= \frac{(k+1)(k+2)(2k+3)}{6}$$

which is exactly the statement for $n = k + 1$. Therefore the formula is true for all natural numbers.

29. For $n = 1$: 2 is a factor of $1^2 + 1 = 2$.

For $n = k$: assume that 2 is a factor of

$n^2 + n \Rightarrow k^2 + k = 2p$ for some integer p.
Then for $n = k + 1$:

$$(k+1)^2 + (k+1) = k^2 + 3k + 2$$

$$= (k^2 + k) + (2k+2)$$

$$= 2p + 2(k+1) = 2(p+k+1)$$

$(p+k+1)$ is an integer, so $2(p+k+1)$ is divisible by 2. Therefore the formula is true for all natural numbers.

31. For $n = 1 : 6$ is a factor of $1(1+1)(1+2) = 6$.

For $n = k$: Assume that 6 is a factor of

$n(n+1)(n+2) \Rightarrow k(k+1)(k+2) = 6p$ for

some integer p.
Then for $n = k + 1$:

$$(k+1)(k+2)(k+3)$$

$$= k(k+1)(k+2) + 3(k+1)(k+2)$$

$$= 6p + 3(k+1)(k+2).$$

Either $k+1$ or $k+2$ is even, so
$3(k+1)(k+2)$ is divisible by 6 and, thus,
$6p + 3(k+1)(k+2)$ is divisible by 6.
Therefore the formula is true for all natural numbers.

33. For $n = 1: (ab)^1 = a^1 b^1$ is true. Assume that it is true for $n = k: (ab)^k = a^k b^k$. Then for $n = k + 1$:

$$(ab)^{k+1} = (ab)^k (ab) = a^k b^k ab = a^{k+1} b^{k+1},$$

which is exactly the statement for $n = k + 1$. Therefore the formula is true for all natural numbers.

8.4 Applying the Concepts

35. For $n = 2: \dfrac{2^2 - 2}{2} = 1$, which is the correct number of hugs for two people. Assume that the number of hugs for k people is $\dfrac{k^2 - k}{2}$. If there are $k + 1$ people, then we can separate one of them. The remaining k have $\dfrac{k^2 - k}{2}$ hugs by the hypothesis. The $k + 1$st person has to hug each of the other k people. So the number of hugs is

$$\frac{k^2 - k}{2} + k = \frac{k^2 - k + 2k}{2} = \frac{k^2 + 2k + 1 - (k+1)}{2}$$

$$= \frac{(k+1)^2 - (k+1)}{2}$$, which is exactly the statement for $n = k + 1$. Therefore the formula is true for all natural numbers $n \geq 2$.

37. a. The number of sides of the nth figure is $3\left(4^{n-1}\right)$. For $n = 1: 3\left(4^0\right) = 3$, which is the number of sides of the triangle. Assume the number of sides of the kth figure is $3\left(4^{k-1}\right)$. When making the $k + 1$st figure, each new small triangle splits the original side into two sides, thus doubling the number of sides. In addition, each new triangle adds on two new sides per original side. So, the number of sides of the $k + 1$st figure is four times the number of sides of the kth figure, i.e., $4(3)(4^{k-1}) = 3(4^k)$, which is exactly the formula for $n = k + 1$. Therefore the formula is true for all natural numbers.

b. The perimeter of the nth figure is $3\left(\dfrac{4}{3}\right)^{n-1}$.

For $n = 1: 3\left(\dfrac{4}{3}\right)^0 = 3$, which is the perimeter of the equilateral triangle with side length 1. Assume that the perimeter of the kth figure is $3\left(\dfrac{4}{3}\right)^{k-1}$. When making the $k + 1$st figure, each new small triangle adds on two new sides and deletes one small side, each with length $1/3$ of the original side. So, each new triangle increases the perimeter by $1/3$ of an original side. Therefore the perimeter of the $k + 1$st figure is $4/3$ times that of the kth figure, i.e., $\dfrac{4}{3}(3)\left(\dfrac{4}{3}\right)^{k-1} = 3\left(\dfrac{4}{3}\right)^k$, which is exactly the formula for $n = k + 1$. Therefore the formula is true for all natural numbers.

39. The smallest number of moves to accomplish the transfer is $2^n - 1$. Prove as follows: for $n = 1: 2^1 - 1 = 1$ is the number of necessary moves for one ring. Assume that the smallest number of moves for k rings is $2^k - 1$. When we move the biggest peg from the bottom to another peg, all the other rings must be stacked on the third peg in decreasing order from bottom to top. This requires $2^k - 1$ moves by the hypothesis. It takes one move to put the biggest ring onto the other peg. Then we have to move all the other rings on top of it, with again requires $2^k - 1$ moves. Altogether, then, there are

$$(2^k - 1) + 1 + (2^k - 1) = 2(2^k) - 1 = 2^{k+1} - 1$$

moves, which is exactly the formula for $n = k + 1$. Therefore the formula is true for all natural numbers.

8.4 Beyond the Basics

41. For $n = 1$, 5 is a factor of $8^1 - 3^1 = 5$. For $n = k$, assume that 5 is a factor of $8^n - 3^n \Rightarrow 8^k - 3^k = 5p$ for some integer p. Then for $n = k + 1$: $8^{k+1} - 3^{k+1} = 8(8^k) - 3(3^k) = 8(8^k) - 8(3^k) + 5(3^k) = 8(8^k - 3^k) + 5(3^k)$

    $8(5p) + 5(3^k) = 5(8p + 3^k)$. Since $8p + 3^k$ is an integer, $5(8p + 3^k)$ is divisible by 5. Therefore the formula is true for all natural numbers.

43. For $n = 1$, 64 is a factor of $3^{2(1)+2} - 8(1) - 9 = 64$. For $n = k$, assume that 64 is a factor of $3^{2n+2} - 8n - 9 \Rightarrow$

    $3^{2k+2} - 8k - 9 = 64p$ for some integer p. Then for $n = k + 1$:

    $3^{2(k+1)+2} - 8(k+1) - 9 = 3^{2k+4} - 8(k+1) - 9 = 3^2(3^{2k+2}) - 8k - 8 - 9 = 9(3^{2k+2}) - 9(8k) - 9(9) + 8(8k) + 64$

    $= 9(3^{2k+2} - 8k - 9) + 64(k+1) = 9(64p) + 64(k+1) = 64(9p + k + 1)$

    Since $9p + k + 1$ is an integer, $64(9p + k + 1)$ is divisible by 64. Therefore the formula is true for all natural numbers.

45. For $n = 1$, 3 is a factor of $2^{2(1)+1} + 1 = 9$.

    For $n = k$, assume that 3 is a factor of $2^{2n+1} + 1 \Rightarrow 2^{2k+1} + 1 = 3p$ for some integer p. Then for $n = k + 1$:

    $2^{2(k+1)+1} + 1 = 2^{2k+3} + 1 = 2^2 2^{2k+1} + 1 = 2^2 2^{2k+1} + 4 - 3 = 4(2^{2k+1} + 1) - 3 = 4(3p) - 3 = 3(4p - 1)$

    Since $4p - 1$ is an integer, $3(4p - 1)$ is divisible by 3. Therefore the formula is true for all natural numbers.

47. For $n = 1$, $a - b$ is a factor of $a^1 - b^1 = a - b$. For $n = k$, assume that $a - b$ is a factor of

    $a^n - b^n \Rightarrow a^k - b^k = (a - b)P(a,b)$ where $P(a,b)$ is a polynomial of a and b. Then for $n = k + 1$:

    $a^{k+1} - b^{k+1} = a\left(a^k - b^k\right) + b^k(a - b) = a(a - b)P(a,b) + b^k(a - b) = (a - b)\left(aP(a,b) + b^k\right)$

    Since $aP(a,b) + b^k$ is a polynomial of a and b, $a - b$ is a factor of $(a - b)\left(aP(a,b) + b^k\right)$. Therefore the statement is true for all natural numbers.

49. For $n = 1$: $\displaystyle\sum_{k=1}^{2} \frac{1}{k+1} = \frac{1}{2} + \frac{1}{3} \leq \frac{5}{6}$ is true. Assume that it is true for $n = m$: $\displaystyle\sum_{k=1}^{m+1} \frac{1}{k+m} \leq \frac{5}{6}$.

    Then for $n = m + 1$: $\displaystyle\sum_{k=1}^{m+2} \frac{1}{k+m+1} = \sum_{k=2}^{m+3} \frac{1}{k+m} = \left(\sum_{k=1}^{m+1} \frac{1}{k+m}\right) + \frac{1}{2m+2} + \frac{1}{2m+3} - \frac{1}{m+1}$.

    In the last expression, $\dfrac{1}{2m+2} + \dfrac{1}{2m+3} - \dfrac{1}{m+1} = \dfrac{(2m+3) + (2m+2) - 2(2m+3)}{2(m+1)(2m+3)} = -\dfrac{1}{2(m+1)(2m+3)} < 0$

    So, $\displaystyle\sum_{k=1}^{m+2} \frac{1}{k+m+1} = \left(\sum_{k=1}^{m+1} \frac{1}{k+m}\right) + \frac{1}{2m+2} + \frac{1}{2m+3} - \frac{1}{m+1} < \left(\sum_{k=1}^{m+1} \frac{1}{k+m}\right) \leq \frac{5}{6}$

    This is exactly the formula for $n = k + 1$. Therefore the formula is true for all natural numbers.

8.4 Critical Thinking/Discussion/Writing

50. The proof is not valid because the first step of the mathematical induction, which is to show that the statement is true for P_1, is missing from this proof.

51. For $n = m = 4$: $2^4 \geq 4^2 \Rightarrow 16 \geq 16$ is true. Now assume that the statement is true for $n = k$: $2^k \geq k^2$. Now we must use P_k to prove that P_{k+1} is true. That is, $P_{k+1} : 2^{k+1} \geq (k+1)^2$.

$2 \cdot 2^k \geq 2k^2 \Rightarrow 2^{k+1} \geq 2k^2 \Rightarrow 2^{k+1} \geq k^2 + k^2$

Since $k \geq 4$, we have $2k \geq 8 \Rightarrow 2k > 1$. Also, since $k \geq 4$, we have $k^2 \geq 4k$.

$2^{k+1} > k^2 + k^2 > k^2 + 4k \Rightarrow 2^{k+1} > k^2 + 4k \Rightarrow 2^{k+1} > k^2 + 2k + 2k \Rightarrow$
$2^{k+1} > k^2 + 2k + 2k > k^2 + 2k + 1 \Rightarrow 2^{k+1} > k^2 + 2k + 1 \Rightarrow 2^{k+1} > (k+1)^2$

Thus, $P_k \Rightarrow P_{k+1}$, and the statement is true for all natural numbers.

52. For $n = m = 6$, $6! > 6^3 \Rightarrow 720 > 216$ is true. Now assume that the statement is true for $n = k$: $k! > k^3$. Now we must use P_k to prove that P_{k+1} is true. That is, $P_{k+1} : (k+1)! \geq (k+1)^3$.

$k! > k^3 \Rightarrow (k+1) \cdot k! \geq (k+1) \cdot k^3 \Rightarrow (k+1)! \geq (k+1) \cdot k^3$.

Since $k+1 \geq 7$, we have $(k+1)! \geq (k+1) \cdot k^3 \Rightarrow (k+1)! \geq 7k^3$.

We need to show that $7k^3 \geq (k+1)^3$ for $k \geq 6$:

$(k+1)^3 = k^3 + 3k^2 + 3k + 1 \Rightarrow 7k^3 \geq k^3 + 3k^2 + 3k + 1 \Rightarrow 6k^3 \geq 3k^2 + 3k + 1 \Rightarrow$
$k^3 \geq \frac{1}{2}k^2 + \frac{1}{2}k + \frac{1}{6} \Rightarrow k^3 \geq \frac{1}{2}\left(k^2 + k + \frac{1}{3}\right) \Rightarrow 7k^3 \geq \frac{7}{2}\left(k^2 + k + \frac{1}{3}\right)$
$k \geq 6 \Rightarrow 7k^2 \cdot k \geq 7k^2 \cdot 6 \Rightarrow 7k^3 \geq 42k^2$

Now compare $\frac{7}{2}\left(k^2 + k + \frac{1}{3}\right)$ with $42k^2$:

$12\left[\frac{7}{2}\left(k^2 + k + \frac{1}{3}\right)\right] = 42k^2 + 42k + 14 \Rightarrow 42k^2 \geq \frac{7}{2}\left(k^2 + k + \frac{1}{3}\right)$, so

$7k^3 \geq \frac{7}{2}\left(k^2 + k + \frac{1}{3}\right) \Rightarrow (k+1)! \geq \frac{7}{2}\left(k^2 + k + \frac{1}{3}\right) \Rightarrow (k+1)! \geq (k+1)^3$.

8.4 Maintaining Skills

53. $(x+y)^2 = (x+y)(x+y) = x^2 + 2xy + y^2$

55. $(x+y)^4 = (x+y)(x+y)^3 = (x+y)\left(x^3 + 3x^2y + 3xy^2 + y^3\right)$
$= x^4 + 3x^3y + 3x^2y^2 + xy^3 + yx^3 + 3x^2y^2 + 3xy^3 + y^4 = x^4 + 4x^3y + 6x^2y^2 + 4xy^3 + y^4$

57. In exercise 53, we see that the sum of the exponents in each term is 2. In exercise 54, the sum of the exponents in each term is 3. In exercise 55, the sum of the exponents in each term is 4, and in exercise 56, the sum is 5. Thus, we can say that the sum of the exponents on x and y in each term in the expansion of $(x+y)^n$ is n.

8.5   The Binomial Theorem

8.5 Practice Problems

1. $(3y-x)^6 = 1(3y)^6 + 6(3y)^5(-x) + 15(3y)^4(-x)^2 + 20(3y)^3(-x)^3 + 15(3y)^2(-x)^4 + 6(3y)(-x)^5 + 1(-x)^6$
$= 729y^6 - 1458y^5x + 1215y^4x^2 - 540y^3x^3 + 135y^2x^4 - 18yx^5 + x^6$

2. a. $\binom{6}{2} = \frac{6!}{2!(6-2)!} = \frac{6 \cdot 5 \cdot 4!}{2!4!} = 15$ **b.** $\binom{12}{9} = \frac{12!}{9!(12-9)!} = \frac{12 \cdot 11 \cdot 10 \cdot 9!}{9!3!} = \frac{12 \cdot 11 \cdot 10}{3 \cdot 2 \cdot 1} = 220$

3. $(3x-y)^4 = \binom{4}{0}(3x)^4 + \binom{4}{1}(3x)^3(-y) + \binom{4}{2}(3x)^2(-y)^2 + \binom{4}{3}(3x)(-y)^3 + \binom{4}{4}(-y)^4$

$$= \frac{4!}{0!4!}(81x^4) - \frac{4!}{1!3!}(27x^3y) + \frac{4!}{2!2!}(9x^2y^2) - \frac{4!}{3!1!}(3xy^3) + \frac{4!}{4!0!}y^4$$

$$= 81x^4 - 108x^3y + 54x^2y^2 - 12xy^3 + y^4$$

4. The term x^3y^9 is the tenth term in the expansion of $(x+y)^{12}$. So its coefficient is $\binom{12}{9}$.

$$\binom{12}{9} = \frac{12!}{9!(12-9)!} = \frac{12 \cdot 11 \cdot 10 \cdot 9!}{9!3!} = \frac{12 \cdot 11 \cdot 10}{3 \cdot 2 \cdot 1} = 220$$

5. $\binom{n}{n-r}x^r(2a)^{n-r} = \binom{15}{15-3}x^3(2a)^{15-3} = \binom{15}{12}x^3(2a)^{12} = \frac{15!}{12!(15-12)!}x^3(2a)^{12} = \frac{15!}{12!3!}x^3(2a)^{12}$

$$= \frac{15 \cdot 14 \cdot 13 \cdot 12!}{12!3!}x^3(2a)^{12} = \frac{15 \cdot 14 \cdot 13}{3 \cdot 2 \cdot 1}x^3a^{12} = 455x^3(2a)^{12} = 1,863,680x^3a^{12}$$

8.5 Basic Concepts and Skills

1. The expansion of $(x+y)^n$ has $\underline{n+1}$ terms.

3. Expanding a difference, such as $(2x-y)^{10}$, results in <u>alternating signs</u> between terms.

5. False. If n is even, then there are an odd number of terms in the expansion, so one coefficient appears just once.

7. $\dfrac{6!}{3!} = \dfrac{6 \cdot 5 \cdot 4 \cdot 3!}{3!} = 120$

9. $\dfrac{12!}{11!} = \dfrac{12 \cdot 11!}{11!} = 12$

11. $\binom{6}{4} = \dfrac{6!}{4!(6-4)!} = \dfrac{6 \cdot 5 \cdot 4!}{4!2!} = 15$

13. $\binom{9}{0} = \dfrac{9!}{0!(9-0)!} = \dfrac{9!}{0!9!} = 1$

15. $\binom{7}{1} = \dfrac{7!}{1!(7-1)!} = \dfrac{7 \cdot 6!}{1!6!} = 7$

17. $(x+2)^4 = x^4 + 4(2x^3) + 6(2^2x^2) + 4(2^3x) + 2^4 = x^4 + 8x^3 + 24x^2 + 32x + 16$

19. $(x-2)^5 = x^5 + 5(-2)x^4 + 10(-2)^2x^3 + 10(-2)^3x^2 + 5(-2)^4x + (-2)^5 = x^5 - 10x^4 + 40x^3 - 80x^2 + 80x - 32$

21. $(2-3x)^3 = 2^3 + 3(2^2)(-3x) + 3(2)(-3x)^2 + (-3x)^3 = 8 - 36x + 54x^2 - 27x^3$

23. $(2x+3y)^4 = (2x)^4 + 4(2x)^3(3y) + 6(2x)^2(3y)^2 + 4(2x)(3y)^3 + (3y)^4$
$\qquad = 16x^4 + 96x^3y + 216x^2y^2 + 216xy^3 + 81y^4$

25. $(x+1)^4 = x^4 + 4(1x^3) + 6(1^2x^2) + 4(1^3x) + 1^4 = x^4 + 4x^3 + 6x^2 + 4x + 1$

27. $(x-1)^5 = x^5 + 5(-1)x^4 + 10(-1)^2x^3 + 10(-1)^3x^2 + 5(-1)^4x + (-1)^5 = x^5 - 5x^4 + 10x^3 - 10x^2 + 5x - 1$

29. $(y-3)^3 = y^3 + 3(-3)y^2 + 3(-3)^2 y + (-3)^3 = y^3 - 9y^2 + 27y - 27$

31. $(x+y)^6 = \binom{6}{0}x^6 + \binom{6}{1}x^5 y + \binom{6}{2}x^4 y^2 + \binom{6}{3}x^3 y^3 + \binom{6}{4}x^2 y^4 + \binom{6}{5}xy^5 + \binom{6}{6}y^6$

$$= \frac{6!}{0!(6-0)!}x^6 + \frac{6!}{1!(6-1)!}x^5 y + \frac{6!}{2!(6-2)!}x^4 y^2 + \frac{6!}{3!(6-3)!}x^3 y^3 + \frac{6!}{4!(6-4)!}x^2 y^4$$

$$+ \frac{6!}{5!(6-5)!}xy^5 + \frac{6!}{6!(6-6)!}y^6$$

$$= x^6 + 6x^5 y + 15x^4 y^2 + 20x^3 y^3 + 15x^2 y^4 + 6xy^5 + y^6$$

33. $(1+3y)^5 = 1^5 + 5(1^4)(3y) + (10)(1^3)(3y)^2 + (10)(1^2)(3y)^3 + (5)(1)(3y)^4 + (3y)^5$

$$= 1 + 15y + 90y^2 + 270y^3 + 405y^4 + 243y^5$$

35. $(2x+1)^4 = (2x)^4 + 4(2x)^3(1) + 6(2x)^2(1)^2 + 4(2x)(1)^3 + 1^4 = 16x^4 + 32x^3 + 24x^2 + 8x + 1$

37. $(x-2y)^3 = x^3 + 3(x^2)(-2y) + 3(x)(-2y)^2 + (-2y)^3 = x^3 - 6x^2 y + 12xy^2 - 8y^3$

39. $(2x+y)^4 = (2x)^4 + 4(2x)^3 y + 6(2x)^2 y^2 + 4(2x)y^3 + y^4 = 16x^4 + 32x^3 y + 24x^2 y^2 + 8xy^3 + y^4$

41. $\left(\dfrac{x}{2}+2\right)^7 = \binom{7}{0}\left(\dfrac{x}{2}\right)^7 + \binom{7}{1}\left(\dfrac{x}{2}\right)^6 (2) + \binom{7}{2}\left(\dfrac{x}{2}\right)^5 (2^2) + \binom{7}{3}\left(\dfrac{x}{2}\right)^4 (2^3) + \binom{7}{4}\left(\dfrac{x}{2}\right)^3 (2^4) + \binom{7}{5}\left(\dfrac{x}{2}\right)^2 (2^5)$

$$+ \binom{7}{6}\left(\dfrac{x}{2}\right)(2^6) + \binom{7}{7}(2^7)$$

$$= \frac{7!}{0!(7-0)!}\left(\dfrac{x}{2}\right)^7 + \frac{7!}{1!(7-1)!}\left(\dfrac{x}{2}\right)^6 (2) + \frac{7!}{2!(7-2)!}\left(\dfrac{x}{2}\right)^5 (2^2) + \frac{7!}{3!(7-3)!}\left(\dfrac{x}{2}\right)^4 (2^3)$$

$$+ \frac{7!}{4!(7-4)!}\left(\dfrac{x}{2}\right)^3 (2^4) + \frac{7!}{5!(7-5)!}\left(\dfrac{x}{2}\right)^2 (2^5) + \frac{7!}{6!(7-6)!}\left(\dfrac{x}{2}\right)(2^6) + \frac{7!}{7!(7-7)!}(2^7)$$

$$= \frac{x^7}{128} + \frac{7x^6}{32} + \frac{21x^5}{8} + \frac{35x^4}{2} + 70x^3 + 168x^2 + 224x + 128$$

43. $\left(a^2 - \dfrac{1}{3}\right)^4 = (a^2)^4 + 4(a^2)^3\left(-\dfrac{1}{3}\right) + 6(a^2)^2\left(-\dfrac{1}{3}\right)^2 + 4(a^2)\left(-\dfrac{1}{3}\right)^3 + \left(-\dfrac{1}{3}\right)^4 = a^8 - \dfrac{4}{3}a^6 + \dfrac{2}{3}a^4 - \dfrac{4}{27}a^2 + \dfrac{1}{81}$

45. $\left(\dfrac{1}{x}+y\right)^3 = \left(\dfrac{1}{x}\right)^3 + 3\left(\dfrac{1}{x}\right)^2 y + 3\left(\dfrac{1}{x}\right)y^2 + y^3 = \dfrac{1}{x^3} + \dfrac{3y}{x^2} + \dfrac{3y^2}{x} + y^3$

47. The term containing x^7 is the fourth term in the expansion: $\binom{10}{7}x^7 y^3 = \dfrac{10!}{7!(10-7)!}x^7 y^3 = 120x^7 y^3$

49. The term containing x^3 is the tenth term in the expansion:

$\binom{12}{3}x^3(-2)^9 = \dfrac{12!}{3!(12-3)!}(-512)x^3 = -112,640x^3$

51. The term containing x^6 is the third term in the expansion:

$\binom{8}{2}(2x)^6(3y)^2 = \dfrac{8!}{2!(8-2)!}(64x^6)(9y^2) = 16,128x^6 y^2$

53. The term containing y^9 is the tenth term in the expansion:

$$\binom{11}{9}(5x)^2(-2y)^9 = \frac{11!}{2!(11-2)!}\left(25x^2\right)\left(-512y^9\right) = -704,000x^2y^9$$

55. $(1.2)^5 = (1+0.2)^5 = 1^5 + 5(1)^4(0.2) + 10(1)^3(0.2)^2 + 10(1)^2(0.2)^3 + 5(1)(0.2)^4 + (0.2)^5 = 2.48832$

8.5 Beyond the Basics

57. The middle term is the sixth term in the expansion:

$$\binom{10}{5}\left(\sqrt{x}\right)^5\left(-\frac{2}{x^2}\right)^5 = \frac{10!}{5!(10-5)!}x^2\sqrt{x}\left(-\frac{32}{x^{10}}\right) = -\frac{8064\sqrt{x}}{x^8}$$

59. The middle term is the seventh term in the expansion:

$$\binom{12}{6}(1^6)(-x^2y^{-3})^6 = \frac{12!}{6!(12-6)!}\left(\frac{x^{12}}{y^{18}}\right) = \frac{924x^{12}}{y^{18}}$$

61. By the Binomial Theorem, we have

$$2^n = (1+1)^n = \binom{n}{0}(1)^n(1)^0 + \binom{n}{1}(1)^{n-1}(1)^1 + \binom{n}{2}(1)^{n-2}(1)^2 + \cdots + \binom{n}{n}(1)^0(1)^n$$

$$= \binom{n}{0} + \binom{n}{1} + \binom{n}{2} + \ldots + \binom{n}{n}$$

63. $\binom{k}{j} + \binom{k}{j-1} = \frac{k!}{j!(k-j)!} + \frac{k!}{(j-1)!(k-j+1)!} = \frac{k!(j-1)!(k-j+1)! + k!\,j!(k-j)!}{j!(k-j)!(j-1)!(k-j+1)!}$

$$= \frac{[k!(j-1)!(k-j)!][j+(k-j+1)]}{j!(k-j)!(j-1)!(k-j+1)!} = \frac{k!(j+(k-j+1))}{j!(k+1-j)!} = \frac{k!(k+1)}{j!(k+1-j)!} = \frac{(k+1)!}{j!(k+1-j)!} = \binom{k+1}{j}$$

65. $(2x-1)^4 + 4(2x-1)^3(3-2x) + 6(2x-1)^2(3-2x)^2 + 4(2x-1)(3-2x)^3 + (3-2x)^4$

$$= \left((2x-1)+(3-2x)\right)^4 = 2^4 = 16$$

67. $(3x-1)^5 + 5(3x-1)^4(1-2x) + 10(3x-1)^3(1-2x)^2 + 10(3x-1)^2(1-2x)^3 + 5(3x-1)(1-2x)^4 + (1-2x)^5$

$$= \left((3x-1)+(1-2x)\right)^5 = x^5$$

69. The constant term is the third term in the expansion: $240 = \binom{6}{2}(kx)^4\left(-\frac{1}{x^2}\right)^2$

$$= \frac{6!}{2!(6-2)!}k^4x^4\left(\frac{1}{x^4}\right) = 15k^4 \Rightarrow k^4 = 16 \Rightarrow k = \pm 2$$

71. The general term in the expansion has the form $\binom{11}{k}(2x^2)^k\left(-\frac{1}{4x}\right)^{11-k} = \binom{11}{k}(2^k)\left(-\frac{1}{4}\right)^{11-k}\frac{x^{2k}}{x^{11-k}}$. In

order to get a constant term, $2k$ must equal $11-k$. However, the solution of this equation is not an integer, so there is no constant term in the expansion.

8.5 Critical Thinking/Discussion/Writing

72. $2^6 = (1+1)^6$

$$= \binom{6}{0}(1)^6(1)^0 + \binom{6}{1}(1)^5(1)^1 + \binom{6}{2}(1)^4(1)^2 + \binom{6}{3}(1)^3(1)^3 + \binom{6}{4}(1)^2(1)^4 + \binom{6}{5}(1)^1(1)^5 + \binom{6}{6}(1)^0(1)^6$$

$$= \binom{6}{0} + \binom{6}{1} + \binom{6}{2} + \binom{6}{3} + \binom{6}{4} + \binom{6}{5} + \binom{6}{6}$$

73. $0 = (1-1)^{10}$

$$= \binom{10}{0}(1)^{10}(1)^0 - \binom{10}{1}(1)^9(1)^1 + \binom{10}{2}(1)^8(1)^2 - \binom{10}{3}(1)^7(1)^3 + \binom{10}{4}(1)^6(1)^4 - \binom{10}{5}(1)^5(1)^5 + \binom{10}{6}(1)^4(1)^6$$

$$- \binom{10}{7}(1)^3(1)^7 + \binom{10}{8}(1)^2(1)^8 - \binom{10}{9}(1)^1(1)^9 + \binom{10}{10}(1)^9(1)^{10}$$

$$= \binom{10}{0} - \binom{10}{1} + \binom{10}{2} - \binom{10}{3} + \binom{10}{4} - \binom{10}{5} + \binom{10}{6} - \binom{10}{7} + \binom{10}{8} - \binom{10}{9} + \binom{10}{10}$$

74. $(x+y)^2 = x^2 + 2x + y^2$. If $x \geq 0$ and $y \geq 0$, then $2xy \geq 0$ and

$x^2 + 2xy + y^2 \geq x^2 + y^2 \Rightarrow (x+y)^2 \geq x^2 + y^2 \Rightarrow x + y \geq \sqrt{x^2 + y^2}$

75. $(x+y)^n = x^n + \binom{n}{1}x^{n-1}y + \binom{n}{2}x^{n-2}y^2 + \cdots + \binom{n}{n-1}xy^{n-1} + y^n$. If $x > 0$ and $y > 0$, then all the

intermediate terms in the expansion are positive. Therefore $(x+y)^n > x^n + y^n$ if $n > 1$.

76. $(1+x)^n = 1 + \binom{n}{1}x + \binom{n}{2}x^2 + \ldots + \binom{n}{n}x^n = 1 + nx + \binom{n}{1}x + \binom{n}{2}x^2 + \ldots + \binom{n}{n}x^n$. If $x > 0$, then all the terms in

the sum are positive, and $1 + nx + \binom{n}{1}x + \binom{n}{2}x^2 + \ldots + \binom{n}{n}x^n > 1 + nx$. Therefore, $(1+x)^n > 1 + nx$ if $n > 1$.

8.5 Maintaining Skills

77. $5 \cdot 6 \cdot 7 \cdot 8 = \dfrac{8!}{4!}$

79. $2 \cdot 4 \cdot 6 \cdot 8 \cdot 10 \cdot 12 = 2 \cdot (2 \cdot 2) \cdot (2 \cdot 3) \cdot (2 \cdot 4) \cdot (2 \cdot 5) \cdot (2 \cdot 6) = 2^6 \cdot 6!$

81. $\dfrac{12!}{10!} = 12 \cdot 11 = 132$

83. $\dfrac{8!}{5!3!} = \dfrac{8 \cdot 7 \cdot 6}{3 \cdot 2 \cdot 1} = 56$

Section 8.6 Counting Principles

8.6 Practice Problems

1.

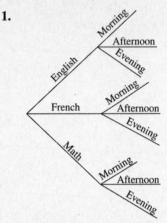

There are 9 ways to choose a course.

2. There are 7 restaurants and 5 movies, so there are $7 \cdot 5 = 35$ different choices.

3. There are $10^6 = 1,000,000$ possibilities.

4. There are $7! = 7 \cdot 6 \cdot 5 \cdot 4 \cdot 3 \cdot 2 \cdot 1 = 5040$ possible arrangements.

5. There are $9 \cdot 8 \cdot 7 \cdot 6 = 3024$ possibilities.

6. a. $P(9,2) = \dfrac{9!}{(9-2)!} = \dfrac{9 \cdot 8 \cdot 7!}{7!} = 72$

   b. $P(n,0) = \dfrac{n!}{(n-0)!} = \dfrac{n!}{n!} = 1$

7. {bears, bulls, lions}, {bears, bulls, tigers}, {bears, lions, tigers}, {bulls, lions, tigers}
   $C(4, 3) = 4$

8. a. $C(9,2) = \dfrac{9!}{(9-2)!2!} = \dfrac{9 \cdot 8 \cdot 7!}{7!2!} = 36$

   b. $C(n,0) = \dfrac{n!}{(n-0)!0!} = \dfrac{n!}{n!} = 1$

   c. $C(n,n) = \dfrac{n!}{(n-n)!n!} = \dfrac{n!}{n!} = 1$

9. $C(12,3) = \dfrac{12!}{(12-3)!3!} = \dfrac{12 \cdot 11 \cdot 10 \cdot 9!}{9!3!}$
   $= \dfrac{12 \cdot 11 \cdot 10}{3 \cdot 2 \cdot 1} = 220$
   There are 220 ways that three of the twelve chocolates can be chosen.

10. $\dfrac{6!}{2!2!2!} = 90$
    There are 90 ways to send six counselors in pairs to three different locations.

8.6 Basic Concepts and Skills

1. Any arrangement of n distinct objects in a fixed order in which no object is used more than once is called a <u>permutation</u>.

3. When r objects are chosen from n distinct objects, the set of r objects is called a <u>combination</u> of n objects taken r at a time.

5. True

7. $P(6,1) = \dfrac{6!}{(6-1)!} = 6$

9. $P(8,2) = \dfrac{8!}{(8-2)!} = 56$

11. $P(9,9) = \dfrac{9!}{(9-9)!} = 362,880$

13. $P(7,0) = \dfrac{7!}{(7-0!)} = 1$

15. $C(8,3) = \dfrac{8!}{(8-3)!3!} = 56$

17. $C(9,4) = \dfrac{9!}{(9-4)!(4!)} = 126$

19. $C(5,5) = \dfrac{5!}{(5-5)!(5!)} = 1$

21. $C(3,0) = \dfrac{3!}{(3-0)!0!} = 1$

23. a. The first letter can be chosen in 26 different ways. The second letter can also be chosen in 26 different ways, so the number of different 2-letter codes is $26 \cdot 26 = 676$.

    b. The first letter can be chosen in 26 different ways. The second letter can be chosen in 25 different ways, so the number of different 2-letter codes is $26 \cdot 25 = 650$.

25. The president can be chosen in 50 ways, while the vice-president can be chosen in 49 ways. So the number of possibilities is $50 \cdot 49 = 2450$.

27. There are four people to be seated in a row of four chairs, so there are
$$P(4,4) = \frac{4!}{(4-4)!} = 24 \text{ different ways to}$$
arrange the people.

29. There are five shirts, three pairs of pants, and four ties, so there are $5 \cdot 3 \cdot 4 = 60$ possible outfits.

31. There are five different toppings. So there are
$$C(5,2) = \frac{5!}{(5-2)!2!} = 10 \text{ different pizzas with}$$
two additional toppings.

33. There are eleven problems, so it is possible to choose nine of them in
$$C(11,9) = \frac{11!}{(11-9)!(9!)} = 55 \text{ different ways.}$$

35. Order is not important, so find the number of combinations. There are $C(11,4) = 330$ different combinations.

37. Order is important, so find the number of permutations. There are 8 seats and six people will be seated, so there are
$$P(8,6) = \frac{8!}{(8-6)!} = 20,160 \text{ different ways to}$$
seat 6 people.

39. Order is important, so find the number of permutations or use the Fundamental Counting Principle. There are
$$P(5,5) = \frac{5!}{(5-5)!} = 120 \text{ ways to visit the five}$$
colleges.

8.6 Applying the Concepts

41. a. The first digit cannot be a zero or a one, so there are eight possibilities for the first digit. There are two possibilities for the second digit and ten possibilities for the third digit. So there are $8 \cdot 2 \cdot 10 = 160$ possible three-digit area codes.

b. There are eight possibilities for the first digit, ten possibilities for the second digit, and ten possibilities for the third digit. So there are $8 \cdot 10 \cdot 10 = 800$ possible three-digit area codes.

43. a. If no repetitions are allowed, there are $24 \cdot 23 \cdot 22 = 12,144$ three-letter fraternity names.

b. If repetitions are allowed, there are $24^3 = 13,824$ three-letter fraternity names.

45. To reach a majority decision, five, six, seven, eight or all nine of the nine justices must agree. Since order is not important, find the sum of the combinations
$$C(9,5) + C(9,6) + C(9,7) + C(9,8) + C(9,9)$$
$$= \frac{9!}{(9-5)!(5!)} + \frac{9!}{(9-6)!(6!)} + \frac{9!}{(9-7)!(7!)}$$
$$+ \frac{9!}{(9-8)!(8!)} + \frac{9!}{(9-9)!(9!)}$$
$$= 126 + 84 + 36 + 9 + 1 = 256$$
There are 256 ways to form a majority of five from the nine justices.

47. Al can choose from $8 \cdot 8 \cdot 8 = 512$ different outfits. Since there are 365 days in a year, he can wear a different outfit every day.

49. a. There are $7! = 5040$ different ways to arrange the houses since any arrangement of the seven houses could be made to correspond to a permutation of the seven designs by agreeing that the first four go on one side of the street and the remaining three go on the other side of the street.

b. There are $7! = 5040$ different ways to arrange the houses since any arrangement of the seven houses could be made to correspond to a permutation of the seven designs by agreeing that the first five go on one side of the street and the remaining two go on the other side of the street.

51. a. There are seven letters in the word CHARITY. If two letters in each four-letter group must be an A and an R, then there are five letters that can be chosen in groups of two: $C(5,2) = \frac{5!}{(5-2)!(2)!} = 10$ possibilities.

b. If one letter in each four-letter group must be an A and none of the letters in the group can be an R, then there are five letters that can be chosen in groups of three:
$$C(5,3) = \frac{5!}{(5-3)!(3)!} = 10 \text{ possibilities.}$$

c. If none of the letters in each four-letter group can be an A or an R, there there are five letters that can be chosen in groups of four: $C(5,4) = \dfrac{5!}{(5-4)!(4!)} = 5$ possibilities.

53. The store will need to stock $5 \cdot 7 \cdot 3 = 105$ shirts to have one of each type.

55. The maximum number of folders is necessary if all of the questionnaires have different rankings. There are $10! = 3,628,800$ possible rankings, so they would need that many folders.

57. In each group of four, if one child is "constant", then the other three children can be chosen in $C(6,3) = \dfrac{6!}{(6-3)!3!} = 20$ ways.

So the maximum number of times any one child will go to the circus is 20. There are $C(7,4) = \dfrac{7!}{(7-4)!4!} = 35$ combinations of four children, so the father will go to the circus 35 times.

59. Because order is not important, we use combinations. Under the "B", there are $C(15,5) = \dfrac{15!}{(15-5)!5!} = 3003$ combinations of numbers. There are the same number under the "I", "G", and "O". Under the "N", there are $C(15,4) = \dfrac{15!}{(15-4)!4!} = 1365$. So there are

$3003^4 \cdot 1365 = 111,007,923,832,370,565$ different bingo cards.

61. There are 10 people and four are needed for the committee. Once the first person is chosen, that person's spouse is eliminated, so there are 8 ways to choose the next person. That person's spouse cannot be chosen next, so there are 6 ways to choose the third person, and 4 ways to choose the fourth person. That gives 1920 possible permutations. However, the order that people are chosen for the committee is not important, so the number of combinations is $\dfrac{1920}{4!} = 80$.

63. If Cory wants to give his sister one coin, there are 4 different combinations of one coin. If he wants to give his sister two coins, there are $C(4,2) = 6$ different combinations. If he wants to give his sister three coins, there are 4 combinations, and if he wants to give his sister four coins, there is one combination. So there are $4 + 6 + 4 + 1 = 15$ different combinations of coins.

65. Out of six letters, there are two "A"s and the rest are singles. So, there are $\dfrac{6!}{2!1!1!1!1!} = 360$ distinguishable ways to arrange the letters.

67. Out of seven letters, there are three "S"s, two "C"s, and the rest are singles. So, there are $\dfrac{7!}{3!2!1!1!} = 420$ distinguishable ways to arrange the letters.

8.6 Beyond the Basics

69. The company needs n workers in combinations of three. So

$C(n,3) = \dfrac{n!}{(n-3)!3!} \geq 20 \Rightarrow$

$\dfrac{n(n-1)(n-2)(n-3)!}{(n-3)!3!} \geq 20 \Rightarrow$

$n^3 - 3n^2 + 2n \geq 120 \Rightarrow$

$n^3 - 3n^2 + 2n - 120 \geq 0$

The possible rational zeros are

$\pm 1, \pm 2, \pm 3, \pm 4, \ \pm 5, \pm 6, \pm 8, \pm 10, \pm 12,$

$\pm 15, \pm 20, \pm 24, \pm 30, \pm 60, \ \pm 120$.

Using synthetic division, we find that 6 is a zero:

```
6| 1   -3    2  -120
        6   18   120
   _____
   1    3   20    0
```

and $(n-6)(n^2 + 3n + 20) = 0$.

The two zeros of $n^2 + 3n + 20 = 0$ are complex, so we reject them. Therefore, six different workers are necessary.

71. There are $C(30,2) = \dfrac{30!}{(30-2)!2!)} = 435$ lines.

73. $\dbinom{m+1}{m-1} = 3! \Rightarrow \dfrac{(m+1)!}{(m-1)!((m+1)-(m-1))!} = 6 \Rightarrow \dfrac{(m+1)!}{(m-1)!2!} = 6 \Rightarrow \dfrac{(m+1)m(m-1)!}{(m-1)!} = 12 \Rightarrow$

$m^2 + m - 12 = 0 \Rightarrow (m+4)(m-3) = 0 \Rightarrow m = 3$ or $m = -4$

Reject the negative solution because factorials are not defined for negative numbers. The solution is $\{3\}$.

75. $2\dbinom{n-1}{2} = \dbinom{n}{3} \Rightarrow \dfrac{2(n-1)!}{2!(n-3)!} = \dfrac{n!}{3!(n-3)!} \Rightarrow \dfrac{(n-1)(n-2)(n-3)!}{(n-3)!} = \dfrac{n(n-1)(n-2)(n-3)!}{6(n-3)!} \Rightarrow$

$6(n-1)(n-2) = n(n-1)(n-2) \Rightarrow n = 6$

77. Using the Binomial Theorem, $\dbinom{n}{0} + \dbinom{n}{1} + \dbinom{n}{2} + \cdots + \dbinom{n}{n} = 64$ is the sum of the coefficients in the nth row

of the expansion of $(x+y)^n$. If $x = 1$ and $y = 1$, then

$(1+1)^n = \dbinom{n}{0} + \dbinom{n}{1} + \dbinom{n}{2} + \cdots + \dbinom{n}{n} = 64 \Rightarrow 2^n = 64 \Rightarrow n = 6$.

79. $\dbinom{m}{4} = \dbinom{m}{5} \Rightarrow \dfrac{m!}{(m-4)!4!} = \dfrac{m!}{(m-5)!5!} \Rightarrow \dfrac{5!}{4!} = \dfrac{(m-4)!}{(m-5)!} \Rightarrow 5 = \dfrac{(m-4)(m-5)!}{(m-5)!} \Rightarrow 5 = m-4 \Rightarrow m = 9$

8.6 Critical Thinking/Discussion/Writing

81. There are $5 \cdot 5 = 25$ terms of the form $x^n y^m$ if n and m are any integers such that $1 \le n \le 5$ and $1 \le m \le 5$.

82. There are two terms in the first factor and three terms in the second factor, so there are $2 \cdot 3 = 6$ terms.

8.6 Maintaining Skills

83. $E = \{2, 5, 7\} \Rightarrow n(E) = 3$

85. $E = \{0\} \Rightarrow n(E) = 1$

For exercises 87–90, $A = \{1, 2, 3, 5\}$ and
$B = \{3, 4, 5, 6, 7\}$.

87. $n(A) = 4, \ n(B) = 5$

89. $A \cap B = \{3, 5\}$
$n(A \cap B) = 2$

For exercises 91–94, $A = \{a, b, c\}$ and
$B = \{2, 4, 6, 8\}$.

91. $n(A) = 3, \ n(B) = 4$

93. $A \cap B = \varnothing$
$n(A \cap B) = 0$

8.7 Probability

8.7 Practice Problems

1. a. $E_1 = \{1, 3, 5\}, \ P(E_1) = \dfrac{3}{6} = \dfrac{1}{2}$

b. $E_2 = \{5, 6\}, \ P(E_2) = \dfrac{2}{6} = \dfrac{1}{3}$

2. $S = \{(H, H), (H, T), (T, H), (T, T)\}$
$E = \{(H, T), (T, H)\}$

$P(E) = \dfrac{2}{4} = \dfrac{1}{2}$

3. $E = \{(6, 1), (5, 2), (4, 3), (3, 4), (2, 5), (6, 1)\}$
There are 36 possible outcomes, so

$P(E) = \dfrac{6}{36} = \dfrac{1}{6}$

4. Since the order in which the numbers are selected is not important, the sample space S consists of all sets of 6 numbers that can be selected from 50 numbers. So,

$n(S) = C(50, 6) = \dfrac{50!}{(50-6)!6!}$

$= \dfrac{50 \cdot 49 \cdot 48 \cdot 47 \cdot 46 \cdot 45}{6 \cdot 5 \cdot 4 \cdot 3 \cdot 2 \cdot 1} = 15{,}890{,}700$

$P(\text{winning the lottery}) = \dfrac{1}{15{,}890{,}700}$

5. $P(\text{jack or king}) = P(\text{jack}) + P(\text{king})$

$= \dfrac{4}{52} + \dfrac{4}{52} = \dfrac{8}{52} = \dfrac{2}{13}$

6. $P(\text{Brandy is not selected}) = \dfrac{C(9,3)}{C(10,3)}$

$$= \frac{9!}{3!6!} \cdot \frac{3!7!}{10!} = \frac{7}{10}$$

$P(\text{Brandy is selected})$

$$= 1 - P(\text{Brandy is not selected}) = 1 - \frac{7}{10} = \frac{3}{10}$$

7. The probability that a student selected at random is a male is $\dfrac{45,143}{82,074} \approx 0.55$. The probability that a student selected at random attends Sacramento College is $\dfrac{20,878}{82,074} \approx 0.25$. The probability that a student selected at random is a male attending Sacramento College is $\dfrac{9040}{82,074} \approx 0.11$. So, the probability that a student selected at random is a male or attends Sacramento College is approximately $0.55 + 0.25 - 0.11 = 0.69$.

8.7 Basic Concepts and Skills

1. If no outcome of an experiment results more often than any other outcome, then the outcomes are said to be equally likely.

3. If it is impossible for two events to occur simultaneously, then the events are said to be mutually exclusive.

5. True (if "between 0 and 1" includes 0 and 1.)

7. $S = \{(\text{Wendy's, McDonald's, Burger King}),$
(Wendy's, Burger King, McDonald's),
(McDonald's, Wendy's, Burger King),
(McDonald's, Burger King, Wendy's),
(Burger King, McDonald's, Wendy's),
(Burger King, Wendy's, McDonald's)$\}$

9. $S = \{\{\textit{Thriller, The Wall}\}, \{\textit{Thriller, Eagles: Their Greatest Hits}\}, \{\textit{Thriller, Led Zeppelin IV}\}, \{\textit{The Wall, Eagles: Their Greatest Hits}\}, \{\textit{The Wall, Led Zeppelin IV}\}, \{\textit{Eagles: Their Greatest Hits, Led Zeppelin IV}\}\}$

11. $S = \{(\text{white, male}), (\text{white, female}), (\text{African-American, male}), (\text{African-American, female}), (\text{Native American, male}), (\text{Native American, female}), (\text{Asian, male}), (\text{Asian, female}), (\text{other, male}), (\text{other, female}), (\text{multiracial, male}), (\text{multiracial, female})\}$

13. $P = 0$.

15. $P = 1$

17. experimental

19. theoretical

21. experimental

23. theoretical

25. experimental

27. An event that is very likely to happen has probability 0.999. An event that will surely happen has probability 1. An event that is a rare event has probability 0.001. An event that is equally likely to happen or not happen has probability 0.5. An event that will never happen has probability 0.

29. $E = \{1, 6\}, P(E) = \dfrac{1}{3}$

31. $E = \{5, 6\}, P(E) = \dfrac{1}{3}$

33. $E = \{2, 4, 6\}, P(E) = \dfrac{1}{2}$

35. $E = \{\text{club queen, spade queen, heart queen, diamond queen}\}, \ P(E) = \dfrac{1}{13}$

37. $E = \{\text{heart ace, heart 2, heart 3, heart 4, heart 5, heart 6, heart 7, heart 8, heart 9, heart 10, heart jack, heart queen, heart king}\}$,

$P(E) = \dfrac{1}{4}$

39. $E = \{\text{heart jack, heart queen, heart king, diamond jack, diamond queen, diamond king, club jack, club queen, club king, spade jack, spade queen, spade king}\}, \ P(E) = \dfrac{3}{13}$

41. $E = \{8 + 3\}, P(E) = \dfrac{1}{10}$

43. $E = \{0 + 3\}, P(E) = \dfrac{1}{10}$

45. $E = \{1 + 3, 3 + 3, 5 + 3, 7 + 3, 9 + 3\}, P(E) = \dfrac{1}{2}$

8.7 Applying the Concepts

47. The probability the Tony will not like his blind date is $1 - 0.3 = 0.7$

49. $P = \dfrac{20}{100} = \dfrac{1}{5}$

51. 6% of the Marines were women, so 94% were men. The probability that a Marine chosen at random is a man is $\dfrac{94}{100} = \dfrac{47}{50}$.

53. If you have 10 tickets and there is only one prize, then the probability of winning is $\frac{10}{125} = \frac{2}{25}$. If there are two prizes, then there are $C(125, 2) = 7750$ possible winning combinations, and you have $C(10, 2) = 45$ possible winning tickets. So the probability of winning is $\frac{45}{7750} = \frac{9}{1550}$.

55. a. There are $2^2 = 4$ possible combinations (bb, bg, gb, gg). The probability of having two boys is $1/4$.

b. The probability of having two girls is $1/4$.

c. The probability of having one boy and one girl is $1/2$.

57. a. Out of the 4440 families, 2370 families own two cars, so the probability of a family owning two cars is $\frac{2370}{4440} = \frac{79}{148}$.

b. Out of the 4440 families, 37 own no cars, 1256 own one car, and 2370 own two cars. So the probability of a family owning at most two cars is $\frac{3663}{4440} = \frac{33}{40}$.

c. Out of the 4440 families, none own six cars, so the probability of a family owning six cars is 0.

d. All of the families own fewer than six cars, so the probability of a family owning at most six cars is 1.

e. Of the 4440 families, all but 37 own at least one car, so the probability of a family owning at least one car is $\frac{4403}{4340} = \frac{119}{120}$.

59. The probability that a non-Hispanic white American has no lactose intolerance is $1 - 0.2 = 0.8 = \frac{4}{5}$. The probability that an African-American, Asian, or Native American has no lactose intolerance is $1 - 0.75 = 0.25 = 1/4$.

61. a. There are $C(10, 2) = 10$ different combinations of people for the committee. There is only one combination of the two oldest members, so the probability of the committee consisting of the two oldest members is $1/10$.

b. There is only one combination of the oldest and youngest members, so the probability is $1/10$.

63.

Number of people who are	Depressed according to the test	Normal according to the test
Actually depressed	90	10
Actually normal	135	765

If 10% of the population suffers from depression, then we expect that 10% of the 1000 people tested = 100 people will suffer from depression and 900 will not.

Of the 100 people, the test will work for 90% of them, or 90 people, and the remaining 10 people are normal according to the test. Of the 900 people who don't suffer from depression, the test works for 85% of them, or 765 people. The remaining 135 are depressed according to the test.

65. The possible combinations are {bbbb, bbbg, bbgg, bggg, gggg}. (Birth order is not important.) So, there are two ways to have three children of one gender and one of the other gender, while there is only one way to have two children of each gender. Therefore, it is more likely to have three children of one gender and one of the other.

67. a. Out of 180 people, 50 people received the placebo, a total of 56 people had no pain relief, while 34 people received the placebo and had no pain relief. So the probability that a person either received the placebo or had no pain relief is $\frac{50}{180} + \frac{56}{180} - \frac{34}{180} = \frac{2}{5}$.

b. Out of 180 people, 70 people received the new medicine, a total of 69 people had complete pain relief, while 40 people who received the new medicine also had complete pain relief. So, the probability that a person either received the new medicine or had complete pain relief is

$$\frac{70}{180} + \frac{69}{180} - \frac{40}{180} = \frac{99}{180}.$$

8.7 Beyond the Basics

69. To find the probability that a number between 1 and 1,000,000 does not contain the digit 3, note that each place value through the hundred-thousand's place can contain one of nine digits. So the probability that a number between 1 and 1,000,000 will not contain the digit 3 is $\dfrac{9^6}{1,000,000} = \dfrac{531,441}{1,000,000}$. So the probability that a number between 1 and 1,000,000 will contain the digit 3 is

$$1 - \frac{531,441}{1,000,000} = \frac{468,559}{1,000,000}.$$

71. Because one of the digits is repeated, there are

$$\frac{4!}{2!1!1!} = 12 \text{ different ways to arrange the}$$

digits. The probability that Maggie will dial the correct number is $1/12$.

8.7 Critical Thinking/Discussion/Writing

72. The probability that a person chosen at random in the survey would report his or her highest level of education as high school graduation is $\dfrac{59,840,000}{187,000,000} = \dfrac{8}{25}$.

73. If there are x dark caramels in the box, then there are $2x$ light caramels, and a total of $3x$ caramels in the box. So the probability of choosing a dark caramel is $\dfrac{x}{3x} = \dfrac{1}{3}$.

74. If a single die is rolled two times, then there are $6 \cdot 6 = 36$ possible outcomes. If the first number rolled is n, then there are $6 - n$ possible rolls in which the second roll will be greater than the first roll. So there are

$$\sum_{n=1}^{5} 6 - n = 15 \text{ possible rolls. The probability}$$

that the second roll will result in a number larger than the first roll is $\dfrac{15}{36} = \dfrac{5}{12}$.

Chapter 8 Review Exercises

1. $a_1 = 2(1) - 3 = -1, a_2 = 2(2) - 3 = 1,$
$a_3 = 2(3) - 3 = 3, a_4 = 2(4) - 3 = 5,$
$a_5 = 2(5) - 3 = 7$

3. $a_1 = \dfrac{1}{2(1)+1} = \dfrac{1}{3}, a_2 = \dfrac{2}{2(2)+1} = \dfrac{2}{5},$
$a_3 = \dfrac{3}{2(3)+1} = \dfrac{3}{7}, a_4 = \dfrac{4}{2(4)+1} = \dfrac{4}{9},$
$a_5 = \dfrac{5}{2(5)+1} = \dfrac{5}{11}$

5. This is an arithmetic sequence with a common difference of -2. $a_n = 32 - 2n$ for $n \geq 1$.

7. $\dfrac{9!}{8!} = 9$

9. $\dfrac{(n+1)!}{n!} = n + 1$

11. $\displaystyle\sum_{k=1}^{4} k^3 = 1^3 + 2^3 + 3^3 + 4^3 = 100$

13. $\displaystyle\sum_{k=1}^{7} \dfrac{k+1}{k}$
$= \dfrac{1+1}{1} + \dfrac{2+1}{2} + \dfrac{3+1}{3} + \dfrac{4+1}{4} + \dfrac{5+1}{5}$
$\qquad + \dfrac{6+1}{6} + \dfrac{7+1}{7} = \dfrac{1343}{140}$

15. $\displaystyle\sum_{k=1}^{50} \dfrac{1}{k}$

17. The sequence is arithmetic. $a_1 = 11, d = -5$

19. The sequence is not arithmetic.

21. $a_n = 3n$

23. $a_n = x + n - 1$

25. $a_3 = a_1 + d(3-1) \Rightarrow 7 = a_1 + 2d$
$a_8 = a_1 + d(8-1) \Rightarrow 17 = a_1 + 7d$
$\begin{cases} a_1 + 2d = 7 \\ a_1 + 7d = 17 \end{cases} \Rightarrow -5d = -10 \Rightarrow d = 2, a_1 = 3$
$a_n = 3 + 2(n-1) = 2n + 1$

27. $d = 2, a_1 = 7, a_n = 37 \Rightarrow$
$37 = 7 + 2(n-1) \Rightarrow n = 16$
$S = 16\left(\dfrac{7+37}{2}\right) = 352$

29. $d = 5, a_1 = 3, n = 40 \Rightarrow$
$a_n = 3 + 5(40-1) \Rightarrow a_n = 198$
$S = 40\left(\dfrac{3+198}{2}\right) = 4020$

31. The sequence is geometric. $a_1 = 4, r = -2$

33. The sequence is not geometric.

35. $a_1 = 16, r = -\dfrac{1}{4}, a_n = 16 \cdot \left(-\dfrac{1}{4}\right)^{n-1} = \dfrac{(-1)^{n-1}}{4^{n-3}}$

37. $a_{10} = a_1 r^{10-1} = 2 \cdot 3^9 = 39,366$

39. $r = \dfrac{1/5}{1/10} = 2; S_{12} = \dfrac{(1/10)\left(1-2^{12}\right)}{1-2} = \dfrac{819}{2}$

41. $r = \dfrac{1/6}{1/2} = \dfrac{1}{3}; S = \dfrac{1/2}{1-(1/3)} = \dfrac{3}{4}$

43. $a_1 = \dfrac{3}{5}, r = \dfrac{3}{5}; \displaystyle\sum_{i=1}^{\infty}\left(\dfrac{3}{5}\right)^i = \dfrac{3/5}{1-3/5} = \dfrac{3}{2}$

45. For $n = 1, \displaystyle\sum_{k=1}^{1} 2^k = 2 = 2^{1+1} - 2$ is true.

Assume that it is true for

$n = m: \displaystyle\sum_{k=1}^{m} 2^k = 2^{m+1} - 2$.

Then for $n = m + 1$,

$\displaystyle\sum_{k=1}^{m+1} 2^k = \left(\sum_{k=1}^{m} 2^k\right) + 2^{m+1} = 2^{m+1} - 2 + 2^{m+1}$

$= 2(2^{m+1}) - 2 = 2^{m+2} - 2$

which is exactly the statement for $n = m + 1$. Therefore the formula is true for all natural numbers.

47. $\dbinom{12}{7} = \dfrac{12!}{7!(12-7)!} = \dfrac{12 \cdot 11 \cdot 10 \cdot 9 \cdot 8 \cdot 7!}{7!5!}$

$= \dfrac{12 \cdot 11 \cdot 10 \cdot 9 \cdot 8}{5 \cdot 4 \cdot 3 \cdot 2 \cdot 1} = 792$

49. $(x-3)^4$
$= x^4 + 4(-3)x^3 + 6(-3)^2 x^2 + 4(-3)^3 x + (-3)^4$
$= x^4 - 12x^3 + 54x^2 - 108x + 81$

51. The term containing x^5 is the eighth term.
$\dbinom{12}{5}x^5(2^7) = \dfrac{12!}{5!(12-5)!}x^5(2^7) = 101,376x^5$

53. If no repetitions are allowed, using the Fundamental Counting Principle, there are $4 \cdot 3 \cdot 2 \cdot 1 = 24$ different numbers that can be written.

55. Since order is important, find the number of permutations of 7 taken 7 at a time:

$P(7,7) = \dfrac{7!}{(7-7)!} = 5040$.

There are 5040 different ways that seven people can line up.

57. The five movies can be listed in $5! = 120$ different ways.

59. Order is not important, so find the number of combinations of three candies from a group of

ten. There $C(10,3) = \dfrac{10!}{3!(10-3)!} = 120$ ways

to choose three candies from a box of ten candies.

61. There are $C(12, 2) = 66$ ways to choose two shirts from 12, and $C(8, 3) = 56$ ways to choose three pairs of pants from eight. So, there are $66 \cdot 56 = 3696$ ways to choose two shirts and three pairs of pants.

63. There are nine letters, so $n = 9$. There are two R's, three E's, and two T's. So there are

$\dfrac{9!}{2!3!2!1!1!} = 15,120$ distinguishable ways to

arrange the letters.

65. Pair the two coins that appear together as one so that there are three positions to fill. Then, (with the paired coins counted as one) there are $3! = 6$ ways to fill these positions. Since the paired coins can be inserted into any position in $2! = 2$ ways, there are $3!2! = 6 \times 2 = 12$ allowable ways to arrange the coins. The total number of ways to arrange the coins is $4! = 24$, so the probability of arranging the coins so that the two most recent

dates will be next to each other is $\dfrac{12}{24} = \dfrac{1}{2}$.

67. There are six letters, so there are $C(6, 2) = 15$ combinations of two letters. There are four consonants, so there are $C(4, 2) = 6$ combinations of two consonants. So the probability of choosing two consonants is $6/15 = 2/5$.

69. There are $C(8, 2) = 28$ ways to choose two people from the group of either. There are five ways to choose one man and three ways to choose one woman, so there are $5 \cdot 3 = 15$ ways to choose one man and one woman. So the probability of choosing one man and one woman is $15/28$.

71. There are 13 clubs, so there are $C(13, 2) = 78$ ways to choose two clubs. There are $C(52, 2) = 1326$ ways to choose two cards from the entire deck, so the probability of choosing two clubs is $\dfrac{78}{1326} = \dfrac{1}{17}$.

73. a. $1/9$ **b.** $4/9$ **c.** $5/9$

Chapter 8 Practice Test A

1. $a_1 = 3(5 - 4(1)) = 3, a_2 = 3(5 - 4(2)) = -9,$
$a_3 = 3(5 - 4(3)) = -21, a_4 = 3(5 - 4(4)) = -33,$
$a_5 = 3(5 - 4(5)) = -45$.
The sequence is arithmetic.

2. $a_1 = -3(2^1) = -6, a_2 = -3(2^2) = -12,$
$a_3 = -3(2^3) = -24, a_4 = -3(2^4) = -48,$
$a_5 = -3(2^5) = -96$. The sequence is geometric.

3. $a_1 = -2, a_2 = 3a_1 + 5 = 3(-2) + 5 = -1,$
$a_3 = 3a_2 + 5 = 3(-1) + 5 = 2,$
$a_4 = 3a_3 + 5 = 3(2) + 5 = 11,$
$a_5 = 3a_4 + 5 = 3(11) + 5 = 38$

4. $\dfrac{(n-1)!}{n!} = \dfrac{(n-1)!}{n(n-1)!} = \dfrac{1}{n}$

5. $a_7 = -3 + 6(4) = 21$

6. $a_8 = 13\left(-\dfrac{1}{2}\right)^7 = -\dfrac{13}{128}$

7. $a_1 = 1, a_{20} = 58$
$\displaystyle\sum_{k=1}^{20}(3k - 2) = 20\left(\dfrac{1+58}{2}\right) = 590$

8. $a_1 = \dfrac{3}{8}, a_5 = \dfrac{3}{128}$

$\displaystyle\sum_{k=1}^{5}\left(\dfrac{3}{4}\right)(2^{-k}) = \dfrac{\dfrac{3}{8}\left(1 - \left(\dfrac{1}{2}\right)^5\right)}{1 - \dfrac{1}{2}} = \dfrac{93}{128}$

9. $a_1 = \dfrac{9}{50}; \displaystyle\sum_{k=1}^{\infty}18\left(\dfrac{1}{100}\right)^k = \dfrac{9/50}{1 - 1/100} = \dfrac{2}{11}$

10. $\dbinom{13}{0} = \dfrac{13!}{0!(13-0)!} = 1$

11. $(1 - 2x)^4$
$= 1^4 + 4(1^3)(-2x) + 6(1^2)(-2x)^2$
$\qquad\qquad + 4(1)(-2x)^3 + (-2x)^4$
$= 16x^4 - 32x^3 + 24x^2 - 8x + 1$

12. The term containing x^1 is the fourth term.
$\dbinom{4}{1}(2x)^1(1)^3 = \dfrac{4!}{1!(4-1)!}(2x)^1(1)^3 = 8x$

13. There are $2^{10} = 1024$ different ways to answer every question on a ten-question true-false test.

14. $P(9, 2) = \dfrac{9!}{(9-2)!} = 72$

15. $C(7, 5) = \dfrac{7!}{(7-5)!5!} = 21$

16. There are $6 \cdot 10 = 60$ ways to fill the positions.

17. There are $26 \cdot 26 \cdot 10^4 = 6,760,000$ possible license plate numbers.

18. $\dfrac{5}{12}$ **19.** $\dfrac{5}{6}$

20. There are 18 coins and $C(18, 2) = 153$ ways to choose two coins. There are $C(7, 2) = 21$ ways to choose two quarters. So the probability of choosing two quarters is $\dfrac{21}{153} = \dfrac{7}{51}$.

Chapter 8 Practice Test B

1. $a_1 = 4(2(1) - 3) = -4, a_2 = 4(2(2) - 3) = 4,$
$a_3 = 4(2(3) - 3) = 12, a_4 = 4(2(4) - 3) = 20,$
$a_5 = 4(2(5) - 3) = 28$. The sequence is arithmetic. The answer is C.

2. $a_1 = 2(4^1) = 8, a_2 = 2(4^2) = 32,$
$a_3 = 2(4^3) = 128, a_4 = 2(4^4) = 512,$
$a_5 = 2(4^5) = 2048$. The sequence is geometric. The answer is D.

3. $a_1 = 5, a_2 = 2a_1 + 4 = 2(5) + 4 = 14,$
$a_3 = 2a_2 + 4 = 2(14) + 4 = 32,$
$a_4 = 2a_3 + 4 = 2(32) + 4 = 68,$
$a_5 = 2a_4 + 4 = 2(68) + 4 = 140$.
The answer is A.

4. $\dfrac{(n+2)!}{(n+2)} = \dfrac{(n+2)(n+1)!}{(n+2)} = (n+1)!$.
The answer is C.

5. $a_8 = a_1 + 7d = -6 + 7(3) = 15$
The answer is A.

6. $a_{10} = 247\left(\dfrac{1}{3}\right)^9 = \dfrac{247}{19,683}$
The answer is B.

7. $a_1 = -3, a_{45} = 4(45) - 7 = 173$
$\displaystyle\sum_{k=1}^{45}(4k - 7) = 45\left(\dfrac{-3+173}{2}\right) = 3825$
The answer is D.

8. $a_1 = \dfrac{8}{3}, r = 2; \displaystyle\sum_{k=1}^{5}\left(\dfrac{4}{3}\right)(2^k) = \dfrac{\frac{8}{3}\left(1 - 2^5\right)}{1 - 2} = \dfrac{248}{3}$.
The answer is A.

9. $a_1 = 8; \displaystyle\sum_{k=1}^{\infty}8(-0.3)^{k-1} = \dfrac{8}{1 - (-0.3)} \approx 6.15$.
The answer is B.

10. $\dbinom{12}{11} = \dfrac{12!}{11!(12-11)!} = 12$
The answer is B.

11. $(3x - 1)^4$
$= (3x)^4 + 4(3x)^3(-1) + 6(3x)^2(-1)^2$
$\quad + 4(3x)(-1)^3 + (-1)^4$
$= 81x^4 - 108x^3 + 54x^2 - 12x + 1$
The answer is A.

12. The term containing x is the third term.
$\dbinom{3}{1}(2x)^1(3)^2 = \dfrac{3!}{1!(3-1)!}(2x)^1(9) = 54x$
The answer is D.

13. There are $4 \cdot 10 \cdot 3 = 120$ ways to choose a necklace, a pair of earrings, and a bracelet. The answer is C.

14. $P(7,3) = \dfrac{7!}{(7-3)!} = 210$
The answer is A.

15. $C(10,7) = \dfrac{10!}{7!(10-7)!} = 120$
The answer is D.

16. The first digit can be chosen in nine ways (note that the first digit cannot be zero), and the second digit can be chosen in nine ways. So there are $9 \cdot 9 = 81$ two-digit numbers that can be formed without using a digit more than once. The answer is B.

17. Since one CD is a "must buy", four CDs must be chosen from the remaining seven. There are $C(7,4) = \dfrac{7!}{4!(7-4)!} = 35$ ways to choose the CDs. The answer is D.

18. The answer is C.

19. There are $6 \cdot 6 = 36$ possible outcomes. The outcomes in which the sum is greater than 9 are $\{(4, 6), (6, 4), (5, 5), (5, 6), (6, 5) (6, 6)\}$, so the probability of getting a sum greater than 9 when two die are rolled is $\dfrac{6}{36} = \dfrac{1}{6}$.
The answer is A.

20. There are $52 - 13 = 39$ non-spades in the deck, so the probability of choosing a card that is not a spade is $\dfrac{39}{52} = \dfrac{3}{4}$.
The answer is D.

Cumulative Review Exercises (Chapters P–8)

1. $|3x - 8| = |x| \Rightarrow 3x - 8 = x$ or $3x - 8 = -x \Rightarrow$ $x = 4$ or $x = 2$. The solution is $\{2, 4\}$.

2. $x^2(x^2 - 5) = -4 \Rightarrow x^4 - 5x^2 + 4 = 0 \Rightarrow$
$(x^2 - 4)(x^2 - 1) = 0 \Rightarrow$
$(x - 2)(x + 2)(x - 1)(x + 1) = 0 \Rightarrow$
$x = 2$ or $x = -2$ or $x = 1$ or $x = -1$
The solution is $\{-2, -1, 1, 2\}$.

3. $\log_2(3x-5)+\log_2 x=1\Rightarrow$

$\log_2 x(3x-5)=1\Rightarrow 3x^2-5x=2\Rightarrow$

$3x^2-5x-2=0\Rightarrow(3x+1)(x-2)=0\Rightarrow$

$x=-\dfrac{1}{3}$ or $x=2$

Reject the negative answer. The solution is $\{2\}$.

4. Let $u=e^x$. Then $e^{2x}-e^x-2=0\Rightarrow$

$u^2-u-2=0\Rightarrow(u-2)(u+1)=0\Rightarrow$

$u=2$ or $u=-1$. If $u=-1$, then

$-1=e^x\Rightarrow\ln(-1)=x$, which is impossible,

so reject -1. If $u=2$, then $2=e^x\Rightarrow x=\ln 2$.
The solution is $\{\ln 2\}$.

5. $\dfrac{6}{x+2}=\dfrac{4}{x}\Rightarrow 6x=4x+8\Rightarrow x=4$

6. $2(x-8)^{-1}=(x-2)^{-1}\Rightarrow\dfrac{2}{x-8}=\dfrac{1}{x-2}\Rightarrow$

$2x-4=x-8\Rightarrow x=-4$

7. $\sqrt{x-3}=\sqrt{x}-1\Rightarrow\left(\sqrt{x-3}\right)^2=\left(\sqrt{x}-1\right)^2\Rightarrow$

$x-3=x-2\sqrt{x}+1\Rightarrow 2=\sqrt{x}\Rightarrow x=4$

8. $x^2+4x\geq 0\Rightarrow x(x+4)\geq 0$

Solving the associated equation, we have $x=0$
or $x=-4$. The intervals to be tested are
$(-\infty,-4],[-4,0],[0,\infty)$.

Interval	Test point	Value of $x(x+4)$	Result
$(-\infty,-4]$	-5	5	$+$
$[-4,0]$	-1	-3	$-$
$[0,\infty)$	1	5	$+$

The solution is $(-\infty,-4]\cup[0,\infty)$.

9. $18\leq x^2+6x\Rightarrow x^2+6x-18\geq 0$

Solving the associated equation, we have

$x=\dfrac{-6\pm\sqrt{36-4(-18)}}{2}=\dfrac{-6\pm\sqrt{108}}{2}$

$=\dfrac{-6\pm 6\sqrt{3}}{2}=-3\pm 3\sqrt{3}$

The intervals to be tested are $\left(-\infty,-3-3\sqrt{3}\right]$,

$\left[-3-3\sqrt{3},-3+3\sqrt{3}\right],\left[-3+3\sqrt{3},\infty\right)$.

Interval	Test point	Value of $x^2+6x-18$	Result
$\left(-\infty,-3-3\sqrt{3}\right]$	-10	22	$+$
$\left[-3-3\sqrt{3},-3+3\sqrt{3}\right]$	0	-18	$-$
$\left[-3+3\sqrt{3},\infty\right)$	3	9	$+$

The solution is

$\left(-\infty,-3-3\sqrt{3}\right]\cup\left[-3+3\sqrt{3},\infty\right)$.

10. $\begin{cases}5x+4y=6\\4x-3y=11\end{cases}\Rightarrow\begin{cases}15x+12y=18\\16x-12y=44\end{cases}\Rightarrow$

$31x=62\Rightarrow x=2$

$5(2)+4y=6\Rightarrow y=-1$

11. $\begin{cases}x-3y+6z=-8\\5x-6y-2z=7\\3x-2y-10z=11\end{cases}\Rightarrow\begin{bmatrix}1 & -3 & 6 & | & -8\\5 & -6 & -2 & | & 7\\3 & -2 & -10 & | & 11\end{bmatrix}$

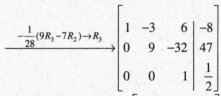

$\xrightarrow[R_3-3R_1\rightarrow R_3]{R_2-5R_1\rightarrow R_2}\begin{bmatrix}1 & -3 & 6 & | & -8\\0 & 9 & -32 & | & 47\\0 & 7 & -28 & | & 35\end{bmatrix}$

$\xrightarrow{-\frac{1}{28}(9R_3-7R_2)\rightarrow R_3}\begin{bmatrix}1 & -3 & 6 & | & -8\\0 & 9 & -32 & | & 47\\0 & 0 & 1 & | & \frac{1}{2}\end{bmatrix}$

$\xrightarrow{\frac{1}{9}(R_2+32R_3)\rightarrow R_2}\begin{bmatrix}1 & -3 & 6 & | & -8\\0 & 1 & 0 & | & 7\\0 & 0 & 1 & | & \frac{1}{2}\end{bmatrix}\Rightarrow$

$y=7,z=\dfrac{1}{2};x-3(7)+6\left(\dfrac{1}{2}\right)=-8\Rightarrow x=10$

The solution is $\left\{\left(10,7,\dfrac{1}{2}\right)\right\}$.

12.

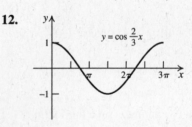

$y=\cos\dfrac{2}{3}x$

13.
$$\sin\left(-\frac{7\pi}{12}\right) = -\sin\left(\frac{7\pi}{12}\right) = -\sin\left(\frac{\pi}{4} + \frac{\pi}{3}\right)$$
$$= -\left(\sin\frac{\pi}{4}\cos\frac{\pi}{3} + \cos\frac{\pi}{4}\sin\frac{\pi}{3}\right)$$
$$= -\left(\frac{\sqrt{2}}{2}\cdot\frac{1}{2} + \frac{\sqrt{2}}{2}\cdot\frac{\sqrt{3}}{2}\right)$$
$$= -\left(\frac{\sqrt{2}+\sqrt{6}}{4}\right) = \frac{-\sqrt{2}-\sqrt{6}}{4}$$

14. $r = \sqrt{2^2 + \left(2\sqrt{3}\right)^2} = 4$

$\tan\theta = \sqrt{3} \Rightarrow \theta = \frac{\pi}{3}$ or $\theta = \frac{4\pi}{3}$

$2 + 2\sqrt{3}i$ lies in Quadrant I, so $\theta = \frac{\pi}{3}$

The polar coordinates for $\left(2 + 2\sqrt{3}i\right)$ are

$4\left(\cos\frac{\pi}{3} + i\sin\frac{\pi}{3}\right)$.

15. Shift the graph of $y = \sqrt{x}$ three units to the right.

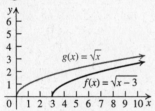

16. The domain is $(-\infty, -1) \cup (-1, 0) \cup (0, \infty)$.

17. $7\ln x - \ln(x-5) = \ln\left(\frac{x^7}{x-5}\right)$

18. $[A\,|\,I] = \begin{bmatrix} 1 & 0 & -2 & 1 & 0 & 0 \\ 4 & 1 & 0 & 0 & 1 & 0 \\ 1 & 1 & 7 & 0 & 0 & 1 \end{bmatrix}$

$\xrightarrow[R_3 - R_1 \to R_3]{-4R_1 + R_2 \to R_2} \begin{bmatrix} 1 & 0 & -2 & 1 & 0 & 0 \\ 0 & 1 & 8 & -4 & 1 & 0 \\ 0 & 1 & 9 & -1 & 0 & 1 \end{bmatrix}$

$\xrightarrow{R_3 - R_2 \to R_3} \begin{bmatrix} 1 & 0 & -2 & 1 & 0 & 0 \\ 0 & 1 & 8 & -4 & 1 & 0 \\ 0 & 0 & 1 & 3 & -1 & 1 \end{bmatrix}$

$\xrightarrow[R_2 - 8R_3 \to R_2]{R_1 + 2R_3 \to R_1} \begin{bmatrix} 1 & 0 & 0 & 7 & -2 & 2 \\ 0 & 1 & 0 & -28 & 9 & -8 \\ 0 & 0 & 1 & 3 & -1 & 1 \end{bmatrix} \Rightarrow$

$A^{-1} = \begin{bmatrix} 7 & -2 & 2 \\ -28 & 9 & -8 \\ 3 & -1 & 1 \end{bmatrix}$

19. Since order is important, find the number of permutations of four people taken two at a time: $P(4,2) = \frac{4!}{(4-2)!} = 12$.

There are 12 arrangements.

20. There are 18 people in total. Three people who are 25 years or older have received speeding tickets, and six people whose ages are between 17 and 24 have received a speeding ticket. The probability that a person chosen at random will be 25 years old or older is $\frac{6}{18}$. The probability that a person chosen at random will have received a speeding ticket is $\frac{9}{18}$. The probability that a person 25 years old or older will have received a speeding ticket is $\frac{3}{18}$. So, the probability that a person chosen at random from the room will be a person who is 25 years or older or has gotten a speeding ticket is $\frac{6}{18} + \frac{9}{18} - \frac{3}{18} = \frac{12}{18} = \frac{2}{3}$.